METAMATERIAL

Edited by **Xun-Ya Jiang**

Metamaterial
Edited by Xun-Ya Jiang

Published by InTech
Janeza Trdine 9, 51000 Rijeka, Croatia

Publishing Process Manager Vana Persen

Technical Editor Teodora Smiljanic

Cover Designer InTech Design Team

Additional hard copies can be obtained from orders@intechopen.com

Metamaterial, Edited by Xun-Ya Jiang

p. cm.

ISBN 978-953-51-0591-6

For EU product safety concerns:
IN TECH d.o.o., Prolaz Marije Krucifikse Kozulić 3, 51000 Rijeka, Croatia,
info@intechopen.com or visit our website at intechopen.com.

Contents

Section 1

The Physics of Metamaterials

1

Novel Electromagnetic Phenomena in Graphene and Subsequent Microwave Devices Enabled by Multi-Scale Metamaterials

Dimitrios L. Sounas and Christophe Caloz
École Polytechnique de Montréal
Canada

1. Introduction

Metamaterials, i.e. artificial materials with properties not found in nature (Caloz & Itoh, 2006; Marqués et al., 2008), have gained a lot of attention in the past decade. The virtually unlimited freedom in tailoring their constitutive parameters created hopes for the realization of unique devices, such as subwavelength lenses and invisibility cloaks, which were essentially impossible to achieve with classical materials. However, the implementation of such devices has been hindered by the inherent limitation of conventional metamaterials: non-perfect homogeneity due to the periodicity being insufficiently smaller than the wavelength, high insertion and thermal losses due to the conduction properties of the constituent elements and narrow bandwidth caused by the strongly resonant nature of the metamaterials constituent elements. Furthermore, the complex 3D structure of bulk metamaterials makes them unsuitable as microwave substrates and superstrates for planar devices where integration is a critical issue. Therefore, there is a need to develop a new generation of metamaterials with higher compactness and more homogeneity, exploiting the recent advances of nanotechnology, for instance in the fields of nanowires, nanopolymers, carbon nanotubes and graphene (Dragoman & Dragoman, 2009; Hanson, 2007).

Graphene is a 2D material consisting of carbon atoms in a 2D honeycomb lattice (Geim & Novoselov, 2007). It was first synthesized from graphite in the form of micro-flakes in 2004 (Novoselov et al., 2004) and since then it has been gaining a continuously increasing interest. Its gap-less linear band-structure results in unique phenomena, such as high electron mobility, ambipolarity, non-zero minimum conductivity and anomalous half-integer quantum Hall effect (Neto et al., 2009). Until now, a lot of research has been devoted to the design of electronic transport devices, such as transistors and non-linear components, mainly exploiting the high mobility and the ambipolar field effect of graphene (Lin et al., 2009; Wang et al., 2009). Recently, it has been demonstrated that graphene also possesses unique electromagnetic properties, such as low loss surface waves (Koppens et al., 2011; Mikhailov & Ziegler, 2007; Vakil & Engheta, 2011), huge Faraday rotation (Crassee et al., 2011; Sounas & Caloz, 2011b) and super-confined edge magnetoplasmons (Mishchenko et al., 2010; Sounas & Caloz, 2011a). These properties, combined with the ambipolar field effect which provides enhanced control capabilities of graphene's conductivity, open a new path towards super-compact and enhanced-functionality microwave devices.

The chapter is organized as follows. Section 2 provides a brief overview of the electronic band structure of graphene, which is responsible for most of its unique properties. Section 3 introduces the surface conductivity model of graphene in the presence of a static magnetic field and gives a physical interpretation of the conductivity expressions. Section 4 constitutes the core of the chapter providing a theoretical analysis of Faraday rotation in graphene. In Section 5 two applications based on the gyrotropic properties of graphene, namely a circular waveguide Faraday rotator and a spatial isolator, are proposed. Section 6 describes a practical implementation of such devices through the multiscale metamaterials concept. Finally, Section 7 summarizes the chapter.

2. Electronic band structure of graphene

The unique properties of graphene originate from its electronic band structure. Using a tight-binding approximation and considering only the nearest-neighbor terms, the energy-wavevector dispersion relation of electrons in graphene is found as (Neto et al., 2009)

$$E_{\pm}(k_x, k_y) = \pm t\sqrt{3 + 2\cos\left(\sqrt{3}k_y a\right) + 4\cos\left(\frac{\sqrt{3}}{2}k_y a\right)\cos\left(\frac{3}{2}k_x a\right)}, \tag{1}$$

where $t = 2.8$ eV is the nearest-neighbor hopping energy, $a = 1.42$ Å is the interatomic distance, k_x and k_y are the x and y wavevector components, respectively, and the plus and minus signs refer to the upper (conduction) and the lower (valence) bands, respectively. Figs 1(a) and 1(b) plot E versus k_x and k_y in the entire Brillouin zone and around the K symmetry point, respectively. The K and K' points are very important, since the energy at these points, which is 0, is the chemical potential[1] of intrinsic graphene. A zero chemical potential means a completely full valence band and a completely empty conduction band, hence minimal conductivity.

The first thing one can observe from Fig. 1 is the absence of an energy gap between the valence and the conduction bands. Because of this, graphene is usually referred to as a semi-metal[2] or a zero-gap semiconductor[3]. The zero gap allows changing the type of conduction charge carriers from electrons to holes and vice versa by shifting the chemical potential from positive (conduction band) to negative (valence band) values, a phenomenon known as ambipolarity. The chemical potential can be easily tuned by an electrostatic voltage between graphene and an electrode parallel to it, as it is shown in Fig. 2. Assuming that the surface carrier density of graphene is n_s, the electric dispacement on both sizes of graphene is oriented perpendicularly to graphene, along the z axis, and it has a value $D_z = (n_s e/2)\mathrm{sgn}(z)$. The electric field in the substrate is then $E_z = D_z/(\varepsilon_r\varepsilon_0) = -n_s e/(2\varepsilon_r\varepsilon_0)$. Integrating E_z from graphene ($z = 0$) to the PEC ($z = -d$) plane, the apllied voltage is found as

$$V_g = \frac{n_s e d}{2\varepsilon_r\varepsilon_0}. \tag{2}$$

[1] The chemical potential is the energy level with 0.5 probability of being fully occupied by electrons (the Fermi-Dirac distribution is 0.5). At zero temperature and under zero applied field (equilibrium), the chemical potential is equal to the Fermi level.

[2] Metals have their conduction band penetrating into their valence band.

[3] Semiconductors have a non-zero gap between the valence and the conduction bands.

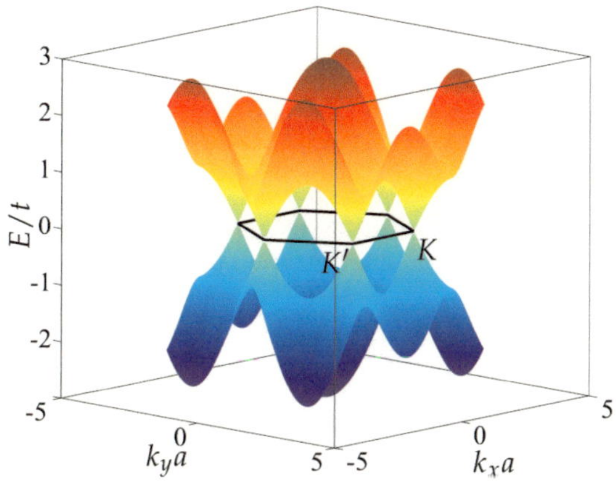

(a)

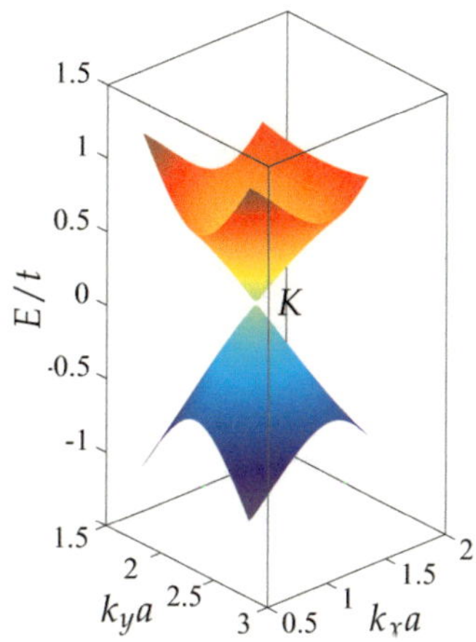

(b)

Fig. 1. Energy-wavenumber dispersion diagram of electrons in graphene. (a) Dispersion diagram in the entire Brillouin zone. (b) Dispersion diagram around the K symmetry point (Dirac point).

Therefore, n_s, and subsequently μ_c, can be controlled via V_g (varying V_g either provides electrons to graphene or depletes graphene from electrons). This way of controlling the carrier density is ofter called electrical doping, in contrast to chemical doping, where the carrier density is controlled by the density of impurity atoms introduced into the graphene lattice.

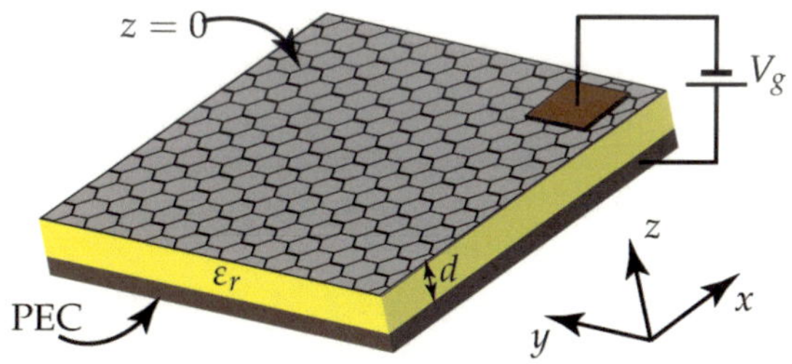

Fig. 2. Structure for controlling the chemical potential of graphene. A conducting electrode is placed at the bottom of graphene, forming with graphene a parallel plate-capacitor. An electrostatic voltage is applied between graphene and the electrode.

As already mentioned, the K and K' points are of significant importance in graphene. Using a Taylor expansion around these points in Equation 1 and keeping only the first order term, the energy-wavevector dispersion relation takes the form

$$E(\mathbf{q}) = \pm \hbar v_F |\mathbf{q}|, \tag{3}$$

where $\mathbf{q} = \mathbf{k} - \mathbf{k}_{\mathbf{K},\mathbf{K}'}$ and $v_F = 3ta/(2\hbar) \approx 10^6$ m/s is the Fermi velocity. Equation 3 is similar to the dispersion relation $E = \hbar ck$ of photons in vacuum and indicates that electrons in graphene behave like zero-mass particles (Dirac fermions) with an energy-independent velocity. As a result electrons in graphene exhibit quantum phenomena, such as the half-integer quantum Hall effect and the Klein paradox, normally only encountered in high energy relativistic particles.

Another important consequence of the graphene band structure, and more specifically of the double valley degeneracy associated with the K and K′ point, is the extremely low phonon scattering of electrons. This induces an upper limit of 200,000 cm^2/Vs in the graphene mobility at room temperature and more than 1,000,000 cm^2/Vs at low temperature! However, the practical mobility is much lower (it can be so low as 1,000 cm^2/Vs), due to various defects in the graphene lattice, such as impurities and wrinkles, and due to scattering from the substrate. Except for phonon scattering, which is related to the nature of the material and therefore imposes an upper bound on mobility, the scattering factors mainly depend on the production process. Therefore, improved production processes are expected to significantly increase the mobility to much higher values.

Biasing graphene with a static magnetic field $\mathbf{B}_0$ perpendicular to its plane results in quantization of the band diagram, as illustrated in Fig. 3. This phenomenon is called Landau quantization and it exists in all semiconductors. However, contrary to semiconductors with a parabolic band dispersion, the energy levels (Landau levels) in graphene are non-uniformly spaced, as a result of the linear dispersion close to the K and K' points. Furthermore, the 0th order Landau level has a zero energy, irrespectively of the applied magnetic field, which is the reason for the observation of the quantum Hall effect in graphene even at room temperature. The energy of the nth Landau level in graphene is

$$E_n = \sqrt{n}L, \tag{4}$$

where $L = \sqrt{2\hbar e B_0 v_F^2}$ is the Landau energy scale.

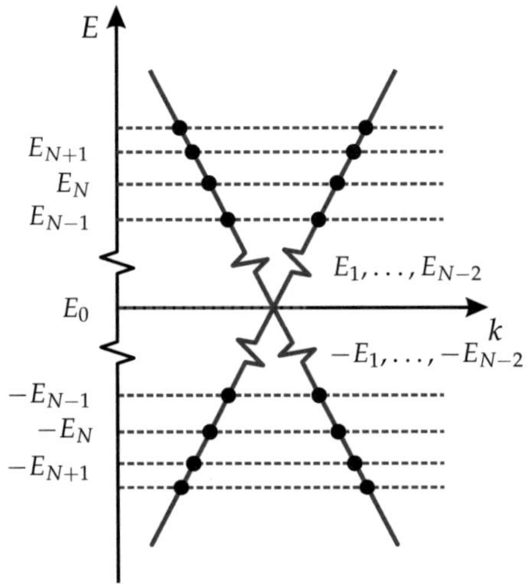

Fig. 3. Energy-wavenumber dispersion diagram in graphene, biased with a static magnetic field perpendicular to its plane. The dispersion diagram is quantized in the discrete energy levels $E_0, E_{\pm 1}, \ldots$.

Before closing this introductory section, let us provide the relation between the carrier density and the chemical potential, since the latter enters in all the conductivity expression that will be given in the next section. It is known from solid state physics that electrons are distributed in the energy levels of any material according to the Fermi-Dirac distribution

$$f_d(E) = \frac{1}{1 + e^{(E-\mu_c)/(k_B T)}}, \tag{5}$$

which represents the probability of finding an electron at the energy level E. In Equation 5 μ_c is the chemical potential, k_B is the Boltzmann constant and T the temperature in K. Fig. 4 plots $f_d(E)$ for $\mu_c > 0$, that is a partially filled conduction band. The electron and hole densities are then

$$n_{se} = \int_0^\infty f_d(E)N(E)dE \tag{6a}$$

$$n_{sh} = \int_{-\infty}^0 \left[1 - f_d(E)\right] N(E)dE, \tag{6b}$$

where

$$N(E) = \frac{2E}{\pi\hbar^2 v_F^2} \tag{7}$$

is the density of states. The density of states is the same for electrons and holes in graphene, due to the band-structure symmetry around the $E = 0$ plane. The effective total carrier density is then $n_s = n_{se} - n_{sh}$. Note that in intrinsic graphene $n_{se} = n_{sh}$ and, therefore, $n_s = 0$. Inserting Equation 5 into Equations 6 and then taking their difference yields

$$n_s = \frac{2(k_B T)^2}{\pi\hbar^2 v_F^2} \left[\mathrm{Li}_2\left(-e^{-\frac{\mu_c}{k_B T}}\right) - \mathrm{Li}_2\left(-e^{\frac{\mu_c}{k_B T}}\right)\right], \tag{8}$$

where Li_2 is the polylogarithm function of second order (Abramowitz & Stegun, 1964). Inserting this relation into (2) provides the relation indicateing the required biasing voltage for a desired chemical potential. If $\mu_c \gg k_B T$, that is for highly doped graphene, Equation 8 takes the asymptotic form

$$n_s = \frac{\mu_c^2}{\pi\hbar^2 v_F^2}. \tag{9}$$

3. Conductivity of magnetically biased graphene

The essentially 2D structure of graphene makes surface conductivity the most natural appropriate quantity to model its electrical properties. When graphene is biased with a static magnetic field $\mathbf{B}_0$ perpendicular to its plane, conductivity takes a tensorial form, which may be deduced by considering the motion of an electron under an electric field $\mathbf{E}$ in the graphene plane, as illustrated in Fig. 5. For simplicity, the electron is considered to be initially at rest. Let us first examine the case $\mathbf{E} = E_x\hat{\mathbf{x}}$. The electric field exerts a force $\mathbf{F}_e = -e\hat{\mathbf{x}}$ on the electron, which accelerates it in the $-x$ direction. As long as the electron acquires a non-zero velocity, a magnetic force $\mathbf{F}_m = -e\mathbf{v} \times \mathbf{B}_0$ along the $-y$ direction is exerted on it and deflects it towards the $-y$ direction. The motion of the electron is, therefore, a combination of two simpler motions, one along the $-x$ axis and one along the $-y$ axis, and the electric current has two components, one along the $+x$ direction and one along the $+y$ direction[4]. For small

[4] The current is opposite to the electron's velocity, due to the negative charge of the electron.

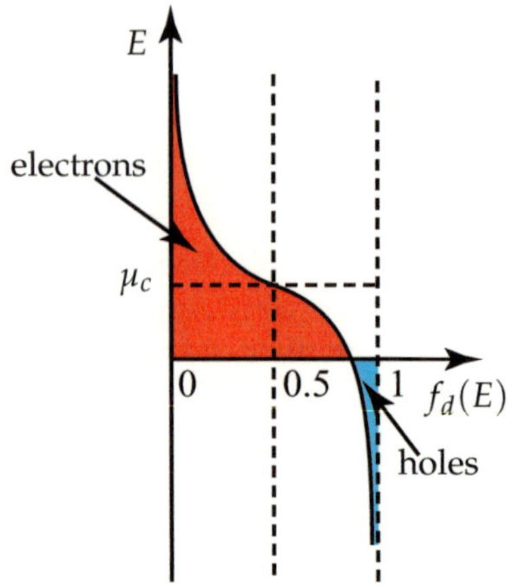

Fig. 4. Fermi-Dirac distribution for $\mu_c > 0$. The red area corresponds to free electrons and the blue to free holes (electrons and holes that can conduct).

E, when nonlinear effects are negligible, the current can thus be related to the electric field through

$$\mathbf{J} = \sigma_{xx} E_x \hat{\mathbf{x}} + \sigma_{yx} E_x \hat{\mathbf{y}}. \tag{10}$$

The proportionality coefficients σ_{xx} and σ_{xy} correspond to the longitudinal (parallel to **E**) and transverse (perpendicular to **E**) conductivities, respectively, which depend on the band structure of the material, the frequency of **E** and $\mathbf{B}_0$. Following a similar analysis for $\mathbf{E} = E_y \hat{\mathbf{y}}$, which is the case depicted in Fig. 5(b), and assuming that graphene exhibits the same properties in all its directions, we derive

$$\mathbf{J} = -\sigma_{yx} E_y \hat{\mathbf{x}} + \sigma_{xx} E_x \hat{\mathbf{y}}. \tag{11}$$

Equations 10 and 11 can be combined into the single equation

$$\mathbf{J} = \bar{\bar{\sigma}} \cdot \mathbf{E}, \tag{12}$$

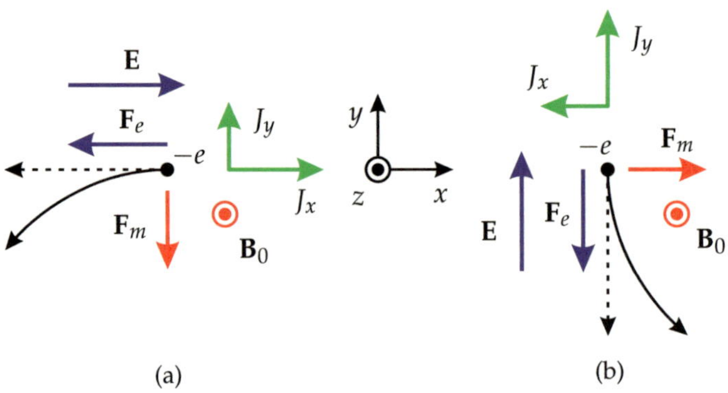

Fig. 5. Motion of an electron on graphene in an electric field and a static magnetic field. The electron, initially accelerated in the direction of the electric field, is deflected in the direction perpendicular to this field, due to the static magnetic field. Two electric current components, parallel and perpendicular to the electric field, are subsequently generated. (a) Case of the electric field along the x-axis. (b) Case of the electric field along the y-axis.

where

$$\bar{\bar{\sigma}} = \sigma_{xx}\bar{\bar{I}}_t + \sigma_{yx}\bar{\bar{J}}_t \tag{13}$$

is the conductivity tensor of graphene and

$$\bar{\bar{I}}_t = \hat{\mathbf{x}}\hat{\mathbf{x}} + \hat{\mathbf{y}}\hat{\mathbf{y}}, \tag{14a}$$

$$\bar{\bar{J}}_t = \hat{\mathbf{y}}\hat{\mathbf{x}} - \hat{\mathbf{x}}\hat{\mathbf{y}}. \tag{14b}$$

Closed-form expressions for σ_{xx} and σ_{yx} have been obtained in (Gusynin et al., 2009) through a quantum mechanical analysis which involves the Kubo formula (Kubo, 1957). Assuming an energy independent scattering mechanism, which is equivalent to considering an average scattering rate Γ, σ_{xx} and σ_{yx} are given by

$$\begin{aligned}\sigma_{xx}(\omega, B_0) = {} & \frac{e^2 v_F^2 |eB_0| \hbar(\omega - j2\Gamma)}{-j\pi} \sum_{n=0}^{\infty} \left\{ \frac{1}{E_{n+1} - E_n} \right. \\ & \times \frac{f_d(E_n) - f_d(E_{n+1}) + f_d(-E_{n+1}) - f_d(-E_n)}{(E_{n+1} - E_n)^2 - \hbar^2(\omega - j2\Gamma)^2} \\ & \left. + (E_n \to -E_n) \right\},\end{aligned} \tag{15a}$$

$$\begin{aligned}\sigma_{yx}(\omega, B_0) = {} & -\frac{e^2 v_F^2 eB_0}{\pi} \sum_{n=0}^{\infty} [f_d(E_n) - f_d(E_{n+1}) - f_d(-E_{n+1}) + f_d(-E_n)] \\ & \times \left[\frac{1}{(E_{n+1} - E_n)^2 - \hbar^2(\omega - j2\Gamma)^2} + (E_n \to -E_n) \right].\end{aligned} \tag{15b}$$

A physical explanation of Equation 15 may be provided with the help of Fig. 6. Each of the summation terms in Equation 15 corresponds to an electron transition between two Landau levels. Two types of transitions exist: intraband, between levels in the same band, and interband between levels of different bands. From all the possible intraband transitions, the only ones allowed by the selection rules are these from level n to $n+1$ in the conduction band and from level $-n-1$ to $-n$ in the valence band. Similarly, the only allowed interband transitions are from level $-n$ to $n+1$ and from $-n-1$ to n. The photon energy needed for a transition between the levels E_{initial} and E_{final} is $E_{\text{final}} - E_{\text{initial}}$. A photon with this energy corresponds to an electromagnetic wave of frequency $(E_{\text{final}} - E_{\text{initial}})/\hbar$. The probability of such a transition is proportional to the difference between the probability $f_d(E_{\text{initial}})$ of the initial level being full and the probability $f_d(E_{\text{final}})$ of the final level being empty. The transitions with the highest probabilities are those crossing μ_c, across which the largest difference in $f_d(E)$ exists, emphasized by thicker arrows in Fig. 6. Therefore, assuming a finite temperature and that μ_c lies between the levels E_N and E_{N+1}, the interband transitions with the lowest energy are the ones that involve the $-N$ and $N+1$ levels or the $-N-1$ and N levels. As a result, interband transitions occur essentially at frequencies $\hbar\omega_{\text{inter}} \geq E_N + E_{N+1}$.

Assume now that $\mu_c \gg L$. For the magnetic field value of 1 T, used in the chapter, $L = 0.036$ eV, and the condition $\mu_c \gg L$ is thus largely satisfied if $\mu_c > 1$ eV. Then $E_{N+1} > \mu_c \gg L$, which through Equation 4 yields $N \gg 1$ and $E_N \approx E_{N+1} \approx \mu_c$ (due to the $\sqrt{n}$ compression factor in E_n), resulting in $\hbar\omega_{\text{inter}} \geq 2\mu_c$. In most of practical situations, $\mu_c \geq 0.05$ eV, so that $\hbar\omega_{\text{inter}} \geq 0.15$ PHz. Therefore, all the results of the chapter, which pertain to microwave and millimeter-wave operation, are derived by considering only the

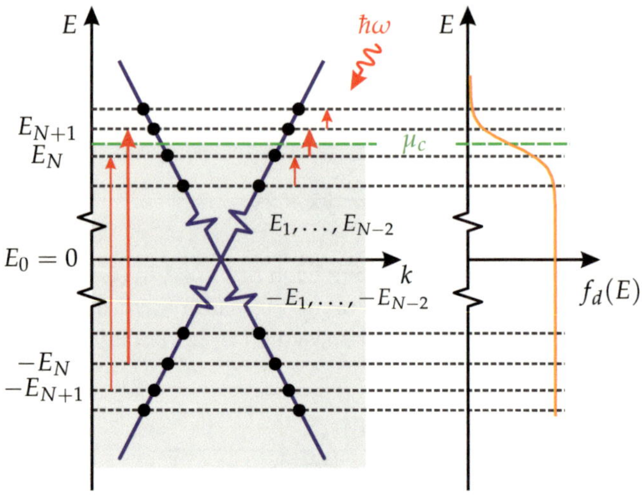

Fig. 6. Electron transitions in graphene when illuminated by an electromagnetic wave of frequency ω. The red arrows represent the allowed intraband (right-hand side) and interband (left-hand side) electron transitions between the energy levels. A photon can be absorbed by graphene if its energy $\hbar\omega$ coincides with the energy of any of the allowed electron transitions. The probability of an electron transition depends on the probability of the initial level to be full and the probability of the final level to be empty. The probability of a level with energy E to be full is given by the Fermi-Dirac distribution $f_d(E)$ (orange curve on the right).

intraband terms in Equation 15. Under the condition $N \gg 1$, the Landau levels around μ_c, which produce non-negligible intraband transitions, lie very close to each other. It can be shown (Kubo, 1957) that in this case σ_{xx} and σ_{yx} follow the Drude model form

$$\sigma_{xx}(\omega, B_0) = \sigma_0 \frac{1 + j\omega\tau}{(\omega_c\tau)^2 + (1 + j\omega\tau)^2}, \tag{16a}$$

$$\sigma_{yx}(\omega, B_0) = \sigma_0 \frac{\omega_c\tau}{(\omega_c\tau)^2 + (1 + j\omega\tau)^2}, \tag{16b}$$

where

$$\sigma_0 = \frac{2e^2\tau}{\pi\hbar^2} k_B T \ln\left(2\cosh\frac{\mu_c}{2k_B T}\right) \tag{17}$$

is the DC conductivity of graphene, $\tau = 1/(2\Gamma)$ is the scattering time and ω_c is the cyclotron frequency. The latter corresponds to the difference between the levels E_N and E_{N+1},

$$\omega_c = \frac{E_{N+1} - E_N}{\hbar} = \frac{L}{\hbar\left(\sqrt{N+1} + \sqrt{N}\right)} \approx \frac{L^2}{2\hbar\mu_c} = \frac{eB_0 v_F^2}{\mu_c}. \tag{18}$$

For the B_0 values used in the chapter, $\mu_c \gg L$ also implies $\mu_c \gg k_B T$ ($k_B T = 0.026$ eV at 300 K), and Equation 17 subsequently simplifies to $\sigma_0 = n_s e \mu$, where $n_s = \mu_c^2/\pi\hbar^2 v_F^2$ is the carrier density and $\mu = e\tau v_F^2/\mu_c$ the mobility. Figure 7 plots σ_{xx} and σ_{yx} versus frequency using both Equation 15 and 16, for $B_0 = 1$ T, $\mu = 5000$ cm^2/Vs, $n_s = 10^{13}$ cm^{-2} and $T = 300$ K, corresponding to values widely used throughout the chapter. Excellent agreement between the two models is observed in the frequency range shown.

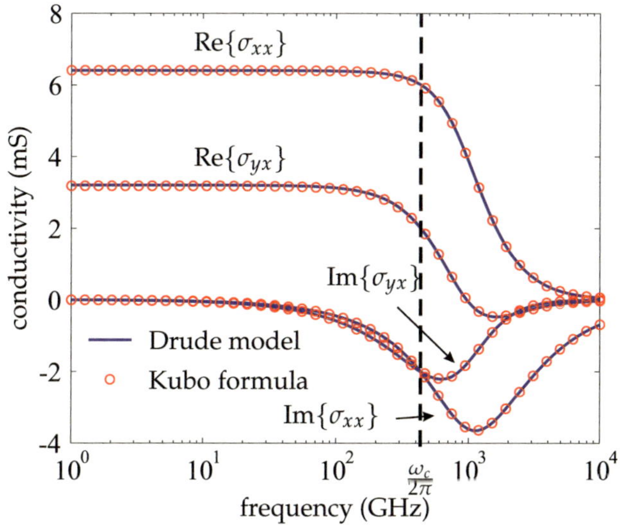

Fig. 7. Graphene conductivity versus frequency computed from the exact expressions 15 (circles) and from the approximate expressions 16 (lines), for $\mu = 5000\ \text{cm}^2/\text{Vs}$, $n_s = 10^{13}\ \text{cm}^{-2}$, $B_0 = 1$ T and $T = 300$ K.

4. Faraday rotation in graphene

When a plasma, such as a semiconductor, is biased with a static magnetic field it exhibits Faraday rotation at frequencies above the cyclotron frequency. The reason why Faraday rotation is only observed above the cyclotron resonance is that below the resonance the plasma is impenetrable. Graphene, being a semiconductor, is also expected to exhibit Faraday rotation (Crassee et al., 2011; Sounas & Caloz, 2011b). However, contrary to semiconductors, Faraday rotation in graphene exists also below the cyclotron resonance, as a result of its super-thin, semi-transparent nature. This section theoretically analyzes Faraday rotation in graphene. The results will be used in the next section where possible applications of this phenomenon will be suggested.

Consider a graphene sheet coinciding with the $z = 0$ plane in free space and biased with a static magnetic field $\mathbf{B}_0 = B_0\hat{\mathbf{z}}$ and a plane wave impinging normally on the sheet from medium 1 towards medium 2, as illustrated in Fig. 8. The electric field of the incident, reflected and transmitted waves then read

$$\mathbf{E}^{\text{inc}} = \mathbf{E}_0^{\text{inc}} e^{-jk_0 z}, \tag{19a}$$

$$\mathbf{E}^{\text{ref}} = \mathbf{E}_0^{\text{ref}} e^{jk_0 z}, \tag{19b}$$

$$\mathbf{E}^{\text{tran}} = \mathbf{E}_0^{\text{tran}} e^{-jk_0 z}. \tag{19c}$$

In order to find $\mathbf{E}_0^{\text{ref}}$ and $\mathbf{E}_0^{\text{tran}}$, we will use the boundary conditions

$$\hat{\mathbf{z}} \times \left(\mathbf{E}^{\text{tran}} - \mathbf{E}^{\text{inc}} - \mathbf{E}^{\text{ref}}\right)\Big|_{z=0} = \mathbf{0}, \tag{20a}$$

$$\hat{\mathbf{z}} \times \left(\mathbf{H}^{\text{tran}} - \mathbf{H}^{\text{inc}} - \mathbf{H}^{\text{ref}}\right)\Big|_{z=0} = \bar{\bar{\sigma}} \cdot \mathbf{E}^{\text{tran}}\Big|_{z=0}, \tag{20b}$$

which hold in the graphene plane. The magnetic fields $\mathbf{H}^{\text{inc}}$, $\mathbf{H}^{\text{ref}}$ and $\mathbf{H}^{\text{tran}}$ are related to the corresponding electric fields through

$$\mathbf{H}^{\text{inc}} = \frac{\hat{\mathbf{z}} \times \mathbf{E}^{\text{inc}}}{\eta_0}, \tag{21a}$$

$$\mathbf{H}^{\text{ref}} = -\frac{\hat{\mathbf{z}} \times \mathbf{E}^{\text{inc}}}{\eta_0}, \tag{21b}$$

$$\mathbf{H}^{\text{tran}} = \frac{\hat{\mathbf{z}} \times \mathbf{E}^{\text{inc}}}{\eta_0}, \tag{21c}$$

where η_0 is the free-space wave impedance. Inserting Equation 19 into 20a yields

$$\mathbf{E}_0^{\text{tran}} - \mathbf{E}_0^{\text{inc}} - \mathbf{E}_0^{\text{tran}} = \mathbf{0}. \tag{22}$$

Similarly, inserting Equation 21 into 20b, one gets

$$-\mathbf{E}_0^{\text{tran}} + \mathbf{E}_0^{\text{inc}} - \mathbf{E}_0^{\text{ref}} = \eta_0 \bar{\bar{\sigma}} \cdot \mathbf{E}_0^{\text{tran}}. \tag{23}$$

Solving Equations 22 and 20b simultaneously with respect to $\mathbf{E}_0^{\text{tran}}$ gives

$$\mathbf{E}_0^{\text{tran}} = 2\left(2\bar{\bar{I}}_t + \eta_0 \bar{\bar{\sigma}}\right)^{-1} \cdot \mathbf{E}_0^{\text{inc}}. \tag{24}$$

Therefore, the transmission dyadic for incidence from medium 1 towards medium 2 is

$$\bar{\bar{T}}_{21} = 2\left(2\bar{\bar{I}}_t + \eta_0 \bar{\bar{\sigma}}\right)^{-1} = \frac{2}{(2+\eta_0\sigma_{xx})^2 + (\eta_0\sigma_{yx})^2}\left[(2+\eta_0\sigma_{xx})\bar{\bar{I}}_t - \eta_0\sigma_{yx}\bar{\bar{J}}_t\right], \tag{25}$$

where the dyadic identity (Lindell, 1996)

$$\left(a\bar{\bar{I}}_t + b\bar{\bar{J}}_t\right)^{-1} = \frac{1}{a^2+b^2}\left(a\bar{\bar{I}}_t - b\bar{\bar{J}}_t\right), \tag{26}$$

has been used. Through a similar analysis, it can be shown that the transmission dyadic for incidence from medium 2 towards medium 1 is

$$\bar{\bar{T}}_{12} = \bar{\bar{T}}_{21}, \tag{27}$$

a condition which, as it will be shown later, indicates non-reciprocity.

Consider now the specific case of a left-handed (LH) circularly polarized incident wave with respect to the z-axis $\mathbf{E}_{\text{LH},0}^{\text{inc}} = \hat{\mathbf{x}} + j\hat{\mathbf{y}}$. The transmitted wave is then, according to Equation 25,

$$\mathbf{E}_{\text{LH},0}^{\text{tran}} = \bar{\bar{T}}_{21} \cdot \mathbf{E}_{\text{LH},0}^{\text{inc}} = \frac{2}{2+\eta_0\sigma_{\text{LH}}}\mathbf{E}_{\text{LH},0}^{\text{inc}}, \tag{28}$$

where $\sigma_{\text{LH}} = \sigma_{xx} - j\sigma_{yx}$. Therefore, in the case of a LH circularly polarized wave the transmission coefficient simplifies to

$$T_{\text{LH}} = \frac{2}{2+\eta_0\sigma_{\text{LH}}}. \tag{29}$$

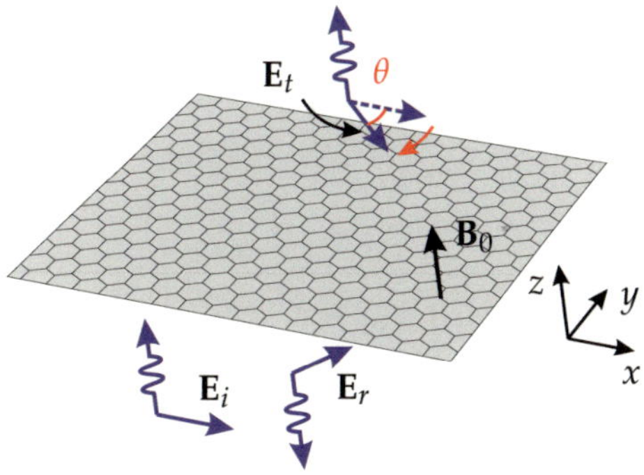

Fig. 8. Magnetically biased graphene sheet in the $z = 0$ plane. A plane wave impinges normally on graphene.

Similarly, the transmission coefficient for a right-handed (RH) circularly polarized wave is

$$T_{\rm RH} = \frac{2}{2 + \eta_0 \sigma_{\rm RH}}, \tag{30}$$

where $\sigma_{RH} = \sigma_{xx} + j\sigma_{yx}$. Equations 29 and 30 reveal that $T_{\rm LH} \neq T_{\rm RH}$, a phenomenon known as circular birefringence, with Faraday rotation being one of its most important consequences.

Fig. 9 plots the amplitude and phase of $T_{\rm LH}$ and $T_{\rm RH}$ versus frequency for a graphene sheet with $\tau = 1.84 \times 10^{-13}$ s, $\mu_c = 0.37$ eV and $B_0 = 1$ T. Two things can be observed: a) the amplitudes of $T_{\rm LH}$ and $T_{\rm RH}$ are almost the same far from the cyclotron resonance and b) there is a huge phase difference between $T_{\rm LH}$ and $T_{\rm RH}$ below the cyclotron resonance. These two facts result in a large Faraday rotation below the cyclotron resonance. To prove this statement, consider an x-polarized incident wave $\mathbf{E}_0^{\rm inc} = \hat{\mathbf{x}}$. Such a wave can be decomposed into a RH circularly polarized wave and a LH circularly polarized wave as

$$\mathbf{E}^{\rm inc} = \frac{1}{2}(\hat{\mathbf{x}} + j\hat{\mathbf{y}}) + \frac{1}{2}(\hat{\mathbf{x}} - j\hat{\mathbf{y}}). \tag{31}$$

The transmitted wave is then

$$\mathbf{E}^{\rm tran} = \frac{T_{\rm LH}}{2}(\hat{\mathbf{x}} + j\hat{\mathbf{y}}) + \frac{T_{\rm RH}}{2}(\hat{\mathbf{x}} - j\hat{\mathbf{y}}). \tag{32}$$

If $|T_{\rm LH}| \approx |T_{\rm RH}|$, as in the case far below and above from the cyclotron resonance, Equation 32 becomes

$$\begin{aligned}\mathbf{E}^{\rm tran} &= \frac{|T_{\rm LH}|}{2}\left[e^{j\varphi_{\rm LH}}(\hat{\mathbf{x}} + j\hat{\mathbf{y}}) + e^{j\varphi_{\rm RH}}(\hat{\mathbf{x}} - j\hat{\mathbf{y}})\right] \\ &= \frac{|T_{\rm LH}|}{2}e^{j\varphi_{\rm av}}\left[\hat{\mathbf{x}}\cos\left(\frac{\Delta\varphi}{2}\right) - \hat{\mathbf{y}}\sin\left(\frac{\Delta\varphi}{2}\right)\right],\end{aligned} \tag{33}$$

where $\varphi_{\rm LH} = \arg\{T_{\rm LH}\}$, $\varphi_{\rm RH} = \arg\{T_{\rm RH}\}$, $\varphi_{\rm av} = (\varphi_{\rm LH} + \varphi_{\rm RH})/2$ and $\Delta\varphi = \varphi_{\rm LH} - \varphi_{\rm RH}$. Since $\Delta\varphi > 0$ (Fig. 9) the field has been rotated by an angle

$$\theta = \frac{\Delta\varphi}{2} \tag{34}$$

in the RH direction with respect to the $+z$-direction (the direction of the magnetic field). For $\omega \ll \omega_c$, where σ_{xx} and σ_{yx} are almost purely real, the phases of T_{LH} and T_{RH} are found from Equations 29 and 30 as

$$\varphi_{\mathrm{LH}} = \tan^{-1}\left(\frac{\eta_0\sigma_{yx}}{2+\eta_0\sigma_{xx}}\right) \tag{35a}$$

$$\varphi_{\mathrm{RH}} = -\tan^{-1}\left(\frac{\eta_0\sigma_{yx}}{2+\eta_0\sigma_{xx}}\right). \tag{35b}$$

Thus, for $\omega \ll \omega_c$, the rotation angle is

$$\theta = \tan^{-1}\left(\frac{\eta_0\sigma_{yx}}{2+\eta_0\sigma_{xx}}\right). \tag{36}$$

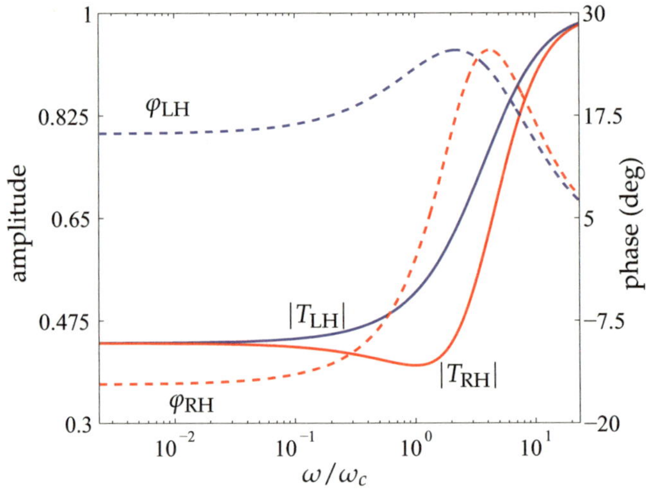

Fig. 9. Transmission coefficient for normal incidence of a LH and a RH circularly polarized wave on a graphene sheet in free space with $\tau = 1.84 \times 10^{-13}$ s, $\mu_c = 0.37$ eV and $B_0 = 1$ T.

Fig. 10 presents θ and the transmission amplitude ($|T_{\mathrm{LH}}| \approx |T_{\mathrm{RH}}|$) versus μ_c and B_0 for a graphene sheet with $\tau = 1.84 \times 10^{-13}$ s at 30 GHz and $T = 300$ K. The rotation angle clearly depends on both μ_c and B_0, providing two degrees of freedom in controlling it. Regarding the μ_c dependence, $\theta \to 0$ as $\mu_c \to 0$. This is expected from the fact that for $\mu_c = 0$ there are very few electrons in the conduction band, hence the conductivity is very small and the interaction between the electromagnetic waves and the material is negligible. As μ_c increases, θ also increases up to a specific μ_c value, which depends on B_0. As μ_c further increases, θ decreases and tends to 0 for very large μ_c, because the conductivity becomes very large and the material behaves like a perfect electric conductor. Concerning the B_0 dependence, Fig. 10 indicates that θ increases along a line of constant transmission amplitude. For μ_c fixed, θ increases linearly with B_0 when $\omega_c\tau \ll 1$, it becomes maximum when $\omega_c\tau \approx 1$ and it decreases inversely proportionally with B_0 when $\omega_c\tau \gg 1$.

In addition to controlling the amount of rotation, we can also control its direction via μ_c, exploiting the ambipolar properties of graphene. Specifically, by inverting μ_c from positive to negative values, the type of charge carriers changes from electrons to holes. Holes have

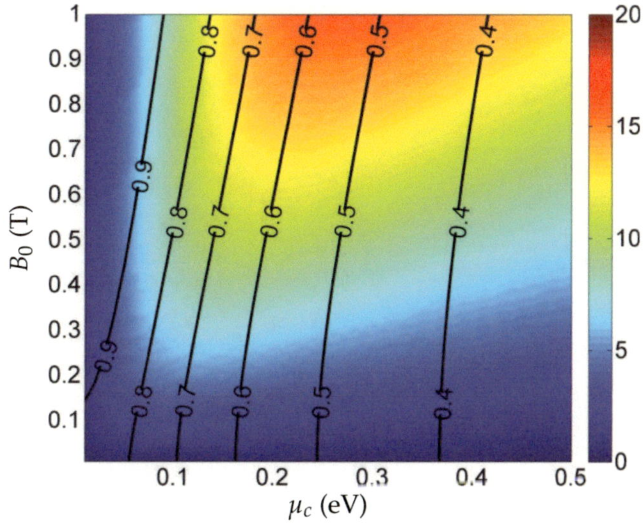

Fig. 10. Rotation angle (color plot) and transmission amplitude (contour plot) versus μ_c and B_0 for a normally incident plane wave on a graphene sheet in free space, for the parameters $\tau = 1.84 \times 10^{-13}$ s, $f = 30$ GHz, and $T = 300$ K.

an opposite charge than electrons and, therefore, provide an opposite transverse current than electrons, hence an opposite rotation direction, as illustrated in Fig. 11. This property is of significant practical importance, since it allows the control of the rotation direction via an electrostatic potential applied to graphene, as shown in Fig. 2, while keeping the magnetic field constant. Note that the static magnetic field is usually provided by permanent magnets and therefore it is extremely difficult to invert.

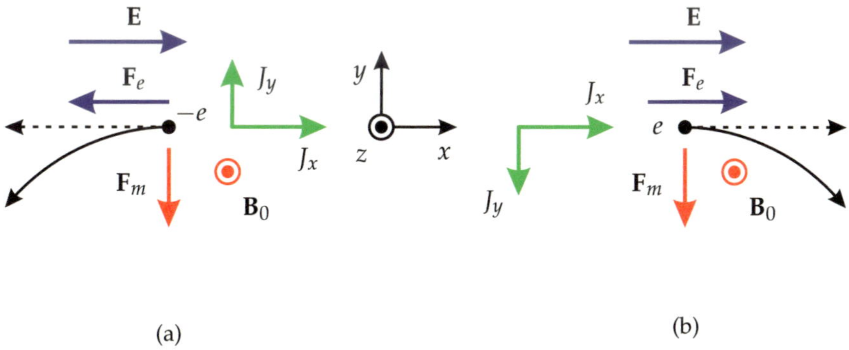

Fig. 11. Reversal of the direction of cyclotron rotation resulting from reversal of the type of charge carriers. (a) Case of electrons carriers. (b) Case of hole carriers, where the the transverse current has been reversed due to the reversal of the charge sign.

Although neglected until now, the fact that σ_{xx} and σ_{yx} are not purely real or purely imaginary numbers results in the transmitted wave being elliptically instead of linearly polarized. Fig. 12 plots θ and the axial ratio of the transmitted wave versus frequency for $\tau = 1.84 \times 10^{-13}$ s, $\mu_c = 0.37$ eV, $B_0 = 1$ T and $T = 300$ K. For $\omega \ll \omega_c$, when σ_{xx} and σ_{yx} are almost purely real, the axial ratio is very high, indicating a linearly polarized wave. As frequency increases, the axial ratio decreases and it becomes minimum close to the cyclotron resonance, where

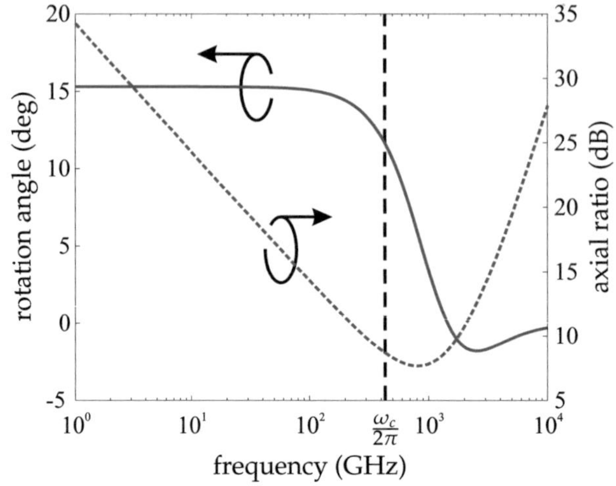

Fig. 12. Rotation angle, computed by Equation 34, and axial ratio, computed from Equation 25 and Balanis (2005), versus frequency for a normally incident plane wave on a graphene sheet in free space, for the parameters $\tau = 1.84 \times 10^{-13}$ s, $\mu_c = 0.37$ eV, $B_0 = 1$ T, and $T = 300$ K.

there is a strong interaction between the material and the circular polarization of the same handedness as the cyclotron motion of electrons. For a further increase of frequency this "resonance" between the circularly polarized waves and the electrons diminishes gradually and the axial ratio increases.

Fig. 12 also shows that Faraday rotation in graphene is an extremely broadband phenomenon below the resonance (in the case of Fig. 12 from 0 to 200 GHz), as a result of the non-dispersive characteristics of conductivity below the resonance (Fig. 7). It has to be stressed that this frequency-independent behavior is different from the corresponding found in ferrites above the ferromagnetic resonance (Lax & Button, 1962), since the former is due to the non-dispersive characteristics of the material while the latter is due to the increase of the electrical length with frequency.

The direction of Faraday rotation in graphene, as in any magneto-optical material, does not depend on the propagation direction but on the direction of the applied static magnetic field. This can be directly seen from $\bar{\bar{T}}_{21} = \bar{\bar{T}}_{12}$ (Equation 27), which states that for an incident wave of a given polarization the polarization of the transmitted wave is always the same, irrespectively of the propagation direction. Thus, graphene is a non-reciprocal material. In order to demonstrate this statement, consider a plane wave impinging on graphene from medium 1 to medium 2 and then reflected back to medium 1 without changing its polarization (e.g. reflection by a PMC wall after graphene), as illustrated in Fig. 13. As the wave passes through graphene from medium 1 towards medium 2, its polarization is rotated by an angle θ in the LH direction with respect to the $+z$ direction (the biasing magnetic field direction). As the wave then passes again through graphene from medium 2 back to medium 1, its polarization is again rotated by an angle θ in the LH direction with respect to the $+z$ axis. Therefore, the polarization of the wave transmitted from medium 1 to medium 2 and then back from medium 2 to medium 1 has undergone an overall rotation of 2θ, which is a manifestation of non-reciprocity.

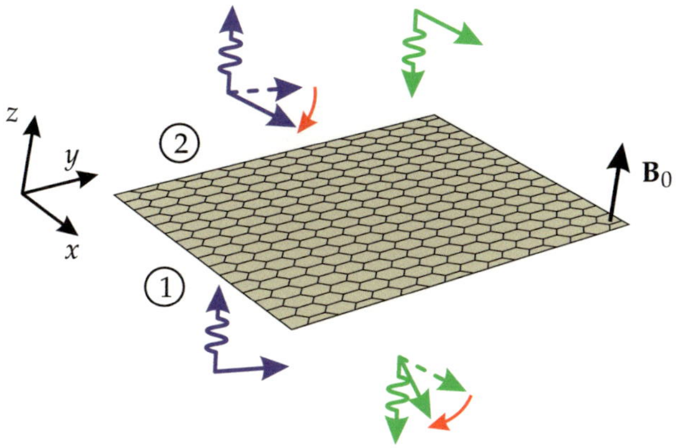

Fig. 13. Non-reciprocity in graphene. A wave normally impinging on graphene from medium 1 to medium 2 and transmitted back to medium 1 undergoes a total rotation angle of 2θ, where θ is the rotation angle for a single pass through graphene.

5. Applications

In this section, two applications of the Faraday rotation effect in graphene will be presented, namely a circular waveguide polarization rotator and a non-reciprocal spatial isolator. Although such devices can be realized with conventional gyrotropic media, such as ferrites, graphene offers increased tunability and electric rotation reversal, and possibly other benefits (such as high heat sink) still to be investigated. The purpose of the section is to show the feasibility of using graphene in such types of devices and not to provide an optimized design. Therefore, the dimensions of the proposed devices are comparable to their ferrite counterparts. However, we believe that by a proper design, e.g. by using a stacked structure composed of alternating graphene sheets and dielectric layers, highly compact devices, with dimensions much smaller than the dimensions of the ferrite devices, can be achieved.

5.1 Circular waveguide faraday rotator

The proposed circular waveguide Faraday rotator consists of a graphene sheet loading the cross section of a circular waveguide, as shown in Fig. 14. The graphene sheet is perpendicular to the waveguide axis and biased with an axial static magnetic field $\mathbf{B}_0$. The waveguide has a radius a and it is excited in its dominant H_{11} (TE_{11}) mode. The incident mode is assumed to be polarized along the x axis and propagates along the $+z$ direction.

The transverse electric field of the mode is

$$\mathbf{E}_{T,11}^{\mathrm{h,inc}}(\rho,\varphi) = \hat{\mathbf{z}} \times \nabla_t \times \psi_{11}(\rho,\varphi), \tag{37}$$

where ∇_t is the transverse (on the $\rho - \varphi$ plane) del operator and $\psi_{11}(\rho,\varphi) = J_1(k_{c,11}^h \rho)\cos\varphi$, with $k_{c,11}^h = x'_{11}/a$ and x'_{11} the first root of $J'_1(x)$. Using $\cos\varphi = (e^{j\varphi} + e^{-j\varphi})/2$, the incident mode can be analyzed into two degenerate orthogonal modes, a LH circularly polarized mode H_{11}^-

$$\mathbf{E}_{T,11}^{h-} = \hat{\mathbf{z}} \times \nabla_t \times \psi_{11}^- \tag{38a}$$

and a RH circularly polarized one H_{11}^+

$$\mathbf{E}_{T,11}^{h+} = \hat{\mathbf{z}} \times \nabla_t \times \psi_{11}^+, \tag{38b}$$

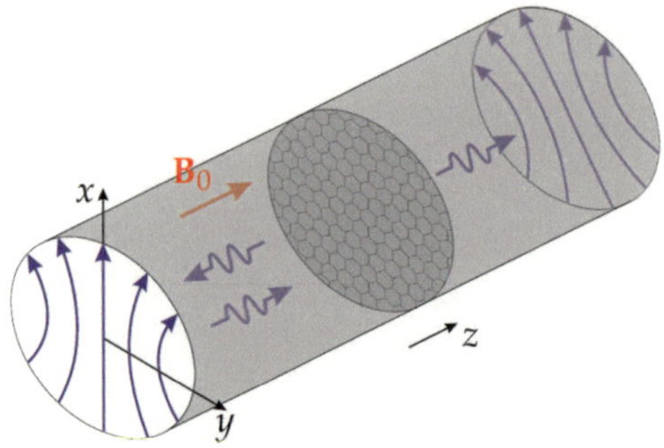

Fig. 14. Circular cylindrical waveguide of radius a loaded by a graphene sheet perpendicular to its axis. The sheet is placed at the $z = 0$ plane and biased by a perpendicular static magnetic field B_0. The waveguide is excited in its dominant H_{11} (TE_{11}) mode.

both with amplitude equal to 1/2. In Equation 38 $\psi_{11}^{\pm} = J_1(k_{c,11}^{h}\rho)e^{\mp j\varphi}$. Since any pair of modes with $e^{-j\varphi}$ and $e^{j\varphi}$ azimuthal dependencies are orthogonal to each other and because of the phase matching condition in the graphene plane, any of these modes creates reflected and transmitted modes with the same azimuthal dependence. Therefore, the H_{11}^{-} mode will excite the H_{1n}^{-} and E_{1n}^{-} modes, while the H_{11}^{+} mode will excite the H_{1n}^{+} ans E_{1n}^{+} modes. Next, we focus on the H_{11}^{-} mode. The analysis for the H_{11}^{+} is completely similar.

The transverse reflected and transmitted fields in the graphene plane ($z = 0$ plane) read

$$\mathbf{E}_T^{-\text{ref}} = \sum_{n=1}^{\infty} \left(R_{1n}^{h-} \mathbf{E}_{T,1n}^{h-} + R_{1n}^{e-} \mathbf{E}_{T,1n}^{e-} \right), \tag{39a}$$

$$\mathbf{E}_T^{-\text{tran}} = \sum_{n=1}^{\infty} \left(T_{1n}^{h-} \mathbf{E}_{T,1n}^{h-} + T_{1n}^{e-} \mathbf{E}_{T,1n}^{e-} \right), \tag{39b}$$

respectively. For the computation of the coefficients R_{1n}^{h-}, R_{1n}^{e-}, T_{1n}^{h-} and T_{1n}^{e-}, the boundary conditions

$$\hat{\mathbf{z}} \times \left(\mathbf{E}_T^{\text{tran}} - \mathbf{E}_T^{\text{inc}} - \mathbf{E}_T^{\text{ref}} \right) \Big|_{z=0} = 0, \tag{40a}$$

$$\hat{\mathbf{z}} \times \left(\mathbf{H}_T^{\text{tran}} - \mathbf{H}_T^{\text{inc}} - \mathbf{H}_T^{\text{ref}} \right) \Big|_{z=0} = \bar{\bar{\sigma}} \cdot \mathbf{E}_T^{\text{tran}} \Big|_{z=0}, \tag{40b}$$

should be used. The reflected and transmitted magnetic fields in the graphene plane are given by

$$\mathbf{H}_T^{-\text{ref}} = -\sum_{n=1}^{\infty} \left(R_{1n}^{h-} \frac{\hat{\mathbf{z}} \times \mathbf{E}_{T,1n}^{h-}}{Z_{1n}^{h}} + R_{1n}^{e-} \frac{\hat{\mathbf{z}} \times \mathbf{E}_{T,1n}^{e-}}{Z_{1n}^{e}} \right), \tag{41a}$$

$$\mathbf{E}_T^{-\text{tran}} = \sum_{n=1}^{\infty} \left(T_{1n}^{h-} \frac{\hat{\mathbf{z}} \times \mathbf{E}_{T,1n}^{h-}}{Z_{1n}^{h}} + T_{1n}^{e-} \frac{\hat{\mathbf{z}} \times \mathbf{E}_{T,1n}^{e-}}{Z_{1n}^{e}} \right), \tag{41b}$$

where $Z_{1n}^{h} = \eta_0(k_0/\beta_{1n}^{h})$ and $Z_{1n}^{e} = \eta_0(\beta_{1n}^{e}/k_0)$ are the wave impedances of the H_{1n} and E_{1n} modes, respectively, and β_{1n}^{h} and β_{1n}^{e} are the corresponding propagation numbers.

Inserting Equations 39 into 40a and using the orthogonality conditions (Collin, 1990)

$$\int_S \mathbf{E}_{T,1n}^{h-} \cdot \mathbf{E}_{T,1m}^{h+} = \int_S \mathbf{E}_{T,1n}^{e-} \cdot \mathbf{E}_{T,1m}^{e+} = 0, \quad n \neq m \tag{42}$$

$$\int_S \mathbf{E}_{T,1n}^{h-} \cdot \mathbf{E}_{T,1m}^{e+} = \int_S \mathbf{E}_{T,1n}^{e-} \cdot \mathbf{E}_{T,1m}^{h+} = 0, \tag{43}$$

with S the waveguide cross section, one obtains

$$R_{1n}^{h-} + \delta_{1n} = T_{1n}^{h-}, \tag{44a}$$

$$R_{1n}^{e-} = T_{1n}^{e-}. \tag{44b}$$

Similarly, introducing Equation 41 into 40b yields

$$-T_{1n}^{h-} + \delta_{1n} - R_{1n}^{h-} = Z_{1n}^{h}\sigma_{xx}T_{1n}^{h-} + Z_{1n}^{h}\sigma_{yx}\sum_{m=0}^{\infty}\left(J_{nm}^{hh}T_{1m}^{h-} + J_{nm}^{he}T_{1m}^{e-}\right), \tag{45a}$$

$$-T_{1n}^{e-} - R_{1n}^{e-} = Z_{1n}^{e}\sigma_{xx}T_{1n}^{e-} + Z_{1n}^{e}\sigma_{yx}\sum_{m=0}^{\infty}\left(J_{nm}^{eh}T_{1m}^{h-} + J_{nm}^{ee}T_{1m}^{e-}\right), \tag{45b}$$

where

$$J_{nm}^{hh} = \frac{\int_S \mathbf{E}_{1n}^{h+} \cdot \left(\hat{\mathbf{z}} \times \mathbf{E}_{1m}^{h-}\right) dS}{\int_S \mathbf{E}_{1n}^{h+} \cdot \mathbf{E}_{1n}^{h-} dS}, \qquad J_{nm}^{he} = \frac{\int_S \mathbf{E}_{1n}^{h+} \cdot \left(\hat{\mathbf{z}} \times \mathbf{E}_{1m}^{e-}\right) dS}{\int_S \mathbf{E}_{1n}^{h+} \cdot \mathbf{E}_{1n}^{h-} dS}, \tag{46a}$$

$$J_{nm}^{eh} = \frac{\int_S \mathbf{E}_{1n}^{e+} \cdot \left(\hat{\mathbf{z}} \times \mathbf{E}_{1m}^{h-}\right) dS}{\int_S \mathbf{E}_{1n}^{e+} \cdot \mathbf{E}_{1n}^{e-} dS}, \qquad J_{nm}^{ee} = \frac{\int_S \mathbf{E}_{1n}^{e+} \cdot \left(\hat{\mathbf{z}} \times \mathbf{E}_{1m}^{e-}\right) dS}{\int_S \mathbf{E}_{1n}^{e+} \cdot \mathbf{E}_{1n}^{e-} dS}. \tag{46b}$$

In the derivation of Equation 45 the identity $\overline{\overline{J}}_t \cdot \mathbf{v} = \hat{\mathbf{z}} \times \mathbf{v}$, which holds for any vector $\mathbf{v}$, has been used. Omitting the proof, which can be found in (Sounas & Caloz, 2011d), we directly give the final expressions of the J coefficients of Equations 46

$$J_{nm}^{hh} = -\frac{j2}{\sqrt{\left[(x'_{1n})^2 - 1\right]\left[(x'_{1m})^2 - 1\right]}}, \qquad J_{nm}^{he} = \frac{2(x'_{1n})^2}{\left[(x'_{1n})^2 - x_{1m}^2\right]\sqrt{(x'_{1n})^2 - 1}}, \tag{47a}$$

$$J_{nm}^{eh} = -\left(J_{mn}^{he}\right)^*, \qquad J_{nm}^{ee} = 0, \tag{47b}$$

where x_{1n} and x'_{1n} are the nth roots of $J_1(x)$ and $J'_1(x)$, respectively. Solving Equations 44 with respect to $R_{1n}^{h/e-}$ and introducing the result into Equations 45 we get

$$\left(2 + Z_{1n}^{h}\sigma_{xx}\right)T_{1n}^{h-} + Z_{1n}^{h}\sigma_{yx}\sum_{m=0}^{\infty}\left(J_{nm}^{hh}T_{1m}^{h-} + J_{nm}^{he}T_{1m}^{e-}\right) = 2\delta_{1n}, \tag{48a}$$

$$\left(2 + Z_{1n}^{e}\sigma_{xx}\right)T_{1n}^{e-} + Z_{1n}^{e}\sigma_{yx}\sum_{m=0}^{\infty}\left(J_{nm}^{eh}T_{1m}^{h-} + J_{nm}^{ee}T_{1m}^{e-}\right) = 0. \tag{48b}$$

Equations 48 constitute a linear system of equations, the solution of which gives $T_{1n}^{h/e-}$. One can calculate $T_{1n}^{h/e+}$ in a similar manner.

If the operation frequency is between the cutoff frequencies of the dominant mode and the first higher order mode, only the dominant mode exists far away from the graphene discontinuity, since all the higher order modes are evanescent. Then, similar to the case of plane wave incidence on graphene in free space, the rotation angle of the transmitted mode is

$$\theta = \frac{\varphi^{-} - \varphi^{+}}{2}, \tag{49}$$

where here $\varphi^{\pm} = \arg\{T_{11}^{h\pm}\}$.

An approximate closed-form expression for θ can be obtained by considering only the dominant mode terms in Equations 48. The solution of Equations 48 reads then

$$T_{11}^{h\pm} = \frac{2}{(2 + Z_{11}^{h}\sigma_{xx}) \pm j\alpha Z_{11}^{h}\sigma_{yx}}, \tag{50}$$

with $\alpha = 2/[(x_{11}')^2 - 1]$. Subsequently, the rotation angle is found as

$$\theta = \tan^{-1}\left(\frac{\alpha Z_{11}^{h}\sigma_{yx}}{2 + Z_{11}^{h}\sigma_{xx}}\right). \tag{51}$$

The frequency dependence of the rotation angle and the transmission amplitude, computed exactly via Equation 49 and approximately via Equation 51, are presented in Fig. 15 for a waveguide with $a = 8$ cm and a graphene sheet with $\tau = 1.84 \times 10^{-13}$ s, $\mu_c = 0.3$ eV and $B_0 = 1$ T. The rotation angle and the transmission amplitude for normal incidence on graphene in free space, computed through Equation 36, are also plotted for comparison. Very good agreement between the exact (Equation 49) and the approximate (Equation 51) formulas is observed.

Fig. 15 shows that θ decreases as frequency increases. This may be understood from the frequency behavior of Z_{11}^{h}, which tends to ∞ at the cutoff of the H_{11} mode and increases towards η_0 as frequency increases. At the H_{11} mode cutoff frequency, θ takes its maximum value

$$\theta = \tan^{-1}\left(\alpha \frac{\sigma_{yx}}{\sigma_{xx}}\right). \tag{52}$$

However, the transmission amplitude close to cutoff is very small. On the other hand, one would expect that at $\omega \to \infty$, when the H_{11} mode propagates almost parallel to the waveguide axis and $Z_{11}^{h} \to \eta_0$, θ would tend to its free space value (Equation 36), whereas, as seen from Equation 51, it tends to a smaller value[5]. A physical explanation can be given as follows. As the H_{11} wave impinges on graphene, it produces a transverse current component, which is responsible for the polarization rotation. However, this current does not perfectly match the transverse pattern of the mode and therefore only partially contributes to the transmitted field amplitude, thus providing a smaller rotation angle than in free space. This mismatch is also responsible for the excitation of the higher order modes in the waveguide environment.

[5] $\alpha < 1$.

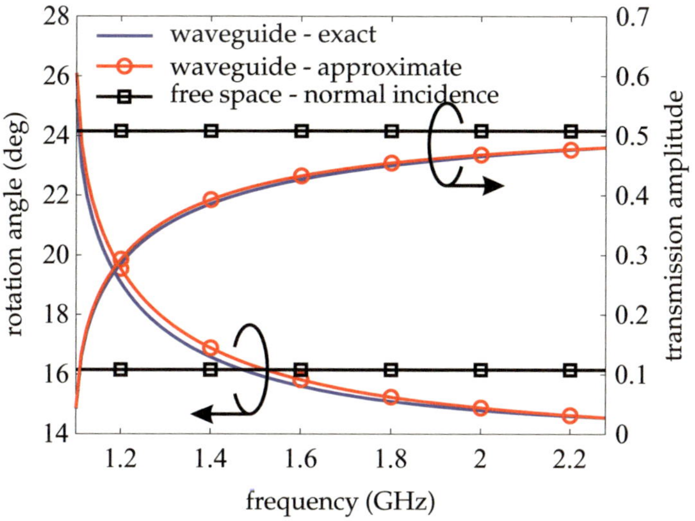

Fig. 15. Frequency dependence of the rotation angle and the transmission amplitude for a waveguide, computed exactly by Equations 48 and 49 and approximately by Equations 50 and 51, for a waveguide with $a = 8$ cm and a graphene sheet with $\tau = 1.84 \times 10^{-13}$ s, $\mu_c = 0.3$ eV, $B_0 = 1$ T and $T = 300$ K.

5.2 Spatial isolator

As a second application of the gyrotropic properties of graphene, we propose the spatial isolator of Fig. 16 (Sounas & Caloz, 2011c). It consists of a magnetically biased graphene sheet in the middle of two wire grids, which are rotated by an angle of 45° with respect to each other. The structure allows propagation of a y-polarized wave from medium 1 towards medium 2, while it blocks any wave propagating from medium 2 towards medium 1.

The operation principle of the structure is the same as of the ferrite-based non-reciprocal spatial isolator presented in (Parsa et al., 2010). A y-polarized wave propagating along the $-z$ direction passes through the wire grid at $z = d_1$, it is then rotated by 45° in the LH direction with respect to the z-axis due to graphene and it eventually passes through the wire grid at $z = d_2$. On the other hand, a wave propagating along the $+z$ direction is directly blocked by the grid at $z = d_2$ if it is polarized parallel to the wires of this grid. If it is perpendicularly polarized to the wires of this grid, it passes through this grid, it is then rotated by 45° in the LH direction as it passes through graphene and it is eventually blocked by the grid at $z = d_1$.

The structure can be analyzed by using the transmission line model of Fig. 17. In this model graphene and the wire grids are represented by shunt dyadic admittances and the space between graphene and the grids by transmission line sections with wave impedance η_0 and propagation number k_0. The shunt admittance which models the graphene sheet is simply the tensorial conductivity of graphene (Equation 13). The shunt admittance which models the grid at $z = d_1$ is

$$\bar{\bar{Y}}_1 = Y_{g\parallel}\hat{\mathbf{x}}\hat{\mathbf{x}} + Y_{g\perp}\hat{\mathbf{y}}\hat{\mathbf{y}}, \tag{53}$$

where

$$Y_{g\parallel} = j\eta_0^{-1}\frac{\lambda}{d}\frac{1}{\ln\left[\sin\left(\frac{\pi w}{2d}\right)\right]} \tag{54a}$$

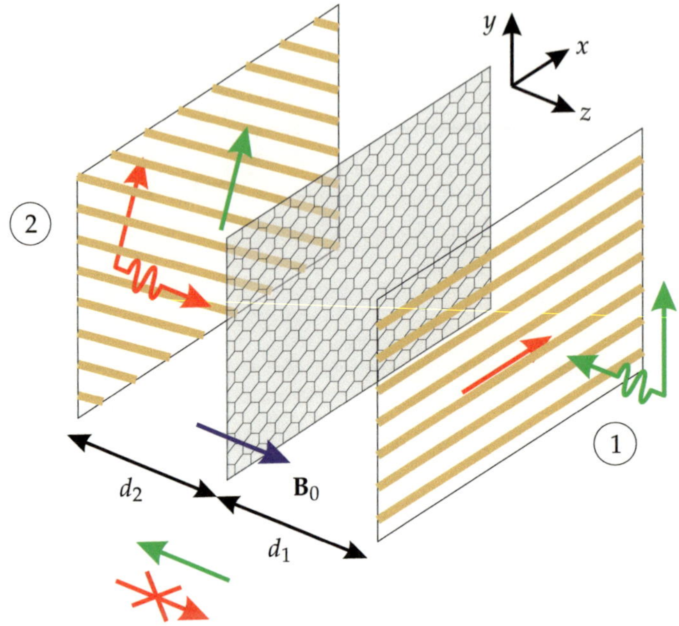

Fig. 16. Graphene-based non-reciprocal spatial isolator, consisting of a graphene sheet and two wire grids, one with wires parallel to the x-axis and the other with wires tilted by $-45°$ with respect to the x-axis. A y-polarized wave propagating along the $-z$-direction passes through the isolator, while any wave propagating along the z-direction does not.

and

$$Y_{g\perp} = -j\eta_0^{-1}\frac{4d}{\lambda}\ln\left\{\sin\left[\frac{\pi(d-w)}{2d}\right]\right\} \tag{54b}$$

are the parallel and the perpendicular to the wires admittances of the grid, respectively (Tretyakov, 2003). In Equation 54 w is the wire width and d is the periodicity of the grid. The shunt admittance of the grid at $z = d_2$ is

$$\bar{\bar{Y}}_2 = Y_{g\|}\hat{\mathbf{t}}\hat{\mathbf{t}} + Y_{g\perp}(\hat{\mathbf{z}}\times\hat{\mathbf{t}})(\hat{\mathbf{z}}\times\hat{\mathbf{t}}), \tag{55}$$

where $\hat{\mathbf{t}} = (\hat{\mathbf{x}}+\hat{\mathbf{y}})/\sqrt{2}$ is the unit vector parallel to the wires.

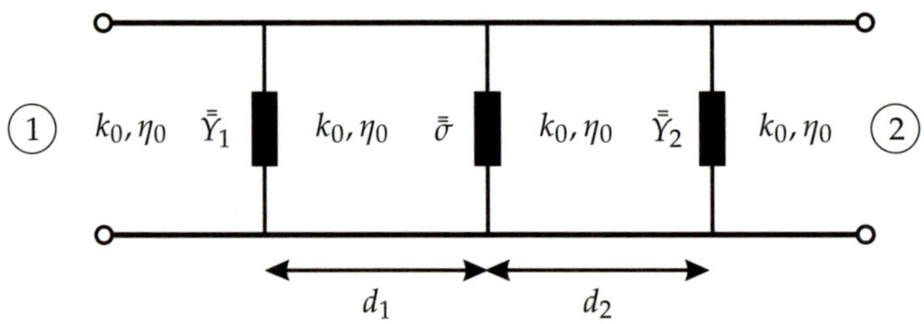

Fig. 17. Transmission line model of the proposed graphene-based non-reciprocal spatial isolator.

The ABCD matrix of the structure, through which the transmission coefficients will be computed, is

$$\begin{bmatrix} \bar{\bar{A}} & \bar{\bar{B}} \\ \bar{\bar{C}} & \bar{\bar{D}} \end{bmatrix} = \begin{bmatrix} \bar{\bar{I}} & \bar{\bar{0}} \\ \bar{\bar{Y}}_1 & \bar{\bar{I}} \end{bmatrix} \cdot \begin{bmatrix} \cos(k_0 d_1) & j\eta_0 \sin(k_0 d_1) \\ j\eta_0^{-1} \sin(k_0 d_1) & \cos(k_0 d_1) \end{bmatrix} \cdot \begin{bmatrix} \bar{\bar{I}} & \bar{\bar{0}} \\ \bar{\bar{\sigma}}_1 & \bar{\bar{I}} \end{bmatrix}$$
$$\cdot \begin{bmatrix} \cos(k_0 d_2) & j\eta_0 \sin(k_0 d_2) \\ j\eta_0^{-1} \sin(k_0 d_2) & \cos(k_0 d_2) \end{bmatrix} \cdot \begin{bmatrix} \bar{\bar{I}} & \bar{\bar{0}} \\ \bar{\bar{Y}}_2 & \bar{\bar{I}} \end{bmatrix}. \quad (56)$$

The S-matrix is then

$$\begin{bmatrix} \bar{\bar{S}} \end{bmatrix} = \begin{bmatrix} \bar{\bar{I}} & -(\bar{\bar{A}} + \eta_0 \bar{\bar{B}}) \\ -\eta_0 \bar{\bar{I}} & -(\bar{\bar{C}} + \eta_0 \bar{\bar{D}}) \end{bmatrix}^{-1} \cdot \begin{bmatrix} -\bar{\bar{I}} & \bar{\bar{A}} - \eta_0 \bar{\bar{B}} \\ -\eta_0 \bar{\bar{I}} & \bar{\bar{C}} - \eta_0 \bar{\bar{D}} \end{bmatrix} \quad (57)$$

and the isolation may be subsequently computed as

$$I = 20 \log_{10} \left| \bar{\bar{S}}_{21} \cdot \hat{\mathbf{y}} \right| - 20 \log_{10} \left| \bar{\bar{S}}_{12} \cdot \frac{\hat{\mathbf{x}} + \hat{\mathbf{y}}}{\sqrt{2}} \right|. \quad (58)$$

Fig. 18 presents isolation in dB (color plot), the transmission loss from medium 1 to medium 2 in dB (continuous lines contour plot) and the rotation angle of graphene in free space (dashed lines contour plot) versus μ_c and B_0 for $\tau = 1.84 \times 10^{-13}$ s. The results are derived for $d_1 = d_1 = \lambda/4$, which has been found to be the condition of the highest isolation. As an example of how to read Fig. 18, the point indicated by the circle corresponds to $B_0 = 0.61$ T, $\mu_c = 0.11$ eV, $I = 20$ dB, a 3 dB transmission loss from medium 1 towards medium 2 and 9° of rotation angle for graphene in free space. One observes that the isolation can be so high as 20 dB even when the rotation angle of graphene in free space is so low as 10°. This contrasts with the ferrite-based isolator, where a 45° rotation angle is always required. This surprising behavior is attributed to the semi-transparent nature of graphene which forces the wave to follow a complex path between the grids, passing several times through graphene. The condition for which this complex path leads to constructive interference for propagation along the $+z$ direction and destructive interference for propagation along the $-z$ direction is $d_1 = d_2 = \lambda/4$. The main reason for the relatively high loss (3 dB) for transmission along the $+z$ direction is the graphene loss, which is increased by the large effective wave path in the space between the grids. This loss can be reduced by using graphene with higher mobility (lower scattering) and properly adjusting the structure parameters.

One of the most significant advantages of the proposed isolator is the control of the isolation direction via the chemical potential, exploiting the ambipolar properties of graphene. In particular, by inverting the chemical potential, the direction of Faraday rotation provided by graphene also inverts, which in turn inverts the isolation direction. Remember that the chemical potential can be easily controlled via an electrostatic voltage applied between graphene and an electrode parallel to graphene (Fig. 2). In such a way it could be possible to fabricate a non-reciprocal antenna radome, which dynamically converts an antenna from transmitting to receiving and vice versa through a proper control signal.

6. Enabling of graphene gyrotropy with metamaterials

As already pointed out, the structures presented in Section 5 are essentially "proofs of concept" and do not follow an optimal design. For example, a practical circular waveguide

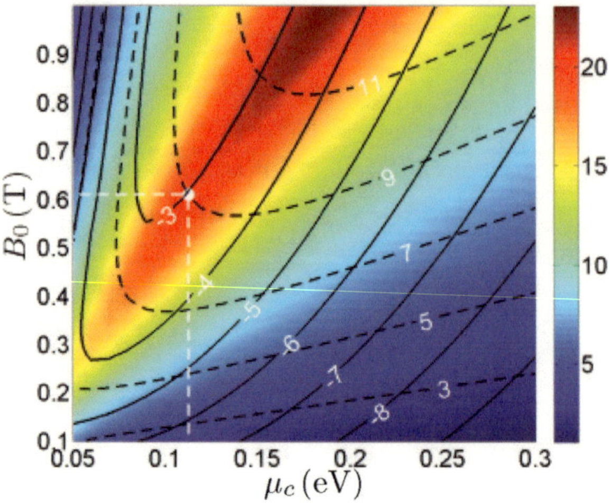

Fig. 18. Color plot: isolation in dB achieved by the proposed graphene-based non-reciprocal spatial isolator versus μ_c and B_0 for $\tau = 1.84 \times 10^{-13}$ s. Solid lines contour plot: power transmission coefficient in dB for the wave allowed to pass through the isolator. Dashed lines contour plot: polarization rotation angle of the graphene sheet in free space.

Faraday rotator, which is a necessary component in a waveguide isolator or a gyrator, requires a rotation angle of at least 90°. Although the rotation angle provided by a single graphene sheet is huge considering its one-atom thickness, it is quite far from the value required in a waveguide isolator. In order to achieve a rotation angle so high as 90° it is necessary to stack several graphene sheets. The material and the spacing between the sheets as well as the rotation angle provided by each sheet must be carefully chosen so that the transmission through such a structure be maximum. In such a way, a microscale metamaterial, similar to this illustrated in Fig. 19(a), can be conceived. Such a metamaterial provides a gradual rotation of the wave polarization, thus mimicking a ferrite medium with, however, a much larger

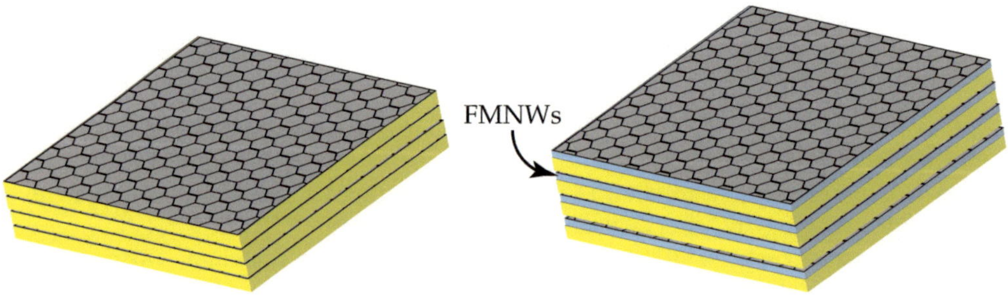

(a) (b)

Fig. 19. Graphene-based multiscale metamaterials, which exhibit a ferrite-type response, however, with a higher rotation power than ferrites. (a) Micro-scale metamaterial consisting of a period stack of graphene sheets and dielectric layers. (b) Multiscale metamaterial consisting of a periodic stack of graphene sheets, ferromagnetic nanowires (FMNW) membranes and dielectric layers. FMNW metamaterial membranes are used for biasing graphene.

Verdet constant (rotation per unit thickness and unit magnetic field). Such an artificial material could be also used in the spatial isolator to reduce significantly its size. In a next nanoscale level, ferromagnetic nanowires membranes (Carignan et al., 2011), which are transparent to normally incident waves, could be introduced between the successive graphene layers, as illustrated in Fig. 19(b) to provide the necessary static magnetic field, making the material self-biased. Finally, boron nitride (BN) could be selected as the separating dielectric between the graphene sheets, since as it has been recently shown, graphene sandwiched between BN layers exhibits extremely high mobility, sometimes approaching that of suspended graphene (Mayorov et al., 2011). It is therefore clear that by exploiting metamaterials concept in the micro, nano and atomic scale simultaneously (Caloz et al., n.d.), novel material with unprecedented properties can be created.

7. Conclusions

The gyrotropic properties of magnetically biased graphene have been analyzed. Graphene exhibits strong non-reciprocity and extremely broadband Faraday rotation at microwave frequencies, below the cyclotron resonance. Two degrees of freedom may be used to control the amount and the direction of rotation, namely the chemical potential (via an applied static voltage) and the static magnetic field. This allows full tuning of Faraday rotation via the chemical potential, while keeping the magnetic field constant, which is usually provided by permanent magnets and is thus difficult to control. Two applications of graphene gyrotropy have been proposed: a circular waveguide Faraday rotator and a spatial isolator. A practical realization of such devices with unprecedented characteristics, such as super compactness, is possible through the multi-scale (micro, nano and atomic scales) metamaterial concept.

8. References

Abramowitz, M. & Stegun, I. A. (1964). *Handbook of mathematical functions with formulas, graphs, and mathematical tables*, Vol. 55, Courier Dover publications.

Balanis, C. A. (2005). *Antenna theory: analysis and design*, John Wiley & Sons.

Caloz, C., Carignan, L.-P., Ménard, D. & Yelon, A. (n.d.). The concept of multi-scale metamaterials.

Caloz, C. & Itoh, T. (2006). *Electromagnetic Metamaterials: Transmission Line Theory, and Microwave Applications: The Engineering Approach*, John Wiley & Sons.

Carignan, L.-P., Yelon, A., Ménard, D. & Caloz, C. (2011). Ferromagnetic nanowire metamaterials: theory and applications, *IEEE Trans. Microwave Theory Tech.* 59(10): 2568–2586.

Collin, R. E. (1990). *Field theory of guided waves*, IEEE Press.

Crassee, I., Levallois, J., Walter, A. L., Ostler, M., Bostwick, A., Rotenberg, E., Seyller, T., van der Marel, D. & Kuzmenko, A. B. (2011). Giant faraday rotation in single- and multilayer graphene, *Nature Phys.* 7: 48–51.

Dragoman, M. & Dragoman, D. (2009). *Nanoelectronics: principles and devices*, Artech House.

Geim, A. K. & Novoselov, K. S. (2007). The rise of graphene, *Nature Materials* 6: 183–191.

Gusynin, V. P., Sharapov, S. G. & Carbotte, J. P. (2009). On the universal ac optical background in graphene, *New J. Phys.* 11: 095013.

Hanson, G. W. (2007). *Fundamentals of Nanoelectronics*, Prentice Hall.

Koppens, F. H. L., Chang, D. E. & García de Abajo, F. J. (2011). Graphene plasmonics: A platform for strong light–matter interactions, *Nano Letters* 11(8): 3370–3377.

Kubo, R. (1957). Statistical-mechanical theory of irreversible processes. I. general theory and simple applications to magnetic and conduction problems, *J. Phys. Soc. Japan* 12(6): 570–586.

Lax, B. & Button, K. J. (1962). *Microwave Ferrites and Ferrimagnetics*, McGraw-Hill.

Lin, Y.-M., Jenkins, K. A., Valdes-Garcia, A., Small, J. P., Farmer, D. B. & Avouris, P. (2009). Operation of graphene transistors at gigahertz frequencies, *Nano Lett.* 9: 422–426.

Lindell, I. V. (1996). *Methods for electromagnetic field analysis*, IEEE Press.

Marqués, R., Martín, F. & Sorolla, M. (2008). *Metamaterials with Negative Parameters: Theory, Design and Microwave Applications*, John Wiley & Sons.

Mayorov, A. S., Gorbachev, R. V., Morozov, S. V., Britnell, L., Jalil, R., Ponomarenko, L. A., Blake, P., Novoselov, K. S., Watanabe, K., Taniguchi, T. & Geim, A. K. (2011). Micrometer-scale ballistic transport in encapsulated graphene at room temperature, *Nano Lett.* 11(6): 2396–2399.

Mikhailov, S. A. & Ziegler, K. (2007). New electromagnetic mode in graphene, *Phys. Rev. Lett.* 99: 016803.

Mishchenko, E. G., Shytov, A. V. & Silvestrov, P. G. (2010). Guided plasmons in graphene *p-n* junctions, *Phys. Rev. Lett.* 104: 156806.

Neto, A. H. C., Guinea, F., Peres, N. M. R., Novoselov, K. S. & Geim, A. K. (2009). The electronic properties of graphene, *Rev. Mod. Phys.* 81: 109–162.

Novoselov, K. S., Geim, A. K., Morozov, S. V., Jiang, D., Zhang, Y., Dubonos, S. V., Grigorieva, I. V. & Firsov, A. A. (2004). Electric field effect in atomically thin carbon films, *Science* 306: 666–669.

Parsa, A., Kodera, T. & Caloz, C. (2010). Ferrite based non-reciprocal radome, generalized scattering matrix analysis and experimental demonstration, *IEEE Trans. Antennas Propag.* . in press.

Sounas, D. L. & Caloz, C. (2011a). Edge surface modes in magnetically-biased chemically-doped graphene strips, *App. Plys. Lett.* . under review.

Sounas, D. L. & Caloz, C. (2011b). Electromagnetic non-reciprocity and gyrotropy of graphene, *Appl. Phys. Lett.* 98: 021911.

Sounas, D. L. & Caloz, C. (2011c). Graphene-based non-reciprocal spatial isolator, *Proc. IEEE International Symposium on Antennas and Propagation, Spokane, WA, (APS 2011)*.

Sounas, D. L. & Caloz, C. (2011d). Gyrotropy and non-reciprocity of graphene for microwave applications, *IEEE Trans. Microw. Theory Tech.* . under review.

Tretyakov, S. (2003). *Analytical modelling in applied electromagnetics*, Artech House, Inc.

Vakil, A. & Engheta, N. (2011). Transformation optics using Graphene, *Science* 332(6035): 1291–1294.

Wang, H., Nezich, D., Kong, J. & Palacios, T. (2009). Graphene frequency multipliers, *IEEE Electron Device Lett.* 30: 547–549.

2

Electrodynamical Analysis of Open Lossy Metamaterial Waveguide and Scattering Structures

L. Nickelson, S. Asmontas, T. Gric, J. Bucinskas and A. Bubnelis
State Research Institute Center for Physical Sciences and Technology, Vilnius, Lithuania

1. Introduction

Large stream of articles devoted to the study of metamaterial waveguide and metamaterial scattering (reflecting) structures points that there is a need for development devices possessing unique characteristics, as multifunctionality, reconfigurability, certain frequency bandwidth, ability to operate at high-powers and high-radiation conditions. The importance of diffraction problems for scattering structures is based on their great practical utility for many applications, such as reflector antennas, the analysis of structures in open space, electromagnetic (EM) defence of structures, the scattering modeling for remote sensing purposes, high frequency telecommunications, computer network, invisibility cloaks technology and radar systems (Li et.al., 2011; Zhou et.al.; 2011, Zhu et al., 2010; Mirza et al., 2009; Abdalla & Hu, 2009; Engheta & Ziolkowski, 2005).

The technological potential of metamaterials for developing novel devices offers a very promising alternative that could potentially overcome the limitations of current technology. The metamaterial waveguide and scattering structures can operate as different devices that possess different specific qualities as well. In order to create a new microwave device it is necessary to know the main electrodynamical characteristics of metamaterial structures on the basis of which the device is supposed to be created.

Here are presented electrical field distributions and dispersion characteristics of open metamaterial waveguides in subsections 1-3 and numerical analysis of the scattered and absorbed microwave powers of the layered metamaterial cylinders in the subsection 4.

2. Analyses of electrical field and dispersion characteristics of square metamaterial lossy waveguides by the SIE method

Here the open (without conductor screen) square metamaterial waveguides with sizes 5x5 mm^2 and 4x4 · mm^2 are investigated by our algorithms that were created on the base of the Singular Integral equations' (SIE) method (Nickelson & Sugurov, 2005; Nickelson et al., InTech2011). Due to the fact that metamaterial is a substance with losses we have determined the complex roots of the dispersion equation by using of the Muller method in our researches (Nickelson et al., InTech2011).

Our computer programmes are written in MATLAB. They let us to investigate open absorptive waveguides with different shapes of cross-section (Gric & Nickelson, 2011). We have used the values of the complex relative permittivity ε_{met} and the complex relative permeability μ_{met} of metamaterial from (Penciu et al, 2006) in suctions 1 and 2.

2.1 Numerical investigations of square 5x5 mm^2 metamaterial waveguide

The dispersion characteristics and the 3D electric field distributions are presented here (Figs 1.1–1.3). The characteristics are shown in Fig. 2.1 (Gric et al., 2010). The main wave (mode) is denoted with black points; the first higher mode is denoted with circles. The values of ε_{met} and μ_{met} are different at every frequency. The real part of the permittivity is always negative at the all frequency range 75-115 GHz. The imaginary part of the permittivity is negative approximately when $90 \leq f \leq 100$ GHz. The real part of the permeability is negative when $100 \leq f \leq 105$ GHz. The imaginary part of the permeability is always equal to zero or a positive number at the all mentioned frequency range. In the Fig. 2.1 are presented dependencies of the complex propagation constant h = h′- i h″ on the frequency.

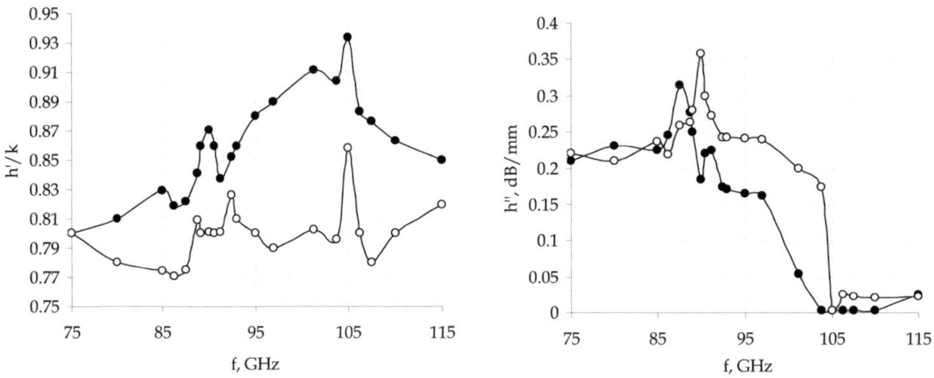

(a) (b)

Fig. 2.1. Dispersion characteristics of square 5x5 mm^2 metamaterial waveguide
(a) – the dependence of the normalized propagation constant, (b) – the dependence of the attenuation constant on frequencies.

In Fig. 2.1(a) we see that dependencies of the normalized propagation constant h'/k on the frequency, where $h'=2\pi/\lambda$ and λ is the wavelength of microwave in the metamaterial waveguide, $k=2\pi f/c$, k is the wavenumber in vacuum, f is an operating frequency, c is the speed of light in vacuum. The curves of the main mode and the first higher mode are not smooth. The magnitudes $h'/k < 1$ for both modes (Fig. 2.1(a)). It means that the main and the first higher modes are the fast waveguide waves. In Fig. 2.1(b) are shown dependencies of the waveguide attenuation constant (losses) h″ on the frequency. We see that the values of the main and higher mode losses are commensurate and the losses are not higher in the frequency range. The loss maximums of the main mode and the first higher mode are slightly shifted. The maximum of main mode losses is 0.32 dB/mm at f = 87.5 GHz and the wavelength of this mode is equal to 4.2 mm. The maximum of the first higher mode is 0.36 dB/mm at f = 90 GHz and the wavelength of this mode is equal to 4 mm (Fig. 2.1). It is important to remark that the losses of the main mode are very low at the frequency range from 105 GHz till 115 GHz. This feature could be used in practice for creation of feeder lines and specific devices that require low distortions in the signal transmission.

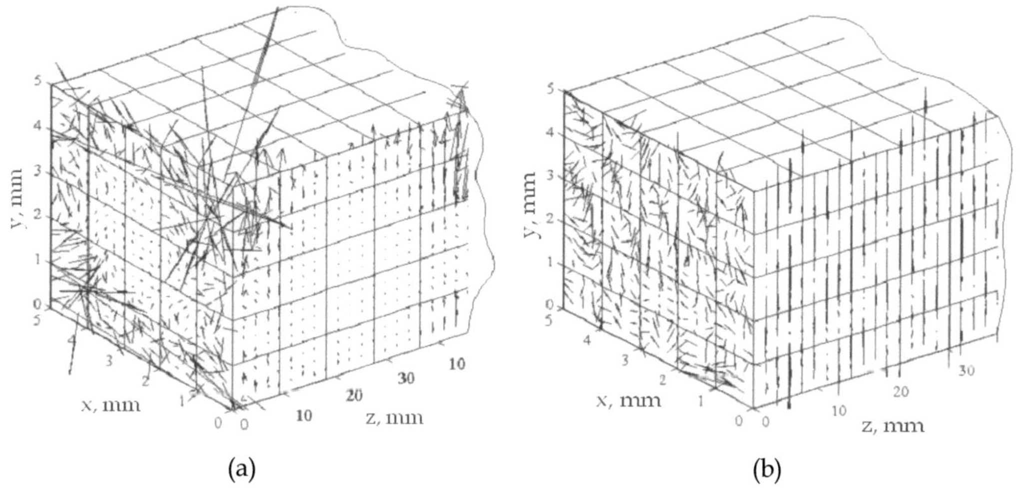

Fig. 2.2. The 3D electric field distribution of the main mode propagating in the square 5x5 mm^2 metamaterial waveguide. (a) - f= 95 GHz, (b) - f=110 GHz.

The 3D electric field distributions of the main mode at f = 95 GHz and 110 GHz are shown in Fig. 2.2. The metamaterial has ε_{met} = -23.75-i 18.75 and μ_{met} = 1.75+i 6.25 at f = 95 GHz. The metamaterial has ε_{met} = -10.83-i0.02 and μ_{met} = 0.5 – i 0.01 at f = 110 GHz.

We see that the waveguide losses are large at f = 95 GHz and they are low at f = 110 GHz (Fig. 2.1(b)). We present here the electric field line distribution at these two frequencies in order to compare how the losses influence on the field picture. The calculations of the electric fields in this section were fulfilled at the approximately 10000 points in every cross-section.

The metamaterial is an epsilon-negative media at f = 95 GHz and f = 110 GHz. In Fig. 2.2(a) we see the electric field is very small in the center of the waveguide cross-section. Such distribution can be explained by the large loss at f = 95 GHz and the EM wave does not deeply penetrate into the metamaterial. The refractive index of single negative metamaterial and the transverse propagation constants of waveguide made of the metamaterial are imaginary numbers. For this reason the electric field concentrates near the waveguide border. The tendency of the field to concentrate at the interface of the single negative metamaterial is the important feature of this kind metamaterial.

The behaviour of the EM field components when approaching to the apex of the waveguide cross-section contour (waveguide edge) are an important point of an electrodynamical solution. We would like to note that the condition on the waveguide edge is satisfied in our solutions of considered electrodynamical problems.

Examining the electric field lines near the upper right corner of the square waveguide we would like to note that the electric field lines are directed counter-clockwise on the right side of the corner and the lines are directed clockwise on the left side of the corner. The electric field lines near the left bottom corner of the square waveguide are distributed in similar way. The lines are directed counter-clockwise on the right side of the corner and they are directed clockwise on the left side of the corner. The electric field lines are diverging of the left bottom corner while the lines are converging to the right upper corner. For this reason, we believe that there is an increased density of electric lines in the upper corner and the weaker density in the bottom corner (Fig. 2.2(a)).

The same effect of no uniformity in the electric field distribution on the waveguide perimeter we can see for the circular waveguide (see below in section 2, Figs 2.2, 2.3).

f = 95 GHz, square waveguide 5x5 mm^2		
E_z [V/m]	E_x [V/m]	E_y [V/m]
$3.765 \cdot 10^{-1} - i\ 6.411 \cdot 10^{-1}$	$1.5099 - i\ 2.1941$	$-0.9943 + i\ 1.9673$
H_z [A/m]	H_x [A/m]	H_y [A/m]
$-3.39 \cdot 10^{-2} + i\ 7.51 \cdot 10^{-2}$	$-7 \cdot 10^{-4} + i1.1 \cdot 10^{-3}$	$2.6 \cdot 10^{-3} - i\ 3.9 \cdot 10^{-3}$
f = 110 GHz		
E_z [V/m]	E_x [V/m]	E_y [V/m]
$-1.2766 \cdot 10^{-5} - i\ 1.6592 \cdot 10^{-5}$	$0.0134 + i\ 0.0211$	$-0.0291 + i\ 0.0702$
H_z [A/m]	H_x [A/m]	H_y [A/m]
$-3.0 \cdot 10^{-3} - i\ 2.9 \cdot 10^{-3}$	$1.3048 \cdot 10^{-4} - i\ 3.0558 \cdot 10^{-4}$	$5.9501\ 10^{-5} + i\ 1.0936 \cdot 10^{-4}$

Table 2.1. The EM field components of the main mode at the point with coordinates $x = 4\ mm$ and $y = 4\ mm$ when f=95 GHz and f=110 *GHz*

f = 95 GHz, square waveguide 5x5 mm^2		
E_z [V/m]	E_x [V/m]	E_y [V/m]
$3.932 \cdot 10^{-1} - i6.181 \cdot 10^{-1}$	$1.3730 - i1.8577$	$-9.535 \cdot 10^{-1} + i\ 1.6598$
H_z [A/m]	H_x [A/m]	H_y [A/m]
$-3.20 \cdot 10^{-2} + i6.36 \cdot 10^{-2}$	$-8 \cdot 10^{-4} + i1.4 \cdot 10^{-3}$	$2.5 \cdot 10^{-3} - i3.4 \cdot 10^{-3}$
f = 110 GHz		
E_z [V/m]	E_x [V/m]	E_y [V/m]
$1.3083 \cdot 10^{-6} - i1.2678 \cdot 10^{-5}$	$-4.6 \cdot 10^{-3} + i2.30 \cdot 10^{-2}$	$-2.63 \cdot 10^{-2} + i4.31 \cdot 10^{-2}$
H_z [A/m]	H_x [A/m]	H_y [A/m]
$-1.9 \cdot 10^{-3} - i2.4 \cdot 10^{-3}$	$1.0715 \cdot 10^{-4} - i1.7759 \cdot 10^{-4}$	$-2.7720 \cdot 10^{-5} + i1.0632 \cdot 10^{-4}$

Table 2.2. The EM field components of the first higher mode at the point with coordinates x = 4 mm and y = 4 mm when f = 95 GHz and f = 110 GHz

In tables 2.1 and 2.2 we demonstrate the values of complex EM field components of the main mode and the first higher mode at two frequencies. On the base of the table data we can say, that the both modes on these frequencies are hybrid modes.

We do not classified the waveguide modes here in the usual way, e.g. the hybrid magnetic HE_{mn} or the hybrid electric EH_{mn} modes because the kind of mode of strong lossy waveguides may change when we change the frequency [Nickelson et al., 2011; Asmontas et al., 2010].

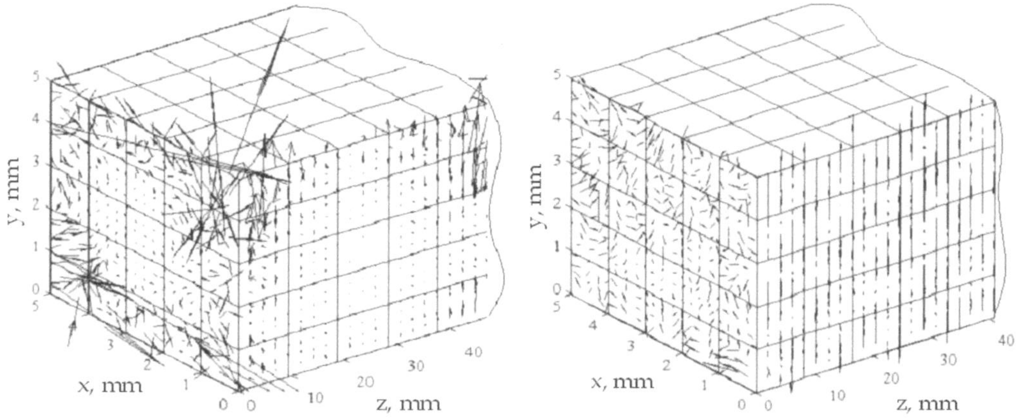

(a) (b)

Fig. 2.3. The 3D electric field distribution of the first higher mode propagating in the square 5x5 mm^2 metamaterial waveguide. a) – f= 95 GHz, b) – f= 110 GHz.

In Fig. 2.3(a) we see that the electric field at f = 95 GHz for the first higher modes is also concentrated near the metamaterial borders and the strongest field is at two diagonal corners of the cross-section, i.e. at the right upper corner and the left bottom corner. There is a strong asymmetry of the electric field distribution on the perimeter of waveguide. It happened probably by reason that the real and imaginary parts of permittivity are negative and relatively large at this frequency.

In Figs 2.2(b)–2.3 (b) we see that the electric field distributions of the main and first higher modes have a more homogeneous picture in the waveguide cross-section at f = 110 GHz. This happened because the electric field penetrates deeper into the metamaterial at this frequency because the waveguide loss is small at f = 110 GHz (Fig. 2.1(b)). We can also see here some asymmetry of electric field lines on the waveguide cross-section. The projections of the vector electric fields on the waveguide sidewalls are depicted along the waveguide (Figs 2.2-2.3).

2.2 Numerical investigations of square 4x4 mm^2 metamaterial waveguide

The dispersion characteristics of the square metamaterial waveguide are presented in Fig.2.4 (Gric et al., 2010). We used the values of ε_r^m and μ_r^m from (Penciu et al, 2006). In Fig. 2.4(a) the normalized propagation constant h'/k is shown. In Fig. 2.4(a) we see that the magnitude h' is less than the wave number *k* in the frequency range 85-108 GHz. It means that at these frequencies the main and the first higher modes are fast waves. The wavelengths of both modes differ slightly. The losses of both propagating modes in the metamaterial square waveguide change in very complicated way when the frequency increases. Mode losses have several maximums and minimums. We see that the losses of the main mode become a very small at the frequency 103.75 GHz.

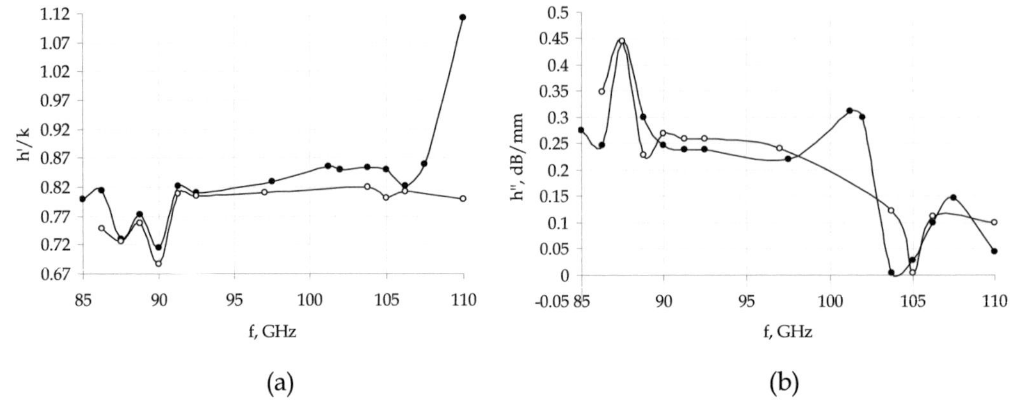

Fig. 2.4. The dispersion characteristics of the square 4x4 mm^2 metamaterial waveguide (a) – the dependence of the normalized propagation constant, (b) - the dependence of the attenuation constant on frequencies.

In Fig. 2.5 we see that the electric field is weak inside of square metamaterial waveguide of size 4x4 mm^2 at f=92.5 GHz. The electric field lines concentrate in the two diagonal waveguide corners in other way in the comparison with the waveguide of size 5x5 mm^2 at f=95 GHz.

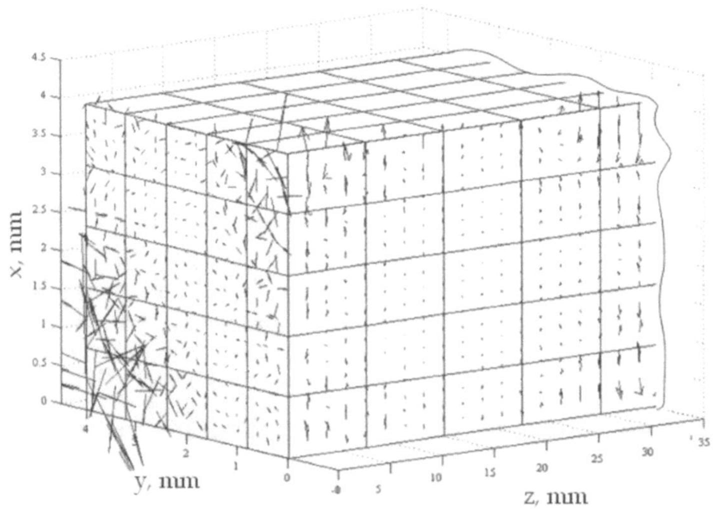

Fig. 2.5. The 3D vector electric field distribution of the main mode propagating in the open square 4x4 mm^2 metamaterial waveguide at f=92.5 GHz.

The 3D electric field distribution of the main mode at the frequency 92.5 GHz is depicted in Fig. 2.5. The permittivity of the metamaterial at f=92.5 GHz is -35-i2.5 and the permeability is 2.25+i0.25. The metamaterial is a single negative matter. The calculations of the electric fields were fulfilled at the approximately 160 points in every cross-section (Fig 2.5).

3. Investigations of the circular waveguides by the partial area method

Here the open circular metamaterial waveguide is investigated by the partial area method (Nickelson et al., 2008). The presentation of longitudinal components of the electric $E_z{}^m$ and magnetic $H_z{}^m$ fields that satisfy to the Maxwell's equations in the metamaterial medium is in the form:

$$E_z{}^m = A_1 J_m(k_\perp^+ r)\exp(im\varphi), \quad H_z{}^m = B_1 J_m(k_\perp^+ r)\exp(im\varphi) \tag{3.1}$$

where A_1, B_1 are unknown arbitrary amplitudes, J_m is the Bessel function of the m−th order, $k_\perp^+$ is the transverse propagation constant of the metamaterial medium, r is the radius of the circular metamaterial waveguide, m is the azimuthal index characterizing azimuthal variations of the field, φ is the azimuthal angle. The presentation the electric field $E_z{}^a$ and the magnetic field $H_z{}^a$ components that satisfy to Maxwell's equations in air are:

$$E_z{}^a = A_2 H_m(k_\perp^- r)\cdot\exp(im\varphi), \quad H_z{}^a = B_2 H_m(k_\perp^- r)\cdot\exp(im\varphi) \tag{3.2}$$

where A_2, B_2 are unknown arbitrary amplitudes, H_m is the Hankel function of the m−th order and the second kind, $k_\perp^-$ is the transverse propagation constant of air medium.

As far as the circular waveguide is researched in the cylindrical coordinate system we have to satisfy the boundary conditions for two components of the electric field (E_φ, E_z) and the magnetic field (H_φ, H_z) on the cylindrical interface metamaterial-air. The condition at infinity also is satisfied. The sign of $k_\perp^-$ changes on the opposite one for the metamaterial hollow-core waveguide, when a hole is surrounded by a metamaterial medium. After substitution of expressions (1) and (2) in the transverse components expressed in terms of the longitudinal components (Kong, 2008) we obtain the expressions of all transverse EM field components. The result of solution is the dispersion equation in the determinant form. We determine complex roots of the dispersion equation by using of the Muller method.

3.1 Investigations of the circular metamaterial waveguide (r=2.5 mm) by the partial area method

We discovered the particularity in the electric field distribution on the cross-section of the open circular metamaterial waveguide at the operating frequency 95 GHz. We find that this waveguide could be used as a narrowband filter at frequencies 102-102.5 GHz.

The circular metamaterial waveguide with r=2.5 mm was researched. The dispersion characteristics and the 3D electric and magnetic field distributions were calculated (see Figs 2.1-2.5). The dispersion characteristics are presented in Fig. 3.1. The main mode is denoted with black points and the first higher mode is denoted with circles.

In Fig. 3.1 (a) we see that the normalized propagation constants of the main and the first higher modes are fairly smooth except only one protrusion at frequencies between 97 GHz and 102 GHz. There is a large peak of the main mode losses at frequency f = 101.25 GHz. At this frequency metamaterial is double negative with $\varepsilon_{r,met}$ = -9.17 -i0.83 and $\mu_{r,met}$ = -0.75. We see that losses of the main mode are very small at the frequency ranges 75-100 GHz and 102.5-115 GHz. While the losses of the first higher mode are significantly higher at this frequency ranges

(Fig. 3.1(b)). Therefore the investigated circular metamaterial waveguide can be used as a filter at the frequencies 100-102.5 GHz and as a one mode lossless waveguide at the frequency ranges 75-100 GHz and 102.5-115 GHz. As we can see the first higher mode is a quickly attenuated wave because it has very large losses in all frequency range.

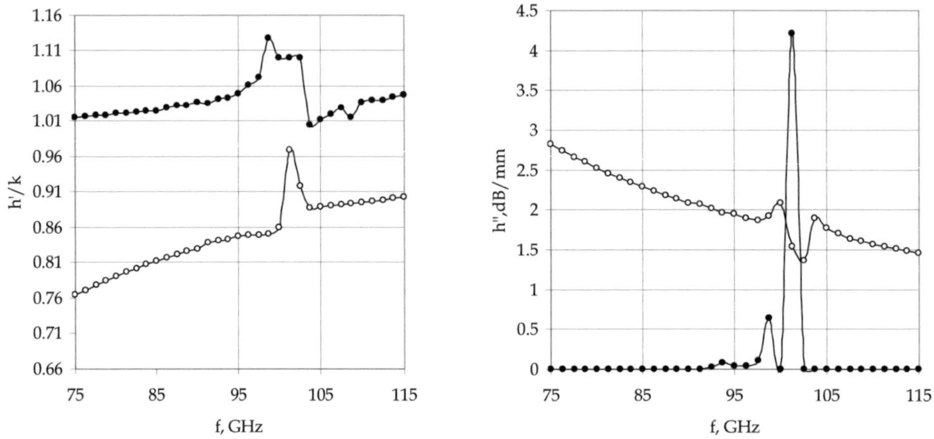

(a) (b)

Fig. 3.1. The dispersion characteristics of the circular metamaterial waveguide with r=2.5 mm (a) – the dependence of the normalized propagation constant, (b) – the dependence of the attenuation constant on frequencies.

The 3D electric field distributions of the main mode were calculated at frequencies 95 GHz and 110 GHz (Fig. 3.2 and 3.3) as well as the first higher mode at the same frequencies (Fig. 3.4 and 3.5). The electric field inside the circular metamaterial waveguide is much smaller than outside of waveguide for this reason we have increased the electric field strength lines inside of the waveguide in order to see them (Figs 3.2(b) – 3.5(b)).

Because the metamaterial has more losses at 95 GHz than at 110 GHz, so the electric field inside of waveguide is weaker at 95 GHz (see tables 3.1, 3.2) compare to the inner electric field at 110 GHz. On this reason we increased the electric field strength lines at 95 GHz in the 143 times and at 110 GHz in the 14 time. The calculations of the electric fields were fulfilled at the approximately 10000 points in every cross-section.

f = 95 *GHz*, waveguide diameter 5mm		
E_z [V/m]	E_x [V/m]	E_y [V/m]
$-2.5897\cdot10^{-5}+i2.7266\cdot10^{-4}$	$-2.8425\cdot10^{-5}+i1.9342\cdot10^{-5}$	$4.5991\cdot10^{-6}-i3.3487\cdot10^{-5}$
H_z [A/m]	H_x [A/m]	H_y [A/m]
$2.5111\cdot10^{-7}-i2.0158\cdot10^{-7}$	$-2.3843\cdot10^{-8}-i2.8570\cdot10^{-8}$	$2.5496\cdot10^{-6}+i1.8364\cdot10^{-7}$
f = 110 *GHz*		
E_z [V/m]	E_x [V/m]	E_y [V/m]
0.0067 +i0.0067	-0.0027 + i0.0027	$-7.8928\cdot10^{-4}-i7.8928\cdot10^{-4}$
H_z [A/m]	H_x [A/m]	H_y [A/m]
$3.1409\cdot10^{-12}-i1.1109\cdot10^{-5}$	$-3.3354\cdot10^{-6}-i3.3354\cdot10^{-6}$	$7.1930\cdot10^{-5}-i7.1930\cdot10^{-5}$

Table 3.1. The electromagnetic field components of the main mode at the point with coordinates r = 2 mm, φ = 45° when f = 95 and f = 110 GHz

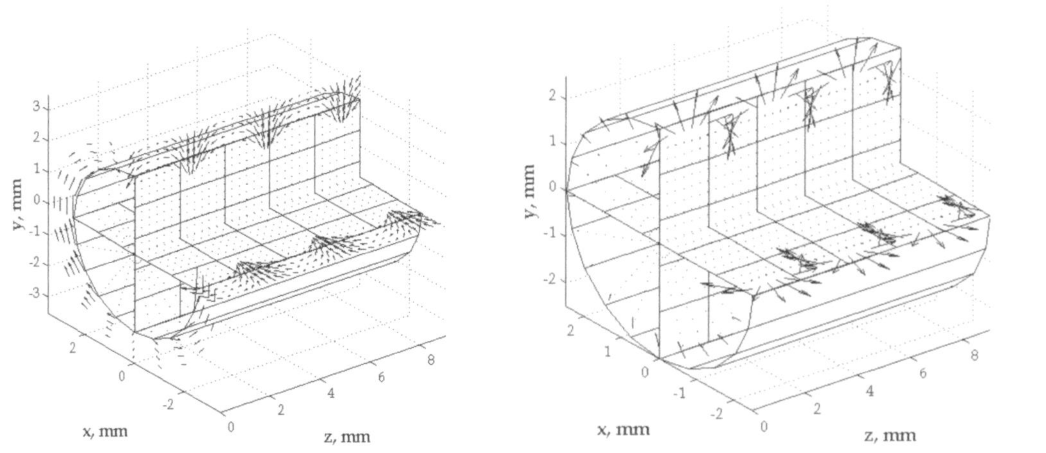

(a) (b)

Fig. 3.2. The 3D electric field distributions of the main mode of circular metamaterial waveguide with r=2.5 mm at f = 95 GHz. (a) – the electric field strength lines outside the waveguide (b) – the 143 times were increased electric field strength lines inside the waveguide.

In In Figs 3.2–3.5 we see that the electric field is irregular on the waveguide perimeter of the cross-section while the cross-section of the waveguide is a circle. We see that and the most part of the electric field localizes on the border and outside of the waveguide. The electric field is strongest outside the waveguide when φ is 0 or π radians. The electric field has the minimum values and the electric field lines are directed clockwise or counter-clockwise to the right and left of the points with φ equal to π/2 or 3π/2 radians (Figs 3.2(a)–3.5(a)). We see that at the points when the electric field outside of the metamaterial waveguide has the maximum value the field inside of the waveguide is minimal. In Figs 3.2(b)–3.5(b) we see that the maximum electric field inside the metamaterial waveguide is when φ is equal to π/2 or 3π/2 radians. The length of the circular waveguide in z-direction (the Figs 3.2– 3.5) is three times longer than wavelength of microwave in the waveguide in our calculations.

f = 95 *GHz*, waveguide diameter 5 mm		
E_z [V/m]	E_x [V/m]	E_y [V/m]
$1.6760 \cdot 10^{-4}$ -i$3.3526 \cdot 10^{-4}$	$3.5399 \cdot 10^{-5}$-i$1.2553 \cdot 10^{-5}$	$-2.7112 \cdot 10^{-5}$-i$1.2042 \cdot 10^{-6}$
H_z	H_x	H_y
$-1.8851 \cdot 10^{-7}$-i$1.9128 \cdot 10^{-7}$	$4.6704 \cdot 10^{-8}$+i$8.1377 \cdot 10^{-8}$	$-3.1731 \cdot 10^{-6}$-i$1.4861 \cdot 10^{-6}$
f = 110 *GHz*		
E_z [V/m]	E_x [V/m]	E_y [V/m]
$-5.3 \cdot 10^{-3}$ -i$1.22 \cdot 10^{-2}$	$4.1 \cdot 10^{-3}$ -i$2.1 \cdot 10^{-3}$	$-3.2013 \cdot 10^{-4}$ +i$9.6358 \cdot 10^{-4}$
H_z	H_x	H_y
$-1.0713 \cdot 10^{-5}$+i$9.0740 \cdot 10^{-7}$	$7.1766 \cdot 10^{-6}$+i9.3919	$-1.3582 \cdot 10^{-4}$+i$5.6938 \cdot 10^{-5}$

Table 3.2. The EM components of the first higher mode with coordinates at the point with coordinates r = 2 mm , φ = 45° when f=95 and f=110 GHz

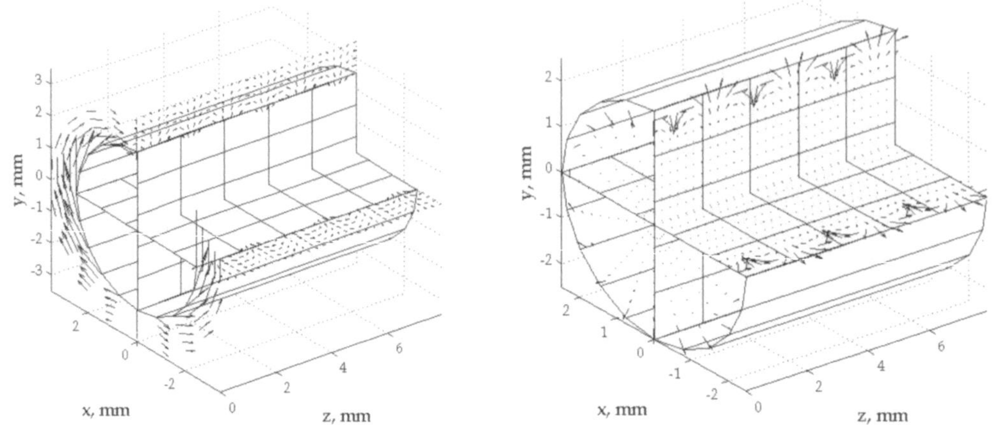

(a) (b)

Fig. 3.3. The 3D electric field distributions of the main of circular metamaterial waveguide with r=2.5 mm at f = 110 GHz. (a) – the electric field strength lines outside the waveguide (b) – the 14 times were increased electric field strength lines inside the waveguide

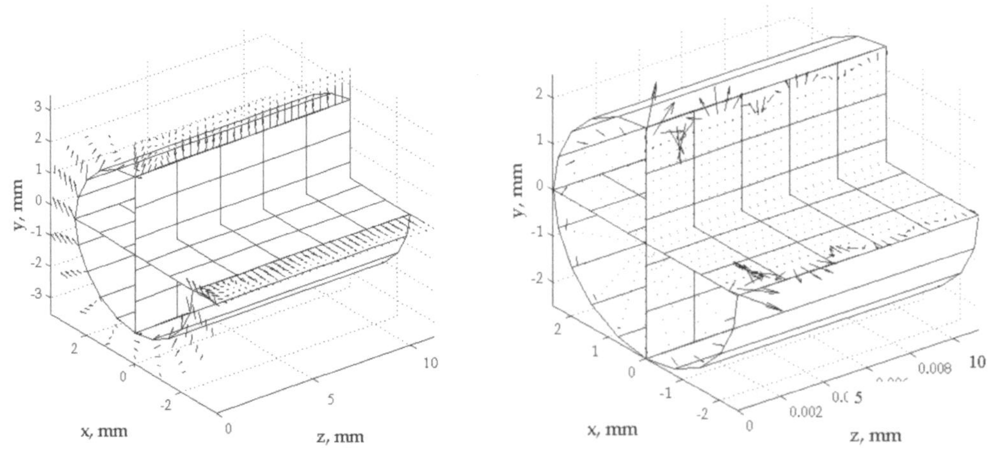

(a) (b)

Fig. 3.4. The 3D electric field distributions of the first higher mode of circular metamaterial waveguide with r=2.5 mm at f = 95 GHz (a) – the electric field strength lines outside the waveguide (b) – the 143 times were increased electric field strength lines inside the waveguide

We can see that the electric field along the waveguide changes periodically (Figs 3.2–3.5). Comparing Figs 3.2(a) and 3.3(a) we see that the main mode' electrical field at 110 GHz is large and have a little different distribution in longitudinal direction in comparison with the electrical field at 95 GHz. Since the waveguide losses of the main mode at frequencies 95

GHz and 110 GHz are small (please, compare with losses of square waveguide (Fig.2.1)), for this reason the electric field amplitudes vary slightly in the longitudinal direction. Comparing Figs 3.4 and 3.5 we see that the larger is the electrical field outside of the waveguide at the point with a certain angle φ the smaller is the electrical field inside of waveguide at the point with the same φ. The last statement is true for all the investigated cases. The electric field amplitude of the first higher mode became smaller in longitudinal direction with increasing of coordinate z (Figs 3.4(b) and 3.5(b)). We observe in this case, the fast wave attenuation. It happened because the losses of the first higher mode at 95 GHz and 110 GHz are enough large in comparison with the main mode (Fig. 3.1(b)).

The electrical field inside of the waveguide is very small at all frequencies. However the observable electric field strength lines appear at the waveguide boundary.

We would like to note that the boundary conditions on the border of waveguide are satisfied, i.e. the tangential components of electric and magnetic fields are equal on the interface air-metamaterial. For this reason when the electric field lines has a tangential character outside the waveguide (Figs 3.2(a), 3.3(a)) the same character of tangential components has to be on the interface of the metamaterial side. The direction of electric field lines changes with removing deeper in the metamaterial from the interface.

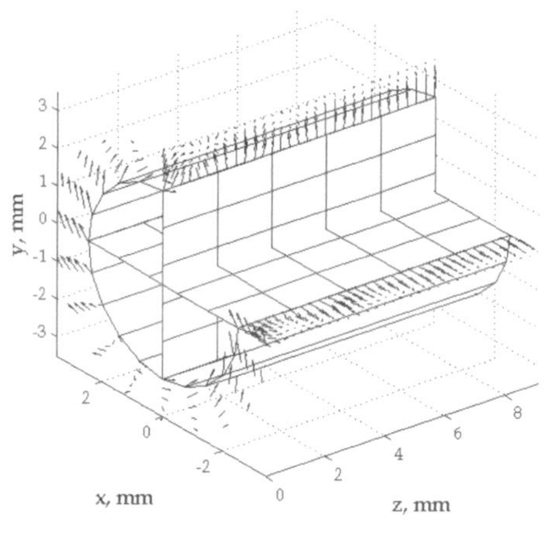

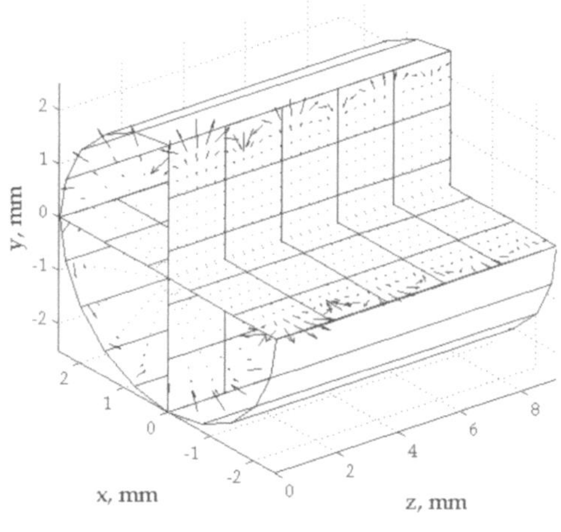

Fig. 3.5. The 3D electric field distributions of the first higher mode of circular metamaterial waveguide with r=2.5 mm at f = 110 GHz (a) – the electric field strength lines inside the waveguide (b) – the 14 times were increased electric field strength lines outside the waveguide.

Comparing dispersion characteristics of square and circular waveguides (Figs 2.1 and 3.1) we see that they are different due to the boundary conditions, which have a strong influence on the dispersion characteristics in our frequency range. As an additional example, the dispersion characteristic ($2\pi f \cdot SQRT(\varepsilon_{r,met}\ \mu_{r,met})$) of plane EM wave propagating in the same metamaterial only when the one has the infinite dimensions is strongly different in comparison with the waveguide dispersion characteristics. The relatively small losses of the EM wave in the infinite metamaterial are only at the frequencies between 102.5 GHz and 105 GHz.

We would like to draw attention to the fact that the feature of the irregular distribution of electric field lines in the cross-section of square and circular metamaterial waveguides at 95 GHz is very similar.

3.2 Investigations of the circular metamaterial waveguide (r = 2 mm) by the partial area method

3.2.1 The metamaterial rod waveguide

We have investigated circular metamaterial waveguides by our algorithm that was created using the partial area method (Nickelson et al., 2008).

In Fig. 3.6 the dispersion characteristics of the metamaterial waveguide are presented. In Fig. 3.6 (a) the normalized propagation constant h'/k of the main mode and the first higher mode is shown. The main mode is denoted with black points and the first higher mode is denoted with circles.

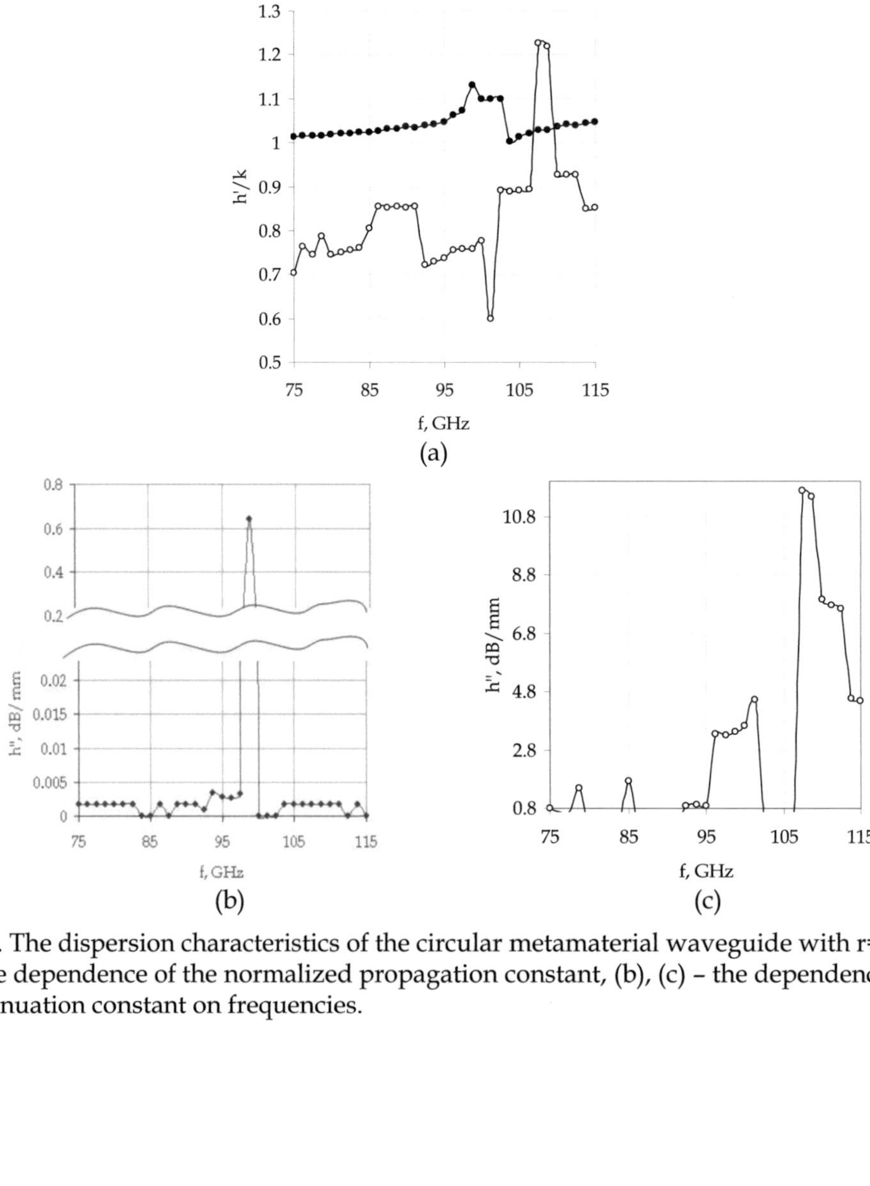

Fig. 3.6. The dispersion characteristics of the circular metamaterial waveguide with r=2 mm (a) – the dependence of the normalized propagation constant, (b), (c) – the dependence of the attenuation constant on frequencies.

In Fig. 3.6 (b) losses of the main mode are shown. In Fig. 3.6 (c) losses of the first higher mode are presented. We show the losses of the modes in the different scales because the losses are not commensurate. We see (Fig. 3.6 (a)) that the main mode is a slow mode and the first higher mode can be a slow mode or a fast mode dependent on the frequency range.

There are two frequency ranges 75-97.5 GHz and 100-115 GHz when losses of the main mode are extremely small. On the other side there are frequency ranges, for example f=95 - 101.25 GHz and f=107.5-115GHz when losses of the first higher mode are large.

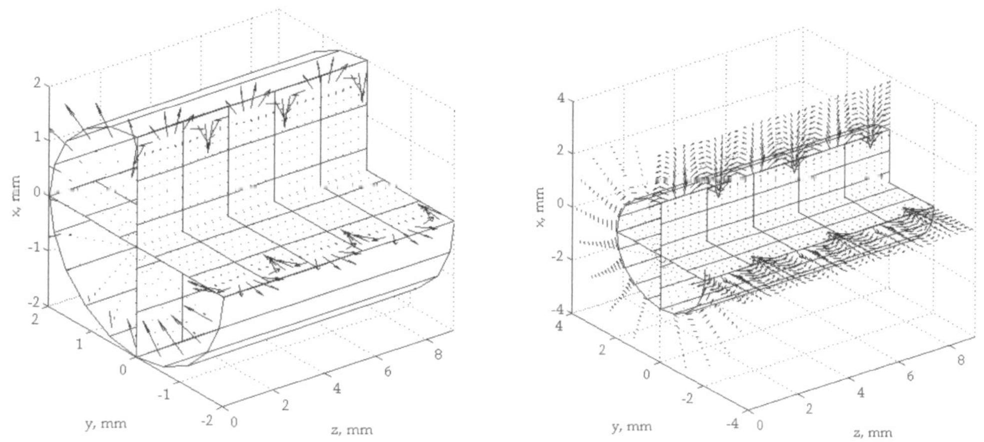

(a) (b)

Fig. 3.7. The 3D vector electric field distributions of the main mode of circular metamaterial waveguide with r=2mm (a) – inside; (b) – inside and outside it.

We have calculated the 3D vector electric field distributions of the main mode propagating in the open circular metamaterial waveguide. The calculation was fulfilled inside and outside the waveguide in 1500 points. The electric field distributions were calculated at frequency f=95 GHz. At this frequency the metamaterial is single-negative. At this frequency $\varepsilon_{r,met}$ = -23.75- i18.75 and $\mu_{r,met}$ = 1.75+ i1.625.

In Fig. 3.7(a) an enlarged picture of the electric field lines inside the metamaterial waveguide is shown. We see that the strongest electric field is in the thin surface layer which is located at the interface metamaterial-air. In Fig. 3.7(b) the electric field lines inside and outside of the metamaterial waveguide are shown. In Fig. 3.7(b) we see that the electric field inside the waveguide is significantly weaker than outside it. We also clearly see that the electric field distributions are periodically repeated in the longitudinal direction.

3.2.2 The metamaterial hollow-core waveguide

The dispersion characteristics of the metamaterial hollow-core waveguide with the radius of the hole (in the metamaterial medium) equal to 2 mm are presented in Fig. 3.8.

In Fig. 3.8 (a) dispersion characteristics of the main and the first higher modes are presented. Both modes are the fast ones. Their electromagnetic energy concentrates in the hollow-core air area. There are the frequency ranges where modes propagate with very small losses. We can see that losses of the main mode in the frequency range 75-90 GHz and 104-115 are very low and they can be large in the frequency range 91.25-103.75 GHz. The first high mode' losses change abruptly. The losses can be very low at some frequencies approximately 75- 87 GHz, 106 and 111 GHz.

Losses of the first high mode are low and this mode can easily propagate and it can modulate the amplitude of the main mode in devices that were created on the base of the metamaterial waveguide with r=2 mm. There is also good possibility to create a devise on the base of the first high mode in the range f=95-105 GHz.

We have calculated the 3D vector electric field distributions of the main mode propagating in the hollow-core metamaterial waveguide. The electric field distributions were calculated at frequency f = 95 GHz. The calculations of the electric fields were fulfilled at the approximately 10000 points in every cross-section.

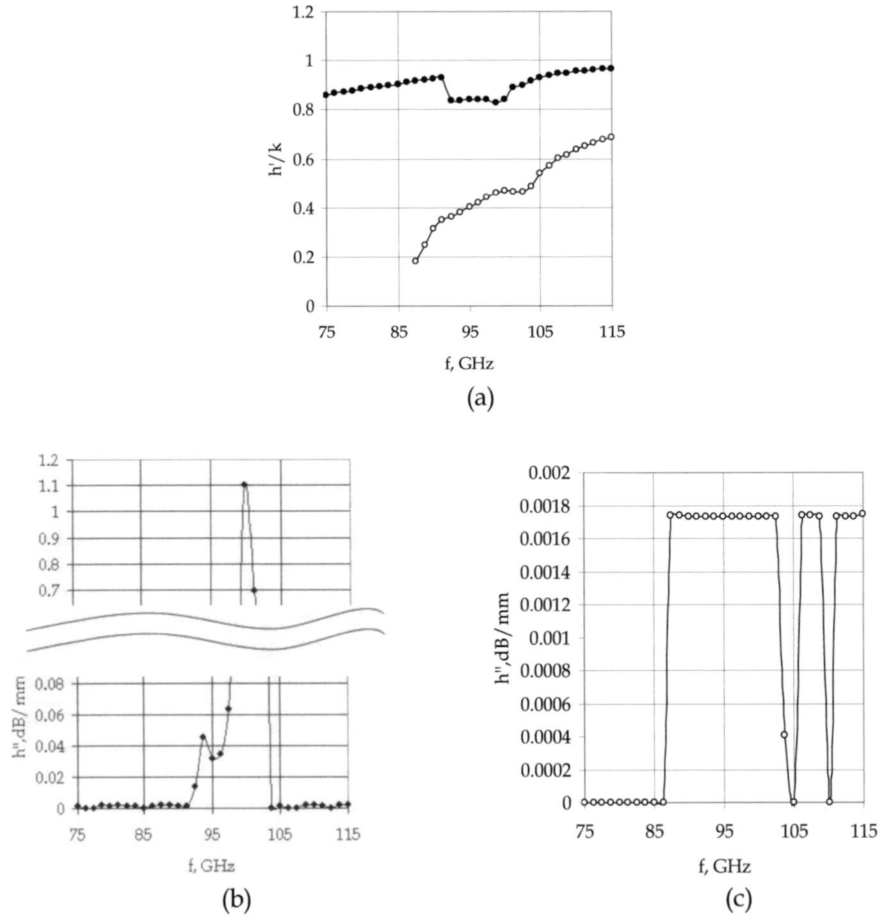

Fig. 3.8. The dispersion characteristics of the hollow-core metamaterial waveguide with r=2 mm (a) – the dependence of the normalized propagation constant, (b)& (c) – the dependence of the attenuation constant on frequencies.

In Fig. 3.9 we see that the electric field inside the air hole is significantly weaker than outside in the metamaterial media. We also clearly see that the electric field distributions are periodically repeated in the longitudinal direction. As the waveguide losses exist only in the metamaterial and the most part of EM energy propagates into air hole then we see no decrease in the amplitude of the electric field with a change in coordinate z (please, compare with Figs 3.4(b), 3.5(b)).

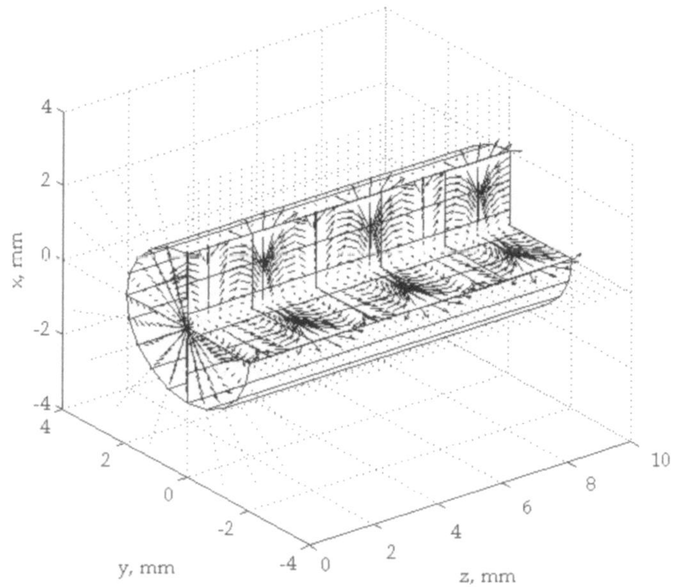

Fig. 3.9. The 3D vector electric field distribution of the main mode of the hollow-core metamaterial waveguide with r=2 mm at frequency 95 GHz.

4. Conclusions on sections 2&3

1. The open lossy metamaterial waveguides with different shapes of the cross-section were investigated by using of our computer programs that have written in MATLAB language. The computer codes were based on the method of singular integral equations and the partial area method.
2. We have calculated the dispersion characteristics (the propagation and attenuation constants) at the frequency range 75-115 GHz as well as the 2D and 3D electromagnetic field distributions. We took the electromagnetic parameters of metamaterial close to practice. Our computer algorithms can be useful working out microwave devices on the base of waveguides made of strong lossy materials.
3. We discovered the special feature of the open lossy metamaterial waveguides. Propagation and attenuation constants depend on the waveguide sizes in an unpredictable complex manner. E.g., the circular metamaterial waveguide with r=2.5 mm (Fig. 3.1) and r=2 mm (Fig. 3.6).
4. The microwave signals propagating on the open metamaterial waveguides (r=2.5 mm) are absorbed at the narrow 2.5 GHz frequency range. This waveguide can be used as a band-stop filter at f_1=100-102.5 GHz when the losses in the passing frequencies 75-100, 102.5-115 GHz are 100 times less than losses at f_1.

5. Investigated EM fields of square and circular waveguides have irregular distributions on the waveguide perimeter. Investigated EM field lines for the main and first higher modes of the squire waveguides are concentrated near the metamaterial borders and the strongest field is at two diagonal corners of the cross-section. The electric field lines are diverging of the one corner while the lines are converging at the other corner.

5. Dispersion characteristic analysis of circular anisotropic metamaterial waveguide with the effective metamaterial permittivity and permeability near to zero

5.1 Introduction

In the past several years many specialist focused on the experimental and theoretical investigations of the zero-refractive index (or zero-index) metamaterials. These metamaterials attracted researches due to their unconventional constitutive parameters and anomalous effects to work out novel electromagnetic devices. Zero- index metamaterial may have the epsilon-near-zero (ENZ) or mu-near-zero (MNZ) properties simultaneously or in turn, one after another at different frequencies.

Zero-index metamaterials are dispersive (electromagnetic parameters dependence on a frequency) media. The constitutive parameters of anisotropic dispersive metamaterials can be described by expressions that involve the plasma frequencies. The metamaterial on frequencies near to plasma resonances is called a plasmonic metamaterial.

Zero-index metamaterials are used in different devices as a transformer to achieve the perfect impedance match between two waveguides with a negligible reflection or to improve the transmission through a waveguide bend as well as for the matching of waveguide structure impedance with the free space impedance (when the metamaterial epsilon and mu are simultaneously very close to zero). Plasmonic metamaterial provides manipulating of the antenna phase fronts and enhancing the antenna radiation directivity. In a Zero-index metamaterial waveguide can be observed a super-tunneling effect. ENZ metamaterials may allow reducing of waveguide sizes and can be used as a frequency selective surface (Bai et al., 2010; Ko&Lee, 2010; Lopez-Garcia et al., 2011; Luo et al., 2011; Wang & Huang, 2010; Oraizi et al. 2009; Zhou et al., 2009, Liu et al., 2008).

5.2 Analysis and simulation of phase constant dependencies

Here we presented the phase constant (real part of the waveguide longitudinal propagation constant) of propagating modes on the circular anisotropic metamaterial waveguide when the metamaterial permittivity and permeability may take values close to zero at certain frequencies. Further we call a plasmonic waveguide.

The solution of Maxwell's equations for the circular anisotropic metamaterial waveguide was carried out by the partial area method (Nickelson et al., 2009). The computer program for the dispersion characteristic calculations has created in MATLAB language. Computer program allows take into account a very large material attenuation (Nickelson et al., 2011; Asmontas et al., 2010). In this section constitutive parameters of the uniaxial electrically and magnetically anisotropic metamaterial were taken from the article (Liu et al., 2007). In the mentioned article was considered an anisotropic dispersive lossless metamaterial slab. For this reason there were given only the real parts of the permittivity $\varepsilon_{r,ij}$ = (ε_{xx}, ε_{xx}, ε_{zz}) and permeability $\mu_{r,ij}$ =(μ_{xx}, μ_{xx}, μ_{zz}) tensor components.

The tensor components of the relative permittivity and the relative permeability are described by following formulae (Liu et al., 2007):

$$\varepsilon_{xx}=1-\omega^2_{epxx}/\omega^2, \quad \varepsilon_{zz}=1-\omega^2_{epzz}/\omega^2, \tag{5.1}$$

$$\mu_{xx}=1-\omega^2_{mpxx}/\omega^2, \quad \mu_{zz}=1-\omega^2_{mpzz}/\omega^2, \tag{5.2}$$

here $\omega=2\pi f$ is the angular frequency of microwaves, f is the operating frequency. The electric plasma frequencies of metamaterial are $\omega_{epxx}= 2\pi f_{epxx}$, $f_{epxx}=(12)^{1/2}$ GHz, $\omega_{epzz}=2\pi f_{epzz}$, f_{epzz}=2.5 GHz. The magnetic plasma frequencies of metamaterial are $\omega_{mpxx}= 2\pi f_{mpxx}$ GHz, $f_{mpxx} = (6)^{1/2}$ GHz and $\omega_{mpzz}=2\pi f_{mpzz}$ GHz, f_{mpzz}=2 GHz. The values of angular frequencies are taken from (Liu et al., 2007). The magnitudes of tensor components ε_{xx}, ε_{zz}, μ_{xx}, μ_{zz} are real numbers. In Fig. 5.1(a,b) are presented their dependencies on the frequency.

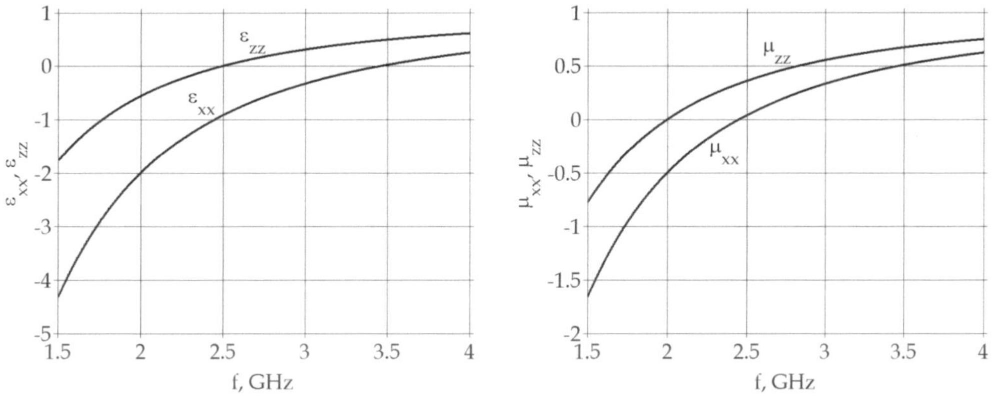

(a) (b)

Fig. 5.1. Dependences of the relative (a) permittivity and (b) permeability tensor components of the metamaterial on the frequency.

We see that the permittivity components ε_{xx} and ε_{zz} have negative values from 1.5 to ~3.5 GHz and from 1.5 to ~2.5 GHz, respectively. The permeability components μ_{xx} and μ_{zz} have negative values from 1.5 to~2.5 GHz and from 1.5 to ~2 GHz, respectively. We realize that all tensor components are negative at the frequency range from 1.5 GHz to ~2 GHz. Absolute values of tensor components are less than 1 at the frequency range from ~2.5 GHz to 4 GHz.

The values of tensor components become equal to zero at the operating frequency f equal to the metamaterial electric f_{epxx}=3.46 GHz, f_{epzz}= 2.5 GHz or magnetic f_{mpxx} =2.45 GHz, f_{mpzz}=2 GHz plasma frequencies.

In Figs. 5.2–5.7 are shown dispersion characteristics (phase constants) of open circular waveguide made of the uniaxial electrically and magnetically anisotropic metamaterial. The calculations are performed for the left-handed (extraordinary) circularly polarized microwaves when exp(+imφ), m=0, 1, 2,…is the wave (mode) azimuthal periodicity index, φ is the azimuthal coordinate. We investigated only modes with the index m=1, because it is known the main mode of open circular dielectric waveguide specify by m=1 and the main mode of a dielectric waveguide is a hybrid mode HE_{11} (Nickelson et al., 2009).

Here is presented the phase constant h′ (the real part of longitudinal propagation constant) dependencies of plasmonic metamaterial waveguides with radii r equal to 0.1 mm, 1 mm, 3 mm, 5 mm, 7 mm, 10 mm and 15 mm. The phase constant h′ is equal to $2\pi/\lambda_w$, where λ_w is the wavelength of certain mode. Our aim is to investigate how an increase in the plasmonic waveguide radius affects on the eigenmode numbers, mode cutoff frequencies and a shape of dispersion characteristics.

The analysis of Figs 5.2-5.5 shows that there are three main frequency areas where localize dispersion curves.

A shape all dispersion characteristics Figs 5.2-5.5 are unusual in the comparison with traditional dispersion characteristics of open cylindrical waveguides made of dielectrics, semiconductors or magnetoactive semiconductor plasma (Nickelson et al., 2011; Asmontas et al., 2009; Nickelson et al., 2009). Because the dispersion characteristic branches of analyzed plasmonic waveguides are vertical.

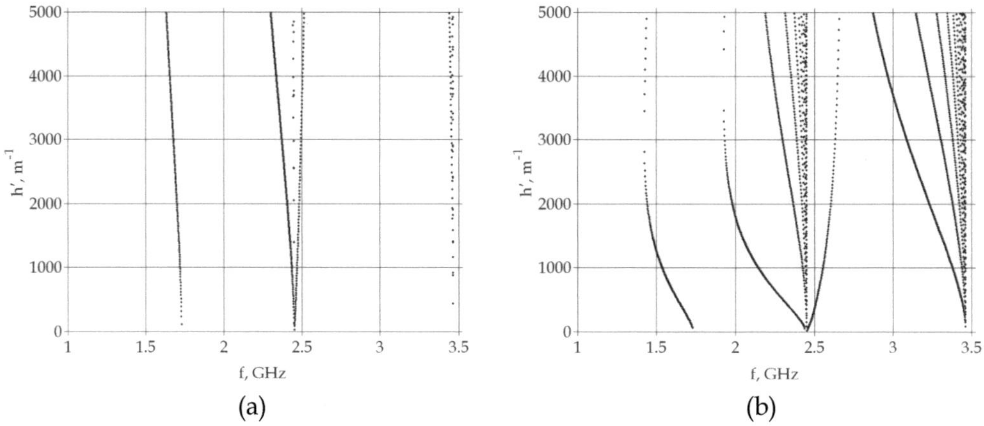

Fig. 5.2. The phase constant dependencies of propagating modes on the anisotropic metamaterial waveguide with (a) r=0.1 mm and (b) r=1 mm.

We see that there is a single mode with the cutoff frequency close to f=1.5 GHz. The cutoff frequency of this mode shifted in the direction of lower frequencies with increasing of the waveguide radius. This first single mode is special one because the mode does not match any of plasma f_{epxx}, f_{epzz}, f_{mpxx}, f_{mpzz} frequencies. We can observe how a shape of the dispersion characteristic changes in the vicinity of the cutoff frequency.

We would like to draw your attention to the fact that the anisotropic metamaterial is described by the negative tensor components ε_{xx}, ε_{zz}, μ_{xx}, μ_{zz} in the frequencies less than 2 GHz (see formulae 5.1 and 5.2). It is mean that the first mode propagates in the waveguide when the metamaterial is double negative (DN). This wave is particularly important because small changes in frequency produce large changes in phase.

We can watch a package of dispersion branches closed to cutoff frequency 2.5 GHz. We see that the left lateral dispersion branch of the package is a special eigenmode, i.e. this one is separated by a larger distance from other eigenmodes. The vertical branch of the left lateral mode is located at the magnetic plasma f_{mpzz} frequency equal to 2 GHz. We can distinguish also the right lateral dispersion branch of the package. The mode with this dispersion characteristic is also more specific one. i.e. this mode is separated by a larger distance from other modes. The vertical branch of this mode is located about 2.7 GHz and shifted on the higher frequency side with increasing of a radius.

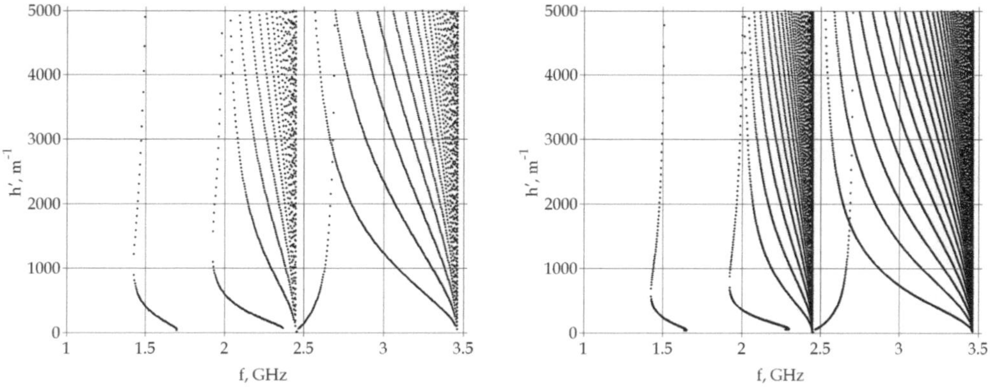

(a) (b)

Fig. 5.3. The phase constant dependencies of propagating modes on the anisotropic metamaterial waveguide with (a) r=3 mm and (b) r =5 mm.

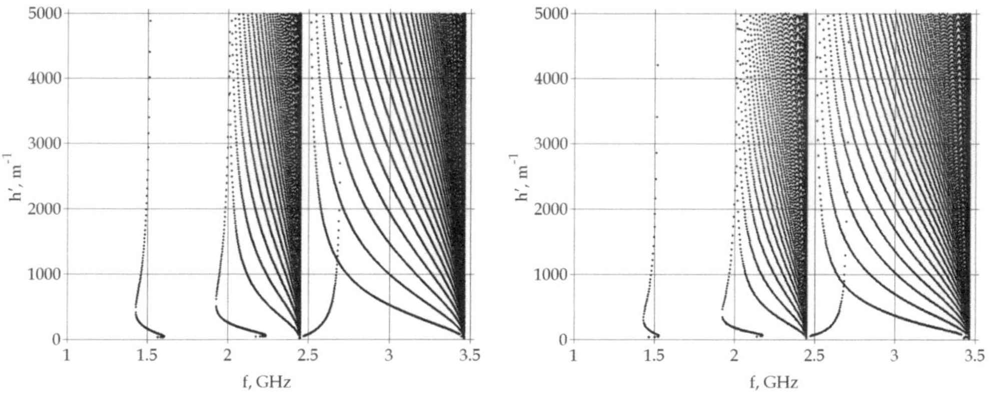

(a) (b)

Fig. 5.4. The phase constant dependencies of propagating modes on the anisotropic metamaterial waveguide with (a) r=7 mm and (b) r=10 mm.

A dense bunch of dispersion curves located between the extreme left and right curves that were previously described. The number of curves increases rapidly at increasing of waveguide radius. It is interesting to note that all dispersion branches of the dense bunch are within the frequency band of 2-2.5 GHz. Apparently the dense bunch of dispersion characteristics related to plasma f_{mpzz} and f_{epzz} frequencies. The cutoff frequencies of all dispersion characteristics of the dense bunch are the same and equal to f~2.46 GHz. The dispersion curves fan out from a point with a value equal to f_{mpxx}.

Second dense bunch of dispersion curves is at the electric plasma frequency f_{epxx}~3.46 GHz. The number of curves increases rapidly at increasing of waveguide radius. All dispersion characteristics are within the frequency band of 2.5 GHz and 3.46 GHz.

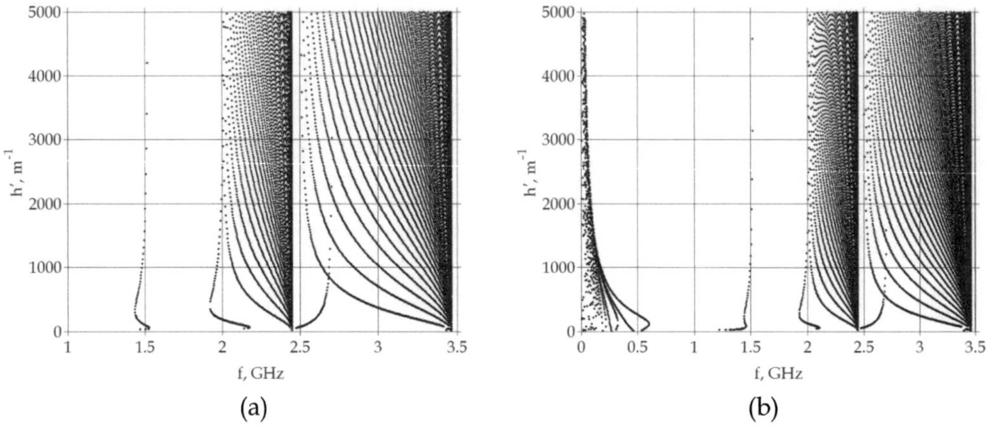

Fig. 5.5. The phase constant dependencies of propagating eigenmodes on the open plasmonic metamaterial waveguide with (a) r=10 mm (b) r=15 mm.

The dispersion characteristics on the right side of the bunch are more vertical. The greatest number of modes can be excited at the electric plasma frequency f_{epxx}~3.46 GHz in the comparison with other plasma frequencies. The cutoff frequencies of dispersion characteristics of this dense bunch are the same and equal to f~3.46 GHz. We did not find the plasmonic metamaterial waveguide eigenmodes in the frequency range from 3.5 GHz till 2000 GHz.

We expanded the searching of eigenmodes on frequencies below f=1 GHz for the plasmonic waveguide with r= 15 mm. In Fig. 5.5 (b) is shown eigenmodes dispersion characteristics of plasmonic metamaterial waveguide in the frequency range from 5 MHz till 3.5 GHz. We present here more detailed calculations of the dispersion characteristics of the plasmonic waveguide eigenmodes with the radius equal to 15 mm (Figs. 5.6 and 5.7). Here are presented the new dispersion characteristic branches of waveguide eigenmodes in the band of frequency from 5 MHz till 600 MHz (Fig. 5.6 (a)). We see that the dispersion curves have the clear expressed cutoff frequencies and they have an opposite slope in the comparison with dispersion curves of open ordinary waveguides (Nickelson et al., 2011; Asmontas et al., 2010).

It should be stressed that these very low frequency dispersion characteristics have obtained by solving of Maxwell's equations with certain boundary conditions. These low frequency modes are also the metamaterial waveguide eigenmodes.

In the Fig. 5.6 (b) is shown the dispersion characteristic of a singular mode on a larger scale. We see that this mode from 1.35 till ~1.45 GHz is a static mode, because no dependence on the frequency. We can observe the anomaly dispersion closed to f~ 1.45 GHz and the very strong dispersion in the frequency band over ~1.47 GHz. We can see the dispersion curves in the area of cutoff frequencies on a larger scale 1.9-2.5 GHz in Figs 5.7(a) and (b). We can note the anomalous dispersion hook of eigenmodes in the band f=1.9-2 GHz.

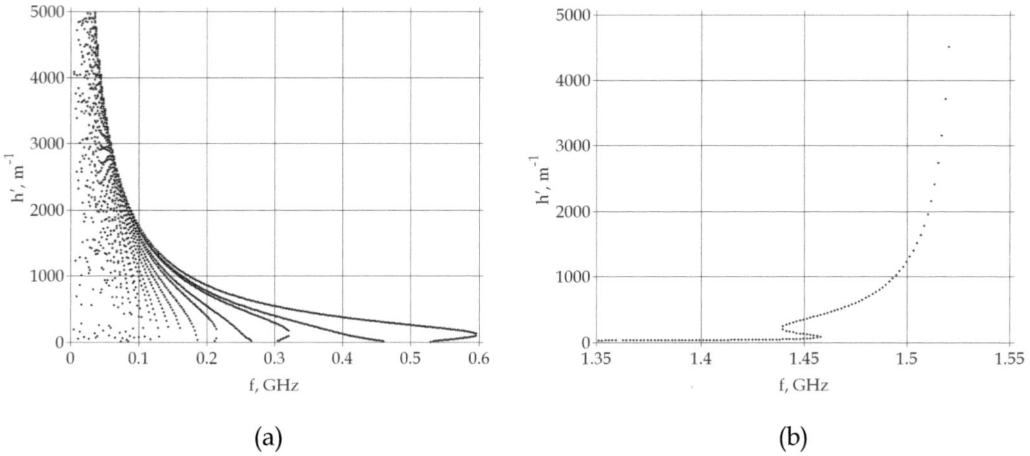

Fig. 5.6. Low frequency branches of phase constants of open plasmonic metamaterial waveguide with r=15 mm at frequencies: (a) f=0.005-0.06 GHz and (b) f=1.35-1.55 GHz.

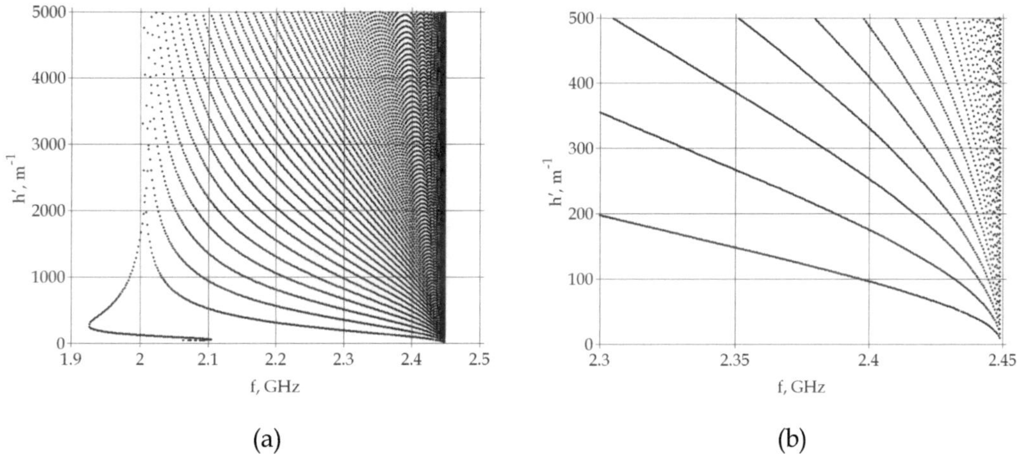

Fig. 5.7. Frequency branches of phase constants of waveguide with r=15 mm at frequencies: (a) 1.9-2.5 GHz and (b) 2.3-2.45 GHz.

6. Conclusions

1. The open anisotropic metamaterial waveguides with seven different radii were investigated by using of our computer programs that have written in MATLAB language. The algorithm is based on the partial area method.
2. We discovered the anomalous dispersion of the analyzed plasmonic waveguide eigenmodes (Figs. 3.2-3.5).
3. We find a mode with the cutoff frequency close to f=1.5 GHz with some anomalous features, e.g. the small changes in the frequency produce the very large changes in phase. This property could be useful in practical realizations.

7. Microwave scattering and absorption by layered metamaterial-glass cylinders

In this section we are going to give our calculation results for a single-layered cylinder and a twelve layered cylinder. The single–layered cylinder consists of a metamaterial core which is coated with an acrylic-glass layer. The twelve-layered cylinder consists of a conductor core which is covered with 12 metamaterial and acrylic-glass alternately layers. The acrylic-glass material is an external layer of each cylinder. The calculation of the scattered (reflected) and absorbed powers are based on the rigorous solution of scattering boundary problem (Nickelson & Bucinskas, 2011). The solution of mentioned electrodynamical problems and expressions of absorbed and scattered powers are given in article (Bucinskas et al., 2010).

The number and thickness of layers is not limited in the presented algorithm. The central core of multilayered cylinder can be made of different isotropic materials as a metamaterial, a ceramic matter or a semiconductor as well as of a perfect conductor. The isotropic coated layers can be of strongly lossy (absorbed) materials.

The signs of the complex permittivity and the complex permeability can be negative or positive in different combinations.

Here are presented the scattered and absorbed power of layered cylinder dependent on the hypothetic metamaterial permittivity and permeability signs and losses. We used for calculations our computer programs which are written in FORTRAN language.

The extern radius of both (single- and twelve-layered) cylinders is the same and equal to 2 mm. We show here our results only in the frequency range from 1 till 120 GHz. We present dependencies of the scattered and absorbed powers by the cylinders at the incident perpendicularly (the angle $\psi = 0^{\circ}$, Fig. 7.1) and parallel (the angle $\psi = 90^{\circ}$, Fig. 7.1) polarized microwaves. An incident angle of microwave is $\theta = 90^{\circ}$ here.

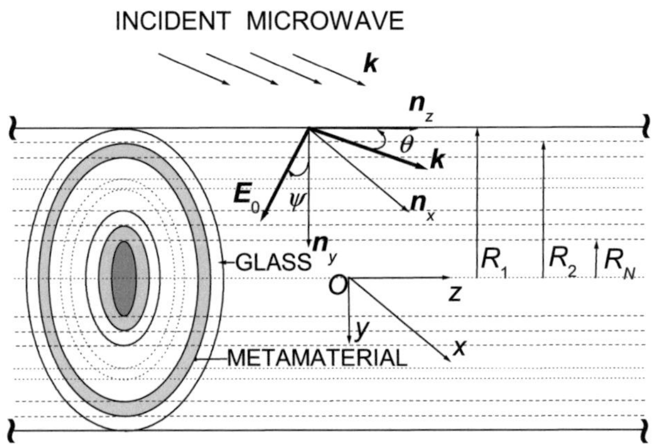

Fig. 7.1. *N*-layered metamaterial-glass cylinder model and designations

We admitted that acrylic-glass material is a non-dispersive and weakly lossy one with the complex permittivity $\varepsilon_g = \varepsilon_g' - i\varepsilon_g'' = |\varepsilon_g| \exp(-i\ \delta_g) = 3.8 - i\ 0.0005$, i.e. the phase of the complex glass permittivity is $\delta_g = \arctan(\varepsilon_g'/\varepsilon_g'') = 1.3 \cdot 10^{-4}$ [rad] and the glass permeability is equal to $\mu_g = 1$.

The complex metamaterial permittivity is $\varepsilon_{met} = \varepsilon'_{met} - i\varepsilon''_{met} = |\varepsilon_{met}| \exp(-i\,\delta_{met})$. The complex permeability is $\mu_{met} = \mu'_{met} - i\mu''_{met} = |\varepsilon_{met}| \exp(-i\,\delta_{met})$. The metamaterial ε_{met} and μ_{met} are the same for the single-layer and twelve-layer cylinders. The module of the metamaterial permittivity is $|\varepsilon_{met}|=20$ and the metamaterial permeability is $|\mu_{met}|=2$. The phase of metamaterial permittivity is $\delta_{\varepsilon,met}=0.7854$ [rad] and the phase of permeability is $\delta_{\mu,met}=0.6981$ [rad] for the single-layer and twelve layer cylinders.

We present here the absorbed power by the acrylic-glass layer (Fig. 7.3) and the absorbed power by the metamaterial (Fig. 7.4) for the single-layer cylinder. The total absorbed power by single-layered cylinder is equal to the sum of absorbed powers by metamaterial and acrylic-glass. And we show here the total absorbed power by all layers of the twelve-layered cylinder.

In Figs 7.2-7.6 are presented the averaged scattered or absorbed power values per oscillation period for the unit length of the metamaterial-glass cylinder.

The absorbed and scattered power calculations were fulfilled using by formula (28) in (Bucinskas et al., 2010)]. The integral of the formula has a positive sign when we calculate the scattered power. And this integral has a negative sign when we calculate the absorbed power. For this reason the scattered power has a positive value and the absorbed power has a negative one (see Figs 7.2-7.6).

We presented here the dependencies of scattered and absorbed powers of the single-layered cylinder on signs of ε'_{met}, ε''_{met}, μ'_{met} and μ''_{met}, i.e. when metamaterial is double positive (DP), ε- single negative (SN) or μ - single negative (SN), double negative (DN).

7.1 Numerical analysis of the scattered and absorbed microwave power of the single-layer cylinder

In this subsection we investigate the scattered and absorbed powers of the incident microwave by the cylinder consists a core of metamaterial which is covered with a single-layer of acrylic-glass. The external cylinder radius is R_1=2 mm and the cylinder core has radius R_2=1.8 mm, so the thickness of the glass layer is 0.2 mm.

Nowadays there is a huge interest to the composite materials with untraditional values of the complex permittivity ε_{met} and the complex permeability μ_{met}. We analyzed here the scattered and absorbed powers for four versions of hypothetic metamaterial parameter signs. The complex metamaterial permittivity $\varepsilon_{met}=s_1|\varepsilon_{met}|\exp(-s_2 i\delta_{met,\varepsilon})$ was taken for two combinations of signs, when $s_1=s_2=\pm1$. It is known that each metamaterial is intended for use in a specific frequency range and has a specific value of the effective permittivity and permeability at the certain frequency. For this reason we took the absolute values of real and imaginary parts of permittivity ε_{met} and permeability μ_{met} constant at all frequencies in our calculations. And the major impact makes the sizes' relation of wavelength and cylinder layers.

In figures 7.2-7.4 of this section are analyzed how the signs of the complex metamaterial permittivity and permeability influence on the scattered and absorbed powers when the plane perpendicularly or parallel polarized microwave impinges on the metamaterial-glass cylinder (Fig. 7.1).

Designations in Figs 7.2-7.4 correspond: the curve 1 is for a DP material when $s_1=s_2=s_3=s_4=+1$ (line with black squares); the curve 2 is for a SN material when $s_1=s_2=+1$, $s_3= s_4=-1$ (line with empty squares); the curve 3 is for a SN material when $s_1=s_2=-1$, $s_3= s_4=+1$ (line with black triangulars); the curve 4 is for a DN material when $s_1=s_2=s_3=s_4=-1$ (line with empty triangulars).

The scattered and absorbed dependences of the metamaterial-glass cylinder when the incident microwave has the perpendicular or parallel polarization are shown in Figs 7.2-7.4. In Fig. 7.2 is presented the dependence of total scattered power W^S on the microwave frequency *f* at two polarizations. We see that the character of curves for all metamaterial sign versions (curves 1-4) is the same at the perpendicular polarization. A comparison of curves 1-4 (Fig. 7.2(a)) shows that only curve 3 that describes by the metamaterial permittivity and permeability signs $s_1=s_2=-1$, $s_3=s_4=+1$ is the most different in comparison with other three cases. At the beginning the scattered power grows till the maximum value after that decreases till the minimum and later increases again with increasing of frequency. The total scattered power maximums of all curves are in the frequency range about 44-53 GHz. Curves 1 and 4 practically coincide with each other. The lowest scattered power is for the curve 3 at the frequencies about 1-35 GHz and the total scattered power minimum exists for the curve 2 approximately at the frequency 80 GHz.

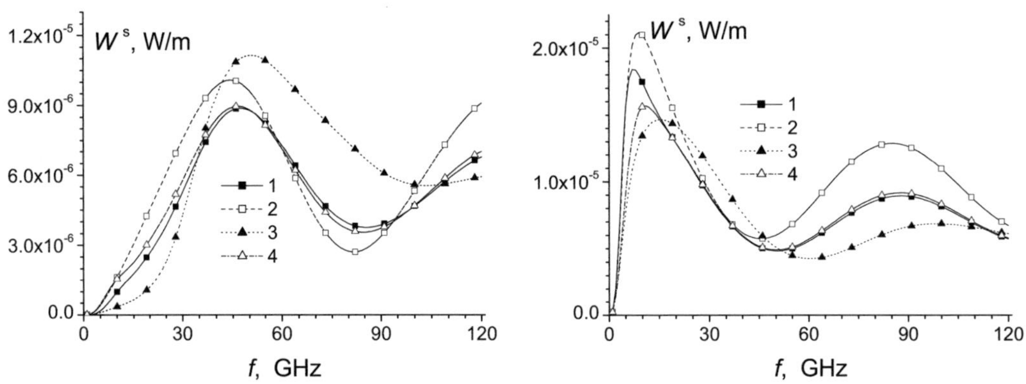

(a) (b)

Fig. 7.2. Scattered power of metamaterial-glass cylinder on the frequency of incident (a) - perpendicular and (b) - parallel polarized microwaves.

The total scattered power of the incident perpendicularly and parallel polarized microwaves (Fig. 7.2(a, b)) are different. The scattered microwave power curves for the incident parallel polarized microwave have two maximums in the frequency range 1-120 GHz. The first position of the scattered power maxima are in the narrow frequency interval about 10-15 GHz and the second position of the maxima is in the interval 85-100 GHz. The largest scattering is at the lower frequencies. The maximum scattering is higher for the incident parallel polarized microwave in comparison with the incident perpendicularly polarized one. While the total scattered power curves for the incident perpendicularly polarized microwave (Fig. 7.2(a)) have only one maximum.

In Fig. 7.3 is presented the absorbed microwave power $W^{a,glass}$ by the coated glass layer at two microwave polarizations. We see that the behavior of the microwave power absorption strongly depends on the signs s_1, s_2, s_3, s_4 of the permittivity and the permeability. The absorption is especially different at the higher frequencies. We see that the absorption power of the glass layer is larger at higher frequencies.

In Fig. 7.3(b) is presented the absorbed power by the coated glass layer for the incident parallel polarized microwave. There are some small distortion "hooks" of the absorbed power at the low frequencies. The absorbed power by glass layer increases with increasing of frequencies (curves 1, 3, 4) for frequencies that are larger than 20 GHz. The absorbed power is approximately constant when the SN metamaterial core permittivity and permeability have signs $s_1=s_2=+1$, $s_3=s_4=-1$.

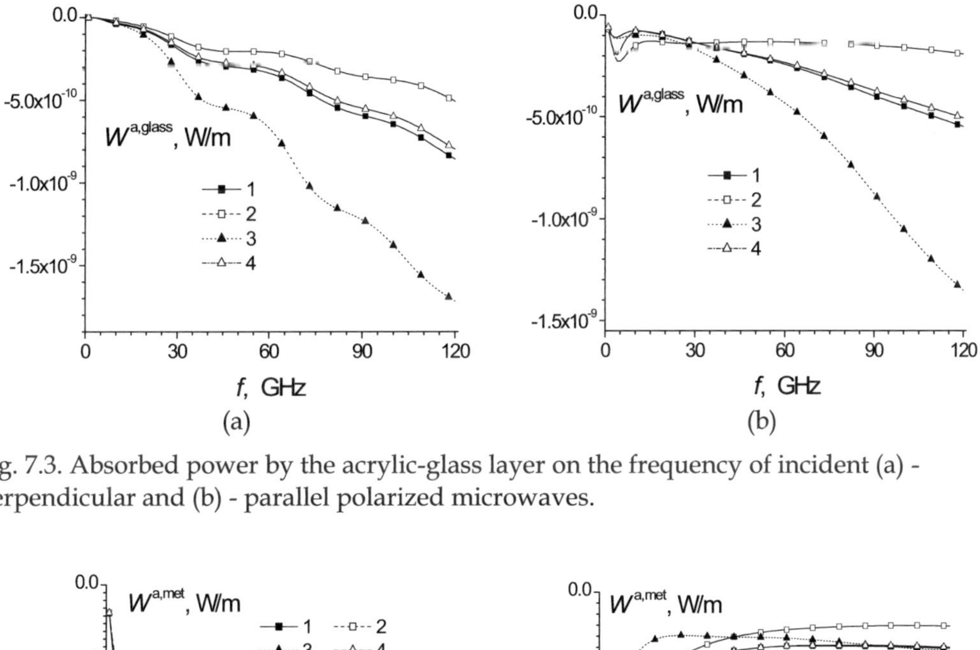

(a) (b)

Fig. 7.3. Absorbed power by the acrylic-glass layer on the frequency of incident (a) - perpendicular and (b) - parallel polarized microwaves.

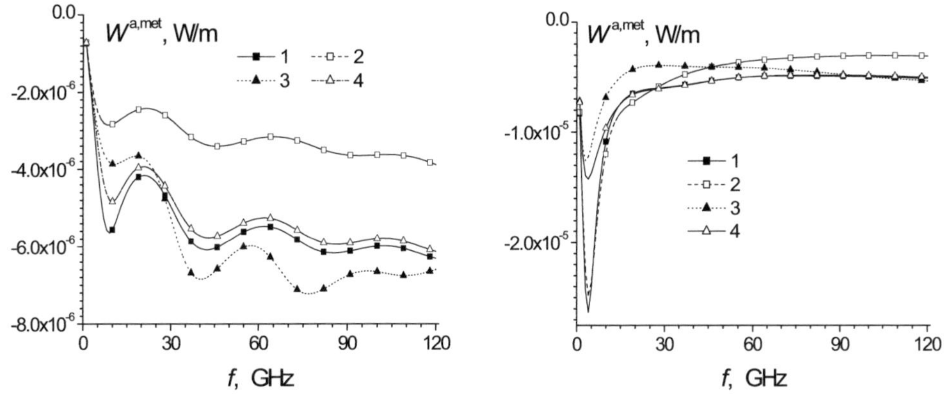

(a) (b)

Fig. 7.4. Absorbed power by metamaterial core on the frequency of incident (a) - perpendicular and (b) - parallel polarized microwaves

In Fig. 7.4(a) is given the absorbed power by the metamaterial core of the cylinder for the perpendicular microwave polarization. We see that the $W^{a,met}$ magnitudes have the pronounced wave-like nature dependent on the frequency. The metamaterial core absorption has the largest absolute value when the metamaterial permittivity has some negative values and the permeability has some positive values (curve 3). In Fig. 7.4(b) is given the metamaterial core absorbed power $W^{a,met}$ for the incident parallel polarized microwave. We see that the absorbed-powers have maximum values at about 5 GHz and their values vary slightly after 20 GHz with increasing of frequency. The absorption by the metamaterial is the largest at the low frequencies (curves 1 and 2) when the metamaterial has the positive permittivity. The comparison of absorbed powers in figures 7.3 and 7.4 shows that dependencies are absolutely different.

7.2 Numerical analysis of the scattered and absorbed microwave power of twelve-layer cylinder

In this subsection we present the total scattered power and the total absorbed power of incident microwave by the cylinder that consists of conductor core covered with 12 metamaterial and glass alternately layers. Designations in Figs 7.5 and 7.6 correspond: the curve 1 is for a DP material when $s_1=s_2=s_3= s_4=+1$ (line with black squares) and the curve 2 is for a DN material when $s_1=s_2=s_3= s_4=-1$ (line with empty triangulars).

In Figs 7.5 and 7.6 the permittivities and permeabilities metamaterial and acrylic-glass are the same as in the previous subsection. We see that the total scattered and absorbed powers strongly dependent on the polarization of incident wave. The characteristics of perpendicular polarized wave (Figs 7.5(a) and 7.6(a)) and the parallel polarized wave (Figs 7.5(b) and 7.6(b)) are completely different. Particularly noticeable the correlation between the total scattered and absorbed powers for the parallel polarized wave of the twelve-layered cylinder. We see that extremums of the total scattered and absorbed powers coincide in the frequency scale. For example, when the scattered power of cylinder has a maximum (Fig. 7.5(b)) then the absorbed power has a minimum (Fig. 7.6(b)) at f~10 GHz.

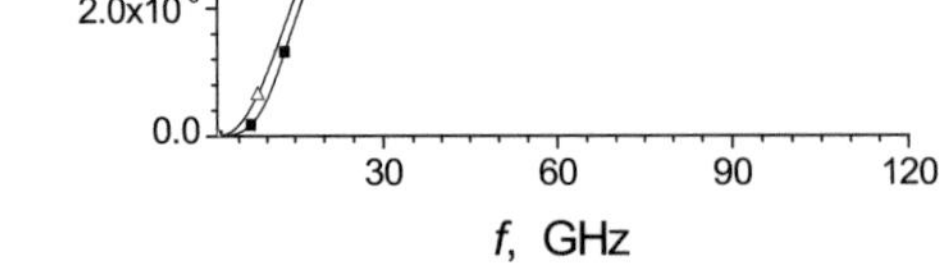

(a)

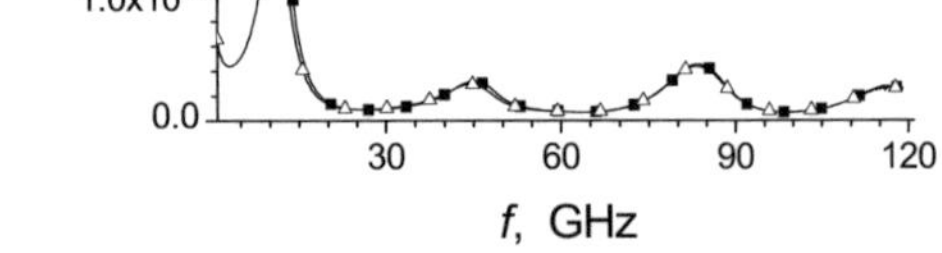

(b)

Fig. 7.5. Total scattered power of twelve layer metamaterial-glass cylinder on the frequency of incident (a) - perpendicular and (b) - parallel polarized microwaves

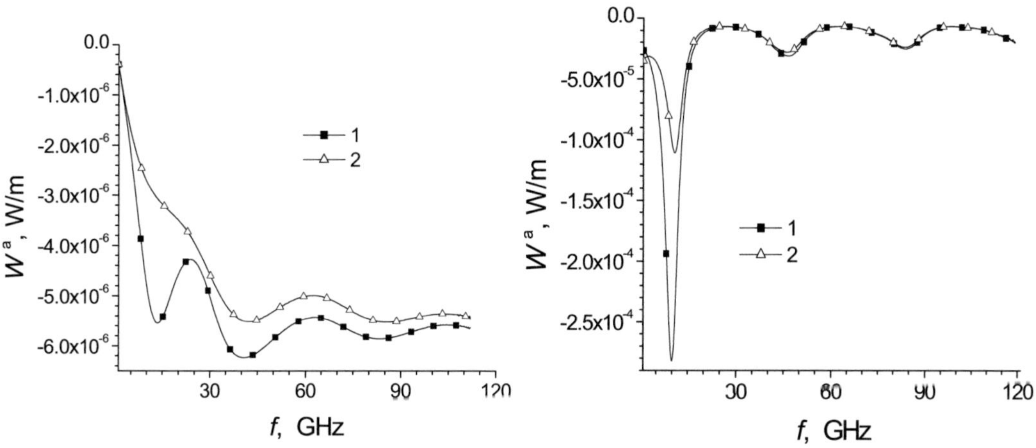

(a) (b)

Fig. 7.6. Total absorbed power by the twelve layered metamaterial-glass cylinder on the frequency of incident (a) - perpendicular and (b) - parallel polarized microwaves.

The comparison of scattered power of single- and twelve-layered cylinders shows that dependencies for the perpendicular polarized microwave are very similar. The scattering power maximum is at ~45 GHz and the minimum is at ~80 GHz (Figs 7.2(a) and 7.5(a)).

The scattered powers of single- and twelve-layered cylinders for the parallel polarized microwave are different (Figs 7.2(b) and 7.5(b)). We see that the scattered power of single-layered cylinder has only two maximums in the considered frequency range while the twelve-layered cylinder dependency has four maximums. The first scattered power maximum of twelve-layered cylinder has a sharp pike and is more than twice larger in comparison with the single-layered cylinder (Figs 7.2(b) and 7.5(b)).

Comparing Figs 7.3(a), 7.4(a) and 7.6(a) for perpendicular polarized microwave we see that the absorbed power determined mainly due to metamaterial losses. The absorbed power extremums of the twelve layered cylinder (Fig. 7.6 (a)) shifted to the higher frequencies in the comparison with single-layered cylinder (Fig. 7.4(a)) for the DP metamaterial (curve 2).

Comparing curves for single- and twelve–layered cylinders (Fig. 7.4(a), curve 4 and Fig. 7.6 (a), curve 2) when the metamaterial is SN we can note that their lineaments are different at the low frequencies.

The dependencies of absorbed parallel polarized microwave power for single-layered and twelve-layered cylinders in general are alike. The first minimum of absorbed power is shifted from ~5 GHz (Fig. 7.4(b)) till 10 GHz (Fig. 7.6 (b)) i.e. with growing of the number of layers the minimum shifted to the side of higher frequencies. The absorbed power dependency of twelve layered cylinder has a wavy behavior in the frequency range from 20 GHz till 120 GHz in a comparison with the single-layered one.

8. Conclusions

1. We found that the scattered power dependences have wave behaviors. The minimal scattering from the metamaterial-glass cylinder are observed for the every metamaterial with some sign combinations of the permittivity and the permeability at the special frequency range (Figs 7.2).
2. We found that the largest absorbed power by the coated acrylic-glass layer is observed for the case when the metamaterial is a single negative material with the negative permittivity. The absorbed power of the glass layer increases with increasing of frequency in the range 1-120 GHz for both microwave polarizations (Figs 7.2).
3. The metamaterial core absorbed power of the parallel polarized incident microwave has the minimum value at low frequencies and slightly dependent on the frequency at the range 20-120 GHz (Fig. 7.4(b)).
4. The comparison of single- and twelve- layered cylinder characteristics shows that the absorbed power extremums shift to the direction of higher frequencies when the number of layers becomes larger (Figs 7.4 and 7.6).

9. References

Abdalla, M. A.& Hu, Z. (2009). Multi-band functional tunable lh impedance transformer, *J. of Electromagn. Waves and Appl.,* Vol. 23, pp. 39–47, ISSN 0920–5071.

Asmontas, S.; Nickelson, L.; Bubnelis, A.; Martavicius, R. & Skudutis, J. (2010). Hybrid Mode Dispersion Characteristic Dependencies of Cylindrical Dipolar Glass Waveguides on Temperatures, *Electronics and Electrical Engineering,* Vol. 106, No. 10, pp. 83-86, ISSN 1392-1215.

Asmontas, S.; Nickelson, L.; Gric, T. & Galwas, B. A. (2009). Solution of Maxwell's Equations by the Partial Area Method for the Electrodynamical Analyses of Open Lossy Metamaterial Waveguides, *in Proceedings of the International Conference Differential Equations and their Applications. DETA–2009,* pp. 17–20, ISBN 978-9955-25-747-9, Panevezys, Lithuania, 10-12 Sept. 2009, TECHNOLOGIJA, Kaunas University of Technology, Kaunas.

Asmontas, S.; Nickelson, L. & Plonis, D. (2009). Dependences of propagation constants of cylindrical n-Si rod on the material specific resistivity, *Electronics and Electrical Engineering,* Vol. 94, No. 6, pp. 57-60, ISSN 1392-1215.

Bai, J.; Shi, S. & Prather, D. W. (2010). Analysis of epsilon-near-zero metamaterial super-tunneling using cascaded ultra-narrow waveguide channels. *Progress In Electromagnetics Research M,* Vol. 14, pp. 113-121, E-ISSN 1559-8985.

Bucinskas, J.; Nickelson, L. & Sugurovas, V. (2010). Microwave scattering and absorption by a multilayered lossy metamaterial-glass cylinder, *Progress In Electromagnetics Research,* PIER105, pp. 103–118, ISSN 1070-4698, E-ISSN 1559-8985.

Bucinskas, J.; Nickelson, L. & Sugurovas, V. (2010). Microwave diffraction characteristic analysis of 2D multilayered uniaxial anisotropic cylinder, *Progress In Electromagnetics Research,* PIER109, pp. 175-190, ISSN 1070-4698, E-ISSN 1559-8985.

Chen, H.; Wu, B.-I. & Kong, J. A. (2006) Review of electromagnetic theory in left-handed materials *J. of Electromagn. Waves and Appl.,* Vol. 20, No. 15, pp. 2137–2151, ISSN 0920–5071.

Engheta, N. & Ziolkowski R.W. (2005). A positive future for double-negative metamaterials. *IEEE Transactions on microwave theory and techniques,* Vol. 53, No. 4, pp. 1535- 1556, ISSN 0018–9480.

Feng, T. H.; Li, Y.H.; Jiang, H. T.; Li W. X., Yang F.; Dong X. P.& Chen H. (2011) Tunable single-negative metamaterials based on microstrip transmission line with varactor diodes loading, *Progress In Electromagnetics Research,* PIER 120, 35-50, ISSN 1070-4698, E-ISSN 1559-8985.

Gric, T. & Nickelson, L. 2011. Electrodynamical investigation of the photonic waveguide structure. *Electronics and Electrical Engineering,* Vol. 111, No. 5, pp. 1-4, ISSN 1392-1215.

Gric, T.; Nickelson, L. & Asmontas, S. (2010). Electrodynamical characteristic particularity of open metamaterial square and circular waveguides. *Progress In Electromagnetics Research,* PIER109, pp. 361-379, ISSN 1070-4698, E-ISSN 1559-8985.

Gric, T.; Ašmontas, S. & Nickelson, L. (2010). 3D vector electric field distributions and dispersion characteristics of open rectangular and circular metamaterial waveguides, *Proceedings of 4th Microwave & Radar Week: 18th Int. Conf. on Microwaves, Radar, and Wireless Communicatins MIKON-2010,* Vol. 2, pp. 578–581, ISBN 978-9955-690-20-7, IEEE Catalog Number CFP10784-PRT, Vilnius, Lithuania, 14-16 June 2010, JUSIDA, Vilnius.

Ko, S.-T. & Lee, J.-H. (2010). Dual Property of Mu-near-zero to Epsilon-near-zero material, *J.of the Korean Physical Society,* Vol. 57, No. 1, pp. 51-54, ISSN 03744884.

Kong, J.A. (2008). *Electromagnetic wave theory,* 1016 p., EMW Publishing, ISBN 0-9668143-9-8, Cambridge, Massachusetts, USA.

Li, M.-H.; Yang, H.-L.; Hou, X.-W.; Tian, Y.& Hou, D.-Y. (2011). Perfect metamaterial absorber with dual bands, *Progress In Electromagnetics Research,* PIER108, 37-49, ISSN 1070-4698, E-ISSN 1559-8985.

Liu, S.-H.; Liang, C.-H.; Ding, W.; Chen, L. & Pan, W.-T. (2007). Electromagnetic wave propagation through a slab waveguide of uniaxially anisotropic dispersive metamaterial, *Progress In Electromagnetics Research,* PIER76, pp. 467–475, ISSN 1070-4698, E-ISSN 1559-8985.

Lopez-Garcia, B.; Murthy, D. V. B. & Corona-Chavez, A. (2011). Half mode microwave filters based on epsilon near zero and mu near zero concepts. *Progress In Electromagnetics Research,* Vol. 113, pp. 379-393, ISSN 1070-4698, E-ISSN 1559-8985.

Luo, J.; Chen, H.; Lai, Y.; Xu, P. & Gao L. (2011) Anomalous transmission properties of epsilon-near-zero metamaterials. *Proceedings of* Progress In Electromagnetics Research Symposium, pp. 1299-1302, ISSN 1559-9450, ISBN 978-1-934142-18-9, Suzhou, China, 12-16 Sept. 2011, The Electromagnetics Academy, Cambridge, MA 02138.

Liu, R.; Cheng, Q.; Hand, T.; Mock J.J.; Cui T. J.; Cummer S. A. &Smit, D. R.(2008) Experimental Demonstration of Electromagnetic Tunneling Through an Epsilon-Near-Zero Metamaterial at Microwave Frequencies, Physical Review Letters, PRL 100, pp. 023903-(1-4), ISSN 0031-9007.

Mirza, O.; Sabas, J. N.; Shi, S. & Prather, D. W. (2009) Experimental demonstration of metamaterialbased phase modulation, *Progress In Electromagnetics Research,* PIER 93, pp. 1–12, ISSN 1070-4698, E-ISSN 1559-8985.

Mirzavand, R.; Honarbakhsh, B.; Abdipour, A. & Tavakoli A. (2009). Metamaterial-based phase shifters for ultra wide-band applications. *J. of Electromagn. Waves and Appl.,* Vol. 23, pp. 1489–1496, ISSN 0920–5071.

Nickelson, L.& Bucinskas, J. (2011). Microwave diffraction dependencies of a conductor cylinder coated with twelve glass and semiconductor layers on the *n*-Si specific resistivity. *Electronics and Electrical Engineering,* Vol. 115, No. 9, pp. 47-50, ISSN 1392-1215.

Nickelson, L.; Gric, T. & Asmontas, S. (2011). Chapter 6: Electrodynamical modelling of open cylindrical and rectangular carbide waveguides, In: *Properties and applications of Silicon Carbide*, (Ed. Rosario Gerhardt), 535 p., pp 115-141, InTech, ISBN 978-953-307-201-2, Rijeka, Croatia.

Nickelson, L.; Bubnelis, A.; Baskys, A. & Navickas, R. (2011). The magnetoactive *p*-Ge rod waveguide loss analysis on the concentration of two component hole charge carriers, *Electronics and Electrical Engineering*, Vol. 110, No. 4, pp. 53-56, ISSN 1392-1215.

Nickelson, L.; Asmontas, S.; Malisauskas, V. & Martavicius, R. (2009). The dependence of open cylindrical magnetoactive *p*-Ge and *p*-Si plasma waveguide mode cutoff frequencies on hole concentrations, *Journal of Plasma Physics*, Vol. 75, No. 1, pp. 35–51, ISSN 0022-3778.

Nickelson, L.; Gric, T.; Asmontas, S. & Martavicius, R. (2009) Analyses of the gyroelectric plasma rod waveguide, *Proceedings of 17th IEEE International Pulsed Power Conference*, pp. 727-730, IEEE Catalog Number: CFP09PPC-DVD, Library of Congress: 2009901215, DC USA, ISBN 978-1-4244-4065-8, Washington, DC, USA, 29 June - 2 July 2009.

Nickelson, L.; Gric, T.; Asmontas, S. & Martavicius, R. (2008). Electrodynamical analyses of dielectric and metamaterial hollow-core cylindrical waveguides. *Electronics and Electrical Engineering*, Vol. 82, No. 2, pp. 3-8, ISSN 1392-1215.

Nickelson, L.; Galwas, B.A.; Gric, T. & Ašmontas, S. (2008). Electric field distributions in the cross sections of the metamaterial hollow-core and rod waveguides, *Proceedings of 17th International Conference on Microwaves, Radar and Wireless Communication MIKON-2008*, Vol. 2, pp. 497-500, ISBN 978-1424-431-22-9, Wrocław, Poland, 19-21 May 2008, MDruk, Warszawa, Poland.

Nickelson, L. & Shugurov, V. (2005). *Singular integral equations' method for the analysis of microwave structures*, 348 p., VSP Brill Academic Publishers, ISBN 90-6764-410-2, Leiden-Boston, The Netherlands.

Oraizi, H. & Abdolali, A. (2009) Some aspects of radio wave propagation in double zero metamaterials having the real parts of epsilon and mu equal to zero, *J. of Electromagn. Waves and Appl.*, Vol. 23, pp. 1957–1968, ISSN 0920–5071.

Penciu, R.S.; Kafesaki, M.; Gundogdu, T.F.; Economou, E.N. & Soukoulis, C.M. (2006). Theoretical study of left-handed behavior of composite metamaterials. *Photonics and Nanostructures - Fundamentals and Applications*, Vol. 4, pp. 12-16, ISSN 1569-4410.

Wang, B. &. Huang, K.-M (2010). Shaping the radiation pattern with mu and epsilon-near-zero metamaterials. *Progress In Electromagnetics Research*, PIER 106, pp. 107-119, ISSN 1070-4698, E-ISSN 1559-8985.

Zhou, H.; Ding, F.; Jin, Y. & He, S. L. (2011). Terahertz metamaterial modulators based on absorption, *Progress In Electromagnetics Research*, PIER119, pp. 449-460, ISSN 1070-4698, E-ISSN 1559-8985.

Zhou, H.; Pei,Z.; Qu, S.; Zhang, S. & Wang, J. (2009) A planar zero-index metamaterial for directive emission, *J. of Electromagn. Waves and Appl.*, Vol. 23, pp. 953–962, ISSN 0920–5071.

Zhu, B.; Wang, Z.; Huang, C.; Feng, Y.; Zhao, J. & Jiang, T. (2010) Polarization insensitive metamaterial absorber with wide incident angle, *Progress In Electromagnetics*, PIER101, pp. 231-239, ISSN 1070-4698, E-ISSN 1559-8985.

3

The Group Velocity Picture of Metamaterial Systems

Xunya Jiang, Zheng Liu, Wei Li, Zixian Liang, Penjun Yao, Xulin Lin, Xiaogang Zhang, Yongliang Zhang and Lina Shi

State Key Laboratory of Functional Materials for Informatics, Shanghai Institute of Microsystem and Information Technology, CAS, Shanghai China

1. Introduction

Metamaterial is a very fresh concept in modern photonics, which are referred to a new class of electromagnetic media whose permittivity or permeability is beyond traditional values. Fifty years ago, a kind of new materials whose permittivity ϵ and permeability μ are simultaneously negative was theoretically predicted to possess a negative refractive index n with many unusual properties[1]. In last decade, negative-n metallic resonating composites and two dimensional (2D) isotropic negative-n material have been constructed[2, 3], and negative light refraction was observed[4]. The unconventional properties of such materials, such as the evanescent wave could be amplified by negative-n so that the sub-wavelength resolution could be achieved[5], have drawn an increasing amount of attention in both science and engineering[6]. After negative-n material, more such unconventional materials are found, so that a new concept "metamaterial" is generated, which termed for the effective medium with very special permittivity ϵ_{eff}, or permeability ϵ_{eff}, or both, over a certain finite frequency band. Such physical media are composed of distinct elements (photonic atoms) which are generally made of sub-wavelength metallic structure and their size scale is much smaller than the wavelengths in the frequency range of interest. Thus, the effective composite media could be considered homogeneous at the wavelengths under consideration. Since their abnormal properties and related totally new phenomena can even go beyond the traditional physical limit, metamaterial becomes one of hottest topics in modern photonics.

However, from the beginning of the metamaterial research, there are many arguments for a lot topics, such as, Pendry's famous pioneer work of superlens[5] was commended several times. One main reason of so many arguments is that the light *beams* in different metamaterials seem to be too strange (even weird) to be acceptable. So it is natural to argue whether these beams could be real. Another main reason is a general weakness of current metamaterial studies which mainly focus on the single frequency properties and neglect the dispersion. Actually these two reasons are related. We know that the dispersion , in the frame of classical electrodynamic, means the electromagnetic response of the material to the external field, and plays the key role in the metamaterial abnormal properties. For these strange beams, such as negative refraction beams, with dispersion we can obtain the group velocity (energy velocity) which determine the beams propagating direction. So the group velocity should be the basic picture for us to understand these strange beams and help us to design related devices. More seriously, the dispersion is related with some very basic limitations of this world, *e.g.* the causality limitation that the group velocity in metamaterial should be less than the vacuum light speed. If our design of devices is based on the metamaterials which violate these basic limitations, the design surely can not work since such metamaterial could not exist in this world. From this view, the group velocity picture is not only needed in understanding and explanation, but also required in research of some topics and design of devices. For instance, in the research of the limitation of the cloak[7, 8] and abnormal phenomena on the interface of the hyperbolic metamaterial[9], if we neglect the dispersion of material, from group velocity picture we will immediately find that we have fallen into a superluminal trap, since the energy velocity in such artificial systems is divergent. So, the group velocity picture can help us avoid such traps.

As a basic value for revealing the complex propagating process, abnormal group velocity has been studied for decades. In the early 1960's the group velocity in material has been studied by Brillouin[10] and the group velocity in strongly scattering media is investigated by J. H. Page, Ping Sheng *et al.* in 1996[11], which indicate that the physical origin of the remarkably low velocities of propagation lies in the renormalization of the effective medium by strong resonant scattering. So far, metamaterials generally are composite of "photonic atoms" which can scatter light coherently. And all abnormal properties of metamaterials, *e.g.* these strange beams, are from these complex coherent scattering. Another byproduct of these scattering is the (abnormal) group velocity. In other words, the strange beam and the abnormal group velocity are two sides of a same coin. Further more, with some abnormal group velocity, such as the extremely slow light, we can design new signal-processing devices or new detecting devices. Hence, exploring the group velocity in metamaterial is very vital for revealing mechanism and the design of the real optical devices.

The numerical simulation takes an important role in research for modern photonics. For metamaterial, since the difficulties of experimental realization, the numerical tools become very essential for researchers. But, in some frequency domain simulation softwares, the dispersion is neglected totally. As we discussed above, we think such softwares can misleading researchers to some imaginary metamaterial which can not exist in this world. Such as for cloaking study, these softwares could present perfect invisibility very easily, but from our study[7, 8], that is misleading one. We strongly recommend the time-domain softwares, such as finite-difference time-domain (FDTD) method or finite-element time-domain(FETD) method. Their simulating results are much more convincible since they are generally with physical dispersion in the simulation and fit for metamaterial studies.

This paper is organized as following. The first section is the introduction in which we generally introduce the group velocity picture of the metamaterial study. As we have discussed, the group velocity is the key for understanding these abnormal properties of metamaterials and also can help us to avoid some traps of basic physical limit. We have also commended the softwares fit for metamaterial studies.

In the second section, the optical properties of the interface between hyperbolic meta-material (with anisotropic hyperbolic dispersion) and common dielectric is investigated. With material dispersion, a comprehensive theory is constructed, and the hyperlens effect that the evanescent wave can be converted into the radiative wave is confirmed. At the inverse process of hyperlens, we find a novel mechanism to compress and stop (slow) light at wide frequency range, which can be used as a removable memory or a light trap. All theoretical results are demonstrated by finite-difference time-domain simulation.

In the third section we propose general evanescent-mode-sensing methods to probe the quantum electrodynamics (QED) vacuum polarization. The methods are based on the phase change and the energy time delay of evanescent wave caused by small dissipation. From our methods, high sensitivity can be achieved even though the external field, realizable in contemporary experiments, is much smaller than the Schwinger critical field.

In the forth section the image field of the negative-index superlens with the quasi-monochromatic random source is discussed, and dramatic temporal-coherence gain of the image in the numerical simulation is observed, even if there is almost no reflection and no frequency filtering effects. From the new physical picture, a theory is constructed to obtain the image field and demonstrate that the temporal coherence gain is from different "group" retarded time of different optical paths. Our theory agrees excellently with the numerical simulation and strict Green's function method. These study should have important consequences in the coherence studies in the related systems and the design of novel devices.

In the fifth section, the dynamical processes of dispersive cloak by finite-difference time-domain numerical simulation are carried out. It is found that there is a strong scattering process before achieving the stable state and its time length can be tuned by the dispersive strength. Poynting-vector directions show that the stable cloaking state is constructed locally while an intensity front sweeps through the cloak. Deeper studies demonstrate that the group velocity tangent component $V_{g\theta}$ is the dominant factor in the process. This study is helpful not only for clear physical pictures but also for designing better cloaks to defend passive radars.

In the sixth section, the limitation of the electromagnetic cloak with dispersive material is investigated based on causality. The results show that perfect invisibility can not be achieved because of the dilemma that either the group velocity V_g diverges or a strong absorption is imposed on the cloaking material. It is an intrinsic conflict which originates from the demand of causality. However, the total cross section can really be reduced through the approach of coordinate transformation. A simulation of finite-difference time-domain method is performed to validate the analysis.

In the last section, we give a summary of our works.

2. Hyper-interface, the bridge between radiative wave and evanescent wave

Many new phenomena are observed at the interfaces between meta-material and common dielectric material, such as the negative refraction which is found at the left-handed material(LHM) surface More interestingly, the evanescent wave (EW) could be amplified at LHM interface so that the super-resolution could be achieved[14]. Besides the LHM, there is another class of anisotropic metamaterial, so called "hyperbolic medium"(HM), in which one of the diagonal permittivity tensor components is negative and results in a hyperbolic dispersion. For convenience, we call the interface between a HM and a common dielectric material as "hyper-interface"(HI). Some surprising electromagnetic properties of HI are intensively studied recently[15, 16]. For instance, HI can convert the EW into the radiative wave (RW) so that the sub-wavelength information could be observed at far-field, which is called "hyper-lens" effect[15]. Very recently it is found that when HI is perpendicular to one asymptote of HM dispersion, abnormal omnidirectional transmission occurs[17]. Although some theoretical and experimental works [16] have demonstrated that the EW really can be converted into RW by HI of the layered cylindrical HM, a full theory involving the "material dispersion"(will be explained later) has not been given so far. For meta-material systems, if without physical dispersion, some abnormal optical properties can not be clearly explained and the dynamical study of wave propagation can not be carried out[20]. Even more seriously, the causality violation because of the superluminal group velocity ($v_g > c$) in HM is pointed out[18], which makes the observed hyper-lens effect doubtful. To solve these problems and predict new phenomena, more robust theory with stronger base is needed.

On the other hand, to compress and to stop (slow) light pulses are very essential for modern optical/photonics research and signal processing. Hence, a new mechanism, which can compress and stop (slow) light pulses and is frequency and direction insensitive, would induce wide interest in related directions.

In this Letter, we theoretically and numerically investigate the novel optical properties of flat HI[22], in which, unlike the cylindrical HI, the translational symmetry guarantees the simple physical picture for intuitive understanding, the quantitative study of the conversion between EW and RW *etc.*

A general theory of HI is constructed with physical dispersion of HM. On the HI, not only the conversion from EW to RW (CER) of hyperlens is confirmed, when RW is incident from HM to dielectric(the inverse process of hyper-lens), but also the almost total conversion from RW to EW (CRE) can occur, *i.e.* there is "no-transmission and no-refelection" (NTNR). More important we find that this is a new mechanism to compress and stop (slow) light pulses in wide frequency and direction range with many potential applications. Theoretically and numerically we demonstrate that the superluminal group velocity in hyperlens is artificial since the HM material dispersion is neglected in previous study[18]. At last, the feasibility to realize these functions on real structures is discussed. All theoretical results are demonstrated by finite-difference-time-domain (FDTD) simulations.

Our model is as follows. Assuming two plane waves are incident to HI from HM and isotropic dielectric, and scattered from HI, as shown in the upper-right insert of Fig.(1). The HI is in the x-z plane, while the incident surface and both HM optical axes lie in the x-y plane. The incident waves are chosen as TM wave with field components (E_x, E_y, H_z) and same "parallel wave-vector" k_x. The HM is with the permittivity tensor as

$$\hat{\epsilon}_p = \begin{pmatrix} \epsilon_1(\omega) & 0 \\ 0 & \epsilon_2 \end{pmatrix} \tag{1}$$

in its principle axes coordinate, where $\epsilon_1 < 0$ and $\epsilon_2 > 0$ are assumed. And the permittivity of isotropic dielectric material is ϵ. The essential point of our model is that the negative diagonal component is dispersive $\epsilon_1 = \epsilon_1(\omega)$, which is called *material dispersion* in our study. It is well known that dispersion is physically required for real meta-materials with abnormal effective constitutive coefficients, such as negative permittivity. We will see that the material dispersion will help us to obtain self-consistent explanation of abnormal optical properties of HI and to avoid causality violation.

For simplicity, the HM is nonmagnetic and Gaussian unit is employed throughout the paper. We define the angle between the HI (or x axis) and the positive-ϵ principle axis of HM is θ,

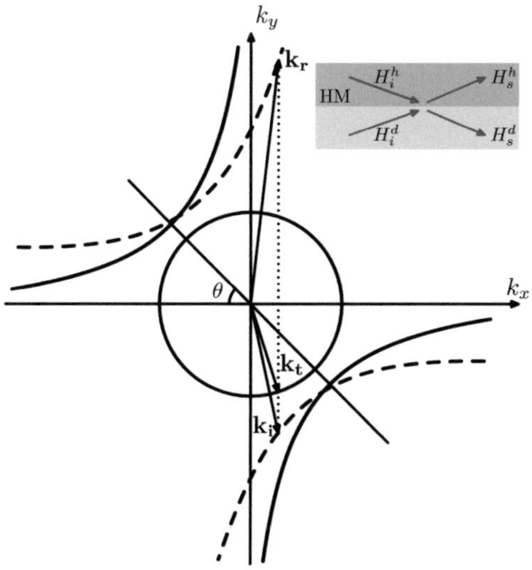

Fig. 1. The frequency contour of HM and isotropic dielectric material in k space. The inset: the schematic figure of our model.

then the most general frequency contour in k space of HM is:

$$\frac{(k_x \cos\theta - k_y \sin\theta)^2}{\epsilon_2} - \frac{(k_y \cos\theta + k_x \sin\theta)^2}{|\epsilon_1|} = (\frac{\omega}{c})^2 \tag{2}$$

In the HM region and the isotropic dielectric region the fields can be expressed uniformly as $\mathbf{H}^\sigma = \mathbf{e_z}(H_{iz}^\sigma + H_{sz}^\sigma)e^{ik_x x - i\omega t}$, $\mathbf{E^{\sigma}} = [\mathbf{e_x}(c_{ix}^\sigma H_{iz}^\sigma + c_{sx}^\sigma H_{sz}^\sigma) + \mathbf{e_y}(c_{iy}^\sigma H_{iz}^\sigma + c_{sy}^\sigma H_{sz}^\sigma)]e^{ik_x x - i\omega t}$ where the coefficients are defined as $c_{\nu\xi}^\sigma = E_{\nu\xi}^\sigma / H_{\nu z}^\sigma$ with $\nu = i, s$ for the incident fields to HI or the scattered fields from HI; $\xi = x, y$; $\sigma = h, d$ for the HM region or dielectric region. Since the translation symmetry of flat HI, the wave-vector parallel component k_x is continuous at both regions. In following discussion, the k vectors are normalized by $\tilde{k}_{\nu\xi}^\sigma = k_{\nu\xi}^\sigma / k_0$ where $k_0 = \omega / c$.

To explore the transmission and reflection properties of HI, we define a scattering matrix associated with the incident fields and out-going fields at HI.

$$(H_{sz}^h, H_{sz}^d)^T = S_{2\times 2}\,(H_{iz}^h, H_{iz}^d)^T \tag{3}$$

where the superscript T means the matrix transpose, $S_{11} = -(c_{ix}^h \epsilon + \tilde{k}_y^d)/(c_{sx}^h \epsilon + \tilde{k}_y^d)$, $S_{12} = (-c_{ix}^h \epsilon + \tilde{k}_{iy}^d)/(c_{sx}^h \epsilon - \tilde{k}_{iy}^d)$, $S_{21} = (c_{sx}^h - c_{ix}^h)\epsilon/(c_{sx}^h \epsilon + \tilde{k}_y^d)$, and $S_{22} = -2\tilde{k}_{iy}^d/(c_{sx}^h \epsilon - \tilde{k}_{iy}^d)$. From the standard boundary conditions, the coefficients $c_{i(s)x}^h$, $c_{i(s)x}^d$ are worked out to be $c_{i(s)x}^h = (-\tilde{k}_{i(s)y}^h \alpha^2 + \tilde{k}_x \alpha\gamma)/(|\epsilon_1|\epsilon_2)$; $c_{i(s)x}^d = -\tilde{k}_{i(s)y}^d / \epsilon$, where factors α, γ are defined as $\alpha = (\epsilon_1 \sin^2\theta - \epsilon_2 \cos^2\theta)^{\frac{1}{2}}$; $\gamma = (|\epsilon_1| + \epsilon_2)\sin 2\theta / 2\alpha$, and the values of $\tilde{k}_{i(s)y}^d = \pm\sqrt{\epsilon - \tilde{k}_x^2}$ and $\tilde{k}_{i(s)y}^h$ and $\tilde{k}_{i(s)y}^h = (\pm(|\epsilon_1|\epsilon_2(\tilde{k}_x^2 + \alpha^2))^{1/2} + \tilde{k}_x \alpha\gamma)/\alpha^2$ are uniquely determined by Eq.(2), respectively. From Eq.(3) we can easily get the reflection and transmission coefficients across the HI from upper to down, or inverse. For the case of hyperlens that the wave is incident from the isotropic medium to the HM, $t_{dh} = S_{22}$; $r_{dh} = S_{12}$. When $\tilde{k}_x^2 > \epsilon$, the incident and the reflected waves in the isotropic dielectric are EWs with y-component wave-vectors as $\tilde{k}_{iy}^d = i\sqrt{\tilde{k}_x^2 - \epsilon} = -\tilde{k}_{sy}^d \equiv i\kappa$. We note that, although single EW can not carry net energy current(time averaged), two EWs, *i.e.* the incident and reflected EWs, *can* carry net energy current $\vec{S}_{iy}$ in isotropic dielectric medium, since the reflected EW gains an extra-phase from complex reflecting coefficient r_{dh}. The energy current $\vec{S}_{iy}$ carried by two EWs can be converted by HI into the RW energy current $\vec{S}_{ty}$ in the HM:

$$|\vec{S}_{iy}| = \frac{\kappa}{\epsilon}\text{Im}(r_{dh}) = -\frac{2c_{sx}^h\kappa^2}{c_{sx}^{h\,2} + \kappa^2}$$
$$= |\vec{S}_{ty}| = -\frac{1}{2}|t|^2 c_{tx} \tag{4}$$

From Eq(4), the hyper-lens effect and the image-improving by CER could be quantitatively studied.

After confirming CER on HI, it is natural to wonder if CRE can occur too, or if there are other novel phenomena on HI. Next we will study the inverse process of hyper-lens, *i.e.* the RW is incident from HM and the transmitted field is in the dielectric. For such inverse processes, there is a critical condition $\theta = \theta_c \equiv \arctan\sqrt{\epsilon_2/|\epsilon_1|}$, which means HI ($x$ axis) perpendicular to one of hyperbola-dispersion asymptotes, or in other words, the asymptote is parallel with y axis now, as shown by the the solid lines in Fig.(1). At this critical condition, especially when the transmitted wave is EW, we will find CRE with NTNR, compressing and stopping light pulses, etc.

Before we get into detailed derivation, for the critical case ($\theta = \theta_c$) we first present two seemingly conflicting conclusions of *reflected wave* from two different arguments, which will clearly show the most tricky point of HI.

The first argument is from the "intuitive way" which is based on Fig.(1). Since there is no reflection wave-vector on the dispersion curve to satisfy the k_x continuity, we intuitively expect that there should be *no* reflected wave with omnidirectional incidence. If the incident angle is large enough $\tilde{k}_x^2 > \epsilon$ so that the transmitted field is EW, and since a single EW can not carry energy current, NTNR is the only possible choice and we expect that CRE will occur on HI. But from the second argument based on Eq.(4), we will obtain a different result. Since $\theta = \theta_c$ is a critical case, we should be more careful and discuss in a more subtle and strict way.

We first suppose the $\theta \neq \theta_c$ as shown by dashed lines Fig.(1), so the finite $\tilde{k}_{ry}$ of reflected field for a fixed $\tilde{k}_x$ can be found. Next we let the angle θ to approach θ_c continuously (which can be realized physically by choosing different direction of HI), then we find that $\tilde{k}_{ry} \to \infty$ when $\theta \to \theta_c$ for a fixed $\tilde{k}_x$.

But surprisingly, when $\theta \to \theta_c$, the reflection coefficient r_{hd}, calculated from Eq.(4) as $\lim_{\theta\to\theta_c} r_{hd} = (\epsilon\tilde{k}_x - \sqrt{|\epsilon_1|\epsilon_2(\epsilon - \tilde{k}_x^2)})/(\epsilon\tilde{k}_x + \sqrt{|\epsilon_1|\epsilon_2(\epsilon - \tilde{k}_x^2)})$ is not zero, and the reflected energy current is not zero too. So the theoretical result seems against our intuition.

To explain the conflicting results, we need to calculate the group velocity inside HM with material dispersion, which will also show that the superluminal group velocity is artificial.

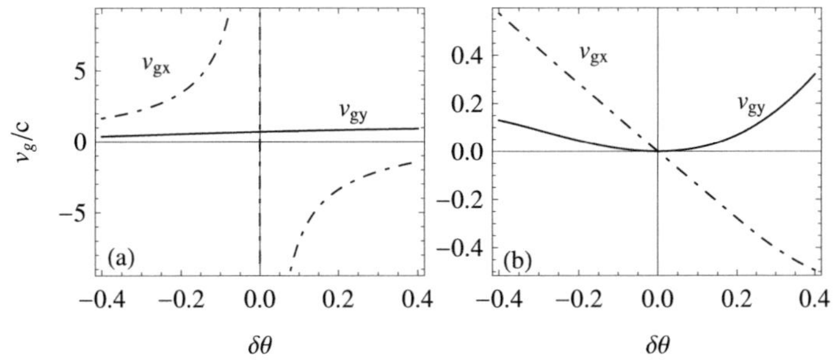

Fig. 2. Two components of $\vec{v}_g$ of reflected field in HM. (a)Without material dispersion of ϵ_1, (b)with material dispersion.

The most general expression of the group velocity of the reflected wave (which is also the group velocity of transmitted field in HM of hyperlens case) can be obtained from Eq.(2) as:

$$v_{gx} = \frac{2c\epsilon_1\epsilon_2(\alpha\gamma\tilde{k}_y^h + \tilde{k}_x\beta^2)}{\epsilon_2^2\epsilon_1(\omega)'\omega\tilde{k}_y^{p2} - \epsilon_1^2\epsilon_2(\omega)'\omega\tilde{k}_x^{p2} - 2\epsilon_1^2\epsilon_2^2} \tag{5a}$$

$$v_{gy} = \frac{2c\epsilon_1\epsilon_2(\alpha\gamma\tilde{k}_x - \tilde{k}_y^h\alpha^2)}{\epsilon_2^2\epsilon_1(\omega)'\omega\tilde{k}_y^{p2} - \epsilon_1^2\epsilon_2(\omega)'\omega\tilde{k}_x^{p2} - 2\epsilon_1^2\epsilon_2^2} \tag{5b}$$

where $\tilde{k}_x^p = (\tilde{k}_x \cos\theta - \tilde{k}_y^h \sin\theta)$ and $\tilde{k}_y^p = (\tilde{k}_y^h \cos\theta + \tilde{k}_x \sin\theta)$ are the "k components" in the principal-axes coordinate of HM and $\beta = (\epsilon_2 \sin^2\theta - \epsilon_1 \cos^2\theta)^{\frac{1}{2}}$. From Eq.(5), we find that, if the material dispersion of HM is neglected $\partial\epsilon_1/\partial\omega = \partial\epsilon_2/\partial\omega = 0$, then we will obtain the superluminal group velocity as shown in Fig.2(a). When $\theta \to \theta_c$, the v_g even diverges.

But with material dispersion, the x and y components of v_g is recalculated, and we find that there is no $v_g > c$ at all cases, as shown in Fig.2(b) in which two components of v_g versus $\theta - \theta_c$ based on Eq.(5), with the parameters $\epsilon_2' = 0$, $\epsilon_2 = 1$, $\epsilon_1(\omega) = (1 - \frac{\omega_p^2}{\omega^2})$[21] and $\omega = \omega_p/\sqrt{2}$, $\theta_c = \pi/4$. When approaching the critical angle $\theta \to \theta_c$, two components can be approximated as:

$$\lim_{\theta\to\theta_c} v_{gx} \sim \frac{1}{\epsilon_1'(\omega)}(k_{ry}^h)^{-1} \tag{6a}$$

$$\lim_{\theta\to\theta_c} v_{gy} \sim \frac{k_x}{\epsilon_1'(\omega)}(k_{ry}^h)^{-2} \tag{6b}$$

. Since $\tilde{k}_{ry}^h \to \infty$ at the critical angle θ_c, the group velocity of the reflected wave should be zero $\mathbf{v}_g = 0$ at the critical angle, as shown in Fig.2(b) too.

What does the zero-group velocity of reflected wave mean? The analysis will give us clear answer. As we have pointed out,since the reflected energy current $\vec{S}_r$ is not zero and $\vec{S}_r = v_g W$ where W is the energy density of reflected wave, hence the energy density W must be infinite large at the critical angle. From Eq.(2), we can obtain that the electric field of reflected wave $|\mathbf{E}_\mathbf{r}^\mathbf{h}|$ is really divergent at the critical angle. The divergent field strength means that it need *infinite long time* to accumulate energy at HI for the reflected field. In other words, there is no reflected wave physically, as our intuition has told us. When the incident angle is large enough

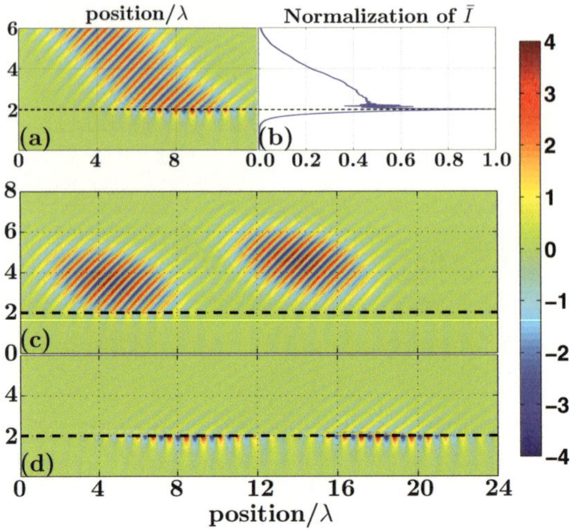

Fig. 3. (a)The magnetic field H_z distributions for a Gaussian beam incident on the interface with $|\tilde{k}_x| > \sqrt{\epsilon}$. (b)The averaged field intensity versus the vertical distance to the HI. (c)Two pulses are arriving at HI at different time. (d) The fields of two pulses, which stay at the incident positions, at 14 periods after the pulse arriving.

$\tilde{k}_x^2 > \epsilon$, since the energy of incident RW can not be transmitted, also can not be reflected, the only answer is that the energy is stored at the HI or CRE occurs. Thus, we can have a self-consistent explanation for our seemingly conflicting results.

To confirm our theoretical discussion at θ_c, the FDTD simulation [79] with strict physical HM dispersion (Drude mode) which satisfies Kramar-Kronig relation, is done. The parameter of HM and dielectric are $\epsilon_1 = -3$, $\epsilon_2 = 3$, and $\epsilon_l = 1.1$. For the case($\tilde{k}_x > \epsilon$), as shown Fig.3(a), a light beam is incident from HM to HI in 45^0 angle,as we predicted, there is no reflection and no transmission, and the field energy is accumulated at HI and stopped there. More detailed observation shows that at the boundary the field energy is mainly at the dielectric side, as shown Fig.3(b).

We also has checked the group velocities of hyperlens cases and and find no violation of the causality. Actually, in FDTD simulation, if there is superluminal group velocity the program will be numerically unstable.

The dynamical study, such as with the pulse incidence, can reveal more interesting phenomena of HI. Since the group velocity along HI is zero at NTNR case as discussed, we expect that the pulse energy will accumulate on HI and stay at the incident position until it is dissipated because of absorption of HM.

The numerical experiments with incident pulses by FDTD are also done. As shown in Fig.3(c) and (d), two pulses arrive at the HI at different time, then they stop at the incident positions on HI. The pulse vertical length is compressed to almost zero, but their width keeps same so that they are still well separated in Fig.3(d). We emphasize at here that this is a novel mechanism to compress and stop (slow) light pulses with special advantages. The first advantage is that this mechanism works at very wide frequency and wide incident-angle range, which is confirmed by FDTD simulation in Fig.3 with incident of pretty short pulses. The frequency and incident-angle insensitivity is because the mechanism is from a simple geometry property,

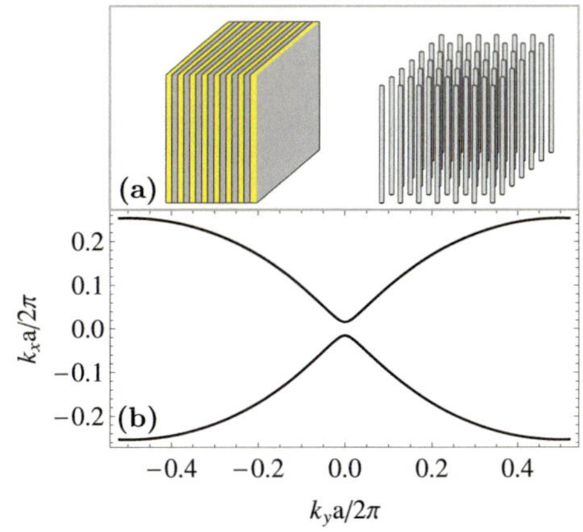

Fig. 4. (a) The structures of real HM: the periodic metal-dielectric layers and the periodic metal nano-wires embedded in a dielectric matrix. (b) The frequency contour of periodic metal-dielectric layers.

i.e. the HI (x axis) perpendicular to one of hyperbola-dispersion asymptotes. The second is that the decay (because of dissipation) of trapped field on HI is much slower than in common metallic material since the trapped field energy is mainly in the dielectric side as shown in Fig.3(b). The third is that the trapped signals are easy to take out (read) since they are on the interface. Because of these advantages, HI could be used as a removable recorder (dynamical memory) in optical/photonic signal processing, or as a wide-frequency wide-angle light trapper in photovoltaic devices.

However, we should point out that the above theoretical and numerical studies are with the assumption of the ideal hyper-dispersion, which is still void when $k_{ry} \to \infty$. In reality, such HM does't exist, so that we need to study the limit of hyper-dispersion of realizable HM. HM can be realized by many structures, i.e. one-dimensional (1D) periodic metal-dielectric binary layers [24, 25] or two-dimensional (2D) periodic metallic lines[26], as shown in Fig.4(a). For these structures, the dispersion relation can be calculated exactly. In Fig.4, the calculated frequency contour of a 1D metal-dielectric binary layers is shown, from which we can see that the effective HM medium is not available anymore when $|k|$ approaches π/a. Based on this limit, we can roughly estimate the slow limit of group velocity by $v_{gx} \sim \frac{1}{\epsilon_1'(\omega)}(k_{ry}^h)^{-1} \propto 1/\gamma_s$ and $v_{gy} \sim \frac{k_x}{\epsilon_1'(\omega)}(k_{ry}^h)^{-2} \propto 1/\gamma_s{}^2$, where $\gamma_s = k_{ry}/k_0$ is the slowing coefficient. For the 2D metallic-line structure, from the modern technical limit we assume the smallest lattice constant as $a = 10$nm. If the incident is the micro-wave $\omega = 5.8$GHz ($\gamma_s \sim 10^7$) and $\epsilon_1'(\omega) = 6.9 \times 10^{-10}$s as in Ref [27], we obtain $v_{gx} \sim 4.6$m/s which means considerably slow light although not totally stopped, and $v_{gy} \sim 7.07 \times 10^{-8}$m/s which means that the strongly-compressed light pulses can be easily achieved.

In conclusion, we have theoretically and numerically investigate the optical properties of HI. The theory with dispersion of meta-material is constructed and the hyperlens effect of CER is confirmed. At the inverse process of hyperlens, the abnormal phenomena of CRE with NTNR and a novel mechanism to compress and stop light in wide frequency range are revealed. Based the calculated group velocity, we demonstrate that the previously-pointed-out superluminal group velocity in HM is artificial since the material dispersion is neglected. FDTD simulations confirm that the HI has potential to be a removable optical/photonic

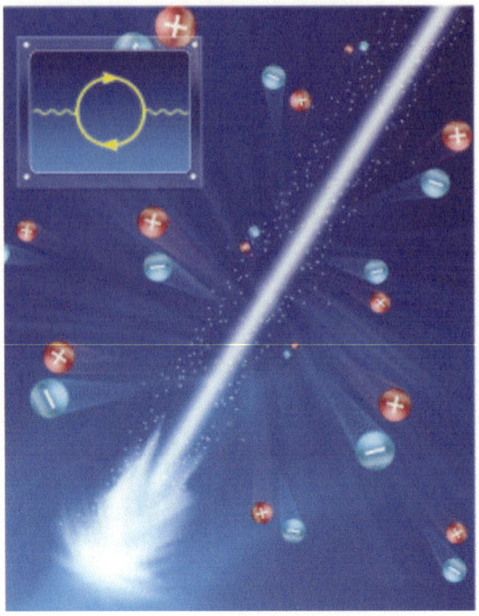

Fig. 5. The schematic picture of vacuum polarization processes with electron-positron pair generation, with which the vacuum becomes dissipative and anisotropic. The insert is the Feynman diagram of the vacuum polarization processes.

recorder, or a wide-frequency wide-angle light trapper. At last the realizability of these phenomena on the real metallic structures is discussed. Obviously, the new mechanism works not only for electromagnetic waves, but also for acoustic or matter waves if hyperbolic dispersion is available, so that more interesting phenomena and applications are waiting for further theoretical and experimental research.

3. The methods to detect vacuum polarization by evanescent modes

Vacuum is one of the most fundamental concepts in all quantum fields[30–32] since all excitations are from the vacuum and determined by vacuum in some way. Modern vacuum concept is started from quantum electrodynamics (QED), which describes the interaction between light and matter (including vacuum), and has been continually studied both experimentally and theoretically[33–38]. According to QED, the vacuum becomes weakly anisotropic, dispersive, dissipative and even nonlinear optical medium, when external electric field is approaching the Schwinger critical value $E_c \simeq 10^{18}V/m$. In other words, the real and imaginary parts of vacuum refractive index could deviate from unit and zero[34, 35], respectively. Physically, the deviation of the imaginary part is mainly from the electron-positron pair generation. However, the electron-positron pair generation, also generally called as *vacuum polarization* (VP) processes[34], which is schematically shown in Fig.1, has not been directly observed for over half century since very high E_c is beyond the contemporary technical limit. Therefore, it is natural to wonder if we can find an approach to probe VP with external field E_{ext} much smaller than E_c.

In this work, we propose evanescent-mode-sensing methods based on new mechanism to detect the QED VP, which is based on the measuring the phase change and the energy time delay of evanescent wave (EW). We find that the required external field could be one order weaker than E_c, which may be realizable by contemporary experiments.

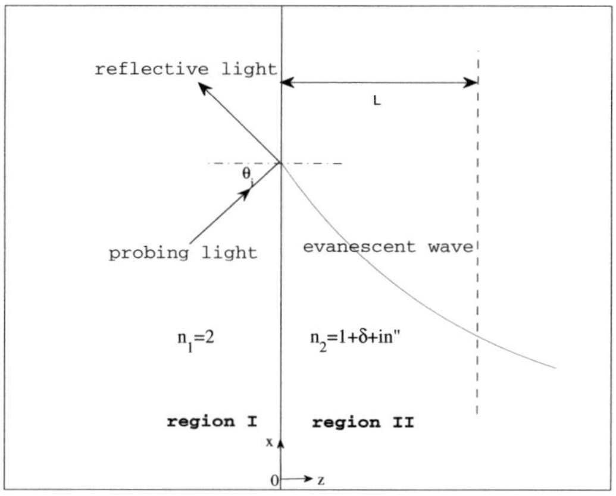

Fig. 6. The schematic diagram of our model.

The idea is from the "dual roles" of real and imaginary parts of refractive index n. Supposing a medium with complex index $n = n' + in''$, our goal is to detect the very tiny change of n' or n''. For radiative waves, since n' determines the real part of wavevector $k \simeq n'\omega/c$ and it is easy to measure the phase change or group delay, so, it is natural to choose the radiative wave to probe small change of n'. On the other hand, for radiative waves, tiny change of n'' causes an extremely small decay change which is very hard to measure in the limited lab space. However, for the evanescent waves, the roles of n' and n'' are totally exchanged, i.e., n' dominates the decay rate, while the n'' introduces a phase change which is much easier to detect. Further more, we will demonstrate that n'' can also introduce the energy propagation for EWs whose energy velocity $v_e \propto n''$ can be extremely slow. Such a slow wave can be detected by measuring the delay time τ at a short distance.

Actually, the tenneling mechanism of EW has been widely studied[39–41]. We would like to emphasize the mechanism difference between ours and that in the previous works. In Ref[41], they are based on "two interfaces" structure (a slab). Such "two-interfaces" structure will generate both evanescent modes $\exp(\pm\kappa x)$ and such two evanescent modes can carry energy current[9], which called as "tunneling mechanism". So, even if the material dissipation is neglected[41] , the energy propagation is still available. However, in our model, since there is only a single interface (Fig.2), obviously if without dissipation there will be no the energy current at all[42], then, no phase change and no the energy delay time. So, our mechanism is based on the dynamical picture and the dissipation is critical.

Here we note that, because the probing light is much weaker than the external field in our model, the nonlinear effect is negligible. For a linear system, all dynamical processes can be solved numerically by sum of multi-frequency componentsčňwhich can be obtained by Green's function methods [42].

Our model is schematically shown in Fig.6, based on the total internal reflection (TIR) at the interface between a dielectric media n_1 (region I) and vacuum (region II). When the incident angle $\theta_i > \theta_c = arcsin(1/n_1)$, the TIR will occur and the transmitted wave in the vacuum is the EW. We choose θ_i is a little larger than θ_c to make sure that almost all frequency components

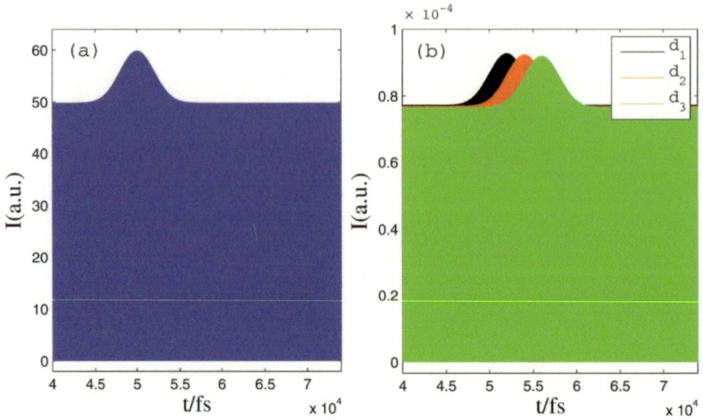

Fig. 7. The irradiance of light I versus time t, (a)the incident light at the interface; (b)the transmitted EW in region II, where the black, the red and the green lines are for the fields at the distance from the interface $d_1 = 0.1\lambda_0$, $d_2 = 0.2\lambda_0$ and $d_3 = 0.3\lambda_0$, respectively, with $\lambda_0 = 600nm$.

are totally reflected when the incidence is the slowly-varying quasi-monochromatic wave. An interferometer or a photon detector is set at distance L from the interface so that the phase and intensity change can be detected.

The time-dependent Maxwell equations are given by $\nabla \times \mathbf{E} = -\mu(z)\mu_0\partial\mathbf{H}/\partial t$ and $\nabla \times \mathbf{H} = \epsilon(z)\epsilon_0\partial\mathbf{E}/\partial t$, where $\epsilon(z)$ and $\mu(z)$ are the relative permittivity and the relative permeability, respectively, and $c = 1/\sqrt{\epsilon_0\mu_0}$. To obtain the concrete results, the system parameters are chosen as following, the incident angle $\theta_i = 0.1667\pi$, the refractive index of region I $n_1 = \sqrt{\epsilon_1} = 2$, and the vacuum refractive index of region II $n_2 = \sqrt{\epsilon_2\mu_2} = 1 + \delta + in''$, where $\delta << 1$ and $n'' << 1$ are the real and imaginary index deviations of vacuum, because of VP processes caused by strong external field. If the incident probing light is a plane wave, the transmitted wave in the vacuum region can be generally written in the form $E(x,z,t) = Eexp(ik_z z + ik_\parallel r_\parallel - i\omega t)$, where $k_\parallel = n_1 \sin\theta_i \omega/c$ and $k_z = \sqrt{(n_2\omega/c)^2 - k_\parallel^2}$ are the wave vectors parallel and perpendicular to the interface. For the EW, k_z is described as:

$$k_z = i\sqrt{(n_1 \sin\theta_i)^2 - (1+\delta)^2}\frac{\omega}{c} + \frac{n''}{\sqrt{(n_1 \sin\theta_i)^2 - (1+\delta)^2}}\frac{\omega}{c}. \quad (7)$$

The physical meaning of k_z is very clear that the imaginary part $Im(k_z) = \kappa_z$ corresponds to the exponential decay of the field, and the real part $Re(k_z)$ causes *a phase change because of VP*. The phase change at distance $z = L$ is

$$\Delta\phi = Re(k_z)L \propto n''L \quad (8)$$

which could be measured by interferometers[43].

Besides the phase change $\Delta\phi$, with the same model as shown in Fig.6, there is another way to detect the tiny n'' by *measuring the time delay of irradiance fluctuation [44] of the evanescent wave*. It is a dynamic process as following. First, we suppose that the incident light is not a plane wave anymore, but with a slow intensity fluctuation, as shown in Fig.7(a). Then, the question is "What will happen for the EW in region II ?" Numerically, from the strict Green's function method with physical dissipation and dispersion, it is found that the fluctuation will propagate on the EW from the interface to far away, as shown in Fig.7(b). So, we can measure the time delay τ of the fluctuation propagation on the EW to detect the VP effect. The propagation speed of irradiance fluctuation can be obtained by the energy velocity v_e which is defined as: $v_e = |\vec{S}_z|/W$, where $\vec{S}_z = \frac{1}{2}Re\,(E \times H^*)\,|_z$ is the averaged Poynting vector along z direction, and $W \simeq \frac{1}{4}(\epsilon_0|E|^2 + \mu_0|B|^2)$ is the local energy density of the electromagnetic wave. In our model, the energy velocity is obtained as:

$$v_e = \chi \cdot n'' \tag{9}$$

with $\chi \simeq c/\left[(n_1 \sin\theta_i)^2\sqrt{(n_1\sin\theta_i)^2 - (1+\delta)^2}\right]$, when the dissipation and dispersion are very weak. The physical meaning of v_e can be understood as the "propagation" speed of irradiance fluctuation of the EW, which can be measured[44].

Hence, experimentally the time delay τ of the irradiance fluctuation at distance L can be measured:

$$\tau = L/v_e \propto 1/n''. \tag{10}$$

Since it is *near field* phenomenon, the detecting should be very near the interface. For the VP effect, since n'' is extremely small, the "propagation" speed of the irradiance fluctuation is so slow that τ gets to peco-second level when the distance is one tenth of the wavelength $L = 60nm$.

Therefore, either the phase change $\Delta\phi$ or the time delay τ are very sensitive for n'', and the EW is a good candidate to probe the VP effect. Here, we note that the famous Kramers-Kronig relations still fit for QED vacuum[34]. Hence, the observation of imaginary part of vacuum index also confirms the dispersion of QED vacuum.

Next, we will quantitatively study the VP detect by our methods. Supposing that an external homogeneous constant electric field E_{ext}, which is perpendicular to the xz plane and smaller than the Schwinger critical electric field E_c, is applied to the vacuum (region II) only, as shown in Fig.6, then, the optical properties of the vacuum can be described by the Euler-Heisenberg Lagrangian L_{eff}[34, 37]. Physically, the imaginary part of Euler-Heisenberg Lagrangian L_{eff} is related to the imaginary part of VP operator, and therefore corresponds to the *electron-positron pair generation*.

Consequently, the vacuum refractive index can be deduced from the Lagrangian L_{eff}[34, 37, 38].

In our model, since the external magnetic field is supposed to be zero, thus the vacuum refractive index is determined only by the external homogeneous constant electric field E_{ext}. We use $n_{\parallel}$ and $n_{\perp}$ to refer the effective refractive index of vacuum when the electric field of probing light are parallel and perpendicular to the field E_{ext}, respectively. $n_{\parallel}$ and $n_{\perp}$ can be obtained from the reference [38]:$n_{\parallel} = 1 + \frac{2\alpha}{45\pi}y^2 + i \cdot \frac{\alpha}{4\pi}\sum_{n=1}^{\infty}\left(\frac{\pi}{y^2} + \frac{1}{n}\frac{1}{y}\right)\exp(-n\pi/y)$, and $n_{\perp} = 1 + \frac{7\alpha}{90\pi}y^2 + i \cdot \frac{\alpha}{4\pi}\sum_{n=1}^{\infty}\left(\frac{2}{3}\pi + \frac{1}{n}\frac{1}{y} + \frac{1}{n^2}\frac{2}{\pi}\right)\exp(-n\pi/y)$, where $y = |E_{ext}|/E_c$, and $\alpha \simeq 1/137$ is the fine-structure constant. Therefore we have $\delta = Re\left(n_{\parallel(\perp)}\right) - 1$, $n'' = Im\left(n_{\parallel(\perp)}\right)$ for $n_{\parallel}$ and $n_{\perp}$ when we solve the equations such as Eq.(7) in this letter.

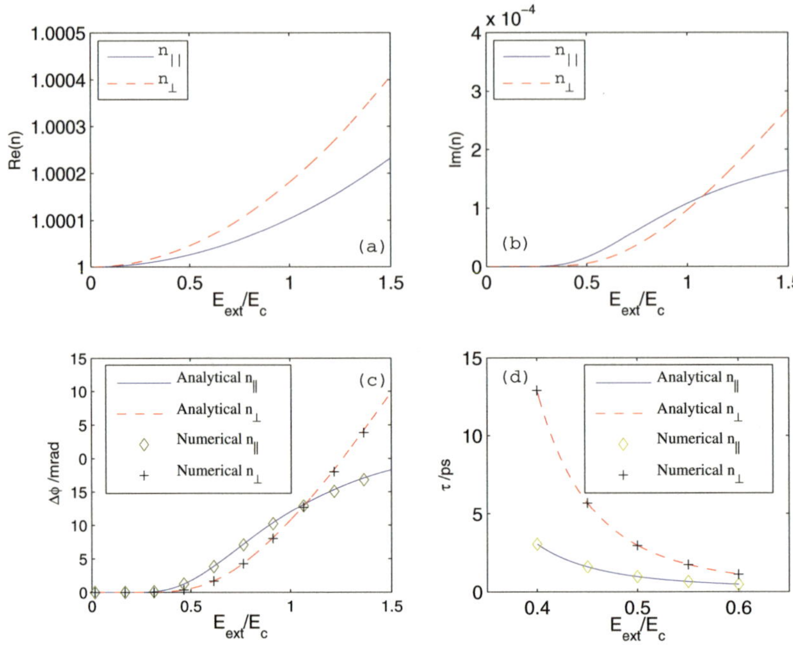

Fig. 8. (a)The real part of vacuum index $n_{\parallel}$ and $n_{\perp}$ versus E_{ext}; (b)The imaginary part of $n_{\parallel}$ and $n_{\perp}$ versus the external electric field strength; (c)$\Delta\phi$ versus with $L_p = 6\mu m$; (d)τ versus E_{ext} with $L_{\tau} = 60nm$. Both the results from theory and from Green's function are shown in (c) and (d).

The parameters of our model in Fig.6 are chosen as following. The wavelength of probing light is $\lambda_0 = 600nm$, the dielectric constant in the region I is $\varepsilon = 4$, and the incident angle is $\theta_{inc} = 0.1667\pi > \theta_c$, so that the field in vacuum is evanescent. The distance L for the phase detecting is $L_p = 6\mu m = 10 \times \lambda_0$, while for τ detecting is $L_{\tau} = 60nm = 0.1 \times \lambda_0$, respectively. The QED theoretical results of real and imaginary part of $n_{\parallel}$ and $n_{\perp}$ are shown in Fig.8(a) and Fig.8(b), respectively. Bring these results into Eq.(8) and Eq.(10), the phase change $\Delta\phi$ and the delayed time τ can be obtained, which are shown in Fig8(c) and Fig.8(d), respectively. Numerically, the phase change with plane wave incidence and the time delay of local amplitude maximum are calculated by Green's function method, which are also shown in Fig.8(c) and Fig.8 (d). Comparing the analytical results from Eq.(8) and Eq.(9) and numerical results, we can find that they agree with each other very well.

Next, we will analyze the possibility to observe the VP effect in experimental conditions. The recent experimental advances[45] have raised hopes that lasers may achieve fields just one or two orders of magnitude below the Schwinger critical field strength. In this case $E_{ext} \sim 0.1E_c$, from our numerical and analytical results in Fig.8, we can see the $\Delta\phi$ can get to $\sim 10^{-1}mrad$ order, which are in measuring limit of contemporary interferometer [43]. Very recently, it is supposed that the electric field E could be effectively amplified 4 times larger by coherent constructive interference of laser beams[36]. If E_{ext} can get to $0.5E_c$ by this method, not only $\Delta\phi$ can be one order larger, but also the delay time τ can get to sub peco-second level and may be measured by contemporary photon detectors.

4. The temporal coherence gain of the negative-index superlens image

Veselago predicted that the negative-index material (NIM) has some unusual properties, such as a flat slab of the NIM could function as a lens for electromagnetic (EM) waves [1]. This research direction was further pushed by works of Pendry and others [4, 5, 46–58] who showed the lens with such NIM (i.e. $\epsilon = \mu = -1 + \delta$) could be a *superlens* whose image resolution can go beyond the usual diffraction limit. After that, several beyond-limit properties of NIM systems are found, such as, the sub-wavelength cavity [59] and the waveguide [60]. Some of the theoretical results are confirmed by experiments[4, 46, 49]. And these beyond-limit properties give us new physical pictures and opportunities to design devices. Recently, new numerical [50, 51] and theoretical Green's function [52] methods are used to understand the phenomena in such systems. But so far almost all studies are done with the strictly single-frequency sources, so that the coherent properties of EM waves (or photons) in the NIM systems have not been studied to the best of our knowledge. Even more seriously, there is no theory for the propagation of coherent functions in NIM systems. The importance of coherence research can not be over-estimated since the coherence is essential in the wave interference , the imaging , the signal processing and the telecommunication [61, 62]. Can we find new frontier to go beyond at the coherent properties in NIM systems? If so, can we develop a simple theoretical method to deal with the image coherence of superlens?

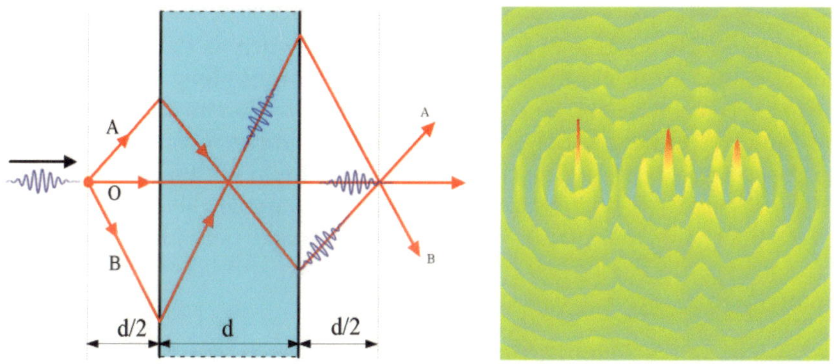

Fig. 9. The schematic diagram of our model with ray paths(left); and the typical snapshot of electric field in our FDTD simulation (right).

In this section, the finite difference time domain (FDTD) method is used in the two-dimensional (2D) numerical experiments to study the temporal coherence of the superlens image with random quasi-monochromatic sources. We observe the dramatic temporal-coherence gain of the superlens image even if the reflection and frequency-filtering effects are very weak. Based on the new physical picture of the signal (the fluctuation of random source) propagation in NIM, we construct a theory to obtain the image field and derive the equation of the temporal-coherence relation between the source and its image. The new mechanism of the temporal-coherent gain can be explained by the key idea that the signals on different paths have different "group" retarded time. Our theory excellently agrees with numerical results and the strict Green's function results.

The setup of the 2D system is shown in Fig.9. The thickness of the infinite-long NIM slab is d. To realize the negative ϵ and negative μ, the electric polarization density $\vec{P}$ and the magnetic moment density $\vec{M}$ are phenomenologically introduced in FDTD simulation [63]. The effective permittivity and permeability of the NIM are $\epsilon_r(\omega) = \mu_r(\omega) = 1 + \omega_P^2/(\omega_a^2 - \omega^2 - i\gamma)$. In our model, $\omega_a = 1.884 \times 10^{13}/s$, $\gamma = \omega_a/100$, $\omega_P = 10 \times \omega_a$. The quasi-monochromatic field is expressed as $E(x,t) = U(x,t)exp(-i\omega_0 t)$, where $U(x,t)$ is a slowly-varying random function, $\omega_0 = \pi/20\delta_t$ is the central frequency of our random sources and $\delta_t = 1.18 \times 10^{-15}s$ is the smallest time-step in FDTD simulation. At ω_0, we have $\epsilon_r = \mu_r = -1.00 - i0.0029$. At here, we emphasize that in our FDTD simulation the smallest space-step $\delta_x = \lambda_0/20$ ($\lambda_0 = 2\pi c/\omega_0$) and the distance ($d/2 = \lambda_0$) of the source from the lens are too large to excite strong evanescent modes of NIM [50, 51, 53]. *Actually the evanescent field in our simulation can be neglected comparing with radiating field, and what we are studying is the property dominated by the radiating field.*

The random source is composed of the randomly generated plane-wave pulses, with the average pulse length t_p and the random starting phase and starting time. In the simulation, we record the field of the source and the image for a duration of $4 \times 10^5\delta_t$ to obtain the data for analysis. For the convenience, we define $E(\omega) = \lim_{T\to\infty} \int_{-T}^{T} E(t)exp(-i\omega t)$ as the $field$ $spectrum$ (FS).

Unusual phenomena.—At first, the FS width of the random source is a little too large ($\Delta\omega_s \simeq \omega_0/20$). When we observe the image temporal-coherence gain, we also find that the FS width of the image is sharper than the source ($\Delta\omega_i < \Delta\omega_s$). It is obvious that there are the frequency-filtering effects because of the NIM dispersion, such as the frequency-dependent interface reflection and focal length. After increasing the pulse-length t_p of the source, we reduce the source FS width to $\Delta\omega_s \simeq \omega_0/100$, then the reflection and focal-length difference are very small [64]. With such source, the FS widths of source and image are almost same $\Delta\omega_i \simeq \Delta\omega_s$, as shown in Fig.10a. The difference between two widths is $< 5\%$, which is our criterion of the $quasi$-$monochromatic$ source. Even so the dramatic gain of temporal coherence is still observed. In Fig.10b, the source field (up) and the image field (down) vs time of FDTD simulation are compared. The *profiles* of them are genically similar, but the image profile is much smoother.

The normalized temporal-coherence function $g^{(1)}(\tau) =< E^*(t)E(t+\tau) > / < E^*(t)E(t) >$ ($<>$ means the ensemble average) of the source (black) and the image (red) from FDTD simulation are shown in Fig. 11. The temporal coherence of the image field is obvious better than the source. From $g^{(1)}$, the image coherent time is obtained $T_i^{co} = \int g_i^{(1)}(\tau)d\tau = 1268\delta_t$, which is about 50% longer than the source coherent time $T_s^{co} = 860\delta_t$.

Although the gain of the spatial coherence only by propagation is well-known[62], the dramatic gain of temporal coherence is generally from the high-Q cavities, contrary to our case, which have strong filtering effects. To reveal the new mechanism of the temporal coherence gain in NIM systems, we also have done more numerical experiments in which *only* the ray near a certain incident angle (shown in Fig.9), such as only paraxial rays ($\theta \simeq 0$), can pass through the superlens. Then the image field profile vs time looks very like the source field and has no gain of coherence anymore. *Therefore, the gain of temporal coherence of the superlens image is not from one ray with certain incident angle, but probably from the interference between the rays with different incident angles.* Then, what is different between the rays with different incident angles? After carefully checking the field profiles of different-incident-angle cases, we find that the profiles have different retarded time. The larger incident angle the longer retarded time.

Physical pictures.—To deeper understand the new mechanism of coherence gain and construct our theory, we need make two physical pictures clear. The first one is about the optical path length (OPL) $\int nds$ which determines the wave phase and the refracted "paths" of rays in Fig.9 according to Fermat's principle (or Snell's law). Based on ray optics, the superlens and traditional lenses have same focusing mechanism, that all focusing rays have same OPL $\int_{paths} nds = const$ ($\int_{paths} nds = 0$ for superlens) from source to image [1]. But this picture is so well-known that it suppresses the other important picture. Because the temporal-coherence information is in the fluctuation signals of random field, the signal propagating picture should be essential for our study. *The optical signals propagate in the group velocity v_g which is always positive.* Obviously, if the path (in Fig.9) is longer (larger incident angle), the signal need a longer propagating time, which is called *group retarded time* (GRT) in this section Inside the NIM, the GRT of a path should be $\frac{d}{cos(\theta)v_g}$ (this is confirmed by our numerical experiments), where θ is the incident angle and $v_g = c/3.04$ is the group velocity of NIM around ω_0 [65]. The total GRT from source to image is $\tau_r = \tau_0/cos(\theta)$ where the $\tau_0 = d/c + d/v_g$ is the GRT of the paraxial ray. Now, the new propagating picture for a signal through superlens is that a signal, generated at at t_s from the source, will propagate on all focusing paths and arrive at

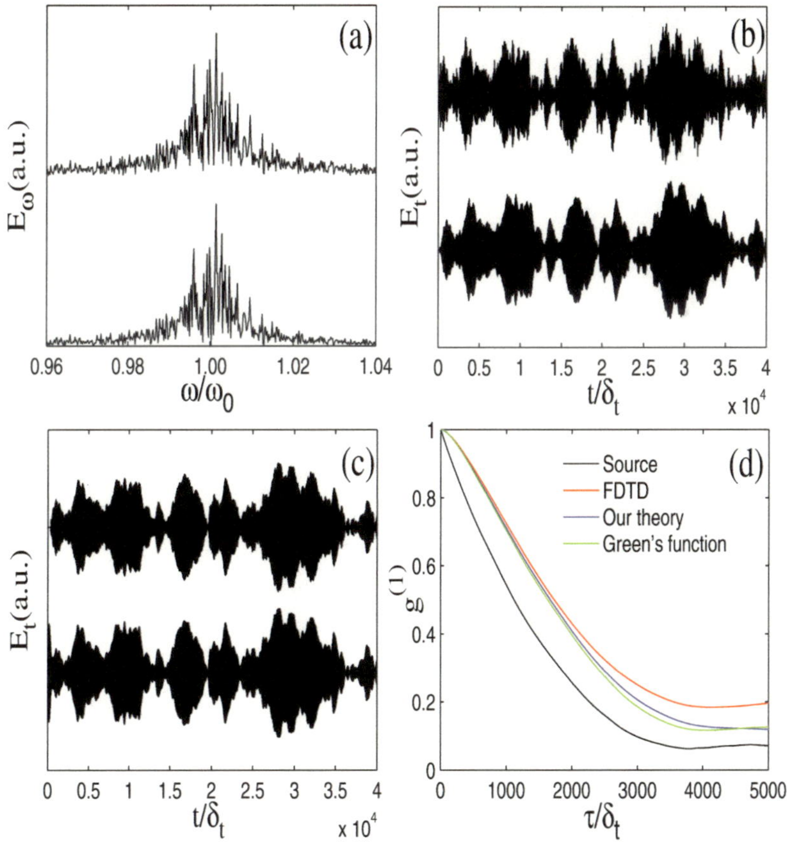

Fig. 10. (a)the FSs of the source(up) and the image(down) . (b)The electric field of the source(up) and its image (down) vs time from FDTD simulation (c) The image field vs time from Eq.(1) (up), and from the Green's function method (down).

image position at very different time $t_s + \tau_0/cos(\theta)$ from different paths(this is schematically shown in Fig.9). This picture is totally different from traditional lenses, whose images don't have obvious temporal-coherence gain because their focusing rays have same OPL *and similar GRT*.

Our theory.—Based on these analysis, we suppose that *the superlens image field of the random quasi-monochromatic source is the sum of all signals from different paths with different GRT*. This is the key point of our theory, and then the image field can be obtained:

$$E_i(t) = \frac{1}{U_0} e^{-i\omega_0 t} \sum_{paths} U_s(t - \tau_r) = \frac{1}{U_0} e^{-i\omega_0 t} \int_{-\frac{\pi}{2}}^{\frac{\pi}{2}} U_s(t - \frac{\tau_0}{\cos(\theta)}) d\theta \tag{11}$$

where $U_s(t)$ is the slowly-varying profile function of the source and U_0 is the normalization factor. In Fig.10c (up), we show the result of the image field based on Eq.(11), we can see it is in excellent agreement with the FDTD result in 2b(down). To show the interference effect of different paths, we assume there are only two paths (such as A and B in Fig.9). Based on Eq.(11) the image field is $E_i = e^{-i\omega_0 t}(U_s(t - \tau_r^A) + U_s(t - \tau_r^B))$, then the temporal coherence of image is $G(\tau) =< E_i^*(t)E_i(t+\tau) >=< U_s^*(t - \tau_r^A)U_s(t - \tau_r^A + \tau) + U_s^*(t - \tau_r^B)U_s(t - \tau_r^B + \tau) + U_s^*(t - \tau_r^A)U_s(t - \tau_r^B + \tau) + U_s^*(t - \tau_r^B)U_s(t - \tau_r^A + \tau) >$. The first two terms are same as the source field (just a time-shift) so they don't contribute to the coherence gain. The last two terms are from interference between two paths. The third (or the forth) term could be very large at the condition $\tau \simeq \tau_r^B - \tau_r^A$ (or $\tau_r^A - \tau_r^B$). This condition can always be satisfied between any two paths since τ is a continuous variable. So the interfering terms between the paths are responsible for the image temporal-coherence gain.

From Eq. (11), after the variable transformation $t_s = t - \tau_0/cos\theta$ and some algebra, the relation of the temporal coherence between the image and the source can be obtained:

$$\begin{aligned} G_i(\tau) &= < E_i^*(t)E_i(t+\tau) > \\ &= \frac{1}{U_0^2} \int_{-\infty}^{-\tau_0} dt_1 \int_{-\infty}^{-\tau_0+\tau} dt_2 h_i^*(t_1)h_i(t_2+\tau)G_s(t_2-t_1) \end{aligned} \tag{12}$$

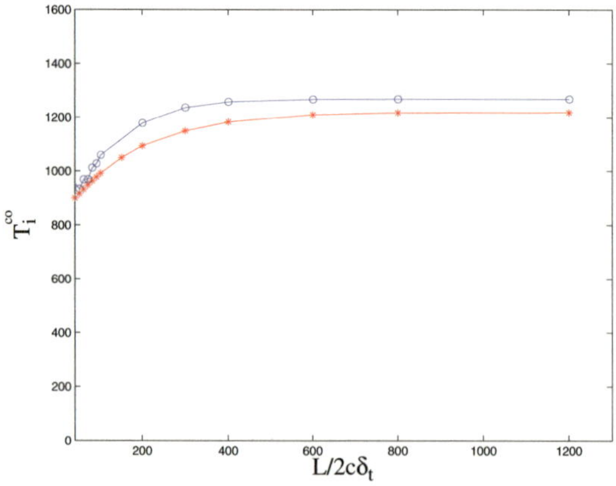

Fig. 11. The normalized temporal-coherence function $g^{(1)}$ vs time of the source field (black), and of the image field which obtained from the FDTD simulation (red), from Eq.(2) (blue) and from the Green's function method(green).

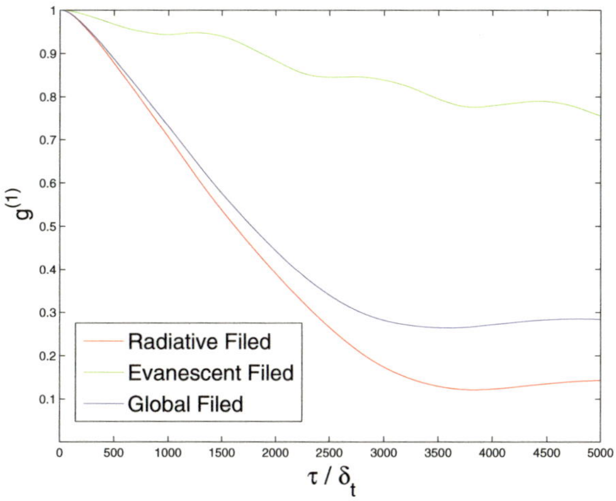

Fig. 12. The coherent time versus the superlens length L, from the FDTD simulation(blue) and from our theory (red).

where $h_i(t) = (\tau_0/t)^2/\sqrt{1-(\tau_0/t)^2}$ is the response function of different incident angles, and $G_s(t_2-t_1) =< E_s^*(t_1)E_s(t_2) >$ is the temporal-coherence function of the source. Eq.(12) can explain the temporal-coherence gain of the image too. Even if the source field is totally temporal *incoherent* $G_s(t_2-t_1) \propto \delta(t_2-t_1)$, based on Eq.(12) we can find that $G_i(\tau)$ is not a δ-function anymore, so the image is partial temporal coherent. The product of $h_i^*(t_1)h_i(t_2+\tau)$ includes the interference between paths. According to our theory, we calculate the image coherence function $g^{(1)}$ vs time (Fig.11 blue) which agree with our FDTD result (Fig.11 red) pretty well (we will discuss the deviation later).

To further confirm our theory and FDTD results, the *strict* Green's function method [52] is engaged to check our results. *We only include the radiating field (no evanescent wave) in Green's function.* The strict image field vs time is shown in Fig.10c(down), and the image temporal-coherence function $g^{(1)}$ vs time is shown in Fig.11 (blue). In Fig.11, we can see that the FDTD result (red) is almost exactly same as the strict Green's function method (green). But our theory (blue) deviates from the strict result at very large $\tau > 3000\delta_t$ which corresponding to very long path(or very large incident angle). This is understandable since in our theory we neglect the dispersion of NIM totally and only use $v_g(\omega_0)$. For the very-large-angle rays a small index difference (from the dispersion of NIM) can cause large focal-length difference. Hence the deviation is from the focus-filtering effect. When we reduce the FS width of source to an even smaller value (*i.e.* $\Delta\omega_s = \omega_0/500$), the deviation of our theory is smaller.

Although our theory is only a good approximation generally, owing to the picture simplicity and clarity the theory can help us to study more complex systems qualitatively and quantitatively. The *finitely-long* 2D superlens is a good example which is hard to deal by Green's function method. In Fig.(11), we plot the coherent time T_i^{co} vs superlens length L of the FDTD simulation (black) and of our theory (red) , respectively. They coincide with each other pretty well (the deviation reason has been discussed). The increase of the T_i^{co} with the increase of L can be explained simply according our theory. Since the image field is $E_i(t) = \frac{1}{U_0}e^{-i\omega_0 t}\int_{\theta_{min}}^{\theta_{max}} U_s(t-\frac{\tau_0}{cos\theta})d\theta$, the large-angle paths ($\theta > \theta_{max}$ and $\theta < \theta_{min}$)and their contribution to the temporal-coherence gain are missed in the short superlens.

Obviously, Eq.(11) is suitable not only for random quasi-monochromatic source, but also for all quasi-monochromatic fields, such as the slowly-varying Gaussian pulses and slowly switching-on process mentioned by [52]. Our theory can be easily extended to 3D systems too. And owing to the fact that what we find is from the radiating field, so the temporal-coherence gain is not the near-field property. Actually, the new mechanism of the temporal-coherence gain is not limited for the $n \simeq -1$ superlens, also applicable to other superlenses, such as the photonic crystal superlens in [46, 49, 58]. But the specialities of $n \simeq -1$ superlens, such as almost no frequency-filtering (no frequency loss) and no reflection (no energy loss), can be used to design novel optical/photonic coherence-gain devices.

In summary, for the first time we have numerically and theoretically studied the temporal coherence of the superlens image with the quasi-monochromatic source. Numerically, we observe that the temporal coherence of the image can be improved considerably even almost without reflection and filtering effects. Based on new physical picture, we construct a theory to calculate the image field and temporal-coherence function, which excellently agree with the FDTD results and strict Green's function results. The mechanism of the temporal coherence gain is theoretically explained by the different GRT of different paths. Although the evanescent wave is very weak in this study, the coherence of evanescent wave in NIM systems is a very interesting topic which will be discussed elsewhere [66]. Other related topics, such as the spatial coherence which is very essential for the image quality of the superlens, can also be studied through the similar methods. Although our study is within the confinement of classic optics, similar investigation can be extended to the quantum optics [62], and interesting results can be expected. Obviously, the temporal-coherence gain of superlens is another evidence that the NIM phenomena are consistent with the causality [49]. We suppose that the temporal-coherence gain phenomena could be observed in micro-wave experiments[4, 46]. Therefore, this study should have important consequences in the future studies of coherence in NIM systems. The no-reflection and no-frequency-filtering coherence gain of the superlens has some potential applications in the imaging, the coherent optical communication, and the signal processing.

5. The physical picture and the essential elements of the dynamical process for dispersive cloaking structures

Recently, the theory[67, 68] has been developed based on the geometry transformation to realize a cloaking structure (CS), in which objects become invisible from outside. Then a two-dimensional (2D) cylindrical CS[69] and a nonmagnetic optical CS[70, 71] are designed. More surprisingly, the experiment[72] demonstrates that such a 2D CS really works with a "reduced" design made of split-ring resonators. These pioneers' works are really attractive and open a new window to realize the invisibility of human dream. However, so far almost all theoretical[67–71, 73–75] studies of the CS are done in the frequency domain and the geometry transformation idea is supposed to work only for a single frequency, so that the effects of the dispersion have not been intensively studied. As pointed out in Ref. [68] and the quantitatively study in our recent work [7], the dispersion is *required* for the cloaking material to avoid the divergent group velocity. For the dispersive CS, new topics, such as the *dynamical process*, can be introduced in. Dynamical study is essential for the cloaking study since without it we can not answer the questions, such as how can the field gets to its stable state?, is there any strong scattering or oscillation in the process?, and how long is the process?, etc. More important, because the real radars generally are pulsive ones, the dynamical process is critical

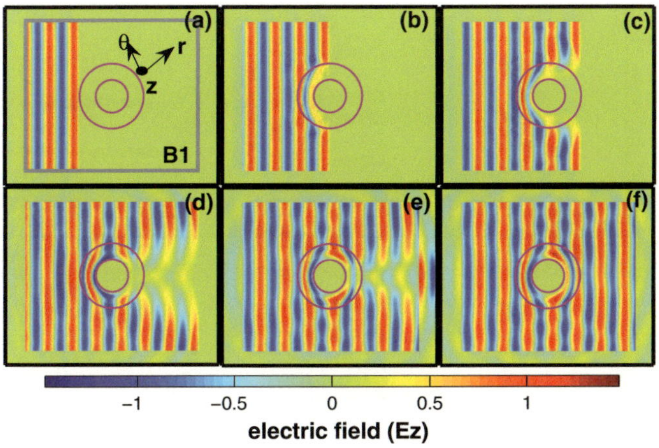

Fig. 13. (Color online) (a) The setup of the system. And the distribution of the electric field at different moments during the process. Parameters are chosen as $A_r = 0.4$, $A_\theta = 1.6$, $A_z = 0.4$ (in the position where $\varepsilon_z < 1$) or $A_z = 1.6$ (where $\varepsilon_z > 1$), and $\gamma = 0.012\omega_0$. (a) $t = 2.28T$. (b) $t = 3.60T$. (c) $t = 4.92T$. (d) $t = 7.20T$. (e) $t = 9.00T$. (f) Stable state. T is the period of the incident EM wave.

for the cloaking effect around the goal frequency. So the dynamical study not only gives us whole physical picture of the cloaking, but also helps us to design more effective cloaks.

In this section, the dynamical process of the electromagnetic (EM) CS is investigated by finite-difference time-domain (FDTD) numerical experiments. In our simulation, the Lorentzian dispersion relations are introduced into the permittivity and the permeability models, then the *real* dynamical process can be simulated[76–78]. Based on numerical simulation, we can follow the details of the dynamical process, such as the time-dependent scattered field, the building-up process of the cloaking effect, and the final stable cloaking state. By tuning the dispersion parameters and observing their effects on the dynamical process and the scattered field, we can find the essential elements which dominate the process. Theoretical analysis of these essential elements can help us to have a deeper physical picture beyond the phenomena and to design more effective cloaks.

The *setup* of the system is shown in Fig. 13(a), similar as the one in Ref. [69]. R_1 and $R_2 = 2R_1$ are the inner and the outer cylindrical radii of the CS, respectively. A perfect electric conductor (PEC) shell is pressed against the inner surface of the CS. The CS is surrounded by the free space with $\varepsilon_0 = \mu_0 = 1$. From the left side, an incident plane wave with working frequency ω_0 is scattered by the CS, the total field and the scattered field can be recorded inside and outside $B1$ respectively by the numerical technique[79]. So the scattering cross-section σ can be calculated easily. Our study is focused on the E-polarized modes, for which only the permittivity and the permeability components ε_z, μ_r, and μ_θ are needed to be considered (For H-polarized modes, considering the corresponding components μ_z, ε_r, and ε_θ, we can obtain the same numerical results in the dynamical process.). All of them are supposed to have the form $1 + F_j(r) \times f_j(\omega)$, where subscript j could be z, r, and θ for ε_z, μ_r, and μ_θ, respectively. The filling factors $F_j(r)$ are only r-dependent, ω_p is the plasma frequency which set to be a constant $\omega_p = 10\omega_0$, and $f_j(\omega) = \omega_p^2/(\omega_{aj}^2 - \omega^2 - i\omega\gamma)$ are the Lorentzian dispersive functions, where

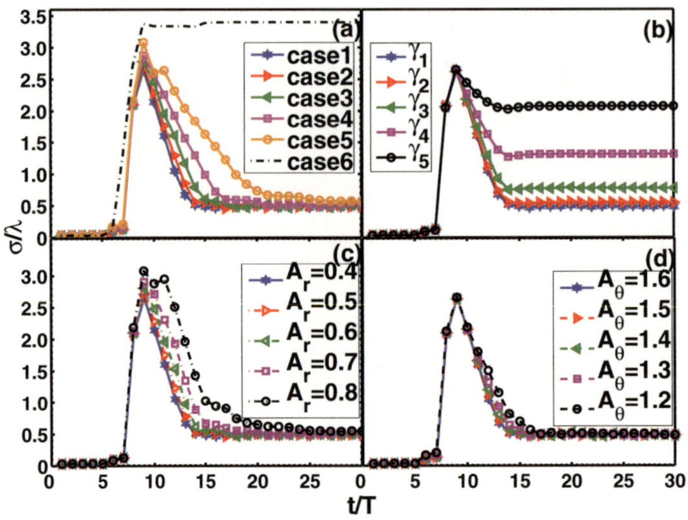

Fig. 14. (Color online) The σ vs t curves. From (a) to (d), $A_z = 0.4$ where $\varepsilon_z < 1$ or $A_z = 1.6$ where $\varepsilon_z > 1$. (a) Keep $\gamma = 0.012\omega_0$ unchanging, choose A_r and A_θ as case 1: $A_r = 0.4$, $A_\theta = 1.6$, case 2: $A_r = 0.5$, $A_\theta = 1.5$, case 3: $A_r = 0.6$, $A_\theta = 1.4$, case 4: $A_r = 0.7$, $A_\theta = 1.3$, case 5: $A_r = 0.8$, $A_\theta = 1.2$. Case 6: only PEC shell without CS. (b) Keep $A_r = 0.4$, $A_\theta = 1.6$ unchanging, choose $\gamma_1 = 0.012\omega_0$, $\gamma_2 = 0.024\omega_0$, $\gamma_3 = 0.048\omega_0$, $\gamma_4 = 0.096\omega_0$ and $\gamma_5 = 0.192\omega_0$. (c) Keep $A_\theta = 1.6$ and $\gamma = 0.012\omega_0$ unvaried, and change A_r. (d) Keep $A_r = 0.4$ and $\gamma = 0.012\omega_0$ unvaried, and change A_θ.

γ is the "resonance width" or called as "dissipation factor," ω_{aj} are the resonant frequency of "atoms" (resonant units) in metamaterials.

For the study of the dispersive CS, we suppose that the real parts of the ε_z, μ_r, and μ_θ always satisfy the geometry transformation of Ref. [69] at ω_0:

$Re[\mu_r(r,\omega_0)] = (r - R_1)/r$, $Re[\mu_\theta(r,\omega_0)] = r/(r - R_1)$, and $Re[\varepsilon_z(r,\omega_0)] = R_2^2(r - R_1)/[(R_2 - R_1)^2 r]$. Then the filling factors $F_j(r)$ at different r can be obtained: $F_r(r) = \{Re[\mu_r(r,\omega_0)] - 1\}/Re[f_r(\omega_0)]$, $F_\theta(r) = \{Re[\mu_\theta(r,\omega_0)] - 1\}/Re[f_\theta(\omega_0)]$, and $F_z(r) = \{Re[\varepsilon_z(r,\omega_0)] - 1\}/Re[f_z(\omega_0)]$.

To investigate the dispersive effect on the dynamical process, we tune the dispersion parameters ω_{aj} in our numerical experiments. We use the working frequency ω_0 as the frequency unit since it is the same for all cases in this section so the ratio $A_j = \omega_{aj}/\omega_0$ represents ω_{aj}. Obviously, for the Lorentzian dispersive relation, the dispersion is stronger when ω_0 and ω_{aj} are closer to each other (the working frequency is near the resonant frequency), or in other words, when A_j approaches *one*. Since there are singular values of real part of ε and μ, in our numerical simulation we have done some approximations,[80] such as we set the maximum and the minimum for ε and μ. Although such approximations will affect the cloaking effect of stable state,[74] we find that the influence of these approximations on the dynamical process is very small and can be neglected.

First, we show an example of evolving electronic field during the dynamical process in Fig. 13 with concrete parameters of A_r, A_θ, A_z, and γ. In Fig. 13(a), the plane wave arrives at the left side of the CS, and is ready to enter the CS. From Fig. 13(b)-13(e), the cloaking effect is

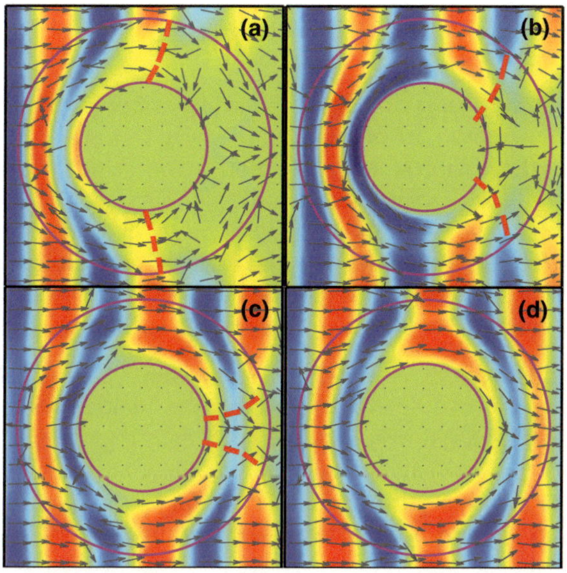

Fig. 15. (Color online) Direction of Poynting vectors and the intensity front (shown by red dashed curves) at moments during the dynamical process. Parameters are chosen as that of Fig. 13. (a) $t = 4.92T$. (b) $t = 7.20T$. (c) $t = 9.00T$. (d) Stable state.

built up step by step, at last, the field gets to the stable state shown in Fig. 13(f). Because of the dispersion, there is an obvious time delay in the cloaking effect and the strong scattered field is observed.

We introduce a time-dependent scattering cross-section $\sigma(t)$ to quantitatively study the dynamical process, which is defined as

$$\sigma(t) = \bar{J}_{scat}(t)/\bar{S}_{inct}, \tag{13}$$

where $t = n \times T$, $n = 0, 1, 2, ...$, T is the period of the incident wave, $\bar{J}_{scat}(t)$ is the one-period-average energy flow of scattered field, and $\bar{S}_{inct}$ is the averaged energy flow density of incident field. To observe the dispersive effect on $\sigma(t)$ during the dynamical process, at the first step, we keep A_z and γ constant and change A_r and A_θ, the results are shown in Fig. 14(a). From the σ versus t curves, we can find the general properties of the dynamical process. First, there is strong scattering in the dynamical process. At the beginning, σ increases rapidly when the wave gets to the CS, then reaches its maximum (at about ninth period). After that, σ starts to decay until it gets to the stable value (of the stable cloaking state). Second, unlike other systems, there is no oscillation in the process. This property will be discussed later. Third, the time length of dynamical process, called as "relaxation time" generally, can be tuned by the dispersion. From Fig. 14(a), we can see that the main dispersive effect is on the decaying process. From case 1 to case 5, A_r and A_θ become closer to *one*, so that the dispersion is stronger. We find that the stronger the dispersion, the longer the relaxation time. For comparing with the cloaking cases, we also show the $\sigma(t)$ of the naked PEC shell in the case 6. From the definition of σ, we know that the area covered by these curve in Fig. 14(a) is proportional to the total scattered energy in the dynamical process. So the CS with *the weaker dispersion will scatter less field* (better cloaking effect) in the dynamical process. But, such a general conclusion is still not enough for us to get a clear physical picture to understand the cloaking dynamical process.

Next, we check whether the absorption of the CS is important in the process. The absorption is determined by the imaginary part of ε and μ. To study this effect, we hold A_r, A_θ, and A_z constant but modify the dispassion factor γ. We modify the filling factors F_j simultaneously, so that the real parts of ε and μ are kept unchanged at ω_0. In such way, we can keep the dispersion strength almost unchanged, but with the imaginary parts of ε and μ changed. Results in Fig. 14(b) show that the stronger absorption only leads to larger stable value of σ, leaving the relaxation time nearly unchanged. Thus, we can exclude the absorption from the relevant parameter list, since it only influences the $\sigma(t)$ of stable state considerably.

To obtain deeper insight of the dynamical process, we need to study the dynamical process more carefully. From Figs. 13(b)-13(e), we can see that the "field intensity" (shown by different color in the figures) propagates slower inside the CS than that in the outside vacuum. And when the inside field intensity "catches up" the outside one [in Fig. 13(f)], the field in the CS gets to the stable state and the cloaking effect is built up. In fact, this catching-up process of the field intensity can be shown more clearly by the direction of Poynting vectors during the dynamical process. From Figs. 15(a)-15(d), we show the direction of Poynting vectors in moments of Figs. 13(c)-13(f), respectively. In the Fig. 15, we see that there is the "intensity front"(shown by red dashed curve) which separates two regions of the CS. At the right side of the front, the field intensity in the CS is much weaker than the outside and the Poynting vector directions are not regular(especially near the front). But at the left-side region which is swept by the intensity front, the Poynting vectors are very regular and nearly along the "cloaking rays" which was predicted at the coordinate transformation[68]. Since the cloaking effect can be interpreted by the mimic picture that the light runs around the cloaking area through these curved cloaking rays, it is not surprising to find that the stable cloaking state is achieved when the intensity front sweeps through the whole CS and these optical rays are well constructed. The surprising thing is that *the stable cloaking state seems to be constructed locally.* We believe this property is related the original cloaking recipe,[68] which makes the cloaking material is almost impedance matched layer by layer. This also explains why there is no oscillation in the cloaking dynamical process generally. This picture also can interpret the strong scattered field in the dynamical process, since these "irregular rays" at the right-side region of the intensity front must be scattered strongly. Further, we can use this picture to analysis the dynamical process of other incident waves, such as the Gaussian beams, which are composed of different plane-wave components.

With these understanding, now we are ready to find the correlation between the relaxation time and the CS dispersion. It is well known that the field intensity (or energy) propagates at the group velocity V_g, which is controlled by the material dispersion. So the intensity front, which determines the dynamical process, should move in V_g. Thus, we can explain the results in Fig. 14, since our modification of the dispersive parameters can cause the V_g changed. But, because the cloaking material is the strong anisotropic material, the V_g at different directions could be very different. Can we predict more precisely which component dominates the relaxation time? The answer is "yes." In Fig. 15(d), we can see that the stable energy flow in the CS is nearly along the θ direction at most regions of the CS. Then it is reasonable for us to argue that it is the component along the θ direction $V_{g\theta}$, not the component along the r direction V_{gr}, that dominates the relaxation time and the total scattered energy in the dynamical process.

For the anisotropic cloaking material, the $V_{g\theta}$ and V_{gr} can be expressed as: $V_{g\theta} = [\nabla_{\mathbf{k}}\omega(\mathbf{k})]_\theta = (\frac{2c}{\sqrt{\varepsilon_z\mu_r}})/(2+\frac{\omega}{\varepsilon_z}\frac{d\varepsilon_z}{d\omega}+\frac{\omega}{\mu_r}\frac{d\mu_r}{d\omega})$ and $V_{gr} = [\nabla_{\mathbf{k}}\omega(\mathbf{k})]_r = (\frac{2c}{\sqrt{\varepsilon_z\mu_\theta}})/(2+\frac{\omega}{\varepsilon_z}\frac{d\varepsilon_z}{d\omega}+\frac{\omega}{\mu_\theta}\frac{d\mu_\theta}{d\omega})$, where c is the velocity of light in vacuum.

In order to illustrate our prediction, the $\sigma(t)$ under different $V_{g\theta}$ and V_{gr} are investigated, respectively. First, we keep the V_{gr} unvaried by holding A_θ, A_z and γ constant [keep $d\varepsilon_z/d\omega$ and $d\mu_\theta/d\omega$ unchanged], only modify A_r to change the $V_{g\theta}$. The results are shown in Fig.14(c), when A_r is closer to *one*, the $V_{g\theta}$ becomes smaller (with larger $d\mu_r/d\omega$), the relaxation time is longer and more energy scattered in the dynamical process. So the larger $V_{g\theta}$ means the better cloaking effect in the dynamical process. On the other hand, when we keep the $V_{g\theta}$ unvaried and change V_{gr} by holding A_r, A_z, and γ constant and modifying A_θ, the results are shown in Fig. 14(d). We find that the relaxation time is almost unchanged with the change of V_{gr}. Obviously, $V_{g\theta}$ is the dominant element in the dynamical process. This conclusion can help us to design a better CS to defend the pulsive radars. In the expression of $V_{g\theta}$, it is also shown how to tune $V_{g\theta}$ by modifying dispersion parameters.

It seems that the larger $V_{g\theta}$, the better cloaking effect in the dynamical process. However, since the V_g (and its components) cannot exceed c generally, there is a minimum limit for the relaxation time of the cloaking dynamical process. We can estimate it through dividing the mean length of the propagation rays by $V_{g\theta}$. In our model, the mean length is $\pi(R_2+R_1)/2$, about three wavelengths. So the relaxation time can not be shorter than three periods. Figure 14 shows that our estimation is coincident with our simulation results. Actually, here we are facing a very basic conflict to make a "better" CS, which is more discussed in our other works.[7, 81] The conflict is from the fact that the *pretty strong* dispersion is required to realize a good *stable* cloaking effect at a certain frequency,[7, 68] but at this research we show that the *weaker* dispersion can realize a better cloaking effect in the *dynamical* process. At real design of the CS, there should be an optimized trade-off.

Based on causality, the limitation of the electromagnetic cloak with dispersive material is investigated in this section The results show that perfect invisibility can not be achieved because of the dilemma that either the group velocity V_g diverges or a strong absorption is imposed on the cloaking material. It is an intrinsic conflict which originates from the demand of causality. However, the total cross section can really be reduced through the approach of coordinate transformation. A simulation of finite-difference time-domain method is performed to validate our analysis.

6. Limitation of the electromagnetic cloak with dispersive material

Through the ages, people have dreamed to have a magic cloak whose owner can not be seen by others. For this fantastic dream, plenty of work has been done by scientists all over the world. For example, the researchers diminished the scattering or the reflection from objects by absorbing screens[82] and small, non-absorbing, compound ellipsoids[83]. More recently, based on the coordinate transformation, J. B. Pendry, et al theoretically proposed a general recipe for designing an electromagnetic cloak to hide an object from the electromagnetic(EM) wave[68]. An arbitrary object may be hidden because it remains untouched by external radiation. Meanwhile, Ulf Leonhardt described a similar method where the Helmholtz equation is transformed to produce similar effects in the geometric limit[67, 75]. Soon, Steven A. Cummer, et al simulated numerically(COMSOL) the cylindrical version of this cloak structure using ideal and nonideal (but physically realizable) electromagnetic

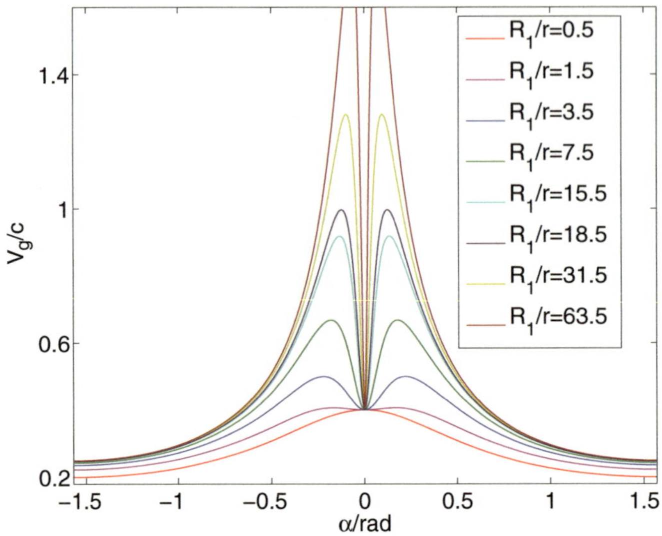

Fig. 16. The group velocity V_g versus α for different R_1/r values

parameters[69]. Especially, Schurig, et al experimentally demonstrated such a cloak by split-ring resonators[72]. In addition, Wenshan Cai , et al proposed an electromagnetic cloak using high-order transformation to create smooth moduli at the outer interface and presented a design of a non-magnetic cloak operating at optical frequencies[70, 71]. According to the general recipe, the electromagnetic cloak is supposed to be perfect or "fully functioned" at certain frequency as long as we can get very close to the ideal design although there is a singularity in the distribution, which has been elucidated further in several literatures [84, 85]. However, in all these pioneering works, the interests are mainly focused on single-frequency EM waves, so that the effects of the dispersion, which is related with very basic physical laws, are not well studied. If the dispersion is introduced into the study, can we have a deeper insight into the cloaking physics?

In this section we will show the ideal cloaking can not be achieved because of another more basic physical limitation—the causality limitation (based on the same limitation, Chen et al obtained a constraint of the band width that limit the design of an invisibility cloak[86]). Starting from dispersion relation and combining with the demand of causality, we will demonstrate that the ideal cloaking will lead to the dilemma that either the group velocity V_g diverges or a strong absorption is imposed on the cloaking material. Our derivation and numerical experiments based on the finite-difference-time-domain(FDTD) methods will show that the absorption cross section will be pretty large and dominate the total cross section for a dispersive cloak, even with very small imaginary parts of permittivity and permeability.

Let's consider a more general coordinate transformation on an initial homogeneous medium with $\varepsilon_i = \mu_i$ in r space: $r' = f(r), \theta' = \theta, \varphi' = \varphi$, following the approach in Ref. [87] and [89], we get the following radius-dependent, anisotropic relative permittivity and permeability: $\varepsilon_{r'} = \mu_{r'} = \varepsilon_i(\frac{r}{f(r)})^2 \frac{df(r)}{dr}$, $\varepsilon_{\theta'} = \mu_{\theta'} = \varepsilon_i / \frac{df(r)}{dr}$ and $\varepsilon_{\varphi'} = \mu_{\varphi'} = \varepsilon_i / \frac{df(r)}{dr}$. We emphasize that since the transformation is directly acted on the Maxwell equations, the above equations are also suited for the imaginary parts of constitutive parameters, and all physical properties of wave propagation in r space should be inherited in r' space, such as the absorption. This

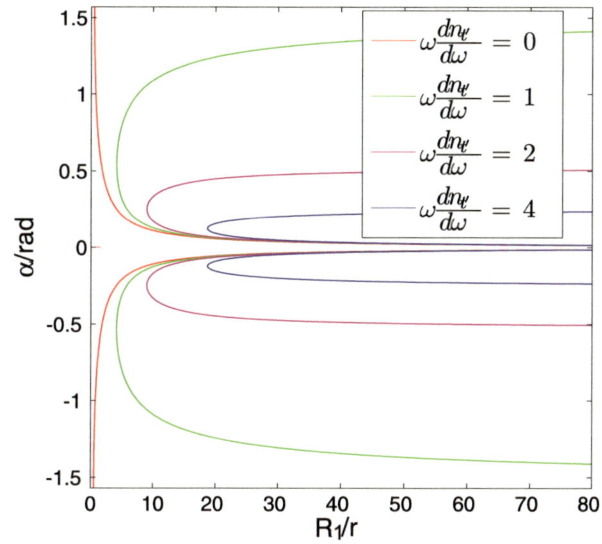

Fig. 17. The relation between R_1/r and α when $V_g = c$ for different $\omega\frac{dn_{t'}}{d\omega}$.

is very important for us to have consistent physical pictures in both spaces. At working frequency ω_0, for a propagating mode with k-vector as $\{k_{r'},k_{\theta'},k_{\varphi'}\}$ inside the cloak, we have the dispersion relation of the anisotropic material[87] as: $k_{r'}^2/n_{r'}^2 + k_{t'}^2/n_{t'}^2 = \omega^2/c^2$, where $k_{t'}^2 = k_{\theta'}^2 + k_{\varphi'}^2$, $n_{r'} = \sqrt{\varepsilon_{\varphi'}\mu_{\theta'}} = n_i/\frac{df(r)}{dr}$, $n_{t'} = n_{\theta'} = n_{\varphi'} = \sqrt{\varepsilon_{r'}\mu_{\theta'}} = n_i r/f(r)$, and $n_i = \sqrt{\varepsilon_i \mu_i}$. Then we can define $k_{r'} = \frac{\omega}{c}n_{r'}cos\alpha$ and $k_{t'} = \frac{\omega}{c}n_{t'}sin\alpha$, the group velocity can be obtained as:

$$V_g = c\sqrt{\frac{(cos\alpha)^2}{n_{r'}^2} + \frac{(sin\alpha)^2}{n_{t'}^2}}/((cos\alpha)^2\frac{m_{r'}}{n_{r'}} + (sin\alpha)^2\frac{m_{t'}}{n_{t'}}) \tag{14}$$

Where $m_{r'} = n_{r'} + \omega\frac{dn_{r'}}{d\omega}$ and $m_{t'} = n_{t'} + \omega\frac{dn_{t'}}{d\omega}$.

If the transformation has the following characteristic: $f(r=0) = R_1$, $f(r=R_2) = R_2$, then when $r' \to R_1$ (or $r \to 0$), $n_{t'}$ will tend to zero, and the group velocity is approximated as:

$$V_g \approx \frac{c}{|sin\alpha|\omega\frac{dn_{t'}}{d\omega}} \tag{15}$$

We will discuss Eq.15 in two cases. The first case is with the finite $\frac{dn_{t'}}{d\omega}$. Obviously, V_g will diverge when $sin\alpha \to 0$ for any finite $\frac{dn_{t'}}{d\omega}$. Such divergence is shown in figure 16 for a concrete example, in which the transformation is $r' = f(r) = (R_2 - R_1)r/R_2 + R_1$ as Ref. [68], $R_2 = 2R_1$, thus $n_{r'} = 2$ and $n_{t'} = 2 - 4/(r/R_1 + 2)$. The dispersion parameters are set as $m_{r'} = 2.5$, $\omega\frac{dn_{t'}}{d\omega} = 4$ at working frequency. In figure 16, the curves of V_g vs α are plotted for different R_1/r values. We can see that, for large $R_1/r(r \to 0)$, the group velocity (more precisely, the tangential component of V_g) will diverge at both peaks around $\alpha = 0$.

Because of the causality limitation, it is well-known that the group velocity can not exceed c except in the "strong dispersion" frequency range (or called "resonant range"). But, if the working frequency is in the "strong-dispersion" range of the cloaking material, the absorption must be very strong and it will destroy the ideal cloaking obviously. So perfect invisibility can not be achieved for the finite $\frac{dn_{t'}}{d\omega}$ because it will lead to superluminal velocity or strong absorption.

In addition, the curves with the criterion condition $V_g = c$ on the plane $[R_1/r, \alpha]$ are plotted for different $\omega\frac{dn_{t'}}{d\omega}$ in figure 17. The region to the left of curves is corresponding to $V_g < c$ and the region to the right is corresponding to $V_g > c$. There exists a maximum $max\{R_1/r\}$ for each curve in order that $V_g \leq c$ can be hold for all the α. Especially, for the no-dispersion case $\omega\frac{dn_{t'}}{d\omega} = 0$, we can see that $V_g > c$ at all R_1/r for large α values, which means the whole cloak is not physical if there is no dispersion. This "dispersion-is-required" conclusion can be generally derived from Eq.14, and it is consistent with the analysis in Ref. [68]. From figure 17, we know that the larger $\omega\frac{dn_{t'}}{d\omega}$, the larger $max\{R_1/r\}$. But anyway, for arbitrary finite $\omega\frac{dn_{t'}}{d\omega}$, $max\{R_1/r\}$ can not be infinite, so that the superluminal range always exists.

The second case of Eq.15 is with divergent $\frac{dn_{t'}}{d\omega}$. From the previous discussion, we know that if the ideal cloak exists, the cloak must be dispersive and $\frac{dn_{t'}}{d\omega}$ must be divergent when $r \to 0$. Actually, when $r \to 0$, since $\sqrt{\varepsilon_{r'}} \propto r$, $\frac{dn_{t'}}{d\omega} \propto \frac{d\varepsilon_{r'}}{d\omega}/\sqrt{\varepsilon_{r'}}$ is really divergent for non-zero $\frac{d\varepsilon_{r'}}{d\omega}$. From Eq.15, we can see that now the $V_g \to 0$ for a finite $\frac{d\varepsilon_{r'}}{d\omega}$ (generally true) at all α values except $\alpha = 0$(or π), so that the group velocity difficulty seems to be overcome. But, since the causality limitation, the non-zero $\frac{d\varepsilon_{r'}}{d\omega}$ means non-zero imaginary part of permittivity (non-zero dissipation). The non-zero dissipation and the almost-zero group velocity will result in very strong absorption. This means that the energy of rays near the inner cloaking radius R_1 is almost totally absorbed by the cloaking material. As we pointed out at the beginning that the absorption in r' space should also appear in r space, because of the consistence between two spaces. The strong absorption in r space can be interpreted in the following way. From the transformation (which is also suited for imaginary part), we can find that when $r \to 0$, the finite imaginary part in r' space corresponds to the infinite imaginary part in r space, which also means very strong absorption in the initial homogeneous medium. So the perfect cloaking is still impossible because of the strong absorption which is enforced by the causality limitation.

For a two-dimensional coordinate transformation: $r' = f(r)$, $\theta' = \theta$, $z' = z$, the same conclusions of the causality limitation can be obtained through the similar analysis, although the coordinate transformation and the singularities are different from the three-dimensional case.

Next, we will discuss the physical meaning of the dilemma that either the group velocity V_g diverges or a strong absorption is imposed on the cloaking material. First, it is an intrinsic conflict which can not be solved by the methods, for example, "the system is imbedded in a medium"[68]. We believe that the ideal cloaking is impossible because of the causality limitation and this conclusion is consistent with the statement of previous studies[89] that the perfect invisibility is unachievable because of the wave nature of light. Second, we have to face the question: " Why the causality is violated for ideal cloaking which is based on the simple coordinate transformation?". Our answer is that the causality is only guaranteed by

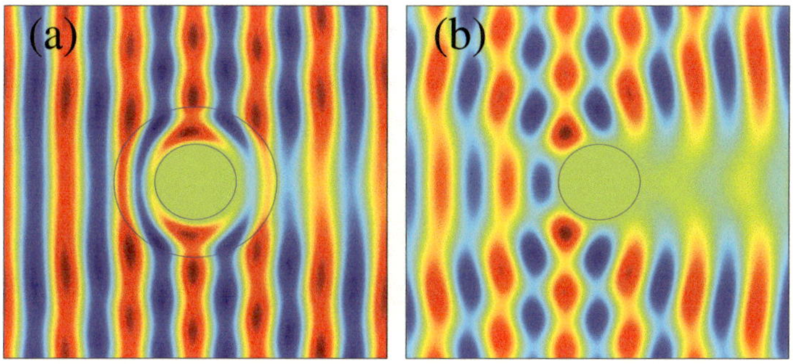

Fig. 18. The snapshots of the electric-field distribution in the vicinity of PEC. (a) the cloaking structure with a PEC at radius R_1, (b) the naked PEC with radius R_1.

the Lorentz co-variant transformation, but the coordinate transformation for ideal cloaking is not Lorentz co-variant. Such violation is obvious if we suppose the initial medium in the r space is not dispersive, such as the vacuum, but as we have pointed out (also mentioned in Ref. [68], the cloaking material (in r' space) must be dispersive to avoid the group velocity over c. Such Lorentz co-variant violation is generally true for ąřtransformation opticsąś since material parameters are non-relativistic, so the causality limitation should be checked widely. Third, from Eq.14, we can find that not only the inner layers of the cloak($r' \to R_1$) but also the other layers($r' > R_1$) must be dispersive. For every layer, a certain dispersive strength is needed to avoid $V_g > c$.

In the following, we will validate that the total cross section can be reduced drastically, and that the perfect cloaking cannot be achieved because of strong absorption by FDTD numerical experiments. Compared with other frequency-domain simulation methods, such as the finite element methods or the transfer-matrix methods, the FDTD simulation can better reflect the real physical process of cloaking. For example, we note that the FDTD calculation will be numerically unstable when the dispersion is not included in the cloak's material. For simplicity, the simulation is limited to two-dimensional cloak[69]. Without lost of generality, only TE modes are investigated in this study (TE modes have the electric field perpendicular to the two-dimensional plane of our model). Thus the constitutive parameters involved here are $\epsilon_{z'}, \mu_{r'}, \mu_{\theta'}$. The dispersion is introduced into our FDTD by standard Lorentz model:

$$\begin{aligned}
\tilde{\epsilon}_{z'}(r',\omega)| &= 1 + F_{z'}(r')\omega_{pz'}^2/(\omega_{az'}(r')^2 - \omega^2 - i\omega\gamma_{z'}) \\
\tilde{\mu}_{r'}(r',\omega) &= 1 + F_{r'}(r')\omega_{pr'}^2/(\omega_{ar'}(r')^2 - \omega^2 - i\omega\gamma_{r'}) \\
\tilde{\mu}_{\theta'}(r',\omega) &= 1 + F_{\theta'}(r')\omega_{p\theta'}^2/(\omega_{a\theta'}(r')^2 - \omega^2 - i\omega\gamma_{\theta'})
\end{aligned} \quad (16)$$

$$r' = \frac{R_2(\theta) - R_1(\theta)}{R_2(\theta)} r + R_1(\theta) \quad (17)$$

$$\theta' = \theta \quad (18)$$

Where $\omega_{pz'}, \omega_{pr'}, \omega_{p\theta'}$ are plasma frequencies, $\omega_{az'}, \omega_{ar'}, \omega_{a\theta'}$ are atom resonated frequencies, $\gamma_{z'}$, $\gamma_{r'}$, $\gamma_{\theta'}$ are damping factors and $F_{pz'}$, $F_{pr'}$, $F_{p\theta'}$ are filling factors. In our simulation, an E-polarized time-harmonic uniform plane wave whose wavelength λ_0 in vacuum is $3.75cm$ is incident from left to right. The real parts of the constitutive parameters at $\omega_0 = 2\pi c/\lambda_0$

satisfy the cloaking coordinate transformation[69, 91], and they are $\mu_{r'} = \frac{r'-R_1}{r'}$, $\mu_{\theta'} = \frac{1}{\mu_{r'}}$, $\varepsilon_{z'} = (\frac{R_2}{R_2-R_1})^2 \frac{r'-R_1}{r'}$, where R_1 is $0.665\lambda_0$, R_2 is $1.33\lambda_0$. And the dispersive parameters are set as follows: if $\varepsilon_{z'} > 1$ then $\omega_{az'} = 1.4\omega_0$, else $\omega_{az'} = 0.6\omega_0$, $\omega_{ar'} = 0.6\omega_0$, $\omega_{a\theta'} = 1.4\omega_0$, $\gamma_{z'} = \gamma_{r'} = \gamma_{\theta'} = \omega_0/100$. $\omega_{pz'} = \omega_{pr'} = \omega_{p\theta'} = 4\omega_0$, $F_{z'}(r) = (\varepsilon_{z'} - 1)\frac{(\omega_{az'}^2-\omega_0^2)^2+\omega_0^2\gamma_{z'}^2}{(\omega_{az'}^2-\omega_0^2)\omega_{pz'}^2}$, $F_{r'}(r) = (\mu_{r'} - 1)\frac{(\omega_{ar'}^2-\omega_0^2)^2+\omega_0^2\gamma_{r'}^2}{(\omega_{ar'}^2-\omega_0^2)\omega_{pr'}^2}$, $F_{\theta'}(r) = (\mu_{\theta'} - 1)\frac{(\omega_{a\theta'}^2-\omega_0^2)^2+\omega_0^2\gamma_{\theta'}^2}{(\omega_{a\theta'}^2-\omega_0^2)\omega_{p\theta'}^2}$. In fact, these parameters have many possible choices. The different groups of parameters correspond to different dynamic processes which we will discuss in another section [92].

Figure 18 shows the snapshots of the electric-field distribution in two cases: the cloak with the perfect electric conductor (PEC) at radius R_1(left), and the naked PEC with radius R_1(right). Obviously, the cloak is very effective. Quantitatively, we calculate the absorption cross section and the scattering cross section of the cloak at the stable state, and they are $0.67\lambda_0$ and $0.24\lambda_0$ respectively, while the scattering cross section of the naked PEC is $3.14\lambda_0$. So, with dispersive cloak, the total cross section is three times smaller, and the absorption cross section dominates as we predicted. To emphasize the huge absorption of the cloak, we use a common homogeneous isotropic media, with $\varepsilon = \mu = 1.1$ but all other parameters are the same as the cloak, to replace the cloaking material. Then we find the absorption cross section is only $0.089\lambda_0$ which is about one order smaller. The reason of strong absorption has been discussed before.

Now we can have a full view of cloaking recipe based on the coordinate transformation. First, the cloaking material must be dispersive, and the strong absorption can not be avoided because of the causality limitation. Thus it is not perfectly invisible. Second, the scattering cross section of the dispersive cloak could be small, so that the scattered field is weak. Although the ideal invisibility is impossible, the cloaking recipe still has a main advantage. The "strong-absorption and weak-scattering" property means that the cloak almost can not be observed except from the forward direction. so that such cloak can well defend the ąřpassive radarsąś which detect the perturbation of the original field. It is well-known that at the Rayleigh scattering case, where the radius of the scatterer is much smaller than the wavelength, the absorption cross section could be larger than the scattering cross section because of the diffraction. The cloaking can be thought as giant Rayleigh scattering case, where the light rays are forced to "diffract" around the cloaked area.

In conclusion, the properties of the dispersive cloak are investigated, and the limitation of causality is revealed. Our study shows that the superluminal velocity or a strong absorption can not be overcome since the intrinsic conflict between the coordinate transformation to obtain the cloaking and the causality limitation. In addition, we validate the results using a numerical simulation which is performed in FDTD algorithm with physical parameters. The numerical experiments show that the absorption cross section is dominant and the scattering cross section can be reduced significantly. The study gives us a full view of the cloaking recipe based on the coordinate transformation, and will have further profound influence on the related topics.

7. Summary

In summary, we have investigate the metamaterial systems from group velocity(energy velocity) picture. From these topics, we demonstrate the importance of group velocity in metamaterial studies. From group velocity, we can find the physical origin of abnormal optical phenomena of metamaterials, such as the "no-transmission no-reflection" on the hyper-medium surface which is from a zero-group-velocity reflecting mode, and the coherence gain of superlens image which is from the different group delay on different paths. From group velocity, we can avoid some traps of violating basic physical limitation, such as the violation of causality limitation in cloaking study. These traps are very serious since the metamaterial from our imagination could exist in this world if violating basic limitations. From group velocity, we can find the key parameter of cloaking dynamical process and help us to optimizing the design of cloak design. From group velocity of evanescent wave, new detecting methods could be found, for example the detecting of QED vacuum polarization by phase change or delay time of evanescent wave. We believe that only with the well-constructed group velocity picture, the deeper understanding of the abnormal optical/photonic properties of metamaterials is possible. All these research works also show that the group velocity study of metamaterials can lead us to many new interesting topics, which are still waiting for further research.

8. Acknowledgement

This work is supported by NKBRPC (Grant No. 2012CB927401), NSFC (Grant Nos. 11004212, 11174309, and 60938004), and the STCSM (Grant Nos. 11ZR1443800 and 11JC1414500).

9. References

[1] V. G. Veselago, Sov. Phys. Usp. 10, 509 (1968).
[2] D. R. Smith, W.J.Padilla, D. C. Vier, S. C. Nemat-Nasser, and S. Schultz, Phys. Rev. Lett. 84, 4184 (2000).
[3] R. A. Shelby, D. R. Smith, S. C. Nemat-Nasser, and S. Schultz, Appl. Phys. Lett. 78, 489 (2001).
[4] R. A. Shelby, D. R. Smith, and S. Schultz, Science 292, 77 (2001).
[5] J. B. Pendry, A. J. Holden, D. J. Robbins, and W. J. Stewart, IEEE Trans. Microwave Theory Tech. 47, 2075 (1999); J. B. Pendry, Phys. Rev. Lett. 85, 3966 (2000).
[6] P. Markos and C. M. Soukoulis, Phys. Rev. B 65, 033401 (2001); Phys. Rev. E 65, 036622 (2002).
[7] P. Yao, Z. Liang, and X. Jiang, Appl. Phys. Lett. 92, 031111 (2008).
[8] Zixian Liang, Peijun Yao, Xiaowei Sun and Xunya Jiang, Appl. Phys. Lett. 92, 131118 (2008)
[9] Zheng Liu, Zixian Liang, Xunya Jiang, Xinhua Hu, Xin Li, and Jian Zi, Appl.Phys. Lett. 96, 113507 (2010)
[10] Brillouin, *Wave Propagation and Group Velocity*, New York:Academic Press Inc., (1960)
[11] J. H. Page, Ping Sheng, H. P. Schriemer, I.Jones, Xiaodun Jing and D. A. Weitzt,Science, 271 634 1996
[12] Hailu Luo, Wei Hu and Zhongzhou Ren, Europhys. Lett. 74 1081(2006).
[13] R.A. Shelby, D.R. Smith, and S. Schultz, Science 92, 792(2001); D.R. Smith and D. Schurig, Phys. Rev. Lett. 90,077405 (2003);C.G. Parazzoli, R.B. Greegor, K. Li, B.E.C. Koltenbah, and M. Tanielian, Phys. Rev. Lett. 90, 107401 (2003).
[14] J. B. Pendry, Phys. Rev. Lett. 85, 3966(2000); N. Fang, H. Lee, C. Sun, X. Zhang, Science 308, 534 (2005); T. Taubner, D. Korobkin, Y. Urzhumov, G. Shvets, R.Hillenbrand, Science 313, 1595 (2006)

[15] J. B. Pendry, Opt. Express 11, 755 (2003); Zhaowei Liu, Hyesog Lee, Yi Xiong, Cheng Sun, and Xiang Zhang, Science 315, 1686 (2007); Igor I. Smolyaninov, Yu-Ju Hung, and Christopher C. Davis, Science 315, 1699 (2007).
[16] Z. Jacob, L. V. Alekseyev, E. Narimanov, Opt. Express 14,8247 (2006); A. Salandrino, N. Engheta, Phys. Rev. B 74, 075103 (2006); Zubin Jacob, Leonid V. Alekseyev, and Evgenii Narimanov, J. Opt. Soc. Am. A 24,A52 (2007); J. B. Pendry, S. A. Ramakrishna, J. Phys. Condens. Matter 14,8463(2002).
[17] Xin Li,Zixian Liang,Xiaohan Liu,Xunya Jiang, and Jian Zi, Appl. Phys. Lett. 93, 171111 (2008).
[18] Hailu Luo, Wei Hu and Zhongzhou Ren, Europhys. Lett. 74 1081(2006).
[19] P. W. Milonni, *Fast Light, Slow Light and Left-Handed Light* Institute of Physics, Great Britain, CRC Press, 2004
[20] As for the explanation of the static solution of the HI it'll be given in another paper.
[21] David R. Smith and Norman Kroll, Phys. Rev. Lett. 85, 2933(2000).
[22] V. A. Podolskiy, and E. E. Narimanov, Phys. Rev. B 71, 201101 (2005); R. Wangberg, J. Elser, E. E. Narimanov, and V. A. Podolskiy, J. Opt. Soc. Am. B. 23, 498 (2006); A. A. Govyadinov, V. A. Podolskiy, Phys. Rev. B 73,115108(2006).
[23] A. Taflove, *Computational Electrodynamics: The Finite-Difference Time- Domain Method* (Artech House, Boston, 1995).
[24] Alessandro Salandrino and Nader Engheta, Phys. Rev. B 74,115108(2006).
[25] N. Fang, H. Lee, C. Sun, and X. Zhang, Science 308, 534 (2005).
[26] Jie Yao,Zhaowei Liu,Yongmin Liu, Yuan Wang,Cheng Sun,Guy Bartal, Angelica M. Stacy and Xiang Zhang, Science 321, 930 (2008).
[27] J. B. Pendry and A. J. Holden and W. J. Stewart, Phys. Rev. Lett. 76, 4773(1996).
[28] A. N. Grigorenko, A. K. Geim, H. F. Gleeson,*et al*, Nature, 438 335(2005)
[29] G. Dolling, M. Wegener, C. M. Soukoulis, and S. Linden, Opt. Lett. 32, 53 (2007).
[30] D. Perkins, Introduction to high energy physics, 4th edition (Cambridge university press, Cambridge, 2000).
[31] N. Ashcroft, N. Mermin, Solid State Physics (Harcourt College Publishers, New York, 1976) ; H. Bruus and K. Flensberg, Many-body quantum theory in condensed matter physics (Oxford university press, Oxford, 2004).
[32] J. Goodman, Statistical Optics (John Wiley & Sons, Inc., New York, 1985).
[33] R. Feynman, QED: the strange theory of light and matter (Princeton university press, Princeton, 1985).
[34] R. Ruffini, G. Vereshchagin and S. Xue, Phys. Rep. 487, 1 (2010).
[35] Y. Ding, A. Kaplan, Phys. Rev. Lett. 63, 2725 (1989).
[36] B. King, A. Piazza and C. Keitel, Nat. photon. 4, 92 (2010).
[37] H. Gies, Phys. Rev. D 61, 085021 (2000).
[38] J. S Heyl and L. Hernquist, J. Phys. A: Math. Gen. 30, 6485 (1997).
[39] R. G. Hunsperger, Integrated Optics: Theory and Technology (Springer-Verlag, New York, 1985).
[40] B.Barwick, D.Flannigan, and A.H. Zewail, Nature 462, 902 (2009).
[41] C. Carniglia and L. Mandel, J. Opt. Soc. Am. 61, 1035 (1971).
[42] J. Kong, Electromagnetic Wave Theory, Electromagnetic Wave Theory (Wiley-Interscience, New York, 1990).
[43] J. Lawall and E. Kessler , Rev. Sci. Instrum. 71, 2669 (2000).

[44] M. Bass, Handbook of Optics, Vol. II (McGraw-Hill, Inc., New York, 1995); W. Su, J. Li and N. Xu, J. Biotechnol. 105, 165 (2003).
[45] T. Tajima, Eur. Phys. J. D 55, 519 (2009).
[46] E. Cubukcu, K. Aydin, E. Ozbay, S. Foteinopoulou, and C. M. Soukoulis Phys. Rev.Lett. 91, 207401 (2003); E. Cubukcu, K. Aydin, E. Ozbay, S. Foteinopoulou, and C.M. Soukoulis, Nature (London) 423, 604 (2003).
[47] J. B. Pendry, Phys. Rev. Lett. 91, 099701 (2003);D. R. Smith, D. Schurig, M. Rosenbluth, S. Schultz, S. A. Ramakrishna, and J. B. Pendry, Appl. Phys. Lett. 82, 1506 (2003).
[48] G. Gomez-Santos, Phys. Rev. Lett. 90, 077401 (2003).
[49] S. Foteinopoulou, E. N. Economou, and C. M. Soukoulis, Phys. Rev. Lett. 90, 107402 (2003); J. B. Pendry and D. R. Smith, Phys. Rev. Lett. 90, 029703 (2003).
[50] X. S. Rao and C. K. Ong, Phys. Rev. B 68, 113103 (2003); X. S. Rao and C. K. Ong, Phys. Rev. E 68, 067601 (2003).
[51] Michael W. Feise, Yuri S. Kivshar, Phys. Lett. A 334 326 (2005)
[52] Lei Zhou, C. T. Chan, Appl. Phys. Lett. 86, 101104 (2005), Y. Zhang, T. M. Grzegorczyk and J. A. Kong, PIER 35,271, (2002)
[53] L. Chen, S. He and L. Shen, Phys. Rev. Lett. 92, 107404(2004).
[54] R. W. Ziolkowski and E. Heyman, Phys. Rev. E 64, 056625 (2001).
[55] S. A. Cummer, Appl. Phys. Lett. 82, 1503 (2003).
[56] P. F. Loschialpo, D. L. Smith, D. W. Forester, F. J. Rachford, and J. Schelleng, Phys. Rev. E 67, 025602(R) (2003).
[57] R. Merlin, Appl. Phys. Lett. 84, 1290 (2004).
[58] C. Luo, S. G. Johnson, J. D. Joannopoulos, and J. B. Pendry Phys. Rev. B 68, 045115 (2003)
[59] Nader Engheta, IEEE Antennas and Wireless Propagation Lett. 1, 10, 2002
[60] Ilya V. Shadrivov, Andrey A. Sukhorukov, and Yuri S. Kivshar, Phys. Rev. E 67, 057602 (2003); A. C. Peacock and N. G. R. Broderick, 11, 2502 (2003).
[61] B. E. A. Saleh, M. C. Teich, *Fundamentals of Photonics* (John Wiley & Sons, New York, 1991)
[62] L. Mandel and E. Wolf *Optical Coherence and Quantum Optics* (Cambridge University, Cambridge, 1995); Marlan O. Scully and M. Suhail Zubairy *Quantum Optics*, (Cambridge University, Cambridge, 1997)
[63] X. Jiang and C. M. Soukoulis, Phys. Rev. Lett. 85, 70 (2000)
[64] In our source frequency range, the index range is about $-1 - i0.0029 \pm (0.006 + i10^{-6})$, so the focal length defference and reflection are very small.
[65] The "group velocity" is not a well-defined value if the working frequency ω_0 is near the resonant frequency ω_a of the NIM. But the GRT is still well-defined.
[66] Xunya Jiang *et. al.* unpublished.
[67] U. Leonhardt, Science 312, 1777 (2006).
[68] J. B. Pendry, D. Schurig, and D. R. Smith, Science 312, 1780 (2006).
[69] S. A. Cummer, B.-I Popa, D. Schurig, and David R. Smith, Phys. Rev. E 74, 036621 (2006).
[70] W. Cai, U. K. Chettiar, A. V. Kildishev, and V. M. Shalaev, Nat. Photonics 1, 224 (2007).
[71] W. Cai, U. K. Chettiar, A. V. Kildishev, V. M. Shalaev, and G. W. Milton, Appl. Phys. Lett. 91, 111105 (2007).
[72] D. Schurig, J. J. Mock, B. J. Justice, S. A. Cummer, J. B. Pendry, A. F. Starr, and D. R. Smith, Science 314, 977 (2006).
[73] H. Chen, B.-I. Wu, B. Zhang, and J. A. Kong, Phys. Rev. Lett. 99, 063903 (2007).

[74] Z. Ruan, M. Yan, C. W. Neff, and M. Qiu, Phys. Rev. Lett. 99, 113903 (2007).
[75] U. Leonhardt, New J. Phys. 8, 118 (2006).
[76] S. A. Cummer, Appl. Phys. Lett. 82, 2008 (2003).
[77] X. Jiang, W. Han, P. Yao, and W. Li, Appl. Phys. Lett. 89, 221102 (2006).
[78] P. Yao, W. Li, S. Feng, and X. Jiang, Opt. Express 14, 12295 (2006).
[79] A. Taflove and S. C. Hagness, *Computational Electrodynamics: The Finite-Difference Time-Domain Method*, 2nd ed. (Artech House, Boston, 2000).
[80] The approximation of permittivity ε and permeability μ in this section are: $Re[\mu_\theta(r,\omega_0)]_{max} = 20$, $Re[\mu_r(r,\omega_0)]_{min} = 1/20$ and $Re[\varepsilon_z(r,\omega_0)]_{min} = 1/5$. unpublished.
[81] H. Chen, Z. Liang, P. Yao, X. Jiang, H. Ma, and C. T. Chan, Phys. Rev. B 76, 241104 (2007).
[82] R. L. Fante, and M. T. McCormack, IEEE Trans. Antennas Propag. 30, 1443(1968).
[83] M. Kerker, J. Opt. Soc. Am. 65, 376(1975). and D. R. Nature Photonics 1, 224 (2007). Graeme W. Milton, Appl. Phys. Lett. 91, 111105 (2007).
[84] Hongsheng Chen, Bae-Ian Wu, Baile Zhang, and Jin Au Kong, Phys. Rev. Lett. 99, 063903 (2007).
[85] Zhichao Ruan, Min Yan, Curtis W. Neff, and Min Qiu, Phys. Rev. Lett. 99, 113903 (2007).
[86] Huanyang Chen, Xunya Jiang, C. T. Chan, arXiv:0707.1126v2.
[87] D. Schurig, J. B. Pendry, and D. R. Smith, Optics Express 14, 9794(2006).
[88] H. Chen and C. T. Chan, Appl. Phys. Lett. 90, 241105 (2007).
[89] Ulf Leonhardt and Thomas G Philbin, New J. Phys. 8, 247(2006).
[90] J. B. Pendry and D. R. Smith, Phys. Rev. Lett. 90, 029703 (2003).
[91] When $r' \rightarrow R_1$, $\mu_{\theta'}$ will tend to infinite. In order that it can be realizable in our numerical simulation, we limit its maximum value to 10^3.
[92] Zixian Liang, Peijun Yao, Xunya Jiang, and Xiaowei Sun, unpublished.
[93] G. W. Milton, M. Briane, and J. R. Willis, New J. Phys. 8, 248 (2006).
[94] A. Greenleaf, Y. Kurylev, M. Lassas, G. Uhlmann, http://www.arXiv:/0703059[math-ph] (2007).
[95] L. D. Landau and E. M. Lifshitz, Electrodynamics of Continuous Media (Pergamon Press, 1975), ch. 11, pp. 315-321
[96] HY Chen and CT Chan, the brief report for the detail of the causality constraint.
[97] T. Koschny, M. Kafesaki, E. N. Economou and C. M. Soukoulis, Phys. Rev. Lett. 93, 107402 (2004).

4

Formation of Coherent Multi-Element Resonance States in Metamaterials

Elena Semouchkina
Michigan Technological University
USA

1. Introduction

Employment of metamaterials in cloaking devices, which could make concealed objects invisible, is based on the capability of resonators used as metamaterial "atoms" to provide effective material parameters, ranging from any positive to any negative values, in the vicinity of the resonance frequency. Metamaterials comprising layers with variously sized resonators (Schurig et al., 2006) or layers with variable density of identical resonators (Semouchkina et al., 2010) have been proposed to obtain the desired spatial distribution of the medium parameters, for example, radial variation of the effective permittivity or permeability in the shell prescribed by the transformation optics relations for cloaking cylindrically shaped objects (Leonhardt, 2006; Pendry et al., 2006). Such approach could be justified, if resonators contribute to the value of the effective parameters by the same way as polarized atoms contribute to the value of the polarization of a dielectric medium, i.e., if the responses of individual atoms are additive so that the medium response exhibits the same dispersion as the response of an individual atom. Then the material can be described by the effective medium theory (EMT), which was adopted for metamaterails description at the onset of metamaterial studies (Pendry et al., 1999; Pendry & Smith, 2003; Smith & Pendry, 2004; Smith et al., 2006). According to the EMT, a polar dielectric material could be represented by a set of parallel dipoles switching their directions by 180^0 in dependence on the phase of the external field. Resonance modes in the "atoms" of metamaterials could also be represented by equivalent dipoles and, similarly to atoms in dielectrics, these dipoles are expected to respond accordingly, i.e. to follow the phase of the external field and demonstrate a coherent response within the half wavelength of incident radiation. However, observation of the performance of the infrared invisibility cloak designed from glass resonators (Semouchkina et al., 2010) has shown that obtaining a coherent response of the cloak structure presents a serious problem, if resonators in the metamaterail are electromagnetically coupled. The problem of coupling between resonators in metamaterials, however, remains to be largely ignored in the literature.

It is well known that metamaterials, as a rule, are designed as periodic structures of close-packed resonators. When the dimensions of resonators are ten times smaller than the wavelength, the EMT is considered to be applicable, so that the waves "do not see" the atoms and propagate in the metamaterial as in a homogenized, i.e. in a uniform medium. Such homogenization is supposed to make metamaterials essentially different from photonic crystals (Pendry & Smith, 2003). It is worth noting that the lattice parameter was not thought to be critical for the EMT application, so that the only ultimate requirement for metamaterials in this attitude is that the lattice parameter should not be equal to the half wavelength of incident radiation, which is typical for photonic crystals. Therefore, for example, the metamaterial for the microwave cloak operating at 8.5 GHz (Schurig et al., 2006), was composed of split ring resonators (SRR) with the planar dimensions of 3 mm x 3 mm packed in concentric arrays with the inter-resonators separation of 0.17 mm, which is negligible compared to the resonator size. Such close packing of SRRs causes questions about possible effects of interaction between resonators and elementary resonance splitting that could deteriorate the EMT applicability.

Our earlier studies (Semouchkina et al., 2004, 2005) indeed pointed out at strong splitting of elementary resonances in close-packed resonator arrays of conventional metamaterials consisting of metal elements. At the modeling of an extended finite metamaterial block, which consisted of 36 SRRs and 12 elongated metal strips arranged in unit cells quite similar to typical arrangements employed in conventional metamaterials, it was shown that at the frequencies corresponding to the metamaterial transmission band, elementary resonances were essentially coupled and integrated in 3D networks, an example of which is presented in Fig. 1. As a sequence of coupling and splitting, most resonators responded resonantly at multiple frequencies and formed specific accociations at different frequencies in the transmission band, making the metamaterial essentially nonuniform with the character of inhomogenities changing in dependence on frequency. No regular wave front propagation across the sample was observed, so that the energy transfer occurred rather due to the hopping mechanism similar to that thought to be inherent in the CROWs (coupled resonator optical waveguides (Yariv et al., 1999).

It is logical to suggest that phenomena similar to the mentioned above should take place in any close-packed metamaterial structures including those used in the microwave cloak described in (Schurig et al., 2006). The reason why these phenomena have not been revealed by the authors can be seen in the employment of simulation models, in which real multi-resonator cloak structures were replaced by layered material structures with prescribed values of the effective permeability for each layer.

An additional reason, why coupling phenomena in metamaterials remain underexplored is seen in the typical approach to consider metamaterial properties to be identical to the properties of a single cell and to ignore interaction between cells. This approach is based on the results of unit cell simulations which could not reveal resonance splitting even at small distances between neighboring resonators defined by the dimensions of the unit cell, because of employment of plane wave excitation source at normal wave incidence and periodic boundary conditions (PBC) at the cell boundaries normal to the direction of wave propagation. Such arangement, however, could only model an infinite 2D array of unit cells and not a 3D metamaterial medium. Our results indicate that drastic changes in electromagnetic (EM) metamaterial response could be observed at stacking 2D arrays in the direction of wave propagation to form a 3D structure, because incident wave has a different phase, when encounters each 2D array. Significant changes also occur in finite 2D arrays and in arrays with some distortions of periodicity that is unavoidable in practice.

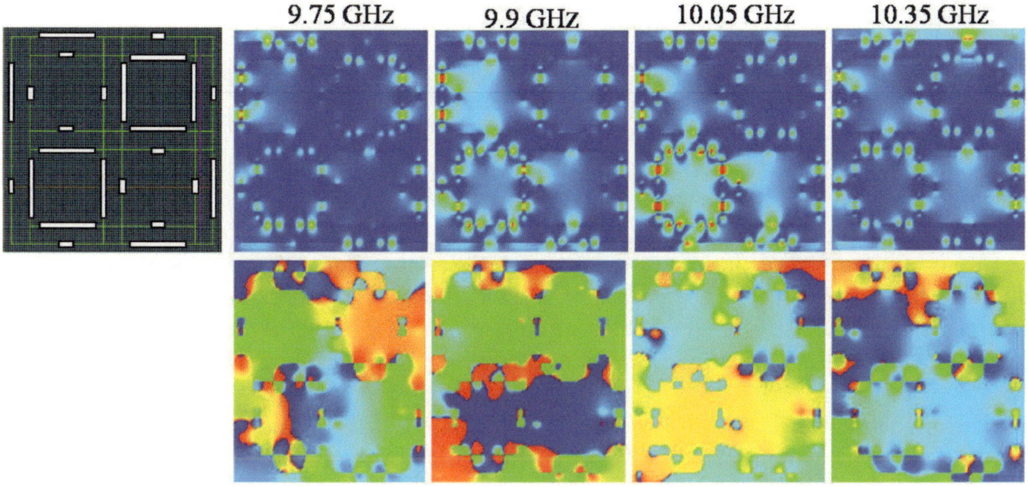

Fig. 1. Cross-section of the metamaterial block depicting SRR sections (longer strips) and cut wires (shorter strips) placed normally to the figure plane (left figure); and distributions of magnetic field amplitude (upper row) and phase (lower row) in the same cross-section at four frequencies within the metamaterial transmission band. Bright spots in the upper row correspond to the areas with highest field magnitude.

This Chapter addresses the listed above underexplored metamaterial problems and describes the effects caused by coupling and splitting of resonances in two types of metamaterial multi-resonator structures, i.e. those composed of SRRs and of dielectric resonators. The results presented here demonstrate that EM responses provided by these structures cannot be reduced to the responses of their constituent single resonators as it is assumed at application of the EMT. The Chapter also shows that splitting of resonances in arrays of dielectric resonators leads to the formation of a bandgap dividing two transmission zones, one at lower, and another one at higher frequncies. While at the gap frequencies elementary resonances in arrays look sporadically flashing at various multiple frequencies, the lower frequency transmission zone is characterized by overlapping of the resonance fields and slow wave propagation. The transmission zone at higher frequencies can be characterized by a coherent response of elementary resonators within half-wavelength distances and by superluminal phase velocities of propagating waves. The obtained results bridge the gap between the properties of metamaterials and photonic crystals employing Mie resonances.

2. Resonance splitting in metamaterial arrays composed of SRRs

2.1 Specific features of SRR characterization

Investigation of the resonance splitting phenomena in SRR, which is presented in the following sub-sections, has been performed on fragments of SRR arrays quite similar to the arrays used in the design of the microwave invisibility cloak (Schurig et al., 2006). This cloaking shell was formed from concentric arrays of unit cells each including one SRR. The size of the sell was of 3.33 mm x 3.33 mm in the SRR plane and 3.18 mm normal to this plane with the SRR located in the central position. The SRRs were placed on the substrate with the thickness of 381 microns and the relative permittivity of 2.33. In the simulations described in the next sub-section, the SRRs with identical parameters (Fig. 2a) were used to build arrays of unit cells similar to various fragments of the cloak described in (Schurig et al., 2006). The only difference between the investigated metamaterial fragments and those used in the publication was that fragments in our case were flat. However, as the radiuses of concentric arrays in (Schurig et al., 2006) were relatively big compared to the cell size, flat arrays with the lengths of up to six cells were a good approaximation of fractions of concentric arrays in the cloak metamaterial. The propagation vector (k-vector) of the incident wave in our studies was always directed along the array rows in the plane of the SRR location, while magnetic field was directed normally to the SRR planes. Therefore, the investigated arrays had to perform similarly to the cloak fragments located close to the diameter of the shell normal to the k-vector of the incident wave. First, the SRR responses in air and at the placement of the resonators on the dielectric substrate have been investigated. Numerical simulations were performed for three arrangements: 1) at the plane wave incidence on the object, 2) at the placement of the SRR in the waveguide of standard dimensions with perfect electric conductor (PEC) walls, and 3) at application of waveguide ports to the standard unit cell and/or to the unit cell increased to incorporate the resonance field "halos" (the following section describes how the "halo" dimensions were found).

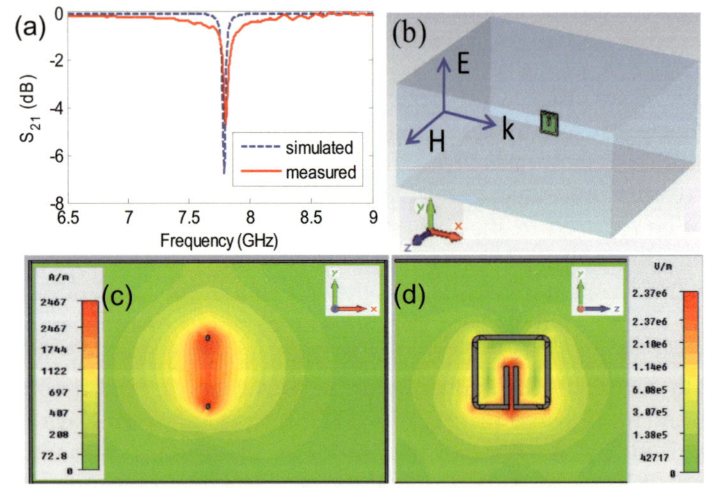

Fig. 2. Resonance in a single SRR placed in the waveguide WR137: (a) measured and simulated transmission spectra S_{21}; (b) arrangement for the measurements; (c) magnetic and (d) electric field patterns.

All simulations were conducted by using the commercial full-wave software package CST Microwave Studio. For the two first arrangements the Transient Solver has been used, while for the third one – the Frequency Domain Solver. Numerical experiments with waveguide ports allowed for obtaining the scattering parameter spectra convenient for controlling the resonance frequency f_{res}, while in experiments with plane wave incidence, the resonance was detected by monitoring the signal of H-field probes placed at the expected maximum field locations in the resonators. Three types of SRRs that differed by the lengths of their inner slot forming strips *s* and the radii of the corners *r*, as summarized in Table 1, have been characterized. The values of *s* and *r* were taken very close to those characteristic for the SRRs of the 1st, 2nd, and 3rd concentric arrays of the microwave cloak described in (Schurig et al., 2006). Small corrections, however, have been introduced in order to provide fitting of the chosen *s* and *r* to the least square dependences which were built to express interrelation between these parameters of SRRs and the effective permeability found in (Schurig et al., 2006) for unit cells with SRRs of proper dimensions.

				Transient solver		Frequency domain solver	
N	*s*	*r*	*substrate*	f_{res} *free space*	f_{res} *waveguide*	f_{res} *unit cell*	f_{res} *halo cell*
	mm	mm		GHz	GHz	GHz	GHz
1	1.654	0.26	yes	7.761	7.745	7.19	7.76
			no	9.16	9.119	8.53	9.17
2	1.677	0.254	yes	7.73	7.67	7.125	7.73
			no	9.125	9.075	8.49	9.185
3	1.716	0.244	yes	7.673	7.657	7.027	7.673
			no	9.061	9.03	8.448	9.068

Table 1. Parameters and resonance frequencies of various SRRs

As seen from Table 1, numerical experiments with plane wave incidence and with the standard waveguide gave quite close results for f_{res}, while in experiments with unit cells used in (Schurig et al., 2006) this frequency appeared to be essentially lower both for SRRs placed in air and on a substrate. The resulting S-parameter spectra in the latter simulations appeared distorted in comparison to the spectra obtained in the standard waveguide. Similar distortions of resonances are known to be observed at too close placement of metal objects or the domain boundaries to the resonator. Therefore, the simulations have been repeated for computational volumes of increased dimensions. Gradual increase of the volume caused better and better fitting of the simulation results to the data obtained in two other sets of numerical experiments. Saturation of changes and coincidence of the data was observed at the distances between resonator and domain boundaries exceeding 1.833 mm in the SRR plane and 4.183 mm in the normal direction. Apparently these dimensions characterize the „halo" of the resonance fields around the SRRs, illustrated by the field patterns presented in Figs. 2a and 2c.

It is obvious that the space occupied by the SRR field „halo" substantially exceeds the volume of unit cells used in (Schurig et al., 2006). As seen in Figs. 2a and 2c, the resonance magnetic and electric fields in the „halo" are not vanishing at the distances from the SRR corresponding to the boundaries of the used unit cells, therefore, the resonance formation and the resonance frequencies f_{res} were affected in respective simulations of such unit cells. In comparison, field magnitudes at the boundaries of unit cells with the size equal to the „halo" size were found to be ten times smaller than those of the cells used in (Schurig et al., 2006) that explains the consistency of the data obtained for the halo-sized cells and the results of other numerical experiments. Only at building the shell of such cells, i.e. at the placement of resonators in arrays of the metamaterial at distances, when „halos" are not overlapped, interaction between resonators could be omitted from consideration. In contrast, the distance between resonators in concentric arrays in (Schurig et al., 2006) did not exceed 0.333 mm, while the distance between arrays (i.e. along the normal to the plane of resonators) was only 3.183 mm.

The data presented in Table 1 do not confirm the expectations for f_{res} to be of about 8.5 GHz as reported in (Schurig et al., 2006) for both the resonance frequency of basic SRRs and the frequency of the cloak response. Resonances at frequencies close to 8.5 GHz were obtained in our experiments only for the unit cells used in (Schurig et al., 2006) at the placement of SRRs in air. It might be that the dielectric substrate was omitted in models used for simulations in (Schurig et al., 2006). The obtained discrepancy does not seem to be critical for the studies of the splitting phenomena in SRR-based metamaterials employed in (Schurig et al., 2006), although it should be taken into account at verification of the cloaking effect, since this effect should not be expected at 8.5 GHz, if the metamaterial perform in accordance to the EMT predictions.

2.2 Resonances in SRR arrays representing fragments of the microwave cloak

In order to additionally verify the results of simulations, individual SRRs and various multi-element arrays of SRRs with the same dimensions as in (Schurig et al., 2006) have been fabricated, and their EM response has been measured in the waveguide WR137 operating at the single TE_{10} mode in the range (5-10) GHz. In experiments, the waveguide WR137 loaded with SRR arrays was connected to the PNA-L Network Analyzer N5230A to measure the transmission spectrum. In order to avoid EM interaction between the arrays and waveguide walls, resonators or arrays were supported by a styrofoam layer inserted into the center region of the waveguide. As seen from Fig. 2d, the resonance in a single resonator placed in the waveguide was observed at f=7.78 GHz and not at 8.5 GHz, i.e. quite close to the results of our simulations for the resonator with the same dimensions.

Next, linear arrays of SRRs arranged along the direction of k-vector of the incident wave have been investigated. Fig. 3a shows that the arrays of 3-5 SRRs demonstrate strong splitting of the resonance, which increases with the number of resonators in the array. It is interesting to note, that instead of multiple dips in the transmission spectrum correlated with the quantity of resonators in the array, only two well expressed drops in the transmission spectra have been observed. The first drop is accompanied by the coherent response of resonators (Fig. 3b), while at the higher frequency drop the resonators in the line respond with opposite phases in two halves of the array. The phase patterns of the wave propagating in free space inside the waveguide at both resonances show strong disturbances of the wave transmission through the waveguide (the curve between bright and dark parts of the image divides the areas corresponding to the half wavelengths of the propagating waves).

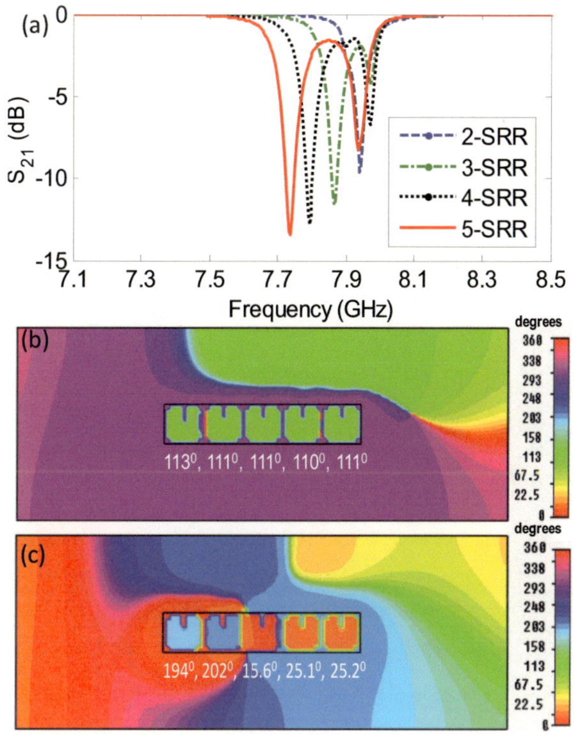

Fig. 3. Resonances in linear arrays arranged in wave propagation direction in the waveguide WR137: (a) S_{21}; (b) coherent response of a 5-SRR array at 7.73 GHz; and (c) assymmetric response at 7.95 GHz.

Much stronger splitting of SRR resonances has been revealed in the SRR columns (Fig. 4) similar to those used in concentric arrays of the SRRs in (Schurig et al., 2006). Placement of such columns in the waveguide WR137 caused some concerns, since at about 1 cm height of the columns their resonance fields could interact with the waveguide top and bottom walls that could cause deterioration of the resonances. Therefore, the simulations for these columns were initially performed in free space at plane wave incidence. As such simulations did not provide an opportunity for deriving the transmission spectra, the resonance field in the columns was monitored by using H-field probes placed in geometrically identical points of the SRR cross-sections (Fig. 4a). As seen from Fig.4e the probes revealed three resonances covering more than 1 GHz of the frequency range (i.e. almost five times wider resonance band than that observed for the linear arrays of SRRs arranged along the direction of the incident wave k−vector). It is worth mentioning that each resonator in the column responded resonantly at all split frequencies while the integrated modes at these frequencies were quite different. As seen from Fig. 4b, at the lower frequency resonance sin-phase oscillations were observed in the two upper resonators, while the third resonator oscillated with a smaller magnitude and with opposite phase. At the higher frequency resonance, the two lower resonators demonstrated sin-phase oscillations while the third resonator oscillated with opposite phase (Fig. 4d). Resonance at the median frequency was supported mainly by sin-phase oscillations in the upper and the lower resonators of the column (Fig. 4c). The third resonator located in the middle of the column did not show strong oscillations, however, the phase of these oscillations was shifted by 180^0 with respect to the oscillations of two other SRRs.

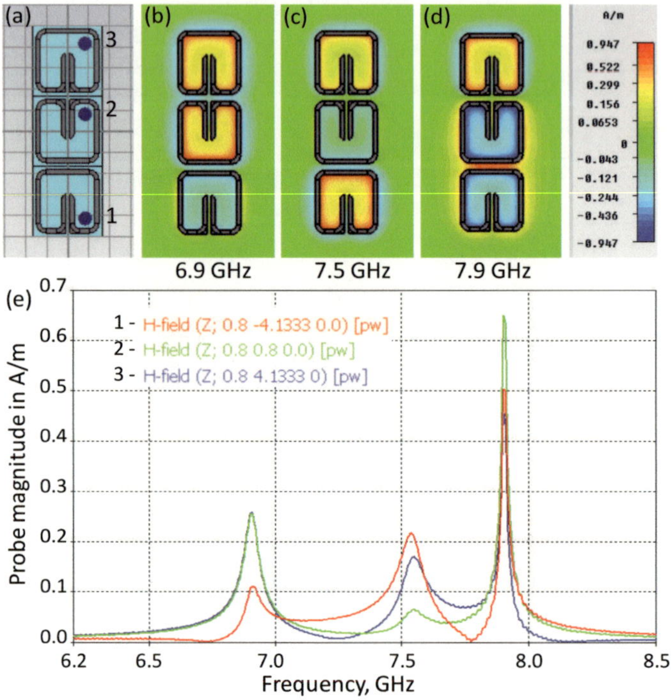

Fig. 4. (a) The model of the SRR column with H-field probes; (b)-(d) H-field resonance responses of the column at signal peak frequencies (colour intensity shows the magnitude of resonance; and (e) the spectra of probe signals.

It is interesting to note that shorted columns of 2 SRRs placed in either "face-to-face" or "back-to-back" arrangement did not demonstrate any splitting of the resonances. As seen from Fig. 5, the former arrangement provided for the unsplit resonance at almost the same frequency as the lower frequency resonance in the column of 3 SRRs, while the latter arrangement provided for the resonance at the frequency close to the frequency of upper resonance in the column of 3 SRRs. It means that the listed above resonances in a 3-SRR column are apparently related to integrated coherent resonances in couples of oppositely arranged SRRs. The coherentness is evident from the phase patterns presented in Fig. 5 on the right.

Essentially more complicated response with multiple resonances of various Q factors was demonstrated by arrays composed of several columns as, for example, the presented by the insert in Fig 6a array of 5 columns. Despite of the complexity, however, it was possible to distinguish three main groups of resonances comparable to resonances revealed in one SRR column (Fig. 4). In fact, the lower frequency group in the 3 x 5 SRR array demontsrated coherent resonances in upper rows of the array (Fig. 6b), while the higher frequency group demontsrated coherent resonances in the lower rows of the array (Fig. 6d). The group of resonances at the median frequencies could be characterized by a symmetry of DR responses with respect to the central row i.e. the resonators located in upper and lower rows responded coherently, even though neighbouring columns demonstrated opposite phases of oscillations. So, it could be said that every other column responded coherently (Fig. 6c).

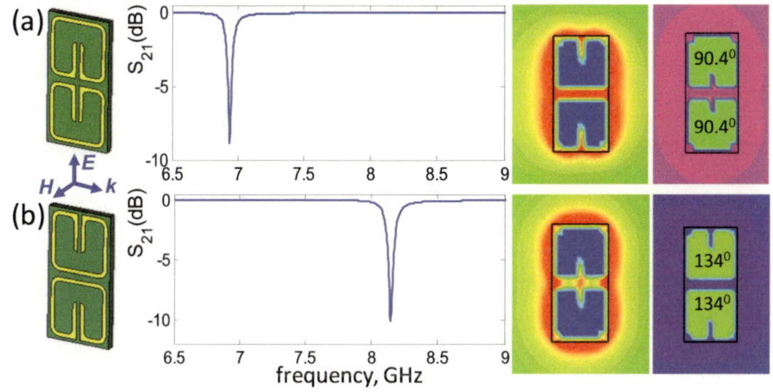

Fig. 5. Geometry, S_{21} spectra, and patterns of H-field magnitude (3d column) and phase (4th column) for SRR pairs arranged: (a) "face-to-face" and (b) "back-to-back ".

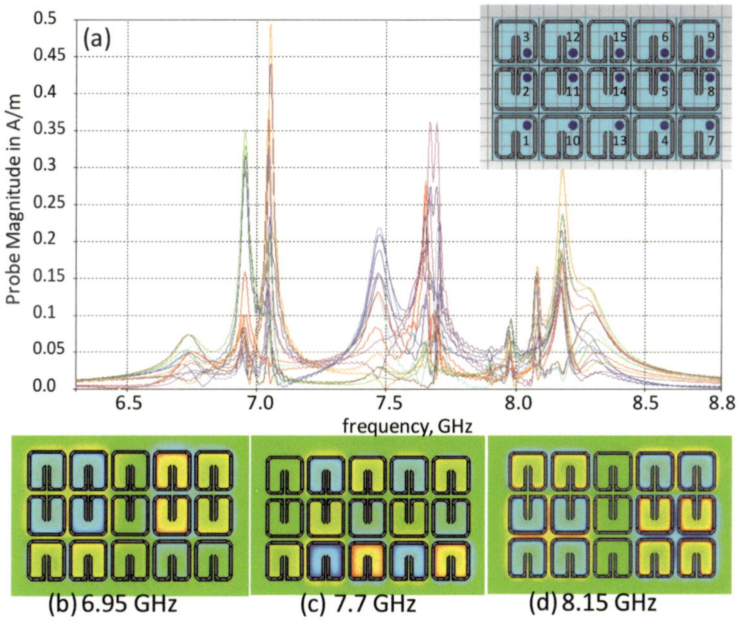

Fig. 6. (a) The spectra of the probe signals; inset shows the SRR array and the H-field probe locations; (b)-(d) typical patterns of the resonance oscillations in the array sampled at 6.64, 7.4 and 8.15 GHz (colour intensity shows the magnitude of resonance oscillations).

Despite of the above mentioned concerns about possible interaction between multi-element arrays and waveguide walls at the measurements we have performed comparison of simulation and measurement results for the transmission spectra at the placement of the 3 x 5 arrays of SRRs in the waveguide WR137. A good matching of the results presented in Fig. 7 is obvious, although both S_{21} spectra demostrate a smaller quantity of the transmission dips compared to that observed in the probe signal spectra (Fig. 7). The three groups of resonances, however, could be clearly distinguished in both spectra presented in Fig. 7.

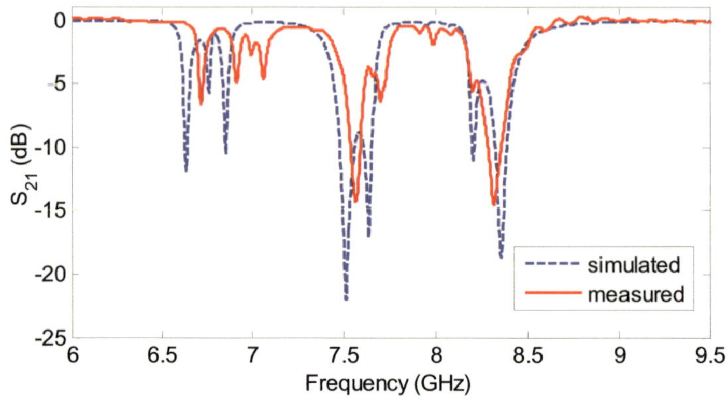

Fig. 7. S_{21} spectra for the 3 x 5 array of SRRs in the waveguide WR137: dashed curve - simulation results, solid curve - the results of measurements.

The responses of multi-layer structures composed of several planar arrays were found to loose the perspicuity of the splitting phenomena observed in the multicolumn arrays. As seen in Fig. 8c, which presents the probe signal spectra for a three-layer array of SRRs (Fig. 8a), it is difficult to distinguish well separated in frequency groups of resonances similar to those observed for 3 x 5 arrays. Instead, split resonances are observed in the essentially extended frequency range of more than 2 GHz. Such character of the multi-layer array response makes it quite doubtful to expect that the proposed in (Schurig et al. 2006) design of cylindrical cloak based on close-packed arrays of SRRs would provide the cloaking effect at the frequency determined from the analysis of the response of a single resonator. The reason for obtaining the invisibility effect in (Schurig et al., 2006) should be rather searched for in some specifics of split resonances, which need to be studied additionally. It is worth noting that some of split modes indeed demonstrated a possibility to support the desired coherent response of SRRs across extended areas of investigated arrays as it is shown in Fig. 8b. It is reasonable to expect that at respective frequencies the participating resonators could contribute to obtaining the desired dispersion of the effective parameters precribed by the transformation optics relations.

3. Resonances in metamaterial arrays composed of dielectric resonators

3.1 Specifics of resonances in dielectric resonators and potential of all-dielectric metamaterials

Dielectric resonators (DRs), especially those of cylindrical shape, are capable of providing EM response quite similar to that of SRRs. In fact, when the incident wave propagates normally to the axis of the cylinder with its H-field directed along this axis, the DR can support the $TE_{01\delta}$ resonance mode, which is equivalent to the formation of a magnetic dipole along the axis of the cylinder, i.e normal to the plane, in which the cross-section of the DR is a circle (Kajfez & Guillon, 1998). The latter plane should host circular displacement currents comparable to circular currents in SRRs.

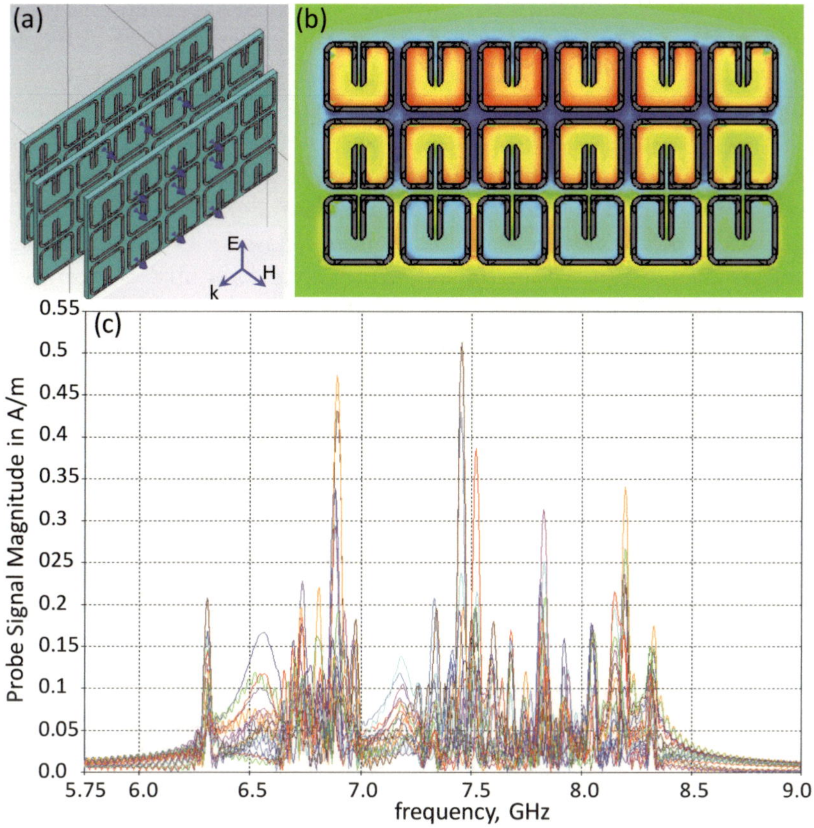

Fig. 8. (a) Geometry of a three- layer SRR array with marked probe locations; (b) coherent response of SRRs at 8.35 GHz; and (c) spectra of probe signals revealing enhanced splitting caused by interaction between SRRs in neighbouring layers.

The publication (Semouchkina et al., 2004) was among the first that started the development of all-dielectric metamaterials. The data presented in Fig. 9 confirm the possibility to use dielectric cylinders of a proper size (the diameter of 6 mm and the height of 3 mm, in this case) and permittivity (dielectric constant of 37.2 in this case) to replace SRRs as magnetic components in metamaterials supporting Lorentz-type resonance necessary for obtaining the effective negative permeability of the medium at microwave frequencies. It was later shown that dielectric resonators can also replace cut metal wires used in conventional metamaterials below their plasma frequency as electric components providing for the effective negative permittivity. Therefore, it was predicted that materials composed of two types of dielectric resonators providing one for electric and another one - for magnetic response could exhibit negative index of refraction (Jylha et al., 2006; Vendik & Gashinova, 2004). In the work (Semouchkina et al., 2005) it was demonstrated that negative refraction can also be provided by dielectric arrays composed of resonators of one type at the placement of the array on a ground plane. Similar effects in arrays composed of DRs of one type with ground planes or in waveguides have been also reported in (Ueda et al., 2007, 2010).

Recently, a metamaterial composed of dielectric resonators has been employed to design an invisisbility cloak for the infrared range (Semouchkina et al., 2009, 2010). In difference from earlier works (Cui et al., 2009, Gaillot et al., 2008, Schurig et al., 2006) on microwave and THz cloaks, the design proposed in (Semouchkina et al., 2009, 2010) utilized identical resonators in the entire cloakling shell. Such design does not require to solve technological problems of fabricating concentric arrays of various nano-sized elements, as well as does not demand a wide band of incident light to excite differently sized resonators. In addition, employment of chalcogenide glass as a resonator material should decrease the loss related limitations for the size of concealed objects recently formulated in (Zhang et al., 2009). The estimates based on the level of the extinction coefficient at 1 micron wavelength of incident light gave more than an order less value for the loss tangent of the glass cloak than values characteristic for previously developed cloaks. The required by the transformation optics dispersion of the effective permeability in the novel cloak design was obtained by using a controlled decrease of the density of resonators from the inner layer of the cloak to the outer layer, i.e. due to radial dispersion of air fractions in the shell. In addition, the requirement to avoid both coupling between resonators and strong resonance splitting as described in previous sections has been satisfied that has led to a spoke-type arrangement of the resonators in the cloak depicted in Fig. 10. The intent to exclude overlapping of the resonance fields "halos" limited minimal angular distances between spokes and minimal gaps between concentric arrays of resonators by the "halo" size, however, it did not deteriorate the cloak performance. The effective parameters of the cloak layers have been characterized by using mixing relations incorporating responses of air fractions and individual resonators (the latter were assigned not to the physical resonator bodies, but to the "halo" volumes). This approach allowed for accurate adjustment of the permeability dispersion, so that the desired performance of the cloak with no distortions of the wave front at frequencies within the cloaking band has been provided.

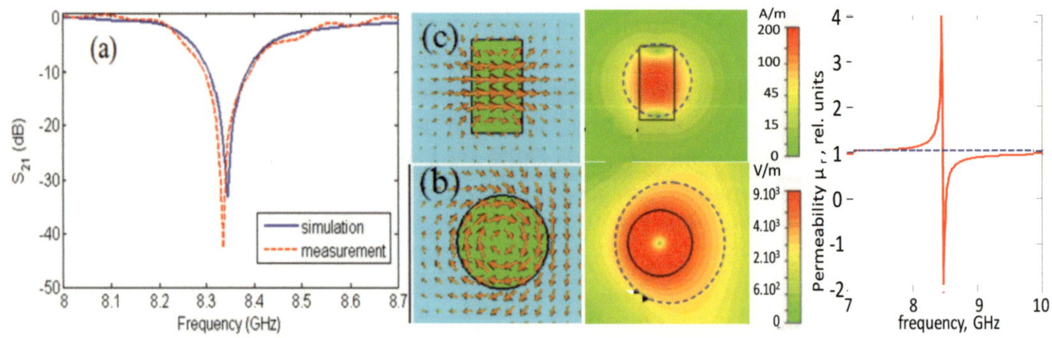

Fig. 9. (a) Simulated and measured S_{12} spectra for a DR in WR137; (b) electric and (c) magnetic field patterns in DR cross-sections; and (c) effective permeability changes at the magnetic-type resonance in the DR (Chen et al., 2011).

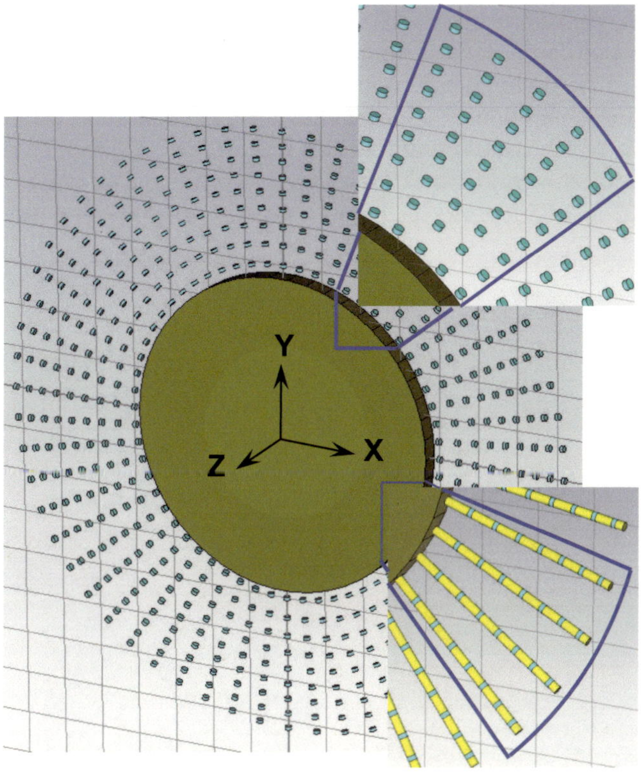

Fig. 10. The design of the infrared invisibility cloak composed of cylindrical glass resonators arranged in concentric arrays; lower inset shows the spokes of resonators with spacers (Semouchkina et al., 2010)

Despite the described advances the application of dielectric metamaterials in cloaking structures has underlined once again the main problem common for all resonance metamaterials - their narrowbandness. According to initial works on cloaking it was thought that application of such metamaterials excudes other than zero bandwidth for the invisibility effects (Pendry et al., 2006). Accordingly, the described above invisibility cloak demonstrated good cloaking effect in the band less than 1.2% or 3 THz. However, it was also revealed that even at a slightly denser packing of resonators than that shown in Fig. 11, the bandwidth could be increased up to 2 times. This effect could be related to the positive consequencies of coupling, which are known since earlier applications of resonator arrays in microwave filters and other devices, where coupling was used for the bandwidth enhancement. The next section will demonstrate the complex nature of coupling related phenomena in dielectric metamaterials and outline the opportunity to use them for the partial lifting of the delay-bandwidth limitations for cloaking devices formulated in (Hashemi et al., 2010).

3.2 Specifics of coupling related phenomena in linear arrays of dielectric resonators

The responses of linear arrays of DRs with the same parameters as those of the DR featured in Fig. 9, at their positioning along the wave propagation direction has been investigated first. In difference from close packed array of SRRs, even one pair of resonators placed at the distance between their bodies of 2 mm demonstrated resonance splitting of about 0.25 GHz (Fig. 11a). Splitting decreased at increasing the distance between the resonators, however, it was still well expressed even at the separation of 10 mm. Only 16 mm separation provided for an unsplit spectrum of S_{21}. It is worth mentioning that this distance is close to the doubled thickness of the electric field „halo" shown in Fig. 9. This confirms the suggestion that resonance coupling and splitting result from overlapping of „halos" created by neighboring resonators.

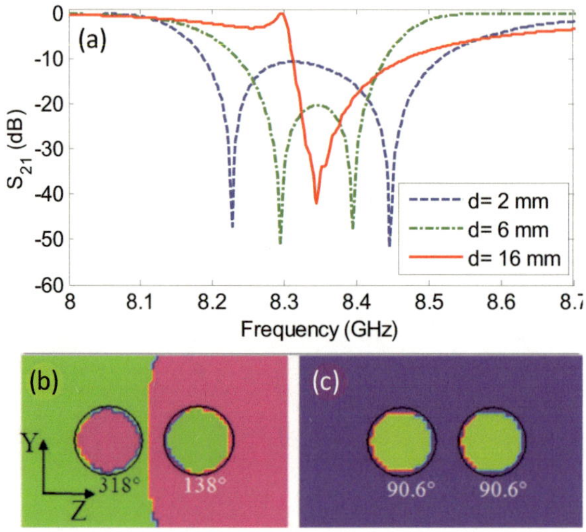

Fig. 11. (a) S_{21} spectra for two DRs arranged along the axis of the waveguide WR137; (b) and (c) phase patterns for H-field distributions in the longitudinal cross-section at 2 split resonances (Chen et al., 2011).

Phase patterns for magnetic field component in the waveguide cross-section presented in Fig. 11 show that at lower frequency, resonance field oscillations in two resonators proceed with opposite phases, while at higher frequency coherent resonance oscillations are observed. Since the resonance modes can be represented by magnetic dipoles, it follows that at lower frequency, magnetic dipoles formed in two resonators remain directed oppositely at any instant, while at higher frequency they always keep parallel orientation. Similar difference between split resonances at transverse interaction of resonators (side-by-side) was observed and discussed in (Liu et al., 2010), where it was concluded that dipole interaction defines the energy difference of two resonance states, and, correspondingly, the degree of their splitting in the frequency scale. In fact, parallel magnetic dipoles are expected to repel each other since the curls of their electric fields tend to cancell each other in the gap between the resonators. Just opposite, counter-directed magnetic dipoles provide for co-directed electric fields in the gap between the resonators that should attract them to each other, thus decreasing the state energy.

At increasing the quantity of resonators in the linear DR array from 2 to 5, the split spectrum of S_{21} demonstrated a better expressed gap of transmission in the frequency range of about 0.45 GHz (Fig. 12). Further addition of resonators to the array led to increasing the quantity of transmission drops in the gap and to its gradual flattening. However, the signals of the probes that measured the magnitude of H-field oscillations in the centers of resonators did not demonstrate any resonance-like field enhancement at corresponding frequencies even though they experienced deep drops of the resonance activity at some other frequencies. Most obvious resonance-like activity was seen only at frequencies corresponding to the boundaries of the transmission gap. Phase patterns in the waveguide cross-section at these frequencies reminded the patterns described above for the pair of resonators, i.e. at the low frequency boundary of the transmission gap all neighboring resonators oscillated with opposite phases, while at the higher frequency boundary all resonators in the array demonstrated coherent responses (Figs. 13a and 13b).

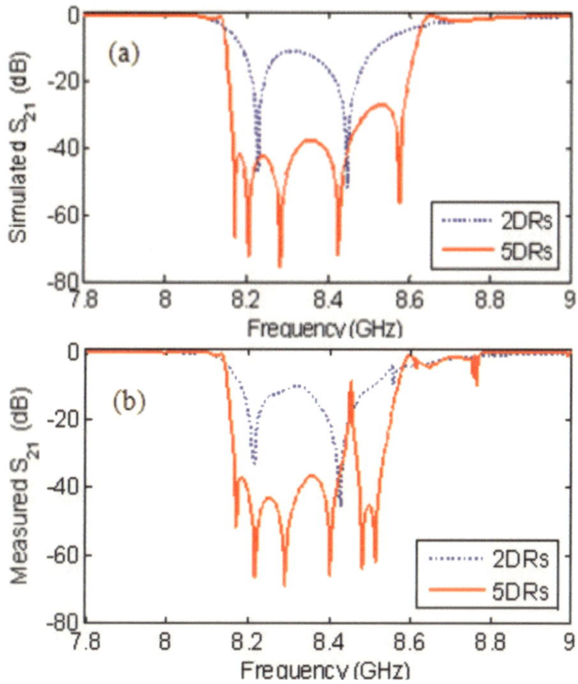

Fig. 12. (a) Simulated and (b) measured spectra of S_{21} for linear DR arrays arranged along k-vector at 2 mm between DR bodies (Chen et al., 2011).

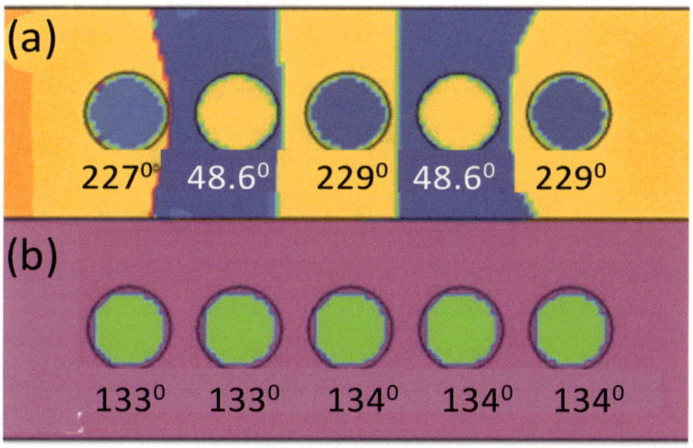

Fig. 13. H-field phase patterns in waveguide at (a) 8.17 GHz and (b) 8.58 GHz (Chen et al., 2011)

It is worth noting that the revealed low frequency and high frequency resonance states in DR arrays remind bonding and anti-bonding states of electrons in molecules and crystals with such difference that instead of electron spins the type of resonance states in DR arrays is defined by respective orientations of formed magnetic dipoles. It is also important to add that a coherent response of resonators presents a special interest for all applications of metamaterials, which depend on the realization of the prescibed by the transformation optics dispersion of the effective medium parameters.

3.3 EM responses of coupled 3D metamaterial arrays of dielectric resonators

It should be pointed out that the described above simulations and experiments with placement the DR arrays in waveguides are, in fact, equivalent to modeling 3D arrays of resonators, due to the mirror reflection effect of the waveguide walls. Since the distance between resonators and the walls of the waveguide WR137 was relatively large compared to inter-resonator separation, the obtained results show that even large lattice parameters in two directions of the modelled 3D arrays (and, repectively, negligible coupling in these directions) does not provide the similarity of the array response to the response of the unit cell with a single resonator inside. However, large lattice parameters are rather unusual for metamaterials applications, therefore, the studies of dense 3D arrays that should present additional interest for practical realization of metamaterials, have also been conducted.

Numerical experiments with 3D arrays having lattice parameters between 8 mm and 12 mm (which corresponded to separations between the resonator bodies ranging from 2 mm to 6 mm, respectively) were carried out in free space. Due to employment of periodic boundary conditions (PBC) in simulations, the 3D arrays under study were infinite in one direction and finite in two other directions. The plane wave incidence was normal to the infinite direction. For simplicity, a cubic structure of the arrays was employed. The individual responses of resonators in two perpendicular to each other linear arrays of resonators were controlled by the probes placed in the centers of corresponding resonators. In additition, more probes were placed at some distances from the array on the way of propagating waves to detect transmission (see schematic in Fig. 14a). Wave propagation through the arrays was controlled by using snap-shots and animation of H-field distribution patterns in XZ cross-section of the array.

As seen in Figs. 14b and 14c, the responses of array elements were characterized by multiple resonances observed even at the frequencies corresponding to the low frequecy half of the transmission gap, between 8.2 GHz and 8.63 GHz (see Fig. 14c). At frequencies below and above the transmission gap the waves passed through the array, however, the specifics of these propagations were quite different. While at lower frequencies the lengths of waves propagating in the array were essentially shorter than in free space so that the wave front movement through the array was lagging the wave front in free space (Fig. 15a), at higher frequencies, *vise versa,* the lengths of waves passing through the array were longer compared to the wavelengths in free space so that the movement of wave front through the array outrided the movement in free space (Fig. 15b). It is worth mentioning that when the frequency below the gap increased to approach its low frequency boundary, the lag in the wave movement through the array became more and more pronounced reaching its maximum at about 8.18 GHz. At frequencies above the transmission gap, the highest velocity of waves passing through the array was observed just near the gap boundary at about 8.66 GHz. At higher frequencies, wave advancing inside the array gradually decreased so that the entire wave front became flat at about 9.12 - 9.15 GHz.

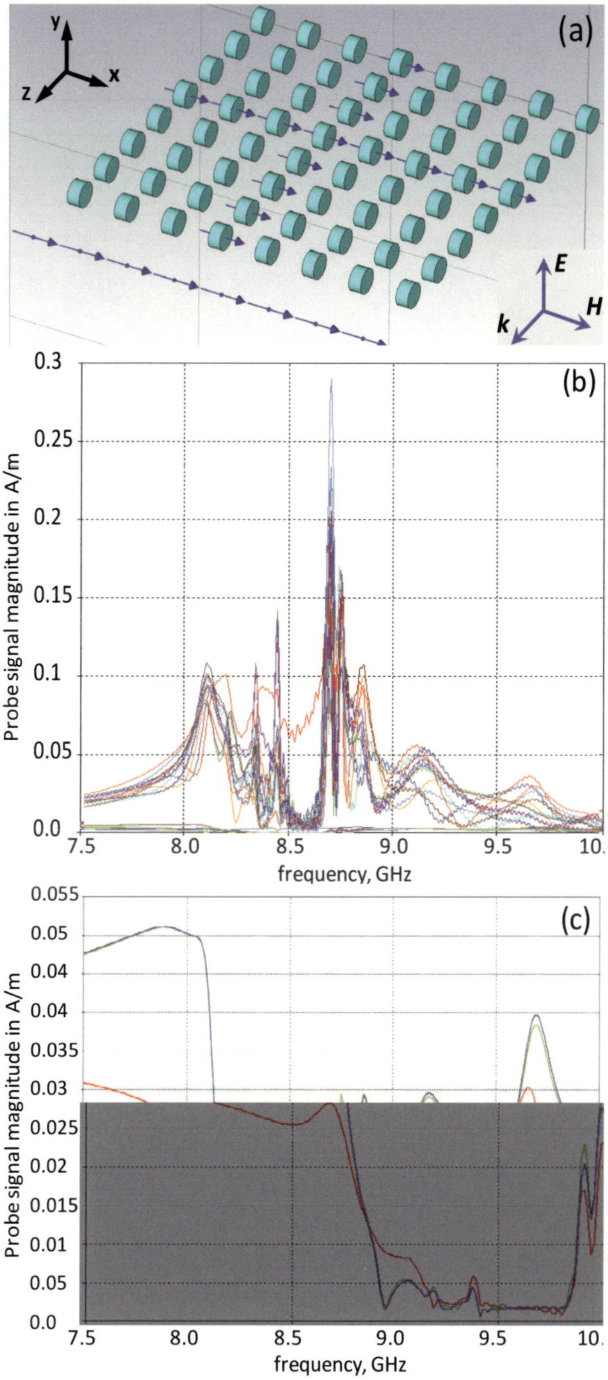

Fig. 14. (a) Basic element of 3D arrays at plane wave incidence in z-and PBC in y-direction; (b) spectra of signals from probes placed in DR centres; and (c) spectra of probes placed outside the sample to detect transmission.

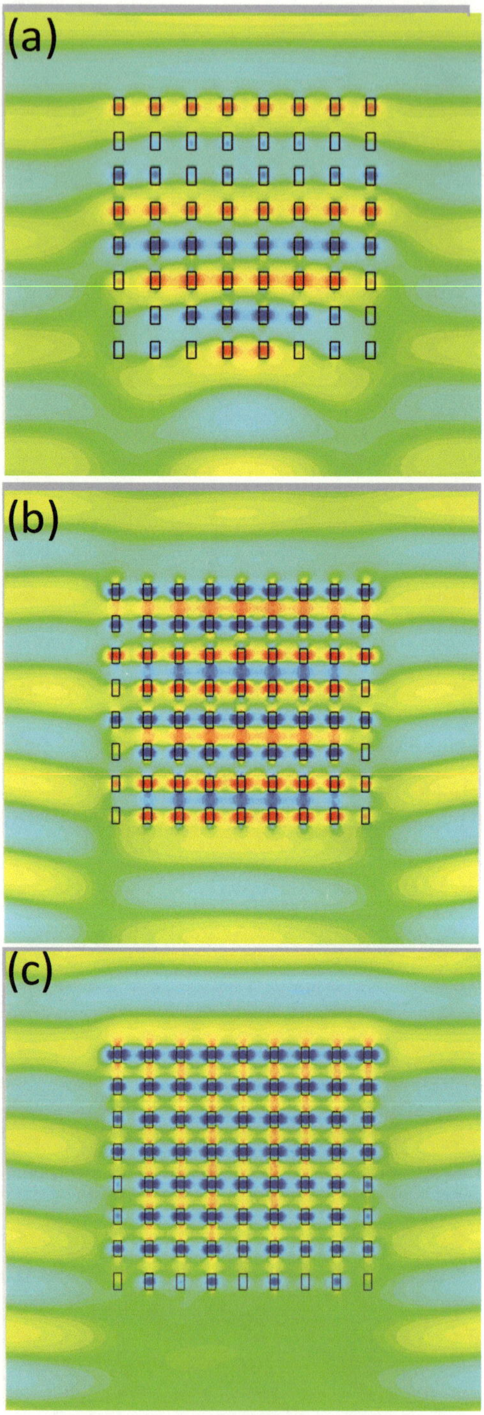

Fig. 15. Wave propagation in 3D DR arrays at frequencies (a) below the gap and (b, c) above the gap at: (a) 8.0 GHz; (b) 8.74 GHz; and (c) 8.64 GHz.

If the investigated array could respond as a uniform medium without any resonance splitting it could be expected that the wave movement inside the array would depend on characteristic for Lorentz-type resonators changes of the effective permeability. As known, these changes have three essential for wave movement ranges: 1) increase of the effective permeability with frequency at approaching the resonance, 2) drop of the effective permeability at the resonance frequency down to negative values and then growth back with crossing the zero level at the so-called magnetic plasma frequency, and 3) continuing increase of the effective permeability in the range of values between 0 and 1, as shown in Fig. 3c. The above changes would define the phase velocity of propagating waves in accordance with the known relation:

$$v_{\mathrm{p}} = \frac{1}{\sqrt{\mu_{\mathrm{r}}\varepsilon_{\mathrm{r}}}}$$

and cause lagging of the wave movement in the array at approaching the resonance, then a transmission gap at frequencies corresponding to the negative values of permeability, and, finally, superluminal wave movement at frequencies exceeding the magnetic plasma frequency. It seems doubtful to expect similar changes of the effective permeability as those observed in the model of a uniform medium (Fig. 3c) from close-packed arrays of coupled resonators with essentially split resonances. Nevertheless, the obtained data testify in favor of some analogy. It follows that despite of the resonance splitting, coupled arrays are capable of responding in the way predicted by the above described changes of the effective permeability near the resonance. It means that coupling and splitting of elementary resonances do not prevent arrays from specific integrated contributions of multiple responses to their effective properties and from exhibiting qualitatively similar to a uniform medium changes of these properties with frequency. Since application of metamaterials in cloaking devices assumes employment of cloaking shells for speeding up waves around the concealed objects, the obtained results show that these expectations could be realized despite of complications caused by the coupling phenomena. It is worth adding that at the "superluminal" phenomena, the resonators in coupled arrays tend to respond coherently within the half wavelengths of the passing waves. Therefore, at frequencies close to the upper boundary of the transmission gap, when the strongest extention of the wavelengths by the resonating array is provided, coherent response can involve even the entire array (Fig. 15c). It should be, however, taken into account that the coherent response formation could occur at frequencies different from the resonance frequency of a single resonator and that the frequency range of the coherent response could depend on the specifics of coupling effects in respective arrays.

The revealed possibility to employ arrays of coupled resonators for obtaining coherent responses within the half wavelengths of the incident wave provides the grounds for the employment of coupling phenomena in metamaterials to enhance the bandwidth of their operation and, in particular, for the enhancement of the cloaking effect bandwidth.

4. Conclusion

The performed studies of coupling and splitting of elementary resonances in arrays of metal split-ring resonators and dielectric resonators have demonstrated that ignoring coupling effects at metamaterial applications can lead to mistakes and malfunctioning. Even the best ideas proposed for designing various devices based on employing resonance metamaterials, can loose an opportunity to be realized properly unless coupling phenomena are taken into account. In particular, the presented data, which disclosed dramatic splitting of the resonances in close-packed arrays of SRRs have shown that these arrays could not be used for demonstrating the effects of invisibility by the cylindrical cloak of concentric resonator arrays at frequencies predicted byusing the analysis of a single SRR.

The arrays of dielectric resonators, which seem perspective due to low loss, especially for employment in cloaking devices, are also found to be a subject of coupling phenomena even at the lattice constants essentially exceeding the dimensions of resonators. However, there are reasons to express an optimism towards their applications as even at strong coupling and resonance splitting phenomena the DR arrays are able to demonstrate integrated responses qualitatively corresponding to the responses expected for the theoretical uniform media with Lorentz-type resonances. In particular, the DR arrays show an ability to support superluminal wave propagation and coherent resonance responses from multiple resonators located within the half wavelength of the propagating wave.

Similar to photonic crystals, DR arrays demonstrate the formation of a transmission gap that separates two transmission bands. However, this bandgap is not related to Bragg resonance and apparently is related to negative values of the effective permeability at elementary Lorentz-type resonances. In this attitude, metamaterial DR arrays could be compared to photonic crystals exploiting Mie resonances. The slow waves, which have been observed in DR arrays at frequencies below the transmission gap, and the superluminal waves observed at frequencies above the gap enhance the similarity with the latter photonic crystals. A detaled comparison of the properties and the physics underlying the observed phenomena in two types of artificial materials could provide a deeper understanding of their specifics.

5. Acknowledgment

This work was supported by the National Science Foundation under Grant No. 0968850. The author wishes to thank graduate students Fang Chen and Xiaohui Wang for performing some simulations and measurements.

6. References

Chen, F., Wang, X. & Semouchkina, E. (2011). Simulation and Experimental Studies of Dielectric Resonator Arrays for Designing Metamaterials, *Proceedings of IEEE International Symposium on Antennas and Propagation,* pp. 2936-2939, ISSN: 1522-3965, Spokane, WA, August 2011

Cui, T. J., Smith, D. R., Liu, R. (2009), *Metamaterials: Theory, Design, and Applications*, Springer, 1st ed., (November 2009), ISBN-10: 1441905723

Gaillot, D. P., Croenne, C. & L ippens, D. (2008). An All-dielectric Route for Terahertz Cloaking. *Optics Express,* Vol.16, No.6, (March 2008), pp. 3986-3992, ISSN: 1094-4087

Hashemi, H., Zhang, B., Joannopoulos, J. D. & Johnson, S. G. (2010). Delay-Bandwidth and Delay-Loss Limitations for Cloaking of Large Objects. *Phys. Rev. Lett.,* Vol.104, (June 2010), 253903, ISSN 0031-9007

Jylha, L., Kolmakov, I., Maslovski, S., & Tretyakov, S. (2006). Modeling of Isotropic Backward-Wave Materials Composed of Resonant Spheres. *Journal of Applied Physics*, Vol.99, 043102, (2006), ISSN:0021-8979

Kajfez, D. & Guillon, P. (1998), *Dielectric Resonators*, Noble Publishing Corp., 2nd ed., Atlanta, (1998), ISBN:1884932053

Leonhardt, U. (2006). Optical Conformal Mapping. *Science*, Vol.312, No. 5781, (June 2006), pp. 1777-1779, ISSN 0036-8075

Liu, N. & Giessen, H. (2010), Coupling Effects in Optical Metamaterials. *Angewandte Chemie Int. Ed.*, Vol.49, No.51, (December 2010), pp. 9838-9852, DOI: 10.1002/anie.200906211, ISSN: 1521-3773

Pendry, J. B., Holden, A., Robbins, D. J, & Stewart, W. J. (1999). Magnetism from Conductors and Enhanced Non-Linear Phenomena. *IEEE Trans. on Microwave Theory & Techniques*, Vol.47, No.11, (November 1999), pp. 2075–2084, ISSN: 0018-9480

Pendry, J. B. & Smith, D. R. (2003). Reversing Light: Negative Refraction. *Physics Today*, Vol.56, (December 2003), pp. 1-8, ISSN:0031-9228

Pendry, J. B., Schurig, D. & Smith, D. R. (2006). Controlling Electromagnetic Fields. *Science*, Vol.312, No.5781, (June 2006), pp. 1780-1782, ISSN 0036-8075

Schurig, D., Mock, J. J., Justice, B. J., Cummer, S. A., Pendry, J. B., Starr, A. F. & Smith, D. R. (2006). Metamaterial Electromagnetic Cloak at Microwave Frequencies. *Science*, Vol.314, No.5806, (November 2006), pp. 977–979, ISSN 0036-8075

Semouchkina, E., Baker, A., Semouchkin, G., Randall, C. & Lanagan, M. (2004). Resonant Wave Propagation in Periodic Dielectric Structures, *Proceedings of the IASTED Int. Conf. ANTENNAS, RADAR AND WAVE PROPAGATION*, pp. 149–154, ISBN: 0-88986-409-8, Banff, Canada, July 2004

Semouchkina, E., Semouchkin, G., Lanagan, M. & Randall, C. A. (2005). FDTD Study of Resonance Processes in Metamaterials. *IEEE Transactions on Microwave Theory & Techniques*, Vol.53, (April 2005), pp. 1477–1487, ISSN: 0018-9480

Semouchkina, E., Werner, D. H. & Pantano, C. (2009). An Optical Cloak Composed of Identical Chalcogenide Glass Resonators, *Proceedings of the Metamaterials'2009 - 3rd Int. Congress on Advanced Electromagnetic Materials in Microwaves and Optics*, , ISBN 978-0-9551179-6-1, London, United Kingdom, September 2009

Semouchkina, E., Werner, D. H., Semouchkin, G. & Pantano, C. (2010). An Infrared Invisibility Cloak Composed of Glass. *Appl. Phys. Lett.*, Vol.96, 233503, (June 2010), ISSN 0003-6951

Smith, D. R., Pendry, J. B. & Wiltshire, M. C. K. (2004). Metamaterials and Negative Refractive Index. *Science*, Vol.305, No.5685, (August 2004), pp. 788–792, ISSN 0036-8075

Smith, D. R. & Pendry, J. B. (2006). Homogenization of Metamaterails by Field Averaging. *JOSA B*, Vol.B23, No.3, (March 2006), pp.391-403, ISSN: 0740-3224

Ueda, T., Lai, A. & Itoh, T. (2007). Demonstration of Negative Refraction in a Cutoff Parallel-Plate Waveguide Loaded with Two-Dimensional Lattice of Dielectric Resonators. *IEEE Transactions on Microwave Theory & Techniques*, Vol.55, No.6, (June 2007), pp.1280–1287, ISSN: 0018-9480

Ueda, T., Michishita, N., Akiyama, M. & and Itoh, T. (2010). Anisotropic 3-D Composite Right/Left-Handed Metamaterial Structures Using Dielectric Resonators and Conductive Mesh Plates. *IEEE Transactions on Microwave Theory & Techniques*, Vol.58, No.7, (July 2010), pp. 1766 –1773, ISSN: 0018-9480

Vendik, O. G. & M. S. Gashinova, M. S. (2004). Artificial Double Negative (DNG) Media Composed by Two Different Dielectric Sphere Lattices Embedded in a Dielectric Matrix, *Proceedings of the 34th European Microwave Conference*, pp. 1209-1212, ISBN/ISSN: 1580-5399-20 9781-5805-39920, Amsterdam, October 2004

Yariv, A., Xu, Y., Lee, R. K. & Scherer, A. (1999).Coupled Resonator Optical Waveguide: a Proposal and Analysis. *Optics Letters*, Vol.24, No.11, (June 1999), pp.711-713, ISSN: 0146-9592

Zhang, B. L., Chen, H. S. & Wu, B.-L. (2009). Practical Limitations of an Invisibility Cloak. *Progress in Electromagnetics Research*, Vol.97, (2009), pp. 407-416, ISSN: 1070-4698

5

Space Coordinate Transformation and Applications

André de Lustrac, Shah Nawaz Burokur, Paul-Henri Tichit,
Boubacar Kante, Rasta Ghasemi and Dylan Germain
Institut d'Electronique Fondamentale,
Univ. Paris-Sud, CNRS UMR 8622, Orsay,
France

1. Introduction

Invisibility or cloaking is an old myth of humanity that must date probably from the time when the success in hunting or war would depend on the ability to hide as much as possible. This invisibility, after being a prolific subject for writers and filmmakers, has become almost a reality in 2006 with the first practical realization of an electromagnetic invisibility cloak.

The invariance of Maxwell's equations in the geometric transformation of coordinates has become a hot topic that year with the first proposal of a cylindrical invisibility cloak by J. B. Pendry and U. Leonhardt. The experimental fabrication and characterization of the first cloak at microwave frequencies have shown that this tool is very effective. After this realization, several applications of this transformation have been proposed for the design of concentrators, waveguides, transitions and bends, directional antennas, and even electromagnetic wormholes.

This space transformation is therefore a powerful tool for the design of devices or components with special properties difficult to obtain from conventional materials and geometries. Theoretically, the method of coordinate transformation involves the generation of a new space derived from an original space where the solutions of Maxwell's equations are known.

The first step is to imagine an initial space and a final space with their topological properties and link them through an analytic transformation. Most of this work is based on a continuous transformation that produces a final space with electromagnetic parameters often complex, heterogeneous and anisotropic.

The challenge is then to effectively design this new space. To make the fabrication easier, simplified parameters were proposed in the early works, with the disadvantage of an impedance mismatch between the material and its environment. More recently, a transformation applied to discrete multi-layer structures has been proposed to further simplify the realization. In this chapter, we present the principles and main applications envisaged for this transformation.

2. Space coordinate transformation

2.1 Principle

In a letter of 1662 Pierre de Fermat established the principle that governs the geometrical optics [Tannery, 1891]. The light follows a stationary optical path between two points. Most of the time, it follows the shortest path. In some cases, it takes the longer one. The optical path is defined by the equation 1, where n is the refractive index of the space, which may depend on the spatial coordinates, and dl a small element of distance:

$$s = \int n dl \tag{1}$$

If n varies with the position in space the path followed by the light can be bent instead of following a straight line. This occurs, for example over a hot road in summer when the index of the air layers above the road varies with the temperature and the height over the road. In this case we can observe a curvature of the path followed by the sunlight that gives the impression that the road is covered with water. Figure 1 shows a schematic of the path of light when the space is not distorted and when this space is distorted (Figure 1a and b).

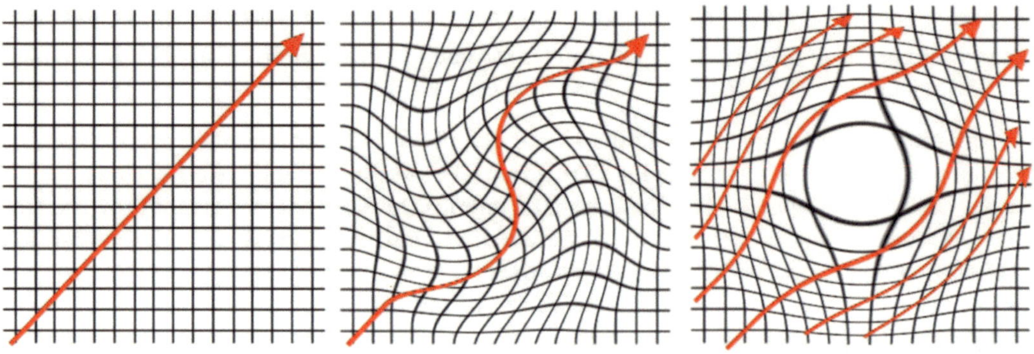

(a) (b) (c)

Fig. 1. (a) Propagation of a light beam in a non distorted space. (b) propagation of the same ray of light in a distorted space. (c) isolation of a region of space by deforming the propagation of light rays around this region.

J. Pendry and U. Leonhardt noted both in their articles published in 2006, the invariance of Maxwell's equations in such a deformed space [Pendry, 2006, Leonhardt, 2006]. J. Pendry has concluded that it was possible to isolate a zone of space by bending light rays around this area (Figure 1c).

2.2 Implementation

The implementation of this transformation is relatively simple. If we consider a Cartesian space where each point is identified by three coordinates (x, y, z), we can define a new space where each point will be identified by three new coordinates (u, v, w). These three new coordinates are based on the original ones.

$$u(x,y,z), v(x,y,z), \text{ and } w(x,y,z) \tag{2}$$

In this case, we use normalized values of the electromagnetic parameters ε and μ that we will define in the new space:

$$\varepsilon'_u = \varepsilon_u \frac{F_u F_v F_w}{F_u^2} \text{ and } \mu'_u = \mu_u \frac{F_u F_v F_w}{F_u^2} \text{ ; etc...} \tag{3}$$

where F_u, F_v et F_w are defined by the following equations:

$$F_u = \left(\frac{\partial x}{\partial u}\right)^2 + \left(\frac{\partial y}{\partial u}\right)^2 + \left(\frac{\partial z}{\partial u}\right)^2 \quad F_v = \left(\frac{\partial x}{\partial v}\right)^2 + \left(\frac{\partial y}{\partial v}\right)^2 + \left(\frac{\partial z}{\partial v}\right)^2$$
$$\text{and } F_w = \left(\frac{\partial x}{\partial w}\right)^2 + \left(\frac{\partial y}{\partial w}\right)^2 + \left(\frac{\partial z}{\partial w}\right)^2 \tag{4}$$

The conventionnal relations of electromagnetics are conserved in the new space :

$$\vec{B}' = \mu_0 \mu' \vec{H}' \quad \text{and} \quad \vec{E} = \varepsilon_0 \varepsilon' \vec{E}' \tag{5}$$

In these relations, μ' and ε' are tensors whose components depend on the spatial coordinates *u*, *v* et *w*. Generally, the new space is anisotropic.

3. Cloaking

The first application of the space coordinate transformation will be the design and the characterization of different electromagnetic invisibility cloaks [2-6].

3.1 Principle

In the case of the first invisibility cloak proposed by Smith and Pendry [Pendry, 2006], the transformation of space concerned a cylindrical space of radius b which is transformed into an annular space between radii a and b (Figure 2). The initial points in space are identified by coordinates r, θ and z. Those of the transformed space are identified by r ', θ' and z' in cylindrical coordinates. Both spaces are assumed infinite in the directions z and z ', which are combined.

The transformation is defined by a relatively simple set of equations (6).

$$r' = \frac{b-a}{b} r + a \text{ ; } \theta' = \theta \text{ ; } z' = z \tag{6}$$

After transformation we obtain :

$$\varepsilon_r = \mu_r = \frac{r-a}{r} \text{ ; } \varepsilon_\theta = \mu_\theta = \frac{r}{r-a} \text{ ; } \varepsilon_z = \mu_z = \left(\frac{b}{b-a}\right)^2 \frac{r-a}{r} \tag{7}$$

The three parameters obtained depend on the distance r and on the geometric parameters of the original and transformed spaces. These are the electromagnetic parameters of the transformed space. Note that the permeability and permittivity have here the same expressions. This guarantees the matching of the wave impedance of the transformed medium with the initial medium.

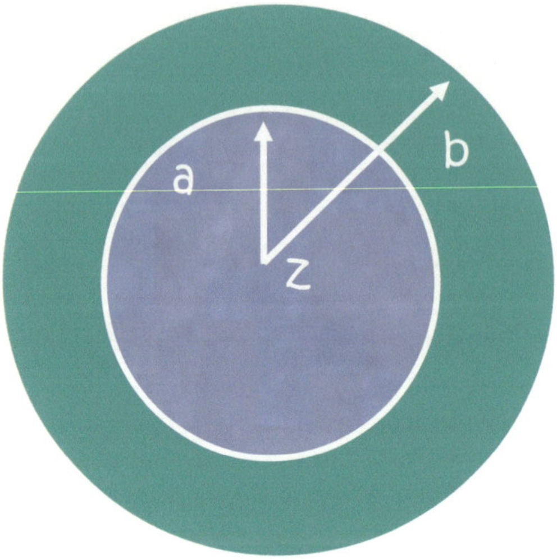

Fig. 2. Initial circular space of radius b, transformed to a ring between a and b.

3.2 Realization of the first microwave cloak

The practical realization of the material forming the transformed space is a difficult task since the three parameters vary simultaneously [Schurig, 2006]. To simplify this material, one solution is to choose a polarization, namely the transverse magnetic (TM) polarization, with the electric field E parallel to the axis z. In this case, only the parameters ε_z, μ_r et μ_θ are important. In this case, a simple set of parameters is possible provided it meets the propagation equation. The next set is one of the possibilities:

$$\varepsilon_z = \left(\frac{b}{b-a}\right)^2 \; ; \; \mu_r = \left(\frac{r-a}{r}\right)^2 \; ; \; \mu_\theta = 1 \tag{8}$$

In these three parameters, two are fixed and one varies; the radial permeability u_r. This material can be realized by the metamaterial concept [Soukoulis, 2011]. Figure 3a shows the basic pattern of the material and the geometric values used with the corresponding values of u_r. The permittivity ε_z is realized using a conventional dielectric.

Figure 3c shows a part of the realized circuit with, in the insert, the evolution of the three parameters μ_r, μ_θ and ε_z. The shape of the elementary patterns depends on the layer of material so that the permeability μ_r also varies and follows the red curve in the figure inset. Figure 4(a) and 4(b) shows the simulation of the cloak with respectively the theoretical parameters of equation 13 and the reduced parameters of equation 14. Figure 4(c) presents the measured electric field cartography of the central bare metallic cylinder where we can observe strong reflections and shadows compared to the case of very low reflections when the cloak is applied around the cylinder (Figure 4(d)).

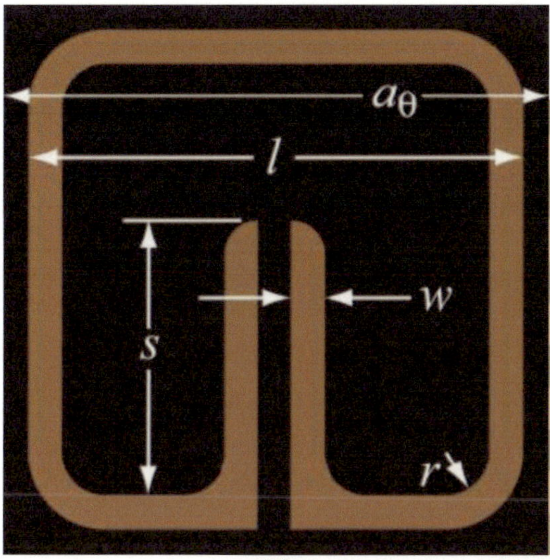

(a)

cyl.	r	s	μ_r
1	0.260	1.654	0.003
2	0.254	1.677	0.023
3	0.245	1.718	0.052
4	0.230	1.771	0.085
5	0.208	1.825	0.120
6	0.190	1.886	0.154
7	0.173	1.951	0.188
8	0.148	2.027	0.220
9	0.129	2.110	0.250
10	0.116	2.199	0.279

(b)

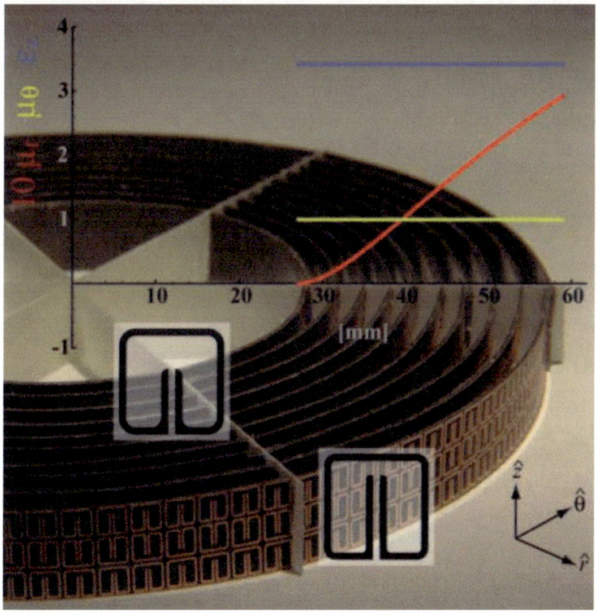

(c)

Fig. 3. (a) Elementary cell of the metamaterial. (b) values of the parameters. (c) Photography of the prototype. The insert shows the variations of μ_r, μ_θ and ε_z.

Figures 4b and 4d are quite similar and clearly show the influence of electromagnetic radiation on the cloak with in particular the cloaking of the central metallic cylinder by the electromagnetic energy. Both figures also illustrate the limits of the exercise as the use of a reduced set of parameters causes reflection of part of the incident energy, and therefore an imperfect reconstruction of the wave after the cloak.

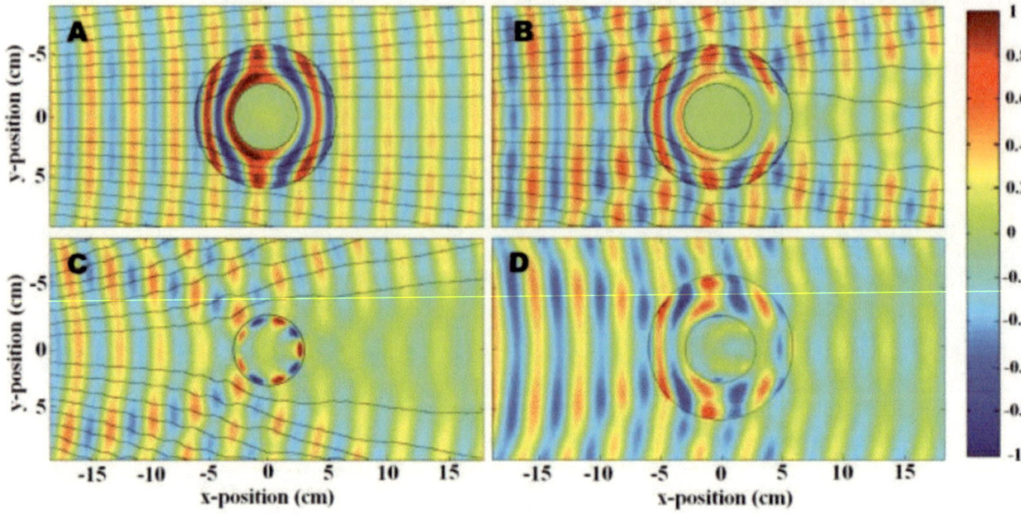

Fig. 4. (a) Simulation of the theoretical cloak. (b) simulation of the cloak with reduced parameters. (c) simulation of the central metallic cylinder. (d) E field measurement.

3.3 Others realizations

The experimental verification of this first cloak has excited the imagination of researchers who have tried to extend to other areas in optics but also in acoustics.

3.3.1 Cloak insensitive to the polarization

In the field of electromagnetics, few achievements have been proposed and tested experimentally. But there are some exceptions [Guven, 2008, Kante, 2008, Kante, 2009]. In reference [Guven, 2008], S. Tretyakov and his team tried to design a cloak based on spiral type resonators and the principle of homogenization. These coils act as a combination of an electric and a magnetic dipole. They can therefore meet the criteria to realize the material of the cloak. Fig. 5a shows the unit spiral resonator cell used by S. Tretyakov and Fig. 5b illustrates the distribution of the resonators in the cloak. The realization of the cloak is simpler than that of Smith and is supposed to work for both TM and TE polarizations of the incident wave with the disadvantage of its large size compared to the cloaked object.

3.3.2 Cloak based on the electric resonance of the SRR

In the references [Kante, 2008, Kante, 2009] the principle is completely different. Instead of using the magnetic resonance of the resonators of Pendry, the authors use the electrical resonance of these resonators. Smith's cloak works for a TM polarization (E-field vertical and H-field in the plane of the cloak). The new cloak works with a TE polarized wave (H-field vertical and E-field in the plane of the cape). For this non-magnetic cloak, the set of parameters is as follows:

$$\mu_z = 1\,,\quad \varepsilon_\theta = \left(\frac{b}{b-a}\right)^2,\quad \varepsilon_r = \left(\frac{b}{b-a}\right)^2 \left(\frac{r-a}{r}\right)^2 \tag{9}$$

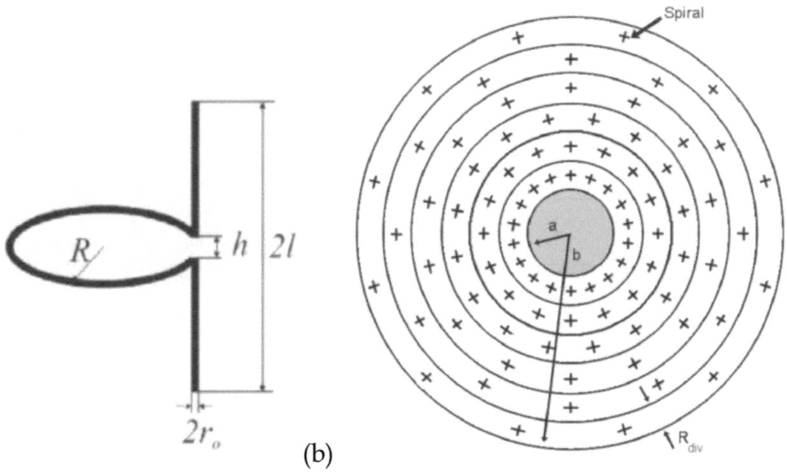

(a) (b)

Fig. 5. (a) Spiral coil used by S. Tretyakov in his cloak. (b) schematic view of the cloak.

This reduced set of parameters holds for a polarization perpendicular to the cylinder axis and satisfies the dispersion relation, but not the equality of the wave impedance between the vacuum and the cloak. A non-zero reflection is then predictable. In our implementation, the radial permittivity profile is designed using the electric response of SRRs by locally changing the dimension of resonators, actually only the SRRs gap size. The typical unit cell and the effective parameters of discrete SRRs at the design frequency (11 GHz) are presented in Fig. 6. While SRRs (embedded in the host medium) are used to achieve the radial variation of the permittivity function, the azimuthal permittivity is mainly implemented by the permittivity of the host medium itself since SRRs have no electric response in this direction. The realized cloak is composed of 15700 elementary SRRs. The cylindrical shell is divided in 20 annular regions of equal thickness (l_r=4.5 mm) with a linear radial variation of the permittivity from 0 to 1 (from the inner to the outer boundary of the cloak) and 157 stripes separated by an angle of about 2.3° (Fig.7 (a)). The inner and outer radius of the cloak are a=6 cm, b=15 cm and the cloak height is 2.25 cm corresponding to $5\times l_z$ i.e. 5 SRR layers. For this set of parameters the reflection coefficient R_p is very weak, equal to 0.0625. The SRRs within a given annular region are identical and designed to have the proper local radial permittivity. The host medium, a commercially available resin is an important design component (closely linked to ε_θ). Its permittivity has been measured and found to be equal to ε_{resin}= 2.75. The SRRs have been printed on a dielectric substrate (as seen in the picture of Fig.1 (c)) with a permittivity close to the resin's one. We chose RO3003 with $\varepsilon_{substrate}$= 3±0.04 and a dielectric loss tangent at 11 GHz of about 0.0013. The 157 stripes were arranged in a moulded water-tight polymeric matrix designed accordingly (Fig. 7 (a)).

In contrast with previously reported structure, the measurements are performed in free space and not in a waveguide configuration(Fig. 7b). A loop antenna, consisting of a circular coil made of the inner conductor of a SMA cable has been designed to map the magnetic field (H_z). The magnetic field is output from the X-band horn antenna. Both antennas are connected to an Agilent 8722ES Vectorial Network Analyzer. The loop antenna position can be controlled via an automated Labview program over a surface of 40 cm*40 cm and getting for each spatial position of the loop antenna the complex (magnitude and phase) scattering parameters. The experimental setup can be seen on Fig. 7(b). Since the resin fills the

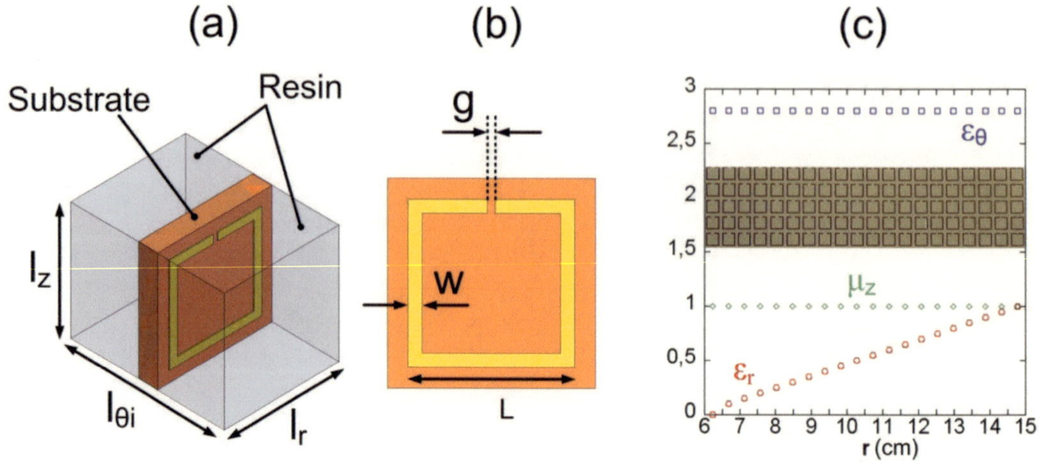

Fig. 6. (a) Unit cell. (b) The dimensions of a typical square SRR are: L=3.6 mm, w=0.3 mm and copper thickness t=35 μm. The SRRs gap g and $l_{\theta i}$, are the only varying parameters. $l_{\theta i}$ linearly decreases from the outer to the inner boundary of the cloak.

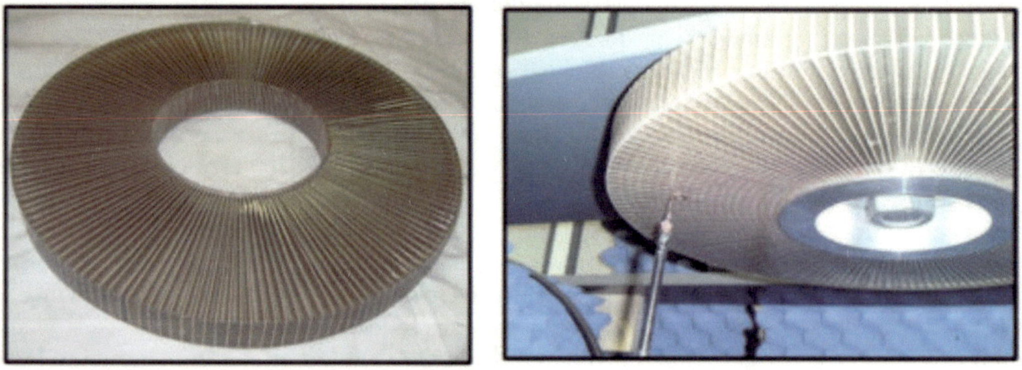

(a) **(b)**

Fig. 7. (a) Realized metamaterial cloaking device (b) Picture of a portion of the experimental setup with the loop antenna mapping the magnetic field at the bottom surface of the cloak.

structure, it is difficult to access the internal field. Instead, the bottom surface of the cloak (see Fig. 8d) has been scanned taking profit of the continuity of field at this boundary in quasi-contact mode. The first measurement maps the magnetic field of the free space radiation from the horn antenna (Fig. 8(a)). The second and third measurements use a metallic cylinder alone (diameter 12 cm) (Fig. 8(b)) and surrounded with the cloak (outer diameter 30 cm) (Fig. 8(d)). The results are presented in Fig. 8 (real part of the complex transmission). The quasi-cylindrical wave output from the horn antenna is nicely resolved in our measurement (Fig. 8(a)). In presence of the metallic cylinder, the scattering and shadowing effects can be clearly observed in Fig. 8(b) as well as interferences between the incident and reflected beams. Fig. 8(d) shows that in the presence of the cloak, the shadowing effect of the metallic cylinder is suppressed and the wave fronts are maintained thus demonstrating the cloaking effect. For comparison, simulation result using commercial finite element code (Comsol Multiphysics) for a cloak with the reduced parameters of equations [Kante, 2009]. is reported in Fig. 8(c). The fact that a non-zero field is detected in the central region of Fig. 8(b) and 8(d) results from radiation leakage below the metallic cylinder in our measurements. More importantly, the bending and redirection of quasi-cylindrical wave fronts inside the cloak can be nicely observed as a change in the radius of the horn antenna waves fronts in Fig. 8(d).

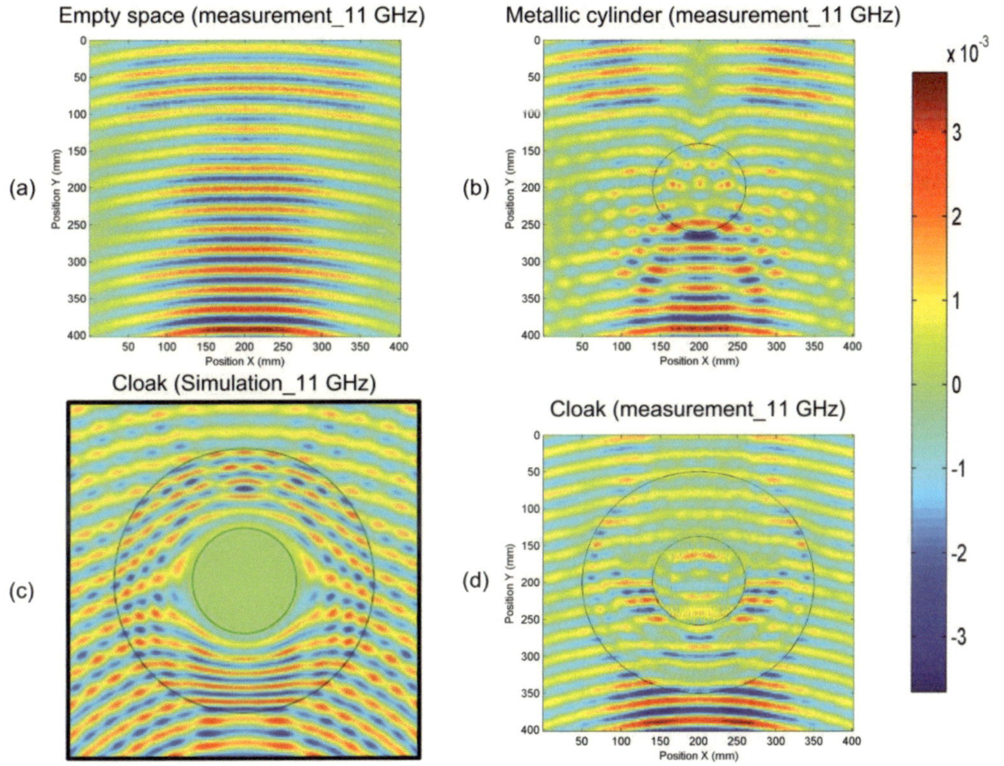

Fig. 8. Real part of the measured magnetic field output from the horn antenna in free space (a) with the metallic cylinder alone (b) and with the cloak surrounding the metallic cylinder (d). Finite element simulation exciting the cloak by the appropriate optical excitation (Comsol Multiphysics) with the reduced set of parameters presented in equations (1) is reported for comparison (c). In all cases, the 11 GHz wave travels from bottom to top.

3.4 Optical cloaks

3.4.1 Propositions of V. Shalaev

In the reference [Cai, 2008], V. Shalaev proposes two possible achievements of cloaks in the infrared and visible domains with a TE and TM polarization (figure 9).

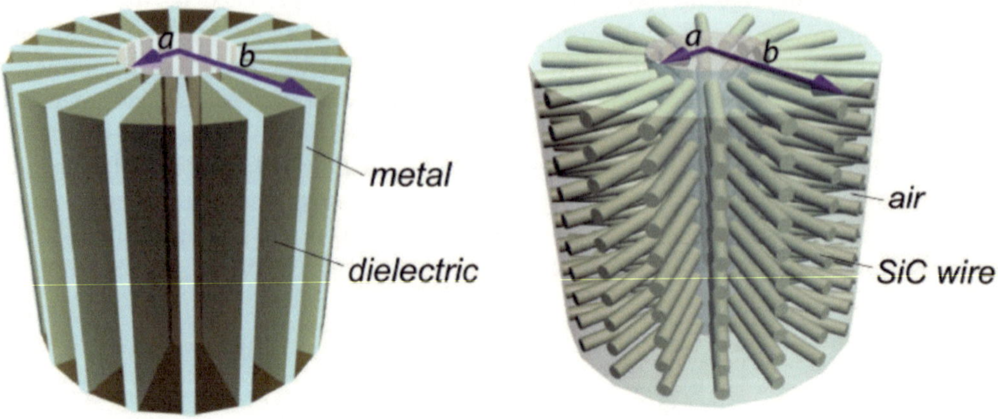

Fig. 9. Proposition of an invisibility cloak for a TE polarization (a) and a TM one (b).

The first one corresponds to a TE polarization. For this polarization, the main parameters are:

$$\varepsilon_r = \left(r'/r \right)^2 : \varepsilon_\theta = \left(\partial g(r') / \partial g(r) \right)^{-2} : \mu_z = 1 \tag{10}$$

Instead for the TM polarization the main parameters are:

$$\mu_r = \left(r'/r \right)^2 \left(\partial g(r') / \partial g(r) \right)^2 : \mu_\theta = 1 : \varepsilon_z = \left(\partial g(r') / \partial g(r) \right)^{-2} \tag{11}$$

In these expressions g(r) is the relation between the intial space and the final one. For example in the case of a cylindrical cloak :

$$r = g(r') = (1 - a / b) r' + a \tag{12}$$

In the case of a TM polarization, where ε must vary between the inside and outside of the cloak, Shalaev proposes to use an effective permittivity given by Wiener relations, varying between the permittivity of a metal (in this case silver or silicon carbide) and that of a dielectric, which can vary with the wavelength of work (like silica or barium fluoride) (Figure 9(a)).

For the TE polarization, where the permeability must vary, the material used can be silicon carbide with the shape of rods embedded in air, which seems to be a bit unrealistic at visible wavelengths. However no practical realization has been proposed.

3.4.2 Other optical cloak for TE polarization

In reference [Kante, 2008] another proposal is to use the electromagnetic properties of metal cut wires. This proposal concerns only the TE polarization, and is directly correlated to the proposed cloak in the microwave region [10] by replacing the resonators of Pendry by gold nanowires.

Indeed, the magnetic resonators of Pendry and the nanowires are equivalent for a TE polarized incident wave, where the electric field is parallel to the long side of the resonator or to the wire. The invisibility cloak that was made in the microwave domain and that operates at 11GHz can be implemented in optics by replacing the resonators used in microwave by gold nanowires (Fig. 10). This figure shows a detail of the cape made using the magnetic resonators of Pendry (Fig. 10 (a)), the simulation of the cloak at 1.5 microns (Fig. 10 (b)), a gold nanowire for a TE polarization of the incident wave (Fig. 10 (c)), the equivalent parameters of the nanowire around its first resonance frequency (Fig. 10 (d)), and a schematic view of the cape in which the resonators are replaced by gold nanowires.

3.5 Cloak with arbitrary shapes

The principles used in the cylindrical cloaks described above can be generalized to a variety of different shapes. The figure below is taken from reference [Nicolet, 2008] where a Fourier expansion is used to access to convex shapes. Reference [Rahm, 2008] proposes a square cloak (Figure 12a), which has been generalized to a polygonal cloak in [Tichit, 2008] (Figure 12b), then to an elliptical one (Figure 12c).

3.6 Broadband cloak

The major disadvantages of the first invisibility cloak were the narrow frequency-operating band and the extreme values of electromagnetic parameters needed. Various attempts have then been proposed to achieve broadband cloaks, or to design cloaks with more realistic parameter values. The most amazing proposition was suggested by U. Leonhardt who proposed to benefit from a non-Euclidean geometry to achieve the broadband [Leonhardt, 2009]. A. V. Kildishev also proposed an approximate solution to achieve a broadband cloak [Kildishev, 2008]. Following the preceding reference, we show the transition from a two-dimensional "conventional" cloak (Figure 13a) to a non-Euclidean one, namely a sphere replacing a single circle (Figure 13b). In the broadband cloak proposed by Shalaev, the principle is simple: light follows a different path depending on the operating frequency (Figure 13c).

Recently other concepts of broadband invisibility cloaks based on the use of broadband non-resonant metamaterials have been proposed [Qiu, 2009, Feng, 2011] (figure 14). These recent works are based on the use of materials made of broadband dielectric multilayer where the electromagnetic parameters are extracted using the relations of Wiener on one-dimensional multi-layer materials. Figure 14 shows an example of a dielectric multi-layer structure used in a cylindrical invisibility cloak. Each concentric layer is constituted by a sub-layer of permittivity ε_A and a sub-layer of permittivity ε_B and η is the thicknesses ratio of the two layers. The effective permittivity parameters of the layer are given by:

$$\varepsilon_\theta = \frac{\varepsilon_A + \eta\varepsilon_B}{1+\eta}, \quad \frac{1}{\varepsilon_r} = \frac{1}{1+\eta}\left(\frac{1}{\varepsilon_A} + \frac{\eta}{\varepsilon_B}\right),$$

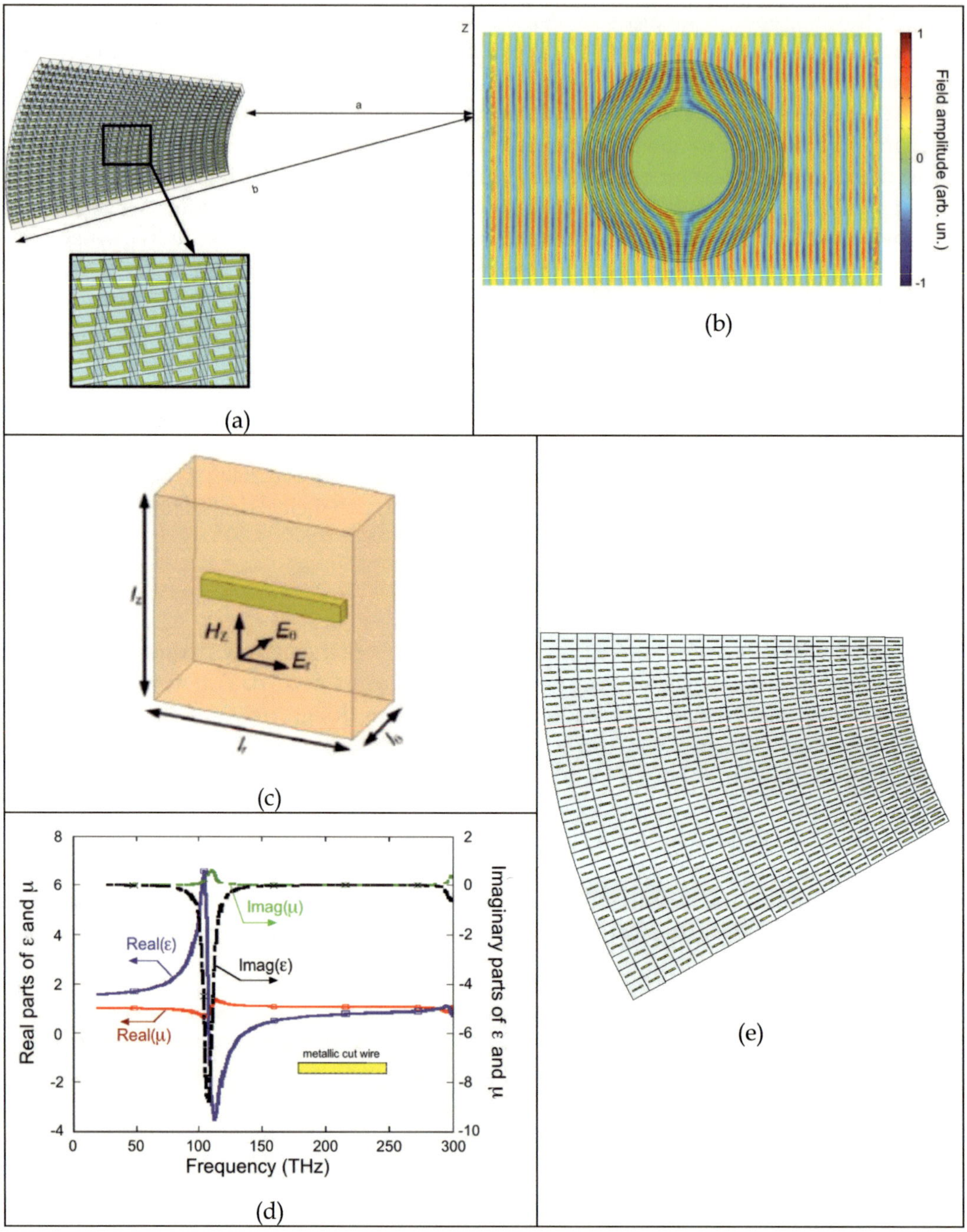

Fig. 10. (a) Cloak made of resonators of Pendry for the TE polarization of the incident field. (b) Simulation of this cloak at 1.5μm. (c) gold nanowire in TE polarization. (d) Variation of the effective permittivity and permeability of the wire as a function of frequency for a nanowire with 300nm length and a width and height of 50 nm on silicon. (e) Portion of the infra-red cloak made by a juxtaposition of gold nanowires on silicon.

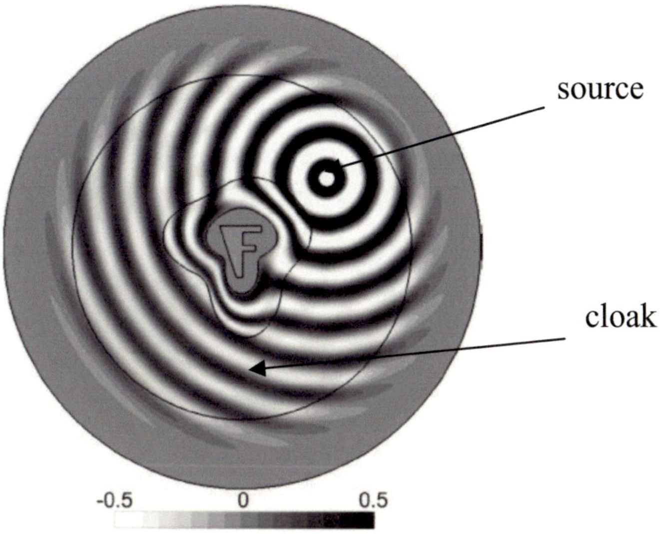

Fig. 11. Electric field radiated by a point source illuminating a cloak whose the shape is obtained by Fourier expansion of a simpler one.

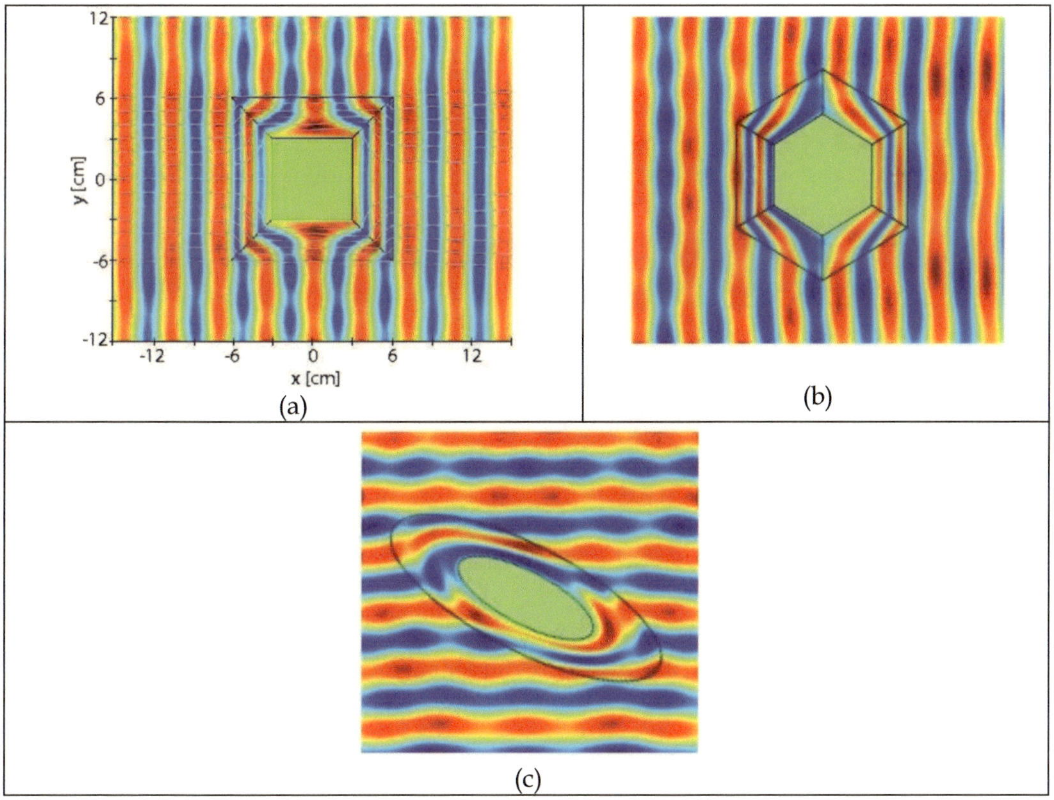

Fig. 12. (a) Squared invisibility cloak [Rahm, 2008]. (b) polygonal cloak [Tichit, 2008]. (c) elliptical cloak [Tichit, 2008].

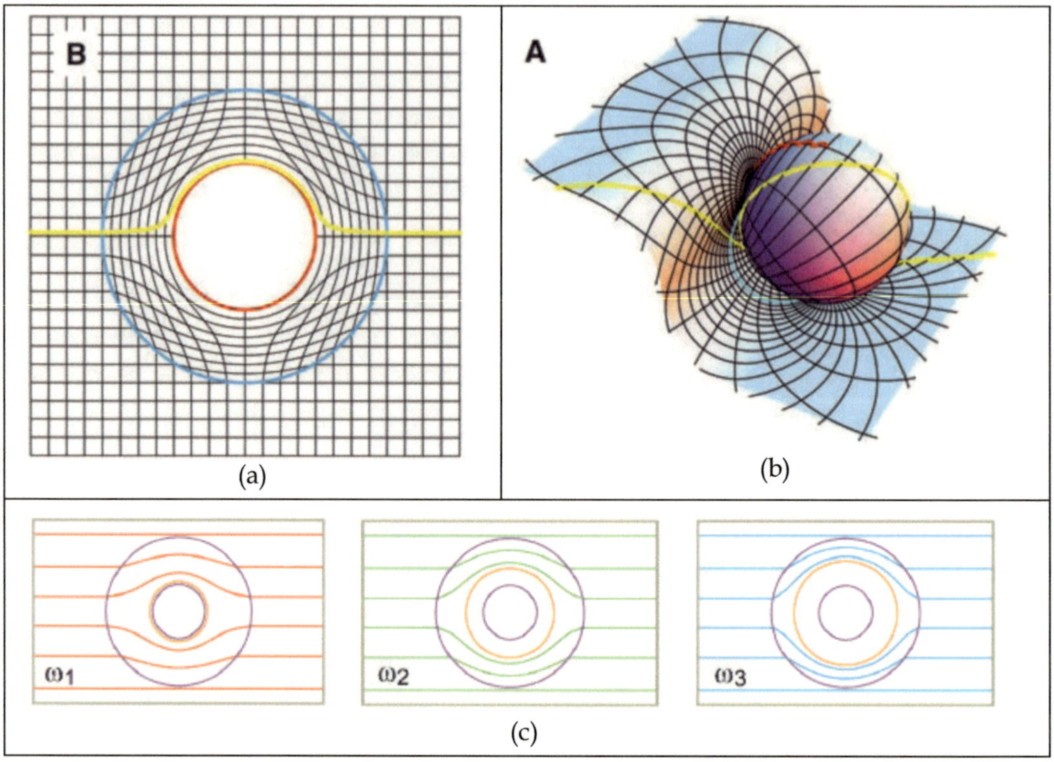

Fig. 13. (a) Classical cylindrical cloak. (b) Cloak in a non-Euclidian space. (c) Principle of the broadband cloak proposed by Shalaev : the light path changes following the frequency.

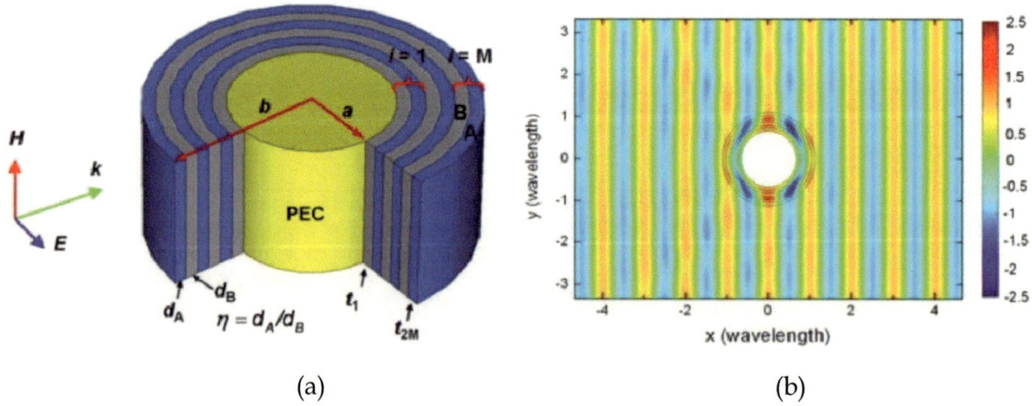

Fig. 14. (a) TM plane wave incident on a PEC cylinder surrounded by concentric multilayers. The inner and outer radii of the shell are a and b, respectively. (b) The total magnetic field distribution for an optimized six-layer cloak.

A TM plane wave is incident on a PEC cylinder surrounded by concentric multilayers as shown in Fig. 14a. The inner and outer radii of the shell are *a* and *b*, respectively. The total magnetic field distribution for an optimized six-layer cloak is presented in Fig. 14b. The main difficulty in this kind of design is the realization of permittivity lower than 1. This 2D approach was also generalized to a 3D cloak [Qiu, 2009].

3.7 Acoustic cloak

The transposition of the concept of electromagnetic cloak in acoustics has been proposed by several research laboratories in 2007 and 2008 [Torrent, 2007, Chen, 2007, Fahrat, 2008, Cummer, 2008], and particularly by the Fresnel Institute at University of Marseille. The variables considered here are the scalar pressure p, the fluid velocity, the density ρ_0, the tensor density and modulus of the fluid density λ. As for the electromagnetic cloak, we have a variation of the above parameters in spherical coordinates as in the set of equations (13) where ρ_r and ρ_ϕ are the components in the plane of the relative bulk density ρ, relative to ρ_0. This example shows the versatility of the concept that can be applied to all media where a wave can propagate.

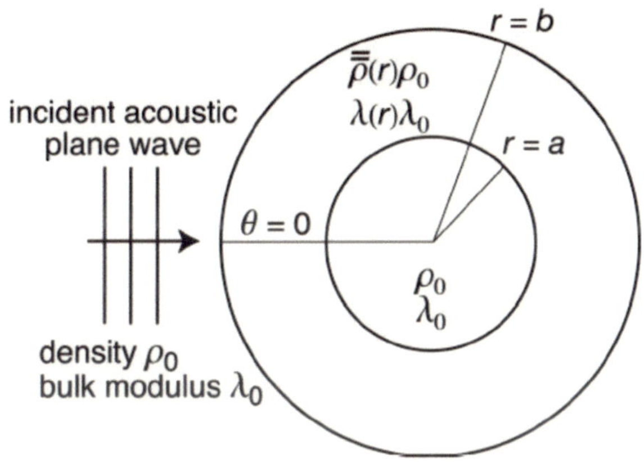

(a)

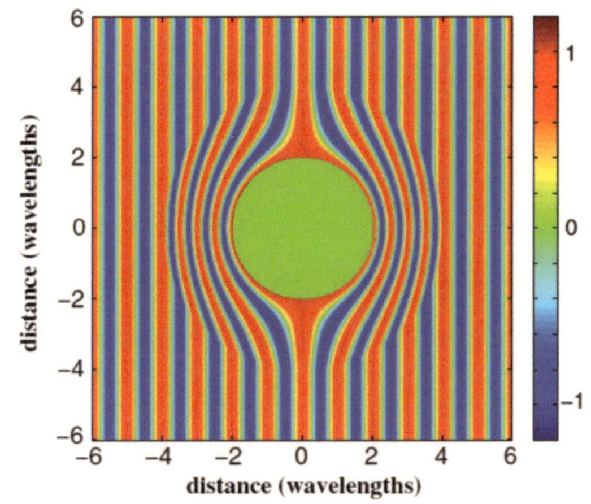

(b)

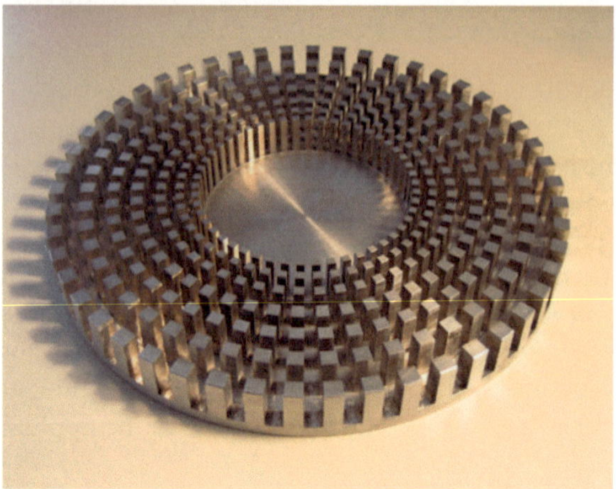

(c)

Fig. 15. (a) The principle of acoustic cloak proposed in [Cummer, 2008]. (b) Real part of the fluid pressure around the object. (c) The cloak proposed in [Fahrat, 2009].

$$\begin{aligned} \rho_\varphi &= \rho_\theta = \frac{b-a}{b} \\ \rho_r &= \frac{b-a}{b}\frac{r^2}{(r-a)^2} \\ \lambda &= \frac{(b-a)^3}{b^3}\frac{r^2}{(r-a)^2} \end{aligned} \tag{13}$$

4. Others applications of the space coordinate transformation

4.1 Antennas

The space coordinate transformation was proposed initially by J. Pendry and U. Leonhardt to design invisibility cloaks. This transformation of space was then used to design new devices including microwave antennas. The principle of these antennas is as follows: we define an initial space in which there is an emitter and a transformed space connected by a geometric transformation to the original space. The transformed space is realized to control the field emitted outside by the antenna. The new coordinates of the transformed space x ', y' and z 'are expressed in terms of x, y, z of the initial space

$$x' = x'(x, y, z),\ y' = y'(x, y, z),\ z' = z'(x, y, z) \tag{14}$$

Then we calculate the electromagnetic parameters of the transformed space

$$\bar{\bar{\varepsilon}}' = \overline{\overline{J\varepsilon J}}^T (\det \bar{\bar{J}})^{-1} \quad \bar{\bar{\mu}}' = \overline{\overline{J\mu J}}^T (\det \bar{\bar{J}})^{-1} \tag{15}$$

where $\bar{\bar{\varepsilon}}$ is the permittivity tensor, $\bar{\bar{\mu}}$ the permeability tensor and $\bar{\bar{J}}$ the Jacobian matrix defined by

$$\bar{\bar{J}} = \frac{\partial(x',y',z')}{\partial(x,y,z)} = \begin{pmatrix} \partial x'/\partial x & \partial x'/\partial y & \partial x'/\partial z \\ \partial y'/\partial x & \partial y'/\partial y & \partial y'/\partial z \\ \partial z'/\partial x & \partial z'/\partial y & \partial z'/\partial z \end{pmatrix} \tag{16}$$

In the initial space, an antenna emits a certain type of radiation. This radiation is then modified by the transformation of the space in which it propagates. Several examples have been proposed in recent papers [Kong,2007, Tichit, 2009, Tichit, 2011, Rui, 2011, Cui 2011].

4.1.1 1st example: Parabolic transformed antenna

An example of a directional antenna is given below [Kong, 2007], where a parabolic space is transformed into a rectangular one (Figure 16a). In this example we are in TM polarization.

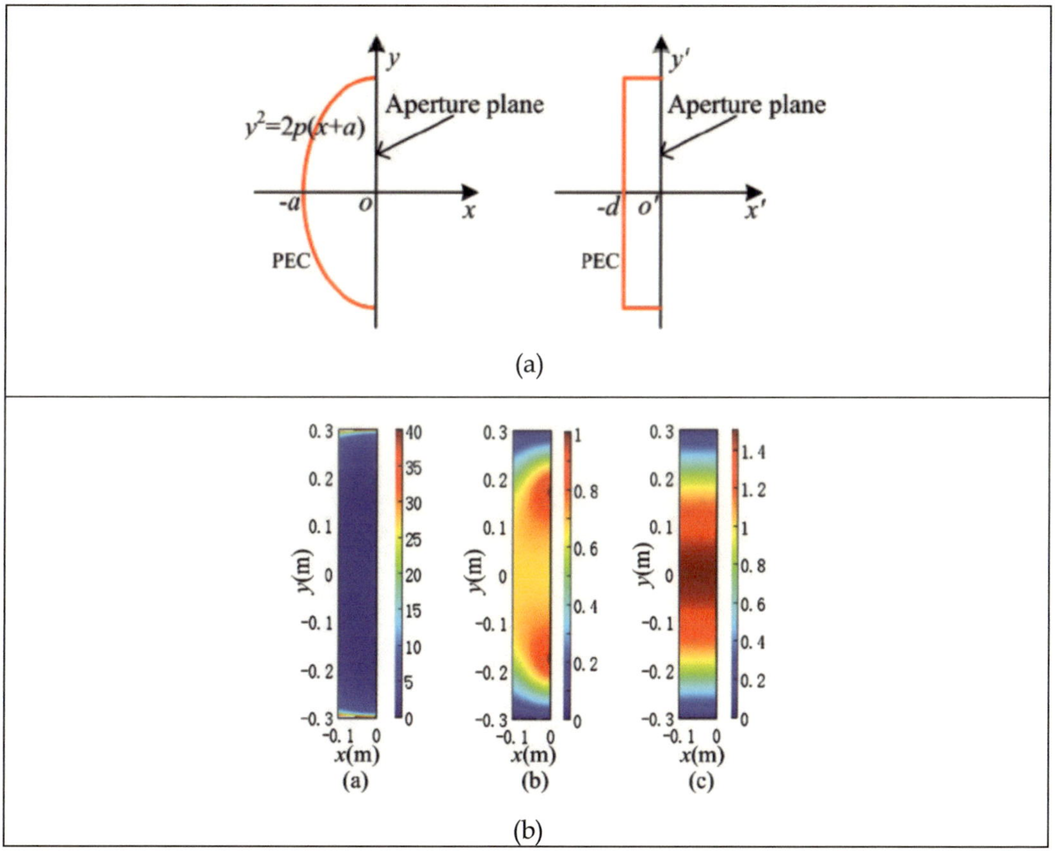

Fig. 16. (a) Initial parabolic space and transformed rectangular space. (b) Variation of the electromagnetic parameters μ_x, μ_y et ε_z of the transformed space.

Figure 16b shows the variations of the calculated electromagnetic parameters. Figure 17 shows the radiation of a horn in the transformed space. In this example, the benefit of the transformation of space is not real in the sense that the obtained antenna has almost the same size as the original antenna. Moreover, in this reference, the antenna has not been realized and the far-field radiation patterns have not been presented. However, space coordinate transformation can be used to design antennas more compact than conventional ones. This is the case of the antenna proposed theoretically in reference 23 and experimentally measured in [Tichit, 2009].

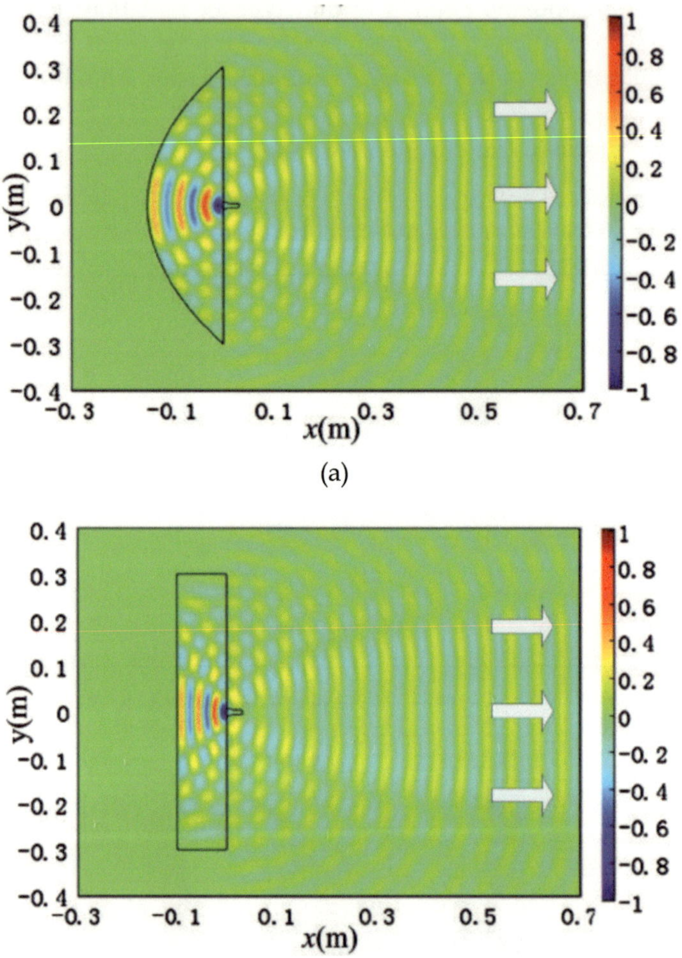

(b)

Fig. 17. (a) Horn antenna emitting in the parabolic space. (b) horn antenna emitting in the transformed space. In both cases the near fields are almost equivalent.

4.1.2 2nd example: Directive antenna

The 2nd example concerns the transformation of an isotropic antenna into a directive one [Tichit, 2009, Tichit, 2011]. This isotropic antenna is taken as an infinite radiating wire. The initial space is then supposed to be the cylindrical space surrounding the wire. The transformed space is a rectangular one as illustrated in Figure 18. After the transformation, the radiating wire in the cylindrical space is then comparable to a plane source radiating in the rectangular space.

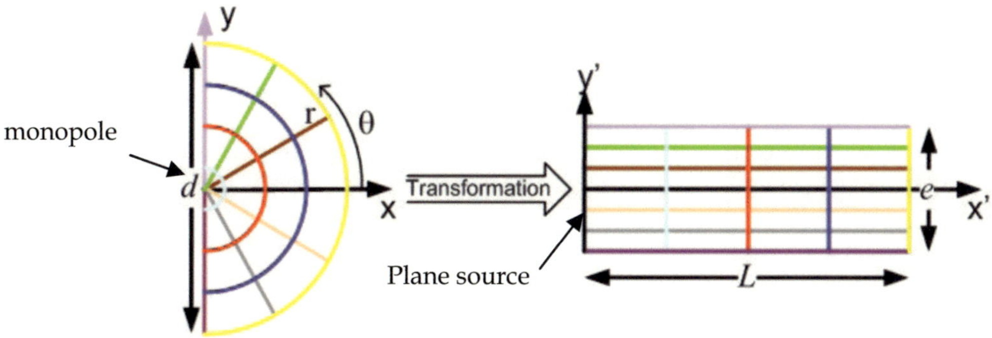

Fig. 18. Initial cylindrical space with the radiating monopole (left) and the transformed space (right) with the transformed plane source.

Figure 19 shows the variations of the electromagnetic parameters ε_{xx}, ε_{yy}, *and* μ_{zz} needed to achieve the space transformation. The expressions of these parameters are as follows:

$$\bar{\bar{\varepsilon}} = \begin{pmatrix} \varepsilon_{xx}(x',y') & 0 & 0 \\ 0 & \varepsilon_{yy}(x',y') & 0 \\ 0 & 0 & \varepsilon_{zz}(x',y') \end{pmatrix} \varepsilon_0 \qquad \bar{\bar{\mu}} = \begin{pmatrix} \mu_{xx}(x',y') & 0 & 0 \\ 0 & \mu_{yy}(x',y') & 0 \\ 0 & 0 & \mu_{zz}(x',y') \end{pmatrix} \mu_0 \quad (17)$$

with

$$\varepsilon_{xx}(x',y') = \mu_{xx}(x',y') = \frac{\pi}{e}x' \qquad \varepsilon_{yy}(x',y') = \mu_{yy}(x',y') = \frac{1}{\varepsilon_{xx}(x',y')} \quad (18)$$

where d, e and L are the geometrical dimensions of the initial and transformed spaces.

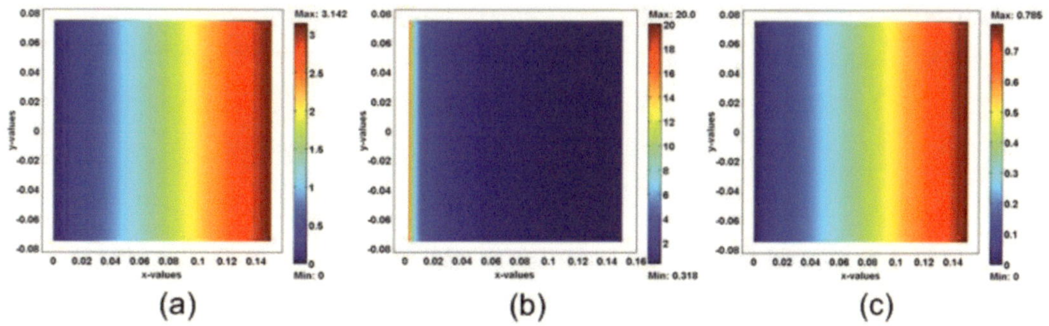

Fig. 19. Variations of the electromagnetic parameters of the transformed space: (a) ε_{xx}, (b) ε_{yy}, et (c) ε_{zz}.

The expressions of the electromagnetic parameters vary continuously, and remain limited to reasonable values. Figure 20 shows the calculated magnetic field at 5, 10 and 40 GHz. The directivity of the antenna increases as the frequency rises. The dimensions of the antenna are shown in Figure 20a. One can observe that there is no reflection between the metamaterial and the air.

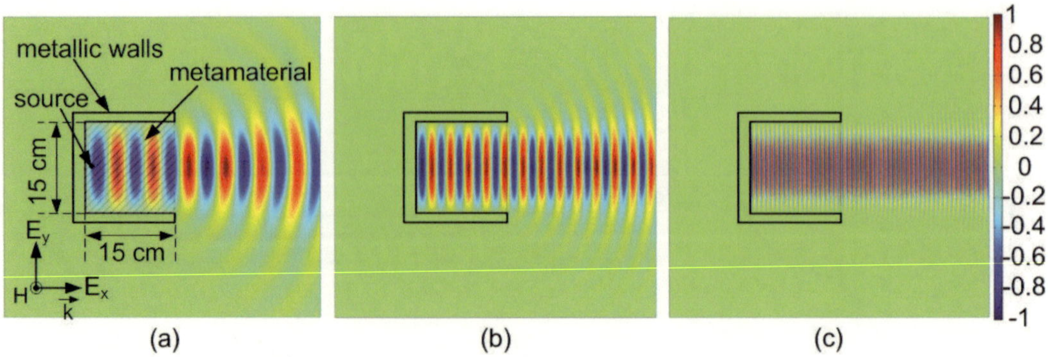

Fig. 20. Magnetic field cartography for a TM wave polarization calculated at (a) 5, (b) 10 and (c) 40 GHz.

The practical realization of this antenna, however, requires a simplification of these parameters. One solution proposed recently is to use a discrete variation of these parameters. The simplification is performed with a conservation of the propagation equation. The following set of parameters is then obtained:

$$\varepsilon_{yy} = \mu_{zz} = 1 \; : \; \varepsilon_{xx} = \left(\frac{\pi x}{e}\right)^2 \tag{19}$$

The material needed must have a variable permittivity in the direction of propagation Ox. The other parameters remain constant. Figure 21a shows a detail of the material used to make the variable permittivity, and the fabricated antenna prototype for an operation near 10 GHz. Figure 21b shows the performances of this antenna. It can be observed that the radiation pattern of the antenna is not affected by the simplification and the discretization of the material.

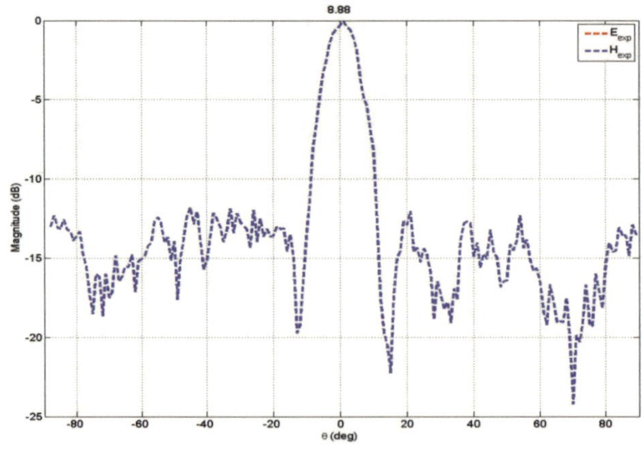

(a)

b

Fig. 21. (a) Normalized measured radiation pattern of the antenna. (b) detail of the metamaterial and of the realized antenna.

4.1.3 3rd example: Broadband Fresnel antenna

The previous antenna made use of resonant metamaterials. Then its bandwidth is inherently narrow. An interesting broadband operation proposal was recently presented by Y. Hao [Rui, 2011]. A broadband Fresnel lens can be realized using a multilayer dielectric structure. The permittivities of the different layers are calculated using space coordinate transformation. Figure 22 shows an example of such antenna. The performances are presented in part II of the figure. We can note the broadband characteristics of the antenna in IId.

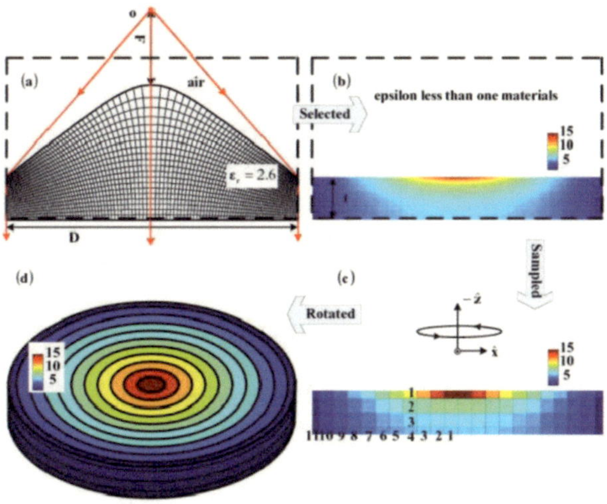

I

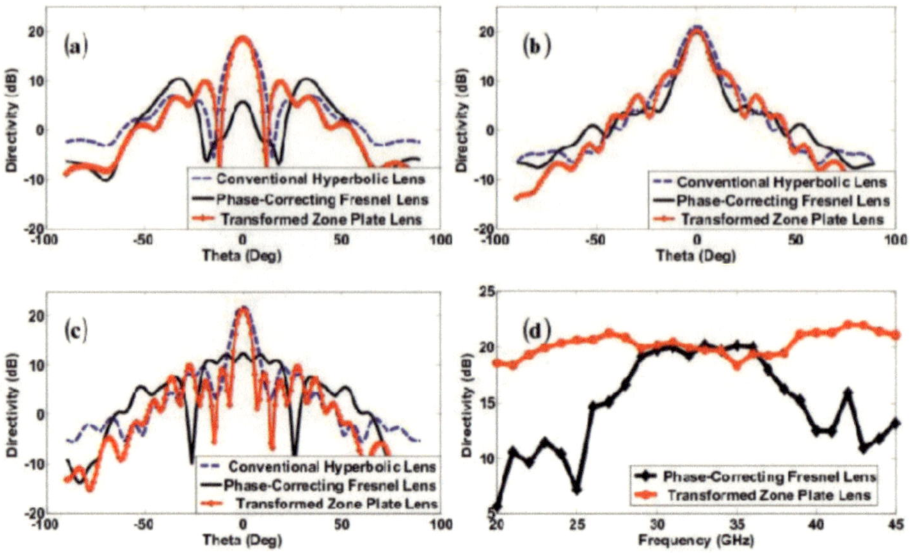

II

Fig. 22. I/ Schematic showing of the transformed zone plate lens antenna. (a) 2D hyperbolic lens with nearly orthogonal mapping. (b) 2D flat lens with the permittivity map consisting of 110×20 blocks. (c) 2D flat lens with the permittivity map consisting of 22×4 blocks. (d) 3D transformed zone plate lens antenna. II/ The radiation patterns of the conventional 3D hyperbolic lens, 3D phase-correcting Fresnel lens and 3D transformed zone plate lens at (a) 20 GHz, (b) 30 GHz, (c) 40 GHz. (d) The comparison of the bandwidth performance of 3D phase-correcting Fresnel lens and 3D transformed zone plate lens from 20 GHz to 45 GHz

4.1.4 4th example: Three-dimensional metamaterial lens antennas

The proposal of T.J. Cui to realize a lens using variable index material so as to focalize the beam of a waveguide is also an interesting application of transformation optics [Cui, 2011]]. Figure 23 shows a photo of the realized prototype and the performances of this lens in X band. The antenna presents two main advantages: the broadband behavior of the dielectric and the easiness of the realization (at microwave frequencies). Indeed the index gradient is realized with an array of variable size closed square rings printed on a dielectric substrate. Remains the classical drawback of the impossibility to realize an index lower than 1.

4.2 Circuits

M. Rahm proposed in [Rahm, 2008] a general method to achieve an invisibility cloak. But he also proposed the implementation of energy concentrator. Figure 24a shows a simulation of such a device. In [Lin, 2008], L. Lin proposed a number of applications such as a phase transformer to transform a cylindrical wavefront to a plane wavefront (Figure 24b) or a power divider (Figure 24c). In [Huangfu, 2008], J. Huangfu proposed a method to achieve wave guiding without reflection at 90° bends (Figure 24d).

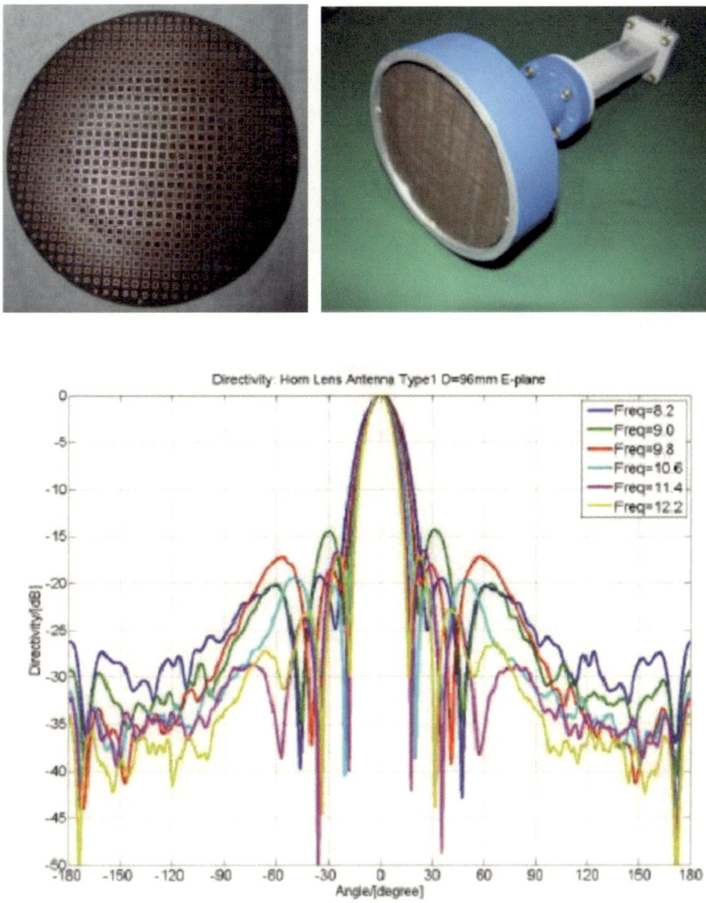

Fig. 23. (a) The 3D flat-lens antenna made of gradient index metamaterial. The aperture size is 9.6 cm. (b) The measured far-field radiation patterns of the 3D metamaterial flat lens antenna in the X band.

Other devices were recently proposed in the domain of the optical waveguiding devices [Ghasemi, 2010, Liu, 2008]. The proposal in [Ghasemi, 2010] tends to answer to a main drawback of use of metallic metamaterials at optical frequencies which is their high losses. A promising approach consists in creating hybrid photonic structures in which metallic parts are coupled with dielectric (and almost lossless) waveguides. In this configuration, useful functionalities are obtained by allowing just enough light to interact with the metallic parts of the system. The remaining part of the energy propagates in the dielectric waveguide, thereby considerably mitigating the losses. Figure 25 shows a mode adapter designed using this approach. The mode adapter allows the transition of the energy flow from a large SOI ridge waveguide to a narrower one. The taper has been achieved using the method of transformation optics. Although the authors simply considered a 2D transformation, they show that this structure can effectively act upon the three dimensional flow of light guided by the SOI structure.

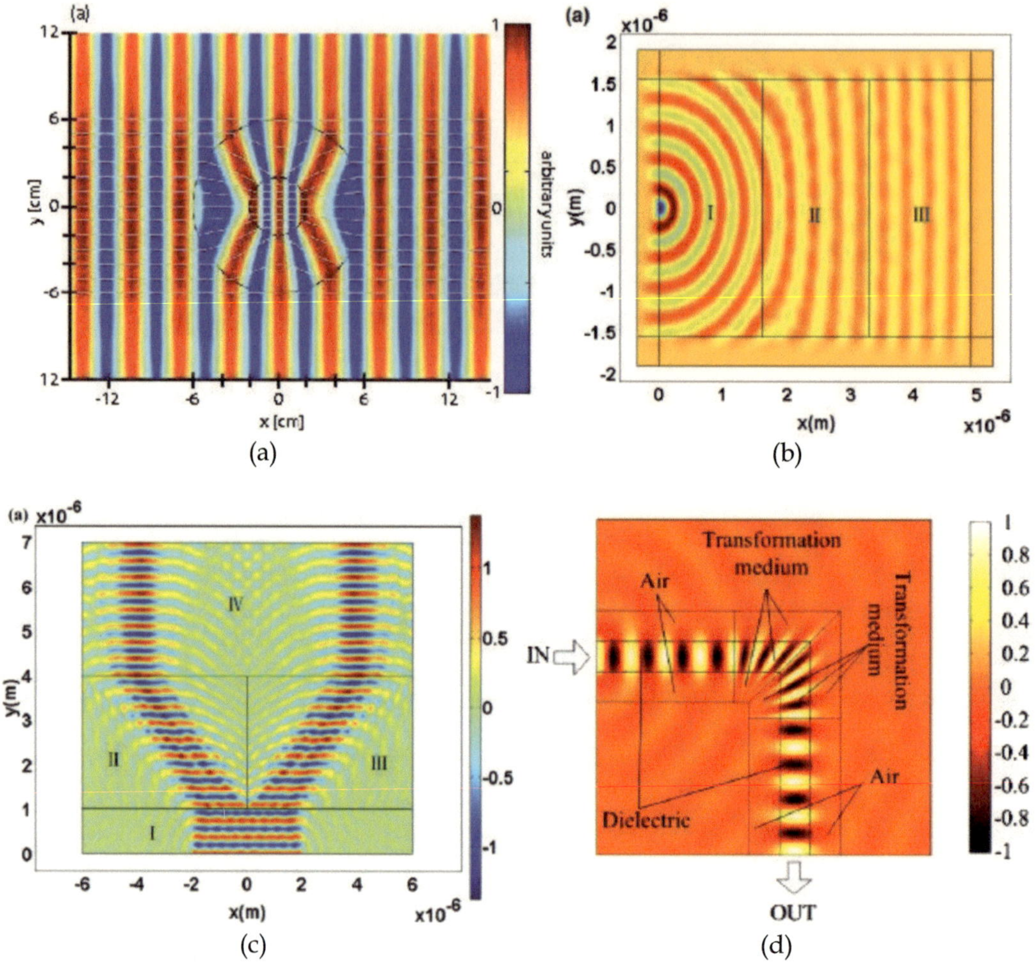

Fig. 24. (a) Energy concentrator proposed in the reference 12. (b) Phase Transformer between 2 regions. (c) Power divider. (d) 90° waveguide bend without loss.

4.3 Broadband carpet cloak

4.3.1 Microwave broadband carpet cloak

D. R. Smith has recently proposed a broadband cloak that can be adjusted to any object placed on the ground [Valentine, 2009]. This cloak allows to reconstruct the reflection of light incident on an object in order to make as if the object was not present. The object must however have dimensions small compared to the dimensions of the cloak. Figure 26a illustrates the operating principle of the cloak. Figure 26b gives a picture of its implementation and shows the pattern of the material permittivity variable used. The idea is to change the optical path followed by the reflected beam. The carpet cloak reconstructs the reflected beam as it is when no object is placed on the ground. This is clearly shown in Figure 26a: in I the ground reflects an incident beam without obstacle, in II the beam is reflected in the presence of an obstacle, in III the reflected beam consists of parallel rays reconstructed by the cloak covering the obstacle.

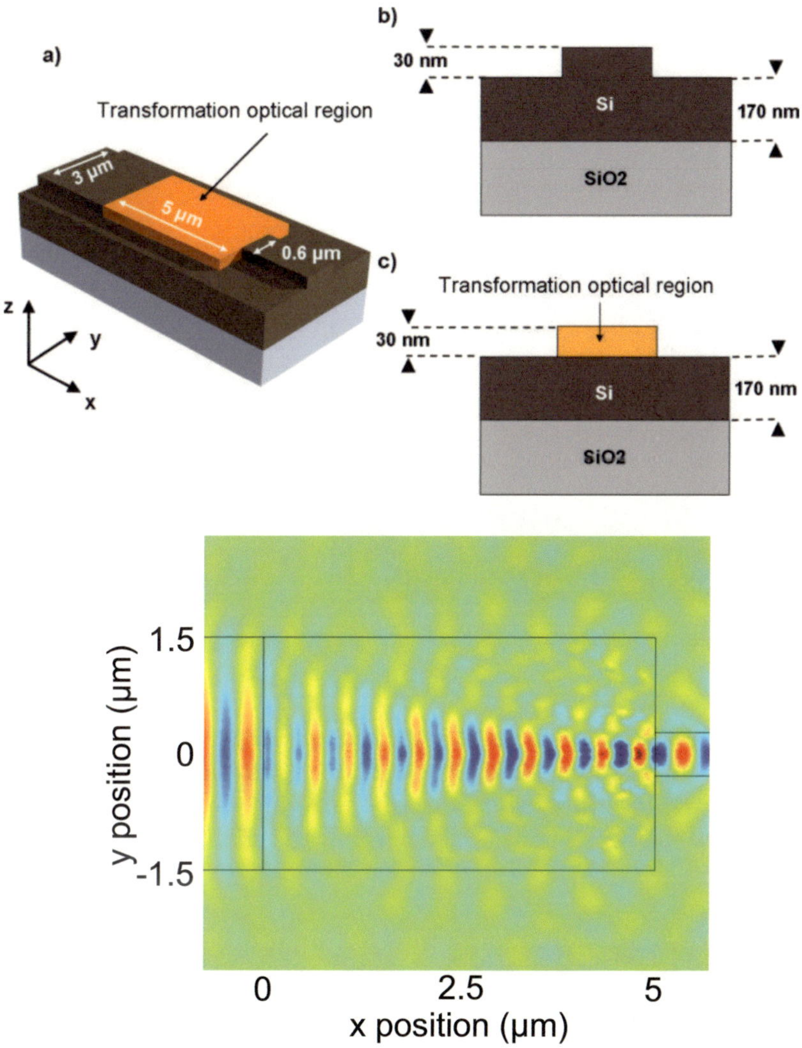

(d)

Fig. 25. (a) Geometry of the mode adapter considered in this study; (b) cross-sectional view of the input SOI ridge waveguide; (c) cross-sectional view of the mode adapter. (d) Transition from the large to the narrow waveguide using a mode adapter. The y-component of the electric field is shown in the x-y plane located halfway through the Si slab.

4.3.2 Optical carpet cloak

The same principle can be applied in infrared and visible domains [Gabrielli, 2009, Greenleaf, 2007, Cheng, 2009]. In reference [Gabrielli, 2009], the authors present the realization and the characterization of a carpet cloak operating in the optical domain. Figure 27 shows a view of the realized carpet on silicon.

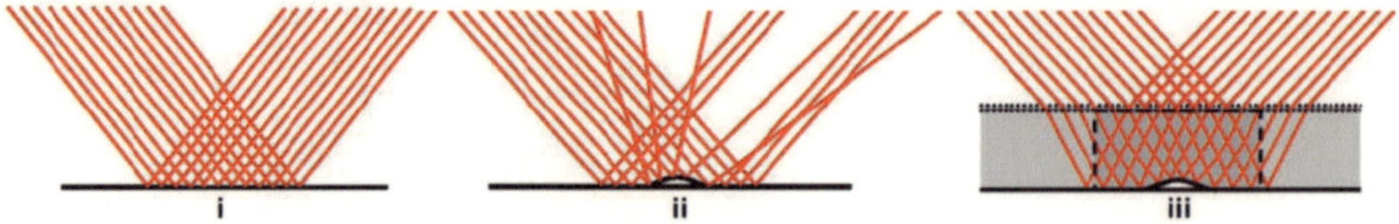

(a)

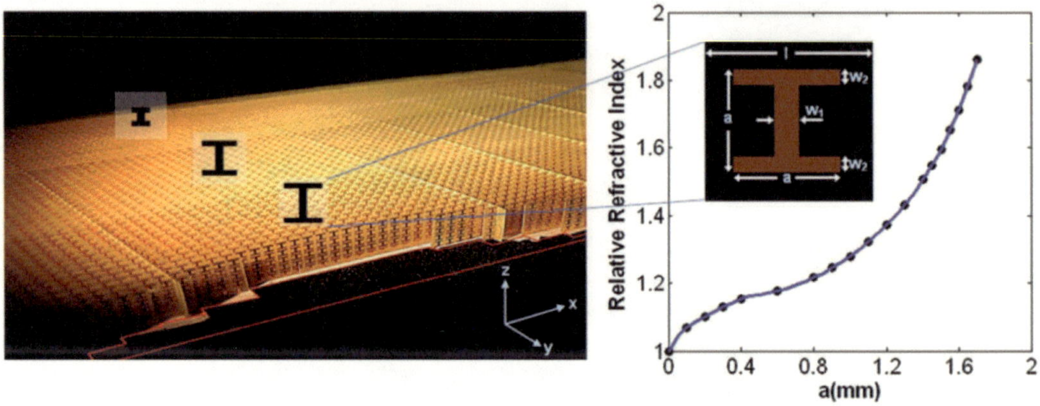

(b)

Fig. 26. Principle of the carpet cloak: in I the ground reflects an incident beam without obstacle, in II the beam is reflected in the presence of an obstacle, in III the reflected beam consists of parallel rays reconstructed by the cloak covering the obstacle. (b) View of the realized carpet cloak and the metamaterial unit cell with variable permittivity used in the carpet.

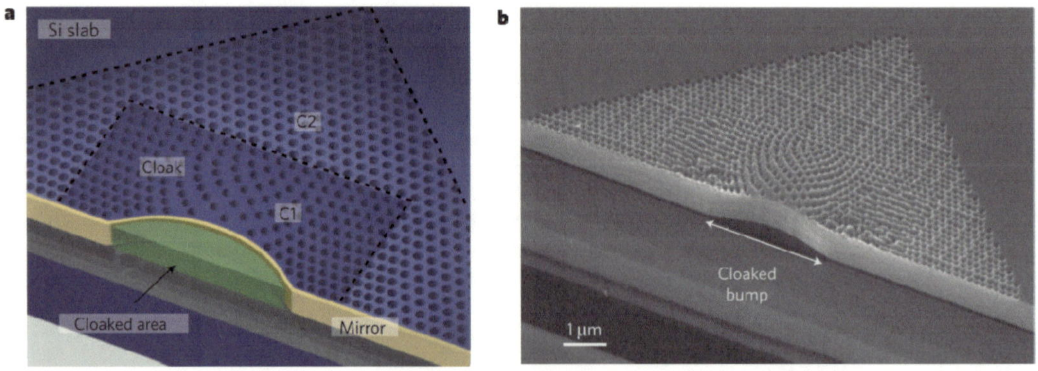

Fig. 27. Principle (a) and realization (b) of an optical carpet cloak on silicon.

Figure 28 shows the carpet cloak operating at a wavelength of 1,540 nm, for an incident Gaussian beam reflected from a curved reflecting surface. Similar reflection characteristics can be observed when compared to a reflection on a flat surface.

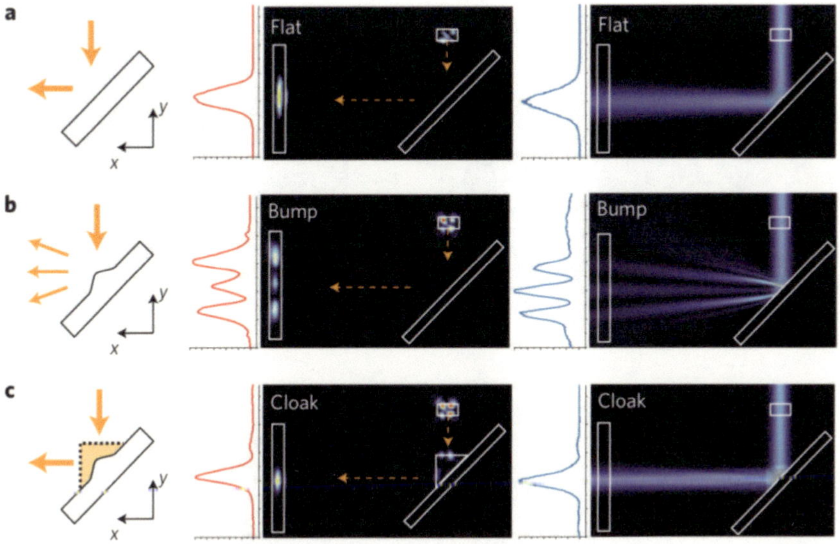

Fig. 28. Optical carpet cloaking at a wavelength of 1,540 nm: The results for a Gaussian beam reflected from a flat surface (a), a curved (without a cloak) surface (b) and the same curved reflecting surface with a cloak (c).

4.4 Electromagnetic wormhole and other cosmological objects

One of the most amazing applications has been proposed by A. Greenleaf [Greenleaf, 2007]. He imagined to create a wormhole using electromagnetic invisibility cloak able to link two remote areas of space and ensures the propagation of an electromagnetic wave between both regions invisible from the outside. Figure 29 shows a schematic illustration of the wormhole where its exterior deflects the incident electromagnetic waves and a section of the wormhole showing a wave propagating inside.

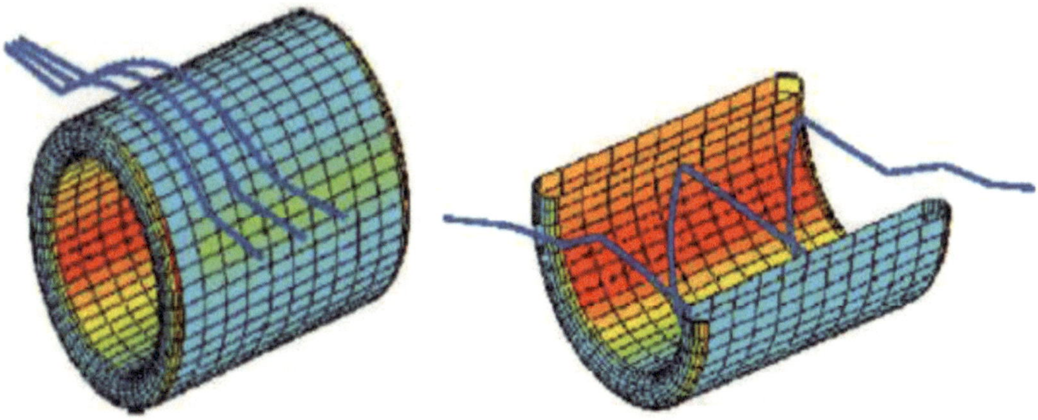

Fig. 29. (a) A schematic illustration of the wormhole whose exterior cloak deflects the incident electromagnetic waves. (b) a section of the wormhole showing a wave propagating inside.

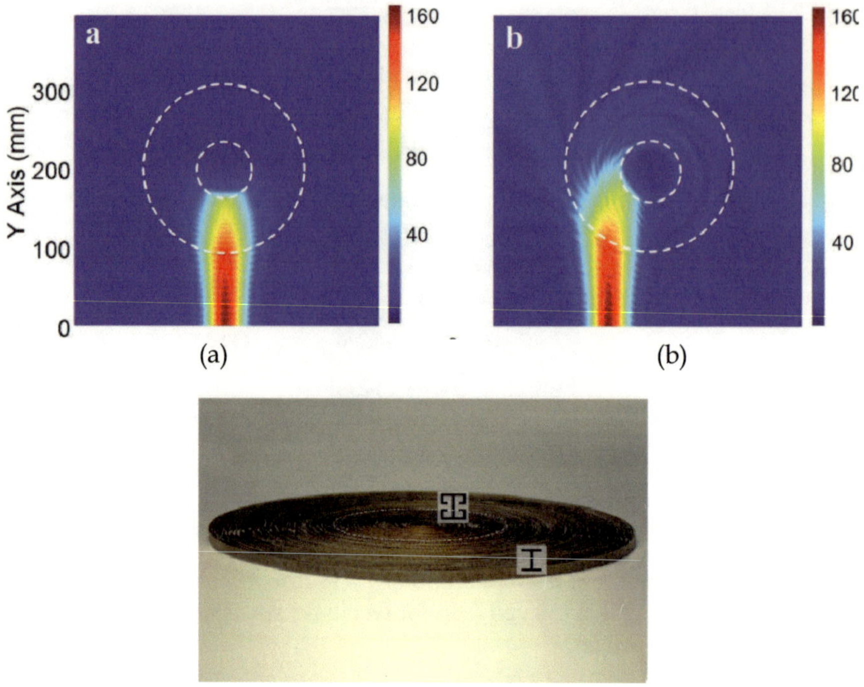

(c)

Fig. 30. (a) Distributions of electric fields |Ez| for the designed black hole at the frequency of 18 GHz: The full-wave simulation result under the on-center incidence of a Gaussian beam. (b) The full-wave simulation result under the off-center incidence of a Gaussian beam. (c) Photograph of the fabricated artificial black hole based on metamaterials, which is composed of 60 concentric layers, with ELC structures in the core layers and I-shaped structures in the shell layers.

5. Conclusion and outlooks

The potential applications of the space coordinate transformation seem to be very various. The examples presented in this chapter show their usefulness, even if they are still far from industrial achievements. Also it appears that these applications can be transposed to any frequency. The conventional metamaterials used in the microwave region are metal-dielectric structures. However metals have different present high losses at infrared and optical frequencies. Therefore the applicability will differ greatly between the microwave domain on one hand, and optical frequencies on the other. In the optical domain, the problem is mainly the achievement of materials with metallic patterns having sizes of about one-tenth of the wavelength (a few hundred nanometers) and the control of their geometry [Soukoulis, 2011]. The other problem is the losses of the metallic metamaterials at optical wavelengths. Innovative approaches have recently been proposed to solve partially this problem. But the problem is not completely solved.

In the microwave field, the achievements seem easier as they involve usually inexpensive materials, and metallic parts of the metamaterials present low losses at these frequencies. But they depend strongly on the complexity of the electromagnetic parameters to achieve. In the rare realizations proposed in literature, these parameters have been simplified, and often the impedance matching has been sacrificed to obtain a feasible material. The metamaterial based design encountered in this case the problem of reflection losses, and are already comparable to existing solutions that have proved their performances, for example in the field of antennas. Another difficulty is the narrow bandwidth of the metamaterials used, in particular those based on resonant structures of the type of split ring resonators of Pendry. In reality this is not really a problem, because broadband metamaterials can be realized from composite metamaterials [Djermoun, 2007], or by using all dielectric structures.

6. References

Tannery P., Henry C. et de Waard C., *Œuvres de Pierre de Fermat*, Gauthier-Villars et Cie, Paris, 1891-1922.

Pendry, J. B., Schurig, D., Smith, D. R., (2006), Controlling Electromagnetic Fields, *Science,* Vol. 312, pp. 1780-1782.

Leonhardt, U., (2006), Optical Conformal Mapping, *Science,* Vol.312, pp.1777-1780.

Cai W., Chettiar U. K., Kildishev A. V., and Shalaev V. M.. (2008), Designs for optical cloaking with high-order transformations, *Optics Express,* Vol. 16, No. 8, pp.5444-5452.

Schurig, D., J. J. Mock, B. J. Justice, S. A. Cummer, J. B. Pendry, A. F. Starr, and D. R. Smith, (2006), Metamaterial Electromagnetic Cloak at Microwave Frequencies, *Science,* Vol.314, pp. 977-980

Soukoulis C.M., Wegener M., (2011), Past achievements and future challenges in the development of three-dimensional photonic metamaterials, *Nature Photonics,* Vol.5, pp.523-530

Guven K., Saenz E., Gonzalo R., Ozbay E. and Tretyakov S., (2008), Electromagnetic cloaking with canonical spiral inclusions, *New Journal of Physics,* Vol.10, 115037

Kanté B., de Lustrac A., Lourtioz J.-M., Burokur S. N., (2008), Infrared cloaking based on the electric response of split ring resonators, *Optics Express,* Vol.16, N° 12, pp.9191-9198,

Kanté B., Germain D., and de Lustrac A., (2009), Experimental demonstration of a nonmagnetic metamaterial cloak at microwave frequencies » *Physical Review B,* Vol. 80, 201104(R).

Nicolet A., Zolla F., Guenneau S., (2008), Electromagnetic analysis of cylindrical cloaks of an arbitrary cross section, *Optics Letters,* Vol.33, No. 14, pp.1584-1586.

Rahm M., Schurig D., Roberts D. A., Cummer S.A., Smith D. R., Pendry J. B., (2008), Design of electromagnetic cloaks and concentrators using form-invariant coordinate transformations of Maxwell's equations, *Photonics and Nanostructures. Fundamental and Applications,* Vol.6, 87.

Tichit P-H., Kante B., de Lustrac A., (2008) Design of polygonal or elliptical invisibility cloak, Nato Workshop Meta 08, Marrakesh, may 07-10 2008.

Leonhardt U., Tyc T., (2009), Broadband Invisibility by Non-Euclidean Cloaking, *Science,* Vol.323, pp.110-112;

Kildishev A. V., Cai W., Chettiar U. K., Shalaev V.M., (2008), Transformation optics: approaching broadband electromagnetic cloaking, *New Journal of Physics,* Vol.10, 115029.

Qiu C.-W., Hu L., Xu X., Feng Y., (2009), Spherical cloaking with homogeneous isotropic multilayered structures, *Physical Review E,* Vol.79, 047602.

Feng Y. J., Xu X. F., Yu Z. Z., (2011), Practical realization of transformation-optics designed invisibility cloak through layered structures, *Antennas and Propagation (EUCAP), Proceedings of the 5th European Conference,* pp. 3456 – 3460, ISBN: 978-1-4577-0250-1, Rome, 11-15 April 2011.

Torrent D. and Sanchez-Dehesa J., (2009), Acoustic metamaterials for new two-dimensional sonic devices, *New Journal of Physics*, Vol.9, 323 (2007).

Chen H., Chan C. T., (2007), Acoustic cloaking in three dimensions using acoustic metamaterials, *Applied Physics Letters,* Vol.91, 183518.

Farhat M. Enoch S., Guenneau S., Movchan A. B., (2008), Broadband cylindrical acoustic cloak for linear surface waves in a fluid, *Physical Review Letters* Vol.101, 134501

Cummer S. A., Popa B. I., Schurig D., Smith D. R., Pendry J., Rahm M., Starr A., (2008), Scattering Theory Derivation of a 3D Acoustic Cloaking Shell, *Physical Review Letters*, Vol.100, 024301.

Kong F., Wu B., Kong J. A., Huangfu J. H., and Xi S., (2007), Planar focusing antenna design by using coordinate transformation technology, *Applied Physics Letters,* Vol.91, 253509.

Tichit P.-H., Burokur S. N., de Lustrac A., (2009), Ultra-directive antenna via transformation optics, *Journal of Applied Physics*, Vol.105, 104912

Tichit P.-H., Burokur S. N., de Lustrac A., (2011), Design and experimental demonstration of a high-directive emission with transformation optics, *Physical Review B,* Vol.83, 155108.

Rui Y., Wenxuan T., Hao Y., (2011), Broadband Dielectric Zone Plate Antenna from Transformation Electromagnetics, *Optics Express,* Vol.19, n°13, pp.12348-12356.

Cui T. J., Zhou X. Y., Ma H. F., (2011) Three-Dimensional Metamaterial Lens Antennas, *Antennas and Propagation (EUCAP), Proceedings of the 5th European Conference,* pp. 3301 - 3303 , ISBN: 978-1-4577-0250-1, Rome, 11-15 April 2011.

Lin L., Wang W., Cui J., Du C., Luo X., (2008), Design of electromagnetic refractor and phase transformer using coordinate transformation theory, *Optics Express,* Vol.16, n°10, pp.6815-6821.

Huangfu J., Xi S., Kong F., Zhang J., Chen H., Wang D., Wu B.J., Ran L., Kong J.A., (2008) Application of coordinate transformation in bent waveguides, *Journal of Applied Physics* Vol.104, 014502.

Tichit P.H., Burokur S.N., de Lustrac A., (2010), Waveguide taper engineering using coordinate transformation technology, *Optics Express,* Vol.18, n°2, pp. 767-772.

Ghasemi R., Tichit P.H., Degiron A., Lupu A. de Lustrac A., (2010), Efficient control of a 3D optical mode using a thin sheet of transformation optical medium, *Optics Express,* Vol.18, n 19, pp. 20305-20312.

Liu R., Ji C., Mock J.J., Chin J.Y., Cui T.J., Smith D.R., (2009), Broadband Ground-Plane Cloak, *Science,* Vol.323, pp.366-369.

Xu X., Feng Y., Hao Y., Zhao J., Jiang T., (2009), Infrared carpet cloak designed with uniform silicon grating structure, *Applied Physics Letters,* Vol.95, pp 184102.

Valentine J., Li J., Zentgraf T., Bartal G., Zhang X., (2009), An optical cloak made of dielectrics, *Nature Materials,* Vol.8, pp.568-571.

Gabrielli L. H., Cardenas J., Poitras C.B., Lipson M., (2009), Silicon nanostructure cloak operating at optical frequencies, *Nature Photonics* Vol.3, pp.461-463.

Greenleaf A., Kurylev Y., Lassas M., and Uhlmann G., (2007), Electromagnetic Wormholes and Virtual Magnetic Monopoles from Metamaterials, *Physical Review Letters,* Vol.99, 183901.

Cheng, Q. & Cui, T. J. (2009), An omnidirectional electromagnetic absorber made of metamaterials, *New Journal of Physics,* Vol.12, 063006

Djermoun A., de Lustrac A., Lourtioz J.M., (2007), A wide band left handed material with high transmission, *Photonics and Nanostructures - Fundamentals and Applications,* Vol.5, n1, pp. 21-28.

6

Effective Medium Theories and Symmetry Properties of Elastic Metamaterials

Ying Wu[1], Yun Lai[2] and Zhao-Qing Zhang[3]

[1]*Division of Mathematical and Computer Sciences and Engineering, King Abdullah University of Science and Technology*

[2]*Department of Physics, Soochow University*

[3]*Department of Physics and William Mong Institute of Nano Science and Technology Hong Kong University of Science and Technology*

[1]*Saudi Arabia*

[2,3]*China*

1. Introduction

Recently, metamaterials have attracted a great deal of attention due to their unusual properties not seen in naturally occurring materials, such as negative refraction (Lezec et al., 2007; Pendry, 2000; Veselago, 1968), superlensing (Grbic & Eleftheriades, 2004) and cloaking (Leonhardt 2006; Pendry et al., 2006), etc. These unusual properties are derived from the resonant structures in their artificial building blocks. The resonant structures interact with the wave, but their small size prevents them from being "seen" individually by the wave with wavelength inside the background much larger than the size of the structures (Pendry & Smith, 2006). Thus, the properties of a metamaterial can be described with homogenized parameters or effective medium parameters. The theory that links the microscopic resonant structures to their effective medium parameters is called the effective medium theory (EMT). For example, the left-handed metamaterial consisting of a periodic array of split ring resonators and conducting thin wires has been successfully demonstrated that, in a frequency regime, it behaves like a homogeneous medium exhibiting negative effective permittivity, ε_{eff}, and negative effective permeability, μ_{eff}, simultaneously (Pendry et al., 1999, 1996). Since the refractive index is defined as $n_{eff}=\sqrt{\varepsilon_{eff}}\sqrt{\mu_{eff}}$, negative refraction is a consequence of double negativity in permittivity and permeability. This example shows how effective medium parameters of particular resonant structures can be used to describe unusual properties of a metamaterial. In turn, a valid and accurate EMT provides an efficient and systematic tool to design and engineer the resonant structures according to certain desired metamaterial properties. During the development of metamaterials, there has always been a continuous effort to find an appropriate EMT for metamaterials such that various novel phenomena can have a theoretical explanation. One can obtain the effective parameters from some phenomenological results, such as transmission and reflections (Baker-Jarvis et al., 1990; Smith et al., 2002), and wave propagation (Andryieuski et al., 2009), but only a theory can give a clear understanding on the physical origin of effecitve parameters. Therefore, EMTs with physical insights can be regarded as theoretical foundations of metamaterials.

The development of EMT has seen a long history accompanied by various approaches. A famous one for electromagnetic (EM) waves is the Maxwell-Garnett theory (Sheng, 2006), which is valid in the quasi-static limit, aka the zero frequency limit. The quasi-static limit requires the wavelengths inside the scatterer, the host, and the effective medium to all be very large compared to the size of the building block (Lamb et al., 1980). However, for metamaterials, the wavelength inside the scatterer could be smaller than the size of the building block and thus lead to resonances at low frequencies. This results in the failure of the widely used quasi-static EMT. Nevertheless, as long as the wavelength in the effective medium is still large compared to the size of the building block, there exists an effective medium description as the wave still cannot probe the fine structures of the building blocks. In this context, another limit is introduced, which is the long wavelength limit. Compared with the quasi-static limit, the long wavelength limit does not have restrictions on the wavelength inside the scatterer, while the wavelengths inside the host and the effective medium should still be large (Lamb et al., 1980). In the study of metamaterials, one aspect is to develop EMTs that are valid in the long wavelength limit. In this chapter, we will focus on the recent developments of EMTs for elastic metamaterials.

1.1 Effective medium theories for electromagnetic metamaterials

Though our focus is on elastic metamaterials, it is necessary to briefly review the EMTs for EM metamaterials to offer a systematic picture of the EMTs. Ever since the birth of EM metamaterials, EMTs have played an integral role in designing metamaterials and explaining their unusual properties. The EMTs for EM metamaterials can be broken down into one of several classes. In one class, the effective parameters are obtained from the average of the computed eigenfields in the unit cell (Chern & Chen, 2009; Chui & L. Hu, 2002; Pendry et al., 1999; Smith & Pendry, 2006). This method gives inherently nonlocal parameters, i.e. parameters that depend on not only frequency but also the Bloch wave vector. For metamaterials with a good effective medium approximation, the nonlocality may be ignored. This method is especially helpful for use with metamaterials with complicated unit structures, such as split rings. Another class of EMTs is called the coherent potential approximation (CPA) method. In this method the effective medium is taken as the background embedded with the scatterer in the unit cell and by implying zero scattering, some elegent formulas of the effective parameters have been obtained (X. Hu et al., 2006; Jin et al., 2009; Wu et al., 2006). This method currently only works for scatterers with isotropic geometry, but it is very accurate in the long wavelength limit even at relatively high frequencies. Interestingly, the obtained effective parameters do not have any imaginary parts if the system does not have any absorption. Similar formulas can also be obtained from the multiple-scattering theory (MST) (Chui & Lin, 2008; Wu & Z. Zhang, 2009). The MST, which will be introduced in Section 2, is capable of producing the dispersion relations of a periodic structure. From dispersion relations, the effective wave speed can be easily calculated while the impedance still remains unknown. Recently, other methods have appeared, such as the quasimode method (Sun et al., 2009) and the first-principles method (Andrea, 2011).

1.2 Effective medium theories for acoustic metamaterials

While the field of EM metamaterial has developed rapidly during the past decade, one of its counterparts, denoted as acoustic metamaterial has also seen fast growth (Ding et al., 2007; Fang et al., 2006; Lee et al., 2010; Yang et al., 2008). The acoustic metamaterial is designed to manipulate acoustic waves (Chen et al., 2010; Lu et al., 2009). Analogous to EM metamaterials, there are two material parameters that describe the wave propagation, which are bulk modulus, κ, and mass density, ρ. In two dimensions (2D), if the constituents of the metamaterials are all fluids, the governing equation can be mapped into 2D EM equations so that the EMT for 2D acoustic metamaterials is the same as that for 2D EM metamaterials. If the scatterers in acoustic metamaterials are solid, the shear modulus of the scatterers can be ignored when the longitudinal velocity contrast between the scatterer and the host is high (Kafesaki & Economou, 1999). In this case, the scattering property is basically the same as the EM cases. The mapping from the EM waves to acoustic waves facilitates the development of EMTs which have also been extensively studied by using various types of methods. MST (Mei et al., 2006; Torrent et al., 2006) and the CPA (Kafesaki et al., 2000; Li & Chan, 2004) represent two classes of them. Exciting news also came from the experimental realizations of acoustic metamaterials, such as acoustic negative refraction (S. Zhang et al., 2009) and acoustic cloaking(S. Zhang et al., 2011).

1.3 Effective medium theories for elastic metamaterials

The term *elastic* metamaterial refers to those metamaterials which are able to sustain not only longitudinal but also shear waves in their effective media. It is well-known that the EMT for an elastic composite in the quasi-static limit is anisotropic in general. The only exception is the hexagonal lattice in two dimensions (Landau & Lifshitz, 1986; Royer & Dieulesaint, 1999; Wu & Z. Zhang 2009). Even for this case, the EMT involves the determination of three effective parameters, i.e., mass density, ρ, bulk modulus, κ, and shear modulus, μ. This is in contrast to the cases of previously mentioned EM and acoustic EMTs which involve only two effective parameters. One more effective medium parameter greatly enriches the types and physics of wave propagation (Chen et al., 2008; Wu et al., 2007), such as mode conversion between longitudinal and transverse waves (Wu et al., 2011); however, it also adds complexity to the EMT. Recently, various EMTs for elastic metamaterials have been proposed, such as those ones based on the plane-wave-expansion method (Krokhin et al., 2003), and integration of fields (Zhou & Hu, 2009).

In this chapter, the symmetry property of an elastic metamaterial is examined based on the MST. It is shown that the elastic metamaterial preserves the quasi-static symmetry properties. For isotropic elastic metamaterials, CPA provides a simple and accurate EMT in the long wavelength limit, which links the scattering properties of the scatterer and the effective medium parameters (Wu et al, 2007). Those formulae can also be derived by using the MST method (Wu & Z. Zhang, 2009). For anisotropic elastic metamaterial, the EMT involves three or more effective elastic moduli (Landau & Lifshitz, 1986). In this case, CPA fails. To tackle this problem, a method based on MST in conjunction with the Christoffel's equation (Royer & Dieulesaint, 1999) has been proposed and the expressions for effective elastic moduli have been obtained (Wu & Z. Zhang, 2009). The combinations of anisotropy and negativities in various effective moduli can give rise to many types of novel wave propagation behaviors that are unseen in normal solids (Lai et al., 2011).

Within the scope of this chapter, all the EMTs mentioned above are limited to linear elastodynamics and do not consider the micro-structure introduced local rotation (Milton & Willis, 2007).

2. Scattering properies of elastic metamaterials

In order to illustrate EMTs and symmetry properties of elastic metamaterials, we start with a simple case where the resonant scatterer in a building block is homogeneous. More complicated scatterers will be discussed in Section 5. The elastic metamaterial considered here is composed of cylindrical inclusions of radius r_s with mass density ρ_s, shear modulus μ_s, and bulk modulus κ_s, embedded in an isotropic matrix, whose material parameters are denoted by $(\rho_0, \mu_0, \kappa_0)$. In two dimensions, the bulk modulus, κ, is related to the shear modulus through the relation $\kappa = \lambda + \mu$, where λ represents the Lamé constant (Royer & Dieulesaint, 1999). Due to the translational symmetry along the cylinder's axis, denoted as *z*-axis, the elastic modes in the system can be decoupled into a scalar part, which is also called shear horizontal mode with vibrations along the *z*-axis, and a vector part, i.e.,*xy*-mode with vibrations in the *x*-*y* plane. *xy*-mode is a mixed polarization of quasi-longitudinal and quasi-shear vertical modes. Since the shear horizontal mode satisfies the scalar wave equation with the same mathematical structure as those for acoustic (Krokhin et al., 2003) and 2D EM cases, this part is skipped. Rather, the focus is on the more complicated case of the *xy*-mode whose wave equation is given by:

$$\rho(\vec{r})\frac{\partial^2 u_i(\vec{r})}{\partial t^2} = \nabla\cdot\left(\mu(\vec{r})\nabla u_i(\vec{r})\right) + \nabla\cdot\left(\mu(\vec{r})\frac{\partial \vec{u}(\vec{r})}{\partial x_i}\right) + \frac{\partial}{\partial x_i}\left[\lambda(\vec{r})\nabla\cdot\vec{u}(\vec{r})\right], \tag{1}$$

where $\vec{u}$ is the displacement field. In general, $\vec{u}$ can be decoupled into a longitudinal part and a transverse part, i.e. $\vec{u} = \nabla\phi_l + \nabla\times(\phi_t\hat{e}_z)$, where ϕ_l and ϕ_t are the longitudinal and transverse gauge potentials, respectively.

2.1 Single-scattering, the scattering coefficients

If there is only one scatterer, the solutions to ϕ_l and ϕ_t can be expanded by using Bessel and Hankel functions. The wave incident on a single scatterer, p, is:

$$\vec{u}_p^{inc}(\vec{r}_p) = \sum_m\left(a_{lm}^p\nabla\left[J_m(k_{l0}r_p)e^{im\theta_p}\right] + a_{tm}^p\nabla\times\left[\hat{z}J_m(k_{t0}r_p)e^{im\theta_p}\right]\right), \tag{2}$$

and the wave scattered by the same scatterer is

$$\vec{u}_p^{sca}(\vec{r}_p) = \sum_m\left(b_{lm}^p\nabla\left[H_m^{(1)}(k_{l0}r_p)e^{im\theta_p}\right] + b_{tm}^p\nabla\times\left[\hat{z}H_m^{(1)}(k_{t0}r_p)e^{im\theta_p}\right]\right), \tag{3}$$

where $J_m(x)$ and $H_m^{(1)}(x)$ are Bessel functions and Hankel functions of the first kind, respectively.

$k_{l0} = \omega\sqrt{\rho_0/(\kappa_0+\mu_0)}$ and $k_{t0} = \omega\sqrt{\rho_0/\mu_0}$ represent the longitudinal and transverse wave vectors in the matrix, respectively. ω is the angular frequency. $\vec{r}_p = (r_p, \theta_p)$ are the polar coordinates originating at the center of the scatterer. The longitudinal and transverse waves in the matrix are coupled by the scatterings of the scatterers, inside which the displacement is given by:

$$\vec{u}_p(\vec{r}_p) = \sum_m \left(c_{lm}^p \nabla\left[J_m(k_{ls}r_p)e^{im\theta_p} \right] + c_{tm}^p \nabla\times\left[\hat{z}J_m(k_{ts}r_p)e^{im\theta_p} \right] \right), \tag{4}$$

where $k_{ls} = \omega\sqrt{\rho_s/(\kappa_s+\mu_s)}$ and $k_{ts} = \omega\sqrt{\rho_s/\mu_s}$ are the longitudinal and transverse wave vectors inside the scatterer, respectively. The coefficients of those Bessel and Hankel functions can be determined by considering the elastic boundary conditions which are the continuities of the radial and tangential component of the displacement field, i.e., u_r and u_θ, and the continuities of the stresses, σ_{rr} and $\sigma_{r\theta}$, at the interface. These continuities on the surface of a cylinder relate b_{lm}^p and b_{tm}^p to a_{lm}^p and a_{tm}^p through:

$$b_{\alpha m}^p = \sum_{\beta=l,t} \sum_{m'} t_{\alpha\beta mm'} a_{\beta m'}^p \quad (\alpha = l,t), \tag{5}$$

where $t_{\alpha\beta mm'} = D_m^{\alpha\beta}\delta_{mm'}$. $D_m^{\alpha\beta}$ are elastic Mie-like scattering coefficients for isotropic scatterers and are functions of k_{ls}, k_{ts}, k_{l0}, k_{t0} and r_s. The explicit expressions for $D_m^{\alpha\beta}$ can be found in (Wu et al., 2007).

2.2 Periodic structures and multiple-scattering

For a collection of scatterers, the MST takes full account of the multiple scatterings between any two scatterers (Liu et al., 2000a; Mei et al., 2003). The wave incident on the scatterer *p* is contributed by two parts: one is the external incident waves from outside the system, and the other part is the scattered waves coming from all the other scatterers inside the system. Thus, the total incident waves on the scatterer *p* are expressed as:

$$\begin{aligned}\vec{u}_p^{inc}(\vec{r}_p) = &\sum_m \left(a_{lm}^{p0} \nabla\left[J_m(k_{l0}r_p)e^{im\theta_p} \right] + a_{tm}^{p0} \nabla\times\left[\hat{z}J_m(k_{t0}r_p)e^{im\theta_p} \right] \right) \\ &+ \sum_{q\neq p}\sum_{m''} \left(b_{lm''}^q \nabla\left[H_{m''}^{(1)}(k_{l0}r_q)e^{im''\theta_q} \right] + b_{tm''}^q \nabla\times\left[\hat{z}H_{m''}^{(1)}(k_{t0}r_q)e^{im''\theta_q} \right] \right),\end{aligned} \tag{6}$$

where (r_q, θ_q) denote $\vec{r}_q$ in the polar coordinates originating at the center of scatterer *q*. Here $\vec{r}_q$ and $\vec{r}_p$ refer to the same spatial point measured from the positions of scatterers *q* and *p*, respectively. For simplicity, the center of scatterer *p* is chosen as the origin and the position of scatterer *q* is denoted by $\vec{\mathrm{R}}_{qp} = (\mathrm{R}_{qp}, \Theta_{qp})$. Thus, $\vec{r}_q = \vec{r}_p - \vec{R}_{qp}$. The relation between $\vec{r}_p$, $\vec{r}_q$ and $\vec{R}_{qp}$ is depicted in Fig. 1.

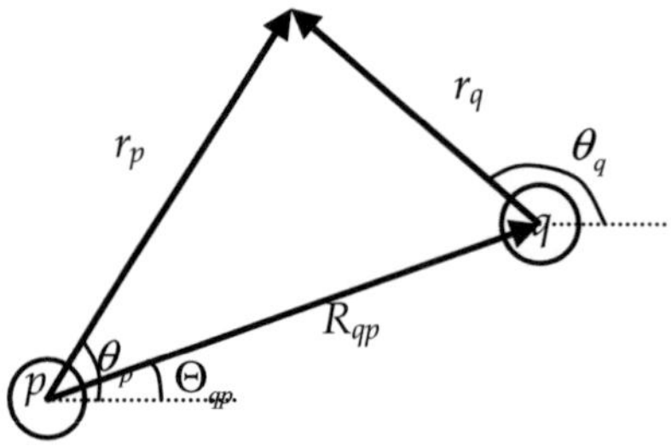

Fig. 1. The spacial relation of $\vec{r}_p$, $\vec{r}_q$ and $\vec{R}_{qp}$.

With the help of Graf's addition theorem (Abramowitz & Stegun, 1972), the Hankel functions depicting the scattered wave coming from the scatterer q can be changed into Bessel functions describing the wave incident on the scatterer p, which is:

$$H_{m''}^{(1)}(k_{\alpha 0} r_q) e^{im''\theta_q} = \sum_m g_{mm''}^{\alpha} J_m(k_{\alpha 0} r_p) e^{im\theta_p} \quad (\alpha = l,t)\,, \tag{7}$$

where

$$g_{mm''}^{\alpha} = H_{m-m''}^{(1)}(k_{\alpha 0} R_{qp}) e^{i(m''-m)\Theta_{qp}} \quad (\alpha = l,t) \tag{8}$$

Substituting Eqs. (7) and (8) into (6), we obtain:

$$\begin{aligned} \vec{u}_p^{inc}(\vec{r}_p) = & \sum_m \left(a_{lm}^{p0} + \sum_{q \neq p} \sum_{m''} b_{lm''}^{q} g_{mm''}^{l} \right) \nabla \left[J_m(k_{l0} r_p) e^{im\theta_p} \right] \\ & + \sum_m \left(a_{tm}^{p0} + \sum_{q \neq p} \sum_{m''} b_{tm''}^{q} g_{mm''}^{t} \right) \nabla \times \left[\hat{z} J_m(k_{t0} r_p) e^{im\theta_p} \right] \end{aligned} \tag{9}$$

Eq. (2) together with Eqs. (5) and (9) leads to the following self-consistent equation:

$$b_{\alpha m}^{p} = \sum_{\beta = l,t} \sum_{m'} t_{\alpha\beta mm'} \left(a_{\beta m'}^{p0} + \sum_{q \neq p} \sum_{m''} g_{m'm''}^{\beta} b_{\beta m''}^{q} \right) \tag{10}$$

This set of self-consistent equations can be written into the standard form of linear equations $Ax = B$, and the multiple-scattering problem is numerically solved.

If the scatterers are arranged in a periodic array, the Bloch theorem, $b_{\alpha m''}^{q} = b_{\alpha m''}^{p} e^{i\vec{K}\cdot\vec{R}_{qp}}$ $(\alpha = l,t)$, can be applied, where $\vec{K}$ is the Bloch wave vector. For an eigenvalue problem, the external incident field is dropped off so that Eq. (10) is converted into:

$$\sum_{\beta=l,t}\sum_{m}\left(\sum_{m'} t_{\alpha\beta mm'}\sum_{q\neq p} g^{\beta}_{m'm} e^{i\vec{K}\cdot\vec{R}_q} - \delta_{mm'}\right) b^{p}_{\beta m} = 0 \qquad (11)$$

Eq. (11) has nontrivial solutions if and only if,

$$\det\left|\sum_{m'} t_{\alpha\beta mm'} S(\beta, m'-m) - \delta_{mm'}\right| = 0\,, \qquad (12)$$

where

$$S(\beta, m'-m) = \sum_{q\neq p} g^{\beta}_{m'm} e^{i\vec{K}\cdot\vec{R}_q}$$

represents lattice sums. The solution of Eq. (12) offers the dependence of frequency, ω, on the Block wave vector, $\vec{K}$, which is known as the dispersion relation. To solve Eq. (12) for dispersion relations, one needs to evaluate the lattice sums first, which can be accomplished through several techniques (Chin et al., 1994; Mei et al., 2003). The expression of $S(\beta, m'-m)$ is given by (Wu and Z. Zhang, 2009):

$$S(\beta,n) = \frac{4i^{n+1}k_{\beta 0}}{\Omega J_{n+1}(k_{\beta 0}a)}\sum_{h}\frac{J_{n+1}(Q_h a)}{Q_h(k_{\beta 0}^2 - Q_h^2)}e^{in\varphi_h} - \left(H_1^{(1)}(k_{\beta 0}a) + \frac{2i}{\pi k_{\beta 0}a}\right)\delta_{n,0}\,,\ (n\geq 0) \qquad (13)$$

$$S(\beta,-n) = -S^{*}(\beta,n)$$

where a denotes the lattice constant, Ω is the volume of the unit cell and (Q_h, φ_h) stands for the vector $\vec{Q}_h = \vec{K} + \vec{K}_h$ in polar coordinates, and $\vec{K}_h$ is the reciprocal lattice vector.

3. Symmetry properties of elastic metamaterials

Since the MST is capable of producing accurate dispersion relations for all frequencies, it is a good approach in the study of the symmetry properties of an elastic metamaterial. The elastic metamaterial is isotropic if its dispersion relation, $\omega(\vec{K})$,does not depend on the direction of $\vec{K}$. In the framework of MST, the dispersion relation can be calculated numerically by solving Eq. (12). In the long wavelength limit, both Ka and $k_0 a$ are all much smaller than unity and appropriate approximations can be made in the Bessel and Hankel functions in the lattice sum $S(\beta, m'-m)$ and the scattering matrix $t_{\alpha\beta mm'}$. This simplifies the secular equation and benefits the derivation of analytic expression of EMT.

Taking $J_n(x) \cong x^n/(2^n n!)$, $H_0^{(1)}(x) \cong 2\ln x/\pi$ and $H_n^{(1)}(x) \cong -2^n(n-1)!/(\pi x^n)$ for $x = k_{l0}r_s$, $k_{t0}r_s$, $k_{l0}a$, $k_{t0}a$ and Ka in Eq. (12), it is easy to find that the leading terms are those with $|m| \leq 2$, where m corresponds to the order of Bessel and Hankel functions and is called the angular quantum number. Thus, we only need to consider the terms with $0 \leq n \leq 4$ in Eq. (13), in which the summation

$$\sum_{h} J_{n+1}(Q_h a)e^{in\varphi_h}\Big/\left[Q_h(k_{\beta 0}^2 - Q_h^2)\right]$$

can be separated into two terms: the first is $\vec{K}_h = 0$ or $\vec{Q}_h \cong \vec{K}$, the second is the sum of all other terms with $\vec{K}_h \neq 0$ or $\vec{Q}_h \cong \vec{K}_h$. Then, the lattice sum takes the following expression:

$$S(\beta,n) \cong \frac{4i^{n+1}}{\Omega}\left(\frac{K^n}{k_{\beta 0}^n(k_{\beta 0}^2 - K^2)}e^{-in\phi_K} + \frac{k_{\beta 0}2^{n+1}(n+1)!}{\left(k_{\beta 0}a\right)^{n+1}}\sum_{h(K_h \neq 0)}\frac{J_{n+1}(K_h a)}{K_h(k_{\beta 0}^2 - K_h^2)}e^{-in\phi_{Kh}}\right), (0 \leq n \leq 4)$$
$$S(\beta,-n) = -S^*(\beta,n), \tag{14}$$

where (K_h, ϕ_{Kh}) denotes the polar coordinates of $\vec{K}_h$, the reciprocal lattice vector of the lattice. The summation of all nonzero reciprocal lattice vectors in the second term in the bracket reveals the dependence of lattice sum on the lattice structure, which influences the symmetry properties of the dispersion relations.

3.1 Isotropic dispersions

For a 2D hexagonal lattice with a lattice constant a, the reciprocal lattice follows:

$$\vec{K}_h = \frac{4\pi}{\sqrt{3}a}\left(h_i\hat{i} + h_j\left(\frac{1}{2}\hat{i} + \frac{\sqrt{3}}{2}\hat{j}\right)\right);\ h_i, h_j \in Z \tag{15}$$

Here, $\hat{i}$ and $\hat{j}$ represent the unit vectors along the x- and y-axes in the reciprocal space. When $n \neq 0$, due to the symmetry of a hexagonal lattice, the summation in the second term of Eq. (14) is zero, which can be proved in the following way. For an arbitrarily chosen reciprocal lattice vector, (K_{h1}, ϕ_{Kh1}), there always exist five other reciprocal lattice vectors at $(K_{h1}, \phi_{Kh1} + N\pi/3)$, $N = 1,2,3,4,5$ such that $\sum_{N=0}^{5} e^{-inN\pi/3} = 0$. Thus, the summation in Eq. (14) vanishes after summing over all the non-zero $\vec{K}_h$ and only the first term of Eq. (14) survives. When $n = 0$, the second term in Eq. (14) no longer sums to zero as $e^{-in\phi_{Kh}} = 1$. However, this term can be ignored in the long wavelength limit because compared to the first term in Eq. (14) which is on the order of ω^{-2}, it is on the order of ω^0. Thus, Eq. (14) is further reduced to:

$$S(\beta,n) \cong \frac{8i^{n+1}K^n}{\sqrt{3}a^2 k_{\beta 0}^n\left(k_{\beta 0}^2 - K^2\right)}e^{-in\phi_K},\ 0 \leq n \leq 4 \tag{16}$$

Substituting Eq. (16) into Eq.(12), we find the following two roots:

$$\left(K_1^{tri}\right)^2 = F_1\left(\tilde{D}_1^{II}\right)F_2\left(\tilde{D}_2^{II}\right) \text{ and } \left(K_2^{tri}\right)^2 = F_1\left(\tilde{D}_1^{II}\right)F_3\left(\tilde{D}_2^{II}, \tilde{D}_0^{II}\right), \tag{17}$$

with

$$F_1\left(\tilde{D}_1^{ll}\right) = -\frac{16i\tilde{D}_1^{ll}(\kappa_0 + \mu_0) - \sqrt{3}a^2\omega^2\rho_0}{\sqrt{3}a^2},$$

$$F_2\left(\tilde{D}_2^{ll}\right)=-\frac{4\tilde{D}_2^{ll}\left(\kappa_0+\mu_0\right)\left(\kappa_0+2\mu_0\right)+i\frac{\sqrt{3}}{2}a^2\omega^2\mu_0\rho_0}{\mu_0\left(4\tilde{D}_2^{ll}\kappa_0\left(\kappa_0+\mu_0\right)-i\frac{\sqrt{3}}{2}a^2\omega^2\mu_0\rho_0\right)}, \tag{18}$$

$$F_3(\tilde{D}_2^{ll},\tilde{D}_0^{ll})=-\frac{\left(4\tilde{D}_0^{ll}\left(\kappa_0+\mu_0\right)+i\frac{\sqrt{3}}{2}a^2\omega^2\rho_0\right)\left(4\tilde{D}_2^{ll}\left(\kappa_0+\mu_0\right)\left(\kappa_0+2\mu_0\right)+i\frac{\sqrt{3}}{2}a^2\omega^2\mu_0\rho_0\right)}{\left(\kappa_0+\mu_0\right)\left(32\tilde{D}_0^{ll}\tilde{D}_2^{ll}\mu_0\left(\kappa_0+\mu_0\right)^2-i\frac{\sqrt{3}}{2}a^2\omega^2\rho_0\left(4\tilde{D}_2^{ll}\kappa_0\left(\kappa_0+\mu_0\right)-i\frac{\sqrt{3}}{2}a^2\omega^2\mu_0\rho_0\right)\right)},$$

where $\tilde{D}_m^{ll}$ is the Mie-like scattering coefficient D_m^{ll} after taking the long wavelength approximation. It is obviously seen that the roots $\left(K_1^{tri}\right)^2$ and $\left(K_2^{tri}\right)^2$ given in Eq. (17) do not depend on the direction of $\vec{K}$, i.e., ϕ_K. This implies that all the dispersion relation for an elastic metamaterial with a hexagonal structure are isotropic near the Γ point in the long wavelength limit.

3.2 Anisotropic dispersions

For the case of a square lattice, the lattice sum is almost the same as that of the hexagonal lattice case except for the $n \neq 0$ case. The reciprocal lattice vector of a square lattice is expressed by:

$$\vec{K}_h=\frac{2\pi}{a}\left(h_i\hat{i}+h_j\hat{j}\right);\ h_i,h_j\in Z \tag{19}$$

For an arbitrarily chosen reciprocal lattice vector (K_{h1},ϕ_{Kh1}), there always exist the other three at

$(K_{h1},\phi_{Kh1}+N\pi/2)$, $N=1,2,3$, which makes the summation $\sum_{N=0}^{3}e^{-inN\pi/2}$

cancel to zero when $1\le n\le 3$, and equals to 4 when $n=4$ respectively. This indicates that the second term of Eq. (14) only vanishes when $1\le n\le 3$. Thus, in the long wavelength limit, the lattice sum can be written, as

$$S(\beta,n)\cong\frac{4i^{n+1}K^n}{a^2k_{\beta0}^n\left(k_{\beta0}^2-K^2\right)}e^{-in\phi_K}\ (0\le n\le 3) \tag{20}$$

for $0\le n\le 3$, and

$$S(\beta,4)\cong\left(\frac{4iK^4}{a^2k_{\beta0}^4\left(k_{\beta0}^2-K^2\right)}+\gamma_\beta e^{i4\phi_K}\right)e^{-i4\phi_K} \tag{21}$$

for $n=4$, where

$$\gamma_\beta = \frac{16 \cdot 2^5 \cdot 5!i}{a^6 k_{\beta 0}^4} \left(\sum_{h_j=0}^{N} \sum_{h_i=1}^{N} \frac{J_5\left(2\pi\sqrt{h_i^2+h_j^2}\right) e^{i4\arctan(h_j/h_i)}}{2\pi\sqrt{h_i^2+h_j^2}\left(k_{\beta 0}^2 - (\frac{2\pi}{a})^2 (h_i^2+h_j^2)\right)} \right) \tag{22}$$

Due to the non-zero γ_β term in Eq. (22), the determinant in Eq. (12) is ϕ_K -dependent, which gives rise to anisotropic dispersion relations. The explicit expressions for K_1^{squ} and K_2^{squ} are very complicated and will be futher discussed in the next Section.

4. Effective medium theory for elastic metamaterials

The MST method is capable of producing the dispersion relations of an elastic metamaterial so that the effective wave speed for elastic waves in the metamaterial can be obtained accordingly. However, it is not able to provide an effective description for each parameters. Knowing the effective parameters will provide a clear theoretical explanation of the unusual phenomenon of a metamaterial and greatly benefit the design of new metamaterials. This Section is devoted to the derivation of EMTs.

4.1 Isotropic media: Coherent potential approximation approach

If the elastic metamaterial is isotropic, i.e., cylinders arranged in a hexogonal lattice, the EMT can be derived by considering the scattering of elastic waves by a coated cylinder embedded in the effective medium with effective parameters (κ_e , μ_e , ρ_e), which is shown in Figure 2 (Wu et al., 2007). The coated cylinder consists of the scatterer surrounded by a layer of the matrix. The inner and outer radii, which are denoted by r_s and r_0 , respectively, satisfy $r_s^2 / r_0^2 = p$, where p is the filling ratio of the scatterer. The effective parameters κ_e , μ_e and ρ_e are determined by the condition that the total scattering of the coated cylinder vanishes which is so-called CPA. This condition together with the boundary conditions on the surface of the coated cylinder at $r = r_0$, provides another two relations: $b_{lm} = D_m^{ll}(e)a_{lm} + D_m^{lt}(e)a_{tm}$ and $b_{tm} = D_m^{tl}(e)a_{lm} + D_m^{tt}(e)a_{tm}$, where $D_m^{\alpha\beta}(e)$ $(\alpha,\beta = l,t)$ can be obtained by replacing λ_s ρ_s μ_s , k_{ls} , k_{ts} and r_s in $D_m^{\alpha\beta}$ $(\alpha,\beta = l,t)$ mentioned in Section 2.1 with λ_e , ρ_e , μ_e , k_{le} , k_{te} and r_0 , respectively, where k_{le} (k_{te}) are the longitudinal (transverse) wave vectors in the effective medium. These relations together with the relations between $b_{\alpha m}$ and $a_{\alpha m}$ $(\alpha = l,t)$ shown in Section 2.1 give the following effective medium condition:

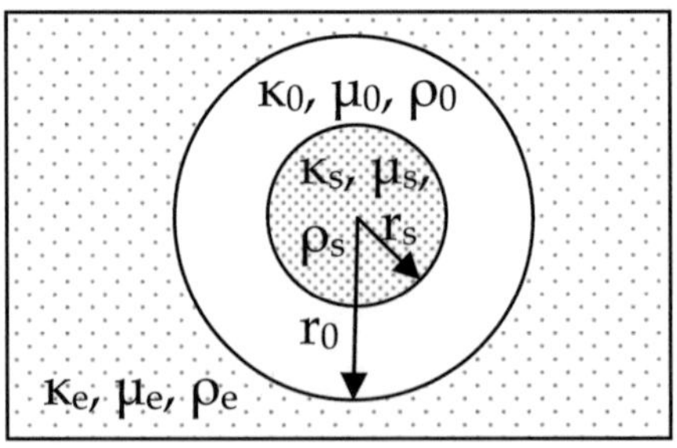

Fig. 2. Micro-sctructure of the effective medium. (Wu et al.,2007)

$$D_m^{\alpha\beta}(e) = D_m^{\alpha\beta} \ (\alpha, \beta = l, t) \tag{23}$$

In the long wavelength limit, where $k_{l0}r_0 << 1$, $k_{t0}r_0 << 1$, $k_{le}r_0 << 1$ and $k_{te}r_0 << 1$, Eq. (23)can be simplified into the following effective medium equations for the elastic metamaterial:

$$\frac{(\kappa_0 - \kappa_e)}{(\mu_0 + \kappa_e)} = \frac{4\tilde{D}_0^{ll}}{i\pi r_0^2 k_{l0}^2}, \tag{24}$$

$$\frac{(\rho_0 - \rho_e)}{\rho_0} = -\frac{8\tilde{D}_1^{ll}}{i\pi r_0^2 k_{l0}^2}, \tag{25}$$

$$\frac{\mu_0(\mu_0 - \mu_e)}{(\kappa_0\mu_0 + (\kappa_0 + 2\mu_0)\mu_e)} = \frac{4\tilde{D}_2^{ll}}{i\pi r_0^2 k_{t0}^2} \tag{26}$$

Obviously, κ_e, ρ_e and μ_e are independently determined by $\tilde{D}_m^{ll}$ of the embedded cylinders alone with angular quantum numbers m = 0, 1 and 2.

The dispersion relation can be reproduced from effective medium parameters. Comparing Eqs. (24)-(26) to Eqs. (17) and (18), it is easy to show that ρ_e/μ_e coincides with $\left(K_1^{tri}/\omega\right)^2$ and $\rho_e/(\kappa_e + \mu_e)$ is identical to $\left(K_2^{tri}/\omega\right)^2$. The equivalence provides a strong evidence that the dispersion relation is isotropic for a hexagonal lattice and its effective medium properties can be evaluated from the EMT derived from CPA.

Eqs. (24)-(26) require the wavelengths in both the host and the effective medium to be much larger than the size of the unit cell, but they do not impose any restriction on the wavelengths inside the scatterer. If the condition of $k_{ls}r_s << 1$, $k_{ts}r_s << 1$ is further considered, the quasi-static limit is reached and Eqs. (24)-(26) becomes:

$$\frac{(\kappa_0 - \kappa_e)}{(\mu_0 + \kappa_e)} = p\frac{(\kappa_0 - \kappa_s)}{(\mu_0 + \kappa_s)}, \tag{27}$$

$$(\rho_0 - \rho_e) = p(\rho_0 - \rho_s), \tag{28}$$

$$\frac{(\mu_0 - \mu_e)}{(\kappa_0\mu_0 + (\kappa_0 + 2\mu_0)\mu_e)} = p\frac{(\mu_0 - \mu_s)}{(\kappa_0\mu_0 + (\kappa_0 + 2\mu_0)\mu_s)} \tag{29}$$

The 3D version was reported by Berryman decades ago (Berryman, 1980). It should be pointed out that elastic EMT cannot recover the acoustic EMT by setting all the shear moduli to be zero, because of the different boundary conditions of elastic and acoustic waves.

Figure 3 shows the equifrequency surface (EFS) of a hexagonal array of silicone rubber cylinders with radii of $0.2a$ embedded in an epoxy host. An EFS is a collection of all states in the $\vec{K}$ space that have the same frequency. The metamaterial is isotropic if its EFS is a circle. Here the dimensionless frequency, $\tilde{f} = (\omega a)/(2\pi c_{t0})$, is used, where c_{t0} is the transverse wave speed inside the host. The silicone rubber's material parameters are $\rho = 1.3\times10^{3} kg / m^{3}$, $\lambda = 6\times10^{5} N / m^{2}$ and $\mu = 4\times10^{4} N / m^{2}$, which means the wave speeds inside the rubber are: 22.87 *m/s* for longitudinal waves and 5.54 *m/s* for transverse waves. The corresponding parameters in the epoxy host are $\rho = 1.18\times10^{3} kg / m^{3}$, $\lambda = 4.43\times10^{9} N / m^{2}$ and $\mu = 1.59\times10^{9} N / m^{2}$, which indicates the wave speeds are $2539.52 \mathrm{m/s}$ $(1160.80 \mathrm{m/s})$ for longitudinal (transverse) waves (Liu et al., 2000b). Apparently, slow wave speeds imply that wavelengths inside the silicone rubber cylinder may be comparable to or even much smaller than the size of the cylinder at low frequencies. Thus, Mie-like resonances may occur, which serve as the built-in resonances required for metamaterials. Here the frequency $\tilde{f}$ is chosen to be 0.03 where both $k_{ls}r_s \simeq 1.9$ and $k_{ts}r_s \simeq 7.9$ are larger than unity indicating it is not in the quasi-static limit. Figure 3(a) shows the corresponding EFS, which exhibits two circular rings, with the inner one denoting the quasi-longitudinal branch and the outer one representing the quasi-transverse branch. The corresponding Ka as a function of ϕ_K is plotted in Fig. 3(b) by open circles. These circles form two horizontal lines, indicating dispersions are isotropic, i.e., effective wave speeds do not vary with directions. Also plotted in Fig. 3(b) are the results of EMT calculated from Eq. (17) or Eqs. (24)-(26), depicted by two solid lines. The complete overlaps between solid lines and circles give a numerical support to the correctness of the EMT in the long wavelength limit.

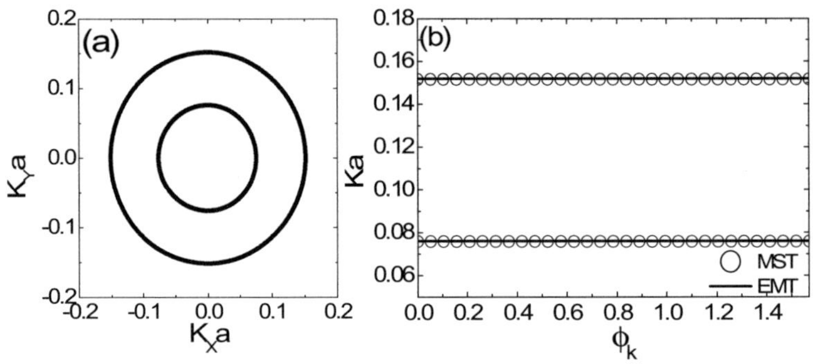

Fig. 3. (a) The equifrequency surface for a hexagonal array. (b) . Ka . as a function of ϕ_K . (Wu & Z. Zhang, 2009)

4.2 Anisotropic media: Christoffel's equation

If the elastic metamaterial is anisotropic, such as cylindrical scatterers arranged in a square lattice, the CPA fails as it only deals with isotropic cases. In this case, the result of MST, i.e., Eq. (14), can give an anisotropic EMT in the form of Christoffel's equation.

Taking the long wavelength limit approximation on Eq. (12) and plug in Eqs. (20) and (22), the expressions for K_1^{squ} and K_2^{squ} can be written as the solutions of the following Christoffel's equation for an anisotropic medium (Royer & Dieulesaint, 1999)

$$\det\begin{vmatrix} \cos^2\phi_K C_{11}\rho_e^{-1} + \sin^2\phi_K C_{44}\rho_e^{-1} - (\omega/K)^2 & \cos\phi_K \sin\phi_K (C_{12}+C_{44})\rho_e^{-1} \\ \cos\phi_K \sin\phi_K (C_{12}+C_{44})\rho_e^{-1} & \cos^2\phi_K C_{44}\rho_e^{-1} + \sin^2\phi_K C_{11}\rho_e^{-1} - (\omega/K)^2 \end{vmatrix} = 0 \quad (30)$$

where ϕ_K denotes the angle between the Bloch wave vector $\vec{K}$ and the x-axis, and ρ_e is the effective mass density derived from CPA. It is convenient to express the three effective moduli in Eq. (32) in terms of the two effective parameters κ_e and μ_e shown in Eqs. (26) and (28) for isotropic media,

$$C_{11} = \kappa_e + \mu_e + \Delta_1, \tag{31}$$

$$C_{12} = \kappa_e - \mu_e - \Delta_1, \tag{32}$$

$$C_{44} = \mu_e + \Delta_2, \tag{33}$$

where $\Delta_{1,2} = \delta(\mu_0 - \mu_e)^2 / [\delta(\mu_0 - \mu_e) \mp 8\rho_0]$, with − and + for Δ_1 and Δ_2, respectively. Here $\delta = i\left(k_{l0}^4 \gamma_l - k_{t0}^4 \gamma_t\right) a^2 / \omega^2$ and $\gamma_{l,t}$ is given by Eq.(22). It is clear that Δ_1 and Δ_2 are responsible for the anisotropy. In the long wavelength limit, δ has the following expression:

$$\delta = -\frac{1920}{\pi^5} \frac{\kappa_0 \rho_0}{\mu_0(\kappa_0 + \mu_0)} \left(\sum_{h_j=0}^{N} \sum_{h_i=1}^{N} \frac{J_5\left(2\pi\sqrt{h_i^2 + h_j^2}\right) e^{i4\arctan(h_j/h_i)}}{\left(\sqrt{h_i^2 + h_j^2}\right)^5} \right) \tag{34}$$

Eq. (30) gives:

$$\left(\frac{\omega}{K_1}\right)^2 = \frac{1}{2\rho_e}\left(C_{11} + C_{44} - \sqrt{\sin^2(2\phi_K)(C_{12} + C_{44})^2 + \cos^2(2\phi_K)(C_{11} - C_{44})^2} \right), \tag{35}$$

$$\left(\frac{\omega}{K_2}\right)^2 = \frac{1}{2\rho_e}\left(C_{11} + C_{44} + \sqrt{\sin^2(2\phi_K)(C_{12} + C_{44})^2 + \cos^2(2\phi_K)(C_{11} - C_{44})^2} \right) \tag{36}$$

Since the origin of anisotropy comes from the term $\gamma_\beta \neq 0$, the isotropy is expected to recover when $\gamma_\beta = 0$ (or $\delta = 0$). In this case, $C_{11} = C_{12} + 2C_{44}$ (Royer & Dieulesaint, 1999) and Eqs. (35) and (36) can be reduced to $(\omega / K_1)^2 = \mu_e / \rho_e$ and $(\omega / K_2)^2 = (\kappa_e + \mu_e) / \rho_e$, which are the square of two known wave speeds. For the case of anisotropic dispersions, Eqs. (35) and (36) give the dispersion relations for the quasi-transverse and quasi-longitudinal bands (Royer & Dieulesaint, 1999). Eq. (35) shows that $(\omega / K_1)^2$ oscillates between two extrema, $(\mu_e + \Delta_2)/\rho_e$ and $(\mu_e + \Delta_1)/\rho_e$ at $\phi_K = 0$ and $\pi/4$, respectively. Similarly, $(\omega / K_2)^2$ oscillates between its two extrema, $(\kappa_e + \mu_e + \Delta_1)/\rho_e$ and $(\kappa_e + \mu_e + \Delta_2)/\rho_e$. If both $|\Delta_1|$ and $|\Delta_2|$ are much smaller than $|\mu_e|$ and $|\kappa_e + \mu_e|$, and the amplitude of the oscillation, $|\Delta_1 - \Delta_2|$, is small, the angle averaged dispersions, $(\omega / K_1)^2$ and $(\omega / K_2)^2$, can be well approximated by μ_e/ρ_e and $(\kappa_e + \mu_e)/\rho_e$, which are the results of isotropic EMT given by Eqs. (24)-(26).

Figure 4(a) is the same as Figure 3(a), but the rubber cylinders are arranged in a square array. The inner ring represents the quasi-longitudinal branch with distinct anisotropy and the outer one is the quasi-transverse branch with weak anisotropy. The corresponding Ka

as a function of ϕ_K is plotted in Fig. 4(b) in open circles, which form two oscillating curves induced by the γ_β term in Eq. (21). In this case, $\Delta_1 = 0.0158$ and $\Delta_2 = -0.0179$ (in the unit of μ_0), which are very small compared to κ_e (2.241) and μ_e (0.733). $(\omega / K_1)^2$ and $(\omega / K_2)^2$ should reach their maximum and minimum at $\phi_K = 0$,respectively. This implies K_1 (K_2) is at its minimum (maximum). Also, K_1 arrives at its maximum at $\phi_K = \pi / 4$, where K_2 takes its minimal value. These are clearly illustrated Fig. 4(b). If we use the ratio $d_i = \left(K_i^{\max} - K_i^{\min}\right) / \langle K_i \rangle, (i = 1,2)$ to characterize the amount of anisotropy (Ni and Cheng, 2005), where $K_i^{\max}$, $K_i^{\min}$ and $\langle K_i \rangle$ are the maximum, minimum and average of K_i, the corresponding quantities are $d_1 = 5.39\%$ and $\langle K_1 a \rangle = 0.1603$ for the transverse waves and $d_2 = 1.20\%$ and $\langle K_2 a \rangle = 0.0793$ for the longitudinal waves. The averaged values of $K_i a$ coincide with the results calculated from the isotropic effective medium, i.e., Eq. (24)-(26), which give $K_t a = 0.1599$ and $K_l a = 0.0794$ as shown in Fig. 4 (b) in two horizontal solid lines. Fig. 4(b) demonstrates that the isotropic EMT can well predict the angle-averaged value of $K_i a$ in the case of anisotropy.

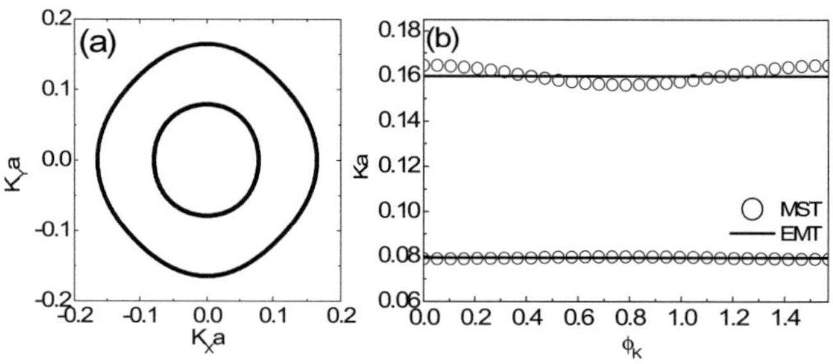

Fig. 4. The same as Figure 3 but the lattice is a square lattice. (Wu & Z. Zhang, 2009)

5. Design of elastic metamaterials

The purpose of deriving EMTs is to reveal the relationship between the resonances of the microstructures and the effective parameters and to provide a guide in the design of new metamaterials with novel properties. Even for isotropic metamaterials, the negativities in three effective parameters as well as their combinations can give rise to various interesting properties unseen in natural materials. For instance, since the effective phase velocities in 2D elastic metamaterials are $c_{le} = \sqrt{\kappa_e + \mu_e}\sqrt{1/\rho_e}$ and $c_{te} = \sqrt{\mu_e}\sqrt{1/\rho_e}$ for longitudinal and transverse waves, respectively, a single negative ρ_e in a frequency regime leads to imaginary c_{le} and c_{te}, which implies the existence of a band gap for both longitudinal and

	$\kappa_e + \mu_e > 0$ $\mu_e > 0$	$\kappa_e + \mu_e > 0$ $\mu_e < 0$	$\kappa_e + \mu_e < 0$ $\mu_e > 0$	$\kappa_e + \mu_e < 0$ $\mu_e < 0$
$\rho_e > 0$	$n_l > 0$; $n_t > 0$	$n_l > 0$; *t*:gap	$n_t > 0$; *l*:gap	*l*&*t*: gap
$\rho_e < 0$	*l*&*t*: gap	$n_t < 0$; *l*:gap	$n_l < 0$; *t*:gap	$n_l < 0$; $n_t < 0$

Table 1. Various wave propagation properties. Positive (negative) *n* indicates positive (negative) propagating bands. *l* and *t* represent longitudinal and transverse waves, respectively.

transverse waves (Liu et al., 2000b). On the other hand, a simultaneous negative ρ_e and μ_e (or $\kappa_e + \mu_e$) induces negative refractive index for the transverse (longitudinal) waves. Table 1 lists eight possible types of wave propagation in 2D elastic metamaterials with different combinations of signs in $\kappa_e + \mu_e$, ρ_e and μ_e. Like acoustic metamaterials (Li & Chan, 2004), the effective bulk modulus and shear modulus κ_e and μ_e determined according to Eqs. (24)-(26) do not satisfy the well-known bounds (Hashin & Shtrikman, 1963; Torquato, 1991) on the effective elastic moduli as these bounds are derived in the quasi-static limit. For anisotropic metamaterials, there exists at least one more effective elastic modulus which can also turn negative. Thus, many more interesting novel wave transport behaviors would be expected. Examples will be shown in a later section.

5.1 Isotropic elastic metamaterials

The simplest isotropic elastic metamaterial which is comprised of silicone rubber cylinders embedded in an epoxy host was exhibited in the last section. If the rubber cylinders' radii are chosen to be $0.3\,a$, it has been shown that various types of resonances were produced (Wu et al., 2007). The displacement fields for three typical resonances are plotted in Fig. 5, which clearly shows in (a), (b) and (c) the dipolar, quadrupolar and monopolar resonances. These resonances are linked to the effective medium parameters in the following manner implied by the EMT. The negative ρ_e arises from a dipolar resonance ($\tilde{D}_1^{ll}$), whereas the quadrupolar ($\tilde{D}_2^{ll}$) and monopolar ($\tilde{D}_0^{ll}$) resonances give rise to negative μ_e and negative κ_e, respectively. We can enlarge the negative regions of effective medium parameters by enhancing these resonances. For instance, the dipolar resonance shown in Fig. 5(a) exhibits the collective motion of the core part of rubber. This mode can be regarded as a simple "mass-spring" harmonic oscillator, with the central part serving as a "mass" and the boundary layer of the rubber serving as "spring". Replacing the inner region of rubber with another heavier cylinder, e.g. lead, will enhance the field oscillation of the silicone rubber, which, in turn, will widen the resonant region of ρ_e. This design was first proposed by Liu *et al* and was named as "locally resonant sonic materials" (Liu et al., 2000b), which were comprised of a cubic array of rubber-coated lead spheres embedded in epoxy. A large low-frequency band gap for both longitudinal and transverse waves, induced by negative mass

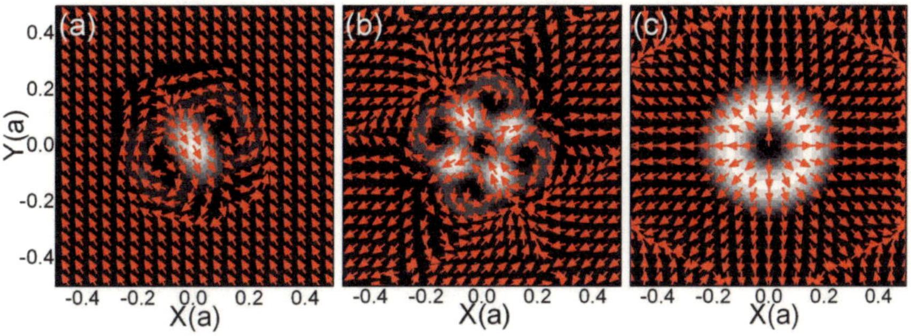

Fig. 5. Displacement fields for different resonances. The arrows represent the direction and the brightness denotes the amplitude, with white indicates larger. (Wu et al., 2007)

density, was found. Figure 5(c) exhibits the field pattern of a monopolar resonance, where the shape of the silicone rubber cylinder remains as a circle, with its cross-sectional area oscillating in time. This suggests that by making the inner core more easily compressed, we would enhance the monopolar resonance. This notion was supported by using air bubbles in water to achieve a large frequency region of negative bulk modulus (Ding et al., 2007). Figure 5(b) shows the relative motion of the rubber. This suggests that by making the core areas easier to deform, we would enhance the quadrupolar resonance so as to enlarge the negative region for shear modulus. An intuitive design is to make the rubber cylinder hollow. A metamaterial based on this design is made of rubber-coated air cylinders embedded in epoxy. The material parameters of air are given by $\rho = 1.23 kg/m^3$ and $\lambda = 1.42\times 10^{10} N/m^2$. The effective medium parameters are evaluated by a generalized EMT, which uses the standard transfer-matrix method to obtain the quantities $\tilde{D}_m^{ll}$ if the scatterers are layered cylinders.

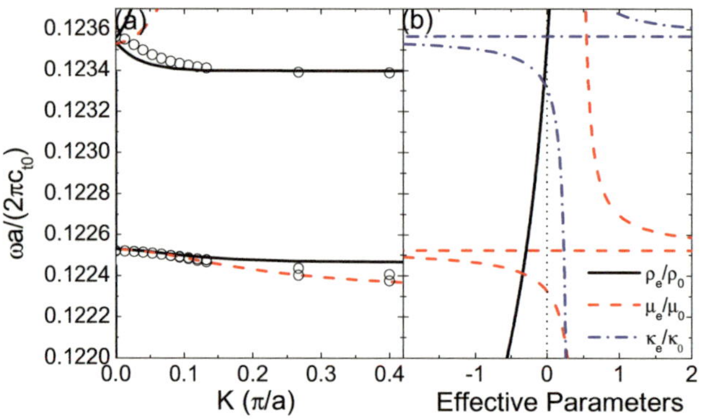

Fig. 6. Band structure and effective medium parameters for an triangular array of hollow rubber cylinders embedded in epoxy. (Wu et al., 2007)

Figure 6(a) shows the band structure of hollow rubber cylinders embedded in epoxy in a hexagonal lattice. The inner and outer radii of hollow rubber cylinders are given as $r_{air} = 0.87 r_s$ and $r_s = 0.3a$, which are carefully chosen so that a negative band for shear wave can be realized. The accurate MST results are plotted in open circles and the EMT predictions are featured by curves, with solid representing the longitudinal branch and dashed corresponding to the transverse branch. In the region of $0.12240 < \tilde{f} < 0.12253$, negative-$n$ bands of both longitudinal and transverse waves are found, which implies that ρ_e, $\kappa_e + \mu_e$ and μ_e are all negative. These negative values are induced only by the resonances of ρ_e and μ_e as shown in Fig. 6(b), in which the individual effective medium parameters are plotted. In another region of $0.12340 < \tilde{f} < 0.12356$, negative-$n$ band purely for longitudinal waves is found, which implies that ρ_e and $\kappa_e + \mu_e$ are both negative. These negative values arise from resonances of both ρ_e and κ_e. Figure 6(a) demonstrates that the isotropic EMT is still a good approximation even for complex scatterers with layered structures. The small discrepancies between the band-structure calculation and the effective medium prediction for the longitudinal branches shown in Fig. 6(a) is due to the less accurate approximation of the Hankel functions when the values of $k_{l0}r_0$ and $k_{t0}r_0$ are not much less than 1.

The structure of the hollow silicone rubber cylinder in epoxy does provide a frequency region of negative band for shear waves, but the bandwidth is too small to be of any practical use. Enlarging the quadrupolar resonance is a challenge task that has yet to be accomplished. Since replacing the inner part of the rubber by an easier deformed material is the direction, another common material, water, becomes a candidate. The material parameters of water are: $\rho = 1.0\times10^3\,\text{kg}/\text{m}^3$ and $\lambda = 2.22\times10^9\,N/m^2$. Figure 7 shows the effective medium parameters for the metamaterial with air core being replaced by water, which exhibits an improvement in the absolute bandwidth for negative shear modulus. However, the bandwidth to mid-frequency ratio (0.00145) is comparable to the previous air core case (0.00183). Moreover, the negative shear band disappears as the region for negative mass density does not overlap with that for negative shear modulus. This example demonstrates that simple replacement of air by water does not improve the negative shear band. Nevertheless, the replacement does enhance the dipolar resonance greatly in the very low frequency regime (which is not plotted here). This fact suggests that water core is a better candidate than air in the context of realizing negative mass density. The difficulty lies in increasing the negative region for shear modulus and adjusting it to overlap with that for the negative mass. This requires optimizing the inner and outer radii of the rubber cylinder. However, it can be shown that only altering the geometry parameters will not make a significant change.

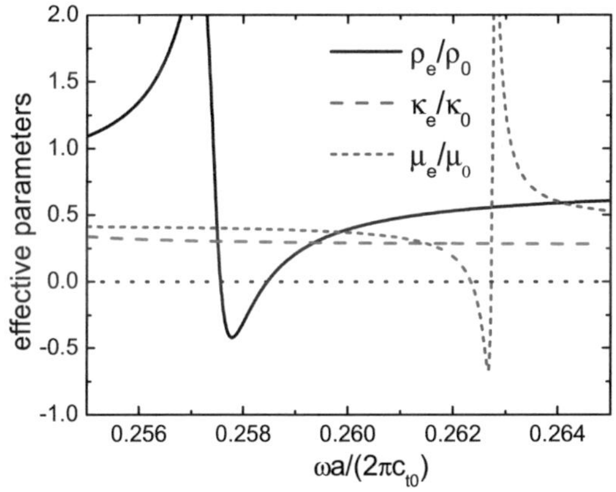

Fig. 7. Effective medium parameters for rubber-coated water cylinders embedded in epoxy in a hexogonal lattice.

Both Figs. 6(b) and 7 show that the μ resonance is very sharp, indicating that the energy is confined in the cylinders locally. To broaden the resonance, one strategy that could be adopted is to make the energy "spread" out of the cylinders so that resonances in different cylinders become "coupled" to each other. This can be realized by reducing the impedance mismatch between the rubber and the host.

Figure 8 shows the effective medium parameters for a rubber-coated water cylinder embedded in a foam host in a hexagonal lattice. The foam is polyethylene foam (HD115) whose material parameters are $\rho = 115kg/m^3$, $\lambda = 6.0\times10^6 N/m^2$ and $\mu = 3.0\times10^6 N/m^2$ (Zhou & Hu, 2009). The light foam makes the host more matched to the rubber than epoxy which will benefit the enhancement of the resonance for shear modulus. It also makes the water-coated rubber core relatively heavier so that the resonance for mass density is also enhanced. By adjusting the geometric parameters to be $0.24a$ and $0.32a$ for inner and outer radii, respectively, a large frequency region, marked by "A" and "B" in Fig. 8, for both negative shear modulus and negative mass density is obtained. The bandwidth to mid-frequency ratio reaches 0.258, which is two orders of magnitude greater than the rubber-coated air cylinders in epoxy. The corresponding band structures as well as the transmission coefficeints of a slab numerically calcuated by MST are plotted in Fig. 9(a), which clearly shows a large negative band for transverse waves denoted by red dots and a narrow negative band for longitudinal waves denoted by blue dots. The polarization of these negative bands is determined through the transmission as shown in Fig. 9(b).

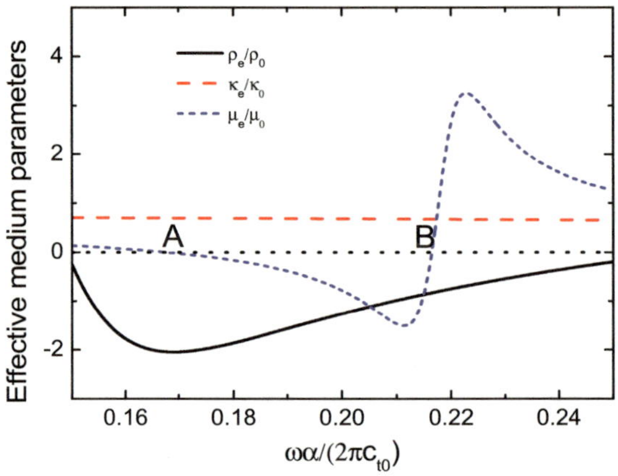

Fig. 8. Effective medium parameters for rubber-coated water cylinders embedded in foam in a hexogonal lattice.

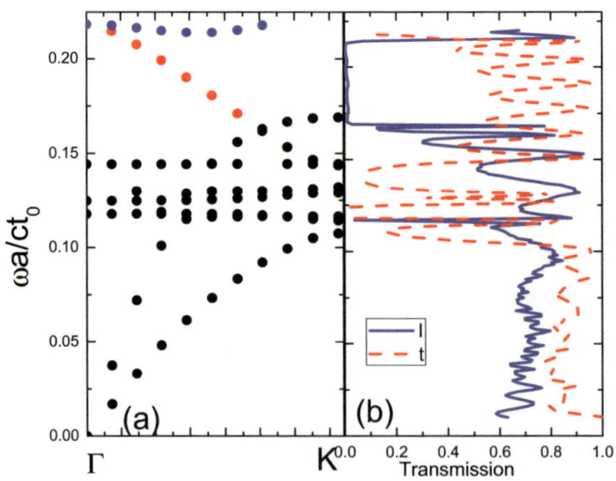

Fig. 9. (a) The dispersion along ΓK direction for the same system as Fig. 8.
(b) Transmission coefficients for longtidinal (solid blue) wave incident and transverse (dashed red) wave incident on a slab of witdth 6a and length 50a. The incident wave is along ΓK direction. (Wu et al., 2011)

5.2 Anisotropic elastic metamaterials and boundary effective medium theory

The rubber-coated water cylinder embedded in foam provides the possiblity of realizing double negative shear bands. The isotropic scatterers and the hexagonal lattice structure result in a simple isotropic effective medium description of the metamaterial, which makes the design of the elastic metamaterial easier. If there is no restriction on the symmetery properties of the scatterer, there would be much more choices at the cost of many more complicated microstructures of the scatterers.

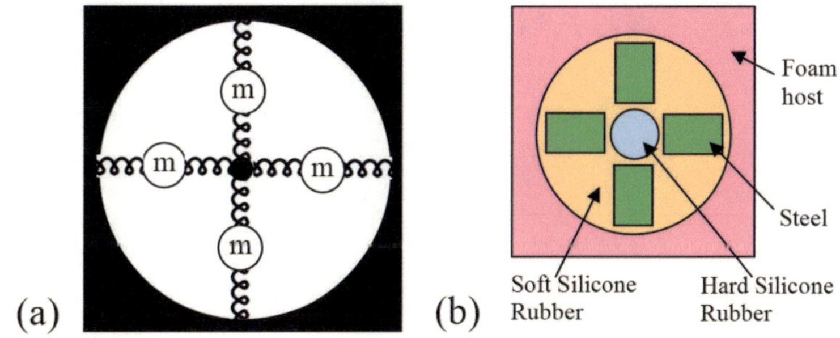

Fig. 10. A schematic figure of the physical model and the practical design. (Lai et al., 2011)

Since Fig. 5(b) exhibits a four-fold symmetry of a quadrupolar resonance, inserting heavier objects into the rubber in a way that is in accordance with the field pattern would help enhance the resonance. Figure 10(a) is a schematic figure of the physical model of the unit cell, which shows four masses connected to their center and the host (Lai et al., 2011). Such a structure is favorable of enhancing the dipolar resonance by the collective motion of the four masses, and the quadrupolar and monopolar resonances by relative motions of the masses. A practical realization of the model is illustrated in Fig. 10(b). The scatterers are composed of four steel rods surrounding a hard silicone rubber cylinder embedded in a soft silicone rubber cylinder. The matrix material is still foam. The lattice structure is a square with lattice constant of $10cm$; the radii of the soft and hard silicone rubber rods are $4cm$ and $1cm$, respectively; the rectangular steel rods are $1.6cm \times 2.4cm$ in size, located at a distance of $2.4cm$ from the center. The material parameters for the foam and soft silicone rubber remain the same as the ones used in the design of rubber-coated water cylinder. The hard siliconerubber and the steel have parameters of: $\rho = 1.415 \times 10^3 kg/m^3$, $\lambda = 1.27 \times 10^9 N/m^2$ and $\mu = 1.78 \times 10^6 N/m^2$ for hard silicone rubber and $\rho = 7.9 \times 10^3 kg/m^3$, $\lambda = 1.11 \times 10^{11} N/m^2$ and $\mu = 8.28 \times 10^{10} N/m^2$ for steel. The four rectangular steel rods serve as the four masses and the soft silicone rubber rods serve as the springs. The insertion of the hard silicone rubber is for the purpose of adjusting the spring constants between the masses.

The band structure of such metamaterial was calculated by using a finite element solver (COMSOL Multiphysics) and is shown in Fig. 11(a). There are two negative bands (red and blue dots), where the lower one (red dots) has a bandwidth about 18 Hz and the higher one (blue dots) has band widths of 18Hz and 10 Hz along ΓM and ΓX direction respectively. These two bands are separated by a small complete gap (178 Hz~198 Hz). The two eigenstates in the lower and upper negative bands at the Γ point are plotted in Figs. 11(b) and 11(c), respectively. The eigenstate in Fig. 11(b) is clearly a quadrupolar resonance, whereas the eigenstate in Fig. 11(c) is a monopolar resonance.

The negative bands can also be understood from an effective medium point of view. Since the scatterer involves a four-fold symmetry, the previously derived formula based on MST for isotropic inclusions does not apply and an EMT based on boundary integration is

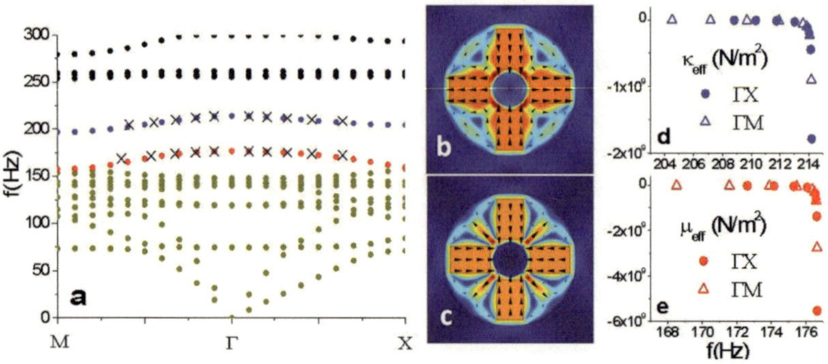

Fig. 11. (a) Band structure of the multi-mass metamaterial. (b) and (c) Displacement field of eigenstates. The color represents the amplitude of displacement (blue/red for small/large values) and the arrows show the displacement vectors directly. (d) and (e) effective medium parameters calculated by a boundary EMT. (Lai et al., 2011)

developed. Though the scatterer is anisotropic, the dispersions and the associated modes can still be obtained from Christoffel's equation, i.e., Eqs. (35) and (36), with three independent effective moduli, C_{11}, C_{12} and C_{44}, and a mass density, ρ. The task is to determine the values of these parameters. The mass density is determined by Newton's law, $\rho^e = -F_x^e / \omega^2 u_x^e a^2$, where both the effective force F_x^e on the unit cell and its effective displacement u_x^e may be obtained from surface integration of the stresses (along the x direction) and the displacements over the unit cell, i.e.,

$$F_x^e = \int T_{xx} dy\Big|_{x=a} - \int T_{xx} dy\Big|_{x=0} + \int T_{xy} dx\Big|_{y=a} - \int T_{xy} dx\Big|_{y=0} \tag{37}$$

and

$$u_x^e = \frac{\int u_x dy\Big|_{x=0} + \int u_x dy\Big|_{x=a}}{2a} \tag{38}$$

The stresses and displacements can be obtained from the COMSOL calculation. Similarly, the effective moduli are evaluated from the effective stress and strain relations: $T_{xx}^e = C_{11}^e S_{xx}^e + C_{12}^e S_{yy}^e$, $T_{yy}^e = C_{12}^e S_{xx}^e + C_{11}^e S_{yy}^e$, and $T_{xy}^e = 2C_{44}^e S_{xy}^e$, where both the effective stresses and the effective strains are evaluated on the unit cell boundary as follows:

$$T_{xx}^e = \frac{\int T_{xx} dy\Big|_{x=0} + \int T_{xx} dy\Big|_{x=a}}{2a}; T_{yy}^e = \frac{\int T_{yy} dy\Big|_{y=0} + \int T_{yy} dy\Big|_{y=a}}{2a};$$

$$T_{xy}^{e} = \frac{\int T_{xy}dx\Big|_{y=0} + \int T_{xy}dx\Big|_{y=a}}{2a}; T_{yx}^{e} = \frac{\int T_{xy}dy\Big|_{x=0} + \int T_{xy}dy\Big|_{x=a}}{2a} \tag{39}$$

and

$$\begin{aligned} S_{xx}^{e} &= \frac{\int u_x dy\Big|_{x=a} - \int u_x dy\Big|_{x=0}}{a^2}; S_{yy}^{e} = \frac{\int u_y dy\Big|_{y=a} - \int u_y dy\Big|_{y=0}}{a^2} \\ S_{xy}^{e} &= \frac{\int u_x dx\Big|_{y=a} - \int u_x dx\Big|_{y=0} + \int u_y dy\Big|_{x=a} - \int u_y dy\Big|_{x=0}}{2a^2} \end{aligned} \tag{40}$$

Though the above equations are presented for calculations along the ΓX direction, the corresponding formula for ΓM direction can be similarly transcribed. Due to the obvious link between the bulk (shear) modulus and monopolar (quadrupolar) resonance, it is more convenient to introduce $\kappa_{\text{eff}} = (C_{11} + C_{12})/2$ and $\mu_{eff} = (C_{11} - C_{12})/2$ as effective elastic bulk modulus and shear modulus. The results for κ_{eff} and μ_{eff} evaluated from the relevant eigenstates are plotted in Figs. 11(d) and 11(e). In the lower negative band, κ_{eff} is positive and finite, while μ_{eff} is negative and diverges at the Γ point, which is in accordance with the quadrupolar resonance. In the higher negative band, μ_{eff} is positive and finite, while κ_{eff} is negative and diverges at the Γ point, which is induced by the monopolar resonance.

Knowing the effective moduli, the corresponding dispersion relations can be calculated by using Christoffel's equation. Along the ΓX direction, compressional wave and shear wave velocities are given by by $\sqrt{C_{11}/\rho}$ and $\sqrt{C_{44}/\rho}$, respectively; whereas along the ΓM direction the compressional and shear wave velocities are $\sqrt{(C_{11} + C_{12} + 2C_{44})/(2\rho)}$ and $\sqrt{(C_{11} - C_{12})/(2\rho)}$. The effective medium results show that the lower negative band supports a longitudinal (transverse) wave along ΓX (ΓM) direction, whereas the upper negative band only allows longitudinal wave in both ΓX and ΓM directions. The corresponding results obtained from the EMT are also plotted in Fig. 11(a) by crosses. Excellent agreements between the finite element results and the EMT prediction are found.

6. Some intriguing properties of elastic metamaterials

Like their EM and acoustic counterparts, elastic metamaterials have shown many intriguing wave transport properties. For example, the total mode conversion and the super-anisotropy are two of them. The total mode conversion can completely convert the incident transverse (longitudinal) wave into a refracted longitudinal (transverse) wave. It is an analogue of the Brewster angle in the EM case (for example, Jackson, 1999), but in a much more stringent and complex manner. It only occurs on the interface between a normal solid and an elastic metamaterial with negative refractive index (Wu et al. 2011). The super-anisotropic behavior has been demonstrated in Section 5.2. Also shown there is the property of sustaining only a longitudinal wave at certain frequencies, which is so-called "fluid-like" solids, blurring the distinction between solids and fluids (Lai et al., 2011).

7. Conclusion

In this chapter, the effective medium properties of 2D elastic metamaterials have been reviewed. Unlike EM or acoustic metamaterials, the elastic metamaterial is in general anisotropic unless the lattice structure is a hexagon with isotropic scatterers. For the isotropic elastic metamaterial, the EMT may be derived from CPA. For the anisotropic metamaterial, the EMT may be obtained from the MST in conjunction with Christoffel's equation, or from the integration of eigenfields on the boundaries. EMT could greatly facilitate the design of new elastic metamaterials, such as rubber-coated water cylinder embedded in foam which gives rise to large negative bands for shear waves and a multi-mass locally resonant structure which results in both negative bands for longitudinal waves and super-anisotropic negative bands.

Elastic metamaterial opens a new research area. The experimental realization would be much more challenging and exciting. The generalization of EMT as well as the symmetry property to more complex lattice structures, such as rectangular lattices, would also be of interest as it will introduce even stronger anisotropy. Meanwhile, finding an EMT that can also treat the rotational modes is a challenging task. Such modes are normally excited at lower frequencies and form flat bands in the band structures.

8. Acknowledgments

The relevant work was supported by Hong Kong RGC Grant No. 605008, and start-up packages from KAUST and Soochow University.

9. References

Abramowitz, M. & Stegun, I. (Eds.). (1972). *Handbook of Mathematical Functions: with Formulas, Graphs, and Mathematical Tables*, Dover, ISBN 0486612724 , New York

Andrea, A. (2011). First-Principle Homogenization Theory for Periodic Metamaterials, *Physical Review B*, Vol.84, No.7, (Aug 2011), pp. 075153, ISSN 1098-0121

Andryieuski, A.; Malureanu, R. & Lavrinenko, A. (2009). Wave Propagation Retrieval Method for Metamaterials: Unambiguous Restoration of Effective Parameters, *Physical Review B*, Vol.80, No.19, (Nov 2009), pp.193101, ISSN 1098-0121

Baker-Jarvis, J.; Vanzura, E. & Kissick, W. (1990). Improved Technique for Determining Complex Permittivity with the Transmission/Reflection Method, *IEEE Transactions on microwave theory and techniques*, Vol.38, No.8, (August 1990), pp. 1096-1103, ISSN 0018-9480

Berryman, J. (1980). Long-wavelength Propagation in Composite Elastic Media. 1. Spherical Inclusions, *Journal of the Acoustical Society of America*, Vol.68, No.6, pp. 1809-1819, ISSN 0001-4966

Chern, R. & Chen, Y. (2009). Effective Parameters for Photonic Crystals with Large Dielectric Contrast, *Physical Review B*, Vol.80, No.7, (August 2009), pp. 075118, ISSN 0163-1829

Chen, H.; Chan, C. & Sheng, P. (2010). Transformation Optics and Metamaterials, *Nature Materials*,Vol.9, No.5, (May 2010), pp.387-396, ISSN 1476-1122

Chen, H.; Fung, K.; Ma, H. & Chan, C. (2008). Polarization Gaps and Negative Group Velocity in Chiral Phononic Crystals, *Physical Review B,* Vol.77, No.22, (June 2008), pp. 224304, ISSN 0163-1829

Chin, S.; Nicorovici, N. & McPhedran, R. (1994). Green's Function and Lattice Sums for Electromagnetic Scattering by a Square Array of Cylinders. *Physical Review E,* Vol.49, No.5, (May 1994), pp. 4590-4602, ISSN 1063-651X

Chui, S. & Hu, L. (2002). Theoretical Investigation on the Possibility of Preparing Left-Handed Materials in Metallic Magnetic Granular Composites, *Physical Review B,* Vol.65, No.14, (March 2002), pp. 144407, ISSN 0163-1829

Chui, S. & Lin, Z. (2008). Long-Wavelength Behavior of Two-Dimensional Photonic Crystals. *Physical Review E,* Vol.78, No.6, (December 2008), pp.065601 ISSN 1539-3755

Ding, Y.; Liu, Z.; Qiu, C. & Shi, J. (2007). Metamaterial with Simultaneously Negative Bulk Modulus and Mass Density, *Physical Review Letters,* Vol.99, No.9, (August 2007), pp. 093904, ISSN 0031-9007

Fang, N.; Xi, D.; Xu, J.; Ambati, M.; Srituravanich, W.; Sun, C. & Zhang, X. (2006). Ultrasonic Metamaterials with Negative Modulus, *Nature Materials,* Vol.5, No.6, (June 2006), pp. 452-456, ISSN 1476-1122

Grbic, A. & Eleftheriades, G. (2004). Overcoming the Diffraction Limit with a Planar Left-Handed Transmission-Line Lens, *Physical Review Letters,* Vol.92, No.11, (March 2004),pp. 117403 ISSN 0031-9007

Hashin, Z. & Shtrikman, S. (1963). A Variational Approach to The Theory of The Elastic Behaviour of Multiphase Materials, *Journal of the Mechanics and Physics of Solids,* Vol.11, No.2, pp. 127-140, ISSN: 0022-5096

Hu, X.; Chan, C.; Zi, J.; Li, M. & Ho, K. (2006). Diamagnetic Response of Metallic Photonic Crystals at Infrared and Visible Frequencies, *Physical Review Letters,* Vol.96, No.22, (June 2006), pp. 223901, ISSN 0031-9007

Jackson, J.; (1999). *Classical Electrodynamics* (Third edition), John Wiley,ISBN 0-471-30932-X, New York

Jin, J.; Liu, S.; Lin, Z. & Chui, S. (2009). Effective-Medium Theory for Anisotropic Magnetic Metamaterials, *Physical Review B,* Vol.80, No.11, (September 2009), pp.115101, ISSN 0163-1829

Kafesaki, M. & Economou, E. (1999). Multiple-scattering Theory for Three-dimensional Periodic Acoustic Composites, *Physical Review B,* Vol.60, No.17, (November 1999), pp. 11993-12001, ISSN 0163-1829

Kafesaki, M.; Penciu, R. & Economou, E. (2000). Air Bubbles in Water: A Strongly Multiple Scattering Medium for Acoustic Waves, *Physical Review Letters,* Vol.84, No.26, (June 2000), pp. 6050-6053, ISSN 0031-9007

Krokhin, A.; Arriaga, J. & Gumen, L. (2003). Speed of Sound in Periodic Elastic Composites, *Physical Review Letters,* Vol.91, No.26, (December 2003), pp. 264302, ISSN 0031-9007

Lai, Y.; Wu, Y.; Sheng, P. & Zhang, Z. (2011). Hybrid Elastic Solids, *Nature Materials,*Vol.10, No.8,(August 2011), pp. 620-624, ISSN 1476-1122

Lamb, W.; Wood, D. & Ashcroft, N. (1980). Long-Wavelength Electromagnetic Propagation in Heterogeneous Media, *Physical Review B,* Vol.21, No.6, (March 1980), pp.2248-2266, ISSN 0163-1829

Landau, L. & Lifshitz, E. (1986). *Theory of Elasticity* (Third edition), Butterworth-Heinemann, ISBN 0 7506 2633 X, Oxford

Lee, S.; Park, C.; Seo, Y.; Wang, Z. & Kim, C. (2010). Composite Acoustic Medium with Simultaneously Negative Density and Modulus, *Physical Review Letters,* Vol.104, No.5, (February 2010), pp. 054301, ISSN 0031-9007

Leonhardt, U. (2006). Optical Conformal Mapping, *Science,* Vol.312, No.5781, (June 2006), pp.1777-1780, ISSN 0036-8075

Lezec, H.; Dionne, J. & Atwater, H. (2007). Negative Refraction at Visible Frequencies, *Science,* Vol.361, No.5823, (April 2007), pp. 430-432, ISSN 0036-8075

Li, J. & Chan C. (2004). Double-negative Acoustic Metamaterial, *Physical Review E,* Vol.70, No.5,(November 2004),pp.055602, ISSN 1063-651X

Liu, Z.; Chan, C.; Sheng, P.; Goertzen, A. & Page, J. (2000). Elastic Wave Scattering by Periodic Structures of Spherical Objects: Theory and Experiment, *Physical Review B,* Vol.62, No.4, (July 2000), pp. 2446-2457, ISSN 1098-0121

Liu, Z.; Zhang, X.; Mao, Y.; Zhu, Y.; Yang, Z.; Chan, C. & Sheng, P. (2000) Locally Resonant Sonic Materials, *Science,* Vol.289, No.5485, (September 2000), pp. 1734-1736, ISSN 0036-8075

Lu, M.; Feng, L. & Chen, Y. (2009). Phononic Crystals and Acoustic Metamaterials, *Materials Today,* Vol.12, No.12, (December 2009), pp. 34-42, ISSN: 1369-7021

Mei, J.; Liu, Z.; Wen, W. & Sheng, P. (2006). Effective Mass Density of Fluid-solid Composites, *Physical Review Letters,* Vol.96, No.2, (January 2006), pp. 024301, ISSN 0031-9007

Mei, J.; Liu, Z.; Shi, J. & Tian, D. (2003). Theory for Elastic Wave Scattering by a Two-Dimensional Periodical Array of Cylinders: An Ideal Approach for Band-Structure Calculations, *Physical Review B,* Vol.67,No.24, (June 2003), pp. 245107, ISSN 1098-0121

Milton, G. & Willis, J. (2007). On Modification of Newton's Second Law and Linear Continuum Elastodynamics,*Proceedings of the Royal Society A-Mathematical Physical and Engineering Sciences,* Vol.463,No.2079,(March 2007), pp.855-880, ISSN1364-5021

Ni, Q. & Cheng, J. (2005). Anisotropy of Effective Velocity for Elastic Wave Propagation in Two-Dimensional Phononic Crystals at Low Frequencies, *Physical Review B,* Vol.72, No.1, (July 2005), pp. 014305, ISSN 0163-1829

Pendry, J. (2000). Negative Refraction Makes a Perfect Lens, *Physical Review Letters,* Vol.85, No.18, (October 2000), pp. 3966-3969, ISSN 0031-9007

Pendry, J.; Holden, A.; Robbins, D. & Stewart, W. (1999). Magnetism from Conductors and Enhanced Nonlinear Phenomena, *IEEE Transactions on Microwave Theory and Techniques,* Vol.47, No.11, (November 1999), pp. 2075-2084, ISSN 0018-9480

Pendry, J.; Holden, A.; Stewart, W. & Youngs, I. (1996). Extremely Low Frequency Plasmons in Metallic Mesostructures, *Physical Review Letters,* Vol.76, No.25, (June 1996), pp. 4773-4776, ISSN 0031-9007

Pendry, J.; Schurig, D. & Smith, D. (2006). Controlling Electromagnetic Fields, *Science*, Vol.312, No. 5781, (June 2006) pp.1780-1782, ISSN 0036-8075

Pendry, J. & Smith, D. (2006). The Quest for the Superlens, *Scientific America*, Vol.295, No.1, (July 2006), pp. 60-67, ISSN 0036-8733

Royer, D. & Dieulesaint, E. (1999). *Elastic Waves in Solids I: Free and Guided Propagation* (First edition), Springer, ISBN 3540659323, Berlin

Sheng, P. (2006). *Introduction to Wave Scattering, Localization, and Mesoscopic Phenomena* (Second edition), Springer, ISBN-10 3-540-29155-5, Berlin Heidelberg

Smith, D. & Pendry, J. (2006). Homogenization of Metamaterials by Field Averaging(Invited Paper), *Journal of the Optical Society of America B-Optical Physics*,Vol.23, No.2 (March 2006), pp.391-403, ISSN 0740-3224

Smith, D.; Schultz, S.; Markos, P. & Soukoulis, C. (2002). Determination of Effective Permittivity and Permeability of Metamaterials from Reflection and Transmission Coefficients, *Physical Review B*, Vol.65, No.19, (May 2002), pp.195104, ISSN 1098-012

Sun, S.; Shui, S. & Zhou, L. (2009). Effective-Medium Properties of Metamaterials: A Quasimode Theory, *Physical Review E*, Vol.79, No.6, (June 2009), pp.066604, ISSN 1063-651X

Torquato, S. (1991). Random Heterogeneous Media: Microstructure and Improved Bounds on Effective Properties, *Applied Mechanics Reviews*, Vol.44, No.2, (Febrary 1991) pp. 37-76, ISSN 0003-6900

Torrent, D.; Hakansson, A.; Cervera, F. & Sanchez-Dehesa, J. (2006). Homogenization of Two-Dimensional Clusters of Rigid Rods in Air, *Physical Review Letters*, Vol.96, No.20, (May 2006), pp. 204302, ISSN 0031-9007

Veselago, V. (1968). The Electrodynamics of Substances with Simultaneously Negative Values of ε and μ, *Soviet Physics Uspekhi*, Vol.10, No.4, pp. 509

Wu, Y.; Li, J.; Zhang, Z. & Chan C. (2006). Effective Medium Theory for Magnetodielectric Composites: Beyond the Long-Wavelength Limit, *Physical Review B*, Vol.74, No.8, (August 2006), pp.085111, ISSN 1098-0121

Wu, Y.; Lai, Y. & Zhang, Z. (2007). Effective Medium Theory for Elastic Metamaterials in Two Dimensions. *Physical Review B*, Vol.76, No.20, (Novemver 2007), pp.025313, ISSN 1098-0121

Wu, Y.; Lai, Y. & Zhang, Z. (2011). Elastic Metamaterials with Simultaneously Negative Effective Shear Modulus and Mass Density, *Physical Review Letters*, Vol.107, No.10, (September 2011), pp. 105506, ISSN 0031-9007

Wu, Y. & Zhang, Z. (2009). Dispersion Relations and Their Symmetry Properties of Electromagnetic and Elastic Metamaterials in Two Dimensions, *Physical Review B*, Vol.79, No.19, (May 2009), pp.195111, ISSN 1098-0121

Yang, Z.; Mei, J.; Yang, M.; Chan, N. & Sheng, P. (2008) Membrane-Type Acoustic Metamaterial with Negative Dynamic Mass, *Physical ReviewLetters*,Vol.101, No.20, (November 2008),pp.204301, ISSN 0031-9007

Zhang, S.; Xia, C. & Fang, N. (2011). Broadband Acoustic Cloak for Ultrasound Waves, *Physical Review Letters*, Vol.106, No.2, (January 2011), pp. 024301, ISSN 0031-9007

Zhang, S.; Yin, L. & Fang, N. (2009). Focusing Ultrasound with an Acoustic Metamaterial Network, *Physical Review Letters,* Vol.102, No.19, (May 2009), pp. 194301, ISSN 0031-9007

Zhou, X. & Hu, G. (2009). Analytic Model of Elastic Metamaterials with Local Resonances, *Physical Review B*, Vol.75, No.19, (May 2009), pp.195109, ISSN 1098-0121

Section 2

The Performance of Metamaterials

7

Resonant Negative Refractive Index Metamaterials

Divitha Seetharamdoo
IFSTTAR, LEOST
France

1. Introduction

Negative refractive index (NRI) media are extensively studied nowadays. The interest in these materials keeps on increasing since the year 2000 when a team at the university of California in San Diego (UCSD) published an experimental demonstration of the existence of a material presenting both a negative permittivity and negative permeability (Shelby et al. (2001); Smith, Padilla, Vier, Nemat-Nasser & Schultz (2000)). They also showed that it is necessary to attribute a negative refractive index to such media (Smith & Kroll (2000)). Novel physical phenomena such as the inversion of Doppler's effect, Cerenkov effect and focusing using flat slabs are then predicted based on the theoretical publication V. G. Veselago dating back to 1967 (Veselago (1968)).

Though different terminologies are used for these media (the actual terms are "left handed media", "double negative metamaterial", "negative refractive index metamaterial"), the concept of backward wave propagation (wave with a phase velocity propagating in the opposite direction with respect to the propagation of energy) dates back to at least 1904 (Moroz (n.d.); Tretyakov (2005)). Indeed, H. Lamb has studied this concept for mechanical systems and A. Schuster in the field of electromagnetism. Independently, H. C. Pocklington (Pocklington (1905)) demonstrated theoretically that in a media supporting backward wave propagation, the phase velocity can be directed in the direction of the source, in the inverse direction of the group velocity. Forty years later (in 1944), L. I. Mandelshtam studied the properties of NRI media (Mandel'shtam (1944)) and more than twenty years later, V. G. Veselago published an exhaustive study on NRI media. The interest in these media then decreased up to the year 2000.

The actual revival of interest for these media can certainly be explained published (Smith, Padilla, Vier, Nemat-Nasser & Schultz (2000)). This demonstration has been performed at microwaves by assembling a medium of periodic metallic wires (for negative permittivity) (Pendry et al. (1996)) and a medium of split ring resonators presenting a negative permeability (Pendry et al. (1999)). These media can be assimilated to a crystalline structure of artificial molecules hence the term metamaterial.

Different technological solutions have been proposed to synthesize negative refractive index media, such as the use of backward wave transmission lines (Eleftheriades et al. (2003); Lai et al. (2004)) and photonic crystals in negative phase velocity regime (Gadot et al. (2003); Gralak et al. (2000); Qiu et al. (2003)). The approach to be chosen depends mainly on the applications and the frequency band of interest. All the approaches have the same aim i.e to synthesize NRI metamaterials for the potential applications and considering the industrial and economic potential of such materials.

The synthesis and study of NRI metamaterials is however difficult because of its heterogeneous characteristics. For an easier study of applications of metamaterials in microwave frequency range, homogenization and macroscopic description of these metamaterials can prove to be very helpful. It can specially allow a higher degree of freedom to overcome the fundamental limitations imposed by natural materials on the performances of microwave devices.

In this chapter, the description of NRI resonant metamaterials in terms of a continuous medium will be analyzed. The electrodynamics of NRI materials will first be described. The effective medium theory as applied to NRI resonant metamaterials as well as the calculation methods will then be detailed. Finally numerical results will be presented together with a thorough analysis and interpretation of the effective parameters calculated with respect to the electrodynamics of negative refractive index materials presented for continuous media.

2. Electrodynamics of negative refractive index materials

2.1 Adequate choice of the sign of the refractive index and wave impedance

For backward wave propagation, an adequate choice of the sign of the refractive index $n(\omega)$ given by (1) is necessary.

$$n(\omega) = \pm\sqrt{\varepsilon(\omega)\mu(\omega)}, \tag{1}$$

where $\varepsilon(\omega)$ and $\mu(\omega)$ represent the effective permittivity and permeability respectively.

2.1.1 Refractive index

The determination of the sign in front of the square root of (1) is done thanks to causal properties which the solutions of wave propagation should respect and energy conservation principles. The choice of this sign allows to define, among other parameters, the direction of the outgoing wave with respect to an interface between a NRI and a conventional material.

To demonstrate that for a material with $(\varepsilon(\omega) < 0, \mu(\omega) < 0)$, the sign of the refractive index should be negative, let us consider a current surface in $x = x_0$ (Smith (2000)). The radiation of this surface current in the medium $(\varepsilon(\omega) < 0, \mu(\omega) < 0)$ is then studied as shown on figure 1.

The wave equation in the medium can be written as fol.:

$$\frac{\partial^2}{\partial x^2}E(x) + k^2E(x) = -j\omega\mu J_0(z), \tag{2}$$

where $E(x)$ is the electric fied component along $\hat{x}$, $\vec{J_0} = i_0\delta(x - x_0)\,\hat{z}$ et $\mu = \mu_0\mu_r$. The solution of this equation is given by:

$$E(x) = \alpha\exp(jk\,|x - x_0|). \tag{3}$$

To determine α, let us first calculate :

$$\frac{\partial^2 E(x)}{\partial x^2} = -\alpha k^2\exp(jk\,|x - x_0|) + 2j\alpha k\delta(x - x_0), \tag{4}$$

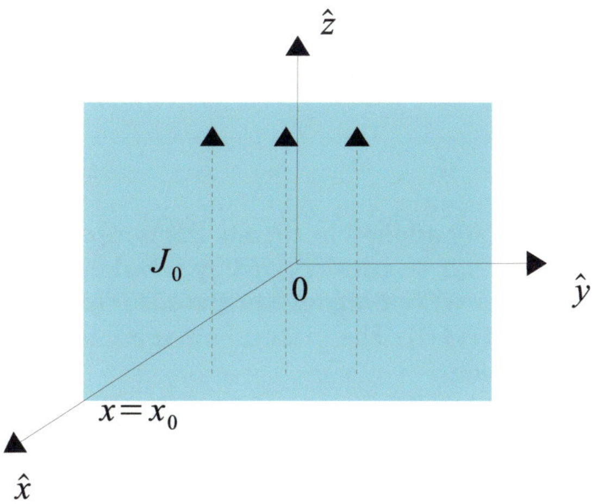

Fig. 1. Surface current J_0 in $x = x_0$ radiating in the medium $(\varepsilon(\omega) < 0, \mu(\omega) < 0)$. The current distribution is considered uniform and infinite in $\hat{y}$ et $\hat{z}$.

and substitute the expressions (4) and (3) in equation (2). α is then given by:

$$\alpha = -\frac{\mu\omega i_0}{2k_0 n} = -\frac{i_0\eta_0}{2}\frac{\mu_r}{n}, \tag{5}$$

and the wave equation becomes:

$$E(x) = -\frac{i_0\eta_0}{2}\frac{\mu_r}{n}\exp(jk\,|x - x_0|). \tag{6}$$

However, if the power P delivered by the current $\vec{J}_0$ to the volume V (Balanis (1989)) is calculated, the fol. equation is obtained:

$$P = -\frac{1}{2}\int_V \vec{E}\cdot\vec{J}_0^{*}dV = \frac{i_0^2\eta_0}{2}\frac{\mu_r}{n}. \tag{7}$$

This equation represents the work done by the source and it must be positive, which implies that $P > 0$ (Balanis (1989)). The ratio μ_r/n must also be positive. *If μ_r is negative, then n must also be negative.* An equivalent demonstration can be done for ε_r. For a propagative medium, the solution retained for the wave equation verifies backward wave propagation.

To determine the constraints with respect to the sign choice of the imaginary part of the refractive index, let us consider the electric field $\vec{E}(\vec{r},\omega)$ in a medium with $n = n' - jn''$ for a time dependence in $\exp(j\omega t)$:

$$\vec{E}(\vec{r},t) = \mathrm{Re}\left[\left|\vec{E}(\vec{r})\right|\exp(-\vec{k_0}\cdot\vec{r}n'')\exp[j(\omega t - \vec{k_0}\cdot\vec{r}n')]\vec{u}_E\right], \tag{8}$$

where $\vec{k_0}$ is the free space wave vector and $\vec{u}_E$ is the unit vector along the direction of the E-field vector $\vec{E}$. If a stable propagation is to be ensured, the magnitude of $\mathrm{Re}[\vec{E}(\vec{r},t)]$ must decrease with time. This implies that the term $\vec{k_0}\cdot\vec{r}n''$ must be positive and:

$$n'' > 0, \tag{9}$$

irrespective of the sign of n' [1]

2.1.2 Wave impedance

Impedance is a concept generally applied to circuits but is also extended to electromagnetic wave propagation. This extension was developped by Schelkunoff and the analogy between the impedance of a medium for wave propagation and the impedance of a transmission line is fully described in (Stratton (1941)). The physical interpretation of the wave impedance Z given here is based on this analogy.

The complex wave impedance of a medium is strongly related to the flux of energy of the wave propagating in the medium. This is why there are fundamental limitations to the values that Z can admit; one of the limitations is directly linked to the passivity of the medium. These limitations apply to both positive and negative refractive index medium.

The passivity or absence of activity within a medium implies that for a plane progressive electromagnetic wave, the mean energy flux must be directed inside the medium in which the wave propagates (Wohlers (1971)). The directions of the vectors $(\vec{E}, \vec{H}, \vec{k})$ and the energy flux $\vec{S}$ for a plane progressive wave at the interface between a conventional material and an NRI is shown in figure 2.

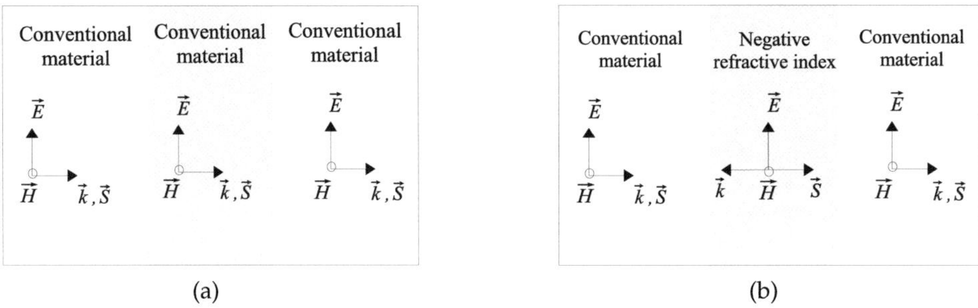

Fig. 2. Direction of the field vectors $(\vec{E}, \vec{H}, \vec{k})$ et $\vec{S}$ for the interaction of a plane wave with at the interface of (a) two conventional material with positive refractive index, and (b) a conventional material and a negative refractive index material.

The wave impedance[2] is defined as the ration of the electric field to the magnetic field component in the propagation plane; the real part is thus given by:

$$\text{Re}[Z(\omega)] = \text{Re}\left[\frac{\bar{E}(\omega)}{\bar{H}(\omega)}\right] = \frac{|\bar{E}(\omega)|}{|\bar{H}(\omega)|}\cos(\varphi_H - \varphi_E), \tag{10}$$

where $\bar{E}(\omega) = |\bar{E}(\omega)|\exp(-j\varphi_E)$ et $\bar{H}(\omega) = |\bar{H}(\omega)|\exp(-j\varphi_H)$. Equation (10) is verified for both positive and negative refractive indexes. The sign of $\text{Re}[Z(\omega)]$ depends only on the term $\cos(\varphi_H - \varphi_E)$.

[1] It can be shown that for the convention $\exp(-j\omega t)$, n'' is positive also but in this case n should be written as $n = n' + jn''$. This is quite similar for the wave impedance, the permittivity and permeability.

[2] The wave impedance is generally defined for a *single* plane wave and in the case of a guided or periodic structure, monomodal wave propagation is assumed and an impedance is assigned to each mode.

The mean of Poynting's vector $\vec{S}_{av}$ is calculated, :

$$\vec{S}_{av}(\vec{r},\omega) = \frac{1}{2}\text{Re}[\vec{E}(\vec{r},\omega) \times \vec{H}^{*}(\vec{r},\omega)] = \frac{1}{2}\left|\vec{E}(\vec{r},\omega)\right|\left|\vec{H}(\vec{r},\omega)\right|\cos(\varphi_H - \varphi_E)\vec{u}_S, \tag{11}$$

where $\vec{E}(\vec{r},\omega) = \left|\vec{E}(\vec{r},\omega)\right|\exp(-j\varphi_E)$, $\vec{H}(\vec{r},\omega) = \left|\vec{H}(\vec{r},\omega)\right|\exp(-j\varphi_H)$ and $\vec{u}_S$ is the unit vector of $\vec{S}_{av}(\vec{r})$.

Knowing that passivity of a medium implies that the energy flux must be directed inside the medium implies that $\vec{S}_{av}(\vec{r},\omega) > 0$. The term $\cos(\varphi_H - \varphi_E)$ is thus always positive for all medium irrespective of the sign of their refractive index(as it can be verified on figure 2). If we apply this restricion to equation (10), the fol. condition is obtained:

$$\text{Re}[Z(\omega)] > 0, \tag{12}$$

for both conventional and NRI materials[3].

There is no particular sign restriction on the imaginary part of the wave impedance. The complex wave impedance provides information non only on wave propagation (as described above) but it also allows physical understanding when there is no wave propagation in a medium (*i.e.* when its imaginary part is much higher that its real part)as to which field component ($\vec{E}$ or $\vec{H}$ fields) is canceled. This information is indeed interesting for the design of artificial magnetic medium such as those based on split-ring resonators. Indeed, if the imaginary part of Z is negative, the medium can be said to be capacitive and there is no wave propagation because of H-field filtering. The response of the medium to an applied magnetic field is thus non-negligible and it can be considered as an artificial magnetic medium.

2.2 Adequate choice of the sign of the effective permittivity and permeability

There are fundamental restrictions limiting the signs that the imaginary part of $\varepsilon(\omega)$ et $\mu(\omega)$ can admit for linear, passive, isotropic homogeneous medium.For conventional material, these restrictions are derived from fundamental theorems of macroscopic electrodynamics (Depine & Lakhtakia (2004); Efros (2004)) and it has been demonstrated in various ways, namely by Callen *et al.* thanks to the fluctuation-dissipation theorems (Callen & Welton (1951)) for arbitrary linear and dissipative systems, and by Landau *et al.* (Landau et al. (1984)) for electromagnetic waves. Based on this last demonstration, we propose to demonstrate the extension of these limitations for NRI materials.

Let us consider a passive, linear, homogeneous, isotropic and dispersive medium of permittivity $\varepsilon(\omega) = \varepsilon'(\omega) - j\varepsilon''(\omega)$ and permeability $\mu(\omega) = \mu'(\omega) - j\mu''(\omega)$. The Poynting vector $\vec{S}(\vec{r},t)$ provides the definition of the power flux density in a medium with variable fields. It can be written in time-domain for dispersive medium as:

$$\vec{S}(\vec{r},t) = \vec{E}(\vec{r},t) \times \vec{H}(\vec{r},t) \tag{13}$$

Using Maxwell-Faraday et Maxwell-Ampere equations in the absence of sources,

$$\nabla \times \vec{E}(\vec{r},t) = -\frac{\partial}{\partial t}\vec{B}(\vec{r},t), \tag{14}$$

[3] This restriction is identical in both conventions $\exp(-j\omega t)$ and $\exp(j\omega t)$

and

$$\nabla \times \vec{H}(\vec{r},t) = \frac{\partial}{\partial t}\vec{D}(\vec{r},t), \tag{15}$$

the divergence of Poynting vector is given as:

$$-\nabla \cdot \vec{S}(\vec{r},t) = \vec{E}(\vec{r},t) \cdot \frac{\partial}{\partial t}\vec{D}(\vec{r},t) + \vec{H}(\vec{r},t) \cdot \frac{\partial}{\partial t}\vec{B}(\vec{r},t). \tag{16}$$

This equation provides an expression of the energy conservation in a dispersive material in time-domain (Balanis (1989)). In frequency domain, the electric and magnetic fields are given, after Fourier transform, by:

$$\vec{E}(\vec{r},t) = \frac{1}{2\pi}\int_{-\infty}^{\infty} \vec{E}(\vec{r},\omega)\exp(j\omega t)\mathrm{d}\omega, \tag{17}$$

$$\frac{\partial}{\partial t}\vec{D}(\vec{r},t) = \frac{j}{2\pi}\int_{-\infty}^{\infty} \omega\varepsilon(\omega)\vec{E}(\vec{r},\omega)\exp(j\omega t)\mathrm{d}\omega. \tag{18}$$

For equation (18), we assume an isotropic material with $\vec{D}(\vec{r},\omega) = \varepsilon(\omega)\vec{E}(\vec{r},\omega)$. Integration of the product of Eq. (17) an Eq. (18) with respect to time gives:

$$\int_{-\infty}^{\infty} \vec{E}(\vec{r},t)\cdot\frac{\partial}{\partial t}\vec{D}(\vec{r},t)\mathrm{d}t = \frac{j}{4\pi^2}\int_{-\infty}^{\infty}\int_{-\infty}^{\infty}\int_{-\infty}^{\infty} \omega\varepsilon(\omega)\vec{E}(\vec{r},\omega)\vec{E}(\vec{r},\omega')\exp\left(j(\omega+\omega')t\right)\mathrm{d}\omega\mathrm{d}\omega'\mathrm{d}t. \tag{19}$$

A first integration with respect to t of the right-hand side of Eq. (19) is done using

$$\int_{-\infty}^{\infty} \exp\left(j(\omega+\omega')t\right)\mathrm{d}t = 2\pi\delta(\omega+\omega').$$

The Dirac distribution is then eliminated by the second integration with respect to ω'. The principle of causality and reality of fields impose (Good & Nelson (1971)):

$$\vec{E}(\vec{r},-\omega) = \vec{E}(\vec{r},\omega)^*,$$

The right-hand side of the equation (19)can finally be written as:

$$\frac{j}{2\pi}\int_{-\infty}^{\infty} \omega\varepsilon(\omega)\left|\vec{E}(\vec{r},\omega)\right|^2\mathrm{d}\omega. \tag{20}$$

After application of the same procedure for magnetic fields $\vec{H}$, we obtain:

$$\int_{-\infty}^{\infty} \vec{H}(\vec{r},t)\cdot\frac{\partial}{\partial t}\vec{B}(\vec{r},t)\mathrm{d}t = \frac{j}{2\pi}\int_{-\infty}^{\infty} \omega\mu(\omega)\left|\vec{H}(\vec{r},\omega)\right|^2\mathrm{d}\omega. \tag{21}$$

Then substituting $\varepsilon(\omega)$ et $\mu(\omega)$ by their complex expression, the energy dissipated (in the period of field variations) in frequency domain is:

$$\int_{-\infty}^{\infty} Q\mathrm{d}t = \frac{1}{2\pi}\int_{-\infty}^{\infty} \omega\left(\varepsilon''(\omega)\left|\vec{E}(\vec{r},\omega)\right|^2 + \mu''(\omega)\left|\vec{H}(\vec{r},\omega)\right|^2\right)\mathrm{d}\omega. \tag{22}$$

The divergence of Poynting vector is expressed as the rate of energy transformation to heat: this dissipated energy depends on $\varepsilon''(\omega)$ et $\mu''(\omega)$. The dependence on $\varepsilon'(\omega)$ et $\mu'(\omega)$ is canceled because the integrand of equation (20) is an odd function of ω [4]. The two terms of the right-hand side of Eq. (22) represent respectively the dielectric and magnetic losses.

The second law of thermodynamics, stating that the entropy of a isolated macroscopic system never decreases, imposes that $Q > 0$. It is thus necessary according to equation (22) that:

$$\varepsilon''(\omega)\left|\vec{E}(\vec{r},\omega)\right|^2 + \mu''(\omega)\left|\vec{H}(\vec{r},\omega)\right|^2 > 0 \tag{23}$$

for positive frequencies ($\omega > 0$). The laws of Thermodynamics also express the irreversible nature of physical processes and the fundamental difference between two types of energy: work and heat (Yavorski & Detlaf (1975)). The energy dissipated by fields into heat is irreversible. In other terms, there can be no exchange between the work done by either the electric field [$\vec{E}(\vec{r},\omega)$ or magnetic field $\vec{H}(\vec{r},\omega)$] and the heat dissipated by the other, implying:

$$\varepsilon(\omega)'' > 0 \quad \text{and} \quad \mu(\omega)'' > 0 \tag{24}$$

This demonstration can be very easily extended to NRI materials. Indeed, the starting point of the demonstration is energy conservation through the expression of the divergence of Poynting vector [Eq. (16)] written thanks to Maxwell-Ampère [Eq. (15)] and Maxwell-Faraday [Eq. (14)] equations as well Poynting theorem [Eq. (13)]. For a NRI material, these equations and theorems are valid (Veselago (1968)). Indeed, only the direction of the vector $\vec{k}$ changes in a NRI material thus giving a negative value for the product $\vec{S}\cdot\vec{k}$ (This product is positive for conventional materials). This is easily verified for a monochromatic wave but can actually also be verified for non-monochromatic wave.

Let us consider for instance, the mean of Poynting vector for a dispersive material excited by the superposition of two monochromatic waves of angular frequency ω_1 et ω_2 such that $\omega_1 \neq \omega_2$ (Pacheco-Jr et al. (2002)):

$$< \vec{S}(\vec{r},t) >= \frac{|E|^2}{2}\left(\frac{\vec{k}(\omega_1)}{\omega_1\mu(\omega_1)} + \frac{\vec{k}(\omega_2)}{\omega_2\mu(\omega_2)}\right) \tag{25}$$

The relation between the direction of Poynting vector $\vec{S}$ and that of the wave vector $\vec{k}$ is clearly shown by this equation. The direction of $\vec{S}$ is independent of the sign of the refractive index and for any propagative medium, *i.e.* when $\vec{k}$ is real, the ratio k/μ is positive. As shown before, if k takes negative values, then μ will be negative too.

3. Effective parameters of resonant NRI metamaterials

3.1 Effective medium theory as applied to metamaterials

The concept of effective medium for the description of heterogeneous systems by a homogeneous one is very attractive in different field of physics. Homogenization procedures allowing the definition of an effective macroscopic response from physical parameters characterizing the heterogeneous system are generally developed. In our case, from the microscopic parameters (geometrical and physical definitions) of the metamaterial, a macroscopic electromagnetic response can be obtained. If this macroscopic definition

[4] Principle of causality imposes that (Good & Nelson (1971)): $\varepsilon(-\omega) = \varepsilon(\omega)^*$ and $\mu(-\omega) = \mu(\omega)^*$.

is performed in accordance with the electrodynamics of continuous medium, they can afterwards be used in Maxwell's equations to predict propagation phenomena and provide physical insight into the design of metamaterial-based microwave and optical devices.

In this chapter, the NRI metamaterials considered are assumed periodic and based on resonant inclusions such as the combination of Split Ring Resonator and wire medium. The general definition of the relevant dimensions for the definition of the effective parameters of such periodic medium is depicted on figure 3 (Baker-Jarvis, Janezic, Riddle, Johnk, Kabos, Holloway, Geyer & Grosvenor (2004)).

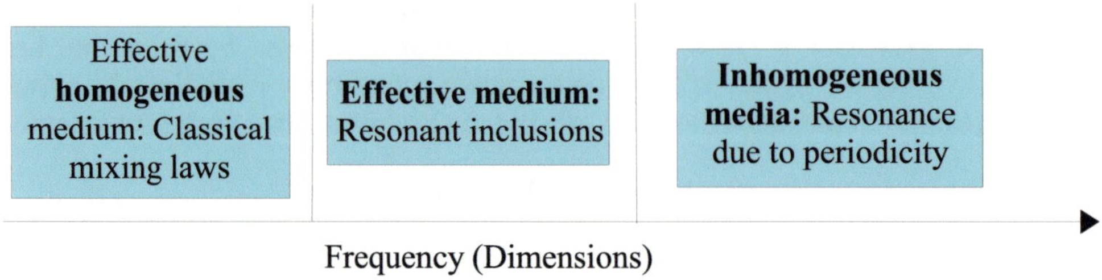

Fig. 3. Dimensions for effective medium of resonant periodic metamaterial.

The left-hand side region represents the quasi-static region where the wavelength is much bigger than the periodicity of the inclusions. The effective parameters of the composite in this zone can be easily calculated using quasi-static solutions or classical mixing rules (Berthier (1993)).

In the right-hand side region, the composite is heterogeneous and the resonances of the medium can be directly linked to the periodicity. Such a composite cannot be considered homogeneous. To study the propagation characteristics of these media, full-wave numerical methods are generally required. The volume under study has to be discretized : a unit-cell is defined and Floquet-Bloch boundary conditions are used. This case is typically the working regime of photonic crystals.

The intermediate region is a region where the inclusions are resonant. The electrical dimensions of the inclusions as well as the periodicity are small compared to the wavelength. Resonant NRI metamaterials belong to this intermediate region. Such a medium is generally considered homogeneous. However, the question which remains to be answered is: how should one study the characteristics of such a medium and how should the associated effective medium be defined?

There is indeed no simple or unique definition to the effective medium concept. The possible approach and definition which will be used in this chapter for NRI metamaterials will be described hereafter.

3.2 Definition of the effective medium concept for composites of the intermediate region

When an EM field is applied to a composite, the fields in the composite results from the interaction between the applied field and the reaction of the inclusions constituting the composite (Baker-Jarvis, Janezic, Riddle, Johnk, Kabos, Holloway, Geyer & Grosvenor (2004); Baker-Jarvis, Kabos & Holloway (2004)). The local field in the composite can be freely propagative, propagative with attenuation or evanescent. The resulting local field is a complicated physical process whereby the applied field polarizes the inclusions which in turn polarize the neighboring inclusions. The group of inclusions then react by creating a modified local field. The presence of inclusions (or perturbations) in a given environment can make an initially evanescent field propagative (de Fornel (1997)). A common example is the insertion of inclusions in a guide under the cut-off frequency. All these complex interactions are visible at the microscopic or local scale. However, the field in a material as expressed in Maxwell's equations is the macroscopic field, usually defined by constitutive equations.

The definition of constitutive parameters require the determination of a relationship between the local field, the applied field and the macroscopic field. the theory of local field of Lorentz (Berthier (1993); Tretyakov (2003)) can be used but it is not always adequate (Baker-Jarvis, Janezic, Riddle, Johnk, Kabos, Holloway, Geyer & Grosvenor (2004); Baker-Jarvis, Kabos & Holloway (2004)). It has been however applied to certain types of composites. The polarizabilities are calculated analytically and the theory of Lorentz then provides the macroscopic parameters. Numerous examples of such calculations are given in ((Tretyakov (2003)) and the papers there cited there. Other methods have also been used in the literature such as those introduced by O. Keller et J. Baker-Jarvis (Baker-Jarvis, Kabos & Holloway (2004); Keller (1996)) but they rely on statistical and quantum approaches. The discussion will be restricted to the context of *classical* electrodynamics.

The definition of effective medium can be mainly performed in two distinct categories of approaches. The first category can be termed *locale* (ğ 3.3) and the second one *global* (ğ 3.4). In the first case, the effective parameters are defined directly from local fields while the second one allows a definition based on global propagation characteristics of the periodic system, for instance from the model of scattering parameters.

3.3 Local approaches

When they are not based on analytical approaches, the input data are the fields or electric and magnetic induction calculated using full-wave numerical methods. The definition of effective parameters from local fields is not straightforward. Three methodologies can be distinguished. The first one consists in defining an equivalence between the local field calculated and the effective parameters of a corresponding homogeneous medium (Pincemin (1995); Silveirinha & Fernandes (2004a;b; 2005a;b)). The second methodology consists in the calculation of the propagation constant from the phase velocity locally determined using time-domain numerical modeling methods (Moss et al. (2002)). Finally, the third methodology consists in the definition of effective parameters by calculation a linear, surface-based or volume-based mean field values on adequately chosen geometries. Several methods are available in the literature (Acher et al. (2000); Bardi et al. (2002); Lerat et al. (2005); Lubkowski et al. (2005); Pendry et al. (1999); Smith (2005); Smith, Vier, Kroll & Schultz (2000); Weiland et al. (2001)). In the method given by Acher *et al.* (Acher et al. (2000)), it is worth noting that a convergence is demonstrated between effective parameters calculated using this type of numerical approach and those obtained by the Bruggeman extended theory of effective medium (taking into account magnetic polarizability) for the asymptotic case of metal-dielectric slab.

3.4 Global approaches

Global approaches provide effective parameters starting from global responses of the periodic system. These responses such as the scattering matrix, the reflection and transmissions coefficients, resonant frequencies are observable or measurable quantities with can be either experimentally determined or numerically calculated from fields defined locally in the unit cell of the periodic NRI metamaterial.

The transformation of the local field to the scattering matrix relies on an analogy between propagation in a periodic structure and in waveguides and circuits. It consists in assimilating the periodic structure in a multiple-access system and to study transmission and reflection between the different accesses. The development of such an analogy requires a few assumptions (Hélier (2001); Richalot (1998); Rivier & Sardos (1982)), namely:

- linearity: the vectors $\vec{E}$ et $\vec{D}$, $\vec{B}$ et $\vec{H}$ are linked to one another by linear relationships,
- stationarity: the properties of the system are invariant with respect to time,
- absence of radiation: the system is closed and energy exchange can only exist between the system accesses,
- existence of pure mode: each access of the system supports a pure mode, i.e. a unique propagation mode characterized by a given propagation constant. If this assumption is not verified, then sufficient supplementary virtual accesses have to be defined to account for higher propagation modes.

Figure 4 depicts an example of a system with three physical accesses modeled using the generalized scattering matrix method described before. Each physical access is artificially decomposed in N virtual accesses, where N represent the number of modes to be taken into account at each physical access.

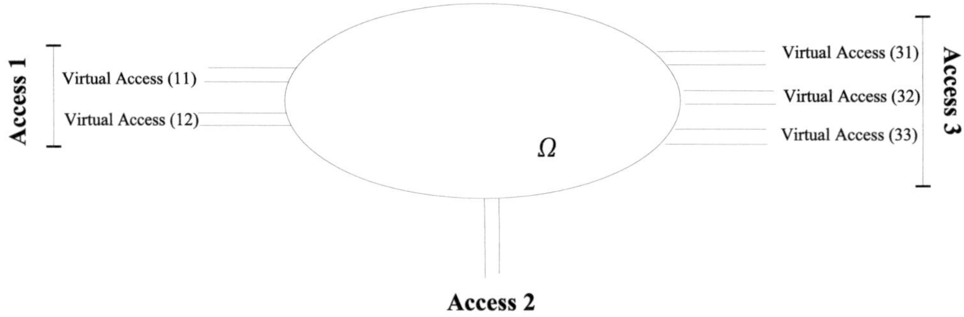

Fig. 4. Example of a system with three physical accesses modeled by a system of N virtual accesses to take inbto account the number of modes present or excited at each physical access.

Such a scattering matrix allows the complete characterization of the structure both in emitting and receiving modes both in near and far fields. The reflection and transmission matrices can thus be directly determined from this matrix followed by the effective parameters; this procedure is further detailed hereafter.

3.5 Calculation of effective parameters of resonant metamaterials

The calculation method described here is belongs to the category of global approaches as defined in section (3.4) and can be divided in two parts, namely for NRI metamaterials structures finite and infinite in the direction of propagation.

3.5.1 NRI metamaterials finite in the propagation direction

For NRI metamaterials finite in the direction of wave propagation, the problem to be solved is the substitution of this periodic structure by a homogeneous slab of same thickness as shown in figure 5.

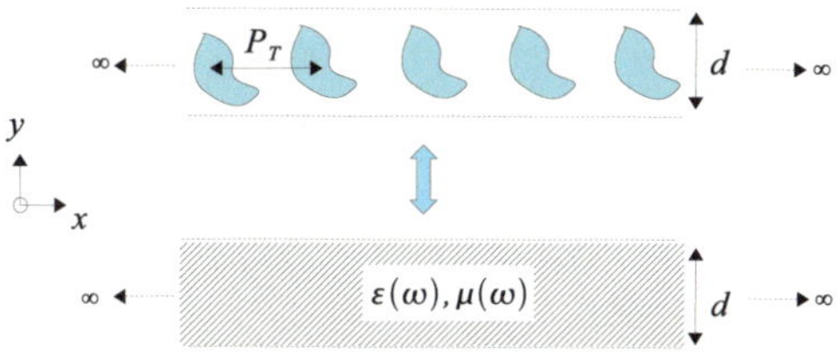

Fig. 5. Equivalence between a periodic composite of transverse periodicity P_T by a homogeneous slab of same thickness d, effective permittivity $\varepsilon(\omega)$ and permeability $\mu(\omega)$.

This substitution or equivalence is only valid under a few assumptions (Lalanne & Hutley (2003); Lalanne & Lalanne (1996)): (i) Only the first mode propagates in the incident medium, transmitted medium and the periodic structure. This condition is given by the fol. equation:

$$|\beta| \leq \frac{\pi}{P_T}. \tag{26}$$

β is the propagation constant in each medium. (ii) Evanescent modes should not be present in the $x0y$ plane. If only the first mode can propagate in the periodic structure with a speed of $c_0/n(\omega)$ where $n(\omega)$ the interference phenomenon is identical to the one which occurs in a homogeneous slab. However, the existence of higher order modes give a much more complex interference phenomenon such that the equivalence with a homogeneous slab is no longer valid.

In the case of a metal-dielectric composite, these equivalence conditions are not automatically satisfied. Indeed, the existing propagation modes and their associated propagation constants depend highly on the nature of the inclusions: their geometry, size and distribution.

The calculation of effective parameters of NRI metamaterials under these assumptions is done in two steps. The first one consists in the determination of the complex reflection and transmission coefficients which can be numerically calculated according to the generalized scattering parameters described in section 3.4. They can also be obtained by experimentally. The second step is then the calculation of the effective permittivity and permeability $(\varepsilon(\omega), \mu(\omega))$ from these reflection and transmission using inversion methods. These methods can either be direct using analytical inversion of Fresnel equations, $(r, t) = f(\varepsilon(\omega), \mu(\omega))$ or performed by an iterative approach. Both approaches are described here.

3.5.1.1 Direct method - Nicholson Ross Weir (NRW) approach

In the NRW method, the wave impedance and refractive index are first calculated. The effective permittivity and permeability are then deduced. The normalized wave impedance of a slab can be described by analogy as the input impedance of a transmission line thus containing information not only on $\bar{E}/\bar{H}$ at the interface of two lines or medium but also of the propagation constant inside the propagating medium. Z is given by:

$$Z = \pm\sqrt{\frac{(1+r)^2 - t^2 e^{-2jk_0 d}}{(1-r)^2 - t^2 e^{-2jk_0 d}}}, \tag{27}$$

where d is the slab thickness, $k_0 = 2\pi/\lambda_0$ is the wave number λ_0 the free-space wavelength. The choice of the sign in front of the square root of Z is done according to the definitions given in section 2.1.

The real part of the refractive index n is given by equation (28):

$$n' = \frac{\arctan\left(Im(Y)/Re(Y)\right) \pm m\pi}{k_0 d}, \tag{28}$$

where $m \in Z$. The variable Y is defined as:

$$Y = e^{-jnkd} = X \pm \sqrt{X^2 - 1}, \tag{29}$$

where

$$X = \frac{e^{jk_0 d}}{2t}\left(1 - r^2 + t^2 e^{-2jk_0 d}\right). \tag{30}$$

The choice of the value of m in equation (28) constitute one of the ambiguities of this method which can be solved in different ways, namely (i) by considering various thicknesses and assuming that there is no coupling between different layers of metamaterials in the direction of propagation (Markos & Soukoulis (2001)), (ii) by comparing the measured group arrival time to the calculated one (Baker-Jarvis, Janezic, Riddle, Johnk, Kabos, Holloway, Geyer & Grosvenor (2004)).

The imaginary part of the refractive index n'' is given by:

$$n'' = \frac{ln\,|Y|}{k_0 d} \tag{31}$$

n'' is calculated using the fundamental limitation described in 2.1, i.e. $n'' > 0$ for both positive or negative refractive index materials.

Using the two independent equations (32 and 33) the effective permittivity and permeability can be deduced.

$$n = \sqrt{\varepsilon(\omega)}\sqrt{\mu(\omega)} \quad \text{and} \quad Z = \sqrt{\mu(\omega)/\varepsilon(\omega)} \tag{32}$$

$$\varepsilon(\omega) = \frac{n}{Z} \quad \text{and} \quad \mu(\omega) = nZ \tag{33}$$

It should be noted that the refractive index is defined as: $n = \sqrt{\varepsilon_{eff}}\sqrt{\mu_{eff}}$ such that when $\varepsilon(\omega)$ and $\mu(\omega)$ are simultaneously negative, the real part of n is also negative. The common formula $n = \sqrt{\varepsilon_{eff}\mu_{eff}}$ should not be used.

3.5.1.2 Iterative method - Optimization approach

This method consists in the minimization of the difference between the functions $F_1(x)$, $F_2(x)$ and the scattering parameters S_{11} et S_{21} according to the fol. cost function:

$$E(x) = |F_1(x) - S_{11}|^2 + |F_2(x) - S_{21}|^2 \qquad \text{with} \quad x = \{\varepsilon(\omega), \mu(\omega)\} \tag{34}$$

The functions $F_1(x)$ et $F_2(x)$ are complex and represent respectively the Fresnel reflection and transmission coefficients defined for a magneto-dielectric slab of thickness d and of infinite transverse dimensions. For a plane wave having an incident angle of θ_i and a polarization TE or TM, the reflection coefficient r and transmission t are defined by:

$$r = \frac{A - B}{A + B} \quad \text{et} \quad t = \frac{2A}{A + B}, \tag{35}$$

$$\text{avec} \quad A = \cos qd + \frac{j \sin qd}{Z_0}, \qquad B = Z_0 \left(\frac{j \sin qd}{Z} + \frac{\cos qd}{Z_0} \right)$$

$$\text{et} \quad q = k_0 \sqrt{1 - \left(\frac{1}{n} \sin \theta_i \right)^2}.$$

For the TE polarization,

$$Z = \frac{\omega \mu_{eff}}{q} \quad \text{and} \quad Z_0 = \frac{\cos \theta_i}{\varepsilon_0 c_0}. \tag{36}$$

For TM polarization,

$$Z = \frac{q}{\omega \varepsilon_{eff}} \quad \text{and} \quad Z_0 = \frac{\mu_0 c_0}{\cos \theta_i}. \tag{37}$$

In our case, the functions $F_1(x)$ and $F_2(x)$ of the equation (34) are replaced by r et t equation (35). The cost function (34) is minimized using the non-linear mean square algorithm of Levenberg-Marquardt. To ensure good results, this algorithm needs to start from a feasible point. This is why a large choice of values for the couple $(\varepsilon(\omega), \mu(\omega))$ is done for the starting point in frequency. Then if the frequency sampling from the numerical simulations is fine enough and the functions considered to be continuous, the starting point chosen is the one for the previous frequency point.

For a few composites, the algorithm can converge to many solutions for large values of the thickness of the slab. These solutions are not local minima but are solutions to Fresnel equations. To choose the right solutions, two physical criteria have been defined:

- If the scattering parameter S_{21} is close to 1, the structure is propagative. The refractive index, being a parameter representative of propagation [5], its imaginary part must be close to zero,
- If S_{11} is close to one, there is no propagation in the structure ; the real part of the refractive index must be close to zero.

These criteria are particularly appropriate for NRI resonant metamaterials and may not be adequate for all type of composites. The principal limitation is that the starting point must be far from resonance and the composite should not present high dissipative losses.

3.5.2 Periodic structure infinite in the propagation direction

If the EM wave propagation is considered in a periodic medium such as the one presented on figure 6(a), the solution of the wave equation provides solution for the propagation constant which are given by $k_n = k + 2m\pi/P$ where $m \in Z$ and $P = P_L$ or P_T.

[5] The imaginary part of the refractive index does not allow to calculate losses by dissipation of a medium (Landau et al. (1984)). It can only represent the presence or absence of propagation in a medium.

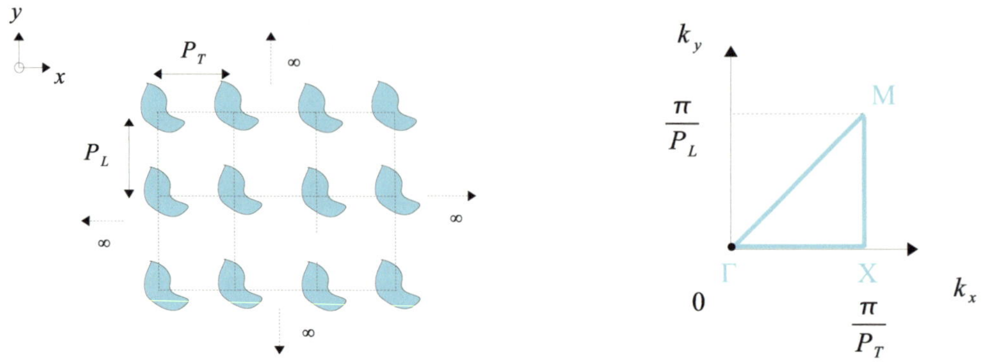

(a) Réseau périodique bi-dimensionnel

(b) Contour de Brillouin associé

Fig. 6. Réseau périodique bi-dimensionnel et zone irréductible de Brillouin associée.

Because the structure is periodic, the analysis of the propagation in only a unit cell of the constant k is enough : the fundamental analysis domain is defined for $-\pi < kP < \pi$. This fundamental domain consists sufficient information for defining propagation inside the whole structure. Figure 6(b) shows the irreducible Brillouin zone associated to the periodic structure for $P_T = P_L$. This Brillouin zone can be briefly described in terms of the propagation constants (k_x, k_y) in the following way:

1. the ΓX contour corresponds to propagation constants $k_yP = 0$ et $k_xP \in [0;\pi]$. for this contour, only normal incidence is considered.
2. the XM contour corresponds to propagation constants $k_xP = \pi$ et $k_yP \in [0;\pi]$. The incidence angles vary from 0°to 45°.
3. the $M\Gamma$ contour corresponds to propagation constants k_xP et $k_yP \in [0;\pi]$. The only incidence angle considered is 45°.

To calculate the two dimensional dispersion diagram of an arbitrary periodic NRI resonant metamaterial, a source-free eigenmode solver of a numerical modeling tool such as Ansoft HFSS (HFSS (2004)) can be used. The calculation volume is sampled by finite elements in the case of the software HFSS and for specific periodic boundary conditions, the eigenvalues of the "periodic cavity" are searched. A couple of propagation constants (k_x, k_y) belonging to the Brillouin zone is imposed as boundary condition and the eigenfrequency is calculated such that the source-free Maxwell equations with the boundary conditions are satisfied. The calculation of each eigenfrequency is performed in an iterative manner (Chang (2005)). To ensure reasonable calculation time, it is thus necessary to impose two parameters which are the lowest eigenfrequency to be calculated and a limited number of eigen frequencies.

4. Numerical results and interpretation of effective parameters of resonant NRI metamaterials

The first unit cell (figure 7a) considered here as resonant NRI metamaterials is based on the metamaterial edge-side coupled split-ring resonators (EC-SRR) proposed in (Greegor et al. (2003)) because the dissipative losses presented by these metamaterials are relatively low. The second NRI metamaterial unit cell considered are based on broad-side coupled

NRI metamaterials (figure 7b) proposed in (Seetharamdoo et al. (2004)) for the reduced bianisotropic properties they present.

(a) EC-SRR unit cell (b) BC-SRR unit cell

Fig. 7. Unit cell of NRI resonant metamaterials constituted of BC-SRR and EC-SRR and metallic lines. These inclusions are printed on the dielectric teflon substrates ($\varepsilon = 2.2$, $\tan\delta = 9 \times 10^{-4}$). The periodicities P_H=4.5 mm, P_T=3.3 mm, and d=3.3 mm.

The unit cell are simulated using Ansoft HFSS and the reflection and transmission coefficients are shown figure 8. A resonance can be observed where the metamaterials are transparent to

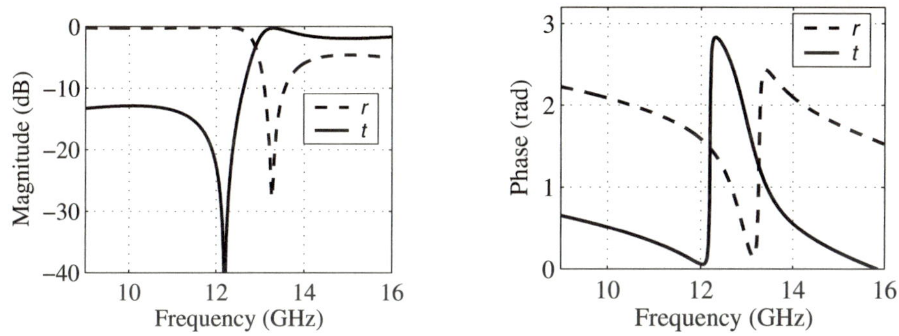

(a) Magnitude of r and t for EC-SRR (b) Phase of r and t for EC-SRR

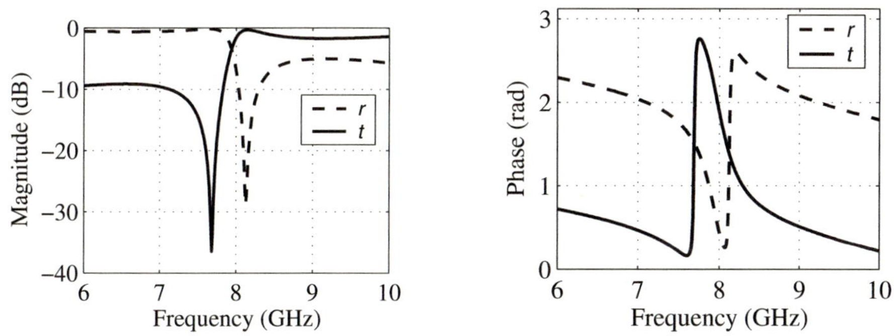

(c) Magnitude of r and t for BC-SRR (d) Phase of r and t for BC-SRR

Fig. 8. Reflection and Transmission coefficients of metamaterials constituted of BC-SRR et EC-SRR the incident wave at the frequencies of 12.3 GHz for the EC-SRR and 8.1 GHz for the BC-SRR respectively.

4.1 Effective parameters calculated by inversion methods

The effective parameters are then calculated by inversion methods presented in the previous section and the refractive index, wave impedance and permittivity and permeability are shown on figure 9.

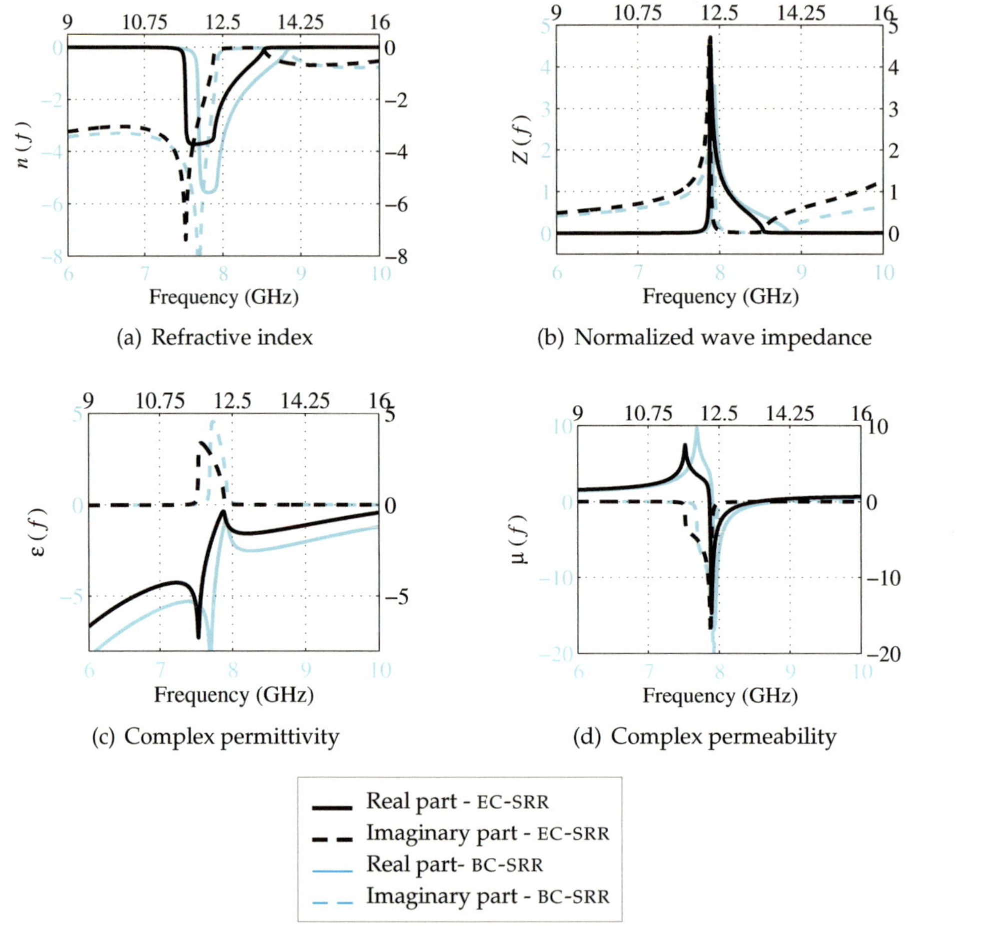

(a) Refractive index (b) Normalized wave impedance (c) Complex permittivity (d) Complex permeability

Fig. 9. Effective paramaters of NRI metamaterials. The upper frequency scale correspond to EC-SRR structures and lower one to BC-SRR.

The NRI metamaterials constituted of EC-SRR and BC-SRR present respectively a negative refractive index from 11.5 GHz - 13.3 GHz and from 7.7 GHz - 8.7 GHz [figure 9(a)]. It should be noted that the refractive index saturates in both cases (11.5 GHz $< f <$ 12.3 GHz for the EC-SRR and 7.7 GHz $< f <$ 7.9 GHz for the BC-SRR). This maximum value can be predicted by equation (26). The effective permeability shown on figure 9(d) is resonant and the imaginary part is positive. The effective permittivity shown on figure 9(c) is anti-resonant and presents also positive values for the imaginary parts. The main frequency bands of interest of these NRI metamaterials are given in table 1.

	BC-SRR	EC-SRR
Negative refractive index	7.7 -8.7 GHz	11.5-13.3 GHz
Negative permeability	7.9 -8.7 GHz	12.3-13.3 GHz
Negative permittivity	6-10 GHz	9-16 GHz
Saturation of the real part of refractive index	7.7-7.9 GHz	11.5-12.3 GHz
$\Im m(\varepsilon) > 0$	7.7-7.9 GHz	11.5-12.3 GHz
$\Im m(\mu) > 0$	7.7-7.9 GHz	11.5-12.3 GHz

Table 1. Frequency bands of interest for NRI metamaterials based on BC-SRR and EC-SRR.

A similar behavior can be observed for both metamaterials but with a shift in frequency. This shift as explained in (Seetharamdoo et al. (2004)) is due to higher capacitive coupling in the BC-SRR compared to the EC-SRR. There is indeed a frequency band for which the real part of the refractive index, effective permittivity and permeability are negative. However, in a part of this frequency band the imaginary parts of $\varepsilon(\omega)$ and $\mu(\omega)$ are positive which not a physically correct as described in section 2.2. This frequency band deserves further analysis and in the next sections for better understanding of these results, a dispersion diagram as well as a multimodal analysis will be proposed for the BC-SRR NRI metamaterial. The choice of this metamaterial for further analysis is justified by the fact that it has also been shown to be 2D-isotropic (Seetharamdoo (2006)).

4.2 Dispersion diagram of NRI metamaterials

The dispersion diagram is calculated using the method described in section **??**. This diagram shown on figure 10(a) gives information on the modes that can propagate in the periodic medium in two dimensions in the irreducible Brillouin zone. The dispersion diagram of the

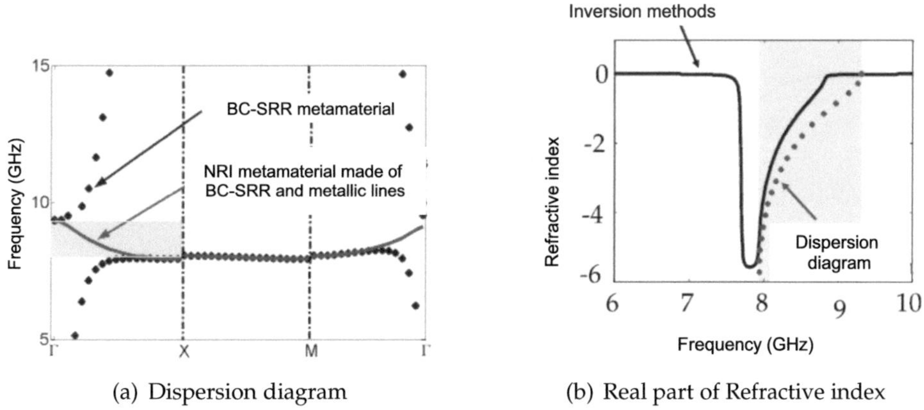

(a) Dispersion diagram

(b) Real part of Refractive index

Fig. 10. (a) Two dimensional dispersion diagram of the medium with BC-SRR only and the NRI metamaterial in the irreducible Brillouin zone. (b) Superposition of the refractive index calculated from the dispersion diagram and the one calculated by the inversion methods. The shaded frequency band represents the frequency band where the refractive index is negative and where there is backward propagation.

metamaterial constituted of only BC-SRR (without the metallic line medium) is also shown. In the shaded frequency band, the metamaterial with BC-SRR only presents a forbidden frequency band while in association with the metallic lines, a propagated frequency band is observed. The phase velocity given by the slope of the curve is negative; the propagation is hence a backward wave propagation. The refractive index can be calculated from this phase velocity and it is compared to the one calculated using inversion methods. As it can be observed, there is indeed a frequency band (7.9 - 8.7 GHz) where both results are in good agreement.

However, in the frequency band (7.7 - 7.9 GHz) where the calculation of effective parameters by inversion methods yield unphysical results, the dispersion diagram shows no propagation. This strongly suggests that the results obtained by the inversion method in this frequency band is not correct and is caused by the finite thickness of the structure. If the structure were large enough in the direction of propagation to represent a periodic or a continuous medium, these unphysical results would not have been obtained. Unfortunately, either in measurements or in the design of NRI metamaterials using numerical modeling, it is not always possible to analyze large structures due to the cost or resources required for the calculation.

4.3 Solution proposed: multi-modal analysis

A simple solution to verify the validity of the results given by inversion methods is to make a multimodal analysis of the periodic NRI resonant metamaterial to detect the existence of higher order modes which would definitely result in incorrect effective parameters calculation by inversion methods using a finite-size structure in the direction of propagation. Figure 11 depicts the modal S_{21} parameters and the associated propagation constants for the first two modes of the periodic structure [6].

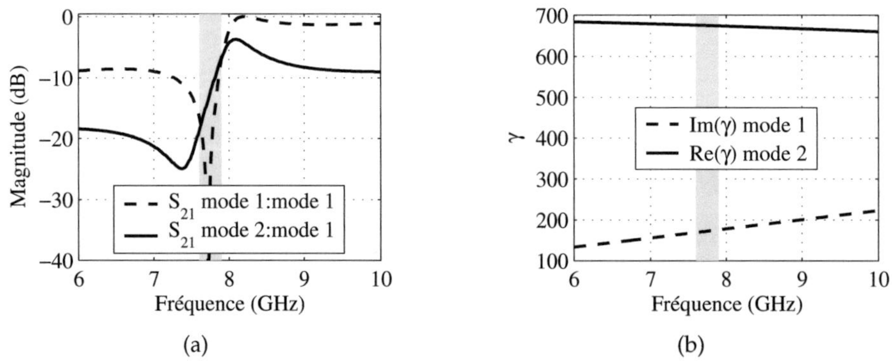

Fig. 11. (a) Modal Scattering parameter S_{21} for the first two modes. (b) Propagation constants of these first two modes. Only $\text{Im}(\gamma)$ is shown for the first mode $\text{Re}(\gamma)$ for the second mode because the corresponding imaginary and real parts are close to zero. The shaded frequency band represents the frequency band where the unphysical results have been observed.

The scattering parameter S_{21} of the fundamental mode presents a resonance at frequency close to 7.8 GHz. Around this frequency and in the shaded frequency band, a second mode can be

[6] S_{21} Mode2:Mode1 represents for instance what is observed from the profile of the second mode on access 2 when only the first mode is excited on access 1.

observed whose magnitude is higher that that of the fundamental mode. In figure 11(b), this second mode can be seen to be evanescent while the first one is propagative. The value of the propagation constant of the evanescent mode is low enough to shown that it can propagate through a few layers of the structure. This implies that the evanescent modes do participate to the interference phenomena is this frequency band and this effect will be more visible with lower dissipative losses in the resonant NRI metamaterial (Seetharamdoo (2006)).

This analysis can prove to be very useful to verify if the assumptions made while using the inversion methods are violated. In this case, one can conclude that the effective parameters calculated in this frequency band is incorrect and non-physical and should thus not be presented or interpreted (Seetharamdoo et al. (2005)).

5. Conclusion

The electrodynamics of NRI materials and the fundamental limitations related to the signs of refractive index, wave impedance, effective permittivity and permeability, both in real and imaginary parts have been fully described. The effective medium theory as it is applied to NRI resonant materials have been detailed with a description of the assumptions linked to this theory for cases of finite thickness in the direction of propagation and infinite dimensions. The methods used for the calculation of effective parameters have been given and applied to numerical models of NRI resonant metamaterials. Unphysical results have been obtained: the imaginary part of the effective permittivity and permeability takes positive values. It has been shown that this is mainly due to the finite size of the structure and that there is a frequency band where the results obtained by the classical inversion methods for the calculation of effective parameters are not correct and this frequency band can be defined thanks to complementary analysis like the calculation of a dispersion diagram and a multimodal analysis.

6. References

Acher, O., Adenot, A. L. & Duverger, F. (2000). Fresnel coefficients at an interface with a lamellar composite material, *Phys. Rev. B* 62: 13748.

Baker-Jarvis, J., Janezic, M. D., Riddle, B. F., Johnk, R. T., Kabos, P., Holloway, C. L., Geyer, R. G. & Grosvenor, C. A. (2004). *Measuring the permittivity and permeability of lossy materials : solids, liquids, metals, building materials and negative-index materials*, Electromagnetics division, Natl. Inst. Stand. Technol. (NIST) : Tech. Note 1536, Boulder, USA.

Baker-Jarvis, J., Kabos, P. & Holloway, C. L. (2004). Nonequilibrium electromagnetics: Local and macroscopic fields and constitutive relationships, *Phys. Rev. E* 70: 036615–1–13.

Balanis, C. A. (1989). *Advanced engineering electromagnetics*, John Wiley and Sons, Canada.

Bardi, I., Remski, R., Perry, D. & Cendes, Z. (2002). Plane wave scattering from frequency-selective surfaces by the finite element method, *IEEE Trans. on Mag.* 38: 641.

Berthier, S. (1993). *Optique des milieux composites*, Polytechnica, Paris.

Callen, H. B. & Welton, T. A. (1951). Irreversibility and generalized noise, *Phys. Rev.* 83: 34–40.

Chang, K. (2005). Encyclopedia of rf and microwave engineering, John Wiley and Sons Inc (USA).

de Fornel, F. (1997). *Les Ondes évanescentes en optique et en optoélectronique*, Eyrolles, Paris.

Depine, R. A. & Lakhtakia, A. (2004). Comment i on: Resonant and anti-resonant frequency dependence of the effective parameters of metamaterials, *Phys. Rev. E* pp. 048601–1.

Efros, A. L. (2004). Comment ii on: Resonant and antiresonant frequency dependence of the effective parameters of metamaterials, *Phys. Rev. E* pp. 048602–1–2.

Eleftheriades, G. V., Siddiqui, O. & Iyer, A. K. (2003). Transmission line models for negative index media and associated implementations without excess resonators, *IEEE Microwave and wireless Comp. Lett.* 13: 51–35.

Gadot, F., Akmansoy, E., Massoudi, S. & de Lustrac, A. (2003). Amplification of anomalous refraction in photonic band gap prism, *Elec. Lett.* 39: 528–529.

Good, R. H. & Nelson, T. J. (1971). *Classical theory of electric and magnetic fields*, Academic press Inc., London.

Gralak, B., Enoch, S. & Tayeb, G. (2000). Anomalous refractive properties of photonic crystals, *J. Opt. Soc. of America : A* 17: 1012–1020.

Greegor, R. B., Parazzoli, C. G., Li, K. & Tanielian, M. H. (2003). Origin of dissipative losses in negative index of refraction materials, *Appl. Phys. Lett.* 82: 2356–2358.

HFSS (2004). *High Frequency Structure Simulator v 9.0, finite element package*, Ansoft Corp.

Hélier, M. (2001). *Les cours de Supélec : Techniques micro-ondes*, Ellipses, Paris.

Keller, O. (1996). Local fields in the electrodynamics of mesoscopic media, *Phys. Rep.* 268: 85–262.

Lai, A., Itoh, T. & Caloz, C. (2004). Composite right/left-handed transmission line metamaterials, *IEEE microwave magazine* pp. 34–50.

Lalanne, P. & Hutley, M. (2003). Artificial media optical properties - subwavelength scale, dans *Encyclopedia of optical engineering* publié par Dekker Encyclopedias - Taylor and Francis Group (USA).

Lalanne, P. & Lalanne, D. (1996). On the effective medium theory of subwavelength periodic structures, *J. Mod. Opt.* 43: 2063.

Landau, L. D., Liftchitz, E. M. & Pitaevskii, L. P. (1984). *Electrodynamics of continuous media*, Butterworth - Heinemann, Oxford.

Lerat, J.-M., Acher, O. & Malléjac, N. (2005). Calcul numérique des paramètres effectifs d'un métamatériau par intégrale des champs, 14èmes *Journées Nationales Microondes (JNM)*, Nantes, France.

Lubkowski, G., Schuhmann, R. & Weiland, T. (2005). Computation of effective material parameters for double negative metamaterial cells based on 3d field simulations, *Negative refraction: Revisiting electromagnetics from microwaves to optics - EPFL Latsis symposium*, Lausanne, Suisse.

Mandel'shtam, L. I. (1944). Lectures on certain problems in the theory of oscillations, *Recueil intégral des travaux, tome 5 - Publié par Leningrad Akademiya Nauk SSRR, 1950*, traduit du Russe par E. F. Kuester, pp. 461–467.

Markos, P. & Soukoulis, C. (2001). Left-handed materials, *arXiv:Cond-mat/0212136* pp. 1–11.

Moroz, A. (n.d.). Some negative refractive index material headlines long before veselago work and going back as far as to 1905..., http://www.wave-scattering.com/negative.html. Lien du 11 mars 05.

Moss, C. D., Grzegorczyk, T. M., Zhang, Y. & Kong, J. A. (2002). Numerical studies of left handed metamaterials, *Progress in Electromagnetics Research* 35: 315.

Pacheco-Jr, J., Grzegorczyk, T. M., Wu, B.-I., Zhang, Y. & Kong, J. A. (2002). Power propagation in homogeneous isotropic frequency dispersive left-handed media, *Phys. Rev. Lett.* 89: 257401–1–4.

Pendry, J. B., Holden, A. J., Robbins, D. J. & Stewart, W. J. (1999). Magnetism from conductors and enhanced non linear phenomena, *IEEE Trans. on Microwave Theory Tech.* 47: 2572.

Pendry, J. B., Holden, A. J., Stewart, W. J. & Youngs, I. (1996). Extremely low frequency plasmons in metallic microstructures, *Phys. Rev. Lett.* 76: 4773.

Pincemin, F. (1995). *Etude de la propagation du rayonnement électromagnétique dans les milieux hétérogènes*, PhD thesis, Ecole centrale Paris.

Pocklington, H. C. (1905). Growth of a wave group when the group velocity is negative, *Nature* 71: 607–608.

Qiu, M., Thylén, L., Swillo, M. & Jaskorzynska, B. (2003). Wave propagation through a photonic crystal in a negative phase refractive index region, *IEEE J. of Sel. Topics in Quantum Electron.* 9: 106–110.

Richalot, E. (1998). *Étude de structures rayonnantes et diffractantes par hybridation de la méthode des élements finis, en couplage direct ou iteratif*, PhD thesis, Institut national polytechnique de Toulouse.

Rivier, E. & Sardos, R. (1982). *La matrice S, du numérique à l'optique*, Masson, Paris.

Seetharamdoo, D. (2006). *Étude des métamatériaux à indice de réfraction négatif : paramètres effectifs et applications antennaires potentielles*, PhD thesis, Université de Rennes 1.

Seetharamdoo, D., Sauleau, R., Mahdjoubi, K. & Tarot, A.-C. (2004). Homogenisation of negative refractive index metamaterials: comparison of effective parameters of broadside coupled and edge coupled split ring resonators, *IEEE Antennas and Propagation Society International Symposium (APS)*, Vol. 4, Monterey, USA, pp. 3761–3764.

Seetharamdoo, D., Sauleau, R., Mahdjoubi, K. & Tarot, A.-C. (2005). Effective parameters of resonant negative refractive index metamaterials : Interpretation and validity, *J. Appl. Phys.* 98: 063505.

Shelby, R. A., Smith, D. R. & Schultz, S. (2001). Experimental verification of a negative index of refraction, *Science* 292: 77–79.

Silveirinha, M. & Fernandes, C. A. (2004a). Effective permittivity of a medium with stratified dielectric host and metallic inclusions, *IEEE Antennas and Propagation Society International Symposium (APS)*, Vol. 4, Monterey, USA, pp. 3777–3780.

Silveirinha, M. & Fernandes, C. A. (2004b). On the homogenization of bulk and sheet metamaterials, *IEEE Antennas and Propagation Society International Symposium (APS)*, Vol. 4, Monterey, USA, pp. 3769–3773.

Silveirinha, M. & Fernandes, C. A. (2005a). Homogenisation of 3d-connected and non-connected wire metamaterials, *IEEE Trans. on Microwave Theory Tech.* 53: 1418–1430.

Silveirinha, M. & Fernandes, C. A. (2005b). Homogenization of metamaterial surfaces and slabs: the crossed wire mesh canonical problem, *IEEE Trans. on Antennas and Propag.* 53: 59–69.

Smith, D. R. (2000). Negative refractive index in left-handed materials, *Phys. Rev. Lett.* 85: 2933 – 2936.

Smith, D. R. (2005). The design of negative index metamaterials : Extending effective medium concepts, *Negative refraction: Revisiting electromagnetics from microwaves to optics - EPFL Latsis symposium*, Lausanne, Suisse, pp. 18–20.

Smith, D. R. & Kroll, N. (2000). Negative refractive index in left-handed materials, *Phys. Rev. Lett.* 85: 2933–2936.

Smith, D. R., Padilla, W. J., Vier, D. C., Nemat-Nasser, S. C. & Schultz, S. (2000). Composite medium with simultaneously negative permeability and permittivity, *Phys. Rev. Lett.* 84: 4184.

Smith, D. R., Vier, D. C., Kroll, N. & Schultz, S. (2000). Direct calculation of permeability and permittivity for a left-handed metamaterial, *Appl. Phys. Lett.* 77: 2246–2248.

Stratton, J. A. (1941). *Electromagnetic theory*, McGraw-Hill book company, New-York.

Tretyakov, S. (2003). *Analytical modeling in applied electromagnetics*, Artech House, USA.

Tretyakov, S. A. (2005). Research on negative refraction and backward-wave media: A historical perspective, *Negative refraction: Revisiting electromagnetics from microwaves to optics - EPFL Latsis symposium*, Lausanne, Suisse, pp. 30–35.

Veselago, V. G. (1968). The electrodynamics of substances with simultaneously negative values of ε and μ, *Sov. Phys. Usp.* 10: 509.

Weiland, T., Schuhmann, R., Greegor, R. B., Parazolli, C. G., Vetter, A. M., Smith, D. R., Vier, D. C. & Schultz, S. (2001). *Ab initio* numerical simulation of left-handed metamaterials: Comparison of calculations and experiments, *J. Appl. Phys.* 90: 5419.

Wohlers, M. (1971). On the passivity and stability of propagating electromagnetic waves, *IEEE Trans. on circuits and systems* 18: 332–336.

Yavorski, B. & Detlaf, A. (1975). *Aide mémoire de physique*, Éditions Mir, Moscou.

8

Nonlinear Left-Handed Metamaterials

Alexander B. Kozyrev and Daniel W. van der Weide
Department of Electrical and Computer Engineering
University of Wisconsin,
Madison
USA

1. Introduction

Metamaterials are artificial structures that are designed to exhibit specific electromagnetic properties required for different applications but not commonly found in nature. The methodology of synthesizing materials composed of micro- and nano-structured components that mimic the electromagnetic response of individual atoms and molecules (meta-atoms and meta-molecules) has proven to be very productive and resulted in the development of metamaterials exhibiting strong magnetic response at microwave and optical frequencies and so-called left-handed metamaterials (LHMs) (both impossible in conventional real-world materials).

LHMs are designed to exhibit simultaneously negative permittivity and permeability (Veselago, 1968; Engheta & Ziolkowski, 2006). In 2000, Smith et al. developed the first experimental left-handed structure, which was composed of metallic split-ring resonators and thin metal wires (Smith et. al., 2000; Shelby et. al., 2001). An alternative transmission line approach for left-handed materials was proposed, almost simultaneously, by several different groups (Belyantsev & Kozyrev, 2002; Caloz & Itoh, 2002; Iyer & Eleftheriades, 2002). This approach, based on nonresonant components, allows for low-loss left-handed structures with broad bandwidth. The unique electrodynamic properties of these materials, first postulated by Veselago in 1968, include the reversal of Snell's law, the Doppler effect, Vavilov-Cherenkov radiation, and negative refractive index, making theses materials attractive for new types of RF and microwave components. The range of applications for LHMs is extensive, and opportunities abound for development of new and powerful imaging and communication techniques. The most tantalizing of these potential applications is the possibility of realizing "perfect" (diffraction-free) lenses based on their inherent negative index of refraction (Pendry, 2000). The slab of LHM can act as an ideal (diffraction-free) lens and thus capable of producing images of objects without any loss of information which is impossible with conventional lenses.

Most studies of LHMs have been concerned with linear wave propagation, and have inspired many applications that were unthinkable in the past (Engheta & Ziolkowski, 2006; Lai et. al., 2004) such as LH phase shifters (Anioniades & Eleftheriades, 2003), LH directional couplers (Caloz et. al., 2004a; Liu et. al., 2002a), and leaky-wave antennas (Lim et. al., 2005;

Liu et. al., 2002b; Grbic & Eleftheriades, 2002). Materials that combine *nonlinearity* with the anomalous dispersion exhibited by LH media (Lapine & Gorkunov, 2004; Lapine et. al., 2003; Powell et. al., 2007; Shadrivov et.al., 2006), however, give rise to a new class of phenomena and promising applications (Zharov et.al., 2003; Shadrivov & Kivshar, 2005; Shalaev, 2007). Here we present a review of the basic *nonlinear* wave propagation phenomena in LH media. We consider left-handed nonlinear transmission lines (LH NLTL) as the simplest systems that would allow us to combine anomalous dispersion with nonlinearity in a controlled fashion. Understanding the nonlinear phenomena in LH NLTL media is important for both the development of new devices and improvement of the performance of recent tunable devices based on LH NLTLs like phase shifters (Kim et. al., 2005a), tunable leaky-wave antennas (Lai et. al., 2002; Sievenpiper, 2005) and notch filters (Gil et. al., 2004).

2. Comparison of conventional RH and LH nonlinear transmission lines

The transmission line approach proves to be a useful description of LH media. It provides insight into the physical phenomena of LH media and is an efficient design tool for LH applications (Lai et. al., 2004). A LH NLTL is the dual of a conventional nonlinear transmission line shown in Fig. 1*b* where inductors are replaced with capacitors and capacitors with inductors. The effective permeability and permittivity of one-dimensional transmission line metamaterials in the lossless case are expressed as follows:

$$\mu_{eff} = -\frac{2d}{\omega^2 C_L}\ ;\ \varepsilon_{eff} = -\frac{d}{\omega^2 L_L}$$

where d is the period of the LH NLTL and ω is the radian frequency. In contrast with RH NLTL where capacitance gives rise to electric nonlinearity, nonlinear capacitances C_L introduce magnetic-type nonlinearity into the LH NLTL (i.e. effective magnetic permeability becomes nonlinear).

Although both the RH and LH NLTLs use the same components arranged in a similar way, the performance of these two circuits is dramatically different. This difference primarily comes from the difference in their dispersion characteristics (see Fig. 1*c*).

A conventional (right-handed) nonlinear transmission line has normal dispersion and frequency increases with the wavenumber. The fundamental wave can travel synchronously with its higher harmonics. In contrast to the RH NLTL, the LH transmission line exhibits anomalous dispersion and frequency decreases with the wave number (see Fig. 1*c*). The waves propagating in such media are also known as backward waves because the direction of group velocity v_g is opposite to phase velocity ($v_p \cdot v_g < 0$). The fundamental wave can travel synchronously with its higher harmonics. The nonlinearity in RH NLTLs provides energy flow to higher frequencies, which results in waveform sharpening and shock wave formation (Gaponov et. al., 1967; Kataev, 1966). Dispersion, however, results in waveform spreading. If a transmission line exhibits both nonlinearity and dispersion, the latter may compensate the nonlinearity, thus resulting in the formation of temporal solitons (Hirota & Suzuki, 1973).

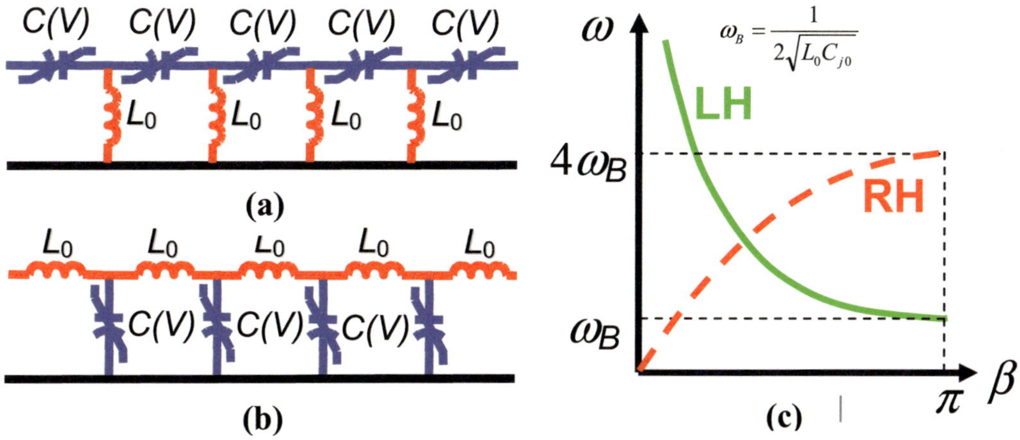

Fig. 1. (a) Equivalent circuit of a LH NLTL; (b) Equivalent circuit of a dual RH NLTL; (c) Typical dispersion curves of LH NLTL (solid line) and RH NLTL (dashed line). Here $\omega_B = 1/(L_0C_0)^{1/2}$.

Nonlinear transmission lines first drew attention in connection with the idea of distributed parametric amplification. It had been predicted that a distributed parametric amplifier or oscillator circuit could exhibit superior stability of operation and efficiency over lumped parametric circuits (Cullen, 1958; Tien, 1958). Lumped parametric amplifiers were popular as very low-noise alternatives to vacuum tubes prior to the widespread use of semiconductor amplifiers (Louisell, 1960). (Parametric resonance responsible for amplification in lumped circuits is similar to the physical mechanism playing on a swing which allows large amplitudes by alternately raising and lowering the center of mass at a certain relation between the frequency of the swing and the frequency of external force.) Their complexity (they require external resonators and matching circuits) and low efficiencies however made them less attractive for widespread use. Conventional NLTLs were thought to be very promising candidates for use in distributed amplifiers because they do not require external resonant circuits and conversion efficiency was claimed to be very high due to a cumulative effect of parametrically interacting waves propagating along NLTLs.

It turned out that parametric interactions (such as three- and four-wave mixing of phase matched waves) in RH NLTLs typically compete with shock wave formation. For instance, parametric generation and amplification in dispersionless RH transmission lines is entirely suppressed by shock wave formation (Landauer, 1960a; Landauer, 1960b). In contrast to conventional NLTLs, both nonlinearity and dispersion present in LH NLTLs (see Fig. 1) lead to waveform spreading (Caloz et. al., 2004b), consequently making shock wave and electronic soliton formation impossible. Anomalous dispersion makes sharp field transients in left-handed NLTL unstable. Once created, they decompose very quickly during propagation of the waveform due to substantial difference in the phase velocities of the propagating waves. This inability to form shock waves enables a variety of parametric processes to occur instead (Kozyrev et. al., 2005; Kozyrev & van der Weide, 2005a). Furthermore, since the parametric interactions no longer compete with shock wave formation, it is possible to use stronger nonlinearities, consequently achieving considerable gain in shorter transmission lines (Kozyrev et. al., 2006).

Both theoretical (Kozyrev & van der Weide, 2005a, 2004) and experimental (Kozyrev et. al., 2005, 2006) investigation demonstrate that nonlinear wave form evolution in a LH NLTL can be understood in terms of competition between harmonic generation, subharmonic generation, frequency down conversion and parametric instabilities.

3. Higher harmonic generation

3.1 Theoretical consideration

In short LH NLTLs, harmonic generation dominates over parametric instabilities (Kozyrev et. al., 2005). The amplitude of the second harmonic in the n-th section of a LH NLTL $V_2(n)$ can be obtained using a small signal approach described in (Kozyrev & van der Weide, 2005):

$$\left|V_2(n)\right| \approx \frac{4K_N V_1^2(0)\sin^2\left(\frac{\beta_1}{2}\right)\cdot\sin^2(\beta_1)}{\sin\left(\frac{\beta_2}{2}\right)\cdot\left[\sin^2\left(\frac{\beta_2}{2}\right)-\sin^2(\beta_1)\right]}\cdot e^{-\alpha n}\cdot\left|\sin\left[(\beta_2-2\beta_1)n/2\right]\right| \qquad (1)$$

where K_N is a "nonlinearity factor" dependent only on diode parameters, $V_1(0)$ is the voltage at the input of the LH NLTL, α is the attenuation constant, n is the section number and β_1 and β_2 are the propagation constants (phase shift per section) for the fundamental wave and its second harmonic, respectively.

The fundamental wave propagating in the LH NLTL is always badly mismatched with its higher harmonics due to inherent anomalous dispersion, yet the generation of higher harmonics can still be very effective. This is possible because of "amplitude singularities". The denominator in (1) has zeros when

$$\sin^2(\beta_2/2)-\sin^2(\beta_1)\rightarrow 0\,. \qquad (2)$$

Due to phase mismatch, the amplitude of the second harmonic varies rapidly with distance. This gives rise to a highly localized energy exchange between the fundamental wave and its second harmonic. It is apparent from (1) that the maximum amplitude of the second harmonic at the end of the N-section line is achieved when

$$(\beta_2-2\beta_1)N=(2k+1)\pi\,,\ k=0,1,2,3... \qquad (3)$$

The same approach applied to RH NLTL predicts linear growth of the second harmonic amplitude (in the lossless case) due to its phase-matching with fundamental wave (Kozyrev & van der Weide, 2005a; Champlin & Singh, 1986). Thus, the theoretical analysis of 2nd harmonic generation in LH NLTLs shows that, despite the large phase mismatch, inherent anomalous dispersion enables the possibility of faster-than-linear growth of the second harmonic amplitude as predicted by (1) in a narrow frequency range where condition (2) is satisfied. This fact explains the dominance of harmonic generation in short LH NLTLs over other parametric instabilities which require long distances of propagation for energy exchange to occur because of long coherence distance (due to phase matching). A somewhat similar singular behavior of the second harmonic amplitude was predicted for the wave reflected from a slab of nonlinear LH medium (Agranovich et. al., 2004).

3.2 Experiment

We fabricated a 4-section LH NLTL having identical sections (shown in Fig. 2*a*) (Kozyrev et.al, 2005). The circuit was realized on a Rogers RT/Duroid 3010 board with ε_r = 10.2 and thickness h = 1.27 mm. The nonlinear capacitance in each section is formed by two back-to-back M/A-COM hyperabrupt junction GaAs flip-chip varactor diodes (MA46H120) with DC bias applied between them. Shunt inductances were implemented with 0.12 mm diameter copper wires connecting the pads to the ground plane on the back side of the board. The pads on the board surface, together with inherent parasitics introduce unavoidable series inductance and shunt capacitance, making the whole circuit a composite right/left-handed transmission line having the equivalent circuit shown in Fig. 2*b*. The dispersion characteristic of a composite right/left-handed transmission line has two passbands divided by the stop band. The low frequency passband exhibits anomalous dispersion (left-handed passband) while the high-frequency one is right-handed. Fig. 3 shows the magnitude of the linear wave transmission (S_{21}) of the LH NLTLs. We measured a –6 dB cut-off frequency at 2.7 GHz for 0 V-bias. The frequency region from 2.7 GHz to 8 GHz for 0 V bias corresponds to the left-handed passband. Parameters of the circuit model in Fig. 2*b* were extracted from the *S*-parameters measured at 0 V bias. They are C_L = 0.99 pF, L_L = 1.695 nH, C_R = 0.05 pF, L_R = 0.966 nH.

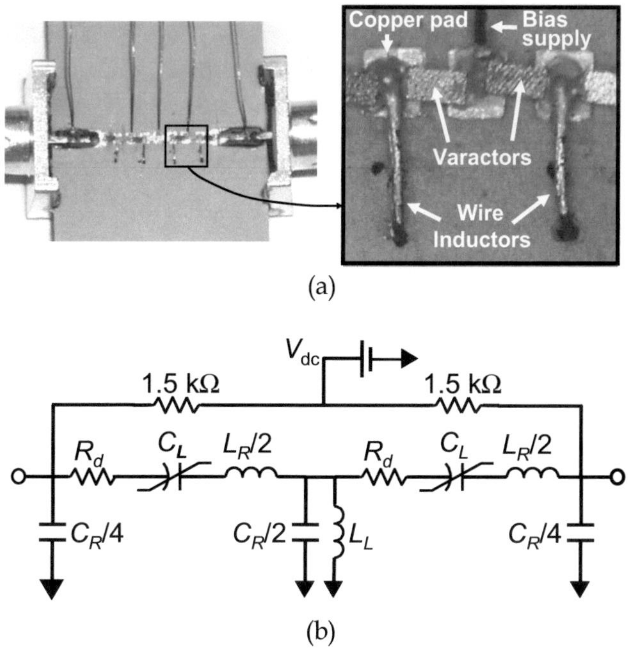

Fig. 2. (a) Fabricated 4-section LH NLTL and (b) equivalent circuit of one stage.

The measured results qualitatively confirm our predictions using small-signal analysis. Figure 4 shows the spectrum of the waveform from the output of 4-section LH NLTL as measured with an Agilent E4448A Spectrum Analyzer, and corresponds to the maximum of the second harmonic conversion efficiency.

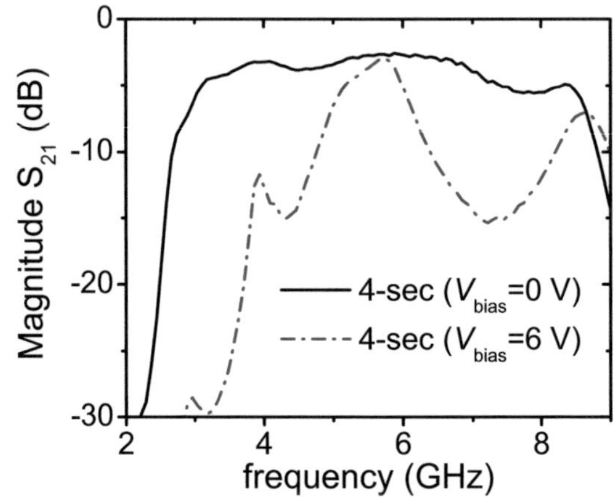

Fig. 3. Measured magnitude of S_{21} parameter for four-section LH NLTL.

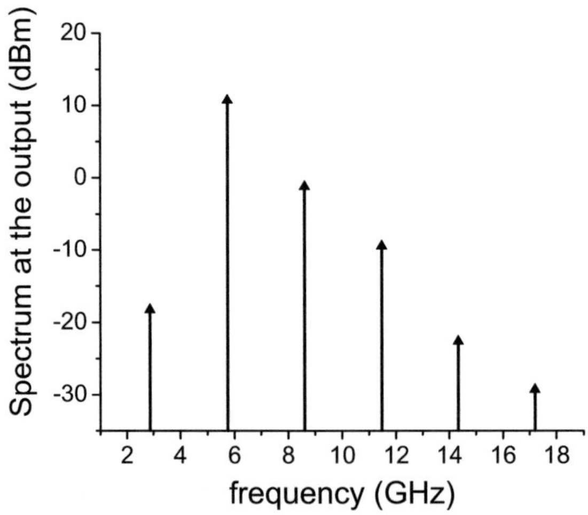

Fig. 4. Spectrum of the output waveform generated by a four-section LH NLTL fed by 2.875 GHz, +17.9 dBm input signal at reverse bias voltage of 6.4 V.

The measured value for the second harmonic conversion efficiency in this 4-section LH NLTL was 19% at 2.875 GHz, using a +17.9 dBm input signal and a reverse bias voltage of 6.4 V. The second harmonic power delivered into a 50 Ω load was +10.72 dBm. The fundamental wave is close to the Bragg cutoff frequency (note the magnitude of S_{21} for the 4-section LH NLTL at bias voltage 6 V as shown in Fig. 2*a*), and thus falls into frequency range for which small-signal analysis predicts amplitude singularity. The second harmonic wave is close to the transmission maximum, which is located in the middle of the left-handed passband. A fundamental of 2.875 GHz generates numerous higher harmonics, with the second harmonic dominating over the fundamental and the other harmonics. Thus, the LH NLTL combines the properties of both a harmonic generator and a bandpass filter, and under certain conditions may provide an almost pure higher harmonic at its output.

The conversion efficiency observed in the LH NLTL is comparable with the per-stage efficiency of a hybrid Schottky-diode RH NLTL operated in a lower frequency range (Duchamp et. al., 2003).

4. Parametric generation and amplification

4.1 Theory

Under certain circumstances, harmonic generation may compete with different parametric processes, resulting in unstable harmonic generation. Effective parametric interaction in medium exhibiting a second-order nonlinearity generally requires phase matching of three waves. The anomalous dispersion of a LH NLTL system enables effective parametric interactions of the type:

$$f_1 + f_2 = f_3, \; \beta_1 - \beta_2 = \beta_3 \tag{4}$$

In the "parametric oscillator configuration", a high-frequency backward pump wave having a frequency f_3 and wavenumber β_3 is excited by the voltage source connected at the input port of a LH NLTL. It generates two other waves having frequencies f_1 and f_2, such that $f_1 < f_2$ and $f_1 + f_2 = f_3$. The wave having frequency f_2 propagates in the opposite direction relative to the pump wave and the wave having frequency f_1 (this is emphasized in (4) with the minus sign). We therefore have a similar situation to backward wave parametric generation (Gorshkov et. al., 1998; Harris, 1966). The backward-propagating parametrically generated wave f_2 enables internal feedback that results in a considerable energy transfer from the pump wave to the parametrically excited waves.

If the amplitude of a high-frequency pump wave exceeds a certain threshold value, it may parametrically generate two other waves. This threshold value depends on the loss present in the LH NLTL, its length and the boundary conditions (matching) at the input and output. No parametric generation occurs when the amplitude of the voltage source is below this value. However, when a weak signal wave is fed into the LH NLTL together with a pump wave having an amplitude below the threshold value, a parametric amplification is observed. In this case, we have two input waves: an intense pump wave and a weak signal wave (Yariv, 1988). The power from the pump wave is transferred to the signal wave, thus amplifying it. A third parasitic idler wave is generated which provides phase matching. From a previous analysis (Gorshkov, et. al. 1998), for the lossless case, the frequencies and powers of these waves also obey the nonlinear Manley-Rowe relations.

4.2 Experiment

Though generation of higher harmonics dominates in short LH NLTLs, in longer transmission lines parametric interactions predominate. We observed efficient parametric amplification in 7-section LH NLTL (Kozyrev et. al., 2006) shown in Fig. 5. The design of the 7-section LH NLTL is similar to the design of four-section LH NLTL described in the previous section.

However, this time, the series nonlinear capacitance has been implemented with Skyworks Inc. SMV 1233 silicon hyperabrupt varactors and shunt inductances with high-Q 10 nH chip inductors (Murata LQW18A_00).

Fig. 6 shows the magnitude and phase of the linear wave transmission (S_{21}) of this 7-section LH NLTL. Parameters of the circuit model in Fig. 2*b* were extracted from the measured *S*-parameters using Agilent ADS software. They are C_L (3.823 V) = 1.34 pF, L_L = 11.43 nH, C_R = 0.62 pF, L_R = 3.18 nH. The dashed line in Fig. 6 shows the magnitude of S_{21} calculated for the circuit model shown in Fig. 2*b* with component values specified above, and it is in a good agreement with measured data. The circuit model of Fig. 2*b* has also been used to calculate the dispersion curve of the LH transmission line as shown in the inset in Fig. 6. As is evident from *S*-parameters and dispersion curve presented in Fig. 6, the transmission line has a left-handed passband (phase velocity is anti-parallel with the group velocity) from 800 MHz to 1.9 GHz at -3.823 V bias.

Fig. 5. Fabricated seven-section LH NLTL.

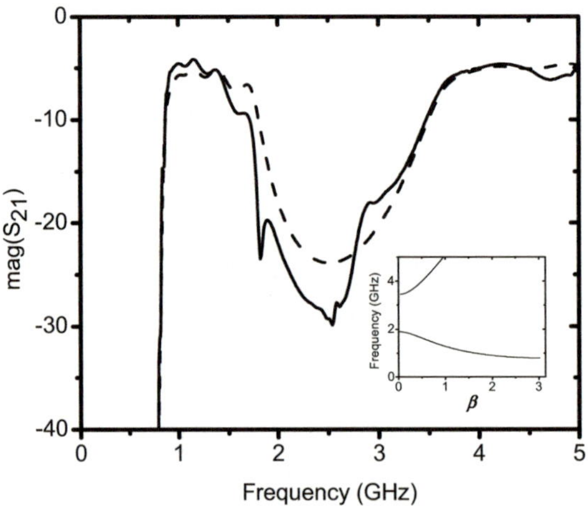

Fig. 6. Measured and simulated (dotted lines) magnitudes of S_{21} parameter for seven-section LH NLTLs for the reverse bias voltage V_B = 3.823 V. Inset shows dispersion curve of the LH NLTL (dependence of the frequency vs relative wave number β).

Figure 7 demonstrates the effect of the intensive pump wave, having frequency $f_P = f_3$, on a weak signal wave ($f_s = f_2$). Figure 7*b* shows the spectrum at the output of the 7-section LH NLTL when only a 1.7279 GHz, 13.96 dBm intensive pump wave is applied at the input. The magnitude of the pump wave was chosen so as to be 0.1 dB below the threshold value required for the occurrence of parametric generation, which manifests itself in distinct, narrow peaks corresponding to the parametrically generated frequencies.

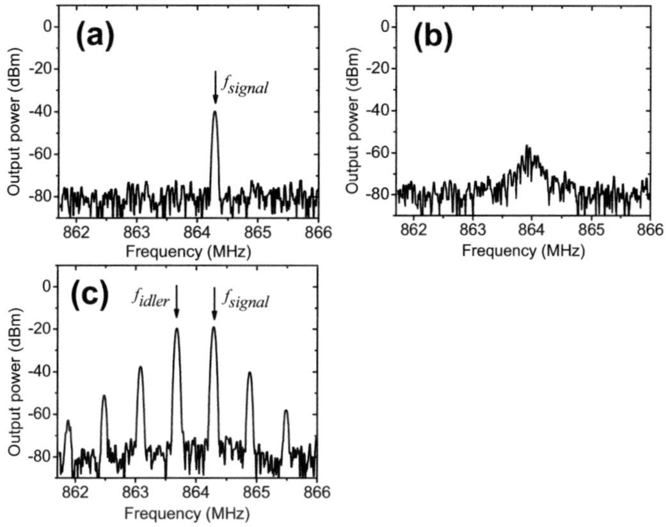

Fig. 7. Spectra of the output waveforms generated by a 7-section LH NLTL fed by: (a) only weak signal source 864.252 MHz, -28 dBm; (b) only pump source 1.7279 GHz, 13.96 dBm; (c) simultaneously signal and pump sources specified in (a) and (b). Reverse bias voltage is 3.87 V.

Figure 7a shows the spectrum at the output when only the 864.252 MHz, -28 dBm signal wave is applied at the LH NLTL input (no pump wave). The graph shows 11.7 dB attenuation of the weak signal wave at the output due loss in the NLTL and power conversion to higher harmonics. And finally, Fig. 7c shows the spectrum at the output when the signal and the pump wave are both applied concurrently at the input of the 7-section LH NLTL. In this spectrum, the components corresponding to the signal wave ($f_s = f_2$), idler wave (f_1), as well as many difference frequencies generated due to the strong nonlinearity in LH NLTL, are evident. Thus, the application of the intensive pump wave results in amplification of the weak signal by 9 dB.

It should be mentioned that parametric amplification of the signal wave propagating in the backward direction with respect to the pump wave has been observed in a round-trip configuration (when both the signal and the pump waves are still applied at the TL input). The parametric interaction of the counter-propagating signal and pump waves becomes possible since both the signal and the pump waves are subject to a strong reflection while propagating along the LH NLTL. This reflection originates from the mismatch at the input and output ports (which is unavoidable due to strong nonlinear variation of capacitance) and results in a standing wave formation. This enables the phase matching of the incident signal wave and the reflected from the output end of the LH NLTL pump wave and hence the parametric interaction of the counterpropagating backward waves.

Figure 8 represents the measured gain of a weak 864.252 MHz signal stimulated by an intense 1.7279 GHz pump wave versus the power of the signal at the input, for fixed values of the pump power. The gain was calculated as the difference between the power of the signal at the output and the power at the input when both are expressed in dBm. Thus, we measured a greater than 10 dB amplification of the signal with power of –32 dBm and below for the power of the pump wave at the input of 13.96 dBm. The measured dependencies of gain verses input signal power becomes flatter with decreasing pump power, thus revealing the potential for amplification in a broad band of the signal power. The results of our measurements in Fig. 8 are in a good agreement with the results of simulations reported in (Kozyrev & van der Weide, 2005b).

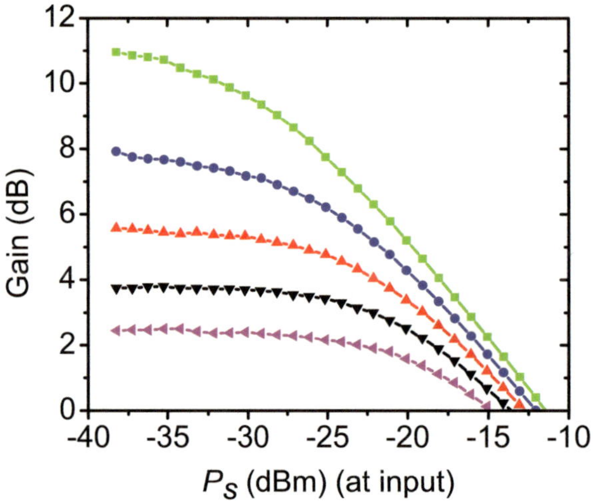

Fig. 8. Measured gain vs power of the signal at the input of LH NLTL for different values of the power of the pump wave at the input $P_{p,in}$.: Green squares - $P_{p,in}$ = 13.96 dBm; Blue circles - $P_{p,in}$ = 13.86 dBm; Red up triangles - $P_{p,in}$ = 13.76 dBm; Black down triangles - $P_{p,in}$ = 13.66 dBm; Magenta left triangles - $P_{p,in}$ = 13.56.

4.3 Higher-order parametric processes

Efficiently generated higher harmonics may also initiate the parametric process. A wave at 2.875 GHz cannot parametrically generate any other waves in the 4-section LH NLTL, shown in Fig. 2, since they would exist below the line's cutoff frequency. The second harmonic at 5.75 GHz excites waves with frequencies of 2.2 GHz and 3.55 GHz depicted in Fig. 9*b* as f_1 and f_2. This basic parametric process then initiates multiple higher-order parametric interactions, resulting in multiple peaks in the spectrum of the output waveform. The progression of this process is shown in Fig. 9*c*, which illustrates conversion of a monochromatic input signal into a wideband output. Further increase in the reverse bias voltage leads to the stabilization of the harmonic generation and suppression of parametric instability (Fig. 9*d*). The variation of bias voltage results in a corresponding change in the dispersion characteristics of the LH NLTL (propagation constants of the interacting waves). This change allows for enabling or disabling of certain nonlinear interactions (phase matching). In our particular case it enables/disables higher-order parametric interactions.

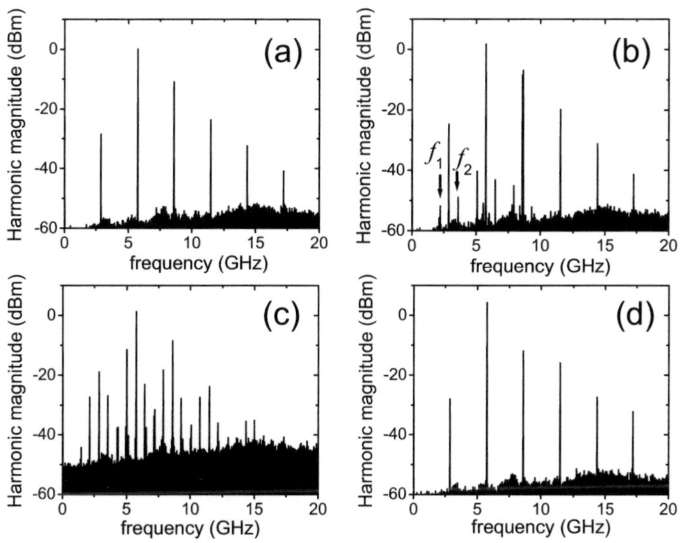

Fig. 9. Spectra of the output waveform generated by a 4-section LH NLTL fed by 2.875 GHz, +19 dBm input signal for different values of the bias voltage (*a* - 4 V, *b* - 4.95 V, *c* - 5 V, *d* - 6.3 V). The results here are decreased by 6 dB due to a protection attenuator.

4.4 Higher-order parametric processes

Parametric amplification can be of interest for building "active" or "amplifying" metamaterials and for providing a means to compensate for inherent LH media loss, a challenge for currently existing metamaterials (Kozyrev et. al., 2006). The primary drawback of current negative-index metamaterials (NIMs) (for example those composed of the arays of metallic wires and split-ring resonators) is their considerable loss, which renders the results ambiguous and the materials all but useless for practical applications. These losses have been overcome to some extent by careful fabrication and assembly techniques (Houck et. al. 2003), but still remain the primary obstacle to using NIMs in imaging applications. It was shown (Tretyakov, 2001) that due to causality requirements, the use of conventional composite NIMs (based on arrays of metallic wires and arrays of split-ring resonators) does not allow for the realization of low-loss NIMs without the incorporation of some active components (transistor amplifiers, etc.) in a composite NIM. The idea of using parametric amplification to compensate for inherent loss in optical left-handed systems has been also discussed in (Popov & Shalaev, 2006).

5. Envelope solitons in LH NLTLs

Besides the nonlinear evolution of a waveform itself, another class of phenomena involving evolution of amplitude and phase of continuous waves is also possible. This type of nonlinear wave propagation phenomena arise in NLTLs having strong frequency dispersion with respect to the average amplitude for amplitude-modulated wave containing a carrier of relatively high frequency and slow (optical-type) nonlinearity. This dispersion may lead to amplitude instability as well as to formation of envelope solitons and periodic modulation of a carrier wave propagating in a stationary manner. The observation of amplitude instability and envelope soliton generation in conventional (RH) NLTLs has already been the subject of many publications (Lonngren & Scott, 1978; Ostrovskii & Soustov, 1972; Yagi & Noguchi, 1976). The experimental observation of the generation of the trains of envelope solitons in LH NLTLs arising from the self-modulational instability under certain conditions of the amplitude and frequency of the pump wave was first reported in (Kozyrev & van der Weide, 2007).

The analysis of LH NLTLs is straightforward when the equations governing envelope evolution can be reduced to the one-dimensional cubic nonlinear Schrodinger equation (NSE), which provides a canonical description for the envelope dynamics of a quasi-monochromatic plane wave (the carrier) propagating in a weakly nonlinear dispersive medium when dissipative processes (including nonlinear damping due to higher harmonic generation and nonlinear wave mixing) are negligible (Gupta & Caloz, 2007; Narahara et. al., 2007). However, in most of the practical situations the parametric decay instabilities and higher harmonic generation can be very significant (Kozyrev et. al, 2005, 2006; Gorshkov et. al., 1998; Lighthill, 1965). The threshold for parametric generation is known to be very low (lower then in conventional RH NLTLs). In order to realize the scenario described by the NSE, the LH NLTL should be operated below this threshold so that the nonlinearity should be very weak and the NLTL impractically long. In contrast, we performed an experimental study of nonlinear envelope evolution and envelope soliton generation in relatively short LH NLTLs and when nonlinear damping is very strong. We are also taking advantage of a fast nonlinearity introduced by Schottky diodes when nonlinear capacitance is a function of the instantaneous value of voltage along the line rather then its amplitude, a type of nonlinearity not described in the framework of the NSE and its modifications developed for slow (retarding) nonlinearity.

Figure 10 shows a typical voltage waveform and its spectrum measured at the output of 7-section LH NLTL in the envelope soliton generation regime. This voltage waveform is a cw signal with carrier at fundamental (pump) frequency and with an envelope representing itself a train of bright solitons appearing as periodic pulses above a cw background. The scenario of the development of modulational instability and/or generation of envelope solitons is very sensitive to the parameters of the signal applied at the input of NLTL. Depending on the amplitude and frequency of the input signal, trains of envelope solitons of different shape and types can be generated. Figure 11 shows envelopes of the measured waveforms at the output of 7-section LH NLTL. These envelopes' functions have been obtained by applying the Hilbert transform to the original voltage waveforms.

Traces (a), (b) and (c) in Fig. 11 show trains of bright envelope solitons of different shapes while traces (d) and (e) show periodic trains of so-called dark solitons (dips in the cw background). The trains of envelope solitons that we observed are also known as cnoidal waves. The interval between individual solitons depends on the amplitude and frequency of the input signal but does not depend on the length of LH NLTL. Comparison of voltage waveforms at the output of 7-, 10- and 17-section LH NLTL for the same input signal parameters shows that the distance between solitons and their shape are preserved during propagation along transmission line and that we deal with the generation of stationary train of solitons. The envelope shape is not smooth since strong nonlinearity gives rise to numerous higher harmonics and subharmonics of carrier frequency. In the spectral domain, generation of envelope solitons manifests itself in the appearance of spectral regions with numerous closely spaced spectral harmonics. The interval between adjacent spectral components is $\Delta f = 1/\tau$, where τ is the period of the train of solitons.

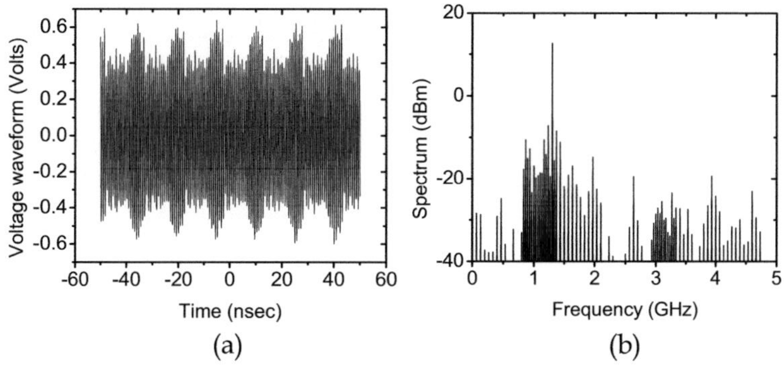

Fig. 10. Voltage waveform (a) and its spectrum (b) measured at the output of 7-section LH NLTL fed by a 1.3125 GHz, +21.6 dBm input signal.

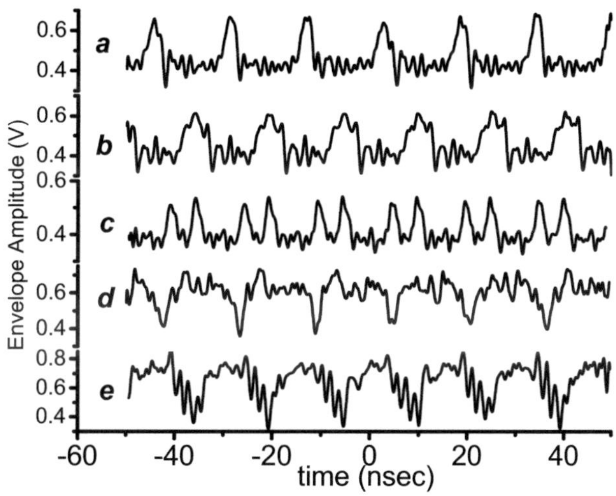

Fig. 11. Measured trains of envelope solitons for different power P_{inp} and the frequency f_{inp} of the input signal. a) f_{inp} = 1.3723 GHz, P_{inp} = 24.66 dBm; b) f_{inp} = 1.3125 GHz, P_{inp} = 21.60 dBm; c) f_{inp} = 1.321596 GHz, P_{inp} = 19.34 dBm; d) f_{inp} = 1.2974 GHz, P_{inp} = 24.64 dBm; e) f_{inp} = 1.102 GHz, P_{inp} = 23.62 dBm

A small variation of the parameters of the input signal leads to switching between the generation of bright and dark solitons [compare traces (a) and (b)] in contrast to the scenario described by the NSE. As is known, systems described by the NSE can be characterized by two main parameters: the nonlinearity parameter $N = \partial\omega/\partial|A|^2$ (ω and A are the carrier frequency and the amplitude) and the dispersion parameter $D = \partial^2\omega/\partial k^2$ (k is the wave number). According to the Lighthill criterion (Lighthill, 1965), either dark or bright solitons are observed depending on the sign of these two parameters. Bright solitons exist when $DN<0$ and dark solitons exits when $DN>0$. The observed switching is enabled by the counterplay of the significant nonlinear damping (due to strong and fast nonlinearity) and strong spatial dispersion exhibited by the periodic LH NLTLs. Neither is taken into account by standard NSE yet both are known to lead to co-existence of bright and dark solitons in other physical systems (Kivshar et. al., 1994; Scott et.al., 2005). For example, somewhat similar processes have recently been observed in the system of an in-plane magnetized single crystal yttrium-iron-garnet (YIG) film in the magnetostatic backward volume wave configuration. However, there is a fundamental issue that distinguishes our work from (Scott et.al., 2005) where the soliton trains have been generated through the nonlinear mode beating of two copropagating magnetostatic backward volume wave excitations in thin YIG film. Thus, a pre-modulated signal was used to achieve soliton generation. In contrast to this work, we applied non-modulated sine wave at the input.

6. Pulse formation in LH NLTL media

As it has already been mentioned in Section 2, both the nonlinearity and dispersion present in LH NLTLs lead to waveform spreading, consequently making shock wave and electronic soliton formation impossible, making them at first blush useless for pulse forming applications. However, this inability to form shock waves enables a variety of parametric processes leading to amplitude istabilitity as well as formation of envelope solitons and periodic modulation of a carrier wave (Kozyrev & van der Weide, 2007) as discussed in the previous section. Here we describe another type of envelope evolution resulting in generation of RF pulses of limited duration with stable amplitude and very short rise/fall times (sharp transients). This type of envelope evolution is primarily enabled by the amplitude-dependent higher harmonic generation rather than self-modulation instability leading to generation of the envelope solitons (Kozyrev & van der Weide, 2010).

Fig. 12 shows a typical dependence of the magnitude of the second harmonics at the output of 7-section LH NLTL shown in Fig. 5 vs magnitude of the input sinusoidal signal. This dependence has three distinct regions. In the first region the power of the generated second harmonic follows a square law as predicted by the small signal analysis (1). When the power of the fundamental wave reaches certain threshold level the second harmonic power jumps by almost 5 dB indicating a bifurcation (multistability region) followed by the saturation region where second harmonic amplitude changes insignificantly with the input power. Step-like dependence of the second harmonic power indicates a bifurcation-type change in the field distribution along the line and formation of field patterns that change dispersion properties of the line resulting in significant increase of the generation efficiency. These field patterns (nonlinear mode build-up) result from the nonlinear interactions and reflection of both fundamental and second harmonic signals from input and output interfaces. An example of such patterns has been investigated in (Kozyrev & van der Weide, 2005a; Kozyrev et. al., 2005) where a significant increase of the 3rd harmonic generation efficiency correlated with self-induced periodicity of the voltage oscillations across nonlinear capacitances on LH NLTL. This self-induced periodicity of the voltage amplitude cross the nonlinear capacitors leads to periodic variation of the capacitance along the line. Due to strong nonlinearity (large capacitance ratio), this periodicity results in a considerable change of the dispersion characteristics and enables quasi-phase matching of the fundamental wave and its higher harmonics.

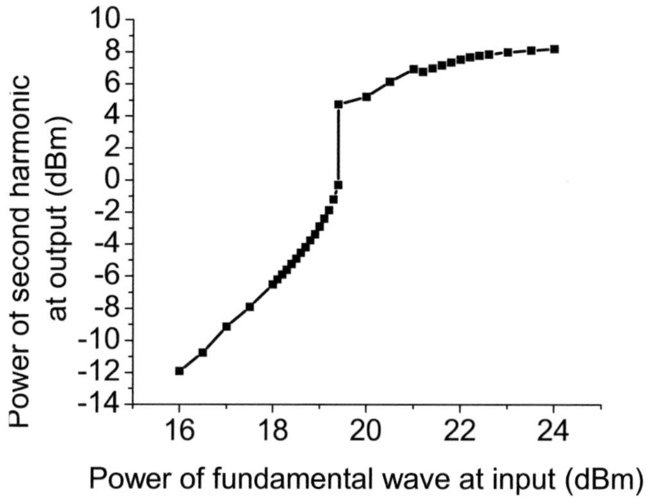

Fig. 12. Dependence of the power of the 2nd harmonic at the output on the power of the fundamental signal at the input in 7-section LH NLTL shown in Fig. 5 measured at 783 MHz and the reverse bias voltage V_B = - 4.1 V.

The step-like dependence of the second harmonic power on the power of the fundamental signal may impact significantly the output waveform if the amplitude in the fundamental wave is modulated around the threshold value. To verify this assumption the LH NLTL was fed by a 783 MHz, +20.5 dBm sinusoidal signal modulated at 100 kHz with the depth of modulation of 50 %. Power of the input signal corresponds to the threshold value in Fig. 12. Figure 13 shows voltage waveforms at the input and output port and spectrum at the output of 7-section LH NLTL. As expected, the voltage waveform at the input is a sinusoidal wave modulated by another sinusoidal signal at 100 MHz. The envelope of the output waveform is dramatically different from the one of the input wave. It represents itself a series of pulses with the shape approaching a rectangular. Furthermore, the carrier frequency of the output signal is the second harmonic of the fundamental signal as revealed by the spectrum presented in Fig. 13*c*. Modulated signal switches second harmonic generation on and off thus enabling generation of a train of RF pulses at the output. Since the fundamental frequency is chosen below the cut-off frequency, it is heavily attenuated in transmission line and only second harmonic is present at the output. Some asymmetry of the shape of the RF pulses at the output is related to the existence of hysteresis and narrow multistability region. The experimental results presented in Fig. 13 clearly demonstrate that a small modulation signal can be used to control the shape, duration and repetition rate of the RF pulses at the output which is very promising to numerous applications.

Our experimental results correlate very well with speculations in (Agranovich et. al., 2004) where authors predicted that the shape of pulses at the output of LH media can be drastically different from those expected from an ordinary nonlinear medium.

Potential applications may include pulse forming circuits, amplifiers of digital signals as well as very efficient modulators at power levels or in frequency ranges not attainable by conventional semiconductor devices.

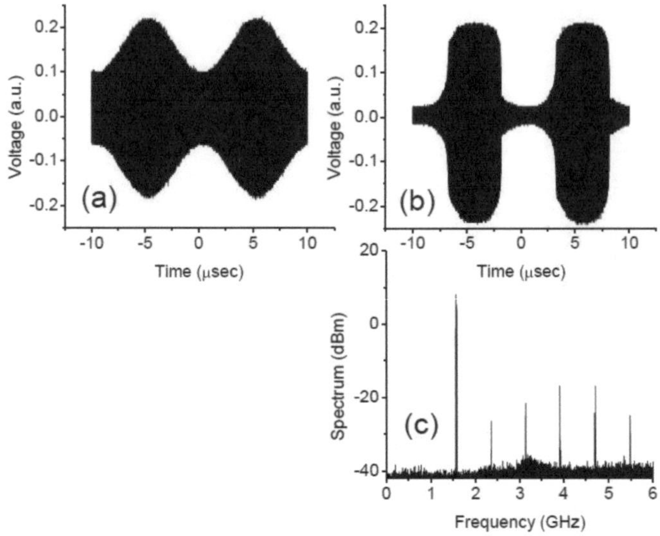

Fig. 13. Voltage waveforms at the input (a) and output (b) port and spectrum at the output (c) of 7-section LH NLTL. Voltage was measured at the coupled output of directional couplers connected at the input and output ports of NLTL.

7. Inverse Doppler effect in nonlinear transmission lines

Some interesting phenomena arise in nonlinear periodic systems (periodically loaded nonlinear transmission lines) supporting propagation of backward spatial harmonics. It is well known that the periodic systems support propagation of the Bloch waves which can be expanded into an infinite set of spatial harmonics (Collin, 1992) so that the field in a periodic structure can be represented as

$$V_B = \sum_{n=-\infty}^{n=\infty} V_{p,n} e^{-j\beta_n z}$$

Each term in this expansion is a spatial harmonic, is periodic in space amplitude

$$V_{p,n}(z+d) = V_{p,n}(z)$$

and has propagation phase constant

$$\beta_n = \beta + 2n\pi / d .$$

All harmonics propagate with the same group velocity; however, some of the spatial harmonics have phase and group velocities that are oppositely directed (backward spatial harmonics) since β_n can be both positive and negative and thus exhibit anomalous dispersion:

$$v_g^{(n)} = d\frac{\omega}{\beta_n} \quad v_p^{(n)} = d\frac{\partial\omega}{\partial\beta_n} = d\frac{\partial\omega}{\partial\beta} .$$

The first experimental observation of an inverse Doppler effect, in which the frequency of a wave is increased upon reflection from a receding boundary, was reported in (Seddon & Bearpark, 2003). They used an experimental scheme based on a magnetic nonlinear transmission line which was suggested recently in (Belyantsev & Kozyrev, 2002, 1999). This scheme falls into a general class of systems that involve the emission of phase matched high-frequency waves by an electromagnetic shock wave propagating along a NLTL with dispersion (Belyantsev et. al., 1995; Belyantsev & Kozyrev, 1998, 2000). The moving boundary that is used to produce a Doppler shift is the discontinuity that is formed between regions of unsaturated and saturated nonlinearity in the transmission line at the leading edge of the pump pulse. Under appropriate conditions, this shock wave (moving discontinuity) generates a Bloch wave propagating in the opposite direction to the moving discontinuity. It occurs when this shock wave is phase matched with a backward spatial harmonic of the exited Bloch wave. Following its reflection from the NLTL input interface, the excited Bloch wave catches up with the moving discontinuity and produces an anomalous Doppler shift (Belyantsev & Kozyrev, 2002). The detailed theory of this phenomenon is presented in (Kozyrev & van der Weide, 2005, 2006).

8. Nonlinear volumetric metamaterials

Nonlinear phenomena similar to ones described in Sections 3-7 have also been observed in volumetric metamaterials (Shadrivov et. al., 2008a, 2008b). For instance, the selective generation of higher harmonics has been observed in metamaterials consisting of split-ring resonantors (SRR) and metal wires shown in Fig. 14. Each SRR contains a variable capacity diode (model Skyworks SMV-1405) which introduces nonlinear current-voltage dependence and resulting nonlinear magnetic dipole moment to each SRR (Shadrivov et. al., 2008a). In terms of effective medium parameters, the manufactured structure has nonlinear magnetization and non- linear effective magnetic permittivity.

Fig. 14. Photograph of the nonlinear tunable metamaterial created by square arrays of wires and nonlinear SRRs. Each SRR contains a varactor (see the inset) which provides power-dependent nonlinear response.

Arrays of SRRs and wires form a square lattice with 29 x 4 x 1 unit cells of the size of 10.5 mm.

To measure the em field scattering on our samples, the metamaterial slab is placed in a parallel plate waveguide. The planes of SRRs are aligned perpendicular to the parallel plate surfaces. We have measured the spectrum of the transmitted signal for different frequencies of the incident em wave. For this purpose, the input antenna (placed at the midpoint of the lower plate, 2 mm from the metamaterial slab, in front of the central unit cell) was fed by the signal generated by an Agilent E4428C ESG vector signal generator and amplified by a 38 dB amplifier. The signal detected by the receiving antenna placed 2 cm behind the metamaterial slab was analyzed using an Agilent E4448A PSA series spectrum analyzer.

Figure 15 shows spectra of the signal detected by the receiving antenna behind the nonlinear LHM slab. Varying the input frequency, we observed efficient selective harmonic generation. Namely, second (Fig. 15*a*), third (Fig. 15*b*), and fourth (Fig. 15*c*) harmonics were selectively generated. Moreover, the generation of a comblike signal was also observed (Fig. 15*d*).

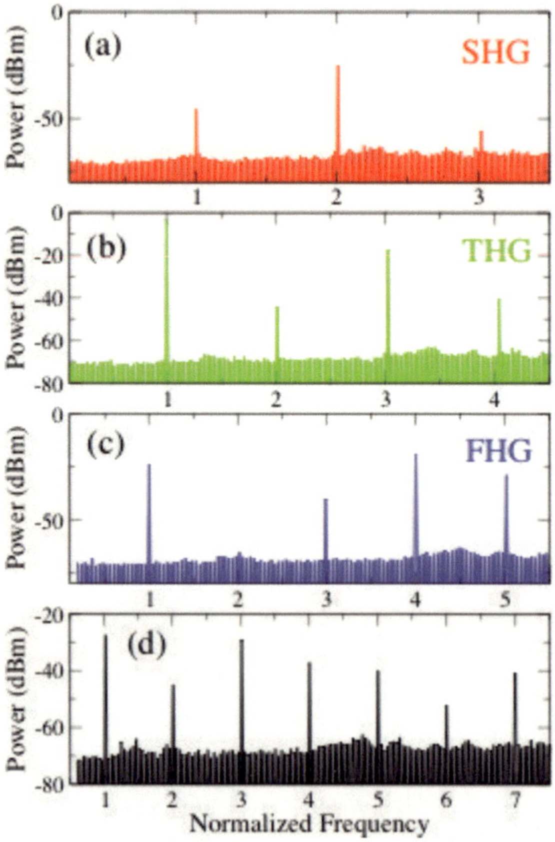

Fig. 15. Spectra of the signals detected by the receiving antenna located behind the nonlinear LHM slab for different source frequencies: (a) 3.415 GHz, (b) 2.29 GHz, (c) 1.733 GHz, and (d) 1.668 GHz. The frequency on each graph is normalized to the corresponding source frequency. Power at the input antenna is +30 dBm.

Selective generation of higher harmonics observed in our experiments is related to the transmission properties of the metamaterial. A particular harmonics dominates over fundamental harmonic and the other higher harmonics when its frequency corresponds to the transparency band. Results of the transmission coefficient measurements performed on our nonlinear LHMs indicate a right-handed transparency band with a maximum transparency at around 7 GHz. This value agrees well with the values of the higher harmonics dominating in our measurements. Furthermore, the presence of very high order harmonics in the spectrum of the trans- mitted signal manifests strong nonlinearity inside the metamaterial which potentially may lead to significant enhancement in nonlinear processes in artificial metamaterials as compared to conventional materials.

9. Conclusion

We have reviewed several nonlinear wave phenomena in LH media, including harmonic generation, parametric amplification and generation of traveling waves, generation of the train of envelope solitons and their competition. Furthermore, LH NLTLs which were considered as a model system in this paper, can be also of interest from the design perspective for development of various compact and robust applications for wireless communications and imaging. LH NLTLs have already been used as the key counterparts of recently designed and implemented tunable phase-shifters, tunable band-pass filters, and the arbitrary waveform generator based on fourier decomposition (by combining broadband power divider, LPF, BPF, HPF, harmonic generator, vector modulator and broadband LNAs on copper board) (Kim et. al., 2005a, 2005b, 2006, 2007). Moreover, extending the results for 1-D LH NLTL to higher dimensions would enable combining harmonic generation in LH NLTL media with focusing (Grbic & Eleftheriades, 2003, 2004), due to the negative refractive index of 2-D or 3-D LH transmission line media. This may lead to the development of highly efficient and powerful frequency multipliers, as well as to building "active" or "amplifying" super lenses. Furthermore, our approach can be also scaled from its current microwave form into terahertz, infrared, and, ultimately, visible form (Goussetis et. al., 2005; Qin et. al., 2007, 2008).

10. Acknowledgment

This work was supported under the Air Force Office of Scientific Research, MURI Grant No F49620-03-1-0420, 'Nanoprobe Tools for Molecular Spectroscopy and Control'.

11. References

Agranovich, V. M.; Shen, Y. R.; Baughman, R. H. & Zakhidov, A. A. (2004). *Physical Review B* Vol.69, p. 165112.

Anioniades, M. & Eleftheriades G. V. (2003). *IEEE Antennas and Wireless Propagation Letters,* Vol.2, p. 103.

Belyantsev, A. M.; Dubnev, A. I.; Klimin, S. L.; Kobelev, Y. A. & Ostrovskii, L. A. (1995). *Technical Physics,* Vol.40, p. 820.

Belyantsev, A. M. & Kozyrev, A. B. (1998). *Technical Physics,* Vol.43, p. 80.

Belyantsev, A. M. & Kozyrev, A. B. (1999). *Materials Science Forum,* Vol.297, p. 349.

Belyantsev, A. M. & Kozyrev, A. B. (2000). *Technical Physics,* Vol.45, p. 747.

Belyantsev, A. M. & A. B. Kozyrev, A. B. (2002). *Technical Physics,* Vol.47, p. 1477.

Caloz, C. & Itoh, T. (2002). *IEEE Antennas and Propagation Society International Symposium Dig.,* Vol.2, p. 412.

Caloz, C.; Sanada, A. & Itoh, T. (2004). *IEEE Transactions on Microwave Theory and Techniques,* Vol.52, p. 980.

Caloz, C.; Lin, I. H. & Itoh ,T. (2004). *Microwave and Optical Technology Letters,*Vol.40, p. 471.

Champlin, K. S. & Singh, D. R. (1986). *IEEE Transactions on Microwave Theory and Techniques,* Vol.MTT-34, p. 351.

Collin, R. E. (1992). *Foundations for microwave engineering,* 2nd ed.: McGraw-Hill, Inc.

Cullen, A. L. (1958). *Nature,* Vol.181, p. 332.

Duchamp, J.-M.; Ferrari, P.; Fernandez, M.; Jrad, A.; Melique, X.; Tao, J.; Arscott, S.; Lippens, D. & Harrison, R. G. (2003). *IEEE Transactions on Microwave Theory and Techniques,* Vol.51, p. 1105.

Engheta, N. & Ziolkowski, R. W. (2006). *Matamaterials: Physics and Engineering Explorations*: John Wiley & Sonc, Inc.

Gaponov, A. V.; Ostrovskii, L. A. & Freidman, G. I. (1967). *Radiophysics and quantum electronics,* Vol.10, p. 772.

Gil, I.; Garcia-Garcia, J.; Bonache, J.; Martin, F.; Sorolla, M. & Marques, R. (2004) *Electronics Letters,* Vol.40, p. 1347.

Gorshkov, A. S.; Lyakhov, G. A.; Voliak, K. I. & Yarovoi, L. A. (1998). *Physica D,* Vol.122, p. 161.

Goussetis, G.; Feresidis, A. P.; Wang, S.; Guo, Y. & Vardaxoglou, J. C. (2005). *Journal of Optics A: Pure and Applied Optics,* Vol.7, p. S44.

Grbic, A. & Eleftheriades, G. V. (2002). *Journal of Applied Physics,* Vol.92, p. 5930.

Grbic, A. & Eleftheriades, G. V. (2003). *Antennas and Wireless Propagation Letters,* Vol.2, p. 186.

Grbic, A. & Eleftheriades, G. V. (2004). *Physical Review Letters,* Vol.92, p. 117403.

Gupta, S. & Caloz, C. (2007). Dark and Bright Solitons in Left-Handed Nonlinear Transmission Line Metamaterials, *IEEE/MTT-S International Symp.,* pp. 979-982.

Harris, S. E. (1966). *Applied Physics Letters,* Vol.9, p. 114.

Hirota, R. & Suzuki, K. (1973). *Proceedings of the IEEE,* Vol.61, p. 1483.

Houck, A. A.; Brock, J. B. & Chuang, I. L. (2003). *Physical Review Letters,* Vol.90, p. 137401.

Iyer, A. K. & Eleftheriades, G. V. (2002). *IEEE MTT-S Int Symp. Dig.,* p. 1067.

Kataev, I. G. (1966). *Electromagnetic Shock Waves.* London: Illife.

Kim, H.; Kozyrev, A. B. & van der Weide, D. W. (2005). *IEEE Microwave and Wireless Components Letters,* Vol.15, p. 366.

Kim, H.; Kozyrev, A. B.; Ho, S.-J. & van der Weide, D. W. (2005). *Microwave Symposium Diges.* p. 4.

Kim, H.; Ho, S. J.; Choi, M. K.; Kozyrev, A. B. & van der Weide, D. W. (2006). *IEEE Transactions on Microwave Theory and Techniques,* Vol.54, p. 4178.

Kim, H.; Kozyrev, A. B.; Karbassi, A. & van der Weide, D. W. (2007). *IEEE Transactions on Microwave Theory and Techniques,* Vol.55, pp. 571-578.

Kivshar, Y. S.; Krolikowski, W. & Chubykalo, O. A. (1994). *Physical Review E,* Vol.50, pp. 5020-32.

Kozyrev, A. B. & van der Weide, D. (2004). *IEEE MTT-S Int. Microwave Symp. Dig.*

Kozyrev, A. B.; Kim, H.; Karbassi, A. & van der Weide, D. W. (2005). *Applied Physics Letters,* Vol.87, p. 121109.
Kozyrev, A. B. & van der Weide, D. W. (2005). *IEEE Transactions on Microwave Theory and Techniques,* Vol.53, p. 238.
Kozyrev, A. B. & van der Weide, D. W. (2005). *IEEE Antennas and Propagation Society International Symposium Dig. P.* 672.
Kozyrev, A. B. & van der Weide, D. W. (2005). *Physical Review Letters,* Vol.94, p. 03902.
Kozyrev, A. B. & van der Weide, D. W. (2006). *Physical Review Letters,* Vol.96, p. 069403.
Kozyrev, A. B.; Kim, H. & van der Weide, D. W. (2006). *Applied Physics Letters,* Vol.88, p. 264101.
Kozyrev, A. B. & van der Weide, D. W. (2007). *Applied Physics Letters,* Vol.91, p. 254111.
Kozyrev, A. B. & van der Weide, D. W. (2010). *Applied Physics Letters,* Vol.96, p. 104106.
Lai, A.; Caloz, C. & Itoh, T. (2004). *IEEE Microwave Magazine,* Vol.5, p. 34.
Landauer, R. (1960). *IBM Journal,* Vol.4, p. 391.
Landauer, R. (1960). *Journal of Applied Physics,* Vol.31, p. 479.
Lapine, M.; Gorkunov, M. & Ringhofer, K. H. (2003). *Physical Review E* Vol.67, p. 065601.
Lapine, M. & Gorkunov, M. (2004). *Physical Review E,* Vol.70, p. 066601.
Lighthill, M. J. (1965). *Journal of the Institute of Mathematics and Its Applications,* Vol.1, p. 269.
Lim, S.; Caloz, C. & Itoh, T. (2005). *IEEE Transactions on Microwave Theory and Techniques,* Vol.53, p. 161.
Liu, L.; Caloz, C.; Chang, C.-C. & Itoh, T. (2002). *Journal of Applied Physics,* Vol.92, p. 5560.
Liu, L.; Caloz, C. & Itoh, T. (2002). *Electronics Letters,* Vol.38, p. 1414.
Lonngren, K. E. & Scott, A. (1978). *Solitons in Action.* New York: Academic Press.
Louisell, W. H. (1960). *Coupled mode and parametric electronics.* New York, London: John
Narahara, K.; Nakamichi, T.; Suemitsu, T.; Otsuji, T. & Sano, E. (2007). *Journal of Applied Physics,* Vol.102, p. 024501.
Ostrovskii, L. A. & Soustov, L. V. (1972). *Izvestiya Vysshikh Uchebnykh Zavedenii, Radiofizika,* Vol.15, p. 242.
Pendry, J. B. (2000). *Physical Review Letters,* Vol.85, p. 3966.
Popov, A. K. & Shalaev, V. M. (2006). *Optics Letters,* Vol.31, p. 2169.
Powell, D. A.; Shadrivov, I. V.; Kivshar, Y. S. & Gorkunov, M. V. (2007). *Applied Physics Letters,* Vol.91, p. 144107.
Qin, C.; Kozyrev, A. B.; Karbassi, A.; Joshkin, V. & van der Weide, D. W. (2007). *IEEE/MTT-S International Symp Dig.* P. 1145.
Qin, C. Kozyrev, A. B. Karbassi, A. & Van Der Weide, D. (2007) *Metamaterials.*
Scott, M. M.; Kostylev, M. P.; Kalinikos, B. A. & Patton, C. E. (2005). *Physical Review B,* Vol.71, p. 174440.
Seddon, N. & Bearpark, T. (2003). *Science,* Vol.302, p. 1537.
Service, R. F. (2003). *Science,* Vol.302, p. 1489.
Shadrivov, I. V. & Kivshar, Y. S. (2005). *Journal of Optics A: Pure and Applied Optics,* Vol.7, p. 68.
Shadrivov, I. V.; Morrison, S. K. & Kivshar, Y. S. (2006). *Optics Express,* Vol.14 .
Shadrivov, I.V.; Kozyrev, A.B.; van der Weide, D.W. & Kivshar, Yu.S. (2008). *Applied Physics Letters,* Vol.93, p. 161903.
Shadrivov, I.V.; Kozyrev, A.B.; van der Weide, D.W. & Kivshar, Yu.S. (2008). *Optics Express,* Vol.16, p. 20266.

Shalaev, V. M. (2007). *Nature Photonics,* Vol.1, p. 41.
Shelby, R. A.; D. R. Smith, D. R. & Schultz, S. (2001). *Science,* Vol.292, p. 77.
Sievenpiper, D. F. (2005). *IEEE Transactions on Antennas and Propagation,* Vol.53, p. 236.
Smith, D. R.; Padilla, W. J.; Vier, D. C.; Nemat-Nasser, S. C. & Schultz, S. (2000). *Physical Review Letters,* Vol.84, p. 4184.
Tien, P. K. (1958). *Journal of Applied Physics,* Vol.29, p.1347.
Tretyakov, S. A. (2001). *Microwave and Optical Technology Letters,* Vol.31, p. 163.
Veselago, V.G. (1968) *Soviet Physics Uspekhi, Vol.*10, p. 509.
Yagi, T. & Noguchi, A. (1976). *Electronics and Communications in Japan,* Vol.59, p. 1.
Yariv, A. (1988). *Quantum electronics*: Wiley.
Zharov, A. A.; Shadrivov, I. V. & Kivshar, Y. S. (2003). *Physical Review Letters,* Vol.91, p. 037401.

9

Non-Unity Permeability in InP-Based Photonic Device Combined with Metamaterial

T. Amemiya[1], T. Shindo[2], S. Myoga[2], E. Murai[2], N. Nishiyama[2] and S. Arai[1,2]
[1]*Quantum Nanoelectronics Research Center, Tokyo Institute of Technology,*
[2]*Dept. of Electrical and Electronic Engineering, Tokyo Institute of Technology,*
Japan

1. Introduction

The relative permeability of every natural material is 1 at optical frequencies because the magnetization of natural materials does not follow the alternating magnetic field of light (see Fig. 1). If we can overcome this restriction and control both the permeability and the permittivity at optical frequencies, we will be able to establish a new field involving optical/photonic devices for future communication technologies. In this paper, we move one step closer to this goal—we demonstrate that in photonic devices, the relative permeability can be controlled by adopting metamaterials.

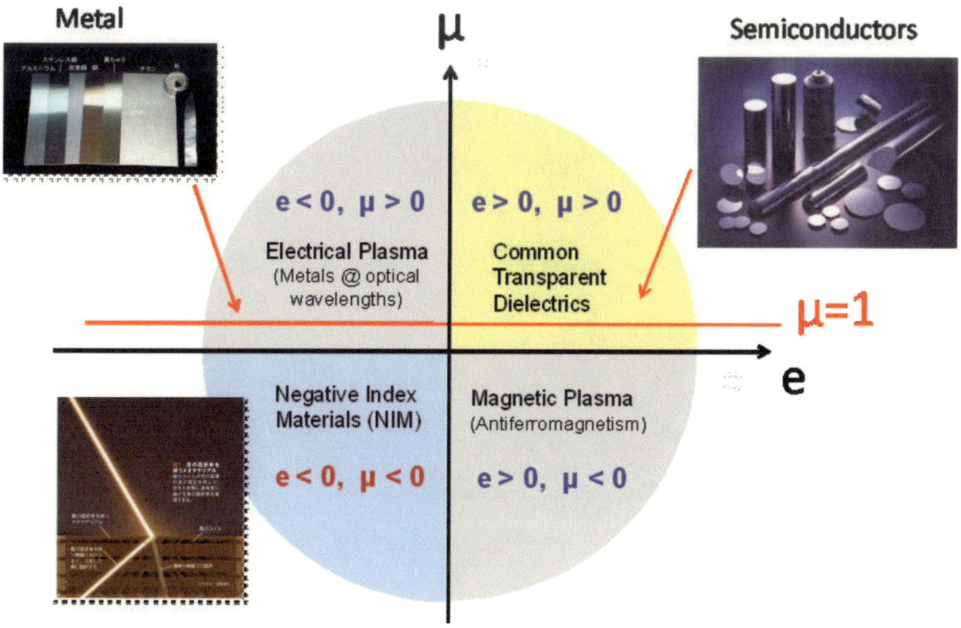

Fig. 1. Constitutive parameters at optical frequencies.

Metamaterials are artificial materials designed to have permittivity and permeability values that are not possible in nature [1]–[4]. They have recently attracted considerable interest because they exhibit unusual properties such as negative refractive indexes and have potential for unique applications such as high-resolution superlenses and invisibility cloaking devices [5, 6].

It is a challenging task to introduce the concept of metamaterials to actual photonic devices (see [7]–[12] for ordinary photonic devices). We hope to apply metamaterials to realize novel optical functionalities that can potentially establish a new field, *meta-photonics*. For this reason, many efforts have been expended in developing advanced optical applications using the concept of metamaterials. Some novel optical functionalities have been realized previously; for example, it has been shown that in theory, it is possible to achieve sophisticated manipulation of light such as slowing, trapping, and storing of light signals [13]–[18].

This paper provides an overview of the present state of research on novel photonic devices with the concept of metamaterials. In Section 2, we outline two promising approaches of making photonic devices combined with metamaterials. One of them is a fiber-based metamaterial device which functions nanoscale light source; another is a Si-based modulator which enables active tuning of metamaterials; the third has the form of III-V semiconductor-based waveguide combined with metamaterials which is compatible with other conventional photonic devices such as lasers and optical amplifier. Although these researches on *meta-photonics* are still in the experimental stage, they will probably reach a level of producing practical devices in the near future. In the succeeding sections, we focus on the III-V semiconductor-based waveguide combined with metamaterials shown in Fig. 2 and make a detailed explanation of the device. In Section 3, we take up the multimode-interferometer (MMI) as an example of photonic devices. First, theoretical investigations of the device are given. Actual devices based on this phenomenon are then developed.

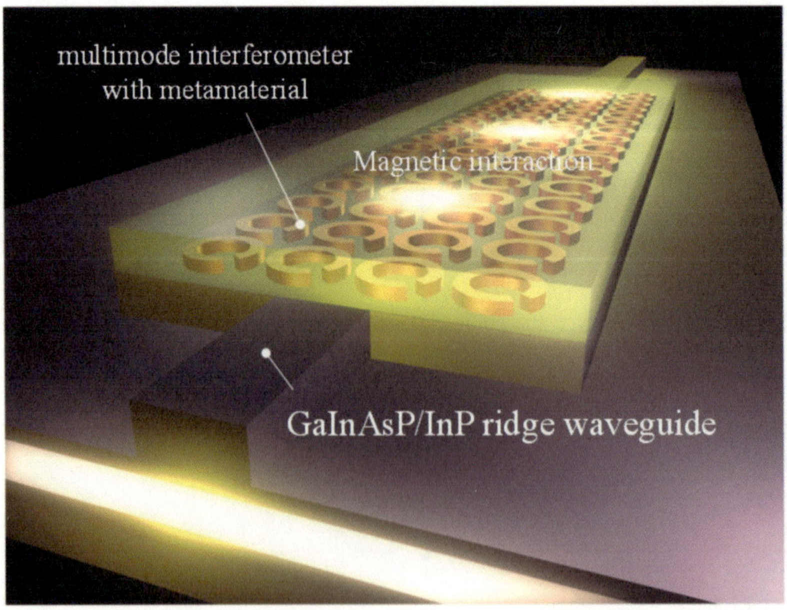

Fig. 2. III-V semiconductor-based waveguide optical device combined with metamaterials

The operating characteristics of the device, i.e., the transmission dependences on the polarization and wavelength of incident light, are also given in this Section. In Sections 4, we report a measurement method for retrieving accurate constitutive parameters (relative permittivity and permeability) from experimental data of the devices. We hope that this paper will be helpful to readers who are aiming to combine photonic integrated devices and metamaterials.

2. Recent progress in photonic devices using the concept metamaterials

2.1 Photonic devices combined with metamaterials

There are several strategies to develop advanced photonic applications which combine with metamaterial. The strategies can be classified into two types. One is based on the optical fiber as in conventional photonic communication system. Transferring the principle of metamaterials to a optical fiber system raises a number of inherent difficulties such as the discoherence of polarization rotation induced by structural birefringence. Therefore new idea is needed to use metamaterials in optical fiber structure. Sophisticated example is the nanoscale light sources consisting of fiber-coupled gold asymmetrically-split ring [19]. This device have attracted attention in recent years because of its compact techniques for producing the device. The other strategy is to combine semiconductor devices with metamaterials. Leading examples are the Si-based modulator which enables active tuning of metamaterial [20-22] and the III-V semiconductor-based waveguide in which relative permeability is not unity [23, 24]. The latter in particular is now the focus of attention because it is compatible with other standard waveguide-based optical devices such as lasers. In the following sections, we give the outline of the fiber-based nanoscale light sources and the Si-based modulator. The III-V semiconductor-based waveguide combined with metamaterials, which has been developed in our laboratory, is explained in detail in Section 3.

2.2 Fiber-based metamaterial device which functions nanoscale light source

Figure 3 shows a schematic of the fiber-based photonic metamaterial devices which act as nanoscale light sources. The device consists of fiber-coupled gold asymmetrically-split ring (ASR) array excited by an electron beam with a trajectory parallel to the surface. Light emission from nanoscale planar photonic metamaterials is induced by beams of free electrons, at wavelengths determined by both the dimensions of metamaterials and the electron beam energy. The ASR resonators are manufactured by focused ion-beam milling through ~70 nm gold films evaporated onto the end faces of standard or tapered optical fibers. Experiments are performed in a scanning electron microscope, which provided simultaneously for imaging and targeted electron beam excitation of samples.

In this device, energy is coupled from incident electrons to the plasmonic modes of the metamaterial structure for which propagating light modes then constitute a decay channel. Low energy beams of free-electrons can act as a broadband excitation sources for the collective plasmonic modes of photonic metamaterials, thereby driving resonant light emission at wavelengths determined by the structural design parameters of the metamaterial, which may be adjusted for operation across the visible to infrared range.

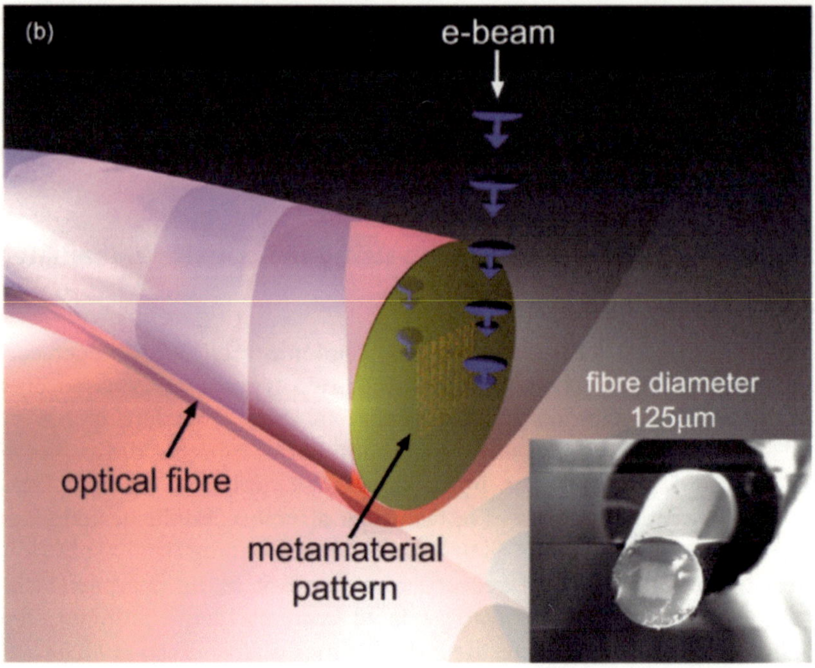

Fig. 3. Fiber-based photonic metamaterial devices which act as nanoscale light sources [19].

2.3 Si-based modulator which enables active tuning of metamaterials

When applying the concept of metamaterials to actual applications, it is indispensable to control properties of metamaterials for tuning electromagnetic responses (i.e., tunable metamaterials). A typical way to create tunable metamaterials is to integrate a reconfigurable material into a metamaterial structure; thereby the active tuning is achieved by applying an external stimulus. GaAs-based modulators with split ring resonators (SRRs) [25] and metamaterial memories on a VO_2 film [26] are such device prototypes operating at 1-10 THz.

Recent researches have shown that one can modulate the optical properties of a metamaterial on a sub-picosecond timescale enabling ultrafast photonic devices [20-22]. In these demonstrations, the authors used a fishnet structure metamaterial [27, 28] with two negative index resonances corresponding to two different periodic wavevectors of the internal gap-mode surface plasmon polaritons. Figure 4 shows one leading example of such device. The device reported here is composed of a BK7 glass substrate and a single metal-dielectric-metal (Ag/α-Si/Ag) functional layer with an inter-penetrating two-dimensional square array of elliptical apertures. In this device, the metamaterial is photoexcited with a visible pump pulse and then the pump-induced, time-resolved change in transmission ($\Delta T/T$) is measured around both the resonances. The longer wavelength resonance has a significantly stronger nonlinear response ($\Delta T/T$~70%) corresponding to its larger absolute value of the negative index and the stronger Drude response of photocarriers at longer wavelengths. These results provide insight into engineering various aspects of the nonlinear response of fishnet structure metamaterials.

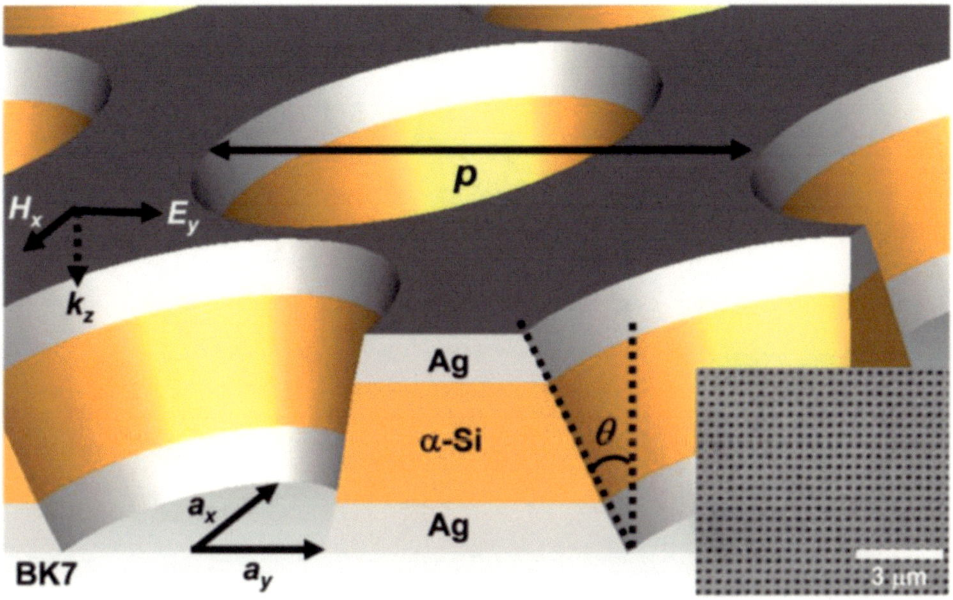

Fig. 4. Si-based modulator which enables active tuning of metamaterials [22].

3. III-V semiconductor-based waveguide optical device with metamaterials

3.1 Waveguide-based photonic devices combined with metamaterials

Encouraged by the results stated in Section 2, we consider introducing metamaterials into conventional photonic devices such as lasers, optical amplifiers, and modulators which have the form of III-V semiconductor-based waveguide. In this Section, the feasibility of employing semiconductor-based photonic devices combined with split-ring resonator (SRR)-based metamaterials is examined both theoretically and experimentally. We used a MMI as the stage of interaction between SRRs and light because the input light dispersed over the whole MMI and therefore gives no saturation to each SRR even if the light was very strong. First, theoretical investigations of the device structures of the SRRs and a MMI for use in the 1.5-μm-wavelength region are given in the former of this Section. Fabrication processes and operating characteristics of the device are then explained in the latter of this Section.

Our metamaterial MMI device is shown in Fig. 5. It consists of a waveguide-based GaInAsP/InP 1 × 1 MMI on which a gold SRR array is attached. If transverse electric (TE)-mode input light for the MMI has a frequency close to the SRR-resonant frequency, magnetic interactions occur between the TE light and SRR array. Therefore, the real part of the macroscopic permeability becomes large positive and negative at frequencies below and above SRR resonance, respectively. Consequently, the SRR array operates as a metamaterial layer to control permeability. The imaginary part of the permeability is not 0 but a finite value under this condition (this causes the absorption loss of light).

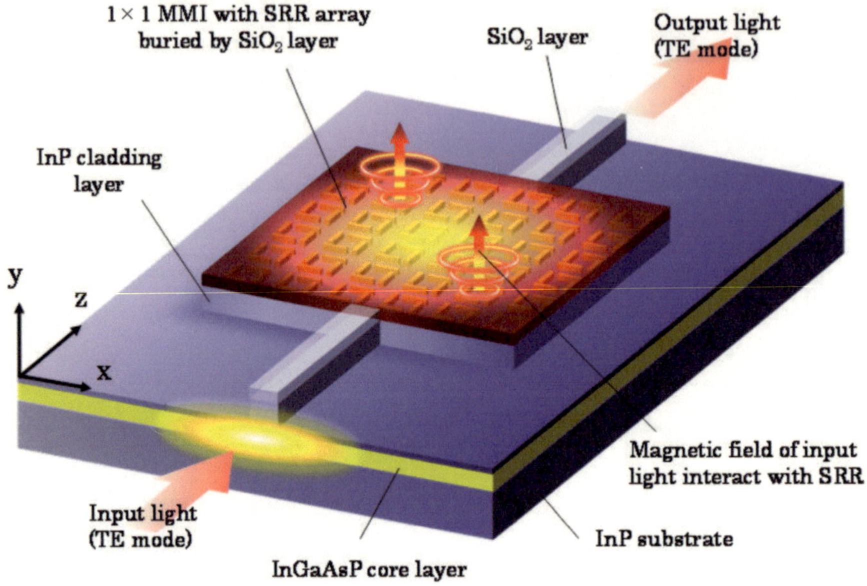

Fig. 5. InP-based 1 × 1 MMI with on which a gold SRR array is attached.

3.2 Theory of waveguide optical devices combined with metamaterials

The key is to create optical metamaterials that can be used to control permeability and obtain *non-unity* values at optical frequencies (note that, on the other hand, permittivity can be controlled more easily than permeability). A promising method for controlling permeability involves the use of a split-ring resonator (SRR). An SRR produces a circular current in response to an incident magnetic flux, thereby producing its own flux to enhance or oppose the incident field. Consequently, an array of extremely small SRRs operates as a metamaterial layer with non-unity permeability [29-33]. We now investigate the optimal structure of a four-cut SRR device for use at an optical frequency of 193 THz (corresponding to 1.55 μm wavelength for low-loss optical fiber communications). The transmission characteristics are obtained by considering the magnetic interactions between the SRRs and light traveling in the MMI.

A. Design of SRR structure for optical frequency

We must first determine the optimal dimensions of the SRR for the 1.5-μm-band frequency. Because the magnetic response of an SRR strongly depends on the conduction characteristics of the metal that forms the SRR, the dispersion of the internal impedance Z of the gold used in our SRRs was calculated. The internal impedance is the ratio of the surface electric field to the total current [34, 35]. *Z* for a unit length and unit width of a metal plane conductor is given by

$$Z(\tau)=\left(\sigma(\omega)\int_0^{\tau}\frac{\exp[ik(\omega)z]+\exp[ik(\omega)(\tau-z)]}{1+\exp[ik(\omega)z]}dz\right)^{-1} \tag{1}$$

Here, $k(\omega)$ is given by

$$k(\omega) = \omega\sqrt{\varepsilon_0\mu_0\left[1 + i\frac{\sigma(\omega)}{\omega\varepsilon_0}\right]}, \tag{2}$$

where ω is the angular frequency of light, ε_0 and μ_0 are the permittivity and permeability of vacuum, respectively, τ is the thickness of the plane conductor, and $\sigma(\omega)$ is the conductivity of the metal as defined by the Drude model.

Figure 6 shows the internal impedance as a function of frequency for a gold layer whose thickness is larger than the penetration depth. As the frequency increases, the real part of the internal impedance first increases sharply and then saturates at around 100 THz; at frequencies higher than 100 THz, it gradually decreases. This dispersion property corresponds to the dielectric behavior of gold. In contrast, the imaginary part changes monotonously with frequency and has large negative values at optical frequencies; this corresponds to ohmic losses in gold.

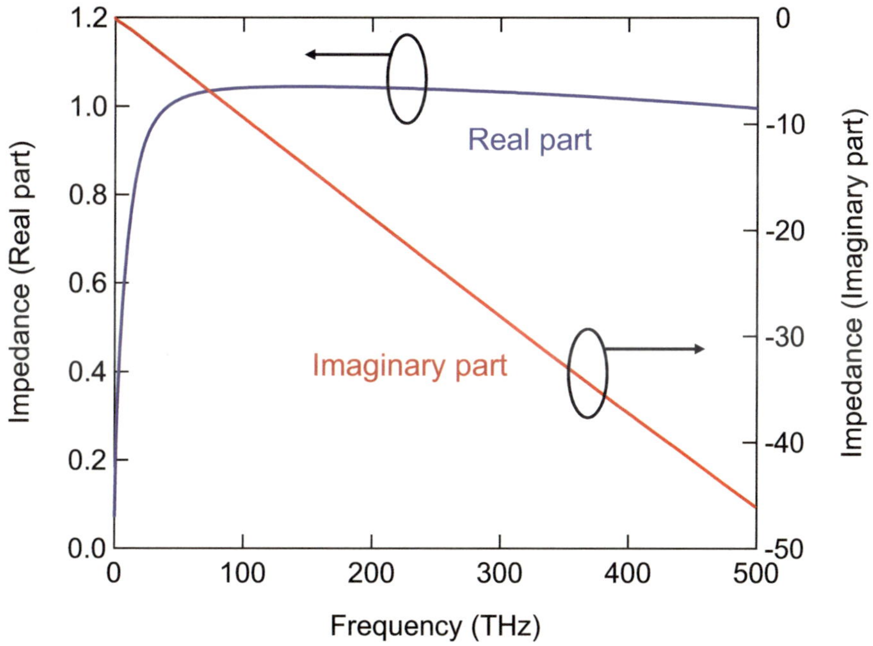

Fig. 6. Internal impedance of gold as a function of frequency (real part is surface resistivity, and imaginary part is internal reactance).

Using these dispersion curves, we calculated the magnetic response of a gold SRR at optical frequencies. A four-cut SRR was considered because it has high resonant frequency due to its small gap capacitance [36, 37]. The SRR was placed in a homogeneous host material (air), as shown in Fig. 5. If an incident AC magnetic field is applied to the SRR, an induced circular current flows in the ring through the gap capacitance. The circular current produces an internal magnetic field, and this produces a magnetic interaction between the SRR and light.

The total electromotive force (emf) induced around the SRR is given by the magnetic flux density B. Therefore, we can equate the potential drop to the emf as follows:

$$\begin{aligned}\partial_t \int_{SRR} B d\sigma &= i\omega\mu_0 \int_{SRR}\left(H_{ext} + \frac{1}{4\pi}\oint \frac{j ds \times \mathrm{r}}{r^3}\right) d\sigma \\ &= V_{ring} + 4V_{gap} = \left[Z(\tau)\cdot\frac{4L}{W_1} - \frac{4W_2}{i\omega\varepsilon_0\varepsilon_m W_1 \tau}\right] j\end{aligned} \quad (3)$$

Here, we have used Biot-Savart's law. In this equation, H_{ext} is the magnetic field of light, ε_m and $Z(\tau)$ are the relative permittivity and internal impedance of gold, respectively, τ is the thickness of the SRR, j is the induced circular current in the SRR; and L, W_1, and W_2 are the dimensions of the SRR (see Fig. 7(a)). The distribution of the magnetic field around the SRR can be calculated using Eq. 3, as illustrated in Fig. 7(b).

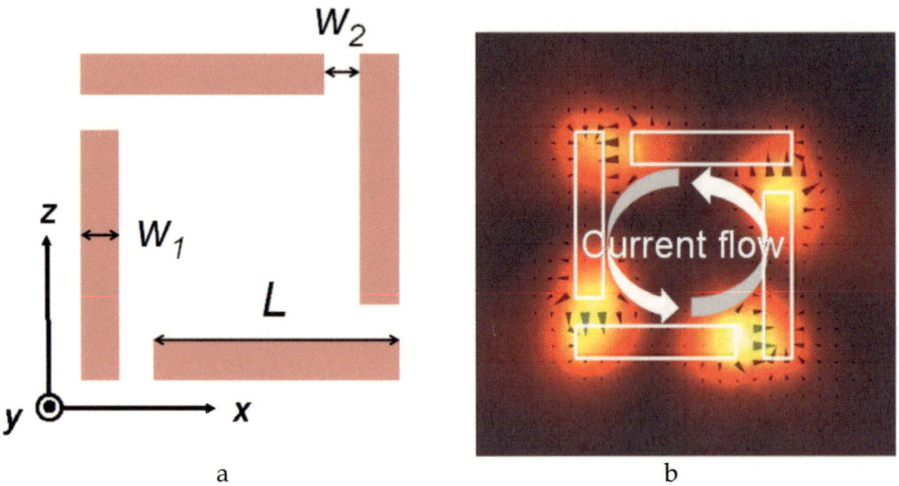

Fig. 7. Four-cut single SRR placed in glass: (a) plane pattern and (b) magnetic field distribution around the SRR at resonant frequency (L = 300 nm). The distribution of field intensity is visualized by a rainbow color map.

B. Macroscopic permeability of SRR array

The results for a single SRR were used to calculate the effective permeability of an SRR array taking into consideration a two-dimensional array layer comprising cubic unit cells, each with an SRR at its center; both the side lengths of each cell and the array pitch of the cells were d. For simplicity, the SRRs were assumed to be placed in air. The macroscopic permeability μ_{yy} of the SRR array layer can be calculated using the field averaging equation [38], and it is given by

$$\bar{\mu}_{yy} = \frac{\bar{B}_y(0, d/2, 0)}{\bar{H}_y(0, d/2, 0)} = \mu_0 \frac{(d)^{-2}\int_{-d/2}^{d/2} dx \int_{-d/2}^{d/2} H_z(x, d/2, z) dz}{(d)^{-1}\int_0^d H_z(0, y, 0) dy}, \quad (4)$$

where B_y and H_y with over lines represent the average values of magnetic flux density and magnetic field, respectively, in the SRR array layer. To get high accuracy in this homogenization approximation, the size of the integration region should be larger than the wavelength of light. In our calculation, we therefore integrated magnetic field over a large cubic tregion that cantained a 3 × 3 cell array (this means that we replaced d with 3d in Eq. 4). Figure 8 shows the real and imaginary parts of permeability as a function of frequency, and the SRR size L is a parameter; here, W_1 and W_2 were set as 100 nm and d was 1.8 μm. The thickness τ of the SRR was set at twice the penetration depth of gold at each frequency. As L decreases, the magnetic resonant frequency increases. Magnetic response could be obtained at the 1.55-μm-band frequency (approximately 193 THz) for the SRR size L of 750 nm (red curves). As L decreases, the magnetic response becomes weaker because the inductance of each SRR decreases; this increases the effective resistance (or decreases the Q factor) of the SRR.

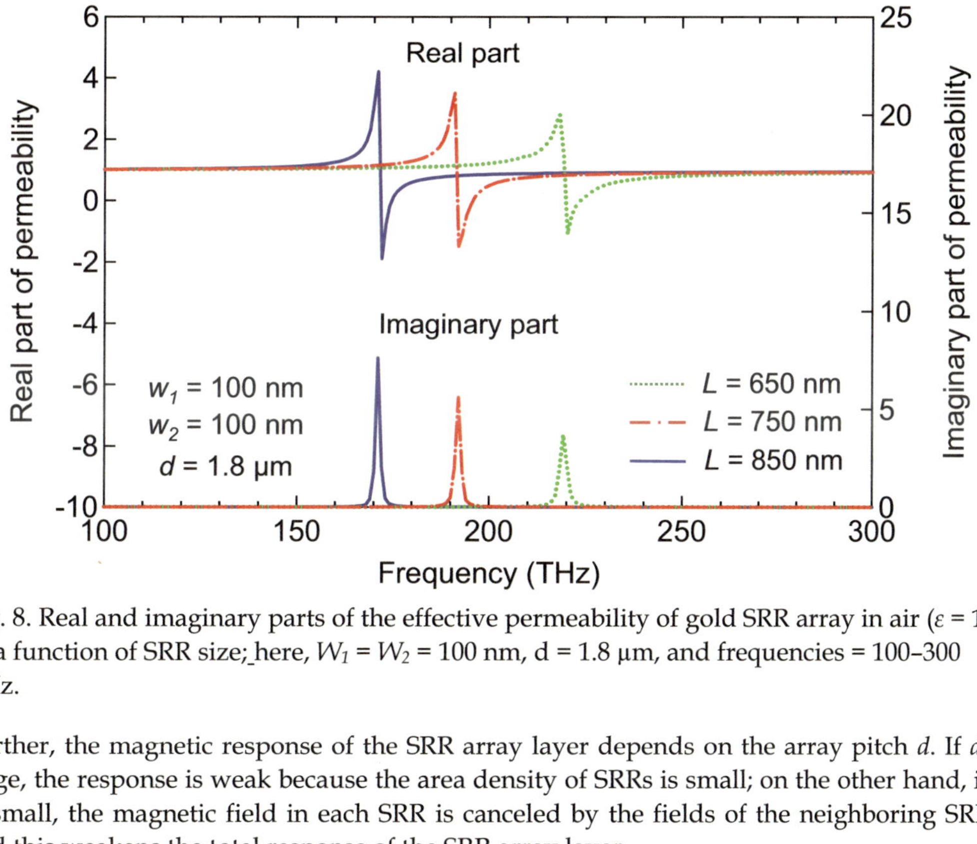

Fig. 8. Real and imaginary parts of the effective permeability of gold SRR array in air ($\varepsilon = 1$) as a function of SRR size;_here, $W_1 = W_2 = 100$ nm, d = 1.8 μm, and frequencies = 100–300 THz.

Further, the magnetic response of the SRR array layer depends on the array pitch d. If d is large, the response is weak because the area density of SRRs is small; on the other hand, if d is small, the magnetic field in each SRR is canceled by the fields of the neighboring SRRs, and this weakens the total response of the SRR array layer.

C. Transmission characteristics of waveguide device with metamaterial

Using the abovementioned results, the transmission characteristics of an MMI with a metamaterial, which is shown in Fig. 5, were estimated using computer simulations based on the transfer-matrix method.

We first designed a device for use at 1.55-µm wavelength. The resultant structure is as follows. The substrate is an InP (refractive index n = 3.16). The constituent layers of the MMI are (i) a core guiding layer: 200-nm-thick $Ga_{0.25}In_{0.75}As_{0.54}P_{0.46}$ (bandgap wavelength λ_g = 1.25 µm; n = 3.38), (ii) InP upper cladding layer (n = 3.16), and (iii) SRR metamaterial layer: 50-nm-thick gold SRR array (L = 750 nm; $W_1 = W_2$ = 100 nm; d = 1.6 µm).

The thickness of the InP cladding layer affects the strength of interaction between the light traveling in the MMI and the SRR array attached to the surface of the cladding layer. Therefore, we determined the optimal thickness of the cladding layer from the following calculations.

After having designed the device structure, we calculated the transmission characteristics of the device as follows. The permeability tensor of the p-th layer, i.e., the InP substrate (p = 1), GaInAsP core layer (p = 2), InP upper cladding layer (p = 3), SRR array layer (p = 4), and air (p = 5), is given by

$$\tilde{\mu}_p = \begin{pmatrix} \mu_{xx}{}^p & 0 & 0 \\ 0 & \mu_{yy}{}^p & 0 \\ 0 & 0 & \mu_{zz}{}^p \end{pmatrix}, \tag{5}$$

where the diagonal elements μ_{xx}, μ_{yy}, and μ_{zz} are 1 at optical frequencies except in the SRR array layer. Using this tensor and the permittivity tensor ε_p for the p-th layer, Maxwell's equations are written as follows:

$$\begin{aligned} \nabla \times \mathrm{H} &= j\omega\varepsilon_0\tilde{\varepsilon}_p \mathrm{E} \\ \nabla \times \mathrm{E} &= -j\omega\mu_0\tilde{\mu}_p \mathrm{H} \end{aligned} \tag{6}$$

We solved Eq. 6 under the condition that the electric and magnetic fields are invariant in the x-direction, that is, $\partial_x = 0$, and their tangential components are continuous at the boundary between the layers. For TE-mode light, the electric field E_x parallel to the z-axis is given by the following differential equation:

$$\partial_y{}^2 E_x + \left(k_0{}^2 \varepsilon_x \mu_z - \frac{\mu_z}{\mu_y}\beta^2 \right) E_x = 0 \tag{7}$$

where ε_x is the diagonal element of the permittivity tensor, and $k_0 = \omega\sqrt{\mu_0\varepsilon_0} = 2\pi/\lambda$ is the free-space propagation constant. The magnetic field H_z parallel to the z-axis (propagation direction) can be calculated using E_x as follows:

$$H_z = -\frac{j}{\omega\mu_0\mu_z}\partial_y E_x \tag{8}$$

An eigenvalue equation can be obtained using the boundary conditions with continuous E_x and H_z. In the calculations, we assumed that E_x and H_z decrease exponentially outside the GaInAsP guiding layer (i.e., in the air and the InP layers). For simplicity, we also assumed that all the layers except the SRR array layer are birefringent. The eigenvalue equation is given by

$$-\frac{j\beta_5}{\omega\mu_0\mu_z}\left(m_{11}-m_{12}\frac{j\beta_1}{\omega\mu_0\mu_z}\right)+\left(m_{21}-m_{22}\frac{j\beta_1}{\omega\mu_0\mu_z}\right)=0\,, \tag{9}$$

where m_{11}–m_{22} are given by

$$\begin{pmatrix} m_{11} & m_{12} \\ m_{21} & m_{22} \end{pmatrix}=\prod_{\text{All}}\begin{pmatrix} \cosh\left(\beta_n d_p\right) & \frac{j\omega\mu_0\mu_z}{\beta_n}\sinh\left(\beta_n d_p\right) \\ \frac{\beta_n}{j\omega\mu_0\mu_z}\sinh\left(\beta_n d_p\right) & \cosh\left(\beta_n d_p\right) \end{pmatrix} \tag{10}$$

$$\beta_n=\sqrt{\left(\mu_z/\mu_y\right)\beta^2-k_0{}^2\varepsilon_x\mu_z} \tag{11}$$

Here d_p is the thickness of the p-th layer. We solved these eigenvalue equations and obtained the effective refractive index β_n/k_0 of each layer. With these results, we calculated the transmission characteristics of the metamaterial MMI using the Fourier expansion method, which is commonly used for MMI propagation analysis [34].

Figure 9 illustrates the example of the distribution profile of light traveling in the metamaterial MMI device. The intensity of the TE electric field at the cross-section (x-z plane) of the device is shown. The MMI is 15-µm wide, 650-µm long, and has a 450-nm-thick InP cladding layer. The wavelength of light is 1565 nm. Figure 9(a) shows our device with a *4-cut* SRR array. The light traveling in the device suffers large propagation losses because of the magnetic interactions between the SRRs and light. Figure 9(b) shows the results for a control device with *no-cut* gold square rings having the same size as a *4-cut* SRR. The *no-cut* ring has no resonant frequency and shows no magnetic interactions with 1.5-µm light; hence, the propagation loss is approximately 10 dB smaller than that of the *4-cut* SRR. These results show that the *4-cut* SRR array can successfully operate as a metamaterial layer at optical frequencies.

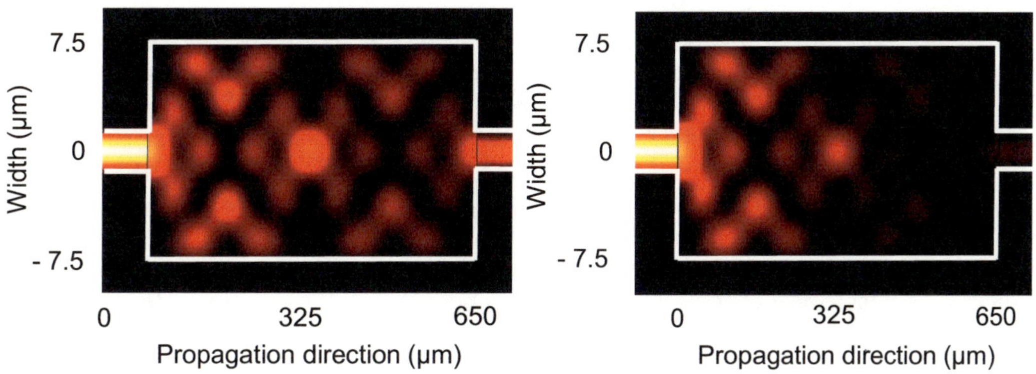

Fig. 9. Distribution profile of light traveling in metamaterial MMI device, calculated for devices with (a) *4-cut* SRR array and (b) *no-cut* gold square rings. Electric field intensity at the cross-section (x-z plane) of the device is visualized by shading. Wavelength of light is 1565 nm.

3.3 Device fabrication and measurement

To move one step closer to the development of actual advanced optical-communication devices using the concept of metamaterials, we fabricated a trial device to confirm the magnetic response of a metamaterial comprising SRRs arrayed on a GaInAsP/InP 1 × 1 MMI coupler. The trial device was fabricated as follows. An undoped $Ga_{0.25}In_{0.75}As_{0.54}P_{0.46}$ core layer (λ_g = 1.22 μm, 200-nm thick) and an undoped InP cladding layer (420-nm thick) were grown on a (100) semi-insulating InP substrate by organometallic vapor phase epitaxy (OMVPE). On the surface of the cladding layer, SRRs consisting of Ti and Au layers were prepared using electron-beam lithography (EBL) and lift-off process.

The fabrication process is illustrated in Fig. 10 with both cross-sectional and plan views. The process flow was as follows: a resist layer of polymethyl methacrylate (PMMA) was first spin coated onto the InP cladding layer. Following spin coating, EBL was used to write a desired SRR array pattern onto the resist. The exposed areas of the resist were dissolved during development with xylene (Fig. 10(b)), resulting in a mask for the subsequent metal-evaporation process (Fig. 10(c)). Subsequently, the unexposed resist was removed along with the metal on top during lift-off with acetone (Fig. 10(d)). Figure 11 shows oblique scanning electron microscope (SEM) images of the SRR array fabricated according to this procedure. The SRRs were made of 5-nm-thick titanium and 20-nm-thick gold, and the dimensions of the individual SRR were designed on the basis of the simulation results shown in Section 3.2. In addition to experimental samples with *4-cut* SRRs (Fig. 11(a)), we also made control samples with SRRs consisting of *2-cut* Au/Ti square rings (Fig. 11(b)) with the same side length as that of the *4-cut* SRR. This *2-cut* SRR has a resonant frequency far lower than 193 THz, about 100 THz, so it does not interact with 1.5-μm light.

After the SRR array was formed, a 75-nm-thick SiO_2 film was deposited on the wafer by using plasma-enhanced chemical vapor deposition. Following the spin coating of PMMA, EBL was used again to write a 1 × 1 MMI pattern onto the resist (Fig. 10(e)). The width and length of the MMI were set to 15 μm and 660 μm. Finally, the exposed regions of the SiO_2 film and InP cladding layer were etched by using buffered HF and reactive ion etching (RIE), respectively, with a mixture gas of CH_4 and H_2 (Fig. 10(f)).

Figures 12(a) and 12(b) show the oblique and cross-sectional SEM images of the MMI region with 300×300-nm SRRs. In this study, the SiO_2 layer was not removed to prevent damages to the nanoscale SRR, which can be observed in Fig. 12(b). A thinner cladding layer (420 nm in Fig. 12(b)) is preferable to obtain large magnetic interactions even though it increases the propagation loss because the optical field coupled to the SRR metal is larger. The light phase is shifted by the magnetic interaction, but this can be neglected compared to the effect of the abovementioned propagation loss. Figure 12(c) shows the magnified plan of the trial device observed using an optical microscope. We made SRRs with different sizes from 300×300 to 550×550 nm (inside size of the square SRR ring). Both the width and gap of the SRR metal region were set to 75 nm.

In the following optical measurements, we observe the magnetic interactions of the propagating light and SRRs in the device. As described in Section 2, if magnetic interactions occur between the SRRs and light, the effective permeability of the SRR array becomes non-unity, i.e., large positive or negative values. At the same time, the imaginary part of the permeability increases from 0 to a finite value, and this implies that light is absorbed in the device. Since it is possible to know the occurrence of the interaction by measuring the propagation loss of light in the device, we primarily measured the transmission and absorption of light in the device.

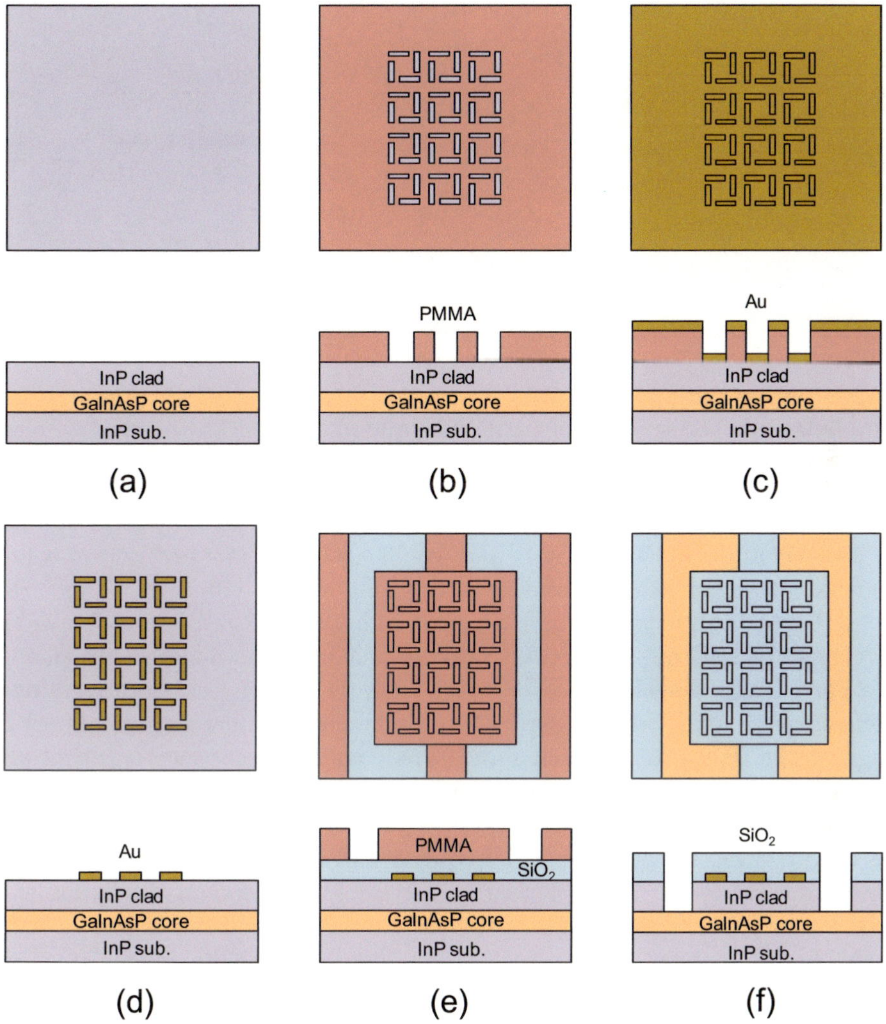

Fig. 10. Fabrication process for metamaterial MMI devices. SRR array was prepared using electron-beam lithography and lift-off process.

In the measurement, light was sent from a tunable laser to the device through a polarization controller. The wavelength was changed in a range of 1420-1575 nm. To clarify the effect of the magnetic interaction, we took the difference between the transmission intensity for the experimental samples (with *4-cut* SRRs) and that for the control samples (with *2-cut* SRRs). This difference shows an intrinsic change in transmission intensity induced by the SRR resonance without including parasitic factors such as wavelength-dependent ohmic loss in SRR metal, lensed-fiber coupling loss, and wavelength-dependent propagation characteristics in the MMI.

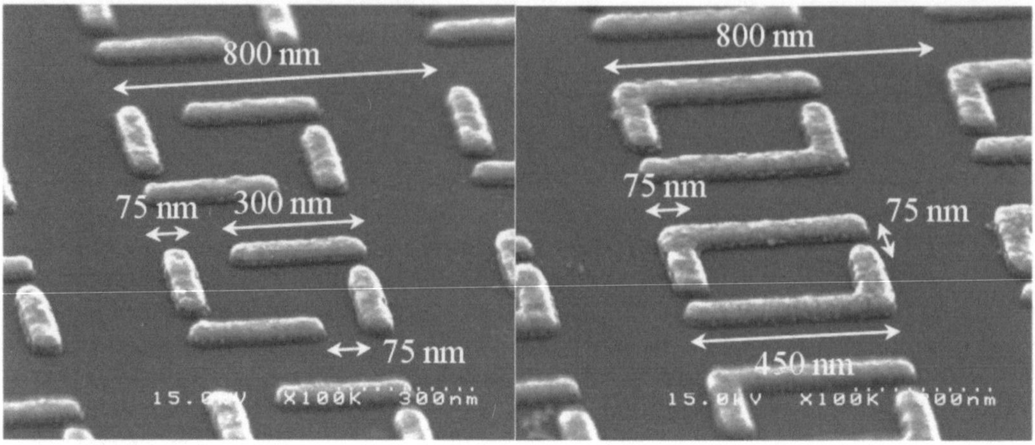

a b

Fig. 11. (a) Enlarged oblique views of 4-cut SRRs, and (b) 2-cut SRRs observed with scanning electron microscopy.

Figure 13 plots the measured intensity difference for devices with a SRR size of (a) 300×300 nm^2, (b) 350×350 nm^2, (c) 400×400 nm^2, and (d) 500×500 nm^2 as a function of wavelength. The magnetic interaction was observed clearly in the device with 350×350-nm^2 SRRs (see Fig. 13(b)). That is, the intensity difference, induced by the SRR resonance, showed its peak at a wavelength of 1500 nm. The peak shifted to a shorter wavelength with smaller SRRs (Fig. 13(a)) and longer wavelength with larger SRRs (Fig. 13(c)), and both were out of this measurement range. With 500×500-nm^2 SRRs, the intensity difference was almost 0 at this wavelength range (Fig. 13(d)), which showed that no SRR resonance occurred at 1.5-μm wavelength. These wavelength-dependent and SRR-size-dependent transmission characteristics show that the magnetic field of light interacted successfully with the SRRs to produce magnetic resonance at optical frequency. In contrast, no intensity change was observed for the TM mode. This polarization-wavelength dependent absorption is positive proof that the magnetic interaction was successfully established in our device for the TE-mode light. In this manner, we can realize *non-unity* permeability in InP-based photonic devices by using the SRR metamaterial

Furthermore, for the TE mode, a magnetic field perpendicular to the axis of the split-ring, by virtue of Ampere's law, created a circulating current via the charge accumulation at the gap (see Section 2 for details). Due to the presence of gaps, the resulting charge distribution was asymmetric; this results in charge accumulation around the capacitive gaps and induces an electric dipole moment. On the other hand, for the TM mode, an electric field is present parallel to the two symmetric sides of the split-ring, whereas in the approximation of the thin metallization, the current in the perpendicular directions was negligible. In addition, the size of the ring was smaller than the incident wavelength, and this ensured that the variations in the electric field between the two sides were also negligible for a first approximation. Consequently, the charge distribution resulting from this incidence was symmetric and did not generate a circulating current.

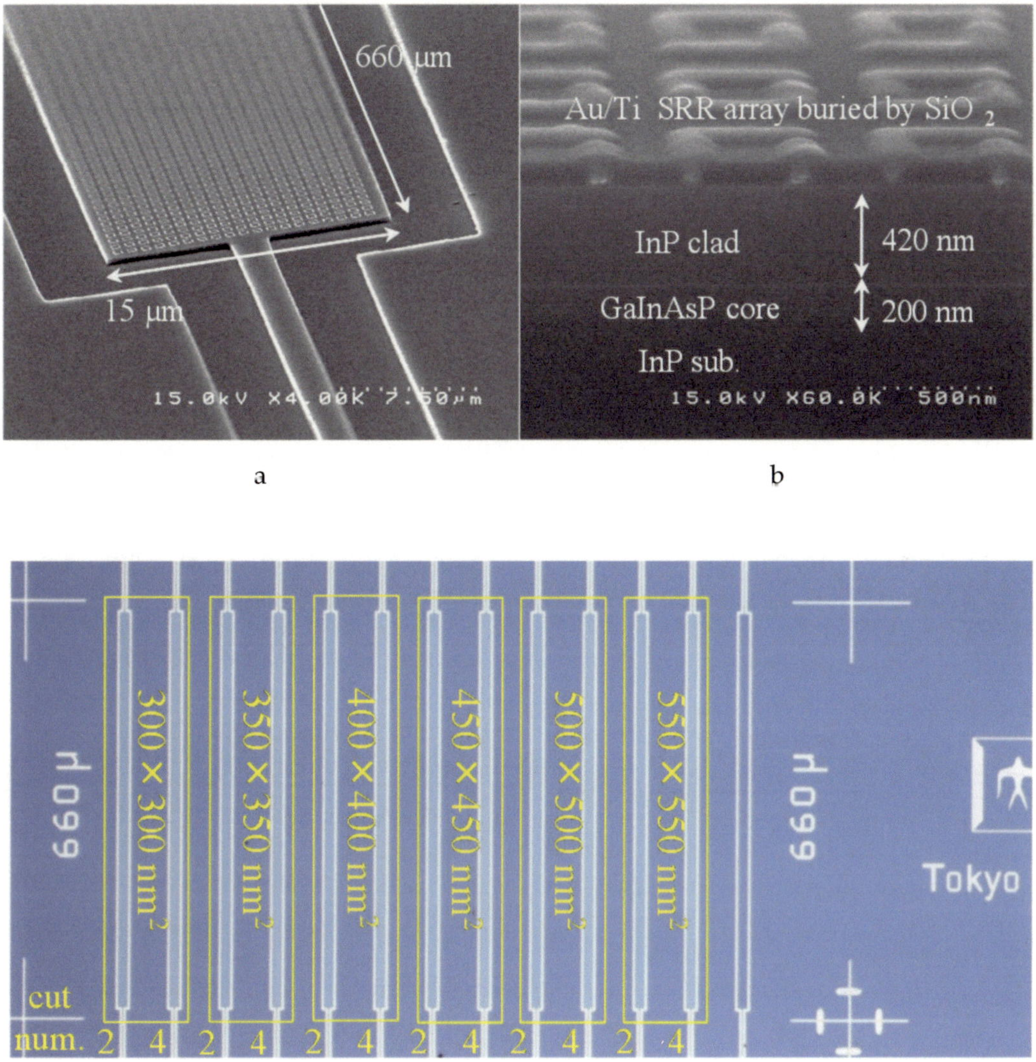

(c)

Fig. 12. (a) Oblique views of completed device and (b) cross-sectional view of 300×300-nm SRRs buried in SiO_2 layer; (c) Plan view of GaInAsP/InP 1 × 1 MMI coupler with SRR array observed with optical microscopy.

The equivalent permeability and permittivity of the SRR array layer are a function of wavelength. To know their values exactly, we must know both intrinsic absorption loss and phase shift of propagating light in the device. However, we cannot extract each separately from the transmission data shown in Fig. 13. So then, to cope with the problem, we have recently proposed a measurement method that uses a Mach-Zehnder interferometer (MZI) and successfully retrieved constitutive parameters for the SRR array. The following Section provides the outline of the results.

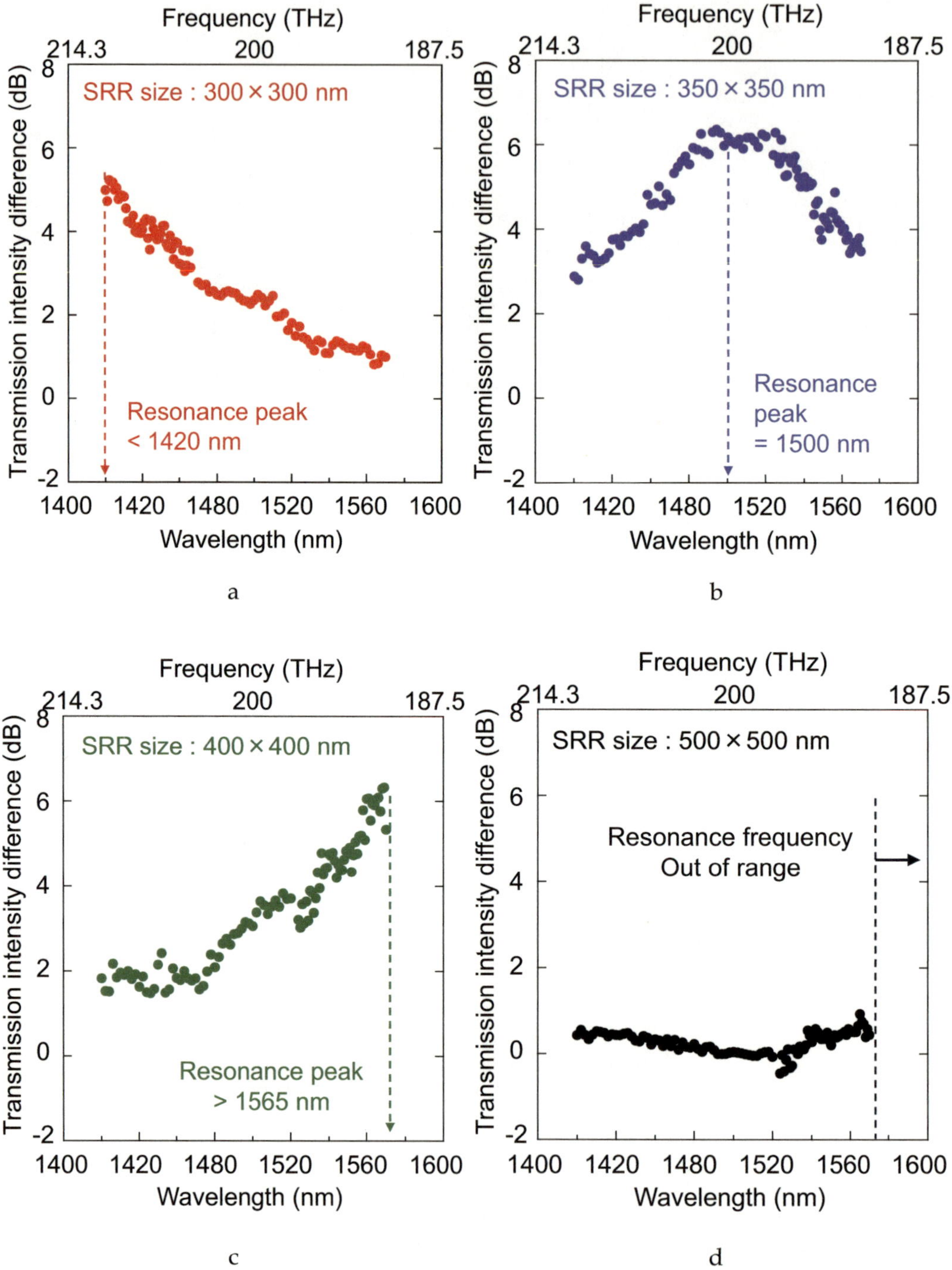

Fig. 13. (Color online) Transmission-intensity difference (or transmission spectra) for devices with SRR size of (a) 300×300 nm^2, (b) 350×350 nm^2, (c) 400×400 nm^2, and (d) 500×500 nm^2 as a function of wavelength from 1420 to 1575 nm, measured for TE-mode light.

4. Permeability retrieval in III-V semiconductor-based waveguide device with metamaterials

The accurate permeability values in waveguide-based photonic devices is a very important factor, since it mostly determines their performance. Accurate permeability data, however, cannot be extracted from the simple transmission data of the previous devices because the transmission data also includes both the effect of SRR's effective permittivity and permeability. In this Section, we have proposed a measurement method that uses a Mach-Zehnder interferometer (MZI) and successfully retrieved constitutive parameters for the SRR array.

Our MZI device is shown in Fig. 14(d). It consists of two 3-dB couplers and two arms made with GaInAsP/InP ridge waveguides, with a metal SRR array attached on one of the arms. For TE-mode input light with the SRR-resonance frequency, the SRR array interacts with the light, thereby behaving as a material with a complex refractive-index. The real part of the refractive-index mainly affects the phase of traveling light, thereby changing the phase difference at the output coupler. The imaginary part mainly causes the propagation loss of light in the arm. Therefore, the transmittance of the MZI is determined by the phase difference and propagation loss.

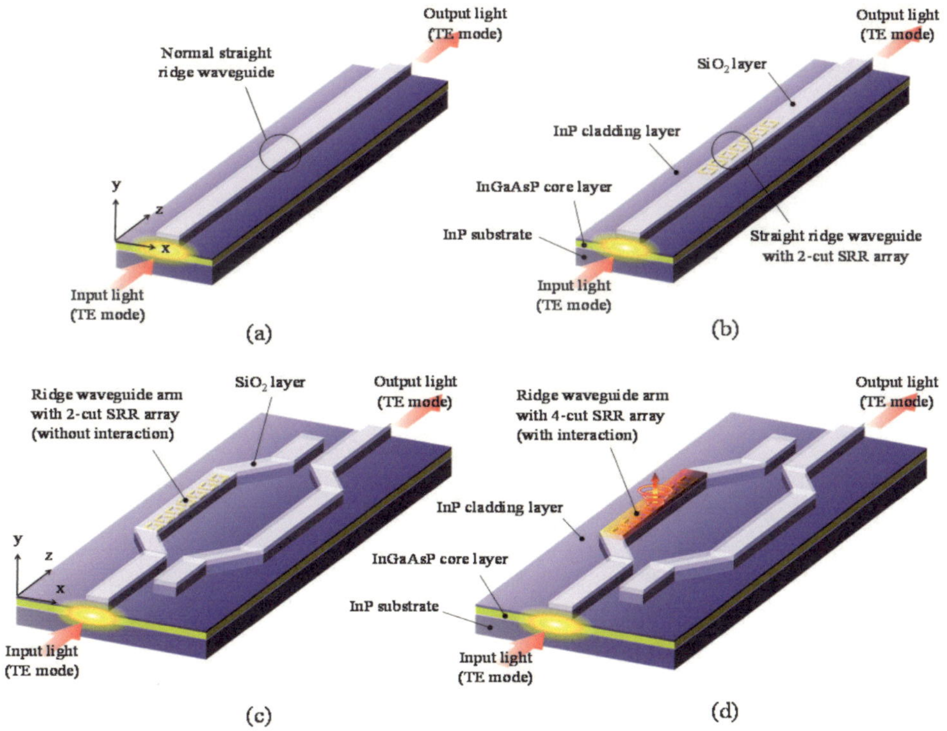

Fig. 14. GaInAsP/InP straight ridge waveguide without SRRs (a) and with 2-cut SRRs (b); Mach-Zehnder Interferometer consisting of GaInAsP/InP waveguides and (c) 2-cut / (d) 4-cut SRR array attached on one arm.

The propagation loss in the arms can be inferred from the transmission data of straight waveguides with/without the SRR array shown in Figs. 14(a) and 14(b). Therefore, the effective permeability and permittivity of the SRR array can be calculated, using measured data for the transmittance of the MZI.

An actual MZI was made for measurement at 1.5-µm optical communication wavelength. Epitaxial layer structures were the same as those of our previous device. On the surface of the device, an SRR array (consisting of 10-nm thick Ti and 40-nm thick Au) was formed using electron-beam lithography (EBL) and a lift-off process. Figure 15(a) shows the oblique images of the SRR array observed with a scanning electron microscope (SEM). The 4-cut SRR was used because it has a high resonant frequency owing to its small gap capacitance as stated in Section 2.

The size of the SRR was designed for use at 1.5-µm band frequency (193 THz). After the formation of the SRR array, a SiO2 mask (100-nm thick) for the MZI pattern was formed on the device with plasma-enhanced chemical-vapor-deposition and EBL. With the SiO2 mask, the MZI structure was formed using CH4/H2 reactive ion etching. Figure 15(b) shows the SEM image of the arms with/without the SRR array. The length of the SRR array along the arm was set to 500 µm. In addition to these experimental samples, straight ridge waveguides with/without the SRRs were made. We also prepared control samples with SRRs consisting of *2-cut* square rings with the same size as that of the *4-cut* SRR. As mentioned in Section 3, the *2-cut* SRR has a resonant frequency far higher than 193 THz, so it has no interaction with 1.5-µm light.

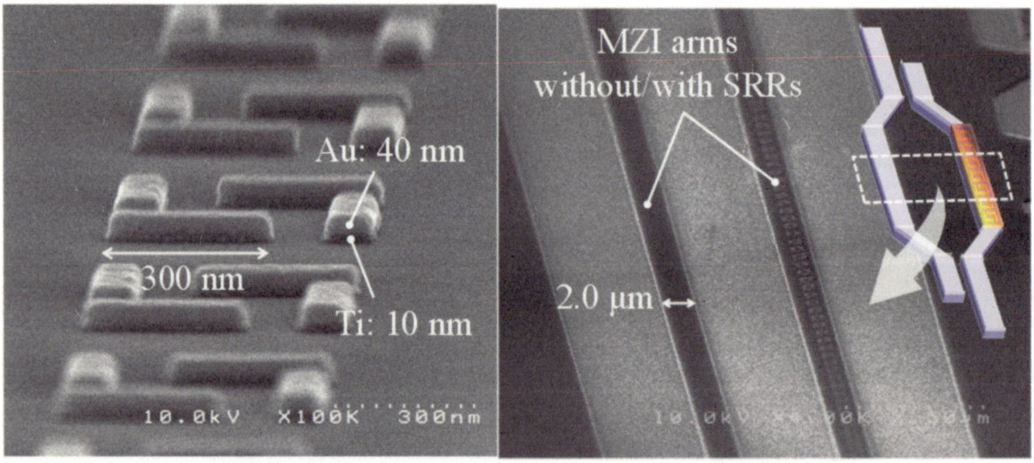

a b

Fig. 15. SRRs and arms of MZI: (a) oblique view of 4-cut SRRs, and (b) two arms, observed with scanning electron microscopy.

Figure 16(a) shows the output intensity for straight waveguides with 4-cut SRR array (blue dots), without SRRs (black dots), and with 2-cut SRRs (red dots). The difference between the curves without SRRs and that with 2-cut SRRs corresponds to a loss caused by light absorption of the SRR metal. The difference between the 2-cut SRRs and 4-cut SRRs shows the loss caused by the magnetic interaction between the SRRs and light. The 4-cut SRRs resonated at about 1510-1520 nm and showed the maximum loss at this wavelength.

Figure 16(b) shows the output intensity for MZIs with 4-cut SRRs and that for without SRRs. Their difference shows the intensity change induced by the phase shift and the absorption loss of light in the MZI. That is, the difference shows an intrinsic change in transmission intensity induced by the SRR resonance without including parasitic factors such as metal absorption loss in the SRRs and lensed-fiber coupling loss in the measurement system.

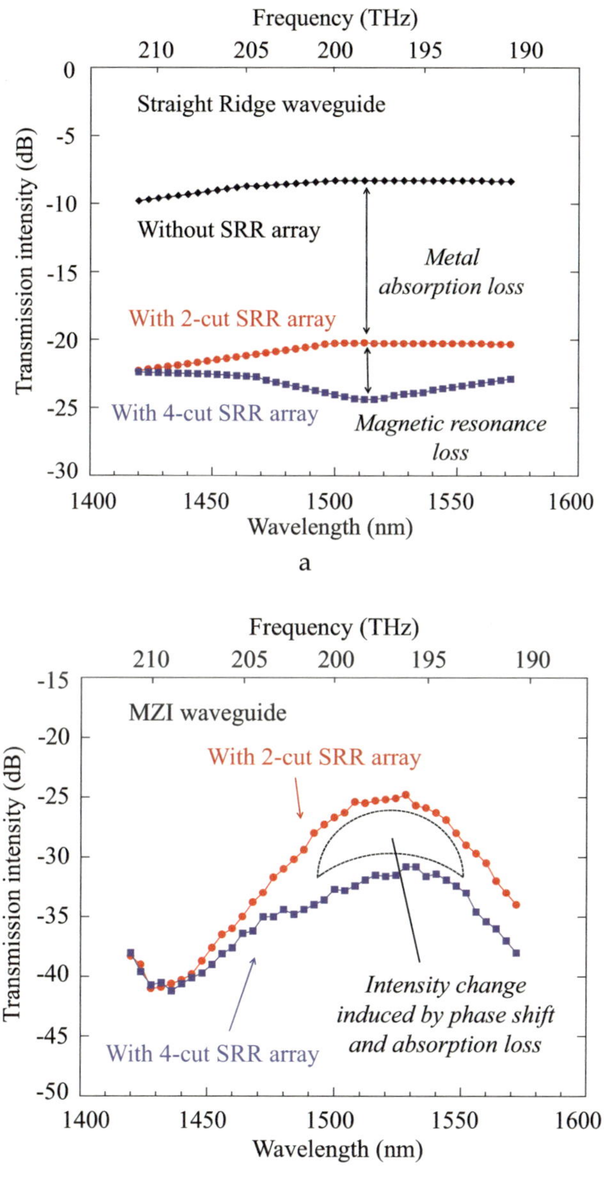

Fig. 16. Output intensity from devices with 4-cut SRRs (blue lines), 2-cut SRRs (res lines) and without SRRs (black lines) as a function of wavelength, measured for (a) straight ridge waveguides and (b) MZIs.

On simple condition that the electric and magnetic fields are constant in the x-direction, that is, ∂x = 0, the wave equation in each layer of the device is given by Eq. (7) in Section 2.

$$\partial_y{}^2 E_x + \left(k_0{}^2 \varepsilon_x \mu_z - \frac{\mu_z}{\mu_y} \beta^2 \right) E_x = 0 \tag{12}$$

To extract the permittivity ε_x of the SRR array without resonance, we used the difference between the transmission intensity of the straight waveguides with *2-cut* SRRs and that of the control device without SRRs. The permittivity ε_x of the SRR array can be calculated, using the transfer-matrix method with Eq. (12) (μ_y and μ_z in Eq. (12) are equal to 1 because the SRRs have no resonance). After that, the real and imaginary parts of permeability μ_y of the SRR array was retrieved, using the obtained permittivity ε_x and the transmission intensity ratio (= (2-cut SRRs)/(4-cut SRRs)) for the straight waveguide and that for the MZI. In calculation, the effective thickness of the SRR layer was set to 350 nm.

Figure 17 shows the retrieved permeability (real and imaginary parts) of the SRR array as a function of frequency. The permeability exhibited a resonance at 200 THz, and the real part of the relative permeability changed from +2.2 to -0.3 in the vicinity of this frequency. This results show the feasibility of semiconductor-based photonic devices combined with metamaterials.

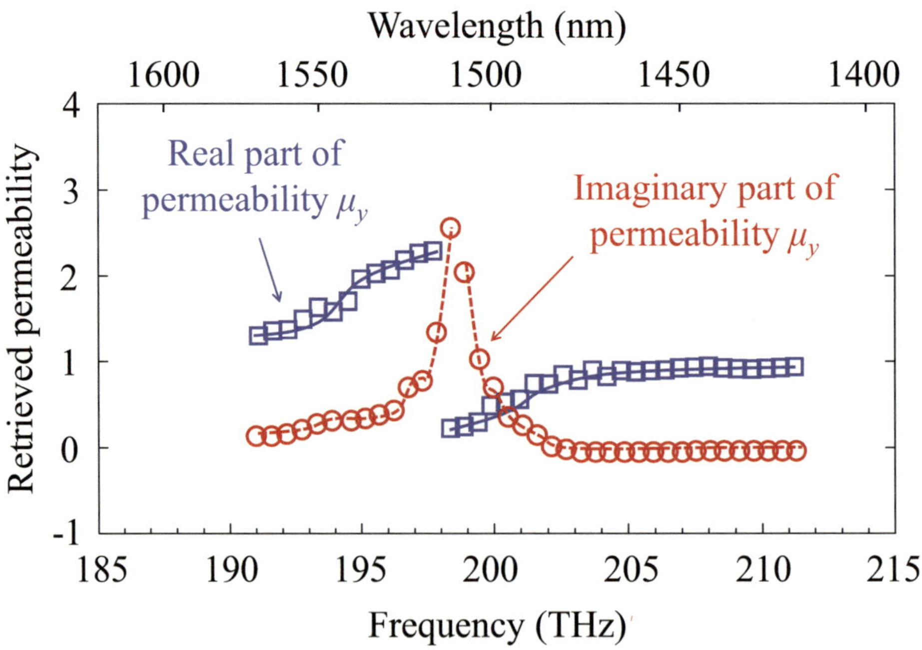

Fig. 17. Retrieved effective permeability (real and imaginary parts) of SRR array on MZI waveguide, plotted as a function of frequency.

5. Conclusion

Realizing *non-unity* permeability at optical frequencies can be expected to lead us to advanced optical-communication devices based on novel operation principles. To move one step closer to this goal, in this paper, we have demonstrated that the permeability in semiconductor photonic devices can be controlled using the concept of metamaterials.

As an actual example, we fabricated a GaInAsP/InP MMI device combined with an SRR array that operated as a metamaterial layer. The operation wavelength was set as 1.55 μm. The transmission characteristics of this metamaterial MMI device strongly depended on the polarization and wavelength of input light. This shows that the SRR array layer interacted with the magnetic field of light and produced magnetic resonance at optical frequencies. After that, to know constitutive parameters in the device exactly, we have proposed a measurement method that uses a MZI and successfully retrieved the accurate permeability value for the SRR array. The permeability exhibited a resonance at 200 THz, and the real part of the relative permeability changed from +2.2 to -0.3 in the vicinity of this frequency. Our results show the feasibility of III-V semiconductor-based waveguide photonic devices combined with metamaterials. This would be useful in the development of novel optical-communication devices.

6. References

[1] R. M. Walser, "Electromagnetic metamaterials," in *Proc.* SPIE, vol. 4467, pp. 1–15 (2001).

[2] H. J. Lezec, J. A. Dionne, and H. A. Atwater, "Negative refraction at visible frequencies," *Science*, vol. 316, no. 5823, pp. 430–432 (2007).

[3] W. Cai, U. K. Chettiar, H. K. Yuan, V. C. de Silva, A. V. Kildishev, V. P. Drachev, and V. M. Shalaev, "Metamagnetics with rainbow colors," *Optics Express*, vol. 15, no. 6, pp. 3333–3341 (2007).

[4] J. Valentine, S. Zhang, T. Zentgraf, E. Ulin-Avila1, D. A. Genov1, G. Bartal1, and X. Zhang, "Three-dimensional optical metamaterial with a negative refractive index," *Nature*, vol. 455, no. 7211, pp. 376–379 (2008).

[5] D. Schurig, J. J. Mock, B. J. Justice, S. A. Cummer, J. B. Pendry, A. F. Starr, and D. R. Smith, "Metamaterial electromagnetic cloak at microwave frequencies," *Science*, vol. 314, no. 5801, pp. 977–980 (2006).

[6] J. B. Pendry, "Negative refraction makes a perfect lens," *Phys. Rev. Lett.*, vol. 85, no. 18, pp. 3966–3969 (2000).

[7] R. Nagarajan, C. H. Joyner, R. P. Schneider, Jr., J. S. Bostak, T. Butrie, A. G. Dentai, V. G. Dominic, P. W. Evans, M. Kato, M. Kauffman, D. J. H. Lambert, S. K. Mathis, A. Mathur, R. H. Miles, M. L. Mitchell, M. J. Missey, S. Murthy, A. C. Nilsson, F. H. Peters, S. C. Pennypacker, J. L. Pleumeekers, R. A. Salvatore, R. K. Schlenker, R. B. Taylor, H. S. Tsai, M. F. Van Leeuwen, J. Webjorn, M. Ziari, D. Perkins, J. Singh, S. G. Grubb, M. S. Reffle, D. G. Mehuys, F. A. Kish, and D. F. Welch, "Large-scale photonic integrated circuits," *IEEE J. Select. Topics Quantum Electron.*, vol. 11, no. 1, pp. 50–65, (2005).

[8] J. Van Campenhout, P. R. Romeo, P. Regreny, C. Seassal, D. Van Thourhout, S. Verstuyft, L. D. Cioccio, J.-M. Fedeli, C. Lagahe, R. Baets, "Electrically pumped InP-based microdisk lasers integrated with a nanophotonic silicon-on-insulator waveguide circuit", *Optics Express*, vol. 15, no. 11, pp. 6744-6749 (2007).

[9] M. J.R. Heck, P. Muñoz, B. W. Tilma, E. A.J.M. Bente, Y. Barbarin, Y. S. Oei, R. Nötzel and M. K. Smit, "Design, Fabrication and Characterization of an InP-based Tunable Integrated Optical Pulse Shaper", *IEEE Journal of Quantum Electron.*, vol. 44, no. 4, p.p. 370-377 (2008).

[10] S. C. Nicholes, M. L. Masanović, B. Jevremović, E. Lively, L. A. Coldren, and D. J. Blumenthal, "World's first InP 8x8 monolithic tunable optical router (MOTOR) operating at 40 Gbps line rate per port," in *Proc. Optical Fiber and Communication Conference*, PDPB1 (2009).

[11] C. R. Doerr, L. Zhang, and P. J. Winzer, "Monolithic InP multi- wavelength coherent receiver," in *Proc. Optical Fiber and Communication Conference*, PDPB1 (2010).

[12] J. J. G. M. van der Tol, Y. S. Oei, U. Khalique, R. Nötzel, and M. K. Smit, "InP-based photonic circuits: Comparison of monolithic integration techniques (review paper)," *Prog. Quantum Electron.*, vol. 34, no. 4, pp. 135–172 (2010).

[13] I. V. Shadrivov, A. A. Sukhorukov, Y. S. Kivshar, "Guided modes in negative-refractive-index waveguides," *Phys. Rev. E*, vol. 67, no. 5, 057602 (2003).

[14] A. C. Peacock and N. G. R. Broaderick, "Guided mode in channel waveguide with a negative index of refraction," *Optics Express*, vol. 11, no. 20, pp. 2502–2510 (2003).

[15] K. L. Tsakmakidis, A. D. Boardman, and O. Hess, "Trapped rainbow storage of light in metamaterials," *Nature*, vol. 450, no. 7168, pp. 397–401 (2007).

[16] A. Figotin and I. Vitebskiy, "Slow-wave resonance in periodic stacks of anisotropic layers," *Phys. Rev. A*, vol. 76, no. 5, 053839 (2007).

[17] K.-Y. Jung, F. L. Teixeira, and R. M. Reano, "Au/SiO2 nanoring plasmon waveguides at optical communication band," *IEEE Journal of Lightwave Technology*, vol. 25, no. 9, pp. 2757-2765 (2007).

[18] K.-Y. Jung and F. L. Teixeira, "Photonic crystals with a degenerate band edge: Field enhancement effects and sensitivity analysis," *Phys. Rev. B*, vol. 77, no. 12, 125108 (2008).

[19] G. Adamo, K. F. MacDonald, F. De Angelis, E. Di Fabrizio, and N. I. Zheludev, "Nanoscale electron-beam-driven metamaterial light sources," *in Proc. 23rd Annual Meeting of the IEEE Photonics Society*, WM2 (2010).

[20] K. M. Dani, Z. Ku, P. C. Upadhya, R. P. Prasankumar, S. R. Brueck, and A. J. Taylor, "Subpicosecond optical switching with a negative index metamaterial," *Nano Lett.*, vol. 9, no. 10, 3565–3569 (2009).

[21] D. J. Cho, W. Wu, E. Ponizovskaya, P. Chaturvedi, A. M. Bratkovsky, S. Y. Wang, X. Zhang, F. Wang, and Y. R. Shen, "Ultrafast modulation of optical metamaterials," *Opt. Express*, vol. 17, no. 20, 17652–17657 (2009).

[22] Keshav M. Dani, Zahyun Ku, Prashanth C. Upadhya, Rohit P. Prasankumar, Antoinette J. Taylor, and S. R. J. Brueck, "Ultrafast modulation of optical metamaterials," *Opt. Express*, vol. 19, no. 5, 3973–3983 (2011).

[23] T. Amemiya, T. Shindo, D. Takahashi, N. Nishiyama, S. Arai, "Magnetic Interactions at Optical Frequencies in an InP-Based Waveguide Device with Metamaterial," *IEEE J. Quantum Electronics*, Vol. 47, No. 5, pp. 736-744 (2011).
[24] T. Amemiya, T. Shindo, D. Takahashi, S. Myoga, N. Nishiyama, S. Arai, "Nonunity permeability in metamaterial-based GaInAsP/InP multimode interferometers," *Optics Lett.*, Vol. 36, No. 12, pp. 2327-2329, Jun. 2011.
[25] H. T. Chen, *et al.*, "Active terahertz metamaterial devices," *Nature*, vol. 444, no. 7119, pp. 597-600 (2006).
[26] T. Driscoll, *et al.*, "Memory Metamaterials," *Science* vol. 325, no. 5947, pp. 1518-1521 (2009).
[27] S. Zhang, W. Fan, N. C. Panoiu, K. J. Malloy, R. M. Osgood, and S. R. J. Brueck, "Experimental demonstration of near-infrared negative-index metamaterials," *Phys. Rev. Lett.* vol. 95, no. 13, 137404 (2005).
[28] Z. Ku, and S. R. J. Brueck, "Comparison of negative refractive index materials with circular, elliptical and rectangular holes," *Opt. Express* vol. 15, no. 8, 4515–4522 (2007).
[29] T. J. Yen, W. J. Padilla, N. Fang, D. C. Vier, D. R. Smith, J. B. Pendry, D. N. Basov, and X. Zhang, "Terahertz magnetic response from artificial materials," *Science*, vol. 303, no. 5663, pp. 1494–1496 (2004).
[30] S. Zhang, W. Fan, B. K. Minhas, A. Frauenglass, K. J. Malloy, and S. R. J. Brueck, "Midinfrared resonant magnetic nanostructures exhibiting a negative permeability," *Phys. Rev. Lett.*, vol. 94, no. 3, 037402 (2005).
[31] C. Enkrich, M. Wegener, S. Linden, S. Burger, L. Zschiedrich, F. Schmidt, J. F. Zhou, Th. Koschny, and C. M. Soukoulis, "Magnetic metamaterials at telecommunication and visible frequencies," *Phys. Rev. Lett.*, vol. 95, no. 20, 203901 (2005).
[32] S. Linden, C. Enkrich, G. Dolling, M. W. Klein, J. Zhou, T. Koschny, C. M. Soukoulis, S. Burger, F. Schmidt, and M. Wegener, "Photonic metamaterials: magnetism at optical frequencies," *IEEE J. Select. Topics Quantum Electron.*, vol. 12, no. 6, pp. 1097–1105 (2006).
[33] C. E. Kriegler, M. S. Rill, S. Linden, and M. Wegener, "Bianisotropic photonic metamaterials," *IEEE J. Select. Topics Quantum Electron.*, vol. 16, no. 2, pp. 367–375 (2010).
[34] S. Ramo, J. R. Whinnery, and T. V. Duzer., *Field and Waves in Communication Electronics* (Wiley 1993), pp. 149–156.
[35] A. Ishikawa, T. Tanaka, and S. Kawata, "Frequency dependence of the magnetic response of split-ring resonators," *J. Opt. Soc. Am. B*, vol. 24, no. 3, pp. 510–515 (2007).
[36] S. A. Ramakrishna, "Physics of negative refractive index materials," *Rep. Prog. Phys.*, vol. 68, no. 2, pp. 449–521 (2005).
[37] A. Ishikawa, T. Tanaka, and S. Kawata, "Negative magnetic permeability in the visible light region," *Phys. Rev. Lett.*, vol. 95, no. 23, 237401 (2005).
[38] D. R. Smith and J. B. Pendry, "Homogenization of metamaterials by field averaging (invited paper)," *J. Opt. Soc. Am B*, vol. 23, no. 3, pp. 391–403 (2006).

[39] L. B. Soldano and E. C. M. Pennings, "Optical multi-mode interference devices based on self-imaging: principles and applications," *IEEE J. Lightwave Technol.*, vol. 13, no. 4, pp. 615–627 (1995).

10

Electromagnetic Response and Broadband Utilities of Planar Metamaterials

Hongqiang Li and Zeyong Wei
Physics Department, Tongji University
China

1. Introduction

Metamaterial (Smith, Pendry et al. 2004) interacts with electromagnetic waves in a resonant manner, affording us new giving rise to a new route for subwavelength photonic devices, such as compact antenna substrate (Li, Hang et al. 2005) and subwavelength resonant cavity (Zhou, Li et al. 2005; Li, Hao et al. 2006). Planar-type configuration,thanks to layer-by-layer fabrication technique, has been recognized as the most efficient and easiest way for the realization of metamaterials from microwave, terahertz, and optical regimes. And it is interesting to note that, in contrast to the common notion that the operational bandwidth of metamaterial is usually very narrow due to the local resonance nature, our recent studies show that planar metamaterial can be broadband in functionality (Wei, Cao et al. 2010; Wei, Cao et al. 2011).

It is worth noting that the studies on plasmonics and metamaterials are heuristic and beneficial to each other. Surface plasmon polaritons (SPPs) modulate light waves at the metal-dielectric interface with wavelength much smaller than that in free space (Raether 1988),which enables the control of light in a subwavelength scale for nanophotonic devices (Barnes, Dereux et al. 2003). SPPs with large coherent length are useful in many areas, including optical processing, quantum information (Kamli, Moiseev et al. 2008) and novel light-matter interactions (Vasa, Pomraenke et al. 2008). The enhancement of local fields by SPPs is particularly crucial to absorption enhancement (Andrew, Kitson et al. 1997), nonlinear optical amplification (Coutaz, Neviere et al. 1985; Tsang 1996) and weak signal probing (Kneipp, Wang et al. 1997; Nie & Emory 1997). Although SPP only exists in the visible and near-infrared regimes where free conduction-band electrons on a metal surface are driven by external fields, its analogue can be found in other frequencies where surface charge-density wave does not exists. With induced surface current oscillations on an array of metallic building blocks (Pendry, Holden et al. 1996; Pendry, Holden et al. 1999; Sievenpiper, Zhang et al. 1999; Yen, Padilla et al. 2004; Hibbins, Evans et al. 2005; Liu, Genov et al. 2006; Lockyear, Hibbins et al. 2009), a metamaterial surface can manipulate electromagnetic waves in a similar way as SPPs. Such spoof SPPs or surface resonance states on a meta-surface can be tuned by geometric parameters.

We will summarize our recent studies on planar metamaterials covering the modal expansion theory (Sheng, Stepleman et al. 1982; Lalanne, Hugonin et al. 2000; Wei, Fu et al. 2010; Wei, Li et al. 2010; Wei, Cao et al. 2010; Wei, Cao et al. 2011), the broadband enhanced transmission (Wei, Cao et al. 2011), negative refraction and subwavelength imaging (Wei, Cao et al. 2010), and the coherent control of spontaneous emission radiations in a wide frequency range (Wei, Li et al. 2010).

2. Broadband response of planar metamaterial

We introduce briefly the modal expansion method developed for multi-layered planar metamaterials in 2.1, and discuss the broadband enhanced transmission through holey metallic multi-layers (Wei, Cao et al. 2011), broadband negative refraction and subwavelength imaging in fishnet stacked metamaterial (Wei, Cao et al. 2010)in 2.2. In Section 2.3, we examine the properties of surface resonance states at a dielectric-metamaterial interface that exhibit magnetic response to the incident waves and strong local field enhancement (Wei, Li et al. 2010). We will show that a thin metamaterial slab, with a thickness much smaller than the operational wavelength, supports delocalized magnetic surface resonance states with a long coherent length in a wide range of frequencies. Operating in a broad frequency range, these spatially coherent SPPs are surface resonance states with quasi-TEM modes guided in the dielectric layer that are weakly coupled to free space, and the coupling strength can be controlled by tuning structural parameters while the frequency can be controlled by varying structural and material parameters. The high fidelity of these surface resonance states results in directional absorptivity or emissivity, which is angle-dependent with respect to frequency. These surface resonance states can give highly directional absorptivity and emissivity, and may thus help to realize interesting effects such as spatially coherent thermal emission, low-threshold plasmon lasing and sensitive photo-electric detection (Cao, Wei et al. 2011).

2.1 A method of modal expansion for planar metamaterials

The MEM is advantageous for analysis of electromagnetic transportation in planar metamaterial with layered geometry. The essence of MEM is to expand local EM fields in each layer as a series of in-plane envelope functions of eigenmodes. Let's demonstrate the formalism of MEM by solving the transmission spectra through one-dimensional metallic lamellar gratings with a thickness of h, as shown in Fig. 1. For simplicity, metals are treated as perfectly electric conductors (PEC) and EM waves only exist in apertures with the metallic layer.

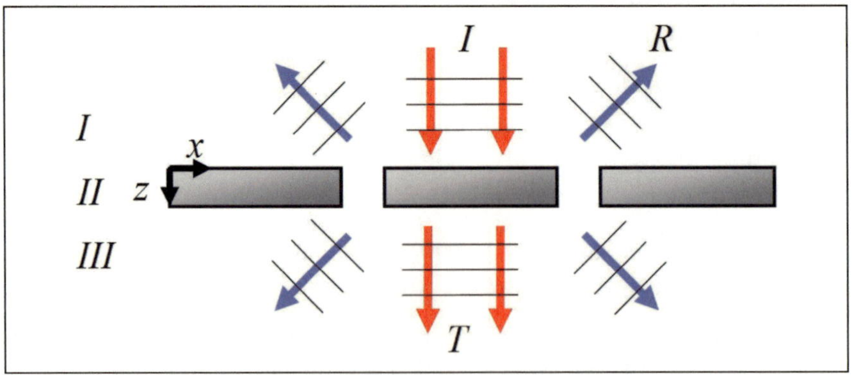

Fig. 1. Schematic of a one-dimensional metallic lamella grating.

For a transverse magnetic (TM) polarized incidence from the free semi-space at $z<0$ (region I), the total magnetic fields in region I and region II ($z>h$), which are along x direction, can be expressed in terms of incidence, reflection and transmission coefficients D0, Rm and Tm, as

$$H_1 = \left[D_0 \exp(jk_{z,0}z) + \sum_{m=-\infty}^{+\infty} R_m \exp(-jk_{z,m}z) \right] \exp(ik_0 \sin\theta x)$$

$$H_3 = \sum_{m=-\infty}^{+\infty} T_m \exp(jk_{z,m}z) \exp(ik_0 \sin\theta x) \quad (1)$$

Where $k_{z,j}^i$ denotes m^{th} order of Bloch wave vector in region i (i=1,2,3), θ is the incident angle. In region II of metallic lamellar gratings, the magnetic field can be written as the expansion coefficients a_l and b_l of forward and backward guided aperture modes, as

$$H_2 = \sum_{l=0}^{\infty} g_l(x) \left[a_l \exp(-jk_{z,l}z) + b_l \exp(jk_{z,l}z) \right] \quad (2)$$

where $g_l(x)$ is in-plane evelope function of waveguide mode of metallic layer. Under the as sumption of PEC for metalis, it can be expressed analytically as

$$g_l(x) = \begin{cases} \cos\left(\dfrac{\pi l}{w_x} x\right), & when \quad -\dfrac{w_x}{2} < x < \dfrac{w_x}{2} \\ 0, & else \end{cases} \quad (3)$$

By applying boundary continuum conditions at $z=0$ and $z=-h$ of the metal-air interfaces, we can obtain

$$\begin{cases} \underline{\Omega}(D_0 + \vec{R}) = \vec{a} + \vec{b} \\ D_0 - \vec{R} = \underline{\chi}(\vec{a} - \vec{b}) \\ u\vec{a} + u^{-1}\vec{b} = \underline{\Omega}(\vec{T} + \vec{R}) \\ \underline{\chi}(u\vec{a} - u^{-1}\vec{b}) = \vec{T} - \vec{R} \end{cases} \quad (4)$$

where $\Omega_{m,l}$, $\chi_{l,m}$ are the overlap integral of the projection between waveguide modes of metallic layers and Bloch modesin dlectric layers or free space, and can be expressed as

$$\begin{cases} \Omega_{m,l} = \dfrac{\int_{-p/2}^{p/2} e^{iG_m x} g_l^*(x) dx}{\int_{-p/2}^{p/2} g_l(x) g_l^*(x) dx} \\ \chi_{l,m} = \dfrac{\int_{-p/2}^{p/2} g_l(x) e^{-iG_m x} dx}{\int_{-p/2}^{p/2} e^{-iG_m x} e^{iG_m x} dx} \end{cases} \quad (5)$$

The coefficients a_l and b_l of forward and backwardwaveguide modecan be derived by Solving EQ.4, as

$$\begin{cases} \vec{a} = \dfrac{M_b^{-1}}{M_b^{-1} \bullet M_a - M_d^{-1} \bullet M_c} 2\underline{\Omega}\vec{D} \\ \vec{b} = \dfrac{M_a^{-1}}{M_a^{-1} \bullet M_b - M_c^{-1} \bullet M_d} 2\underline{\Omega}\vec{D} \end{cases} \quad (6)$$

where

$$\begin{cases} M_a = I + \underline{\Omega}\underline{\chi} \\ M_b = I - \underline{\Omega}\underline{\chi} \\ M_c = [I - \underline{\Omega}\underline{\chi}]u \\ M_d = [I + \underline{\Omega}\underline{\chi}]u \end{cases} \quad (7)$$

One advantage of MEM is that the problem can be solved without solving the inverse matrix of $\Omega_{m,l}$, $\chi_{l,m}$ so that the order of plane waves and that of waveguide modes are not necessarily be the same. At wavelength much larger than the array period, an $l \neq 0$ high order mode is evanescent and contribute little to transmission and reflection as the z component of wavevector is a large imaginary number.And the calculation is quickly convergent by adopting only a few waveguide modes.We can also see from Fig.2 that the calculated transmission spectra converge qucikly with only 7 plane waves considered as well.

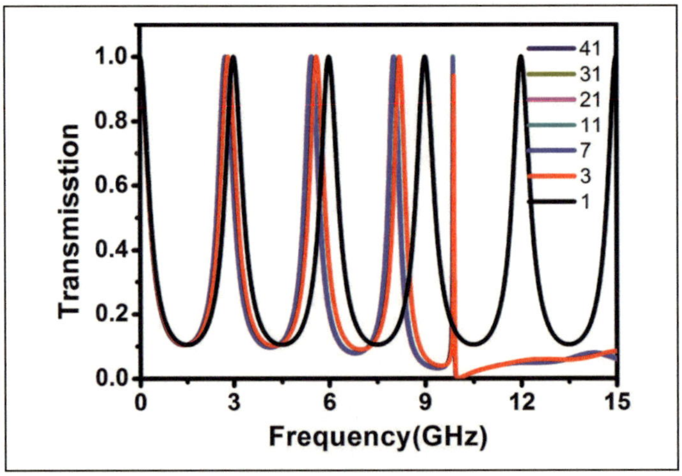

Fig. 2. Calculated 0th-order transmittance through the grating with different numbers of plane wave considered.

It is noticeable that the inter-layer coupling of EM waves can be analytically dealt with projection integral between the in-plane eigenmode functions of two adjacent layers. Under the treatment of MEM, a three-dimensional EM calculation will be simplified to a problem in two dimension. Thus the semi-analytical method is much faster than the conventional numerical simulations such as finite-difference-in-time-domain (FDTD) method, finite-element method etc. For metallic gratings, holey mesh, coaxial and split-ring structures, the method is quickly convergent by adopting only one or a few guided modes of metallic layers. The results shown in Fig.2 can be accomplished on an ordinary PC within a second, which is several orders faster than any other numerical methods. The method can, in principle, be generalized to a layered metallo-dielectric structure perforated with arbitrary shaped apertures as the in-plane functions of eigenmodes within each layer can always be solved by a standard algorithm for eigenvectors of a two-dimensional system.

2.2 Broadband transparency from stacked metallic multi-layers perforated with coaxial annular apertures

Extraordinary optical transmission (EOT) through metallic film perforated with subwavelength hole arrays has attracted considerable attentions since the pioneering study by T.W. Ebbesen and his coworkers (Ebbesen, Lezec et al. 1998; Ghaemi, Thio et al. 1998). Substantial efforts have been devoted to exploring the physical origin of EOT, both theoretically and experimentally, due to the appealing prospect in related applications (Martin-Moreno, Garcia-Vidal et al. 2001; de Abajo & Saenz 2005; Gay, Alloschery et al. 2006; Liu & Lalanne 2008; Xiao, Jinbo et al. 2010; Bahk, Park et al. 2011). Previous studies extensively investigated the EOT effects arising from the resonant tunneling of surface plasmon polaritons (SPPs) (Raether 1988; Barnes, Dereux et al. 2003) through the perforated metallic film. The frequency of such an EOT transmission peak is not only scaled to the period of hole arrays, but also very sensitive to the incident angle as the resonant tunneling occurs via the in-plane Bragg-scattering channels. Very recently, similar phenomena of the EOT through cascaded metallic multi-layers, which are perforated with one-dimensional gratings or two-dimensional hole arrays, have also been brought into attention (Miyamaru & Hangyo 2005; Ye & Zhang 2005; Chan, Marcet et al. 2006; Tang, Peng et al. 2007; Ortuno, Garcia-Meca et al. 2009; Marcet, Hang et al. 2010; Zhou, Huang et al. 2010). The resonant coupling among the SPP modes on different layers can be tuned by the spacing distance and lateral displacement of hole arrays at different layers, leading to tunable transmission peaks and zeros in spectra. It is worth noting that, when the slit size is very large or some kinds of specific apertures are adopted, the waveguide resonant modes of a slit or aperture can also give rise to the phenomena of EOT by allowing electromagnetic waves to propagate through the metallic slab. The cut-off wavelength of guided resonance modes (Baida, Van Labeke et al. 2004; Fan, Zhang et al. 2005; Fan, Zhang et al. 2005; van der Molen, Klein Koerkamp et al. 2005; Wen, Zhou et al. 2005; Wei, Fu et al. 2010) is primarily determined by the geometry of slits or apertures, and thus can be much longer than the array period. Under this circumstance, the EOT can also occur at a rather low frequency which is not scaled to the array period,and is robust against the structure disorder (Ruan & Qiu 2006). To the best of our knowledge, the EOT of metallic multi-layers arising from guided resonance modes has not yet been investigated before.

Here, we investigate the enhanced transmission of metallic multi-layers perforated with periodic arrays of coaxial annular apertures (CAAs). Modal expansion method (MEM) is developed to semi-analytically deal with the electromagnetic properties of the multilayered system. We show that the hybridization of guided resonance modes of CAAs in adjacent layers dramatically extends an enhanced transmission peak into a broad passband that is nearly reflectionless. The passband gets more and more broadened with sharper edges when the system contains more metallic layers. In contrast, these results can not be observed when the wave propagation is dictated by evanescent coupling of SPP modes (Miyamaru & Hangyo 2005; Ye & Zhang 2005; Chan, Marcet et al. 2006; Tang, Peng et al. 2007; Ortuno, Garcia-Meca et al. 2009; Marcet, Hang et al. 2010; Zhou, Huang et al. 2010). Measured transmission spectra are in good agreement with calculations for the model systems with different metallic layers. The broadening and varied fine structures of the EOT passband with the increase of metallic layers,can be understood intuitively by a physical picture of mode splitting of coupled atoms. The passband of the enhanced transmission for a system with only two or three metallic layers, covering a wide frequency range with sharp band-edges, can be estimated by calculated dispersion diagram under the assumption of infinite metallic layers.

A model system with n metallic layers perforated with square arrays of CAAs is of our interest. Figure 3 presents the front-view photo and schematic configuration of a sample with three thin metallic layers ($n=3$) and two sandwiched dielectric space layers. The aperture arrays deposited on different layers are aligned with no displacement in xy plane. The geometric parameters are the lattice constant $p=10\text{mm}$ of square arrays, the outer radius $R=4.8\text{mm}$ and inner radius $r=3.8\text{mm}$ of CAAs, and the thickness $t=0.035\text{mm}$ of metallic layer respectively. Each dielectric layer has a thickness of $h=1.575\text{mm}$ and a permittivity of $\varepsilon_r=2.65$.

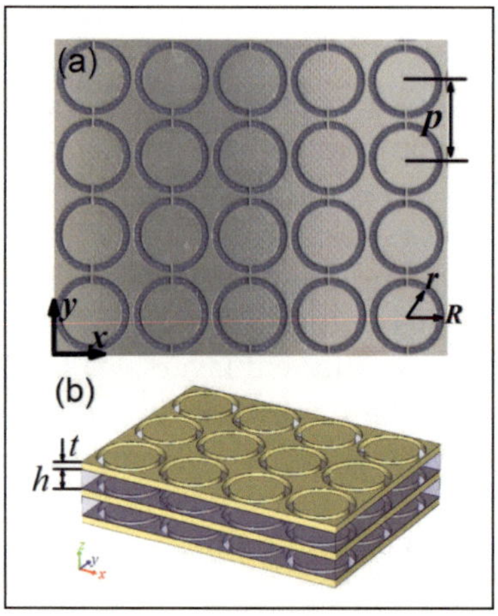

Fig. 3. (a) Top-view photo and (b) 3D schematic of our sample with three metallic layers (n = 3). The metallic layers are perforated with coaxial annular apertures (CAAs).

Under assumption of perfect electric conductor (PEC) for metals, the electromagnetic wave fields within a metallic layer only exist in apertures. In cylindrical coordinate system, the radial and angular field components E_ρ and E_ϕ inside an aperture of the metallic layer can be analytically expanded by the superposition of guided resonance modes of the aperture, as

$$E_\rho(\rho,\phi,z)=\sum_{l=1}^{\infty}(a_l e^{-i\beta_l z}+b_l e^{i\beta_l z})g_l(\rho,\phi)$$

$$E_\phi(\rho,\phi,z)=\sum_{l=1}^{\infty}(a_l e^{-i\beta_l z}+b_l e^{i\beta_l z})f_l(\rho,\phi)\,, \tag{8}$$

where a_l and b_l are the coefficients of forward and backward guided waves inside the CAAs,

$$g_l(\rho,\phi) = \frac{j\omega\mu l}{\rho}\left[N'_l(T_l r)J_l(T_l\rho) - J'_p(T_l r)N_l(T_l\rho)\right]\sin(l\phi)$$

and $$f_l(\rho,\phi) = j\omega\mu T\left[N'_l(T_l r)J'_l(T_l\rho) - J'_p(T_l r)N'_l(T_l\rho)\right]\cos(l\phi)$$

are the lth order modal functions of radial and angular components in aperture with $J_l(x)$ and $N_l(x)$ being the lth order Bessel and Neumann functions, T_l refers to the root of the equation $J'_l(TR)N'_l(Tr) - J'_l(Tr)N'_l(TR) = 0$. By adopting EQ. (8) as expressions of EM fields in metallic layers and plane-waves as those in dielectric layers, we perform MEM to resolve the electromagnetic problems in the multilayered system. The method is quickly convergent by considering only 2 or 3 lowest guided resonance modes of CAAs. A higher order resonance mode contributes little to the interlayer coupling as its wave vector β_l is a large imaginary number. Three guided modes ($l = 1,2,3$) in CAAs and 11×11 orders of plane-wave basis in dielectric layers are adopted in our calculations. The results are very accurate (solid lines in Fig.4) and in good agreement with the measurements (circular dots in Fig.4).

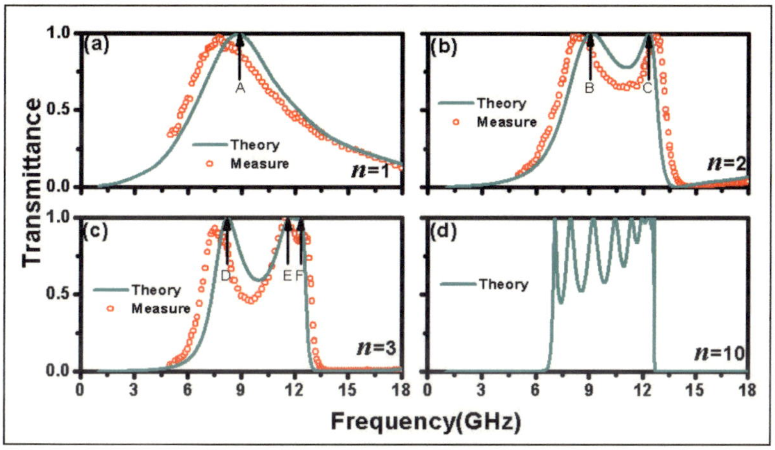

Fig. 4. Transmission spectra through the models with (a) n=1, (b) n=2, (c) n=3, (d) n=10 metallic layers. Solid lines for calculated results by Modal expansion method (MEM), circular dots for measured results in microwave regime.

We see from Fig. 4 (a) that there exists a transmission peak for the n=1 sample at $f_A = 8.7\text{GHz}$ due to the excitation of guided TE_{11} resonance mode in CAAs. We also see from Figs. 4 (b) and 4 (c) that there are two transmission peaks at $f_B = 9.1\text{GHz}$, and $f_C = 12.3\text{GHz}$ for the $n = 2$ sample, three peaks at $f_D = 8.2\text{GHz}$, $f_E = 11.64\text{GHz}$ and $f_F = 12.35\text{GHz}$ for the $n = 3$ sample. Figure 4 (d) presents the calculated transmission spectra of an n=10 model system. It means that, with the increase of metallic layers, more transmission peaks emerge, giving rise to a broad transparent band.

More calculations show that, for the $n = 2$ sample, at an on-resonance frequency $f_B = 9.1\text{GHz}$ or $f_C = 12.3\text{GHz}$ where transmissivity is nearly unity, the spatial distribution of electric fields [see Figs. 5 (a) and 5 (b)] are symmetric or anti-symmetric about the xy plane. And the transmitted waves possess a phase difference of 0 (in phase) or π (out phase) with respect to the incident waves. Therefore the peaks at f_B and f_C, derived from the peak at f_A of the $n=1$ model, come from the excitation and hybridization of the TE_{11} guided resonance modes in apertures at different metallic layers as a results of mode splitting of coupled apertures (or meta-atoms). Further more, the anti-symmetric mode at $f_C = 12.3$GHz of the $n=2$ model splits into two modes of the $n=3$ model: spatial field distribution of the one at $f_E = 11.64$GHz reveals that the incident and outgoing waves are out phase to each other [Fig. 5 (d)] and it is on the opposite for the other at $f_F = 12.35$GHz [Fig.5 (e)], while the resonant mode at the lowest frequency $f_D = 8.2$GHz retains a symmetric feature in field distribution [Fig. 5 (c)] and inherits the in-phase signature from the symmetric mode at f_B of the $n=2$ model.

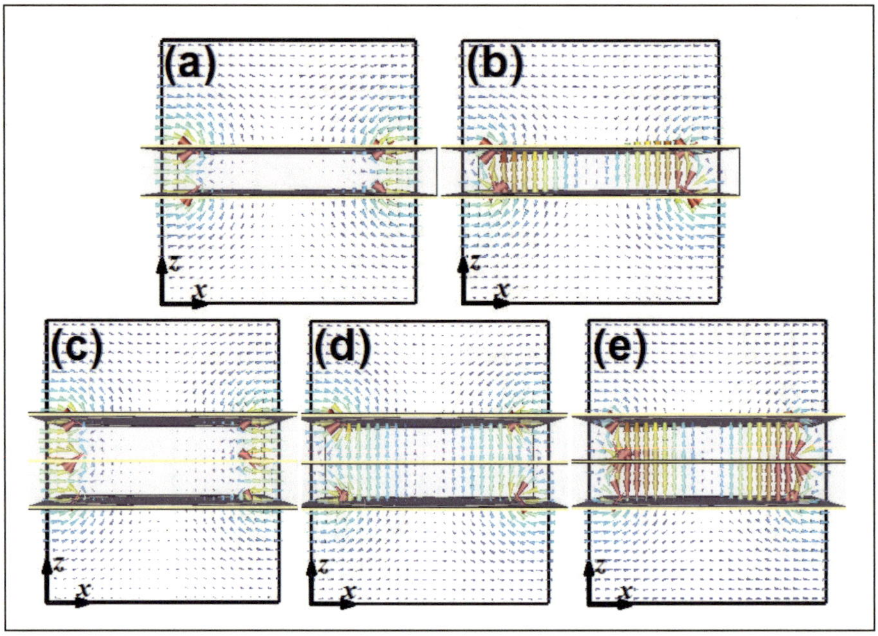

Fig. 5. Spatial distribution of electric fields in the xz plane at on-resonance frequencies of (a) f_B = 9.1GHz, (b) f_C = 12.3GHz for the n = 2 model , and (c) f_D = 8:2GHz, (d)f_E = 11.64GHz, (e) f_F = 12.35GHz for the n = 3 model.

Figure 6 (a) presents the dispersion relation of bulk material periodically constructed with layered CAAs. The band structure is calculated with MEM algorithm assuming periodic boundary conditions along the z axis. The process of mode splitting from $n=1$ to $n=3$, as shown in Fig. 6 (b), depicts the evolution of the enhanced transmission feature from a single transmission peak to a broad passband. It is interesting that the passband between $f_b = 6.77$GHz and $f_t = 12.7$GHz shown in Fig. 6 (b), predicting the passband of the n=10 model quite well, is also a good measure of the bandwidth of the n=3 sample. The total bandwidth is about 60% of the central frequency. In contrast, the EOT observed in multilayered systems of previous studies demonstrates a peak lineshape in spectra as it arises from the resonant tunneling of SPP modes among metallic films instead of guided resonance modes. And the broad passband we observed is not sensitive to the incident angle (not shown), while it is on the contrary when the SPP modes dominate.

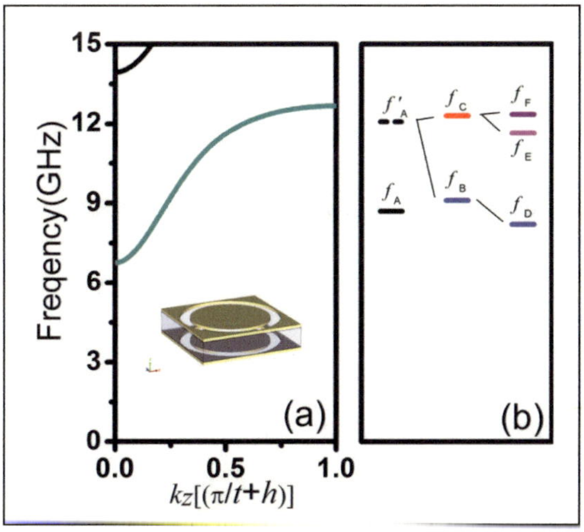

Fig. 6. Transmission spectra through the models with (a) n=1, (b) n=2, (c) n=3, (d) n=10 metallic layers. Solid lines for calculated results by Modal expansion method (MEM), circular dots for measured results in microwave regime.

This work reports for the first time that enhanced transmission peak can be broadened through stacked metallic multi-layers perforated with CAAs. Taking advantage of the excitation of guided resonance modes of CAAs and interlayer coupling, the enhanced transmission of such a system with only three metallic layers can span a wide frequency range covering about 60% of the central frequency. The broadband utility shall have enormous potential applications in optoelectronics, telecommunication and image processing.

2.3 Broadband negative refraction from stacked fishnet metamaterial

Since J.B. Pendry proposed perfect lens (Pendry 2000) using left-handed materials (Veselago 1968), sustained attentions have been drawn to the negative-index metamaterial (NIM) with simultaneously negative permittivity and permeability. The NIM, comprising of subwavelength metallic resonant units, has been designed and realized in both the microwave (Shelby, Smith et al. 2001) and optical regime (Liu, Guo et al. 2008). Negative refraction and subwavelength imaging with NIMs have great application potentials in photonic devices (Grbic & Eleftheriades 2004; Belov, Hao et al. 2006; Wiltshire, Pendry et al. 2006; Freire, Marques et al. 2008; Silveirinha, Fernandes et al. 2008; Silveirinha, Medeiros et al. 2010). Among various types of NIMs, one most promising candidate is the so-called fishnet NIM which comprises of alternating metal/dielectric layers perforated with two-dimensional array of holes (Beruete, Campillo et al. 2007; Beruete, Sorolla et al. 2007; Beruete, Navarro-Cia et al. 2008; Navarro-Cia, Beruete et al. 2008; Navarro-Cia, Beruete et al. 2009). The simple structure also provides a feasible solution for optical NIM (Dolling, Enkrich et al. 2006; Zhang, Fan et al. 2006; Dolling, Wegener et al. 2007; Valentine, Zhang et al. 2008; Ku & Brueck 2009; Ku, Zhang et al. 2009).

In most of the previous studies on fishnet NIMs (Dolling, Enkrich et al. 2006; Zhang, Fan et al. 2006; Beruete, Campillo et al. 2007; Beruete, Sorolla et al. 2007; Dolling, Wegener et al.

2007; Beruete, Navarro-Cia et al. 2008; Navarro-Cia, Beruete et al. 2008; Valentine, Zhang et al. 2008; Ku & Brueck 2009; Ku, Zhang et al. 2009; Navarro-Cia, Beruete et al. 2009), the light waves are incident on the top interface of the metal/dielectric multi-layers. Thus the light waves can not penetrate into the structure below the cut-off frequency of air holes, and the negative index was retrieved only within in a narrow frequency range above the cut-off. In this section, a different incidence configuration is employed by impinging the light waves on the sidewall interface of fishnet NIM that is perpendicular to the metal/dielectric multi-layers. As the uniformly spaced holey metallic layers of fishnet NIM constitute a multiple of slab waveguide channels filled with dielectric spacer layers, the incidence configuration of this kind enables us to fully exploit the optical properties of the fishnet NIM in the long wavelength limit. We show that the evanescent coupling between the slab waveguides gives rise to all-angle negative refraction and sub-wavelength imaging in a wide frequency range starting from zero.

Figure 7 schematically illustrates the structure of our stacked fishnet metamaterial and the incidence configuration. The metal/dielectric layers are lying in $\hat{x}\hat{y}$ plane. The square arrays of air holes perforated on metallic layers are aligned along z axis without lateral displacement in $\hat{x}\hat{y}$ plane. The period of the hole array, the thickness of metallic layer and dielectric layer are *p=6.0mm*, *t=0.035mm* and *h=1.575mm* respectively. The line width of metallic strips along x direction *w=0.2mm* is the same as that along y direction, and the size of square holes is *a=p-g=5.8mm*. The dielectric constant of the dielectric layer is ε_r =2.55. The EM incidence waves are propagating in the $\hat{x}\hat{z}$ plane with an incident angle of θ .

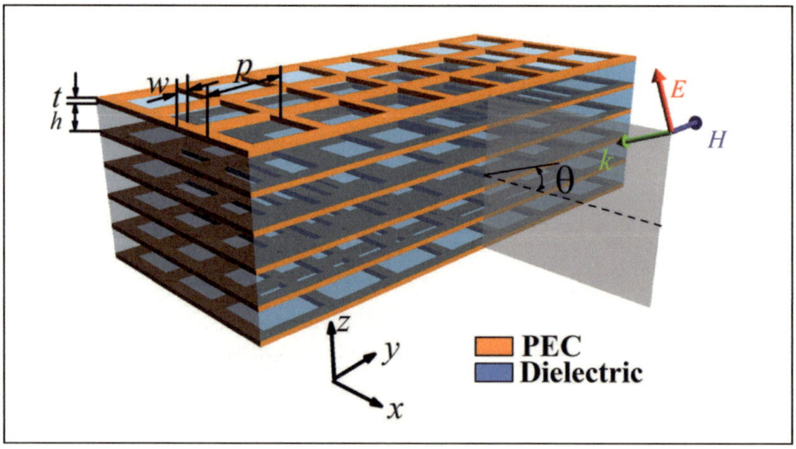

Fig. 7. The schematic of stacked fishnet metamaterial. The red and blue arrows refer to the directions of electric field $\vec{E}$ and magnetic field $\vec{H}$. The plane in gray color denotes the incident plane.

Figure 8 presents the calculated dispersion diagram along the Γ (0,0,0)→X (0.5,0,0), Γ (0,0,0)→M (0.5,0,0.5) and M (0.5,0,0.5)→X (0.5,0,0) directions. The blue dashed line refers to the light line in dielectric. We notice that, the lowest branch along the ΓX direction ($k_z = 0$) precisely reproduces the light line in dielectric. This is understandable with the help of modal expansion method. Detailed calculations show that these $k_z = 0$ states only contain the 0th order Bloch component which is the transverse electromagnetic (TEM) mode that is always orthogonal to the local modes of air hole. In this situation, the free photons in the dielectric are the only choice for the $k_z = 0$ states as no evanescent couplings happen via the breathing air holes. However the lowest branch ($k_z \neq 0$) along the ΓM and XM directions (red solid line in Fig.8) evidently deviates from the light line in dielectric. The $k_z \neq 0$ states on this branch originate from the evanescent coupling between the adjacent slab waveguides via the TE_{10} mode of holes (noting that the overlap integral between a high order of guided Bloch mode and local mode of air hole is not zero). Figure 9 (a) presents the charts of equi-frequency surface (EFS) analysis for the $k_y = 0$ states to further reveal the characteristics of this band. All curves in Fig. 9 (a) are in a hyperbolic-like lineshape, which indicates that all-angle negative refraction occurs in the $\hat{x}\hat{z}$ plane at low frequency regime starting from zero. At lower frequency the curves in Fig. 9 (a) become much more flat, which means that the waves are also strongly collimated inside the structure along the direction parallel to the metal/dielectric layers. A numerical proof of negative refraction is shown in Fig. 9 (b). In our FDTD simulations, a monochromatic one-way Gaussian beam in the $\hat{x}\hat{z}$ plane with a frequency at 11GHz is incident from upside at an incident angle of 30°. The fishnet model is stacked with 500 metal/dielectric layers along z direction. Given the periodicity of hole arrays and the incidence configuration, 60 periods along x direction and one period along y direction are adopted for the metal/dielectric layers in space domain. The negative refraction is clearly shown in Fig. 9 (b) with the magnetic field distribution in xz plane. The black arrows denote the directions of energy flow in the free space and fishnet structure. A refraction angle of -16.2°, retrieved from the refracted direction of the energy flow or the negative Goos-Hanchen shift alternatively, is in good agreement with the estimate in EFS analysis. We also see from Fig. 9 (b) that almost no reflection occurs as the incidence can easily propagates inside the structure by coupling with the guided Bloch modes in the slab waveguide channels.

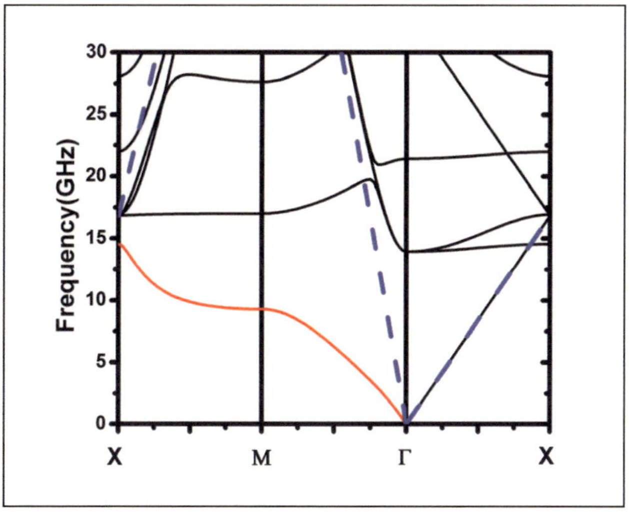

Fig. 8. The dispersion diagram of the stacked fishnet metamaterial.

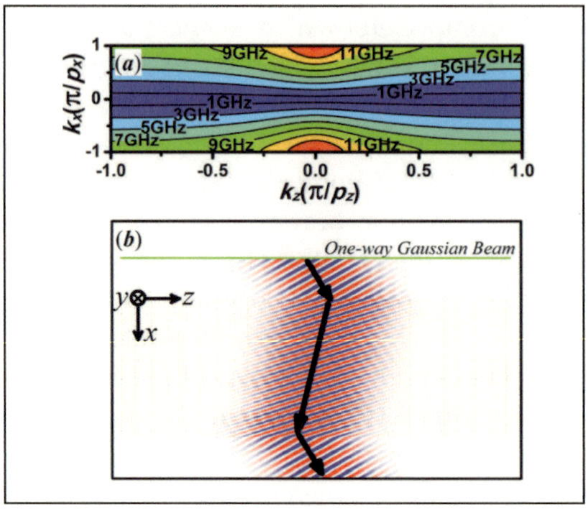

Fig. 9. (a) The charts of EFS analysis for ky=0 states respect to k_z and k_x. (b) The FDTD simulations on the magnetic field distribution in the problem domain. The one-way Gaussian beam is about 70mm away from the top interface of our fishnet model.

The optical properties of such a system can be described with the coupled wave equation (Haus & Molter-Orr 1983; Eisenberg, Silberberg et al. 2000; Pertsch, Zentgraf et al. 2002) by considering the coupling between the nth waveguide channel and its nearest neighbors, the $(n-1)^{th}$, $(n+1)^{th}$ waveguide channels, as:

$$i\frac{da_n(x)}{dx}+\beta a_n(x)+C[a_{n+1}(x)+a_{n-1}(x)]=0 \tag{9}$$

Where $a_n(x)$ denotes the wave fields in the nth slab waveguide, C is the coupling coefficient, and β is the propagation constant of free photons in dielectric. Under the periodic boundary condition along z direction, the dispersion of the system takes the form as

$$k_x=\beta+2C\cos(k_z p) \tag{10}$$

where k_x and k_z are the vector components along the x and z directions. At $k_z=0$, the coupling coefficient $C=0$ is zero (as aforementioned no evanescent coupling occurs) and we have $k_x=\beta$ which is rightly the light line in the dielectric. While C is always negative when $k_z\neq 0$ in the limit of long wavelength [which can be deduced from the charts in Fig. 8 (a)], giving rise to all-angle negative refraction. We note that the silver/dielectric multi-layered structure also supports all-angle negative refraction in a certain optical frequency regime under the same incidence configuration of our study (Fan, Wang et al. 2006). The long range SPPs play an important role for the negative refraction. We also note that, at long wavelength limit, a holey metallic surface can be homogenized into a single-negative medium with electric response in the form of Drude model (Pendry, Martin-Moreno et al. 2004). The plasmon frequency is rightly the cut-off frequency of air holes. Thus it is reasonable for us to consider the stacked fishnet metamaterial as an artificial plasmonic waveguide array. The negative coefficient C implies that for the eigenstates on the lowest branch, the spatial field distributions are anti-symmetric with respect to the plane of air holes. If our findings are applicable in optical regime, the most field energy shall propagate outside the lossy metal film with anti-symmetric field distribution, and low loss is expected. The picture may be helpful as well to explain the low loss measured in a recent experiment about fishnet optical NIMs (Zhang, Fan et al. 2006; Valentine, Zhang et al. 2008).

One important application of all-angle negative refraction is flat lens. The imaging performance of our stacked fishnet metamaterial is examined by the brute-force FDTD numerical simulations. As shown in Fig.10, a monochromatic point source with a frequency at 11GHz is located 15mm away from the surface at the left side of the fishnet structure. The snap shot shown in Fig. 10 (a) clearly indicates a high-quality image achieved at another side of the structure about 15mm away from the interface. The image resolution can be checked by the normalized magnetic field profile at image plane. As illustrated in Fig. 10 (b), along z axis, the full width at half maximum (FWHM) of the field profile is 10mm about one-third of the wavelength. The FWHM at a lower frequency still remains at about 10mm, leading to a better resolution in subwavelength scale along the z direction. But a longer structure is required due to strong collimation effect at lower frequency.

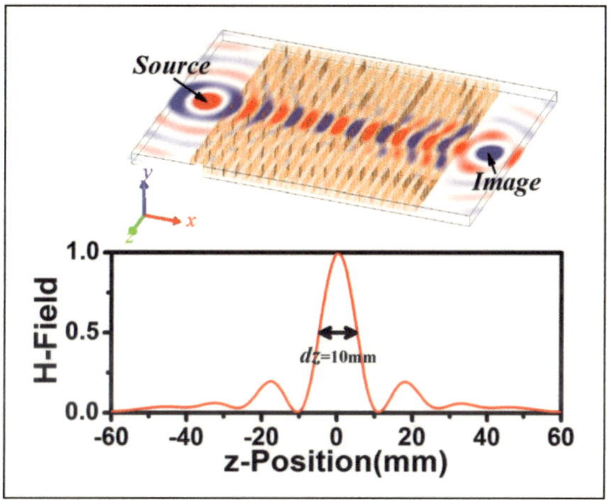

Fig. 10. (a) The snap shot of magnetic field distribution in the incident plane. The model, stacked with 80 metal/dielectric multi-layers along z axis, has 20 periods along x direction and one period along y direction. (b) Normalized magnetic field profile at the source plane (black solid line) and the image plane (red solid line) as a function of the z aixs

In conclusion, the fishnet metamaterials can operate as plasmonic waveguide arrays. The broadband negative refraction, subwavelength imaging in the long wavelength limit have great potentials for photonic devices in microwave, THz and even in the optical regimes.

2.4 Spatially coherent surface resonant states derived from magnetic meta-surface

Surface plasmon polaritons (SPPs) can modulate light waves at the metal-dielectric interface with wavelength much smaller than that in free space (Raether 1988),which enables the control of light in a subwavelength scale for nanophotonic devices (Barnes, Dereux et al. 2003). SPPs with large coherent length are useful in many areas, including optical processing, quantum information (Kamli, Moiseev et al. 2008) and novel light-matter interactions (Vasa, Pomraenke et al. 2008). The enhancement of local fields by SPPs is particularly important as it opens a new route to absorption enhancement (Andrew, Kitson et al. 1997), nonlinear optical amplification (Coutaz, Neviere et al. 1985; Tsang 1996) as well as weak signal probing (Kneipp, Wang et al. 1997; Nie & Emory 1997). As the properties of SPP are pretty much determined by the natural (plasmon) resonance frequency, there is not much room for us to adjust the SPP response for practical applications. With induced surface current oscillations on an array of metallic building blocks (Pendry, Holden et al. 1996; Pendry, Holden et al. 1999; Sievenpiper, Zhang et al. 1999; Yen, Padilla et al. 2004; Hibbins, Evans et al. 2005; Liu, Genov et al. 2006; Lockyear, Hibbins et al. 2009), metamaterial surfaces can manipulate electromagnetic waves in a similar way as SPPs. Such SPPs or surface resonance states on structured metallic surfaces are tunable by geometric parameters.

Here, we examine the properties of surface resonance states at a dielectric-metamaterial interface that exhibit magnetic response to the incident waves and strong local field enhancement. We will see that these surface resonance states can give highly directional absorptivity and emissivity, and may thus help to realize interesting effects such as spatially coherent thermal emission. As the structure is very simple, it can be fabricated down to the IR and optical regime (Grigorenko, Geim et al. 2005; Shalaev 2007; Boltasseva & Shalaev 2008).

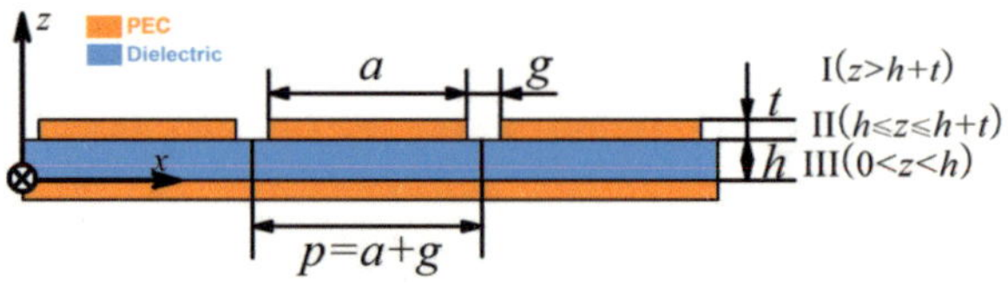

Fig. 11. Schematic picture of the magnetic metamaterial slab

We will show that a thin metamaterial slab, with a thickness much smaller than the operational wavelength, supports delocalized magnetic surface resonance states with a long coherent length in a wide range of frequencies. Operating in a broad frequency range, these spatially coherent SPPs are surface resonance states with quasi-TEM modes guided in the dielectric layer that are weakly coupled to free space, and the coupling strength can be controlled by tuning structural parameters while the frequency can be controlled by varying structural and material parameters. The high fidelity of these surface resonance states results in directional absorptivity or emissivity, which is angle-dependent with respect to frequency. Finite-difference-in-time-domain (FDTD) simulations verify that the highly directional emissivity from the slab persists in the presence of structural disorder in the grating layer.

Such metal-dielectric-metal (MDM) structures were recognized as artificial magnetic surfaces with high impedance by the end of last century [12]. The incident waves induce surface current solenoids on the unit cells of the ultra-thin high-impedance surface, giving rise to magnetic susceptibilities. The magnetic response can be described with an effective permeability in Lorentz type (Sievenpiper, Zhang et al. 1999; Zhou, Wen et al. 2003). After the concept of metamaterial being proposed (Engheta & Ziolkowski 2006), P. Alastair and his co-workers numerically and experimentally proved that the ultra-thin MDM structures can resonantly absorb or transmit radiations at low frequency limit (Hibbins, Sambles et al. 2004). They addressed that the central frequencies of absorption peaks are independent from the incident angle with an interpretation of Farby-Perrot resonant mode (EQ. 1 in Ref. 22). The same group further explored the angle-independent absorption, as the main scenario of the incremental work, by measuring the flat bands of surface wave dispersion in the visible (Hibbins, Murray et al. 2006) as well as the microwave region (Brown, Hibbins et al. 2008). In contrast, we find that the structures with proper design also supports very narrow absorption peaks which are sensitiveto the incident angle and obviously do not satisfy to the Fabry-Perot resonance condition suggested in the previous studies.

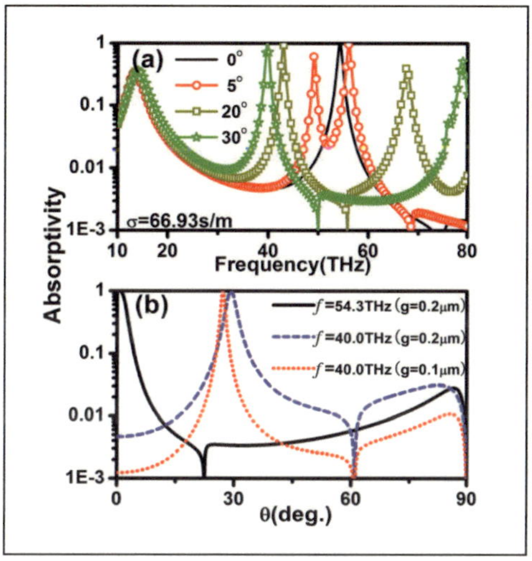

Fig. 12. Absorption spectra under TM-polarized incidence (a) as a function of frequency at incident angles of $\theta = 0^o, 5^o, 20^o, 30^o$ (g=0.2μm) and (b) as a function of incident angle at 54.3THz (solid line, g=0.2μm), 40THz (dashed line, g=0.2μm) and 40THz (doted line, g=0.1μm)

It is worth noting that an angle-independent peak is quite different from an angle-dependent one in physics origin. The former, investigated in Refs. 22-24, comes from the localized surface resonance states, while the latter, found by us, comes from the collective surface resonance states. Mode analysis presents An an intuitive picture for the formation of these collective surface states. When high order quasi-TEM modes are dominant components of the guided waves inside the dielectric layer they will assign phase correlation to the outgoing waves emitted from the air slits of grating, thus are very crucial to the formation of collective response. Weak enough both the leakage from dielectric layer to air slits and the material absorption, the spatial coherence of surface resonant states will survive. As the interaction between the structure and the incident waves will excite quasi-TEM modes inside the dielectric layer, the magnetic induction must be parallel to the MDM surfaces if it exists. Thus a surface resonant state on a MDM structure is usually magnetic in nature. Our findings about spatially coherent surface resonance states are original compared to the common knowledge, and have great potentials in coherent control of SPPs as well as thermal emission radiations.

Our model system is schematically illustrated in Fig. 11. Lying on the $\hat{x}\hat{y}$ plane, the slab comprises an upper layer of a metallic lamellar grating with thickness *t*, a dielectric spacer layer as a slab waveguide with thickness *h* and a metallic ground plane. The metallic strips are separated by a small air gap *g*, giving rise to a period of *p=a+g* for the lamellar grating. The geometric parameters of our model are $t = 0.2\mu m$, $h = 0.8\mu m$, $a = 3.8\mu m$, $g = 0.2\mu m$ and $p = a + g = 4.0\mu m$. Each metallic strip together with the ground plane beneath it constitutes a planar resonant cavity as the building block that gives magnetic responses at cavity resonances (Sievenpiper, Zhang et al. 1999; Lockyear, Hibbins et al. 2009). As the metallic grating is along the $\hat{x}$ direction, the guided waves in the dielectric layer (at $0 < z < h$ in region III) shall always couple to the incident waves with a non-zero component $E_x \neq 0$ of electric field.

As a first step, we consider a transverse magnetic (TM) polarized incident plane wave in the free semi-space (at $z > h + t$ in region I). The electric field $\vec{E}$ lies in $\hat{x}\hat{z}$ plane, the magnetic field $\vec{H}$ is along y axis and the in-plane wave vector is $\vec{k}_{0_\parallel} = k_x \hat{e}_x$ ($k_y = 0$). The total magnetic fields in region I and in region III can be written in terms of the reflection coefficients r_m and the guided Bloch wave coefficients t_m , as

$$\begin{aligned} H_y^I(\vec{r},z) &= \delta_{m,0} e^{-ik^I_{zm} z} e^{ik_x x} + \sum_m r_m e^{ik^I_{zm} z} e^{i(k_x + 2m\pi/p)x} \\ H_y^{III}(\vec{r},z) &= \sum_m [t_m e^{ik^{III}_{zm} z} + t_m e^{-ik^{III}_{zm}(z-2h)}] e^{i(k_x + 2m\pi/p)x}, \end{aligned} \tag{11}$$

where the term $\delta_{m,0} e^{-ik^I_{zm} z} e^{ik_x x}$ denotes the incident plane wave with $\delta_{m,0}$ being the Kronecker function and m being the Bloch order; $e^{i(k_x + 2m\pi/p)x}$ denotes wave component of the mth Bloch eigenmode in the semi-free space (region I) and the dielectric layer (region III) with respect to $\vec{k}_m = \hat{x}(k_x + 2m\pi/p)$. k_m is the in-plane wave vector and $\vec{G}_m = \hat{x} \cdot 2\pi m/p$ is the mth reciprocal lattice vector.

$$k^I_{z_m} = \sqrt{\varepsilon_0 \mu_0 \omega^2 - |\vec{k}_m|^2} \text{ and } k^{III}_{z_m} = \sqrt{\varepsilon_{III} \mu_0 \omega^2 - |\vec{k}_m|^2}$$

are the z components of wave vector for the mth order Bloch eigenmode in region I and region III respectively. ε_0 and ε_{III} are the permittivity of the vacuum and the dielectric , μ_0 is the vacuum permeability. In general, we also derived the method for a plane wave incidence with any specific wavevector and any specific polarization.

We shall mainly consider infrared frequencies, at which the metals can be well approximated as perfectly electric conductors (PEC). The EM fields at $h \leq z \leq h + t$ in region II are squeezed inside the air gaps, in which the magnetic fields can be expressed in terms of the expansion coefficients a_l and b_l of forward and backward guided waves, as:

$$H_y^{II}(\vec{r},z) = \sum_l [a_l e^{-iq_l(z-h-t)} + b_l e^{iq_l(z-h)}] g_l(x), \tag{12}$$

where $g_l(x) = \cos[l\pi/g(x + g/2)]$, $(l = 0,1,...,n,...)$ is the in-plane distribution of guided mode α_l running over all air gaps. $q_l = \sqrt{\varepsilon_0 \mu_0 \omega^2 - (l\pi/g)^2}$ is the z component of wave vector for the lth guided mode α_l .

We can obtain the coefficients $t_m(f,\vec{k}_{0_{//}})$ and $r_m(f,\vec{k}_{0_{//}})$ of the mth guided and reflected waves by applying the boundary continuity conditions for the tangential components of electromagnetic wave fields (over the slits) at the interfaces $z=h$ and $z=h+t$. Given that surface resonance modes are intrinsic response, we can also assign zero to the incident plane wave and apply the boundary continuity conditions for the tangential components of wave fields to derive the eigen-value equations. A surface resonance state can be determined by searching a zero value / minimum of eigen-equation determinant in the reciprocal space provided that it is non-radiative/radiative with infinite/finite life time below/above light line in free space.

We derived the absorption spectra of the slab

$$A(\vec{k}_0,\omega)=1-\sum_m \mathrm{Re}(\frac{k^I_{z_m}}{k^I_{z_0}})\,|\,r_m(\vec{k}_0,\omega)\,|^2$$

which includes the contributions from all Bloch orders of reflected waves. As a consequence, $A(\vec{k}_0,\omega)$ gives information about the surface resonance states as well as the emissivity properties as governed by Kirchhoff's law (Greffet & Nieto-Vesperinas 1998). We shall assume that the dielectric spacer layer is slightly dissipative by assigning a complex permittivity $\varepsilon_{III}=\varepsilon_r\varepsilon_0+i\sigma/\omega$ with $\varepsilon_r=2.2$ and $\sigma=66.93S/m$ $[\,\mathrm{Im}(\varepsilon_{III})\approx 10^{-2}\varepsilon_r\varepsilon_0\,]$ in the calculated frequency regime. In Fig. 12 (a), we present the absorption spectra at various incident angles. The spectra exhibit a low and broad peak at *13.2*THz which is almost independent of the incident angle, while the other absorption peaks at higher frequencies are narrow and sensitive to the incident angle with a maximum absorption approaching 100%. The slab thus acts as an all-angle absorber at *13.2*THz, but exhibits sharp angle-selective absorption peaks at higher frequencies. Shown as solid and dashed lines in Fig. 12 (b), the sharp angular dependence of absorption coefficients (note that the vertical axis is in log-scale) at 40.0*THz* and 54.3THz implicitly implies the existence of spatially coherent surface resonance states. The angle-dependent absorption peaks become lower and disappear gradually with the increase of the material loss. This presents a way to realize nearly perfect absorption with weakly absorptive materials by coherent surface resonance states. The coherent length of a surface resonance state can be estimated by the ratio of the wavelength λ and the full width at half maximum (FWHM) $\Delta\theta$ of the absorption peak (Greffet, Carminati et al. 2002). For example, for the Γ_4 state at 54.3THz and $\vec{k}_{0_{||}}=0$, the angular FWHM of the corresponding absorption peak $\Delta\theta=4.6^o$ (from $\theta=-2.3^o$ to $\theta=2.3^o$) gives rise to a coherent length $\lambda/\Delta\theta=68.5\mu m\approx 12.4\lambda$. The coherent length is about 220λ for the surface resonance state at 50.22THz and $\vec{k}_{0_{||}}=0.01\pi/p$ with $\Delta\theta=0.26^o$ (not shown in figure). The angular FWHM is reduced if the gap size is smaller, as shown with the dashed and dotted lines in Fig.12 (b) for $g=0.2\mu m$ and $g=0.1\mu m$ at 40THz, which means that the coherent length of the surface resonant modes can be controlled by the gap-period ratio g/p. It is worth noting that, although k_y=0 is assumed for the calculated results shown in Fig. 12, the angle-dependent absorption peaks are readily obtained for any specific incident angle.

To quantitatively characterize the formation of these spatially coherent surface resonance states, we employ the eigenmode expansion method to calculate the surface resonance dispersion (in the limit of no material loss) as shown in Fig. 13 (b).The B_1 surface resonance states lie below the light line L_2 (magenta dashed line), and thus are non-radiative as evanescent modes. The surface resonances labeled as B_2 originate from the coupling of the fundamental magnetic resonance modes of the metal strip structure with the free space light line L_2. The surface resonances B_3 and B_4 are harmonic modes of the magnetic resonances that hybridizes with the guided mode inside the dielectric layer. The calculated reflection phase difference between the 0th order reflected and incident electric field, as shown in Fig. 13 (a) for normal incidence (red line), and 2° incidence (blue line), clearly shows that the resonances are magnetic in nature when the surface resonances intersect the zone center at Γ_2 (13.2THz) and Γ_4 (54.3THz) as the reflection phase is zero like what a magnetic conductor surface does to the incident waves. The state Γ_3, invisible in the reflection phase spectrum under normal incidence [red solid line in Fig. 13 (a)], is a dark state as its eigenmode is in mirror symmetry about the *yz* plane and can not couple with free space photons. While the other B_3 states can couple with external light under oblique incidence [see the blue line in Fig. 13 (a)]. For example, there exists in-phase reflection at frequency 50.22THz under an incident angle of 2°, corresponding a B_3 state at frequency 50.22THz and $k_{0_{\parallel}} \approx 0.02\pi / p$.

The angle-independent absorption peak at 13.2THz is due to the B_2 mode, which is only weakly dispersive near the zone center. The more dispersive B_3 and B_4 modes are accountable for the incident-angle sensitive absorption in the higher frequencies in Fig. 12 (a). The field patterns in Figs. 14 (a)-14 (c) present the spatial distribution of the real part of magnetic fields excited by the incident plane waves with incident angles 0°, 2° and 0° for the three surface resonance states on B_2, B_3 and B_4 respectively, and the corresponding vector diagrams of electric fields are shown in Figs 14 (d)-14 (f). We can see clearly that the electric fields reach maximum in strength at the slab upper surface, and exponentially decay along the surface normal into the free space. This is precisely a picture of SPP modes. The field patterns comes from the coincidence of the evanescent wave components in high Bloch orders at both sides of metallic grating.

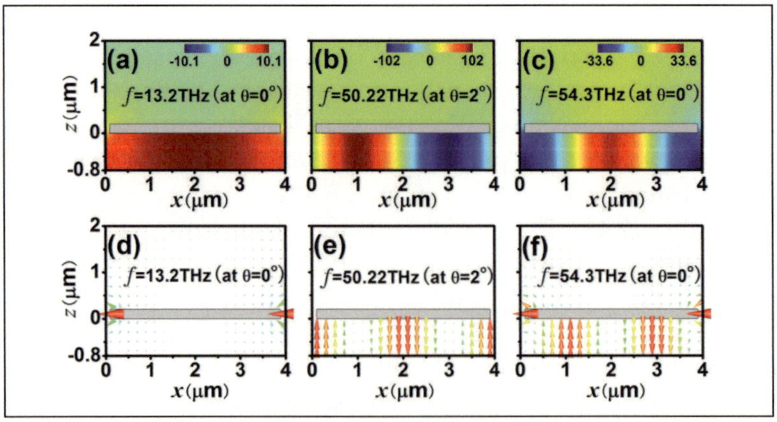

Fig. 14. Spatial distributions of magnetic fields and electric fields in the xz plane for Γ_2 state at f_{Γ_2} =13.2THz, $\theta = 0°$ [(a) and (c)], a state on B_3 at f=50.22THz, $\theta = 2°$ [(b) and (e)]and Γ_4 state at f_{Γ_4} =54.3THz, $\theta = 0°$ [(c) and (f)]

Although only TEM guided modes are allowed to be excited in the thin MDM slab within the frequency of our interest, the B_2 states are quite different from the B_3 and B_4 states in field patterns inside the dielectric layer. We see from Figs. 14 (b), 14 (c) ,14 (e)and 14 (f) that for a B_3 or B_4 state, there are nodes and anti-nodes in field patterns, while for the B_2 state, the magnetic field is almost uniformly distributed. Calculations on local field enhancement inside the dielectric slab resolve the puzzle. Black solid line in Fig. 15 presents the normalized magnetic field $|H|$ inside the dielectric with respect to that of incidence $|H_0|$ under an incident angle of $\theta = 2^o$. The Fourier component in *m=0* order [blue solid line] contributes the most at 13.2THz and the least at 50.22THz and 54.3THz; while it is just opposite for the contributions in combination from the two high order Fourier components with *m=±1* [red solid line]. Figure 15 also indicates that the enhancement of local field of an excited B_3 or B_4 state can be ten times larger than that of an excited B_2 state, the enhancement factor at 50.22THz is about 100 times, while it is only 10 times at 13.2THz.

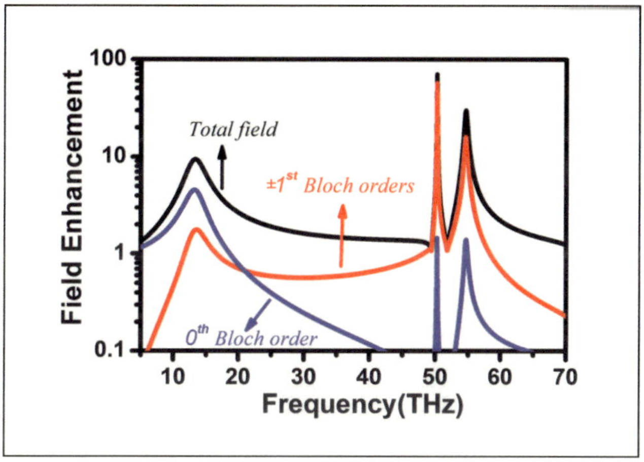

Fig. 15. Magnetic field H inside the dielectric layer normalized to that of incidence H under an incident angle of $\theta = 2^\circ$. Black line: all Bloch orders of TEM guided modes included; Red line: only the 0th Bloch order considered; Blue line: summation of -1st and +1st Bloch orders of TEM guided modes.

We see from Fig. 13 (b) that the surface resonance dispersion of the slab comes from the interaction between the magnetic resonances and the (folded) light lines L_1 (for dielectrics) and L_2 (for air) grazing on the interfaces. In the limit of a small gap-period ratio ($g/p = 0.05$ for example), our system is weakly Bragg-scattered, and as such, when a surface resonance state on branches B_3 or B_4 is excited, the induced wave fields inside the dielectric of region III are guided quasi-TEM modes dominated by $\pm 1^{st}$ Bloch orders. For that reason, the B_3 and B_4 states have high fidelity even though they are leaky modes, as most of their Bloch wavefunction components lying outside the free space light line. As the air gaps of the metallic grating serve to couple the electromagnetic waves of region I and region III, the quality factor of a resonance state can be estimated with the overlap integral between the fundamental waveguide mode $|\alpha_0>$ in the air gap and the dominant Bloch waves $|\vec{k}_m^i>$ ($i = I, III$) in region I or region III for the coupling coefficients

$$C_m^i = \frac{1}{w_x}\int_0^p \frac{k_{zm}^i}{\sqrt{\varepsilon_r}\sqrt{k_x^2 + k_{zm}^{i\ 2}}} e^{-i(k_x+G_m)x} g_0(x)dx$$
$$= \frac{1}{\sqrt{\varepsilon_r}} \frac{k_{zm}^i}{\sqrt{k_x^2 + k_{zm}^{i\ 2}}} \sin c[\frac{(k_x+G_m)g}{2}], \tag{13}$$

when the air gap width $g \ll p$ is satisfied. For the B_2 states, the major Fourier component of the wavefunction is $|\vec{k}_0^i>$ in zero order, and as k_{z0}^{III} is generally not small, C_0^{III} is usually very large according to Eq. 3, and the B_2 states leak out easily. The states on branches B_3 and B_4 have major Fourier components in *m*=±*1* order, and as they are asymptotic to the (folded) dielectric light lines L1, the absolute value of k_{zm}^{III} ($m = +1 \text{ for } k_x < 0$ or $m = -1 \text{ for } k_x > 0$) is very small, resulting in the small coupling coefficients C_{-1}^{III} or C_{+1}^{III}. The B_3 and B_4 modes have to travel a long distance before they leak out. They have a long life time and a good spatial coherence. It also explains why the state Γ_3, a state precisely superposing on folded light line L'_1 in dielectric layer, is dark to the incident plane wave as $k_{zm}^{III} = 0$.

Different from B_3 and B_4 states, the B_2 states have a major Fourier component in *m=0* order which directly couples to the free space photons. As a consequence, the B_2 states, forming a flat band far away from the light line L_2 when $\vec{k}_{0_{\parallel}}$ is small, are localized with resonant frequency scaled by local geometry of unit cell. The high mode fidelity of a B_3 or B_4 state also gives rise to much more intense local field compared to the B_2 states. As shown in Fig. 15, the induced local field is 100 times stronger than the incident field for the state on B_3; while it is only 10 times stronger for Γ_2, and this is consistent with the absorptivity shown in Fig. 12 (a). In addition, the coherent length can be adjusted by the gap width as the kernel $\langle \alpha_0 | \vec{r} | k_m^i \rangle$ is proportional to the gap-period ratio g/p. More calculations demonstrate that the angular FWHM of the absorption peak is reduced from 0.26^o to 0.16^o when the gap is decreased from $0.2\mu m$ to $0.1\mu m$, corresponding to a coherent length of 358λ.

We note that most of the attentions in previous studies have been devoted on the localized B2 states (Hibbins, Sambles et al. 2004; Hibbins, Murray et al. 2006; Brown, Hibbins et al. 2008; Diem, Koschny et al. 2009). While the spatially coherent surface resonance states will lead us into a new vision about coherent control of emission radiations. J.-J. Greffet and co-workers showed that highly directional and spatially coherent thermal emission can be obtained by etching a periodic grating structure into a SiC surface (Le Gall, Olivier et al. 1997; Carminati & Greffet 1999; Greffet, Carminati et al. 2002; Marquier, Joulain et al. 2004). The magnetic resonant modes in our system can do the same, as will be demonstrated below. Our system has the advantage that the operational frequency is tunable by changing the structural parameters, and the operational bandwidth is wide. In addition, our structure supports all-angle functionality for some specific range of frequencies as shown in Fig. 16, although it is periodic only in one direction.

We performed finite-difference-in-time-domain (FDTD) simulations to emulate the emissions from a slab containing point sources with random phases using the same configuration parameters aforementioned. We purposely put disorder in structure to test the robustness of the phenomena. We assigned two Gaussian distributions (they can be uniform distributions or other types as well) independently to the width of metallic strips and the center positions of air gaps to introduce a 4% (standard deviation) structural disorder. The slab has a lateral size of 60 periods along the $\hat{x}$ direction. A total of 1200 point sources with random phases are placed at the mesh points inside the dielectric layer. Directional emissions of a wide range of frequencies above 34THz are confirmed by the simulation. The 4% structural disorder has little impact on the directional emissivity. Fig. 16 (a), 16 (c) and 16 (e) show the far-field emission patterns in the $\hat{x}\hat{z}$ plane (H-plane) at 40.0THz, 54.3THz and 58.0THz. The inset in Fig. 16 (c) is a control calculation in which the top metal gratings are removed, so that there is just a dielectric layer with random phase sources above a metal ground plane. The directivity of emission from the random sources is lost. Fig. 16 (b), 16 (d) and 16 (e) present the absorptivity (under plane wave incidence) as a function of in-plane wave-vector (solid angle) at these frequencies. The strong angle selectivity of the absorption is evident, and by Kirchhoff's law, the thermal emission should also be highly directional, which is a direct consequence of the good spatial coherence of the surface resonance states. As shown in Figs. 16 (b), 16 (d) and 16 (f), the absorption/emission peaks generally trace out an arc in the k_x-k_y plane, but near 54.3*THz* [Fig. 16 (d)], the dominant emission beam is restricted to a small region near the zone center. This is because the Γ_4 state is a minimum point if we consider the band structure in the k_x-k_y plane. That means that at 54.3*THz*, we can obtain a directional emission beam not just in the H-plane, but in all directions, although the structure is periodic in only one direction.

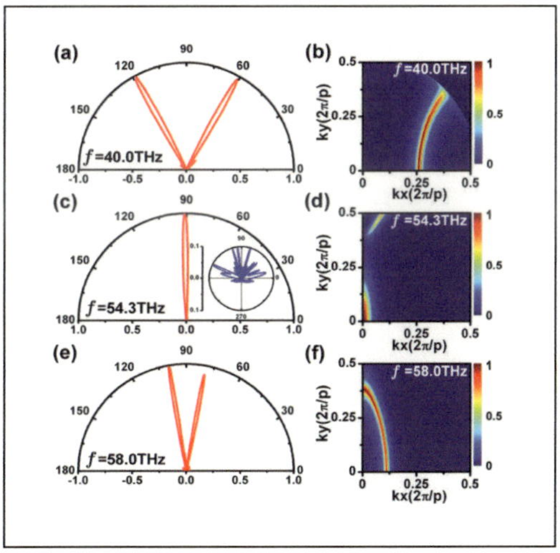

Fig. 16. Radiation patterns in the H-plane (calculated by FDTD) and absorptivity (calculated by mode expansion method) as a function of in-plane wavevectors at f=40.0THz[(a) and (b)], f=54.3THz[(c) and (d)] and f=58.0THz[(e) and (f)]. In FDTD simulations, 1200 point sources with random phases are placed at the mesh points inside the dielectric layer. A 4% structural disorder is included in the80μmsimulation cell, which accounts for the slight asymmetry of the radiation patterns, but also demonstrates the robustness of the angle selectivity with respect to disorder.The inset in (c) is a control calculation in which the top metal gratings are removed, so that there is just a dielectric layer with random phase sources above a metal ground plane. The directivity of emission from the random sources is lost.

We note that there are other schemes to realize coherent thermal radiations, such as by utilizing three-dimensional photonic crystals (Laroche, Carminati et al. 2006) or one-dimensional photonic crystal cavities (Lee, Fu et al. 2005). Our metamaterial slab presents a route to achieve linearly polarized coherent thermal emission radiations in a wide frequency range which can be tuned by adjusting structural parameters and material parameters.

In summary, we proposed a simple metamaterial slab structure that possesses spatially coherent magnetic surface resonance states in a broad range of frequencies. These states facilitate nearly perfect absorption in a thin metamaterial slab containing slightly absorptive materials. As the absorption spectrum is very narrow and sensitive to incident angle, the slab should support directional thermal emission. Direct FDTD simulation with random-phase sources corroborates the existence of strong angular emissivity even in the presence of structural disorder. As the surface resonances originate from artificial resonators, the operational frequency and the response can be tuned by varying the structural configurations. Our findings constitute a simple solution for coherent control of thermal emissions, optical antennas, infrared or THz spectroscopy as well as photon detector.

3. Conclusion

We have shown that a strategy of stacking a multiple layers of holey metallic slabs can give rise to wide transparency band. The multi-layered structure, well known as fishnet metamaterial, also supports broadband negative refraction and sub-wavelength imaging provided that the light waves are incident on its sidewall interface. We also show that coherent control of spontaneous emission radiations can be realizing in a wide frequency range by utilizing spatially coherent magnetic surface resonance states of a magnetic meta-surface. The modal expansion method, developed for magnetic meta-surface and multi-layered holey metallic slabs, is very fruitful for semi-analytical interpretation on the behind physics picture of planar metamaterials.

4. Acknowledgment

H. Li thanks Prof. C.T. Chan, Prof. H. Chen and Prof. D. Z. Zhang for the collaboration and helpful discussion. This work was supported by NSFC (No. 11174221, 10974144, 60678046, 10574099), CNKBRSF (Grant No. 2011CB922001), the National 863 Program of China (No.2006AA03Z407), NCET (07-0621), STCSM and SHEDF (No. 06SG24).

5. References

Andrew, P.; Kitson, S. C. & Barnes, W. L. (1997). Surface-plasmon energy gaps and photoabsorption. *Journal of Modern Optics* Vol. 44 No. 2, (FEB 1997) pp. 395-406,ISSN 0950-0340

Bahk, Y. M.; Park, H. R.; Ahn, K. J.; Kim, H. S.; Ahn, Y. H.; Kim, D. S.; Bravo-Abad, J.; Martin-Moreno, L. & Garcia-Vidal, F. J. (2011). Anomalous Band Formation in Arrays of Terahertz Nanoresonators. *Physical Review Letters* Vol. 106 No. 1, (Jan 5) pp. 013902-013904,ISSN 0031-9007

Baida, F. I.; Van Labeke, D.; Granet, G.; Moreau, A. & Belkhir, A. (2004). Origin of the super-enhanced light transmission through a 2-D metallic annular aperture array: a study of photonic bands. *Applied Physics B: Lasers and Optics* Vol. 79 No. 1, (JUL 2004) pp. 1-8,ISSN 0946-2171

Barnes, W. L.; Dereux, A. & Ebbesen, T. W. (2003). Surface plasmon subwavelength optics. *Nature* Vol. 424 No. 6950, (Aug 14) pp. 824-830,ISSN 0028-0836

Belov, P.; Hao, Y. & Sudhakaran, S. (2006). Subwavelength microwave imaging using an array of parallel conducting wires as a lens. *Physical Review B* Vol. 73 No. JAN 2006) pp. 033108-033104,ISSN 1098-0121

Beruete, M.; Campillo, I.; Rodriguez-Seco, J. E.; Perea, E.; Navarro-Cia, M.; Nunez-Manrique, I. J. & Sorolla, M. (2007). Enhanced gain by double-periodic stacked subwavelength hole array. *Ieee Microwave and Wireless Components Letters* Vol. 17 No. 12, (Dec) pp. 831-833,ISSN 1531-1309

Beruete, M.; Navarro-Cia, M.; Sorolla, M. & Campillo, I. (2008). Polarization selection with stacked hole array metamaterial. *Journal of Applied Physics* Vol. 103 No. 5, (Mar) pp. 053102-053104,ISSN 0021-8979

Beruete, M.; Sorolla, M.; Navarro-Cia, M.; Falcone, F.; Campillo, I. & Lomakin, V. (2007). Extraordinary transmission and left-handed propagation in miniaturized stacks of doubly periodic subwavelength hole arrays. *Optics Express* Vol. 15 No. 3, (Feb) pp. 1107-1114,ISSN 1094-4087

Boltasseva, A. & Shalaev, V. M. (2008). Fabrication of optical negative-index metamaterials: Recent advances and outlook. *Metamaterials* Vol. 2 No. 1, (pp. 1-17,ISSN 1873-1988

Born, M. & Wolf, E. (1980). Principles of Optics: Electromagnetic Theory of Propagation, Interference and Diffraction of Light. Oxford, Angleterre, Pergamon Press.

Brown, J.; Hibbins, A.; Lockyear, M.; Lawrence, C. & Sambles, J. (2008). Angle-independent microwave absorption by ultrathin microcavity arrays. *Journal of Applied Physics* Vol. 104 No. 4, (AUG 15 2008) pp. 043105,ISSN 0021-8979

Cao, J. X.; Liu, H.; Li, T.; Wang, S. M.; Li, T. Q.; Zhu, S. N. & Zhang, X. (2009). Steering polarization of infrared light through hybridization effect in a tri-rod structure. *Journal of the Optical Society of America B-Optical Physics* Vol. 26 No. 12, (Dec) pp. B96-B101,ISSN 0740-3224

Cao, R.; Wei Z.Y.; Li W.; Fang, A.A. Li, H.Q.; Jiang, X.Y.; Chen, H. & Chan, C.T. (2011). Low-threshold directional plasmon lasing assisted by spatially coherent surface plasmon polaritons. *arXiv:1106.3002v1*

Carminati, R. & Greffet, J.-J. (1999). Near-field effects in spatial coherence of thermal sources. *Physical Review Letters* Vol. 82 No. 8, (FEB 22 1999) pp. 1660-1663,ISSN 0031-9007

Chan, H. B.; Marcet, Z.; Woo, K.; Tanner, D. B.; Carr, D. W.; Bower, J. E.; Cirelli, R. A.; Ferry, E.; Klemens, F.; Miner, J.; Pai, C. S. & Taylor, J. A. (2006). Optical transmission through double-layer metallic subwavelength slit arrays. *Optics Letters* Vol. 31 No. 4, (Feb 15) pp. 516-518,ISSN 0146-9592

Chin, J. Y.; Gollub, J. N.; Mock, J. J.; Liu, R. P.; Harrison, C.; Smith, D. R. & Cui, T. J. (2009). An efficient broadband metamaterial wave retarder. *Optics Express* Vol. 17 No. 9, (Apr 27) pp. 7640-7647,ISSN 1094-4087

Chin, J. Y.; Lu, M. Z. & Cui, T. J. (2008). Metamaterial polarizers by electric-field-coupled resonators. *Applied Physics Letters* Vol. 93 No. 25, (Dec 22) pp. 251903,ISSN 0003-6951

Coutaz, J. L.; Neviere, M.; Pic, E. & Reinisch, R. (1985). Experimental study of surface-enhanced second-harmonic generation on silver gratings. *Physical Review B* Vol. 32 No. 4,pp. 2227-2232

de Abajo, F. J. G. & Saenz, J. J. (2005). Electromagnetic surface modes in structured perfect-conductor surfaces. *Physical Review Letters* Vol. 95 No. 23, (Dec 2) pp. 233901-233904,ISSN 0031-9007

Diem, M.; Koschny, T. & Soukoulis, C. (2009). Wide-angle perfect absorber/thermal emitter in the terahertz regime. *Physical Review B* Vol. 79 No. 3, (JAN 2009) pp. 033101,ISSN 1098-0121

Dolling, G.; Enkrich, C.; Wegener, M.; Soukoulis, C. & Linden, S. (2006). Simultaneous negative phase and group velocity of light in a metamaterial. *Science* Vol. 312 (MAY 12 2006) pp. 892-894,ISSN 0036-8075

Dolling, G.; Wegener, M. & Linden, S. (2007). Realization of a three-functional-layer negative-index photonic metamaterial. *Optics Letters* Vol. 32 No. 5, (Mar) pp. 551-553,ISSN 0146-9592

Ebbesen, T. W.; Lezec, H. J.; Ghaemi, H. F.; Thio, T. & Wolff, P. A. (1998). Extraordinary optical transmission through sub-wavelength hole arrays. *Nature* Vol. 391 No. 6668, (FEB 12 1998) pp. 667-669,ISSN 0028-0836

Eisenberg, H.; Silberberg, Y.; Morandotti, R. & Aitchison, J. (2000). Diffraction management. *Physical Review Letters* No. AUG 28 2000) pp. 1863-1866,ISSN 0031-9007

Engheta, N. & Ziolkowski, R. W. (2006). *Metamaterials: physics and engineering explorations,* Wiley & Sons.

Fan, W. J.; Zhang, S.; Malloy, K. J. & Brueck, S. R. J. (2005). Enhanced mid-infrared transmission through nanoscale metallic coaxial-aperture arrays. *Optics Express* Vol. 13 No. 12, (JUN 13 2005) pp. 4406-4413,ISSN 1094-4087

Fan, W. J.; Zhang, S.; Minhas, B.; Malloy, K. J. & Brueck, S. R. J. (2005). Enhanced infrared transmission through subwavelength coaxial metallic arrays. *Physical Review Letters* Vol. 94 No. 3, (JAN 28 2005) pp. 033902,ISSN 0031-9007

Fan, X.; Wang, G.; Lee, J. & Chan, C. T. (2006). All-angle broadband negative refraction of metal waveguide arrays in the visible range: Theoretical analysis and numerical demonstration. *Physical Review Letters* Vol. 97 No. AUG 18 2006) pp. 073901-073904,ISSN 0031-9007

Fan, Y. C.; Han, J.; Wei, Z. Y.; Wu, C.; Cao, Y.; Yu, X. & Li, H. Q. (2011). Subwavelength electromagnetic diode: One-way response of cascading nonlinear meta-atoms. *Applied Physics Letters* Vol. 98 No. 15, (Apr 11) pp. 151903,ISSN 0003-6951

Fedotov, V. A.; Schwanecke, A. S.; Zheludev, N. I.; Khardikov, V. V. & Prosvirnin, S. L. (2007). Asymmetric transmission of light and enantiomerically sensitive plasmon resonance in planar chiral nanostructures. *Nano Letters* Vol. 7 No. 7, (Jul) pp. 1996-1999,ISSN 1530-6984

Freire, M.; Marques, R. & Jelinek, L. (2008). Experimental demonstration of a mu=-1 metamaterial lens for magnetic resonance imaging. *Applied Physics Letters* Vol. 93 No. DEC 8 2008) pp. 231108-231101,ISSN 0003-6951

Gay, G.; Alloschery, O.; de Lesegno, B. V.; Weiner, J. & Lezec, H. J. (2006). Surface wave generation and propagation on metallic subwavelength structures measured by far-field interferometry. *Physical Review Letters* Vol. 96 No. 21, (Jun 2) pp. 213901-213904,ISSN 0031-9007

Ghaemi, H. F.; Thio, T.; Grupp, D. E.; Ebbesen, T. W. & Lezec, H. J. (1998). Surface plasmons enhance optical transmission through subwavelength holes. *Physical Review B* Vol. 58 No. 11, (SEP 15 1998) pp. 6779-6782,ISSN 1098-0121

Grbic, A. & Eleftheriades, G. (2004). Overcoming the diffraction limit with a planar left-handed transmission-line lens. *Physical Review Letters* Vol. 92 (MAR 19 2004) pp. 117403-117404,ISSN 0031-9007

Greffet, J.-J.; Carminati, R.; Joulain, K.; Mulet, J.-P.; Mainguy, S. & Chen, Y. (2002). Coherent emission of light by thermal sources. *Nature* Vol. 416 No. 6876, (Mar 7) pp. 61-64,ISSN 0028-0836

Greffet, J.-J. & Nieto-Vesperinas, M. (1998). Field theory for generalized bidirectional reflectivity: derivation of Helmholtz's reciprocity principle and Kirchhoff's law. *Journal of the Optical Society of America A* Vol. 15 No. 10 pp. 2735-2744

Grigorenko, A. N.; Geim, A. K.; Gleeson, H. F.; Zhang, Y.; Firsov, A. A.; Khrushchev, I. Y. & Petrovic, J. (2005). Nanofabricated media with negative permeability at visible frequencies. *Nature* Vol. 438 No. 7066, (NOV 17 2005) pp. 335-338,ISSN 0028-0836

Han, J.; Li, H. Q.; Fan, Y. C.; Wei, Z. Y.; Wu, C.; Cao, Y.; Yu, X.; Li, F. & Wang, Z. S. (2011). An ultrathin twist-structure polarization transformer based on fish-scale metallic wires. *Applied Physics Letters* Vol. 98 No. 15, (Apr 11) pp. 151908,ISSN 0003-6951

Hao, J. M.; Yuan, Y.; Ran, L. X.; Jiang, T.; Kong, J. A.; Chan, C. T. & Zhou, L. (2007). Manipulating electromagnetic wave polarizations by anisotropic metamaterials. *Physical Review Letters* Vol. 99 No. 6, (Aug 10) pp. 063908,ISSN 0031-9007

Haus, H. & Molter-Orr, L. (1983). Coupled multiple waveguide systems. *Quantum Electronics, IEEE Journal of* Vol. 19 No. 5, (pp. 840-844,ISSN 0018-9197

Hibbins, A.; Murray, W.; Tyler, J.; Wedge, S.; Barnes, W. & Sambles, J. (2006). Resonant absorption of electromagnetic fields by surface plasmons buried in a multilayered plasmonic nanostructure. *Physical Review B* Vol. 74 No. 7, (AUG 2006) pp. 073408,ISSN 1098-0121

Hibbins, A.; Sambles, J.; Lawrence, C. & Brown, J. (2004). Squeezing millimeter waves into microns. *Physical Review Letters* Vol. 92 No. 14, (APR 9 2004) pp. 143904,ISSN 0031-9007

Hibbins, A. P.; Evans, B. R. & Sambles, J. R. (2005). Experimental verification of designer surface plasmons. *Science* Vol. 308 No. 5722, (APR 29 2005) pp. 670-672,ISSN 0036-8075

Kamli, A.; Moiseev, S. A. & Sanders, B. C. (2008). Coherent Control of Low Loss Surface Polaritons. *Physical Review Letters* Vol. 101 No. 26, (DEC 31 2008) pp. 263601,ISSN 0031-9007

Kneipp, K.; Wang, Y.; Kneipp, H.; Perelman, L. T.; Itzkan, I.; Dasari, R. R. & Feld, M. S. (1997). Single molecule detection using surface-enhanced Raman scattering (SERS). *Physical Review Letters* Vol. 78 No. 9, (MAR 3 1997) pp. 1667-1670,ISSN 0031-9007

Ku, Z. & Brueck, S. R. J. (2009). Experimental demonstration of sidewall angle induced bianisotropy in multiple layer negative index metamaterials. *Applied Physics Letters* Vol. 94 No. 15, (Apr) pp. 153107-153103,ISSN 0003-6951

Ku, Z. Y.; Zhang, J. Y. & Brueck, S. R. J. (2009). Bi-anisotropy of multiple-layer fishnet negative-index metamaterials due to angled sidewalls. *Optics Express* Vol. 17 No. 8, (Apr) pp. 6782-6789,ISSN 1094-4087

Lalanne, P.; Hugonin, J. P.; Astilean, S.; Palamaru, M. & Moller, K. D. (2000). One-mode model and Airy-like formulae for one-dimensional metallic gratings. *Journal of Optics a-Pure and Applied Optics* Vol. 2 No. 1, (Jan) pp. 48-51,ISSN 1464-4258

Laroche, M.; Carminati, R. & Greffet, J.-J. (2006). Coherent thermal antenna using a photonic crystal slab. *Physical Review Letters* Vol. 96 No. 12, (MAR 31 2006) pp. 123903,ISSN 0031-9007

Le Gall, J.; Olivier, M. & Greffet, J.-J. (1997). Experimental and theoretical study of reflection and coherent thermal emission by a SiC grating supporting a surface-phonon polariton. *Physical Review B* Vol. 55 No. 15, (APR 15 1997) pp. 10105-10114,ISSN 0163-1829

Lee, B. J.; Fu, C. J. & Zhang, Z. M. (2005). Coherent thermal emission from one-dimensional photonic crystals. *Applied Physics Letters* Vol. 87 No. 7, (Aug) pp. 071904,ISSN 0003-6951

Li, H. Q.; Hang, Z. H.; Qin, Y. Q.; Wei, Z. Y.; Zhou, L.; Zhang, Y. W.; Chen, H. & Chan, C. T. (2005). Quasi-periodic planar metamaterial substrates. *Applied Physics Letters* Vol. 86 No. 12, (MAR 21 2005) pp. 121108,ISSN 0003-6951

Li, H. Q.; Hao, J. M.; Zhou, L.; Wei, Z. Y.; Gong, L. K.; Chen, H. & Chan, C. T. (2006). All-dimensional subwavelength cavities made with metamaterials. *Applied Physics Letters* Vol. 89 No. 10, (SEP 4 2006) pp. 104101-104103,ISSN 0003-6951

Li, T.; Liu, H.; Wang, S. M.; Yin, X. G.; Wang, F. M.; Zhu, S. N. & Zhang, X. A. (2008). Manipulating optical rotation in extraordinary transmission by hybrid plasmonic excitations. *Applied Physics Letters* Vol. 93 No. 2, (Jul 14) pp. 021110,ISSN 0003-6951

Li, T. Q.; Liu, H.; Li, T.; Wang, S. M.; Wang, F. M.; Wu, R. X.; Chen, P.; Zhu, S. N. & Zhang, X. (2008). Magnetic resonance hybridization and optical activity of microwaves in a chiral metamaterial. *Applied Physics Letters* Vol. 92 No. 13, (Mar 31) pp. 131111,ISSN 0003-6951

Liu, H.; Genov, D.; Wu, D.; Liu, Y.; Steele, J.; Sun, C.; Zhu, S. & Zhang, X. (2006). Magnetic plasmon propagation along a chain of connected subwavelength resonators at infrared frequencies. *Physical Review Letters* Vol. 97 No. 24, (DEC 15 2006) pp. 243902,ISSN 0031-9007

Liu, H.; Genov, D. A.; Wu, D. M.; Liu, Y. M.; Liu, Z. W.; Sun, C.; Zhu, S. N. & Zhang, X. (2007). Magnetic plasmon hybridization and optical activity at optical frequencies in metallic nanostructures. *Physical Review B* Vol. 76 No. 7, (Aug) pp. 073101,ISSN 1098-0121

Liu, H. T. & Lalanne, P. (2008). Microscopic theory of the extraordinary optical transmission. *Nature* Vol. 452 No. 7188, (Apr 10) pp. 728-731,ISSN 0028-0836

Liu, N.; Guo, H. C.; Fu, L. W.; Kaiser, S.; Schweizer, H. & Giessen, H. (2008). Three-dimensional photonic metamaterials at optical frequencies. *Nature Materials* Vol. 7 No. 1, (Jan) pp. 31-37,ISSN 1476-1122

Lockyear, M. J.; Hibbins, A. P. & Sambles, J. R. (2009). Microwave Surface-Plasmon-Like Modes on Thin Metamaterials. *Physical Review Letters* Vol. 102 No. 7, (FEB 20 2009) pp. 073901,ISSN 0031-9007

Mansuripur, M. (2002). Classical Optics and its Applications, Cambridge University Press.

Marcet, Z.; Hang, Z. H.; Chan, C. T.; Kravchenko, I.; Bower, J. E.; Cirelli, R. A.; Klemens, F.; Mansfield, W. M.; Miner, J. F.; Pai, C. S. & Chan, H. B. (2010). Optical transmission through double-layer, laterally shifted metallic subwavelength hole arrays. *Optics Letters* Vol. 35 No. 13, (Jul 1) pp. 2124-2126,ISSN 0146-9592

Marquier, F.; Joulain, K.; Mulet, J.-P.; Carminati, R.; Greffet, J.-J. & Chen, Y. (2004). Coherent spontaneous emission of light by thermal sources. *Physical Review B* Vol. 69 No. 15, (APR 2004) pp. 155412,ISSN 1098-0121

Martin-Moreno, L.; Garcia-Vidal, F. J.; Lezec, H. J.; Pellerin, K. M.; Thio, T.; Pendry, J. B. & Ebbesen, T. W. (2001). Theory of extraordinary optical transmission through subwavelength hole arrays. *Physical Review Letters* Vol. 86 No. 6, (FEB 5 2001) pp. 1114-1117,ISSN 0031-9007

Maystre, D. (2001). Photonic crystal diffraction gratings. *Optics Express* Vol. 8 No. 3, (JAN 29 2001) pp. 209-216,ISSN 1094-4087

Menzel, C.; Helgert, C.; Rockstuhl, C.; Kley, E. B.; Tunnermann, A.; Pertsch, T. & Lederer, F. (2010). Asymmetric Transmission of Linearly Polarized Light at Optical Metamaterials. *Physical Review Letters* Vol. 104 No. 25, (Jun 22) pp. 253902,ISSN 0031-9007

Miyamaru, F. & Hangyo, M. (2005). Anomalous terahertz transmission through double-layer metal hole arrays by coupling of surface plasmon polaritons. *Physical Review B* Vol. 71 No. 16, (Apr) pp. 165408-165405,ISSN 1098-0121

Mocella, V.; Dardano, P.; Moretti, L. & Rendina, I. (2007). Influence of surface termination on negative reflection by photonic crystals. *Optics Express* Vol. 15 No. 11, (MAY 28 2007) pp. 6605-6611,ISSN 1094-4087

Navarro-Cia, M.; Beruete, M.; Sorolla, M. & Campillo, I. (2008). Negative refraction in a prism made of stacked subwavelength hole arrays. *Optics Express* Vol. 16 No. 2, (Jan) pp. 560-566,ISSN 1094-4087

Navarro-Cia, M.; Beruete, M.; Sorolla, M. & Campillo, I. (2009). Converging biconcave metallic lens by double-negative extraordinary transmission metamaterial. *Applied Physics Letters* Vol. 94 No. 14, (Apr) pp. 144107-144103,ISSN 0003-6951

Nie, S. & Emory, S. R. (1997). Probing single molecules and single nanoparticles by surface-enhanced Raman scattering. *Science* Vol. 275 No. 5303, (FEB 21 1997) pp. 1102-1106,ISSN 0036-8075

Ortuno, R.; Garcia-Meca, C.; Rodriguez-Fortuno, F. J.; Marti, J. & Martinez, A. (2009). Role of surface plasmon polaritons on optical transmission through double layer metallic hole arrays. *Physical Review B* Vol. 79 No. 7, (Feb) pp. 075425-075427,ISSN 1098-0121

Pendry, J. B. (2000). Negative refraction makes a perfect lens. *Physical Review Letters* Vol. 85 No. 18, (Oct 30) pp. 3966-3969,ISSN 0031-9007

Pendry, J. B.; Holden, A. J.; Robbins, D. J. & Stewart, W. J. (1999). Magnetism from conductors and enhanced nonlinear phenomena. *Ieee Transactions on Microwave Theory and Techniques* Vol. 47 No. 11, (Nov) pp. 2075-2084,ISSN 0018-9480

Pendry, J. B.; Holden, A. J.; Stewart, W. J. & Youngs, I. (1996). Extremely low frequency plasmons in metallic mesostructures. *Physical Review Letters* Vol. 76 No. 25, (Jun 17) pp. 4773-4776,ISSN 0031-9007

Pendry, J. B.; Martin-Moreno, L. & Garcia-Vidal, F. J. (2004). Mimicking surface plasmons with structured surfaces. *Science* Vol. 305 No. 5685, (Aug 6) pp. 847-848,ISSN 0036-8075

Pertsch, T.; Zentgraf, T.; Peschel, U.; Brauer, A. & Lederer, F. (2002). Anomalous refraction and diffraction in discrete optical systems. *Physical Review Letters* Vol. 88 (MAR 4 2002) pp. 093901-093904,ISSN 0031-9007

Plum, E.; Fedotov, V. A.; Schwanecke, A. S.; Zheludev, N. I. & Chen, Y. (2007). Giant optical gyrotropy due to electromagnetic coupling. *Applied Physics Letters* Vol. 90 No. 22, (May 28) pp. 223113,ISSN 0003-6951

Plum, E.; Zhou, J.; Dong, J.; Fedotov, V. A.; Koschny, T.; Soukoulis, C. M. & Zheludev, N. I. (2009). Metamaterial with negative index due to chirality. *Physical Review B* Vol. 79 No. 3, (Jan) pp. 121104 (R),ISSN 1098-0121

Raether, H. (1988). Surface Plasmons on Smooth and Rough Surfaces and on Gratings. Berlin, Springer-Verlag.

Rayleigh (1907). *Proc. R. Soc. London, Ser. A* Vol. 79 No. pp. 399

Rogacheva, A. V.; Fedotov, V. A.; Schwanecke, A. S. & Zheludev, N. I. (2006). Giant gyrotropy due to electromagnetic-field coupling in a bilayered chiral structure. *Physical Review Letters* Vol. 97 No. 17, (Oct 27) pp. 177401,ISSN 0031-9007

Ruan, Z. C. & Qiu, M. (2006). Enhanced transmission through periodic arrays of subwavelength holes: The role of localized waveguide resonances. *Physical Review Letters* Vol. 96 No. 23, (JUN 16 2006) pp. 233901,ISSN 0031-9007

Shalaev, V. M. (2007). Optical negative-index metamaterials. *Nature Photonics* Vol. 1 No. 1, (JAN 2007) pp. 41-48,ISSN 1749-4885

Shelby, R. A.; Smith, D. R. & Schultz, S. (2001). Experimental verification of a negative index of refraction. *Science* Vol. 292 (APR 6 2001) pp. 77-79,ISSN 0036-8075

Sheng, P.; Stepleman, R. S. & Sanda, P. N. (1982). Exact eigenfunctions for square-wave gratings: Application to diffraction and surface-plasmon calculations. *Physical Review B* Vol. 26 No. 6,pp. 2907-2916,ISSN

Sievenpiper, D.; Zhang, L. J.; Broas, R. F. J.; Alexopolous, N. G. & Yablonovitch, E. (1999). High-impedance electromagnetic surfaces with a forbidden frequency band. *Ieee Transactions on Microwave Theory and Techniques* Vol. 47 No. 11, (Nov) pp. 2059-2074,ISSN 0018-9480

Silveirinha, M.; Fernandes, C. & Costa, J. (2008). Superlens made of a metamaterial with extreme effective parameters. *Physical Review B* Vol. 78 (NOV 2008) pp. 195121-195127,ISSN 1098-0121

Silveirinha, M.; Medeiros, C.; Fernandes, C. & Costa, J. (2010). Experimental verification of broadband superlensing using a metamaterial with an extreme index of refraction. *Physical Review B* Vol. 81 No. JAN 2010) pp. 033301-033304,ISSN 1098-0121

Singh, R.; Plum, E.; Menzel, C.; Rockstuhl, C.; Azad, A. K.; Cheville, R. A.; Lederer, F.; Zhang, W. & Zheludev, N. I. (2009). Terahertz metamaterial with asymmetric transmission. *Physical Review B* Vol. 80 No. 15, (Oct) pp. 153104,ISSN 1098-0121

Smith, D. R.; Pendry, J. B. & Wiltshire, M. C. K. (2004). Metamaterials and negative refractive index. *Science* Vol. 305 No. 5685, (Aug 6) pp. 788-792,ISSN 0036-8075

Taflove, A. & Hagness, S. C. (2000). Computational Electrodynamics: The Finite-Difference Time-Domain Method, 2nd ed. Norwood,MA, Artech House.

Tang, Z. H.; Peng, R. W.; Wang, Z.; Wu, X.; Bao, Y. J.; Wang, Q. J.; Zhang, Z. J.; Sun, W. H. & Wang, M. (2007). Coupling of surface plasmons in nanostructured metal/dielectric multilayers with subwavelength hole arrays. *Physical Review B* Vol. 76 No. 19, (Nov) pp. 195405-195408,ISSN 1098-0121

Tsang, T. Y. F. (1996). Surface-plasmon-enhanced third-harmonic generation in thin silver films. *Optics Letters* Vol. 21 No. 4, (FEB 15 1996) pp. 245-247,ISSN 0146-9592

Valentine, J.; Zhang, S.; Zentgraf, T.; Ulin-Avila, E.; Genov, D.; Bartal, G. & Zhang, X. (2008). Three-dimensional optical metamaterial with a negative refractive index. *Nature* Vol. 455 (SEP 18 2008) pp. 376-379,ISSN 0028-0836

van der Molen, K. L.; Klein Koerkamp, K. J.; Enoch, S.; Segerink, F. B.; van Hulst, N. F. & Kuipers, L. (2005). Role of shape and localized resonances in extraordinary transmission through periodic arrays of subwavelength holes: Experiment and theory. *Physical Review B* Vol. 72 No. 4, (JUL 2005) pp. 045421,ISSN 1098-0121

Vasa, P.; Pomraenke, R.; Schwieger, S.; Mazur, Y. I.; Kunets, V.; Srinivasan, P.; Johnson, E.; Kihm, J. E.; Kim, D. S.; Runge, E.; Salamo, G. & Lienau, C. (2008). Coherent exciton-surface-plasmon-polariton interaction in hybrid metal-semiconductor nanostructures. *Physical Review Letters* Vol. 101 No. 11, (SEP 12 2008) pp. 116801,ISSN 0031-9007

Veselago, V. G. (1968). The electrodynamics of substances with simultaneously negative values of ε and μ. *Sov. Phys. Usp.* Vol. 10 No. 4,pp. 509–514

von Freymann, G.; Koch, W.; Meisel, D. C.; Wegener, M.; Diem, M.; Garcia-Martin, A.; Pereira, S.; Busch, K.; Schilling, J.; Wehrspohn, R. B. & Gosele, U. (2003). Diffraction properties of two-dimensional photonic crystals. *Applied Physics Letters* Vol. 83 No. 4, (JUL 28 2003) pp. 614-616,ISSN 0003-6951

Wei, Z. Y.; Cao, Y.; Han, J.; Wu, C.; Fan, Y. C. & Li, H. Q. (2010). Broadband negative refraction in stacked fishnet metamaterial. *Applied Physics Letters* Vol. 97 No. 14, (Oct) pp. 141901,ISSN 0003-6951

Wei, Z. Y.; Cao, Y.;Fan, Y. C.; Yu, X.; & Li, H. Q. (2011). Broadband enhanced transmission through the stacked metallic multi-layers perforated with coaxial annular apertures. *arXiv:1107.4604*

Wei, Z. Y.; Fu, J. X.; Cao, Y.; Wu, C. & Li, H. Q. (2010). The impact of local resonance on the enhanced transmission and dispersion of surface resonances. *Photonics and Nanostructures - Fundamentals and Applications* Vol. 8 No. 2,pp. 94-101

Wei, Z. Y.; Li, H. Q.; Cao, Y.; Wu, C.; Ren, J. Z.; Hang, Z. H.; Chen, H.; Zhang, D. Z. & Chan, C. T. (2010). Spatially coherent surface resonance states derived from magnetic resonances. *New Journal of Physics* Vol. 12 (SEP 15 2010) pp. 093020,ISSN 1367-2630

Wei, Z. Y.; Li, H. Q.; Wu, C.; Cao, Y.; Ren, J. Z.; Hang, Z. H.; Chen, H.; Zhang, D. Z. & Chan, C. T. (2010). Anomalous reflection from hybrid metamaterial slab. *Optics Express* Vol. 18 (JUN 7 2010) pp. 12119-12126,ISSN 1094-4087

Wen, W. J.; Zhou, L.; Hou, B.; Chan, C. T. & Sheng, P. (2005). Resonant transmission of microwaves through subwavelength fractal slits in a metallic plate. *Physical Review B* Vol. 72 No. 15, (Oct) pp. 153406-153404,ISSN 1098-0121

Wiltshire, M.; Pendry, J. & Hajnal, J. (2006). Sub-wavelength imaging at radio frequency. *Journal of Physics-Condensed Matter* Vol. 18 (JUN 7 2006) pp. L315-L321,ISSN 0953-8984

Wood, R. W. (1935). Anomalous Diffraction Gratings. *Physical Review* Vol. 48 No. 12, (pp. 928-936,ISSN

Xiao, X. A.; Jinbo, W.; Sasagawa, Y.; Miyamaru, F.; Zhang, M. Y.; Takeda, M. W.; Qiu, C. Y.; Wen, W. J. & Sheng, P. (2010). Resonant terahertz transmissions through metal hole array on silicon substrate. *Optics Express* Vol. 18 No. 18, (Aug 30) pp. 18558-18564,ISSN 1094-4087

Ye, Y. H. & Zhang, J. Y. (2005). Enhanced light transmission through cascaded metal films perforated with periodic hole arrays. *Optics Letters* Vol. 30 No. 12, (Jun 15) pp. 1521-1523,ISSN 0146-9592

Ye, Y. Q. & He, S. (2010). 90 degrees polarization rotator using a bilayered chiral metamaterial with giant optical activity. *Applied Physics Letters* Vol. 96 No. 20, (May 17) pp. 203501,ISSN 0003-6951

Yen, T. J.; Padilla, W. J.; Fang, N.; Vier, D. C.; Smith, D. R.; Pendry, J. B.; Basov, D. N. & Zhang, X. (2004). Terahertz magnetic response from artificial materials. *Science* Vol. 303 No. 5663, (Mar 5) pp. 1494-1496,ISSN 0036-8075

Zhang, S.; Fan, W.; Panoiu, N.; Malloy, K.; Osgood, R. & Brueck, S. (2006). Optical negative-index bulk metamaterials consisting of 2D perforated metal-dielectric stacks. *Optics Express* Vol. 14 (JUL 24 2006) pp. 6778-6787,ISSN 1094-4087

Zhou, L.; Huang, C. P.; Wu, S.; Yin, X. G.; Wang, Y. M.; Wang, Q. J. & Zhu, Y. Y. (2010). Enhanced optical transmission through metal-dielectric multilayer gratings. *Applied Physics Letters* Vol. 97 No. 1, (Jul 5) pp. 011905-011903,ISSN 0003-6951

Zhou, L.; Li, H. Q.; Qin, Y. Q.; Wei, Z. Y. & Chan, C. T. (2005). Directive emissions from subwavelength metamaterial-based cavities. *Applied Physics Letters* Vol. 86 No. 10, (MAR 7 2005) pp. 101101,ISSN 0003-6951

Zhou, L.; Wen, W.; Chan, C. & Sheng, P. (2003). Multiband subwavelength magnetic reflectors based on fractals. *Applied Physics Letters* Vol. 83 (OCT 20 2003) pp. 3257-3259,ISSN 0003-6951

Zhu, B.; Feng, Y. J.; Zhao, J. M.; Huang, C. & Jiang, T. A. (2010). Switchable metamaterial reflector/absorber for different polarized electromagnetic waves. *Applied Physics Letters* Vol. 97 No. 5, (Aug 2) pp. 051906,ISSN 0003-6951

11

Design and Characterization of Metamaterials for Optical and Radio Communications

André de Lustrac, Shah Nawaz Burokur, Boubacar Kanté, Alexandre Sellier and Dylan Germain
Institut d'Electronique Fondamentale, Univ. Paris-Sud, CNRS UMR 8622, Orsay, France

1. Introduction

Metamaterials have attracted considerable interests (Shelby, 2001, Yen, 2004, Smith, 2004, Linden, 2004, Zhang, 2004) because of their unusual electromagnetic properties (Veselago, 1968) and because of their potential applications such as invisibility cloaks (Leonhardt, 2006, Pendry, 2006, Schurig, 2006, Cai, 2007, Gaillot, 2008, Kante, 2008), the so-called perfect lenses (Pendry, 2000) and gradient index (GRIN) lenses. For example, perfect lenses require the use of Left-Handed (LH) metamaterials (Smith, 2000) having a negative refractive index, which can be produced by a simultaneously negative electric permittivity ε and magnetic permeability μ. Invisibility cloaks require adjustable positive permeability and permittivity from near zero values to several tenths. Traditionally, these properties are achieved by the use of a combination of split-ring resonators (Pendry, 1999) and metallic wires (Pendry, 1996), with periods much smaller than the wavelength of the electromagnetic wave, such that the medium can be considered homogeneous. Lately, pairs of finite-length wires (cut wires pairs) (Shalaev, 2005) have been proposed not only to replace the conventional split-ring resonators (SRRs) to produce a negative magnetic permeability under normal to plane incidence, but also lead to a negative refractive index n in the optical regime. However in a recent review paper (Shalaev, 2007), Shalaev stated that it is very difficult to achieve a negative refractive index with exclusively wire pairs and that the negative index value observed in the ref. (Shalaev, 2005), was accomplished in part because of the significant contribution from the imaginary part of the permeability. Nevertheless, the negative index from only cut wire and plate pairs has never been verified elsewhere (Dolling, 2005). Instead, continuous wires have been combined to the cut wire pairs to produce simultaneously a negative permittivity to lead to a negative index in the microwave domain (Zhou, 2006a). Zhou *et al.* also theoretically proposed a left-handed material using only cut wire pairs by increasing the equivalent capacitance between two consecutive short wire pairs so as to adjust the electric resonance frequency (Zhou, 2006b) This increase of capacitance can only be obtained by strongly reducing the spacing between two consecutive wires, which is quite difficult to achieve at high frequencies. These cited results concern mainly the microwave domain. In the optical regime, infrared and visible domains, the main problem concern the design and the characterization of metamaterials made of unit cells with nanometric dimensions. At this scale, the control and the engineering of the electromagnetic properties of metamaterials are closely linked to the easiness of the fabrication. This easiness is more and more important with the simplicity of the geometrical shapes of the unit cell of the metamaterials. We show that we can change the conventional shape of the split ring resonator of J. Pendry to simple coupled nanowires, keeping the same electromagnetic properties (Linden, 2004, Enkrich, 2005, Kante, 2008a, Burokur, 2009a).

In the first part of this chapter we investigate numerically and experimentally the electromagnetic properties of cut wire pairs metamaterials where the symmetry between the wires on opposite faces is voluntarily broken along the **E**-field direction (Sellier, 2009, Kante, 2009a, Burokur, 2009a). It is reported that in this case the electric resonance of the cut wire pairs can occur at lower frequencies than the magnetic resonance, which is in contrast with the symmetrical configuration. This lower electrical resonance frequency allows realizing a common frequency region where the permeability and the permittivity are simultaneously negative. This claim is verified numerically and experimentally in the microwave domain and indications on designing negative refractive index from structures composed of only cut wire pairs are given.

Then, we investigate numerically and experimentally the reflection and transmission spectra for an obliquely incident plane wave on the asymmetric structure. It is reported that a diffraction threshold appears in E-plane (plane containing vectors E and k), that is the (-1,0) mode starts to propagate (Burokur, 2009b). Besides, resonances in E-plane shift in frequency with increasing oblique incidence. However in H-plane (plane containing vectors H and k), the structure is diffractionless and independent of the incidence and therefore the negative index is maintained in a wide angular range. These statements are verified numerically and experimentally in the microwave domain.

On the other hand, we show that it is possible to engineer the resonances of metamaterial in the infrared domain (Kante, 2008b, Kante, 2009b). We present an experimental and numerical analysis of the infrared response of metamaterials made of split ring resonators (SRR) and continuous nanowires deposited on silicon when the geometry of the SRRs is gradually altered. The impact of the geometric transformation of the SRRs on the spectra of the composite metamaterial is measured in the 1.5-15 μm wavelength range for the two field polarizations under normal to plane propagation. We show experimentally and numerically that tuning the SRRs towards elementary cut wires translates in a predictable manner the wavelength response of the artificial material. We also analyze coupling effects between the SRRs and the continuous nanowires for different spacing between them. The results of our study are expected to provide useful guidelines for the design of optical devices using metamaterials on silicon.

2. Cut wire pairs metamaterials with broken symmetry at microwave frequencies

2.1 Plasmon hybrization

In this chapter, we show that negative dielectric permittivity and negative magnetic permeability can be simultaneously achieved by appropriately controlling the coupling strength between paired cut-wires of adjacent layers. The coupling strength is itself controlled by adjusting either the spacing or the alignment of paired cut-wires. Using an asymmetric alignment, an inverted hybridization scheme, where the asymmetric mode is at a higher frequency than the symmetric mode, is predicted and thus more favorable for obtaining negative refraction. The first experimental demonstration of a negative refraction metamaterial exclusively based on paired cut-wires in the microwave range is reported.

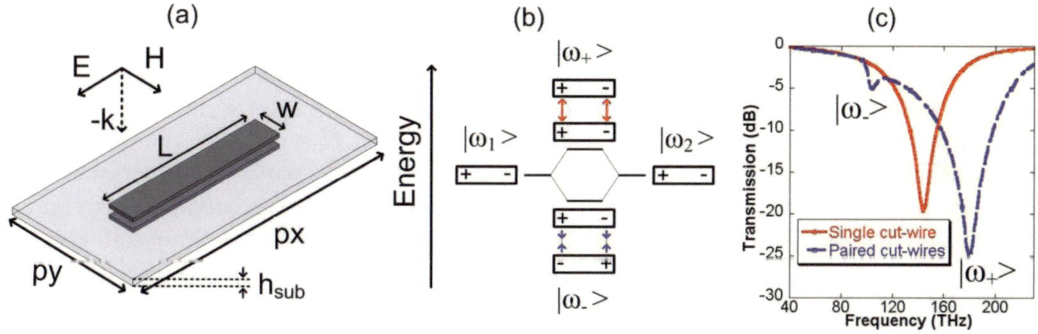

Fig. 1. (a) Schematic of the symmetric cut-wire pair. (b) Hybridization scheme of the two coupled dipoles. (c) Transmission spectra calculated at normal incidence for a periodic array of cut-wires (red) and of paired cut-wires (blue), respectively (px= 1.2 μm; py= 200 nm; w= 30 nm; L= 600 nm; hsub= 100 nm). The 30 nm thick gold cut-wires are described using a Drude model whose parameters can be found in [Kante, 2008a]. The dielectric spacer (SiO2) permittivity is ε_{sub}=2.25.

The plasmon hybridization scheme was introduced by Prodan et al. [Prodan, 2003] who gave an intuitive electromagnetic analogue of molecular orbital theory. Such a scheme has largely been used by the metamaterial community especially for simplifying metamaterial designs at optical wavelengths [Shalaev, 2005, Liu, 2008]. The coupling of two electric dipoles facing each other has thus been exploited to mimic magnetic atoms and alter the effective magnetic permeability of metamaterials in the optical range. While a magnetic activity was indeed obtained from metamaterials comprised of metallic dipoles [Shalaev, 2005, Liu, 2008], negative refraction was reported only in the pioneering demonstration by Shalaev et al. [Shalaev, 2005] who used a periodic array of cut-wire pairs. In order to unambiguously achieve negative refraction, the magnetic activity must actually occur within a frequency band in which the electric permittivity is negative. For this purpose, one solution consists of associating magnetic "atoms" (coupled metallic dipoles) to a broadband "electric plasma" (continuous wires) in the same structure. Many authors have used this solution either in the microwave [Zhou, 2006, Guven, 2006] or in the optical regime [Liu, 2008], thereby contributing to the development of the so-called fishnet structure. We propose another solution based on the control of the coupling between metallic dipoles in such a way that the symmetric and anti-symmetric bands have a sufficient overlap. The coupling strength is varied either by changing the distance between coupled dipoles or by breaking the symmetry of the structure.

Fig. 1(a) shows the rectangular unit cell of the studied 2D structure in the case where the coupled metallic dipoles (cut-wires) are vertically aligned. This structure is henceforth referred to as the symmetric cut-wire structure to distinguish it from the asymmetric structure discussed later (see fig. 2). Both structures consist of 2D periodic arrays of metallic cut-wires separated by a dielectric spacer [Shalaev, 2005]. The electromagnetic wave should propagate normally to the layers with the electric field parallel to the longest side of dipolar elements. The structure can be described in terms of effective parameters as long as the cut-wire width w and spacer thickness hsub are much smaller than the wavelength [Shalaev, 2005]. Two series of calculations were carried out using a finite element simulation package (HFSS from Ansys), one for symmetric structures and the other for asymmetric structures. The effective parameters were obtained from the calculated transmission and reflection coefficients [Gundogdu, 2008].

The first series of calculations were performed to compare the electromagnetic response of a symmetric cut-wire bi-layer (blue curve in Fig. 1(c)) to that of a cut-wire monolayer (red curve in Fig. 1(c)). As it is evident in the figure, only one resonance is observed for the single-face cut-wire structure in the frequency range of interest. This resonance corresponds to the fundamental cut-wire dipolar mode, which in the optical regime can also be interpreted in terms of a localized plasmon resonance [Smith, 2002]. Collective electronic excitations, also called surface plasmons, are indeed the main mechanism at short wavelengths. For the double-face cut-wire structure, the coupling between paired cut-wires lifts the degeneracy of the single cut-wire mode, which hybridizes into two plasmon modes. One mode is symmetric and corresponds to in-phase current oscillations, while the other is anti-symmetric and corresponds to out-of-phase current oscillations. For a symmetric cut-wire pair with a vertical alignment of the two cut-wires, the anti-symmetric mode is the low-energy (-frequency) mode since attractive forces are present in the system. Conversely, repellent forces are produced in the case of the symmetric mode that is therefore the high-frequency mode. The stronger the coupling (the smaller the spacing between the dipoles), the larger the frequency difference between the two modes. The evolution of the transmission spectra with the thickness of the dielectric spacer (or substrate) h_{sub} is illustrated in Fig. 3(a) in the case of a structure designed to operate in the microwave regime. Similar results were obtained for the structure in Fig. 1(c).

A second series of calculations was performed to analyze the influence of a vertical misalignment of metallic dipoles at a fixed spacer (or substrate) thickness. For this purpose, the cut-wire layers were shifted from each other in the horizontal XY plane (Fig. 2(a)) thus breaking the symmetry of the cut-wire structure. The relative displacements dx and dy in the X and Y directions respectively were used as parameters. In the example in the microwave regime, the substrate thickness was chosen to be equal to that of commercially available epoxy dielectric boards (1.2 mm). For this thickness and a vertical alignment of paired cut-wires (dx = dy = 0), the calculated transmission spectrum in Fig. 3(a) revealed a pronounced frequency separation between the symmetric (electric) and anti-symmetric (magnetic) modes. Figs. 3(b) and 3(c) show the evolution of the transmission spectrum for non-zero values of the longitudinal (dx) and lateral (dy) displacements, respectively. Quite surprisingly, as previously reported by A. Christ et al. [Christ, 2007] for the control of Fano resonances in a plasmonic lattice of continuous wires, symmetry breaking can invert the hybridization scheme due to modified Coulomb interactions (Fig. 2(b)) resulting in the symmetric resonance occurring at a lower frequency than the anti-symmetric one. The Coulomb forces in our system result from the interaction of charges located at the cut-wire ends. When the longitudinal shift (dx) is progressively increased, the signs of the charges in close interaction change. As a result the repulsive force becomes attractive and vice versa. Correspondingly, the symmetric mode becomes the low-energy mode while the asymmetric mode is shifted to higher frequencies. It is evident that this inversion process is impossible in the case of a lateral dy displacement of the dipoles (Fig. 3(c)).

Controlling the coupling between metallic dipoles thus allows the two plasmon resonances to be engineered. When the magnetic and electric modes are very close together, the bands of negative permeability and negative permittivity overlap, and a negative refraction material is obtained. More generally, the design of true negative index metamaterials can be achieved by appropriate design of the three degrees of freedom h_{sub}, dx and dy.

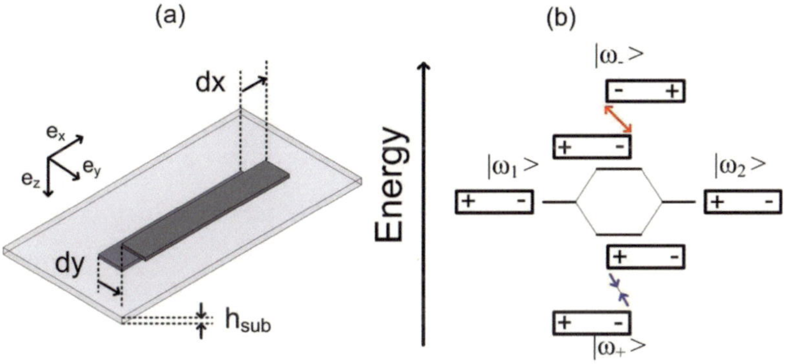

Fig. 2. (a) Asymmetric cut-wire pair with the three degrees of freedom for the control of the coupling strength: hsub, dx and dy. (b) Inverted hybridization scheme.

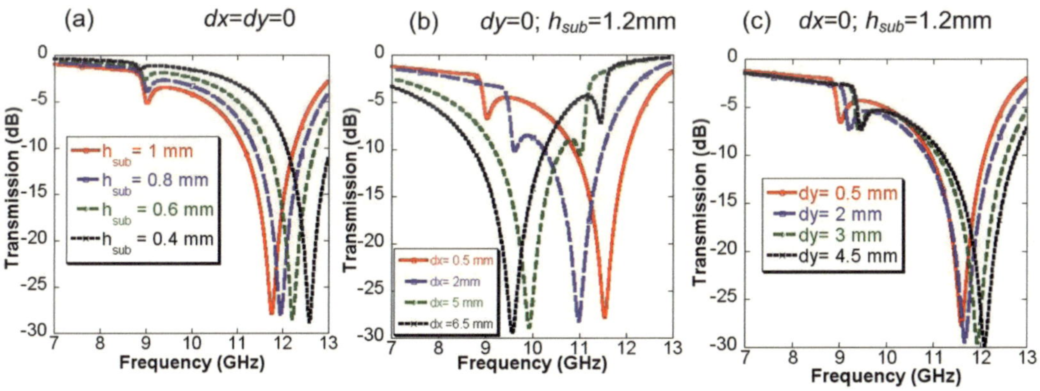

Fig. 3. Influence of the coupling strength on the transmission spectra of a bi-layer structure (px = 19 mm; py = 9.5 mm; w = 0.3 mm; L = 9.5 mm; h_{sub} = 1.2 mm). The substrate permittivity is ε_{sub} = 3.9. (a) variation of the dielectric spacer (or substrate) thickness h_{sub}; (b) variation of the longitudinal shift dx; (c) variation of the lateral shift dy.

2.2 Monolayer double-face structure

In this part, we present a systematic study of the cut wires structure presented in Fig. 4, derived from the previous structures, under normal-to-plane incidence in the microwave domain. Numerical simulations performed using the FEM based software HFSS are run to show and understand the electromagnetic behavior of the design. A single layer of the metamaterial is characterized by reflection and transmission measurements. The retrieved parameters show simultaneous resonances in the permittivity and permeability responses leading to a negative index of refraction.

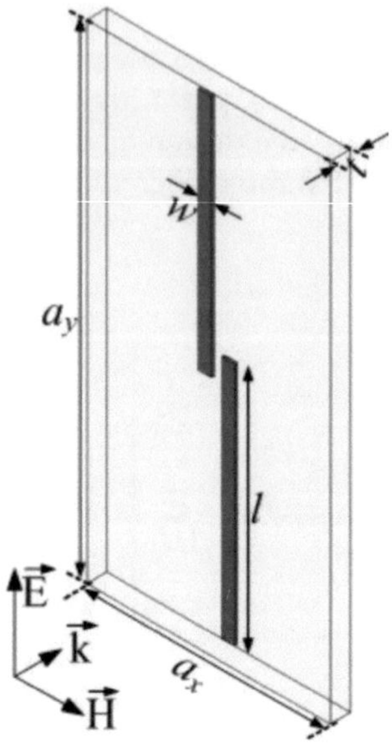

Fig. 4. Unit cell of: cut wires structure under normal-to-plane propagation (ax = 9.5 mm, ay = 19 mm, w = 0.3 mm, l = 9.5 mm). The inserts show the direction and the polarization of the wave.

The cut wires metamaterial illustrated by its unit cell in Fig. 4 is employed to operate in the microwave regime. It consists of a double-face structure composed of periodic cut wires of finite length. The structure is printed on both faces of an epoxy dielectric board of thickness t = 1.2 mm and of relative permittivity $\varepsilon_r = 3.9$. For the different samples reported here, the width of the cut wires denoted by w is 0.3 mm. The length of the cut wires is l = 9.5 mm and the unit cell size in the x and y direction is respectively ax = 9.5 mm and ay = 19 mm. These dimensions have been optimized to operate around 10 GHz and remain the same throughout the section.

The reflection and transmission spectra of the metamaterial are calculated using HFSS by applying the necessary periodic boundary conditions on the unit cell. Several samples of the structure consisting of 10 × 5 cells on a 120 mm × 120 mm epoxy surface are fabricated using conventional commercial chemical etching technique. Measurements are done in an anechoic chamber using an Agilent 8722ES network analyzer and two X-band horn antennas. In the transmission measurements, the plane waves are incident normal to the prototype surface and a calibration to the transmission in free space (the metamaterial sample is removed) between the two horn antennas is done. The reflection measurements are done by placing the emitting and receiving horn antennas on the same side of the prototype and inclined with an angle of about 5° with respect to the normal on the prototype surface. The calibration for the reflection is done using a sheet of copper as reflecting mirror.

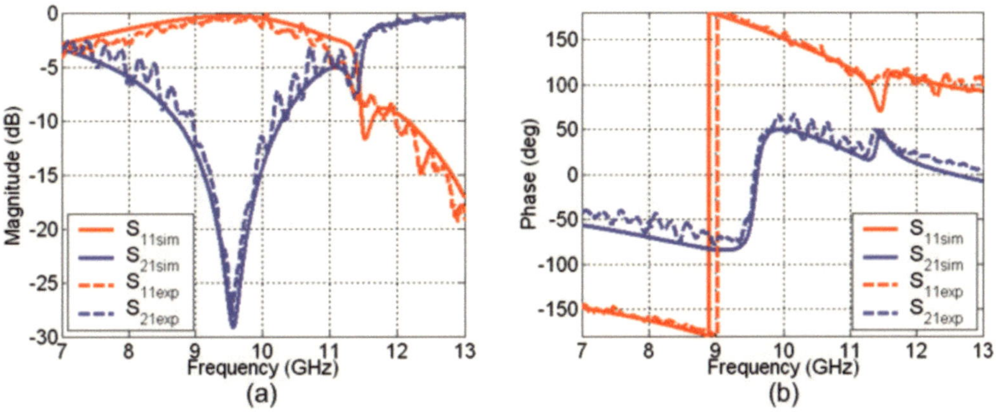

Fig. 5. Calculated and measured reflection (S11) and transmission (S21) responses of the metamaterial for a single layer: (a) magnitude, and (b) phase.

Figure 5 shows the calculated (continuous lines) and measured (dashed lines) S-parameters of the metamaterial for a monolayer configuration. There is a very good qualitative agreement between simulations and measurements. The calculated and measured magnitudes of S21 presented in Fig. 5(a) show clearly two resonance dips, the first one at 9.58 GHz and a second one at 11.39 GHz. We can note in Fig. 5(b) that a change in sign occurs for the transmission phase at the first resonance dip. At the second resonance dip, a peak and a dip is respectively observed in the transmission and reflection phase.

Using the retrieval procedure described in [Nicholson, 1970], based on the inversion of the reflection and transmission coefficients, the effective parameters of the double-face metamaterial structure are extracted. The metamaterial has a period very small compared to the wavelength λ (less than $\lambda/20$) in the propagation direction. The propagation of the electromagnetic wave travelling along this direction is dominated by this deep sub-wavelength period and not by the in-plane period ax or ay. There is only a single propagating mode in the negative-index frequency region, justifying the description of the cut wires metamaterial with an effective index [Valentine, 2008].

The extracted permittivity ε, permeability μ and refractive index n are shown in the various parts of Fig. 6. Two extraction procedures have been performed: the first one uses the calculated S-parameters and the second one is based on the measured S-parameters. As illustrated by the extracted parameters from the calculated and measured S-parameters, the cut wires structure shows firstly an electric resonance at the first resonance dip observed at 9.58 GHz in Fig. 6. This electric resonance exhibits values going negative for the real part of the permittivity in the vicinity of the resonance. Secondly a magnetic resonance with negative values appears at the right hand side of the second resonance dip at 11.5 GHz. Around the same frequency, the real part of the permittivity is still negative. The extracted real part of the refractive index is therefore negative around 11.5 GHz which is the frequency of the LH peak. However, we can also notice that the zero value for the ε response is very close to 13 GHz where a full transmission band is observed in Fig. 6(a). This frequency constitutes the frequency of the RH transmission peak. We can therefore deduce that this RH transmission peak is due to an impedance matching between the structure and vacuum.

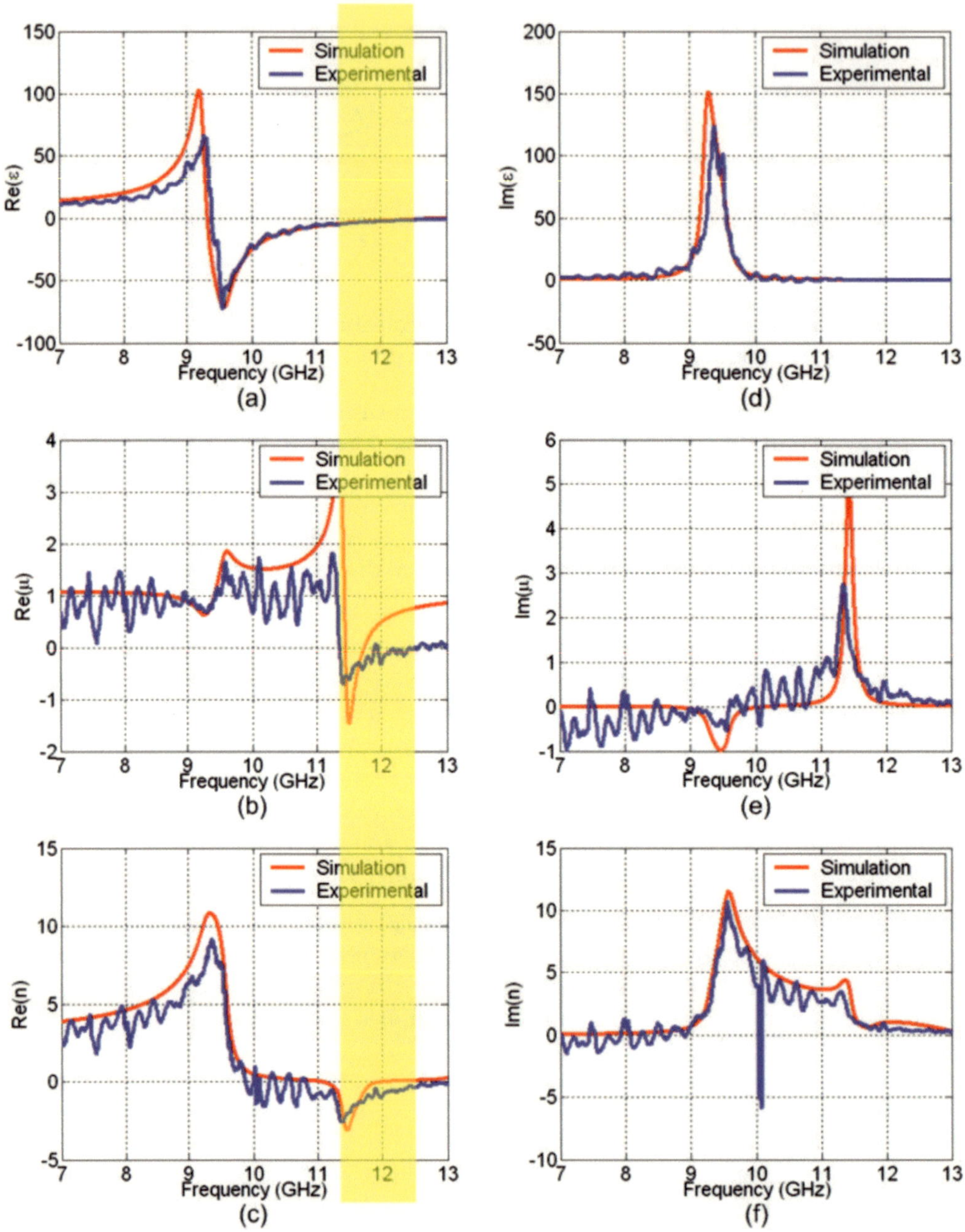
Simulation
Experimental
Re(ε)
Im(ε)
Re(μ)
Im(μ)
Re(n)
Im(n)
Frequency (GHz)
(a)
(b)
(c)
(d)
(e)
(f)

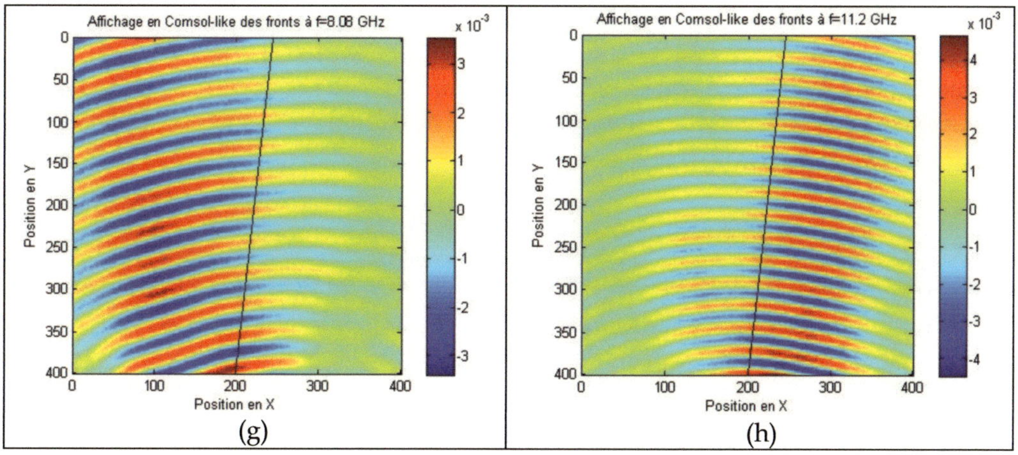

Fig. 6. Extracted electromagnetic properties of the cut wires metamaterial using the simulated and experimental data of Fig. 4: (a)-(c) real parts, and (d)-(f) imaginary parts of the permittivity ε, of the permeability μ and the refraction index n. The shaded yellow area delineates the frequency region where the measured real parts of ε and μ are simultaneously negative. Measurement of the wave propagation trough a prism of 6.3°: (g) E field cartography at 8.08GHz: the optical index is positive, around 5.6 (h) E field cartography at 11.2GHz: the optical index is negative, around -2.3.

Since the real part of n (n′) is given by $n' = \varepsilon' z' - \varepsilon'' z''$ from $n = \varepsilon z$ and $z = \sqrt{(\mu/\varepsilon)}$, the imaginary parts of the permittivity (ε″) and the permeability (μ″) also accounts for n′. Therefore, a negative real part of n can be accomplished without having ε′ and μ′ simultaneously negative. This can happen only if ε″ and μ″ are sufficiently large compared to ε′ and μ′. A wider negative n′ frequency band is observed due to the dispersion of the fabricated prototype. The shaded yellow area in Fig. 6 highlights the frequency region where the measured real parts of the permittivity (ε′) and the permeability (μ′) are simultaneously negative to emphasize the desired measured negative values of n′. Concerning the imaginary parts, a very good qualitative agreement is observed between calculations and experiments. We shall note that the imaginary part of n (n″) is very low in the negative n′ frequency region. Figures 6(g) and 6(h) show the measurements of the near electric field through a prism with an angle of 6.3°. Figure 6(g) shows the E field cartography at 8.08 GHz: the optical index is positive and found to be around 5.6. Figure 6(h) shows the E field cartography at 11.2 GHz: the optical index is negative and calculated to be around -2.3. These values agree very well with the extracted values of the optical index calculated from the measurements of the reflexion and transmission coefficients of the figure 6(c) using the retrieval procedure described in [Nicholson, 1970].

2.3 Stacking of layers

Stacking multiple layers of LH materials may be useful in many practical applications such as subwavelength imaging [Wang, 2007, Ziolkowski, 2003] and directive antennas [Burokur, 2005, Shelby, 2001]. It is obvious that the effective properties obtained from the inversion method on a monolayer give a good idea about the effective properties of the metamaterial. However, other effect such as inter-layer coupling must be taken into account because it affects the material properties of the structure. Therefore, two, three and four layers of the designed bi-layered metamaterial are stacked with a 1 mm air spacing between each layer as presented in Fig. 7(a). Numerical simulations are run to show the expected performances of a bulk metamaterial composed of multiple layers. The transmission spectra for the different number of layers are presented in Fig. 7(b).

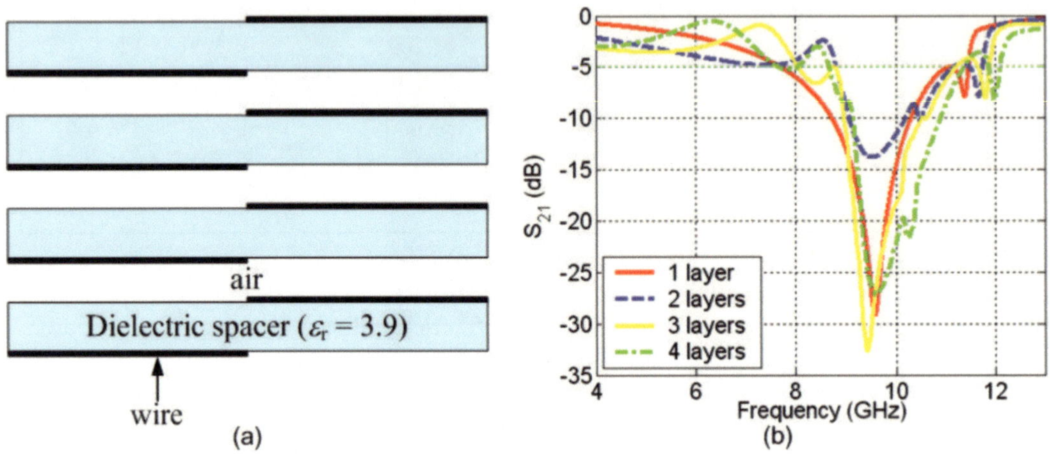

Fig. 7. (a) A bulk metamaterial composed of four layers interleaved with 1 mm air spacing, and (b) transmission spectra for different number of layers.

From the spectra of Fig 7(b), we can note that the frequency of the first transmission dip remains constant with an increasing number of layers while the second dip shifts slightly towards higher frequencies. However, peaks and valleys appear at lower frequencies suggesting a coupling mechanism between consecutive layers. The number of these peaks and valleys increases with an increasing number of layers as shown in Fig. 7(b). The transmission spectra together with the corresponding reflection spectra are used for the extraction of the material properties presented in Fig. 8. It should be noted that the first transmission dip in the single layer case corresponds to an electric resonance where ε' exhibits negative values. However, other $\varepsilon' < 0$ frequency bands can be observed in Fig. 8(a) for the multiple layers cases due to the valleys noted in the transmission spectra. And, since the magnitude of the transmission dips decreases with the number of layers, the magnitude of ε' also decreases as shown in Fig. 8(a). At higher frequencies near 12 GHz, a magnetic resonance is also observed for multiple layers as for the single layer case. However the magnitude tends to decrease while the number of layers increases (Fig. 8(b)). For more than two layers, μ' exhibits only positive values at the resonance near 12 GHz. Besides, another magnetic resonance with $\mu' < 0$ can be observed at lower frequencies with simultaneously $\varepsilon' < 0$ when more than one layer is used. So even if the $\mu' < 0$ frequency band disappears at the second transmission dip due to the $\mu' > 0$, a negative index band is observed at lower frequencies as shown in Fig. 8(c). This negative refractive index results from the coupling mechanism created when several layers of the double-face structure are stacked. The negative index frequency band widens when the number of layers increases.

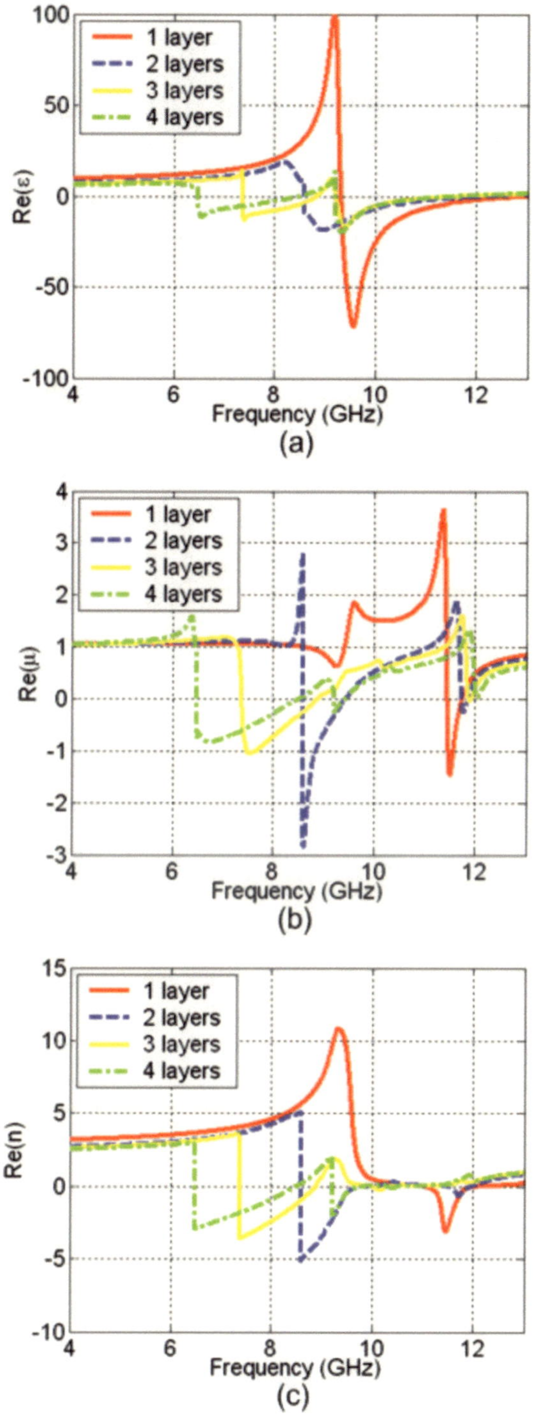

Fig. 8. Extracted material properties for different number of layers. (a) Re(ε), (b) Re(μ), and (c) Re(n).

2.4 Incidence dependence of the negative index

Finally, we investigate numerically and experimentally the reflection and transmission spectra for an obliquely incident plane wave on the asymmetric structure. Three different angles, namely 15°, 30° and 45° in both H- and E-planes of the square lattice are studied in simulations and experiments. Measured reflection and transmission coefficients are compared to simulated ones in Fig. 9 for the H-plane. There is qualitative agreement between simulations and measurements. Calculated and measured magnitudes of S_{21} show clearly two resonance dips, an electric one at 9.5 GHz and a magnetic one at 11.5 GHz. These two resonances are found to be independent of the incidence angle in the H-plane as shown in Figs. 9(a)-(c).

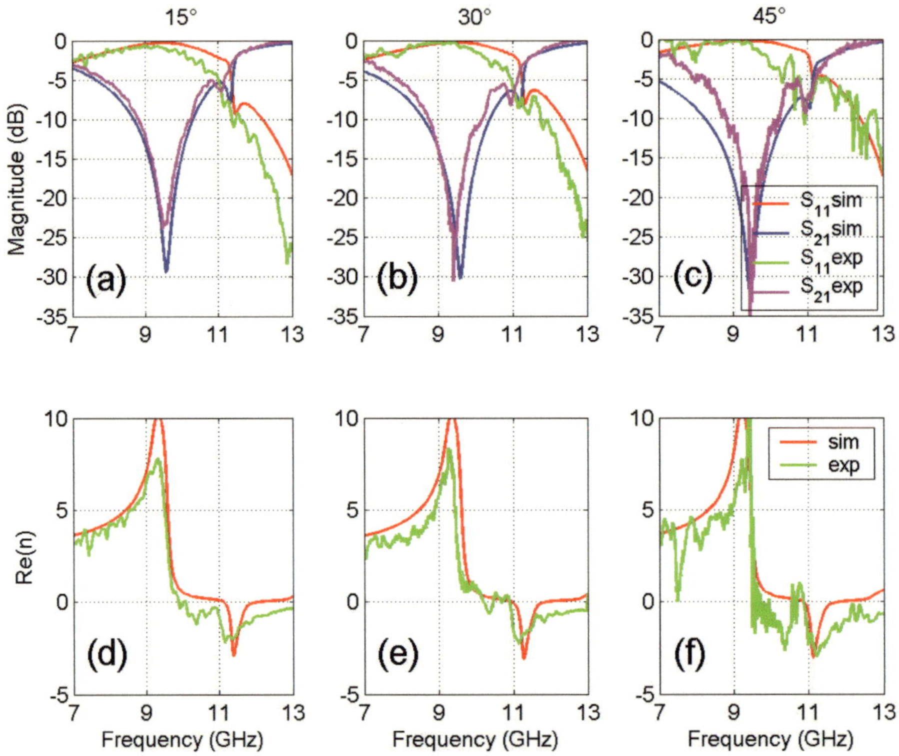

Fig. 9. Oblique incidence (15°, 30° and 45°) in H-plane. (a)-(c) Computed and measured reflection and transmission coefficients. (d)-(f) Real part of effective index n.

To retrieve effective parameters at oblique incidence, the retrieval procedure in [Smith, 2002] has to be modified and anisotropy has to be addressed [Burokur, 2009b]. Indeed cut wire pairs represent a biaxial anisotropic media whose principal axis are along x, y and z. Consequently 2x2 transfer matrices used in normal incidence are no longer sufficient and the full 4x4 transfer matrix accounting for coupling of s- and p-waves should be considered [Yeh, 1998]. However, since the electric field in our case is always along y independently of θ, cross-polarization terms do not arise and therefore we only use one 2x2 matrix for s-waves. In this case the effective index is given by

$$\frac{\varepsilon_y \mu_x \mu_z}{\mu_z \cos^2\theta + \mu_x \sin^2\theta} \tag{1}$$

Figures 9(d)-(f) show that n responses remain mostly unchanged compared to the 0° case studied in [Burokur, 2009, Sellier, 2009].

Measured reflection and transmission coefficients are compared to simulated ones in Fig. 10 for the E-plane. Apart from one sharp feature on each spectrum, simulations and measurements agree qualitatively. This peak which is much sharper than either the anti-symmetric or the symmetric resonance, shifts with incidence angles from 12.47 GHz at 15° to 10.5 GHz at 30° and 9.22 GHz at 45°. It is the manifestation on the specular order (0,0) of a diffraction threshold, namely that of the (0,-1) diffracted order. At these frequencies the (0,-1) diffracted order transits from evanescent to propagating, appearing at grazing incidence. Diffraction thresholds frequencies are calculated in two ways, with HFSS and using (2) from grating theory taking into account the fact that there is no wave vector component along x in E-plane.

$$k_\perp = \pm\sqrt{\frac{\omega^2}{c^2}\varepsilon_r - k_{//}^2} = \pm\sqrt{\frac{\omega^2}{c^2}\varepsilon_r - \left(m\frac{2\pi}{a_x}\right)^2 - \left(\frac{\omega}{c}\sin\theta + n\frac{2\pi}{a_y}\right)^2} \quad m,n\in Z \tag{2}$$

Due to respective values of ax = 9.5 mm and ay = 19 mm, it can be seen that the (0,-1) order is the first diffracted order to become propagating. Results for diffraction thresholds frequencies are summarized in Table I. By comparison, in the H-plane even at an angle of 45° the (0,-1) diffracted order is above 18 GHz. To overcome the appearance of a diffraction threshold, a triangular lattice as shown in the inset of Fig. 10(d) is proposed to replace the square one used in the figure 10(a-c). In the triangular lattice, every other cell along x is laterally displaced by 4.75 mm along y. In this lattice, diffraction threshold frequencies are given by (3):

$$k_\perp = \pm\sqrt{\frac{\omega^2}{c^2}\varepsilon_r - k_{//}^2} = \pm\sqrt{\frac{\omega^2}{c^2}\varepsilon_r - \left((2m-n)\frac{2\pi}{a_x}\right)^2 - \left(\frac{\omega}{c}\sin\theta + n\frac{2\pi}{a_y}\right)^2} \quad m,n\in Z \tag{3}$$

Computed reflection and transmission spectra presented in Figs. 10(d)-(f) confirm the rejection of the diffraction threshold above 13 GHz. Besides, a shift in frequency can be noted for both anti-symmetric and symmetric resonances. This shift is seen to be much stronger than for any resonances in the H-plane and the detuning of both electric and magnetic resonances with respect to the incidence angle θ leads to a loss of the frequency overlap, hence the negative index, above 20°. As it can be noted particularly for θ = 15°, the magnetic resonance shifts towards lower frequencies for the triangular lattice (10.43 GHz) compared to the square one (11.02 GHz). This is most probably due to extra capacitive coupling between y-displaced wires on same face of the dielectric board.

In summary, we presented the dependence of resonances and retrieved effective index on the incident angle in recently proposed asymmetric cut wire pairs. No change has been observed for oblique incidence in the H-plane. However for the E-plane, a diffraction threshold appears for the square lattice rendering the introduction of effective parameters meaningless. A triangular lattice has therefore been proposed to avoid having the diffraction threshold below the electric and magnetic resonances. A detuning of both resonances has been observed leading to a lack of resonance frequencies overlap above 20°.

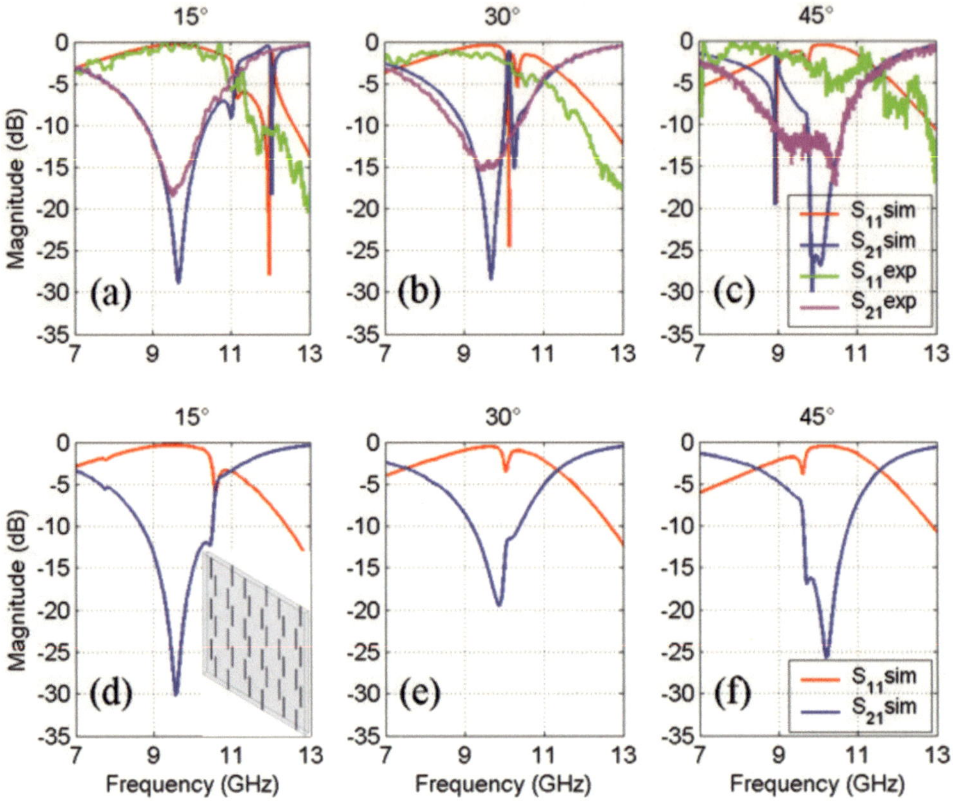

Fig. 10. Oblique incidence (15°, 30° and 45°) in E-plane. (a)-(c) Computed and measured reflection and transmission coefficients for the square lattice. (d)-(f) Computed and measured reflection and transmission coefficients for the triangular lattice.

	HFSS (square lattice) Mode (0,-1)	Grating (square lattice) Mode (0,-1)	HFSS (triangular lattice) Mode (0,-1)	Grating (triangular lattice) Mode (0,-1)
$\theta = 0°$	15.78	15.79	22.31	22.33
$\theta = 15°$	12.58	12.54	19.15	19.15
$\theta = 30°$	10.54	10.53	17.31	17.32
$\theta = 45°$	9.24	9.25	16.35	16.35

Table 1. Computed numerical and analytical diffraction threshold frequencies (GHz)

3. Infrared metamaterials and plasmons hybridization

3.1 Engineering resonances in infrared metamaterials

Recently, a theoretical study showed that optical resonances in both SRR arrays and cut wire arrays could be interpreted in terms of plasmon resonances [Rockstuhl, 2006]. A gradual shift in the SRR resonance frequencies was predicted when reducing the length of SRR legs to the point where each SRR was transformed into a single wire piece. The magnetic and electric properties of these modified versions of SRRs were theoretically investigated in [Zhou, 2007]. Following these theoretical studies, we present here an experimental and numerical analysis of the infrared response of metamaterials made of continuous nanowires and split ring resonators where the geometry is gradually altered. The metamaterial structure is fabricated on low-doped silicon. The impact of the geometric transformation of the SRRs on the spectra of the composite metamaterial is measured in the 20-400 THz frequency range (i.e., in the 1.5 - 15 µm wavelength range) for the two field polarizations under normal to plane propagation. Coupling effects between the SRRs and the continuous nanowires are analyzed for different spacings between them. The results of our study are expected to provide useful guidelines for the design and engineering of negative index metamaterials on silicon.

3.1.1 Design, fabrication, characterization and modeling of metamaterial structures

Four structures consisting of a two-dimensional periodic array of gold nanowires and gold SRRs were fabricated on a 280µm thick silicon substrate (Fig. 11). The fabrication steps included e-beam lithography, high vacuum electron beam evaporation of 5 nm thick titanium and 40 nm thick gold films, and a lift-off process. It is worthwhile noticing that all structures were fabricated in the same run, thereby allowing a meaningful comparison of their optical characteristics. As seen in Fig. 11, the four structures only differ in the shape of SRRs, which are gradually transformed into simple cut wires from structure 1 to structure 4. In the intermediate cases of structures 2 and 3, SRRs appear to be U-shaped with smaller legs than in the standard case of structure 1. Except for this resonator shape, all the other geometrical parameters of the four structures are identical. In each case, the lattice period is ~ 600nm, the width of all wires (continuous and discontinuous) is ~ 50nm, and the continuous wires are parallel to the SRR bases with a separation of 115 ± 20 nm between each continuous wire and the closest SRR base. The SRR gap width in structure 1 is ~100 nm while the length of the two SRR legs is ~ 280nm. This length is reduced to ~190 and 110 nm in structures 2 and 3, respectively. The scanning electron microscope (SEM) images reported in Fig. 11 (middle row) illustrate the excellent regularity of the four fabricated structures.

The transmission and reflection spectra of the fabricated structures were measured under normal-to-plane incidence with a FTIR (Fourier Transformed InfraRed spectrometer) BioRad FTS 60 equipped with a Cassegrain microscope. The FTIR beam was polarized using a KRS5 polarizer adapted to the wavelength region from ~ 1.5 to 15µm (i.e. to frequencies varying from 20 to 200 THz). A diaphragm was used in such a way as to produce a light spot smaller than 100 × 100 µm2 onto the sample (i.e. smaller than the surface of each periodic structure). Measurements were performed for two field polarizations of the incident beam, the parallel polarization with the illuminating electric field parallel to both the continuous wires and the SRR gaps and the perpendicular polarization with the electric field perpendicular to the continuous wires and SRR gaps. The measured transmission spectra were normalized versus the transmission of an unprocessed part of the silicon substrate. The measured reflection spectra were normalized versus the reflection of a 40 nm thick gold film deposited on silicon.

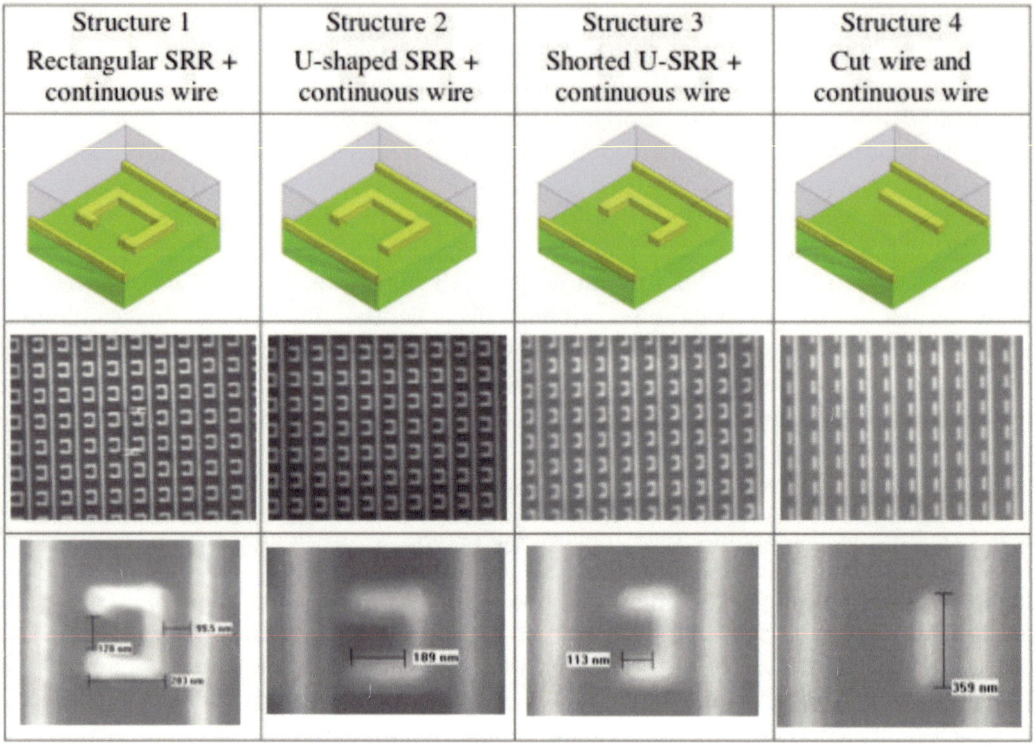

Fig. 11. Schematic representations (top) and scanning electron microscope images (middle and bottom) of the four metamaterial structures fabricated on silicon. The top and bottom pictures show the elementary unit cell of each structure. From left to right, the length of SRR legs is reduced to the point where the SRR is transformed into a single wire piece. The middle pictures show the regularity achieved in the fabrication of the periodic arrays of nanowires and split ring resonators

Numerical simulations of the spectral responses of the four structures were performed with a finite element software (HFSS, 2006). Periodic boundary conditions were applied to the lateral sides of the elementary lattice cell (Fig. 11). The silicon substrate was assumed to be lossless with a constant permittivity equal to 11.9. A Drude model was used to simulate the permittivity and loss tangent of the gold wires:

$$\varepsilon(\omega) = 1 - \frac{\omega_p^2}{\omega(\omega + i\omega_c)} \tag{4}$$

where ω_p and ω_c are the plasma and collision frequency of the gold film, respectively. The values of ω_p and ω_c chosen in the simulations were: $\omega_p = 1.367 \times 10^{16} s^{-1}$ (f_p =2176 THz) and ω_c

= 6.478 x 10^{13} s^{-1} (f_c =10.3 THz). Actually, the collision frequency can be considered to a certain extent as a fit parameter. An increase of the collision frequency results in higher absorption losses of the structures, while it does not change the spectral positions of resonances. The value reported above for ω_c is 2.6 times larger than in bulk gold. This increase is supposed to account for additional scattering experienced by electrons at the metal surfaces.

3.1.2 Results of measurements and simulations

The results of our measurements and simulations are shown in Fig. 12 for the four structures. Results for the parallel polarization are gathered in the series of figures from (a) to (d). Those for the perpendicular polarization are gathered in the series of figures from (e) to (h). In all these figures, resonances manifest themselves as reflection maxima correlated with transmission minima. A slow decrease (resp. increase) of the transmission level (resp. reflection level) is also observed at low frequencies for the parallel polarization. This latter evolution can be readily attributed to the plasmon-like band associated to the periodic array of continuous wires [Pendry, 1998].

For the parallel polarization and the structures with the SRRs of larger sizes (structures 1 and 2), two resonances are observed within the spectral window of measurements [Figs. 12(a) and 12(b)]. Only the first resonance is experimentally observed for structures 3 and 4. The SRR resonances actually shift towards higher and higher frequencies as the whole SRR length is decreased. In the same time, their amplitude becomes smaller and smaller. For the first resonance, the maximum reflection R_{max} decreases from ~0.63 to ~0.5 while the minimum reflection R_{min} remains close to ~0.35 [Fig. 11(b)]. Accordingly, the minimum transmission T_{min} increases from ~0.12 to ~0.25 while the maximum transmission T_{max} remains close to ~0.6 [Fig. 11(a)]. It is worthwhile noticing that the values of R and T out of resonance correspond to those expected for a single face of silicon wafer partially covered with ~10-15% of highly reflecting metal. Whereas (R+T) approaches unity in this case (R_{min}+T_{max}~0.95), its smaller value at resonance (R_{max}+T_{min}~0.75) clearly indicates the presence of dissipative losses in metallic elements. The frequency positions of resonances, the values of R and T out of resonance, the values of (R+T) in general as well as the shapes of experimental curves in Figs. 12(a) and 12(b) are very well reproduced by numerical simulations [Figs. 12(c) and 12(d)]. The only discrepancy between experiments and simulations stems from the smaller amplitudes of resonances measured in experiments, especially those at high frequencies. The second resonance predicted at ~180 THz (λ ~ 1.7µm) for structure 3 is even not resolved in the experiments. Actually, minute deviations of the geometry from unit cell to unit cell and particularly residual surface roughness of the SRRs and continuous wires can explain the damping and inhomogeneous broadening of resonances as well as the increasing importance of these effects at high frequencies.

For the perpendicular polarization and for structures 1, 2 and 3, a single resonance is observed within the spectral range of measurements [Figs. 12(e) and 12(f)]. In contrast, no resonance is detected for the structure with cut wires. The evolution of the SRR resonance with the SRR length is actually similar to that observed for the parallel polarization. Smaller SRR lengths simultaneously lead to higher resonance frequencies and smaller resonance amplitudes. A good agreement is found between experimental results and numerical simulations [Figs. 12(g) and 12(h)]. Previous remarks made for the parallel polarization apply to the perpendicular polarization. Minute deviations from unit cell to unit cell and surface roughness of metallic elements are likely to explain the broader resonances with smaller amplitudes observed in the experiments. The second resonances predicted for structures 1, 2 and 3 in the frequency region from 170 to 200 THz [Figs. 12(g) and 12(h) manifest themselves only as smooth maxima (resp. minima) in the measured reflection (resp. transmission) spectra of Fig. 12(f) (resp. Fig. 12(e)]. Supplementary measurements between 200 and 250 THz (not shown here) did not reveal any other resonance.

Figure 13 shows the distribution of the electric field calculated at the bottom surface of metallic elements for each of the resonant modes observed in Fig. 12. These results are in agreement with previous calculations reported by Rockstuhl et al. [Rockstuhl, 2006]. When the incident field is polarized parallel to the SRR gap, resonant modes possess an odd number of nodes along the entire SRR. This number is equal to one for the first resonance, while it is equal to three for the second resonance. The first resonance also identifies to the so-called LC resonance as defined in previous studies at microwave and far infrared frequencies [Katsarakis, 2004, Katsarakis, 2005]. It simply identifies to the dipolar mode in the case of the cut wire. When the incident field is polarized perpendicular to the gap, resonant plasmon modes possess an even number of nodes. The first resonant mode in this polarization thus exhibits one additional node as compared to the first resonance in the parallel polarization. This in turn requires higher energies of the light field to excite this mode. The frequency of the first resonant mode in the perpendicular polarization is typically two times higher than that of the first resonant mode in the parallel polarization.

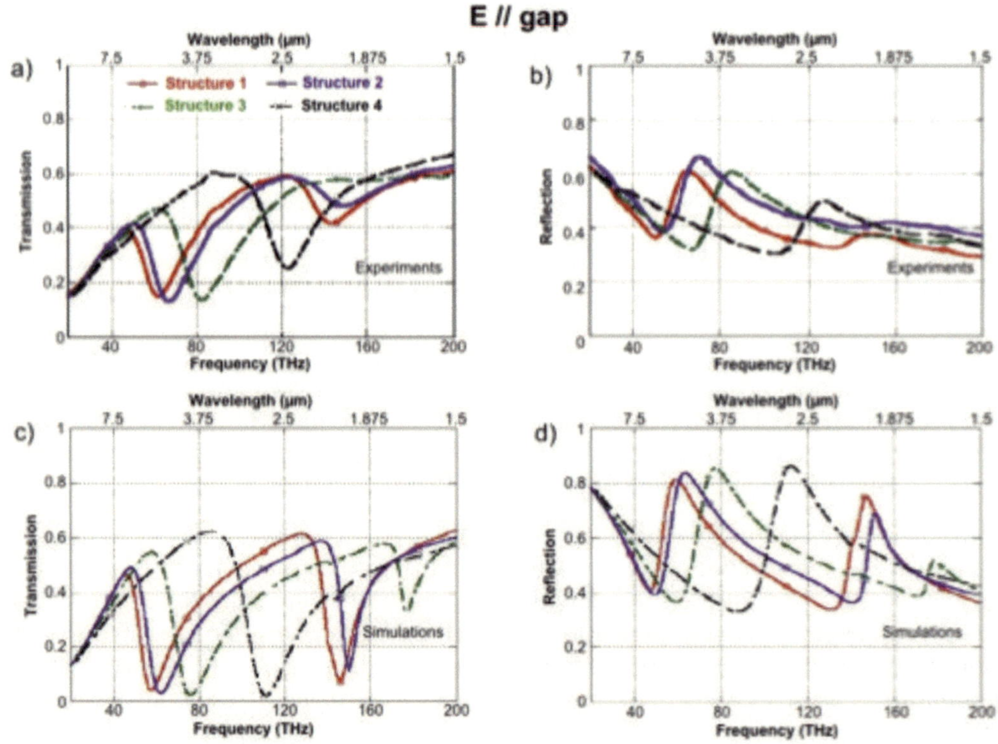

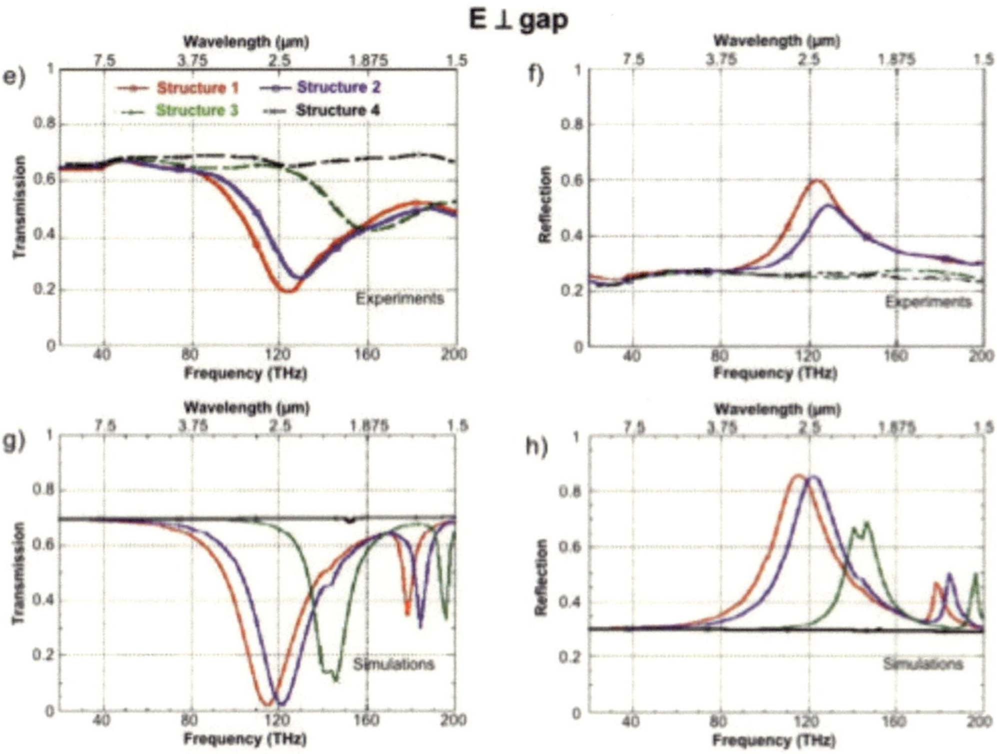

Fig. 12. Measured and simulated transmission/reflection spectra of the four structures depicted in Fig. 10. Curves in red, blue, green and black are for the 1st, 2nd, 3rd and 4th structures, respectively. - (a) and (b): transmission and reflection spectra measured for the parallel polarization (incident electric field parallel to the SRR gap). - (c) and (d): numerical simulations corresponding to (a) and (b), respectively. - (e) and (f): transmission and reflection spectra measured for the perpendicular polarization (incident electric field perpendicular to the SRR gap). - (g) and (h): numerical simulations corresponding to (e) and (f), respectively.

3.1.3 Frequency positions of resonances

At this stage, it is interesting to compare the frequency positions of resonances reported in Figs. 12 to those reported in previous works for similar structures with gold SRRs on glass substrate. For instance, the first resonance calculated in [Rockstuhl, 2006] for the parallel polarization and U-shaped SRRs with 400 nm long base and 190 nm long legs was found to be close to 3800 cm^{-1}, i.e. close to 115 THz instead of 65 THz measured in our experiments for U-shaped SRRs with similar sizes [Figs. 12(a), 12(b), and Fig. 13(b)]. The first resonance calculated in [15] for the parallel polarization and U-shaped SRRs with the same base but with 110 nm legs was found to be near 4800 cm^{-1}, i.e. near 144 THz instead of 80 THz measured in our experiments [Figs. 12(a), 12(b) and Fig. 13(c)]. Actually, all the mode frequencies calculated in [15] are 1.7- 1.9 times higher than those reported in this work, whatever the resonance order and the field polarization are. Approximately the same ratio is obtained when comparing the first resonance measured in [10] for SRRs of standard shape

(~100 THz) to that reported in Fig. 12(a) (~60 THz) for standard SRRs with the same total length (lm ~960 nm). This ratio is actually of the same order of magnitude than the ratio between the refractive index of glass and that of silicon. This suggests that in our case the electromagnetic field at resonance largely extends into the silicon substrate.

3.1.4 Coupling effects between continuous wires and SRRs

In the previous sections, it has been implicitly assumed that the presence of continuous wires had no influence on the resonant response of the structures except for a slow decrease of the transmission observed at low frequencies for the parallel polarization. However, a careful examination of the field distributions in Fig. 13 indicates that, at least for the second resonance, the electromagnetic field extends well in the region comprised between the SRR and the closest wire. We thus performed a numerical analysis to investigate in more detail the possible existence of coupling effects between the SRRs and continuous wires. For this purpose, the spectral responses of the four structures were calculated for different distances d between the SRR base and the closest wire. They were also compared to the spectral response of a periodic array of SRRs only.

Results of our calculations are shown in Figs. 14(a) and 14(b) for structure 1 with three values of d, for a periodic array of SRRs without continuous wires and for a periodic array of wires without SRRs. The dimensions of SRRs are the same for the first four structures. The lattice period is identical for all the structures. Calculations are performed within the same spectral range as in Fig. 12, and the incident electric field is polarized parallel to the continuous wires and/or to the SRR gaps. Figure 14(a) represents the transmission spectra calculated for the different structures. Figure 14(b) shows the electric field distributions calculated for the different resonances observed in Fig. 14(a). As seen in Fig. 14(a), the position of the first SRR resonance is not modified by the presence of the continuous wires whatever the separation between SRRs and wires is. Only the shape of the resonance is modified, and it becomes asymmetric with the presence of the wires.

This asymmetry mainly results from the fact that the optical response of the wires [pink dashed curve in Fig. 14(a)] adds to that of the SRRs. A weak coupling between SRRs and wires only occurs at the smallest separations between the two metallic elements as shown from the calculated distribution of the electric field at the first SRR resonance (Fig. 14(b), second column).

The evolution of the second SRR resonance is quite different. For a sufficiently large separation between the SRRs and wires ($d \geq 100$ nm), the frequency position of this second resonance is still rather independent of the presence of the wires. This justifies our previous interpretations concerning the results of Fig. 12, where the different spectra were obtained for $d \approx 130$ nm. However, for small separations between the SRRs and wires, strong coupling effects exist, which lead to a splitting of the second SRR resonance into two components [Fig. 14(a)]. These components are well separated for the smallest value of d (black dashed curve in Fig. 14(a), $d = 10$ nm). As seen in Fig. 14(b) (second and third columns), the modal field of the low frequency component is found to be concentrated in the region between the SRR base and the closest wire. That of the high frequency component is rather concentrated in the SRR legs.

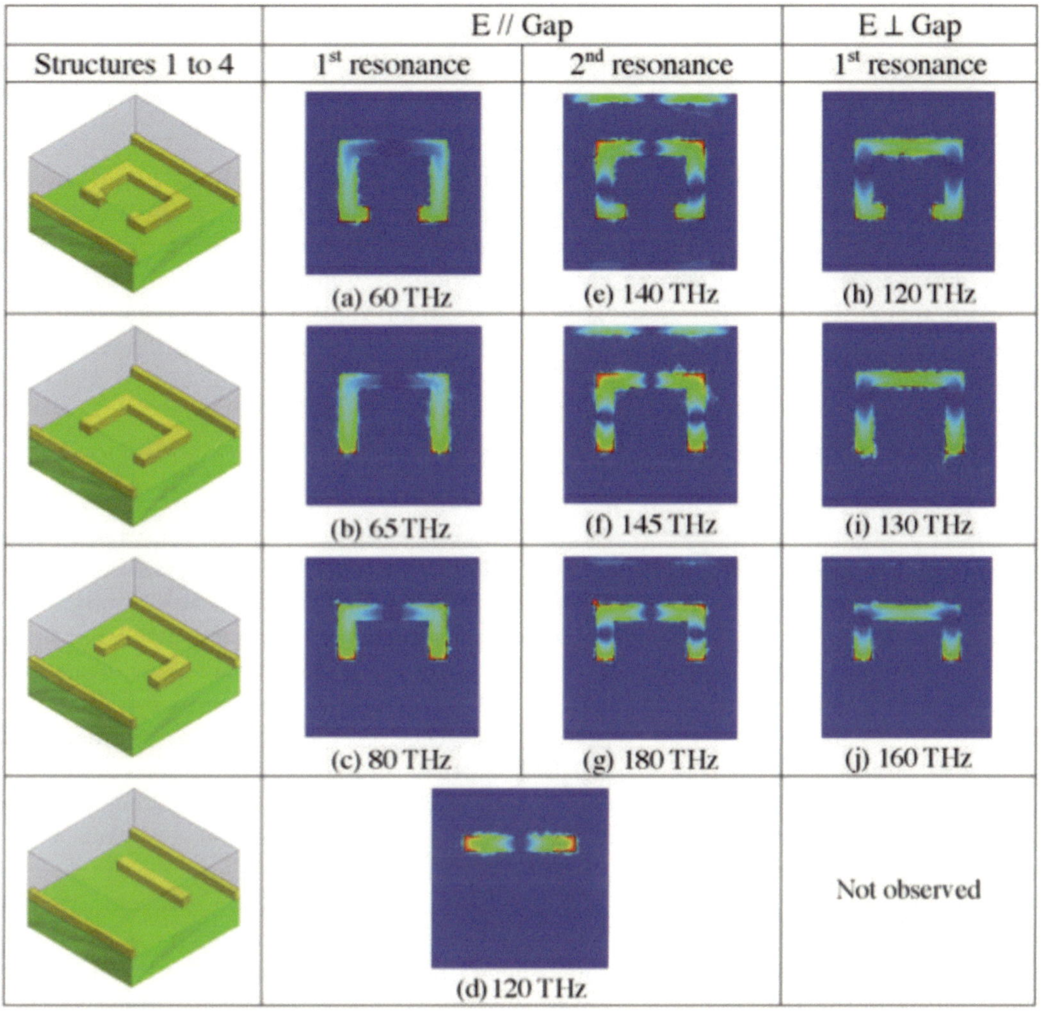

Fig. 13. Magnitude of the normal electric field component (|Ez|) calculated at the bottom surface of the metallic elements for each of the resonant plasmon modes observed in the different spectra of Fig. 11. The different colours, from blue to red correspond to increasing magnitudes of the field component. Modes are classified according to the resonance order and to the polarization of the incident electric field. As expected, both the energy and the number of field nodes increase with the resonance order.

Calculations above were repeated for structures with U-shaped SRRs as structures 2 and 3 and for structures with cut wires as structure 4 (Fig. 11). The same behavior was found in the case of structures with U-shaped SRRs. The frequency position of the first resonance was not modified when the distance between SRRs and continuous wires was varied. The second resonance split into two components for small values of d. In contrast, the structures with cut wires did not exhibit the same "robustness" of the first resonance against coupling effects.

The resonant mode split into two distinct components when the separation between cut wires and continuous wires was smaller than 50 nm. This different behavior can be simply explained by the fact that unlike structures with true SRRs, the field of the first resonant mode is obviously concentrated in the close neighbourhood of the continuous wires, i.e. in the cut wires themselves.

Coupling effects were also investigated from numerical simulations for the perpendicular polarization. We only considered the first SRR resonance since it was the only one clearly observed in the experiments (Fig. 12). The frequency position of this resonance was found to be rather independent of the presence of continuous wires whatever the separation between the SRRs and wires was. However, a small splitting of the resonance was observed for very small separations (d = 10 nm) in the case of U-shaped SRRs with small legs (structure 3).

3.1.5 Metamaterials with negative refraction on silicon

It is now well established that the use of an array of continuous metallic wires allows obtaining a negative permittivity over the whole plasmon-like band when the electric field is polarized parallel to the wires [Pendry, 1998]. Figures 12 and 14(a) presently show that this band can extend well up to near-infrared frequencies for a sufficiently small period of the wire lattice. On the other hand, it has been demonstrated that an array of metallic SRRs can exhibit a magnetic response in the optical domain with a negative permeability at certain frequencies [Enkrich, 2005]. However, this situation only occurs at SRR resonances and for an oblique or grazing incidence, i.e. for an incident magnetic field with a non-zero component along the SRR axis. An additional condition is that the resonant plasmon mode must possess an odd number of field nodes along the SRR [Shalaev, 2005]. Concerning this latter aspect, our experimental results confirm that the first resonant mode, the so-called LC resonance, is by far the most exploitable due both to its strength and to its robustness against parasitic coupling effects. They also show that its frequency position can be finely tuned by adjusting the total length of SRRs and using for instance U-shaped SRRs. One solution to achieve a magnetic response at normal incidence with respect to the sample plane consists in using a stack of SRR layers [Liu, 2008] or simpler, a stack of cut-wires as originally proposed in [Shalaev, 2005]. Coupling between adjacent SRRs or between adjacent cut-wires leads to the formation of hybridized plasmon modes of opposite symmetry. Anti-symmetric plasmon modes can exhibit a magnetic response, and lead to a negative permeability in certain frequency regions. Our experimental results in Fig. 12 show that the resonance associated to the dipolar mode of cut-wires is well pronounced for the fabricated structures. Coupling between two such modes in a multilayer stack should thus allow obtaining a magnetic response at normal incidence. One advantage in using stacked cut-wires instead of stacked SRRs stems from the possibility of achieving more easily a magnetic response at (high) near- infrared frequencies. This is all the more true when metallic nanostructures are fabricated on a high permittivity substrate such as silicon. All the plasmon resonances are shifted to low frequencies, and the realization of very-small-size SRRs operating at telecommunication wavelengths on silicon would require pushing the lithographic techniques to their present limits.

3.2 Optical asymmetric cut-wire pairs

3.2.1 Negative refractive index in optical asymmetric cut-wire pairs

In the previous structures, SRRs were associated with continuous wires to obtain negative index at infrared wavelengths. In this chapter we will study asymmetric cut-wire pairs to obtain also negative index. Metallic nanostructures can be regarded as elementary circuits including nano-capacitors, inductors or resistors [Engheta, 2007]. The simplest resonator that can be imagined is a dipole consisting of a simple metallic cut-wire. Coupling two such oscillators lead to two eigenmodes with opposite symmetry. The virtual current loop of the anti- symmetric mode is now recognized as a mean to create artificial magnetism at optical frequencies [Grigorenko, 2005, Shalaev, 2005]. Fig. 15 shows the structure under consideration. It consists of a periodic array of paired cut-wires separated by a dielectric layer. For simplicity the spacer has been taken to be silicon dioxide (SiO2) with a dielectric permittivity $\varepsilon_r = 2.25$ and a thickness of 100 nm. The surrounding medium is air with $\varepsilon_r = 1$. The relevant polarization of the impinging light is given in Fig. 15(a) with the electric field parallel to the longest side of the cut- wires. Transmission spectra are presented in Fig. 15(b). All the simulations are done using a commercial finite element code (HFSS, 2006). An array of isolated cut-wires is actually found when the separation distance between cut-wires is large: only one resonance (the dipolar mode) is observed in this case (red curve in Fig. 15(b)). This response can be interpreted in terms of a localized plasmon resonance [Rockstuhl, 2006, Kante, 2008]. When the separation distance between cut-wire pairs is progressively diminished, the response of the paired system is modified due to the interaction between its elementary constituents.

Following the plasmon hybridization concept [Kreibig, 1981, Prodan, 2003], coupling effects lift the degeneracy of the single cut-wire mode, thus leading to two distinct plasmons modes [Liu, 2007, Kante 2009], the anti-symmetric and symmetric modes as shown in Fig. 15(b) (blue curve). The symmetric mode with larger restoring force is at a higher energy than the anti-symmetric mode. The main idea of our work is to invert this process. The symmetric mode, which easily couples to incident light, corresponds to a wide rejection band with negative permittivity. The anti-symmetric mode, which is difficult to excite due to the opposite contributions of the two dipoles, manifests itself as a small transmission dip with negative permeability. Achieving such a resonance at a higher frequency than the symmetric mode will ease the overlap condition as proposed recently in [Kante, 2009, Sellier, 2009]. For this purpose, one solution is to break the symmetry of the cut-wire pair. This is achieved in Fig. 15(d) by displacing one of two cut-wires in the direction of the electric field.

Fig. 16(a) and (b) shows the evolutions of the resonant modes and transmission spectra with the longitudinal shift dx between the two cut-wires of each pair. Simulations are performed in the infrared domain. As a major result, for a sufficiently large displacement dx (here dx > 350 nm), the hybridization scheme is inverted with the symmetric mode at a lower frequency than the anti-symmetric mode. The two hybridization schemes are presented in Fig. 15(c) and (e), respectively. Fig. 16(d) clearly shows that a negative refraction regime is obtained (between 145 and 160 THz for dx = 600 nm) for the inverted scheme while the index of refraction remains always positive (dx = 0 nm) for the normal hybridization case (Fig. 16(c)). This extends to infrared frequencies previous results reported by the authors at microwave frequencies [Kante, 2009, Sellier, 2009]. As seen in Fig. 16(d), negative refraction corresponds to an overlap between the region with negative permittivity and that with negative permeability. It is worthwhile noticing that the domain of overlap with negative permittivity and permeability can be controlled through the different degrees of freedom of the structure [Kante, 2009].

3.2.2 Hybridization of the localized plasmons of SRRs

The recipe proposed above for obtaining a negative index with plasmon hybridization can be applied to any structure supporting localized plasmons. The inversion process resulted from a radical change in near field Coulomb interactions between cut-wires in each pair [Kante, 2009, Christ, 2008]. Let us consider for instance a periodic array of paired SRRs, which are fundamental building blocks in the design of metamaterials. Indeed, the use of SRRs has allowed the achievement of negative magnetic permeability, which is impossible with natural materials at high frequencies. However, obtaining a negative magnetic permeability requires the incident wave to possess a magnetic component along the SRR axis. Such a requirement is not easily fulfilled in optics [Dolling, 2006, Liu, 2009]. Moreover, the electromagnetic response of SRRs has been shown to saturate at optical frequencies [O'Brien, 2002]. In most experimental works reported so far in the infrared domain, normal incidence has been used instead of a grazing or oblique incidence [Kante, 2008]. In this situation, only the electric field can couple to the structure. SRR resonances are nothing but plasmonic resonances which can be classified into even and odd modes depending on the polarization the exciting light with respect to the structure [20]. While even modes are excited for an incident electric field perpendicular to the SRR gap, odd modes are excited for an electric field parallel to this gap. It is straightforward to see that only odd modes can lead to a magnetic moment or eventually to a negative permeability under oblique incidence.

(a)

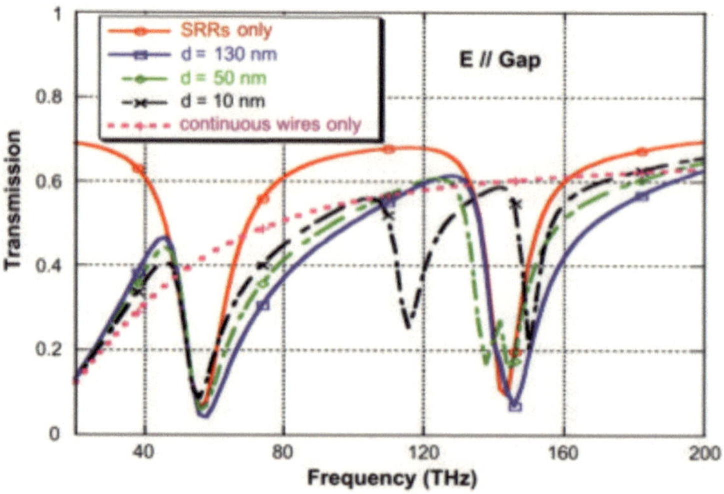

(b)

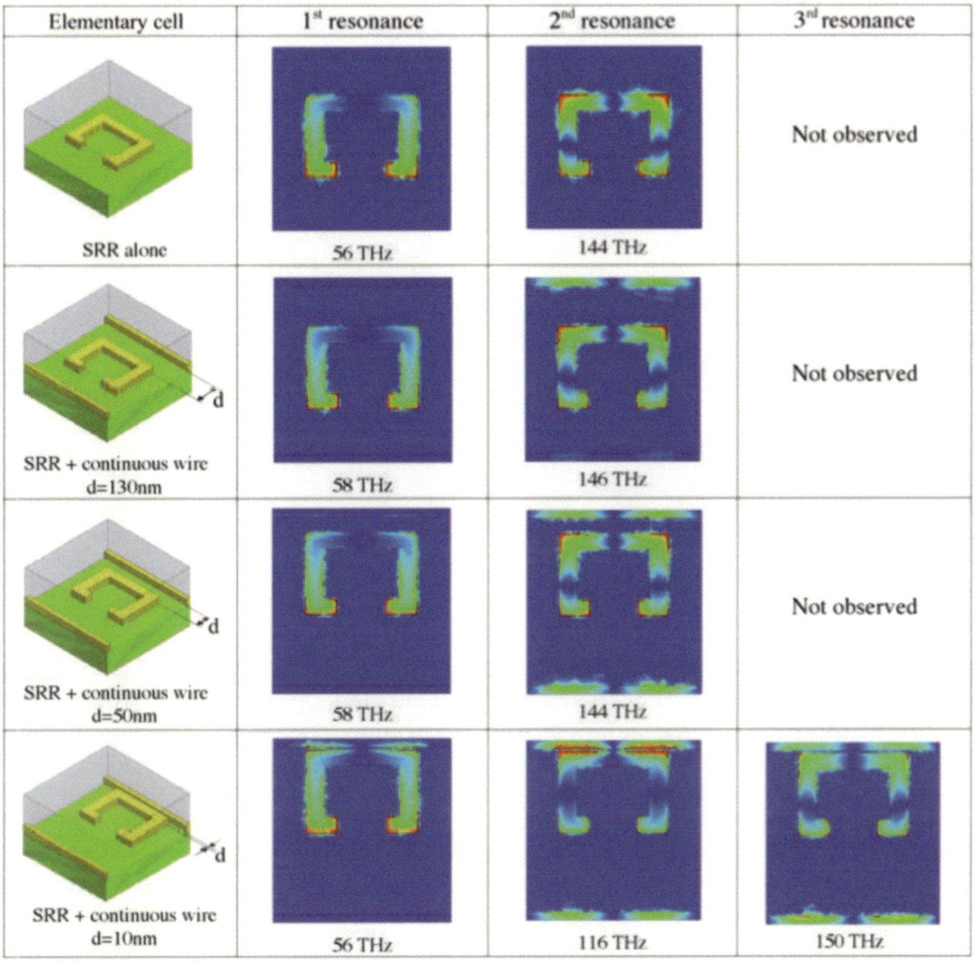

Fig. 14. (a) Transmission spectra calculated for a periodic array of SRRs (red curve), a periodic array of continuous wires (pink dashed curve) and periodic arrays of SRRs and wires with different separations between SRRs and wires: d=130 nm (blue curve), d=50 nm (green dashed curve), d=10 nm (black dashed curve). In each case, the incident electric field is polarized parallel to the gap, the SRR dimensions and lattice period as the same as for structure 1 in Fig. 10. (b) Magnitude of the normal electric field component ($|Ez|$) calculated at the bottom surface of the metallic elements for each of the resonant modes observed in the transmission spectra of Fig. 14 (a). The different colours, from blue to red correspond to increasing magnitudes of the field component. Modes are classified according to the resonance order and to the separation d between the SRR and the closest wire.

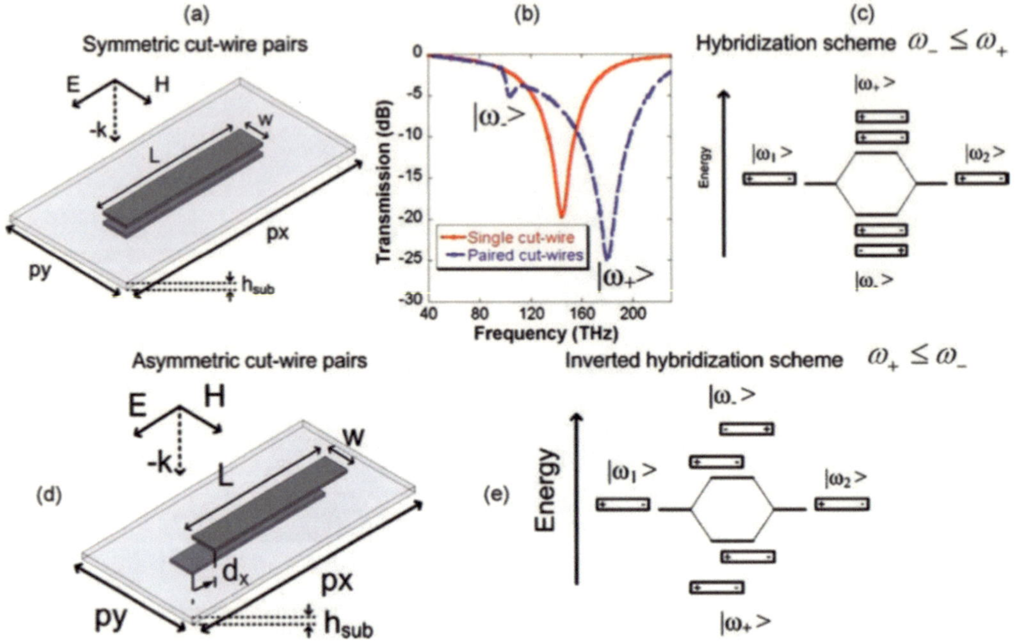

Fig. 15. (a and d) Schematics of the symmetric and asymmetric cut-wire pairs, respectively. (b) Transmission of an array of un-coupled or coupled cut-wires structure with p_x=1.2mm, p_y=200nm, p_z=600nm, w=30nm, L=600nm, h_{sub} =100nm and ε_{sub} =2.25. The pieces of gold metals are described by a Drude model whose parameters can be found in [Kante, 2008]. (c and e) Hybridization scheme and inverted hybridization scheme, respectively.

Fig. 17 (red curves) presents transmission spectra of a single SRR layer designed to operate in the infrared range for two polarizations at normal incidence. When the electric field is perpendicular to the gap, only one resonance (the fundamental even mode) is observed in the frequency range of interest while two resonances are observed for the parallel polarization. These last two resonances respectively correspond to the first and second odd plasmonic modes [Rockstuhl, 2006, Kante, 2008], the fundamental mode being also called LC resonance. When two such SRR structures are brought close to each other, the localized plasmons can hybridize according to a plasmon hybridization scheme similar to the one previously described for cut-wire pairs (Fig. 15). In what follows, the hybridization scheme of an SRR pair is analyzed for both parallel and perpendicular polarization as well as for resonances of different orders. As it will be shown, a negative index of refraction can be obtained at normal incidence for a periodic structure, which is exclusively made of SRRs and based on an inverted hybridization scheme. The SRR pair forming the elementary motif of the periodic structure is depicted in Fig. 17 (left graph). As seen, the lower SRR of the pair is shifted in the x and y directions with respect to the upper one. Following our previous work on cut-wire pairs, this configuration will be simply called "asymmetric SRR pair". A system of coupled SRRs has been recently investigated by Liu et al. [Liu, 2009], but in a twisted configuration. These authors showed that the twist angle between vertically coupled SRRs could modify either the electric or magnetic response of the system resulting in what they called a "stereometamaterial". Our structure can thus be regarded as a particular "stereometamaterial". In contrast, rotating one SRR with respect to another did not produce any inverted hybridization scheme, while this scheme is the most appropriate one for obtaining negative refraction. In fact, split ring resonators are complex structures regarding their responses to an electromagnetic field since they support localized plasmons addressable either by the electric field or the magnetic field. To our knowledge, using the two degrees of freedom (dx, dy) in the design of the SRR pair (Fig. 17 (left graph)) to achieve a negative index of refraction has never been reported so far in the context of SRR-based metamaterials.

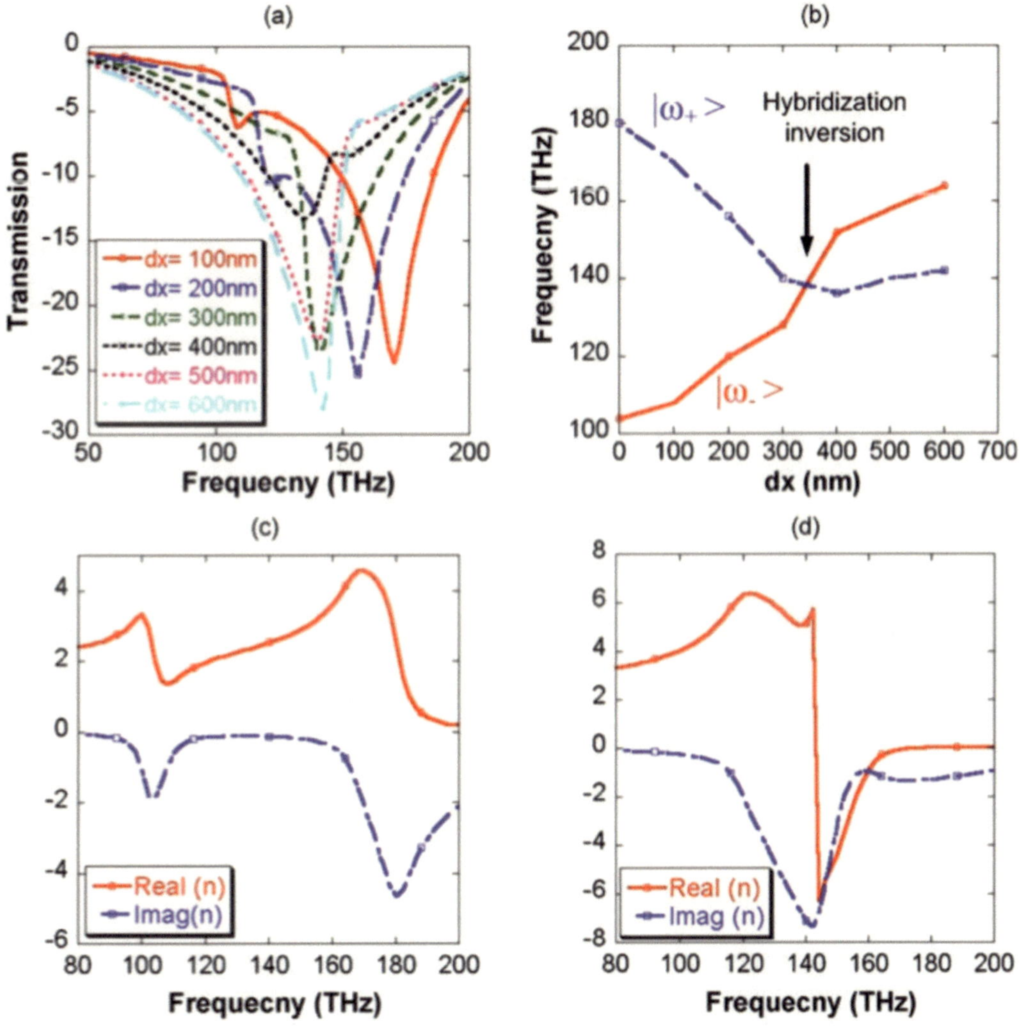

Fig. 16. (a) Transmission spectra of the two-dimensional array of asymmetric cut-wire pairs calculated for different values of the longitudinal shift (dx) between the two cut-wires in each pair. (b) Evolutions of the symmetric and anti-symmetric mode frequencies as a function of the dx shift. (c) Effective index of refraction [30] calculated for dx = 0 with a de-embedding up to the metamaterials interfaces. (d) Same type of calculations for dx = 600 nm.

3.2.3 Hybridization of the LC resonance

The LC resonance (i.e. the fundamental odd mode) is excited at normal incidence when the electric field is parallel to the gap. This resonance has been experimentally observed from microwaves to optics. Electric charges in metallic SRR arms are mainly located near the SRR gap (insert of Fig. 17, middle graph). Fig. 18 shows the modifications of the transmission spectrum when coupling vertically two SRR layers. Curves in the left graph correspond to different values of dx at a fixed dy (dy = 0). Curves in the right graph correspond to different values of dy at fixed dx (dx = 0). As expected, in all cases, coupling between SRRs splits the LC resonance in two eigenmodes. The symmetric mode, which has the largest amplitude, is found at the highest frequency for dx = dy = 0. In principle, a longitudinal magnetic coupling also exists [Liu, 2009], but can be neglected for a qualitative understanding of the hybridization scheme in presence of electric coupling. As seen in the left graph of Fig. 17, a shift of one of the two SRRs in the direction perpendicular to the gap (dx≠0) has a weak influence. One just observes a small decrease of the mode amplitudes. The evolution is radically different when the shift occurs in the direction parallel to the gap (dy ≠ 0). An inversion of the hybridization scheme is observed for sufficiently large values of dy. The overall results obtained for the LC mode of SRRs are then very comparable to those reported for the dipolar mode of cut-wires. The LC mode exhibits indeed a dipole-like behavior.

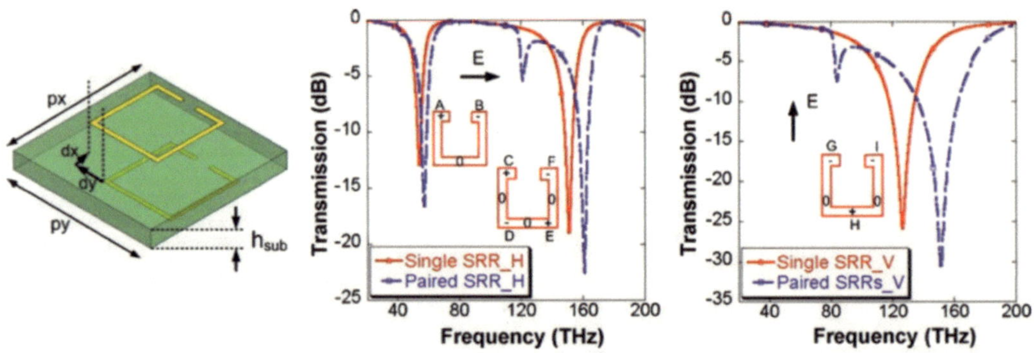

Fig. 17. (Left) Schematics of the asymmetric SRR pair. (Middle) Transmission spectra of periodic SRR arrays at normal incidence for a field polarization parallel to the SRR gap. (Right) Transmission spectra of periodic SRR arrays for a field polarization perpendicular to the SRR gap. In the middle and right graphs, red curves are for one-SRR-layer array while blue curves are for a two-layer array of paired SRRs. Inserts show the distributions of charges and electric field nodes for the different resonances in the one-layer array. Squared SRRs are used with 700 nm side length and 200 nm gap width. The 100 nm wide conducting elements are described using the Drude model reported in [Kante, 2008]. Other parameters are: px = py = 1.4 mm, pz = 1 mm, hsub = 100 nm, εsub = 2.25.

3.2.4 Hybridization of the second odd plasmonic mode

Let us now consider the second odd mode. The corresponding distribution of electric charges and field nodes in SRR arms is shown in the middle graph of Fig. 17 (right insert) for a one-layer SRR array. This picture shows that this mode has a "dipole activity" in both the x and y directions. Charges at the extremities of CD and EF arms produce the dipolar activity along the x direction while those at the extremities of the DE arm produce the dipolar activity along the y direction. Correspondingly, in a two-layer SRR array, an inversion of the hybridization occurs whether one of the two layers is displaced along the x or the y direction (Fig. 18). In each case, the inversion stems from the near field interaction between the active dipoles in the corresponding direction. In the case of a displacement along x, the inversion results from the interaction of dipoles CD and EF with dipoles C'D' and E'F'. The primes referring to the second SRR. In the case of a displacement along y, it results from the interaction between DE and D'E'.

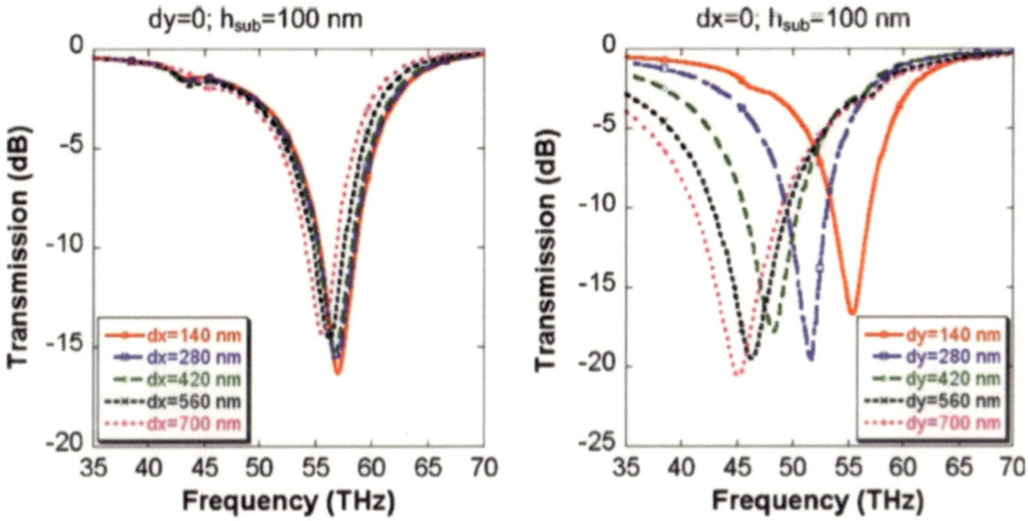

Fig. 18. Transmission spectra of a periodic array of asymmetric SRRs pairs around the LC resonance for different values of dx at dy=0 (left) and for different values of dy at dx = 0 (right)

3.2.5 Hybridization of the fundamental even plasmonic mode (perpendicular polarization)

When the incident electric field is perpendicular to the gap, the SRR can be regarded as two cut-wires in parallel since the charges in the two arms GH and IH oscillate in phase (see insert in the right graph of Fig. 17). In consequence, the hybridization of localized plasmons evolves a priori as in the case of cut-wires pairs [Kante, 2009]. Therefore, only a displacement along the x direction (parallel to the dipoles GH and HI) for one of the SRRs of the asymmetric pair can lead to an inverted hybridization scheme with the possibility of a negative index of refraction.

The inversion of the hybridization scheme for dx $\neq 0$ is illustrated from calculated transmission spectra in Fig. 20(a). Results of calculations for the effective index of refraction are presented in Fig. 20(c) and (d) for a symmetric SRR pair (dx = dy = 0) and an asymmetric pair (dx = 700 nm, dy = 0), respectively. Results obtained for the symmetric pair are comparable to those reported in [Liu, 2008]. There with no overlap between the regions of negative epsilon and negative mu. In contrast, a negative index of refraction is clearly obtained for the asymmetric SRR pair in Fig. 20(d).

Results obtained for a transversal displacement ($dy \neq 0$) of one SRR of the pair may appear to be quite surprising since three peaks are observed in the transmission spectra instead of two (Fig. 20(b)). However, a careful analysis reveals that these peaks have different origins. Three types of dipole–dipole interactions are indeed involved: GH with G'H', GH with I'H' and IH with G' H'. For an intermediate value of dy (i.e. dy = py/2 = 350 nm), the dipole-dipole interactions GH - G'H' and I'H' - IH become degenerate, thereby leading to the disappearance of one of the three peaks.

In conclusion we have numerically demonstrated that the coupling between localized plasmons in periodic arrays of paired cut-wires or SRRs can be controlled by modifying the symmetry of each individual pair. It has been shown that breaking the symmetry of cut-wire or SRR stacks can lead to a negative index of refraction. The scheme proposed here contrasts with previous designs of negative index metamaterials where two kinds of meta-atoms were mixed. Only one type of meta-atom supporting localized plasmons is used. A true negative index band is achieved provided that the coupling between localized plasmons is appropriately controlled.

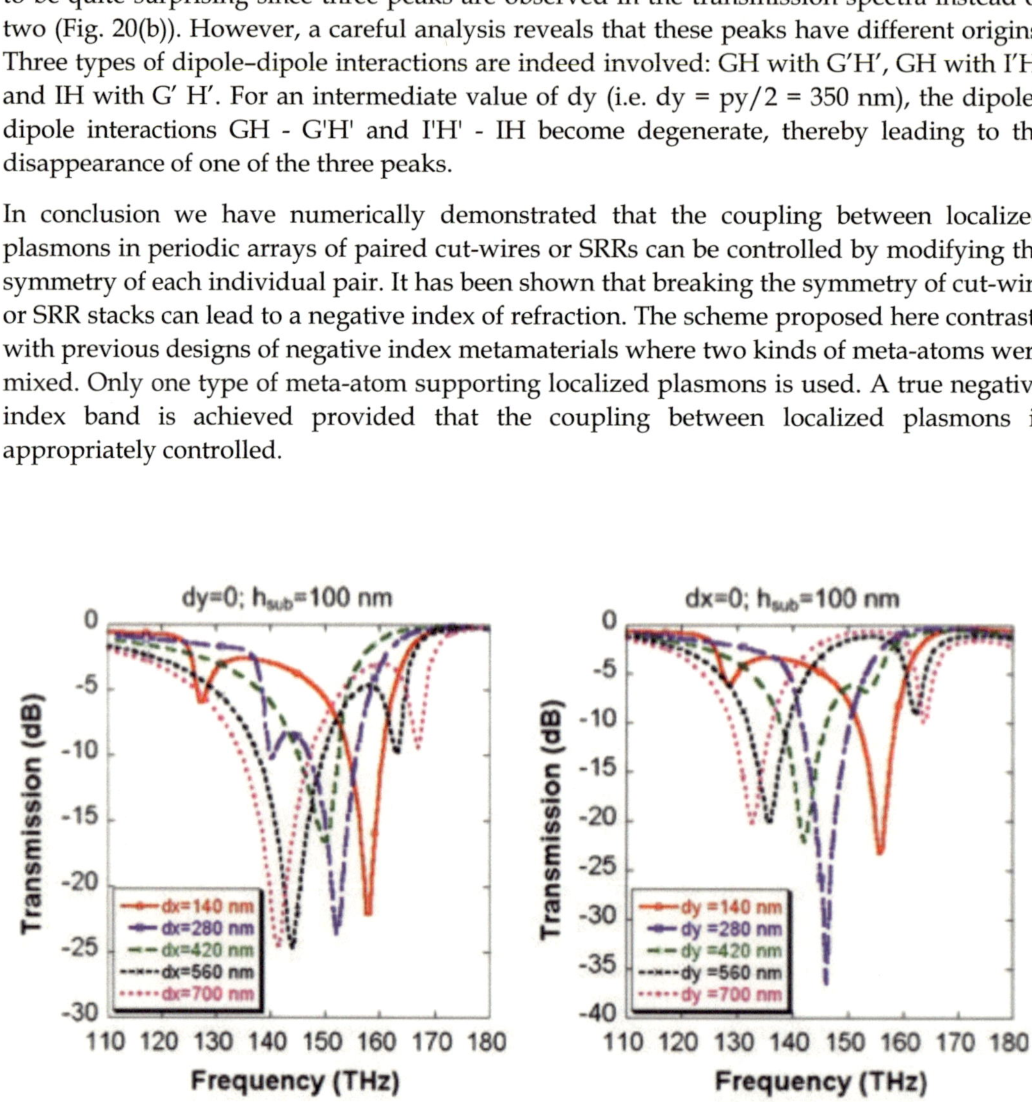

Fig. 19. Transmission spectra of a periodic array of a symmetric SRRs pairs around the second odd SRR mode for different values of dx at dy=0 (left) and for different values of dy at dx = 0 (right).

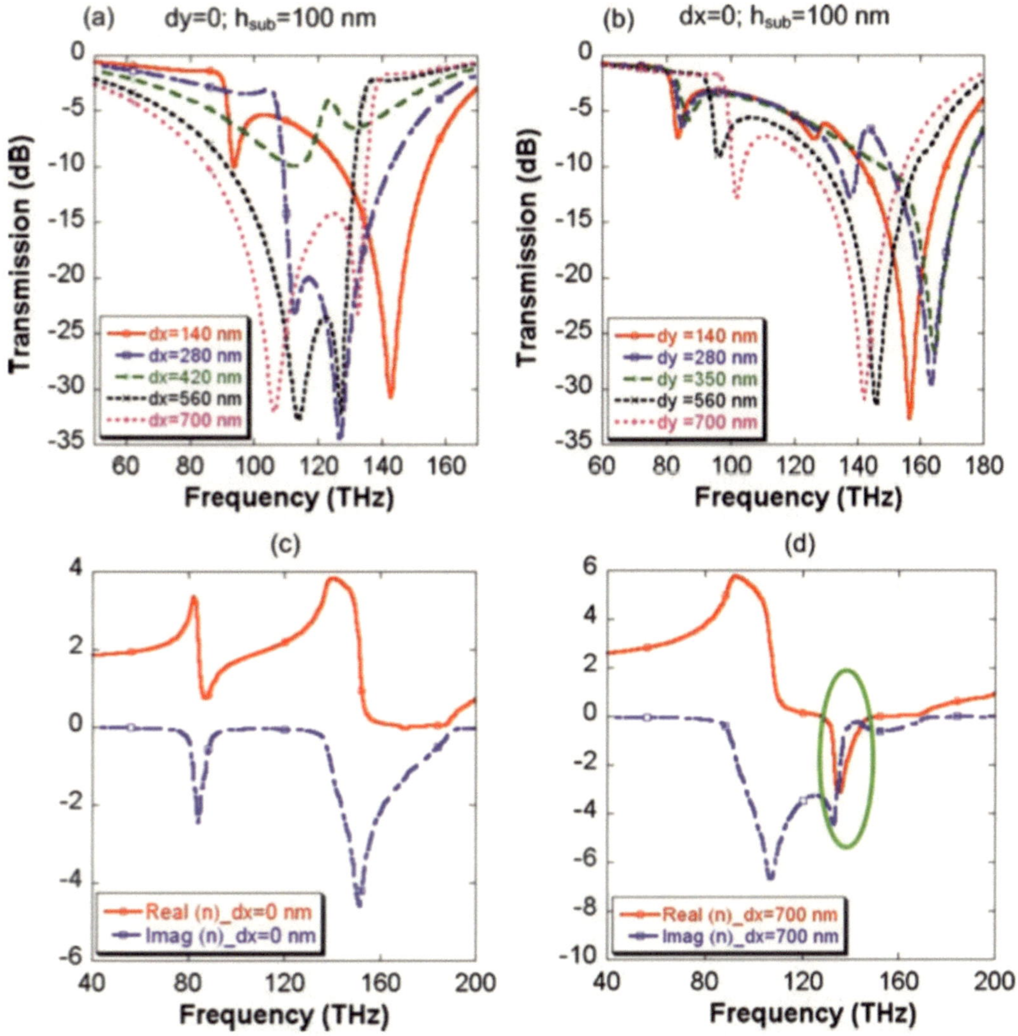

Fig. 20. (Top) Transmission spectra of a periodic array of asymmetric SRRs pairs for vertical polarization and different values of dx at dy=0 (a) and for different values of dy at dx = 0 (b). (Bottom) Effective index of refraction (real and imaginary parts) calculated for dx = 0 and dy = 0 (c) and for dx = 700 nm, dy = 0 (d), pz = 1 mm with a de-embedding up to the metamaterials interfaces.

4. Conclusions

In this chapter, we have presented different metamaterials with a negative index at microwave and optical frequencies. Through numerical simulations and measurements, we have shown that it was possible to obtain a negative index by optimizing the coupling between different layers of metamaterials. We have also shown that this concept can be implemented from microwaves to optics. These works are believed to open a new way for the design of negative refraction metamaterials from microwaves to optics.

5. References

Aydin, K., Bulu, I., Ozbay, E., (2005), Focusing of electromagnetic waves by a left-handed metamaterial flat lens, *Optics Express,* Vol.13, n°22, pp.8753-8759.

Burokur, S. N. , Latrach, M., Toutain, S., (2005), Theoretical investigation of a circular patch antenna in the presence of a left-handed medium, *IEEE Antennas and Wireless Propagation Letters* Vol.4, pp.183-186.

Burokur, S. N., Sellier, A., Kanté, B., de Lustrac, A., (2009a), Symmetry breaking in metallic cut wire pairs metamaterials for negative refractive index, *Applied Physics Letters,* Vol.94, 201111.

Burokur, S. N., Lepetit, T., de Lustrac, A., (2009b), Incidence dependence of negative index in asymmetric cut wire pairs metamaterials, *Applied Physics Letters* Vol.95, 191114.

Cai, W. , Chettiar U. K., Kildishev A. V., Shalaev V. M., (2007), Optical cloaking with metamaterials, *Nature Photonics* Vol.1, pp.224-227.

Christ, A., Ekinci Y., Solak H. H., Gippius N. A., Tikhodeev S. G., Martin O. J. F., (2007), Controlling the Fano interference in a plasmonic lattice, *Physical Review B,* Vol.76, 201405□R.

Christ, A., Marin O.J.F., Ekinci Y., Gippius N. A., Tikhodeev S. G., (2008), Symmetry Breaking in a Plasmonic Metamaterial at Optical Wavelength, *Nano Letters,* Vol.8, pp.2171-2175.

Dolling, G. , Enkrich C., Wegener M., Zhou J. F., Soukoulis C. M., Linden S., (2005), Cut-wire pairs and plate pairs as magnetic atoms for optical metamaterials, *Optics Letters,* Vol.30, n°23, pp.3198-3200.

Dolling, G., Enkrich C., Wegener M., Soukoulis C. M., Linden S., *(2006),* Simultaneous Negative Phase and Group Velocity of Light in a Metamaterial, *Science,* Vol.312, pp. 892-894.

Enkrich, C. , M. Wegener, S. Linden, S. Burger, L. Zschiedrich, F. Schmidt, J. F. Zhou, Koschny, T., Soukoulis, C. M., (2005) , Magnetic Metamaterials at Telecommunication and Visible Frequencies, *Physical Review Letters* Vol.95, 203905.

Gaillot, D. P., Cröenne, C., Lippens, D., (2008), An all-dielectric route for terahertz cloaking, *Optics Express,* Vol.16, n°6, pp.3986-3992

Grigorenko, A.N., Geim A. K., Gleeson H. F., Zhang Y., Firsov A. A., Khrushchev I. H., Petrovic J., (2005), Nanofabricated media with negative permeability at visible frequencies, *Nature,* Vol.438, pp.335–338.

Gundogdu, T., F., Katsarakis N., Kafesaki M., Penciu R. S., Konstantinidis G., Kostopoulos A., Economou E. N., and Soukoulis C. M., (2008), Negative index short-slab pair and continuous wires metamaterials in the far infrared regime, *Optics Express,* Vol.16, n°12, pp. 9173-9180.

Guven, K., M. D. Kaliskan, and E. Ozbay, (2006), Experimental observation of left-handed transmission in a bilayer metamaterial under normal-to-plane propagation, *Optics Express,* Vol.14, n°19, pp. 8685-8693.

HFSS, High-frequency structure simulator version 10.1 Finite-element package, Ansoft Corporation, Pittsburgh, PA (2006).

Kanté, B., Ourir A., Burokur, S. N., Gadot, F., de Lustrac, A., (2008a), Metamaterials for optical and radio communication, *Comptes Rendus Physique,* Vol.9, n°1, pp.31-40.

Kanté, B., de Lustrac, A., Lourtioz, J. M., Burokur, S. N., (2008b), Infrared cloaking based on the electric response of split ring resonators, *Optics Express,* Vol.16, n°12, pp. 9191-9198.

Kanté, B., de Lustrac A., Lourtioz, J.M. , Gadot, F., (2008c), Engineering resonances in infrared metamaterials, *Optics Express*, Vol.16, n°10, pp. 6774-6784.

Kanté, B., Burokur, S. N., Sellier, A. , de Lustrac, A., Lourtioz, J.-M., (2009a), Controlling Plasmon Hybridization for Negative Refraction Metamaterials, *Physical Review B*, Vol.79, 075121.

Kanté, B., de Lustrac, A., Lourtioz, J. M., (2009b), In-plane coupling and field enhancement in infrared metamaterial surfaces, *Physical Review B,* Vol.80, 035108.

Katsarakis, N., Koschny T., Kafesaki M., Economou E. N., and C. M. Soukoulis, (2004), Electric coupling to the magnetic resonance of split ring resonators, *Applied Physics Letters,* Vol.84, pp.2943-2945.

Katsarakis, N., Konstantinidis, G., Kostopoulos, A., Penciu, R. S., Gundogdu, T. F., Kafesaki, M., Economon, E. N., Koschny, T., and Soukoulis, C.M., (2005), Magnetic response of split-ring resonators in the far- infrared frequency regime, *Optics Letters* Vol.30, pp.1348-1350.

Kreibig, U., Althoff, A., Pressmann, H., (1981), Veiling of optical single particle properties in many particle systems by effective medium and clustering effects, *Surface Science* Vol.106 , n°1-3, pp. 308-317.

Leonhardt, U., (2006), Optical Conformal Mapping, *Science,* Vol.312, pp.1777-1780.

Linden, S., Enkrich, C., Wegener, M., Zhou J., Koschny, T., Soukoulis, C. M., (2004), Magnetic Response of Metamaterials at 100 Terahertz, *Science,* Vol.306, pp.1351-1353.

Liu, N., Guo H., Fu L., Kaiser S., Schweizer H., Giessen H., (2007), Plasmon Hybridization in Stacked Cut-Wire Metamaterials, *Advanced Materials*, Vol.19, pp.3628-3632.

Liu, N., Guo H., Fu L., Kaiser S., Schweizer H., and Giessen H., (2008), Three-dimensional photonic metamaterials at optical frequencies, *Nature Materials* Vol.7, pp.31-37.

Liu, N., Liu H., Zhu S., Giessen H., (2009), Stereometamaterials, *Nat. Photon*. Vol.3, pp.157–162

Nicholson, A. M., Ross G.F., (1970), Measurement of the intrinsic properties of materials by time-domain techniques, *IEEE Transactions on Instrumentation And Measusrements* Vol.19, pp.377-382.

O'Brien S, Pendry J.B., (2002), Magnetic Activity at Infra Red Frequencies in Structured Metallic Photonic Crystals, *Journal of Physics Condensed Matter* Vol.14, pp.6383–6394.

Pendry, J. B. , Holden A. J., Stewart W. J., Youngs I, (1996), Extremely Low Frequency Plasmons in Metallic Mesostructures, *Physical Review Letters,* Vol.76, pp.4773-4776.

Pendry, J. B. , Holden A. J., Robbins D. J., Stewart W. J., (1999), Magnetism from conductors and enhanced nonlinear phenomena, *IEEE Trans. Microwave Theory Tech.,* Vol.47, pp.2075-2084.

Pendry, J. B. , (2000), Negative Refraction Makes a Perfect Lens, *Physical Review Letters,* Vol.85, pp.3966-3969.

Pendry, J. B., D. Schurig, and D. R. Smith, (2006), Controlling Electromagnetic Fields, *Science* Vol.312, pp. 1780-1782.

Prodan, E., C. Radloff, N. J. Halas, and P. Nordlander, (2003), A Hybridization Model for the Plasmon Response of Complex Nanostructures, *Science* Vol.302, pp. 419-422.

Rockstuhl, C., F. Lederer, C. Etrich, T. Zentgraf, J. Kuhl, and H. Giessen, (2006), On the reinterpretation of resonances in split-ring-resonators at normal incidence, *Optics Express,* Vol.14, n°19, pp. 8827-8836.

Sellier, A., Burokur S. N., Kanté B., de Lustrac A., (2009), Negative refractive index metamaterials using only metallic cut wires, *Optics Express,* Vol.17, 8, pp. 6301-6310.

Shalaev, V. M., Cai W., Chettiar U. K., Yuan H. K., Sarychev A. K., Drachev V. P., Kildishev A. V., (2005), Negative index of refraction in optical metamaterials, *Optics Letters* Vol.30, pp.3356-3358.

Shalaev, V. M. , (2007), Optical negative-index metamaterials, *Nature Photonics* Vol.1, pp.41-48.

Shelby, R. A. , D. R. Smith, and S. Schultz, (2001), Experimental Verification of a Negative Index of Refraction, *Science* Vol.292, pp.77-79.

Schurig, D., J. J. Mock, B. J. Justice, S. A. Cummer, J. B. Pendry, A. F. Starr, and D. R. Smith, (2006), Metamaterial Electromagnetic Cloak at Microwave Frequencies, *Science,* Vol.314, pp. 977-980.

Smith, D. R., Padilla W.J., Vier D. C., Nemat-Nasser S. C., Schultz S., (2000), Composite Medium with Simultaneously Negative Permeability and Permittivity, *Physical Review Letters,* Vol.84, pp.4184-4187.

Smith, D. R., Schultz S., Markos P., and Soukoulis C. M., (2002), Determination of effective permittivity and permeability of metamaterials from reflection and transmission coefficients, *Physical Review B* Vol.65, 195104.

Smith, D. R. , Pendry J. B., Wiltshire M. C. K., (2004), Metamaterials and Negative Refractive Index, *Science* Vol.305, pp.788-792.

Valentine, J., Zhang S., Zentgraf T., Ulin-Avila E., Genov D.A., Bartal G., Zhang X., (2008), "Three Dimensional Optical Metamaterial Exhibiting Negative Refractive Index," *Nature* Vol.455, pp.376-379.

Veselago, V. G. , (1968), The electrodynamics of substances with simultaneously negative values of e and μ, *Soviet Physics Uspieky* Vol.10, 509.

Wang, G., Fang J. R., Dong X. T., (2007), Refocusing of backscattered microwaves in target detection by using LHM flat lens, *Optics Express,* Vol.15, n°6, pp.3312-3317.

Yeh, P. , (1998), *Optical waves in layered media* (Wiley 1998).

Yen, T. J., Padilla W. J., Fang N., Vier D. C., Smith D. R., Pendry J. B., Basov D. N., Zhang X., (2004), Terahertz Magnetic Response from Artificial Materials, *Science* Vol.303, pp.1494-1496.

Zhang, S. , Fan W., Minhas B. K., Frauenglass A., Malloy K. J., Brueck S. R. J., (2005), Midinfrared Resonant Magnetic Nanostructures Exhibiting a Negative Permeability, *Physical Review Letters,* Vol.94, 37402.

Zhou, J., Zhang L., Tuttle G., Koschny T., Soukoulis C. M., (2006), Negative index materials using simple short wire pairs, *Physical Review B.* Vol. 73, 041101.

Zhou, J., Economon E., Koschny T., Soukoulis C. M., (2006), Unifying approach to left-handed material design, *Optics Letters* Vol.*31,* pp.3620-3622.

Zhou, J., Koschny T., and Soukoulis C. M., (2007), Magnetic and electric excitations in split ring resonators, *Optics Express,* Vol.15, pp.17881-17890 .

Ziolkowski, R.W., Kipple A., (2003), Application of double negative metamaterials to increase the power radiated by electrically small antennas, *IEEE Transaction Antennas Propag*ation Vol.51, pp.2626-2640.

12

Characterization of Metamaterial Transmission Lines with Coupled Resonators Through Parameter Extraction

Francisco Aznar-Ballesta[1], Marta Gil[2], Miguel Durán-Sindreu[3],
Jordi Bonache[3] and Ferran Martín[3]
[1]*Universidad Politécnica de Madrid,*
[2]*Universidad de Castilla-La Mancha,*
[3]*Universitat Autònoma de Barcelona,*
Spain

1. Introduction

Since resonant-type metamaterial transmission lines were proposed (Martín et al., 2003), this kind of transmission lines have been of significant importance in the development of new and innovative microwave devices. The small size and novel characteristics of these transmission lines based on sub-wavelength resonators allows the miniaturization and improvement of existing devices (Bonache et al., 2006a, 2006b; Gil et al., 2007a, 2007b), as well as the design of components with new functionalities (Sisó et al., 2008, 2009). Due to the complicated layouts that these designs usually involve, having an accurate equivalent circuit model is an important assist during the design process. Besides, the application of parameter extraction methods (Bonache et al., 2006c) to obtain the values of the electrical parameters of the circuit model makes possible the characterization of both the transmission line and, therefore, the microwave device. The circuit models and parameter extraction methods presented in this chapter have been widely verified and their accuracy permits even their application for automatic layout generation based on space mapping techniques (Selga et al., 2010), which is a large and useful advance in the design of such structures.

2. Metamaterial transmission lines based on the resonant approach

In this section, several structures of metamaterial transmission lines based on the resonant approach are shown. The considered structures are implemented in planar technology and consist in a host microstrip line or in coplanar wave guide (CPW) loaded with sub-wavelength resonant particles. Each structure requires the combination of resonators and other loading elements in order to achieve the intended propagation. In the following sections, several structures based on different kinds of resonators are presented and discussed.

2.1 Transmission lines based on Split-Ring Resonators (SRRs)

The metamaterial transmission lines based on SRRs were proposed in 2003 by Martín et al. (Martín et al., 2003). They designed a transmission line exhibiting left-handed transmission by loading a CPW structure with SRRs. Left-handed transmission is achieved when the effective permittivity and permeability of a medium are both negative, providing negative values of the phase velocity and refraction index, among other peculiarities (Veselago, 1968). This first left-handed SRR-based CPW was inspired on the medium of Smith et al., 2000. By etching SRRs in the back substrate side of the CPW, beneath the slots, and shunt connected metallic strips between the central strip and ground plane. A one dimensional effective medium with simultaneous negative permeability (due to the presence of the SRRs) and permittivity (thanks to the shunt strips) in a narrow band was achieved (Martín et al., 2003).

The resonators used in this kind of transmission lines based on the resonant approach can be SRRs or other similar resonators with different topologies based on the SRR. The layout for the SRR is shown in Fig. 1(a). In Fig. 1, other examples of these resonators, like the spiral resonator (SR, Fig. 1b) with only one metal layer, are shown. Resonators implemented with two metal layers, like the broadside coupled non-bianisotropic split ring resonator (BC-NB-SRR, Fig. 1c), the broadside coupled spiral resonator with two turns (BC-SR(2), Fig. 1d) and the broadside coupled spiral resonator with four turns (BC-SR(4), Fig. 1e) are also depicted (Marqués et al., 2003; Aznar et al., 2008b).

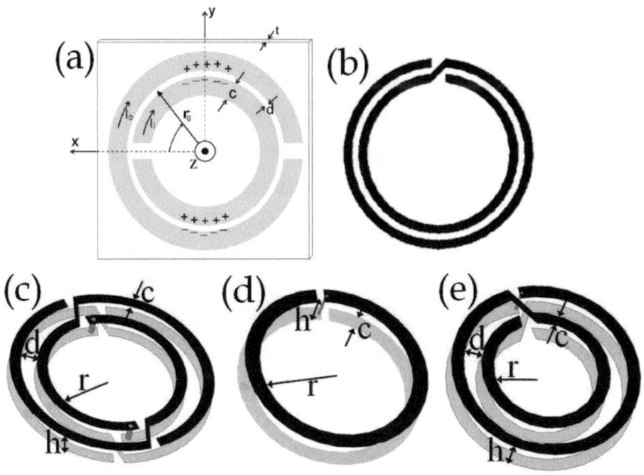

Fig. 1. Examples of topologies of different resonators based on the SRR (a): SR (b), BC-NB-SRR (c), BC-SR(2) (d) and BC-SR(4) (e).

The equivalent circuit model for the SRR is shown in Fig. 2(a) (Baena et al., 2005). The capacitance $C_0/2$ is related with each of the two SRR halves, whereas L_s is the resonator self-inductance. C_0 can be obtained as $C_0=2\pi r C_{pul}$, where C_{pul} represents the per unit length capacitance between de rings forming the resonator. Regarding L_s, it can be approximated to the inductance of a single ring with the average radius of the resonator and the width of the rings, *c*. Taking into account the circuit model of the resonator, its resonance frequency can be calculated as:

$$\omega_0 = \frac{1}{\sqrt{L_S C_S}} \tag{1}$$

The resonators based on the SRR, like the examples represented in Fig. 1 can also be modelled by a simple *L-C* resonant tank (see Fig. 2b). As long as the inductance and the capacitance of the resonator can be increased (within the technology limits), the resonance frequency of the SRR can be decreased, reducing its electrical size (Aznar et al., 2008b). This is the case of the resonators shown in Fig. 1, which are electrically smaller than the SRR.

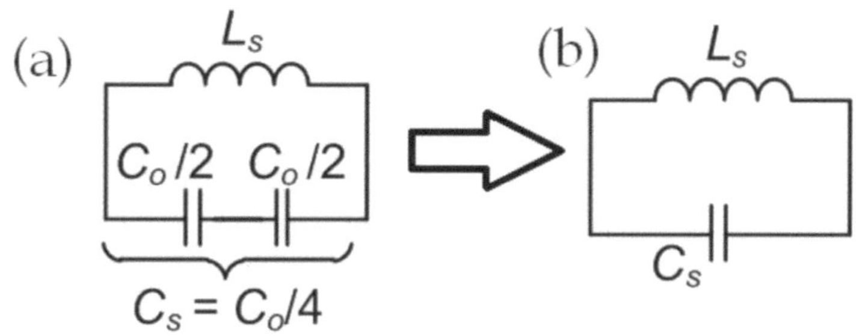

Fig. 2. Equivalent circuit models for a SRR (a) and simplified equivalent circuit model for a resonator based on SRR (b).

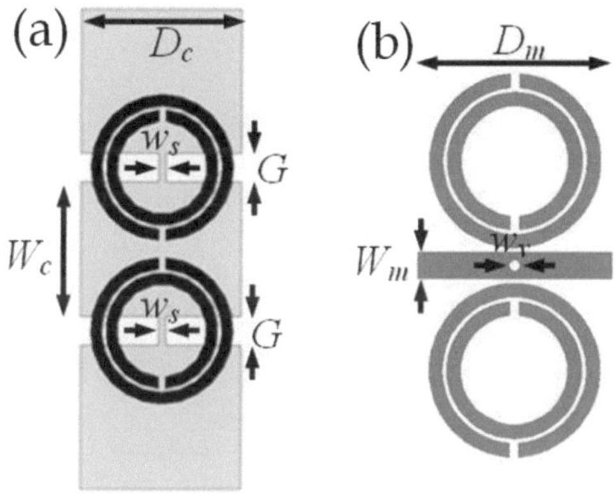

Fig. 3. Layouts of metamaterial transmission lines loaded with resonators based on SRR in CPW (a) and microstrip (b) technology.

The layout of two unit cells of metamaterial transmission lines loaded with resonators based on SRR are shown in Fig. 3(a) for CPW and Fig. 3(b) for microstrip technology. In the CPW configuration, the SRRs are paired on the lower substrate side (where they are etched), beneath the slots of the structure and centred with the shunt strips. The resonators are responsible for the negative effective permeability, whereas negative effective permittivity is achieved by means of the shunt connected strips. In case the shunt strips are eliminated, the CPW loaded with resonators provides only negative permeability, showing stop-band behaviour.

In the case of microstrip lines, the SRRs can be etched in pairs on the upper substrate side, adjacent to the conductor strip. The metallic vias are responsible for the negative permittivity of the structure in this case as well.

2.2 Transmission lines based on Complementary Split-Ring Resonators (CSRRs)

The layout of the CSRR is shown in Fig. 4(a). This resonator results from the application of the Babinet principle to the structure of the SRR, which leads to its complementary counterpart (Falcone et al., 2004). In the CSRR the rings are etched on a metallic surface and its electric and magnetic properties are interchanged with respect to the SRR: the CSRR can be excited by an axial time-varying electric field and exhibits negative values of the dielectric permittivity.

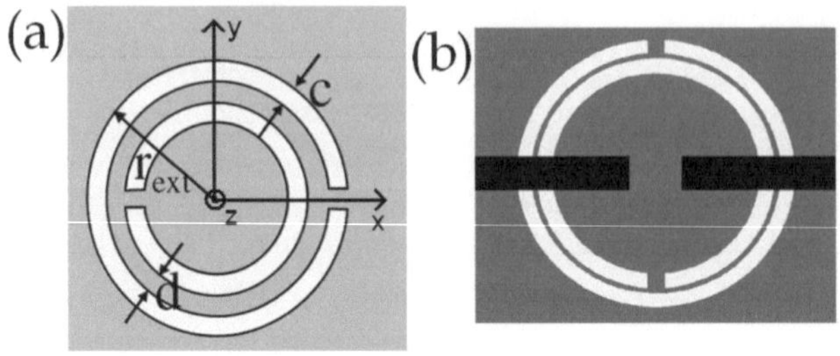

Fig. 4. (a) Representation of a complementary split ring resonator (CSRR); in this case the resonator is etched on a metallic surface; metallic part is represented in grey. (b) Scheme of a unit cell of a CSRR-based resonant-type metamaterial transmission line. CSRRs are etched on the ground plane (in grey) of a microstrip line just below the capacitive gaps etched on the signal strip (black).

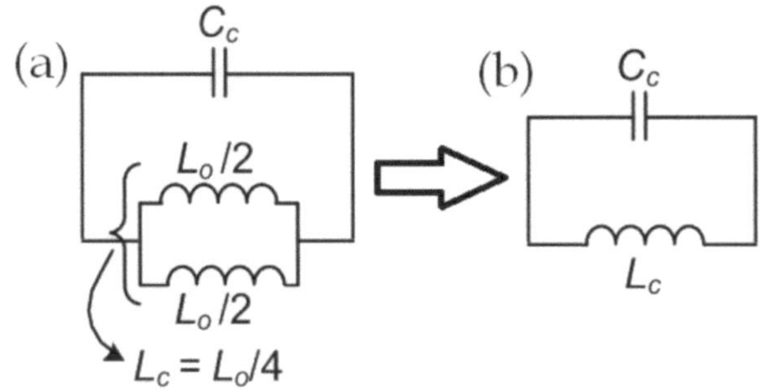

Fig. 5. Equivalent circuit models for a CSRR (a) and simplified equivalent circuit model for a resonator based on CSRR (b).

In this case, the CSRRs are etched on the ground plane of a microstrip transmission line. The CSRRs, which provide negative permittivity, are combined with capacitive gaps etched on the signal strip just above the resonators (see Fig. 4b). The gaps are in this case responsible for the negative permeability. The equivalent circuit model of the CSRR is shown in Fig. 5(a) (Baena et al., 2005). The resonance frequency of the CSRR is roughly the same of the frequency of a SRR with the same dimensions. Many other resonators admit a complementary counterpart, like, for example the SR shown in Fig. 1(b).

2.3 Transmission lines based on open resonators: Open Split-Ring Resonators (OSRRs) and Open Complementary Split-Ring Resonators (OSRRs)

A different kind of resonators consists in open resonators. Fig. 6 shows the layouts and equivalent circuit models of the open SRR (Martel et al., 2004) and the open complementary SRR (Vélez et al., 2009). As can be seen in the layout, the OSRR is based on the SRR and is obtained by truncating the rings forming the resonator and elongating them outwards. The OCSRR can be obtained as the complementary particle of the OSRR, in a similar way as the

CSRR is obtained from the SRR. These resonators, as is shown in Figs. 6 and 7, can be implemented either in microstrip or in coplanar technology (Aznar et al., 2008b, 2009a, 2009b; Durán-Sindreu et al., 2009; Vélez et al., 2010). The equivalent circuit models of the resonators are also shown in Fig. 6. The model of the OSRR is a series LC resonator (Martel et al., 2004). The inductance L_s can be obtained as the inductance of a ring with the average radius of the resonator and the same width, c, of the rings forming the OSRR. The capacitance C_0 is the distributed edge capacitance that appears between the two concentric rings. In a similar way, the OCSRR can be modelled by means of a parallel LC resonant tank (Aznar et al., 2008b; Vélez et al., 2009), where the inductance L_0 is the inductance of the metallic strip between the slot hooks and the capacitance is that of a disk with radius r_0-$c/2$ surrounded by a metallic plane separated by a distance c. According to this, it follows that, for identical dimensions and substrate, the resonance frequency of the OSRR or OCSRR is half the resonance frequency of the SRR or CSRR, respectively.

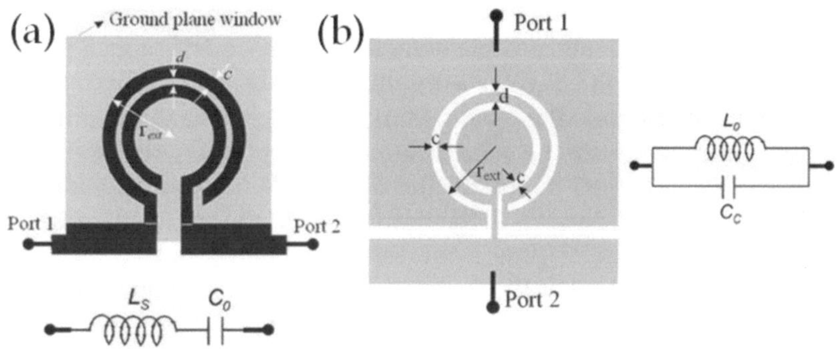

Fig. 6. Examples of microstrip structures loaded with open resonators. (a) Open split-ring resonator (OSRR) and its equivalent circuit model. (b) Open complementary split-ring resonator (OCSRR) and its equivalent circuit model.

As example, Fig. 7(a), shows a CPW transmission line loaded with a pair of OCSRRs. Fig. 7(b) represents an OSRR-based CPW unit cell.

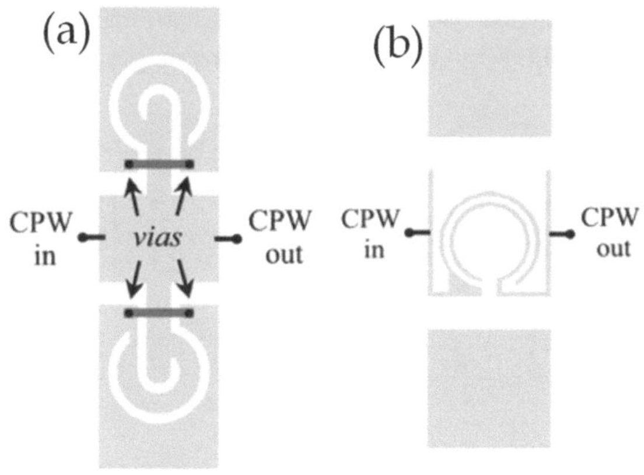

Fig. 7. Layout of a CPW based on OCSRR (a) and based on OSRR (b). In (a) the backside strips (in dark grey) connecting the different ground plane regions are necessary to prevent the slot mode of the CPW and the second resonance of the OCSRRs.

3. Equivalent circuit models for metamaterial transmission lines based on the resonant approach

In this section, we present the equivalent circuit models for the resonant-type metamaterial transmission lines presented in section 2. These models have been widely studied and confirmed and are able to model the composite behaviour of the considered structures (Aznar et al., 2008a; Gil et al., 2006). For a given kind of structure, the circuits are independent on the employed resonator. This means that the model for structures based on SRRs would be the same in case that the employed resonators were, for example, spiral resonators, which are electrically smaller but have the same equivalent circuit as the SRRs.

3.1 Equivalent circuit model for metamaterial transmission lines based on Split-Ring Resonators (SRRs)

The behaviour of the SRR-based structures (with and without shunt connected strips) can be interpreted to the light of the lumped element equivalent circuit models of the unit cells (Aznar et al., 2008a) (see Fig. 8a). Fig. 9 shows the typical behavior of two structures based on SRRs. The first one (left-handed) includes shunt strips, whereas the second one (negative-permeability) does not. In the circuit model, L and C account for the line inductance and capacitance, respectively, C_s and L_s model the SRR, M is the mutual inductive coupling between the line and the SRRs, and L_p is the inductance of the shunt strips (in case they are included). From the transmission line approach of metamaterials (Caloz & Itoh, 2005; Marqués et al., 2008), it follows that the structure exhibits left-handed wave propagation in those regions where the series reactance and shunt susceptance are negative, whereas in case they are positive, the propagation is conventional. According to this, the model in Fig. 8(a) perfectly explains the composite behavior of the structures shown in Fig. 3 (Aznar et al., 2008c). The inductance of the shunt inductive strips L_p is located between the two inductances ($L/2$) that model each line section, to the left and right of the position of the shunt strips. This reflects the location of the inductive strips. The resulting model is neither a π circuit nor a T circuit. Consequently, the transmission zero frequency and the frequency where the phase shift nulls cannot be directly obtained from it. It can be demonstrated that the model of Fig. 8(a) can be transformed to a π circuit, more convenient for its study and formally identical to that of Fig. 8(b) with modified parameters.

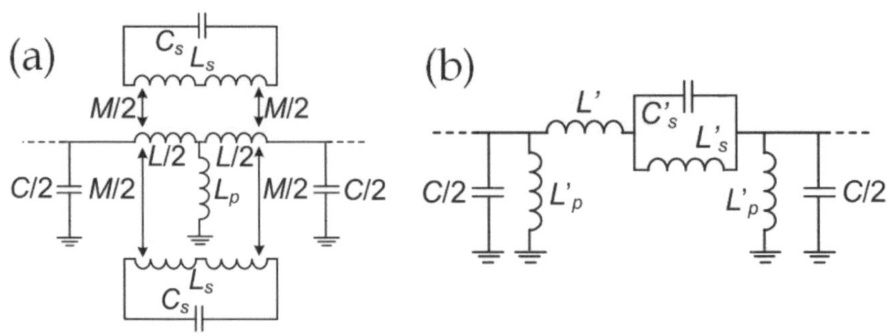

Fig. 8. Proposed circuit model for the basic cell of the left handed CPW or microstrip structure with loaded resonators based on the SRR (a). Transformation of the model to a π circuit (b).

Due to symmetry considerations and reciprocity, the admittance matrix of the circuit of Fig. 8(a) (which is a biport) must satisfy $Y_{12}=Y_{21}$ and $Y_{11}=Y_{22}$. From these matrix elements, the series (Z_s) and shunt (Z_p) impedances of the equivalent π-circuit model can be obtained (Pozar, 1990)

$$Z_S(\omega) = (Y_{21})^{-1} \tag{2}$$

$$Z_P(\omega) = (Y_{11} + Y_{21})^{-1} \tag{3}$$

Y_{21} is inferred by grounding port 1 and obtaining the ratio between the current at port 1 and the applied voltage at port 2. Y_{11} is simply the input admittance of the biport, seen from port 1, with a short circuit at port 2. After a straightforward but tedious calculation, the elements of the admittance matrix are obtained, and by applying Eq. 2 and Eq. 3, we finally obtain

$$Z_S(\omega) = j\omega\left(2 + \frac{L}{2L_P}\right)\left[\frac{L}{2} + M^2 \frac{1 + \frac{L}{4L_P}}{L_S\left(\frac{\omega_0^2}{\omega^2} - 1\right) - \frac{M^2}{2L_P}}\right] \tag{4}$$

$$Z_P(\omega) = j\omega\left(2L_P + \frac{L}{2}\right) \tag{5}$$

with $\omega_0 = (L_S C_S)^{-1/2}$. Expression 4 can be rewritten as

$$Z_S(\omega) = j\omega\left(2 + \frac{L}{2L_P}\right)\left[\frac{L}{2} + L'_S + \frac{L'_S}{1 - L'_S C'_S \omega^2}\right] \tag{6}$$

with

$$L'_S = 2M^2 C_S \omega_0^2 \frac{\left(1 + \frac{L}{4L_P}\right)^2}{1 + \frac{M^2}{2L_P L_S}} \tag{7}$$

$$C'_S = \frac{L_S}{2M^2\omega_0^2}\left(\frac{1 + \frac{M^2}{2L_P L_S}}{1 + \frac{L}{4L_P}}\right)^2 \tag{8}$$

These results indicate that the circuit model of the unit cell of the left-handed lines loaded with SRRs and shunt inductors (Fig. 8a) can be formally expressed as the π circuit of Fig 8(b). These parameters are related to the parameters of the circuit of Fig. 8(a), according to Eq. 7, Eq. 8 and

$$L' = \left(2 + \frac{L}{2L_P}\right)\frac{L}{2} - L'_S \tag{9}$$

$$L'_P = 2L_P + \frac{L}{2} \tag{10}$$

The transmission zero frequency ω_z for the circuit of Fig. 8(a) can be obtained forcing $Z_S(\omega)=\infty$. This gives

$$\omega_Z = \omega_0 \sqrt{\frac{1}{1+\dfrac{M^2}{2L_PL_S}}} \tag{11}$$

It can be observed, that the transmission zero frequency is always located below the resonance frequency of the SRRs ω_0 ($\omega_z<\omega_0$). On the other hand, the frequency where $\phi=0$, ω_S, is obtained by forcing $Z_S(\omega)=0$. This gives

$$\omega_S = \frac{1}{\sqrt{C_S\left(L_S - 2\dfrac{M^2}{L}\right)}} \tag{12}$$

Despite that $Z_S(\omega)=0$ is a function of L_P, unexpectedly, ω_S does not depend on the shunt inductance.

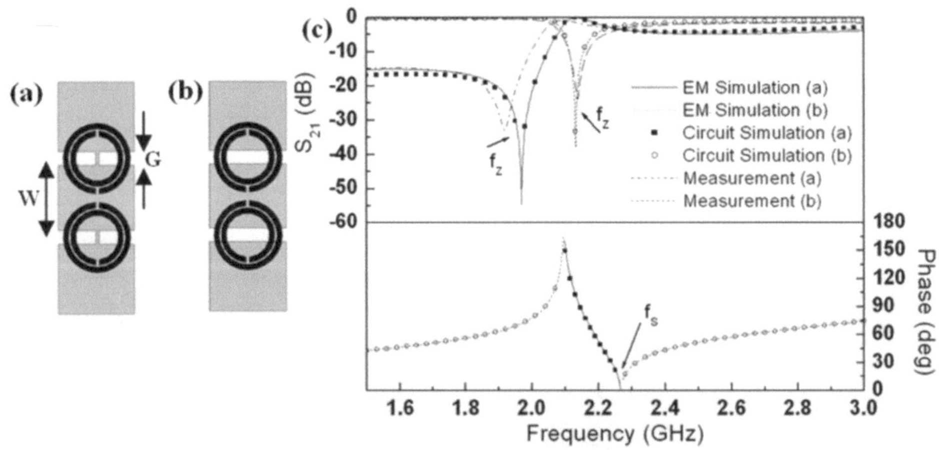

Fig. 9. Layouts of the considered CPW structures with SRRs and shunt strips (a) and with SRRs only (b); simulated and measured transmission coefficient S_{21} and simulated dispersion relation (c). The considered substrate is Rogers RO3010 with thickness h=1.27 mm and dielectric constant ε_r=10.2. Relevant dimensions are rings width c=0.6 mm, distance between the rings d=0.2 mm, and internal radius r=2.4 mm. For the CPW structure, the central strip width is W=7 mm and the width of the slots is G=1.35 mm. Finally, the shunt strip width is 0.2 mm. The results of the electrical simulation with extracted parameters are depicted by using symbols. We have actually represented the modulus of the phase since it is negative for the left-handed line. Discrepancy between measurement and simulation is attributed to fabrication related tolerances.

In Fig. 9 we can see two configuration examples for a CPW with loaded SRRs, with and without shunt strips. Electromagnetic simulation, circuit simulation and measurement show very good agreement in both cases. Therefore, we are able to confirm that the proposed circuit model is correct and accurate. This circuit model can also be applied to the microstrip structure shown in Fig. 3(b), which is in this sense, equivalent to the CPW structure.

3.2 Equivalent circuit model for metamaterial transmission lines based on Complementary Split-Ring Resonators (CSRRs)

The equivalent circuit model for the CSRR-based structure shown in Fig. 4(b) is the circuit shown in Fig. 10(a), which provides an accurate description of the behaviour of the structure (Aznar et al, 2008c, 2008d) and can be transformed into the circuit shown in Fig 10(b).

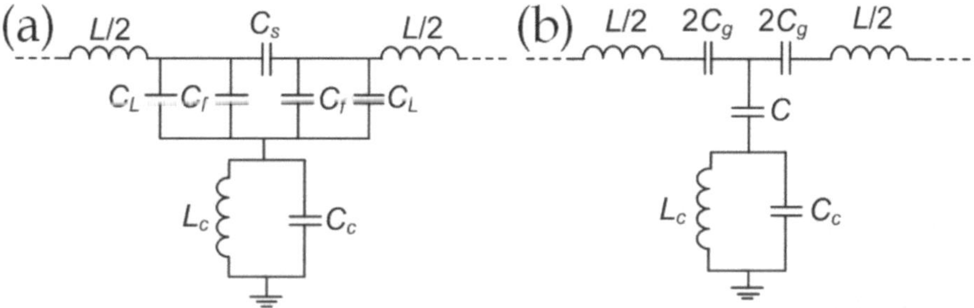

Fig. 10. Equivalent circuit model for the structure based on CSRR shown in Fig. 4(b) (a). Modified circuit model for the structure based on CSRR shown in Fig. 4(b) (b).

In the equivalent circuit (Fig. 10 a), the resonator is modelled by the resonant tank formed by L_c and C_c. The line parameters are L and C_L and the gap is modelled by the π-structure formed by C_s and C_f, which take into account the series and the fringing capacitances due to the presence of the capacitive gap. The modified circuit (Fig. 10b) is perfectly able to reproduce the behaviour of the structure. Nevertheless, the equivalent circuit (Fig. 10a) can be transformed into the modified circuit (Fig. 10b), by means of the following equations:

$$C_{par} = C_f + C_L \tag{13}$$

$$2C_g = 2C_S + C_{par} \tag{14}$$

$$C = \frac{C_{par}\left(2C_S + C_{par}\right)}{C_S} \tag{15}$$

so that the modified circuit (Fig. 10b), much simpler, can substitute the equivalent circuit (Fig. 10a) for a more straightforward work. The excellent agreement between electromagnetic and electrical simulation (employing the proposed model) can be observed in Fig. 11 (a). The notable frequencies (ω_z, ω_0) of this circuit model are indicated below in section 4.1.

In case the host transmission line is loaded just with CSRRs, the resulting structure shows negative permittivity and, thus, stop-band behaviour. This is the case of the structure shown in Fig. 11 (b). For these structures, the circuit model would be the same as in Fig. 10(b) except for the capacitances C_g, which would be eliminated.

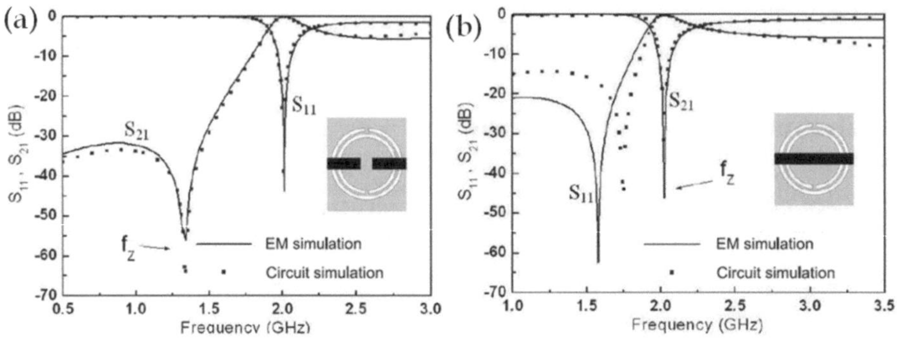

Fig. 11. Simulated (through the Agilent Momentum commercial software) frequency responses of the unit cell structures shown in the insets: (a) microstrip line loaded with CSRRs and series gaps, (b) microstrip line only loaded with CSRRs. The response that has been obtained from circuit simulation of the equivalent model with extracted parameters is also included. For the structures (a) and (b), the dimensions are: the strip line width W_m = 1.15 mm, the length D = 8 mm, and the gap width w_g = 0.16 mm. In both cases, the dimensions for the CSRR are: the outer ring width c_{out} = 0.364 mm, the inner ring width c_{inn} = 0.366 mm, distance between the rings d = 0.24 mm, and the internal radius r = 2.691 mm. The considered substrate is Rogers RO3010 with the dielectric constant ε_r = 10.2 and the thickness h = 1.27 mm.

3.3 Equivalent circuit models for metamaterial transmission lines based on open resonators: Open Split-Ring Resonators (OSRRs) and Open Complementary Split-Ring Resonators (OSRRs)

We will now consider the model of a CPW transmission line loaded with a pair of OCSRRs shown in Fig. 7(a). The resonator is represented by the elements C_p and L_p in Fig. 12(a). Although the particle is electrically small, it has been found that the structure exhibits certain frequency shift at resonance with respect to the frequency theoretically predicted by the equivalent circuit shown in Fig. 6. This frequency shift is expected if access lines are present. However, in the absence of access lines, we still obtain a small (although non negligible) phase shift. This means that the OCSRR-loaded CPW cannot be merely modeled as a two-port network with a shunt connected parallel resonator. To properly model the structure, we must introduce additional elements to account for the phase shift. That is, we must introduce phase shifting lines at both sides of the resonator. Such transmission line sections can be modeled through series inductances (L) and shunt capacitances (C), as depicted in Fig. 12(a). For design purposes, we can also use the simplified model depicted in Fig 12(b).

For a CPW loaded with a series connected OSRR represented in Fig. 7(b), a similar phenomenology results. Thus, to take into account these parasitic effects, we must introduce additional elements in the two port network describing the structure. A typical topology and the circuit model of these OSRR-loaded CPW transmission line sections are depicted in Fig. 12(c) and Fig. 12(d).

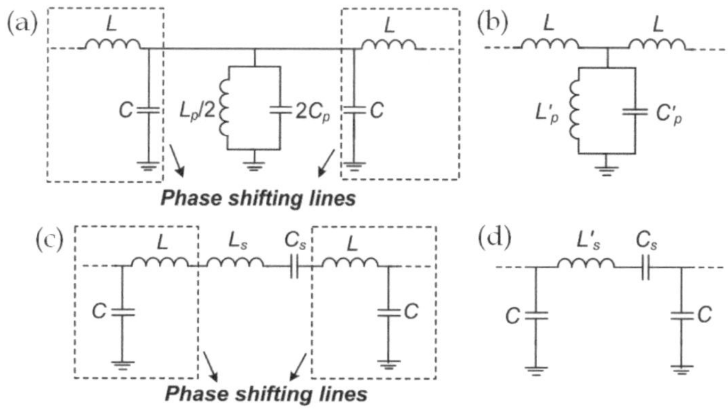

Fig. 12. (a) Circuit model, and (b) simplified circuit model of a CPW transmission line loaded with a pair of OCSRRs. (c) Circuit model, and (d) simplified circuit model of a CPW transmission line loaded with a series connected OSRR.

The transformation to simplify the circuit based on OSRR (Figs. 7b, 12 c and 12 d) is:

$$L'_S = L_S + 2L \tag{16}$$

and the transformations to simplify the circuit based on OCSRR (Figs. 7a, 12 a and 12 b) are:

$$C'_P = 2C_P + 2C \tag{17}$$

$$L'_P = \frac{L_P}{2} \tag{18}$$

4. Parameter extraction technique for metamaterial transmission lines based on the resonant approach

In previous sections we have presented different metamaterial transmission lines loaded with resonators and their equivalent circuit models. In this section we present a parameter extraction technique which, properly modified, can be applied to all of them. With this technique we can determinate the parameters of the circuit model of the structure. This represents an important assist in the design process, which becomes easier and faster.

The parameter extraction method consists in the imposition of several conditions obtained either from the simulated or measured response of the considered structure. The number of imposed conditions must be enough to obtain the values of all the parameters of the circuit.

4.1 Parameter extraction technique metamaterial transmission lines based on Split-Ring Resonators (SRRs)

The parameter extraction for the metamaterial transmission line based on SRRs is focused on the simplified equivalent circuit of the Fig. 8(b). This method was proposed in the reference (Aznar et al., 2008e). Since the number of parameters of the circuit model is five, we also need five conditions to univocally determine all parameters. From the representation of the reflection coefficient of a single unit cell, S_{11}, in the Smith chart, two conditions are obtained. Firstly, we can determine the frequency that nulls the series reactance, f_s, from the intercept of S_{11} with the unit conductance circle (see Fig. 13 a). This is obvious since at this frequency, the real part of the admittance seen from the ports is simply the admittance of the opposite port, that is, $Y_0=(Z_0)^{-1}=(50\Omega)^{-1}=0.02$ S. Hence, S_{11} must be allocated in the unit conductance circle at f_s, as illustrated in the example provided in Fig. 13. This frequency is given by the following expression:

$$f_S = \frac{1}{2\pi}\sqrt{\frac{1}{L'_S C'_S} + \frac{1}{L' C'_S}} \tag{19}$$

Secondly, the susceptance of the unit cell seen from the ports at f_s, whose value can be inferred from the Smith chart, is

$$B(\omega_S) = \frac{C L'_P \omega_S^2 - 2}{L'_P \omega_S} \tag{20}$$

with $\omega_S = 2\pi f_S$. The next condition concerns the parallel resonator of the series branch. Namely, the resonance frequency of this resonator is given by

$$f_Z = \frac{1}{2\pi}\sqrt{\frac{1}{L'_S C'_S}} \tag{21}$$

Notice that this frequency does not coincide with the intrinsic resonance frequency of the magnetically driven resonator, f_o (which is the resonance frequency of the tank formed by L_s and C_s). The transmission zero frequency f_z (Eq. 21) can be easily obtained from the transmission coefficient S_{21} of the unit cell since at this frequency the series branch is opened and the whole power injected from the input port is reflected back to the source. Thus, the transmission coefficient nulls (zero transmission) and f_z can easily be identified from the representation of the transmission coefficient in a decibel scale (see example in Fig. 13 b).

Another condition can be deduced from the phase of the transmission coefficient, ϕ_{S21}. At the frequency where $\phi_{S21}=90^\circ$, $f_{\pi/2}$, the electrical length of the unit cell, $\phi=\beta l$ (β being the phase constant and l the length of the unit cell), is $\phi\left(f_{\pi/2}\right) = -90^\circ$. Since the dispersion relation of a periodic structure consisting of cascaded unit cells, as those in Fig. 8(b) is

$$\cos\phi = 1 + \frac{Z_S(\omega)}{Z_P(\omega)} \tag{22}$$

with Z_s and Z_p being the series and shunt impedances, respectively, of the π-circuit model, it follows that

$$Z_S\left(\omega_{\pi/2}\right) = -Z_P\left(\omega_{\pi/2}\right) \tag{23}$$

with $\omega_{\pi/2} = 2\pi f_{\pi/2}$. Expressions (19-21) and (23) are four of the five conditions needed to univocally determine the circuit parameters in Fig. 8(b). Now, by removing the shunt connected vias or strips in the layouts in Fig. 3, we can represent the corresponding reflection coefficient on a Smith chart and read the susceptance seen from the ports at that frequency where S_{11} intercepts the unit conductance circle. Since this is simply the susceptance corresponding the line capacitance (provided L_p has been removed), we can thus univocally determine C. Hence, this is the fifth condition that is required to extract the circuit parameters of the circuit model in Fig. 8(b).

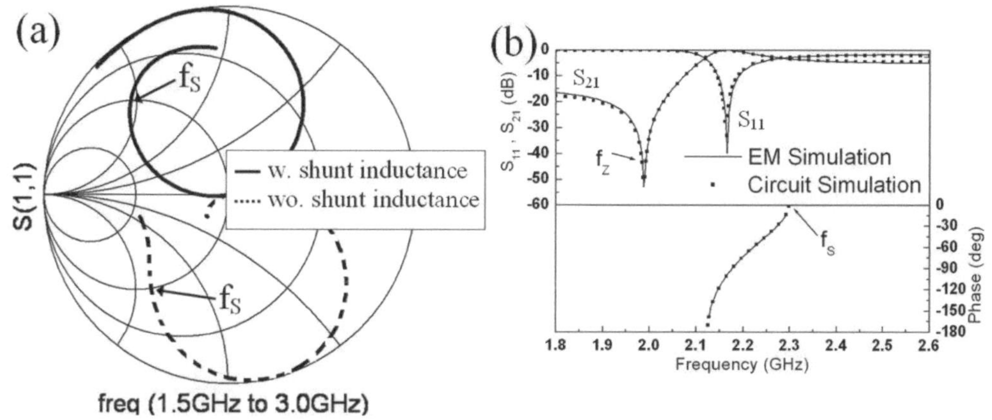

Fig. 13. Reflection coefficient on the Smith chart (a); frequency response (reflection, S_{11}, and transmission, S_{21}, coefficients) depicted in a decibel scale and the dispersion relation (b) for a left handed cell based on a CPW structure.

4.2 Parameter extraction technique metamaterial transmission lines based on Complementary Split-Ring Resonators (CSRRs)

The technique is based on the equivalent circuit model of a CSRR-loaded transmission line shown in Fig. 10(b). This method was proposed in 2006 (Bonache et al., 2006c). The considered structures are a negative permittivity as well as a left handed microstrip line (gap capacitors are required in the latter case). In view of the models, if losses are neglected (this is reasonable in a first order approximation), two characteristic frequencies can be identified: the frequency that nulls the shunt impedance (transmission zero frequency, f_z) and the frequency that nulls the shunt admittance (which obviously coincides with the intrinsic resonance frequency of the CSRR, f_0). These frequencies are given by the following expressions:

$$f_Z = \frac{1}{2\pi\sqrt{L_C(C+C_C)}} \tag{24}$$

$$f_0 = \frac{1}{2\pi\sqrt{L_C C_C}} \tag{25}$$

and they can be either experimentally determined, or obtained from the simulated response of the structure. At f_z a notch in the transmission coefficient is expected and this frequency can be accurately measured. To obtain f_0, a representation of the transmission coefficient on a Smith chart is required (see example in Fig. 14). At this frequency the shunt path to ground is opened, and the input impedance seen from the ports is solely formed by the series elements of the structure (L, for the negative permittivity line, and L and C_g for the left handed line) and the resistance of the opposite port (50 Ω). Therefore, f_0 is given by the intersection between the measured (or simulated) S_{11} curve and the unit normalized resistance circle. From this result we can also obtain the impedance of the series elements at that frequency. This gives directly the value of L for the negative permittivity line. For the left handed line, L can be independently estimated from a transmission line calculator, or from the value extracted for the negative permeability line, corrected by the presence of the gap, whereas C_g can be determined by adjusting its value to fit the impedance value read from the Smith chart at f_0 from the simulation or experiment.

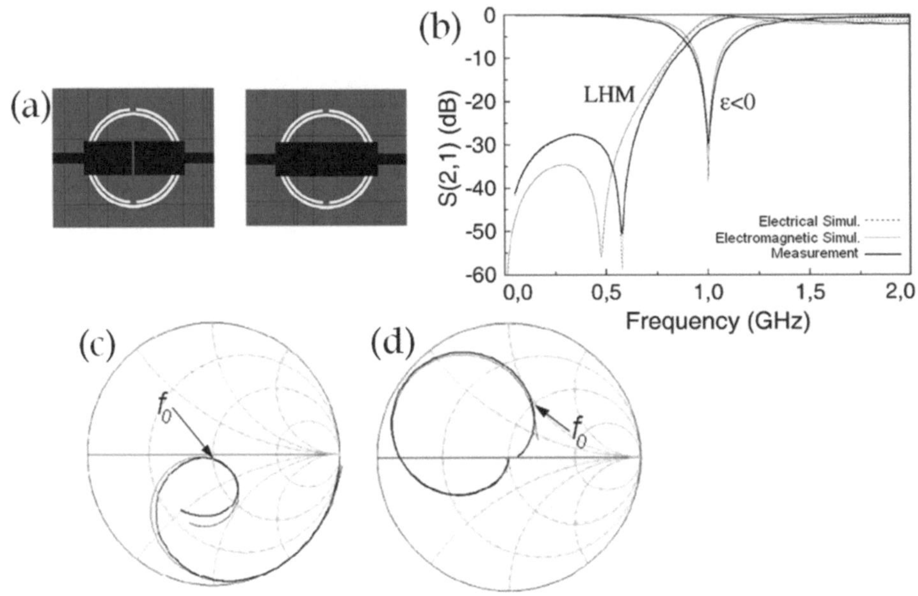

Fig. 14. (a) Layouts of the structures employed in the parameter extraction method for the CSRR-based unit cell. (b) Frequency responses of the measurement, the electromagnetic simulation and the electric simulation employing the extracted parameters. Representation of the S_{11} parameter for the identification of the CSRR resonance frequency (c) for the complete (LH) structure (d) for the structure without gap (ε<0). Measurement and electric simulation.

Expressions (24) and (25) are dependent on three parameters. Therefore, we cannot directly obtain the element values of the CSRR (as desired) and the coupling capacitance. To this end, we need an additional condition, namely:

$$Z_S\left(j\omega_{\pi/2}\right) = -Z_P\left(j\omega_{\pi/2}\right) \tag{26}$$

where $Z_S(j\omega)$ and $Z_P(j\omega)$ are the series and shunt impedance of the T-circuit model of the structure, respectively, and $\omega_{\pi/2}$ is the angular frequency where the phase of the transmission coefficient (which is a measurable quantity) is $\phi(S_{21}) = \pi/2$. Thus, from (23)-(25) we can determine the three reactive element values that contribute to the shunt impedance.

4.3 Parameter extraction technique for metamaterial transmission lines based on open resonators: Open Split-Ring Resonators (OSRRs) and Open Complementary Split-Ring Resonators (OSRRs)

This subsection is focused on the parameter extraction for the structures represented in Fig. 7. The parameters of the circuit model of a CPW loaded with an OSRR (Fig. 12 d) can be extracted from the measurement or the electromagnetic simulation of the structure following a straightforward procedure (Durán-Sindreu et al., 2009a, 2009b). First of all, from the intercept of the return loss curve with the unit conductance circle in the Smith chart, we can directly infer the value of the shunt capacitance according to

$$C = \frac{B}{2\omega_{Z_S=0}} \tag{27}$$

where B is the susceptance in the intercept point. The frequency at this intercept point is the resonance frequency of the series branch

$$\omega^2\Big|_{Z_S=0} = \frac{1}{C_S L'_S} \tag{28}$$

To determine the other two element values of this branch, one more condition is needed. This condition comes from the fact that at the reflection zero frequency ω_z (maximum transmission) the characteristic impedance of the structure is 50Ω. In this π-circuit, the characteristic impedance is given by

$$Z_0(\omega) = \sqrt{\frac{Z_P(\omega)^2 Z_S(\omega)}{2Z_P(\omega) + Z_S(\omega)}} \tag{29}$$

Thus, by forcing this impedance to 50Ω, the second condition results. By inverting (27) and (28), we can determine the element values of the series branch. The following results are obtained:

$$C_S = \left[\frac{\omega_Z^2}{\omega^2\Big|_{Z_S=0}} - 1\right]\left\{\frac{1}{2Z_0^2\omega_Z^2 C} + \frac{C}{2}\right\} \tag{30}$$

$$L'_S = \frac{1}{\omega^2\Big|_{Z_S=0} C'_S} \tag{31}$$

The parameters of the circuit model of a CPW loaded with an OCSRR (Fig. 12 b) can be extracted following a similar procedure. In this case, the intercept of the return loss curve with the unit resistance circle in the Smith chart gives the value of the series inductance

$$L = \frac{X}{2\omega\Big|_{Z_P\to\infty}} \tag{32}$$

where X is the reactance at the intercept point. The shunt branch resonates at this frequency, that is,

$$\omega^2\Big|_{Z_P\to\infty} = \frac{1}{L'_P C'_P} \qquad (33)$$

Finally, at the reflection zero frequency (ω_z), the characteristic impedance, given by

$$Z_0(\omega) = \sqrt{Z_S(\omega)\left[Z_S(\omega) + 2Z_P(\omega)\right]} \qquad (34)$$

must be forced to be 50Ω. From these two latter conditions, we finally obtain

$$L'_P = \left[\frac{\omega_Z^2}{\omega^2\Big|_{Z_P\to\infty}} - 1\right]\left\{\frac{Z_0^2}{2\omega_Z^2 L} + \frac{L}{2}\right\} \qquad (35)$$

$$C'_P = \frac{1}{\omega\Big|_{Z_P\to\infty} L'_P} \qquad (36)$$

and the element values are determined.

5. Conclusion

In this chapter, different kinds of resonant-type metamaterial transmission lines based on subwavelength resonators have been presented and studied. There are several types of resonators which allow their use in the implementation of this kind of artificial transmission lines. SRR-, CSRR- and open resonator-based structures have been presented and studied. The equivalent circuit models for each of the kinds of structures presented have been exposed, together with their parameter extraction methods. These methods provide the circuit model parameters, extracted from the frequency response of the structure, being a very useful design tool and allowing the corroboration of the proposed circuits as correct models for the considered structures. Furthermore, thanks to their accuracy, the parameter extraction methods have already been applied in automation procedures which allow the automatic generation of layouts for this kind of structures (Selga et al., 2010). After the imposition of certain conditions on the frequency response, which implies certain values for the circuit model parameters, the layout of the structure satisfying the required conditions can be automatically generated by means of space mapping techniques. This is a good example of the usefulness and importance of the equivalent circuit models and the parameter extraction methods.

6. References

Aznar, F.; Bonache, J. & Martín, F., Improved circuit model for left-handed lines loaded with split ring resonators, *Applied Physics Letters*, vol. 92, February 2008, pp. 043512-3, ISSN 0003-6951.

Aznar, F.; García-García, J.; Gil, M.; Bonache J. & Martín, F. Strategies for the miniaturization of metamaterial resonators, *Microwave and Optical Technology Letters*, vol. 50, May 2008, pp. 1263-1270, ISSN 1098-2760.

Aznar, F.; Gil, M.; Bonache J. & Martín, F., Revising the equivalent circuit models of resonant-type metamaterial transmission lines, *Proceedings of the IEEE MTT-S Int'l Microwave Symposium*, pp. 323-326, ISBN: 978-1-4244-1780-3 Atlanta (USA), June 2008.

Aznar, F.; Gil, M.; Bonache, J. & Martín, Modelling metamaterial transmission lines: a review and recent developments, *Opto-Electronics Review*, vol. 16, September 2008, pp. 226-236, ISSN 1230-3402.

Aznar, F.; Gil, M.; Bonache, J.; Jelinek, L.; Baena, J. D.; Marqués, R. & F. Martín, Characterization of miniaturized metamaterial resonators coupled to planar transmission lines through parameter extraction, *Journal of Applied Physics*, Vol 104, December 2008, 114501, ISSN 0021-4922.

Aznar, F.; Vélez, A.; Bonache, J.; Menés J. & Martín, F. Compact lowpass filters with very sharp transition bands based on open complementary split ring resonators, *Electronics Letters*, Vol. 45, No. 6, March 2009, pp. 316-317, ISSN 0013-5194.

Aznar, F.; Vélez, A.; Durán-Sindreu, M.; Bonache J. & Martín, F., Elliptic-function CPW low-pass filters implemented by means of open complementary split ring resonators (OCSRRs, *IEEE Microwave and Wireless Components Letters*, vol. 19, November 2009, pp. 689-691, ISSN 1531-1309.

Aznar, F.; Vélez, A.; Durán-Sindreu, M.; Bonache J. & Martín, F., Open complementary split ring resonators: physics, modelling, and analysis, *Microwave and Optical Technology Letters*, vol. 52, July 2010, pp. 1520-1526, ISSN 1098-2760.

Baena, J. D.; Bonache, J.; Martín F.; Sillero, R. M.; Falcone, F.; Lopetegi, T.; Laso, M. A. G.; García-García, J.; Gil, I.; Portillo, M. F. & Sorolla, M. Equivalent-circuit models for split-ring resonators and complementary split-ring resonators coupled to planar transmission lines, *IEEE Transactions on Microwave Theory and Techniques*, vol. 53, April 2005, pp. 1451-1461, ISSN 0018-9480.

Bonache, J.; Gil, I.; García-García, J. & Martín, F. Novel microstrip band pass filters based on complementary split rings resonators, *IEEE Transactions on Microwave Theory and Techniques*, vol. 54, January 2006, pp. 265-271, ISSN 0018-9480.

Bonache, J.; Martel, J.; Gil, I.; Gil, M.; García-García, J.; Martín, F.; Cairó, I. & Ikeda, M. Super compact (<1cm2) band pass filters with wide bandwidth and high selectivity at C-band, *Proceedings of the 36th European Microwave Conference*, pp. 599-602, ISBN 2-9600551-6-0, Manchester, UK, September 2006.

Bonache, J.; Gil, M.; Gil, I.; Garcia-García, J. & Martín, F., On the electrical characteristics of complementary metamaterial resonators, *IEEE Microwave and Wireless Components Letters*, vol. 16, October 2006, pp. 543.545, ISSN 1531-1309.

Caloz, C. & Itoh, T. (2005). *Electromagnetic Metamaterials: Transmission Line Theory and Microwave Applications*, Wiley Interscience, ISBN 978-0-471-66985-2.

Durán-Sindreu, M.; Aznar, F.; Vélez, A.; Bonache, J. & Martín, F. New composite, Right/Left handed transmission lines based on electrically small, open resonators, *Proceedings of IEEE MTT-S International Microwave Symposium*, pp. 45-48, ISBN 978-1-4244-2804-5, Boston (MA), USA, June 2009.

Durán-Sindreu, M.; Vélez, A.; Aznar, F.; Sisó, G; Bonache, J. & Martín, F., Application of open split ring resonators and open complementary split ring resonators to the synthesis of artificial transmission lines and microwave passive components, *IEEE Transactions on Microwave Theory and Techniques*, vol. 57, December 2009, pp. 3395-3403, ISSN 0018-9480.

Falcone, F.; Lopetegi, T.; Laso, M. A. G.; Baena, J. D.; Bonache, J.; Beruete, M.; Marqués, R.; Martín, F. & Sorolla, M. Babinet principle applied to the design of metasurfaces and metamaterials, *Physical Review Letters*, vol. 93, November 2004, 197401, ISSN 1079-7114.

Gil, M.; Bonache, J.; Gil, I.; García-García, J. & Martín, F., On the transmission properties of left handed microstrip lines implemented by complementary split rings resonators, *Int. Journal Numerical Modelling: Electronic Networks, Devices and Fields*, vol. 19, March 2006 , pp 87-103, ISSN 0894-3370.

Gil, M.; Bonache, J.; Gil, I.; García-García, J. & Martín, F., Miniaturization of planar microwave circuits by using resonant-type left handed transmission lines, *IET Microwave Antennas and Propagation*, Vol.1, February 2007, pp. 73-79, ISSN 1751-8725.

Gil, M.; Bonache, J.; García-García, J.; Martel, J. & Martín, F. Composite right/left-handed metamaterial transmission lines based on complementary split-rings resonators and their applications to very wideband and compact filter design", *IEEE Transactions on Microwave Theory and Techniques*, Vol. 55, No. 6, June 2007, pp. 1296-1304, ISSN 0018-9480.

Martel, J.; Marqués, R.; Falcone, F.; Baena, J. D.; Medina, F.; Martín F. & Sorolla, M. A new LC series element for compact bandpass filter design, *IEEE Microwave and Wireless Components Letters*, vol. 14, May 2004, pp. 210-212, ISSN 1531-1309.

Marqués, R.; Mesa, F.; Martel, J. & Medina, F. Comparative analysis of edge- and broadside-coupled split ring resonators for metamaterial design - Theory and experiments, *IEEE Transactions on Antennas and Propagation*, vol. 51, October 2003, pp. 2572-2581, ISSN 0018-926X.

Marqués, R.; Martín, F. & Sorolla, M. (2008). *Metamaterials with Negative Parameters*, John Wiley & Sons, Inc., ISBN 978-0-471-74582-2.

Martín, F.; Bonache, J.; Falcone, F.; Sorolla M. & Marqués, R. Split ring resonator-based left-handed coplanar waveguide, *Applied Physics Letters*, vol. 83, December 2003, pp. 4652-4654., ISSN 0003-6951.

Pozar, M, (1990). *Microwave engineering*, Addison Wesley, ISBN 978-0-471-44878-5.

Selga, J; Rodríguez, A.; Gil, M.; Carbonel, J.; Voria, V.E. & Martín, F., Towards the automatic layout synthesis in resonant-type metamaterial transmission lines, *IET Microwave Antennas and Propagation*, Vol.4, August 2010, pp. 1007-1015, ISSN 1751-8725.

Sisó, G.; Gil, M.; Aznar, F.; Bonache, J. & Martín, F. Generalized Model for Multiband Metamaterial Transmission Lines, *IEEE Microwave and Wireless Components Letters*, Vol. 18, Num. 11, November 2008, pp 728-730, ISSN 1531-1309.

Sisó, G.; Gil, M.; Aznar, F.; Bonache, J. & Martín, F. Dispersion engineering with resonant-type metamaterial transmission lines, *Laser & Photonics Review*, vol. 3, Issue 1-2, February 2009, pp. 12-29, ISSN 1863-8880.

Smith, D. R.; Padilla, W. J.; Vier, D. C.; Nemat-Nasser S. C. & Schultz, S. Composite medium with simultaneously negative permeability and permittivity, *Physical Review Letters*, vol. 84, May 2000, pp. 4184-4187, ISSN 0031-9007.

Vélez, A.; Aznar, F.; Bonache, J.; Velázquez-Ahumada, M. C.; Martel J. & Martín, F. Open complementary split ring resonators (OCSRRS) and their application to wideband cpw band pass filters, *IEEE Microwave and Wireless Components Letters*, vol. 19, April 2009, pp. 197-199, ISSN 1531-1309.

Vélez, A.; Aznar, F.; Durán-Sindreu, M.; Bonache, J. & Martín, F., Stop-band and band-pass filters in coplanar waveguide technology implemented by means of electrically small metamaterial-inspired open resonators, *IET Microwave Antennas and Propagation*, Vol.4, June 2010, pp. 712-716, ISSN 1751-8725.

Veselago, V. G. The electrodynamics of substances with simultaneously negative values of ε and μ, *Soviet Physics Uspekhi*, January 1968, pp. 509-514, ISSN 0038-5670.

13

Synthesis of Planar EBG Structures Based on Genetic Programming

Luisa Deias, Giuseppe Mazzarella and Nicola Sirena
University of Cagliari
Italy

1. Introduction

1.1 Electronic Band–Gap (EBG) structures

Frequency selective surfaces (FSS) consist of two-dimensional periodic arrays of metal patches patterned on a dielectric substrate or apertures etched on a metal screen (Hosseini et al., 2006; Wu, 1995; Yang et al., 1999).

These periodic structures resonate at certain frequencies, thus ensuring filtering characteristics, exploited both in the microwave and optical region of the electromagnetic spectrum. Frequency selective surfaces have been thoroughly studied over the years (Mittra et al., 1988; Munk, 2000), and they have found new life in the past decade when *electromagnetic bandgap* (EBG) (Sievenpiper et al., 1999; Yang et al., 1999) structures were introduced, firstly under the name of photonic bandgap (PBG) materials, in analogy to the bandgaps present in electric crystals, even though no photons were involved. Some well–known EBG structures are the Uniplanar Compact Photonic Band-Gap (UC–PBG)(Yang et al., 1999) (Fig. 1) and the Sievenpiper "mushroom" high–impedance surface (Sievenpiper et al., 1999).

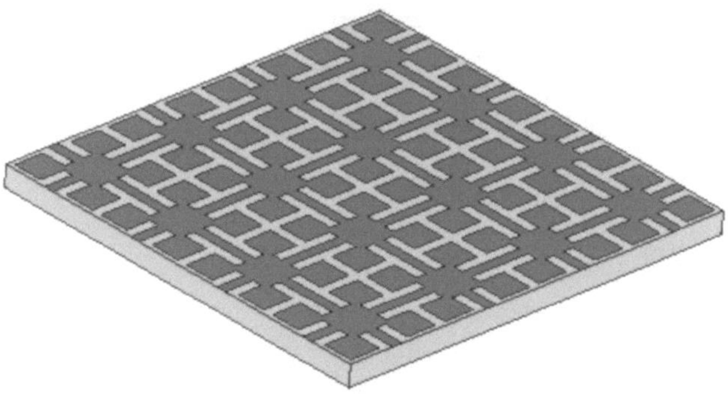

Fig. 1. Uniplanar Compact Photonic Band-Gap (UC–PBG)(Yang et al., 1999)

This entirely new class of structures, encompassing FSS as one of its subclasses (planar EBG) display some very interesting new electromagnetic properties. The presence of a stopband for this structure has been theoretically and experimentally verified and exploited in different realizations, i.e. TEM waveguide, slow–wave planar structure and low–loss conductor–backed coplanar waveguide.

By choosing the proper geometry of the periodic surface we can shape the electromagnetic behavior of these structures and they can be made to act as a so–called electromagnetic crystal, exhibiting frequency bands inside which the propagation of electromagnetic waves is not allowed or is highly attenuated. The concept of suppressing surfaces waves on metals is not new. Surface waves can be eliminated from a metal surface over a finite frequency band by applying a periodic texture. It has been done long before EGB structures were introduced using several geometries, such as a metal sheet covered with small bumps or a corrugated metal slab.

The novelty of EBGs is the application of an array of periodic patches or apertures, i.e. lumped circuit elements, to produce a thin two–dimensional structure that must generally be described by band structure concepts, even though the thickness and periodicity are both much smaller than the operating wavelength.

These periodic structures can also be designed to act as an *artificial magnetic conductor* (AMC) or high–impedance electromagnetic ground plane over a desired (quite) narrow frequency range, corresponding to the forbidden frequency band. Hence, the key feature of these structures is the reflection of an incident plane wave with no phase reversal, unlike normal metal surfaces (Sievenpiper et al., 1999).

High–impedance surfaces are widely studied now as promising antenna substrates (Feresidis et al., 2005; Gonzalo et al., 1999; Hosseini et al., 2006). A possibility of realizing a magnetic wall near the resonant frequency of a very thin structure is very attractive, since this allows one to design low–profile antennas and enhance the performance of printed antennas. The main drawback of this strategy is the reduced bandwidth of the complete antenna, since the frequency range over which these EBG surfaces behave as an AMC is usually narrowband and fixed by their geometrical configuration. The ultimate goal is then to design and incorporate such metamaterial–substrates in antenna structures in order to improve antenna performance.

A key issue in the research field of metamaterials is then represented by the design and optimization of EBG structures. Different techniques have been investigated, among which methods of global optimization such as genetic algorithm and particle swarm optimization (PSO),(Bray et al., 2006; Ge et al., 2007; Kovács et al., 2010; Tavallaee & Rahmat-Samii, 2007; Yeo et al., 2002).

1.2 Genetic programming

Genetic Programming (GP) (Koza, 1992) falls into the larger class of evolutionary computations (Fogel, 2006; Michalewicz, 1992), including genetic algorithms, evolution strategies and evolutionary programming, which can be described as highly parallel probabilistic search algorithms imitating the principles of natural evolution for optimization problems, based on the idea that most real–world problems cannot be handled with binary representations. Such algorithms allow task specific knowledge emerge while solving the problem. Genetic Programming, a variant of genetic algorithms yet with marked differences, is an especially interesting form of computational problem solving.

Genetic Algorithms (GA), already widely used in antenna design (Jones & Joines, 2000; Lohn et al., 2001) and more recently also applied to EBG design and optimization (Bray et al., 2006; Ge et al., 2007; Yeo et al., 2002) iteratively transform populations of mathematical objects (typically fixed–length binary character strings), each with an associated fitness value, into new populations using the Darwinian principle of natural selection.

In GP, unlike GA chromosomes need not be represented by bit-strings and the alteration process includes other "genetic" operators appropriate for the given structure and the given problem. We can say that GA works on the "nucleotide" (i.e. bit) level, in the sense that the antenna or EBG structure is completely defined from the beginning and only a handful of parameters remains to be optimized. The approach proposed by Koza assumes no "a priori" structure. Instead, it builds up the structure of the individuals as the procedure evolves. As a consequence, its solution space has the power of the continuum, while the GA solution space is a discrete one, so it is a very small subspace of the former. The goal of genetic programming is not simply to evolve a bit–string representation of some problem but the computer code that solves that problem.

A fixed–length coding such as GA is rather artificial. As it cannot provide a dynamic variability in length, such a coding often causes considerable redundancy and reduces the efficiency of genetic search. In contrast, GP uses high–level building blocks of variable length. Their size and complexity can change during breeding. Moreover, the typical evolution operators work on actual physical structures rather than on sequences of bits with no intuitive link to the EGB surface shape. The enormous power of this strategy fully allows the exploration of more general shapes.

Getting machines to produce human–like results is the reason for the existence of the fields of artificial intelligence and machine learning. Genetic programming addresses this challenge by providing a method for automatically creating a working computer program from a high–level description of the problem. This is the reason why, unlike Genetic Algorithms, Genetic Programs often deliver elegant human–like solutions not anticipated by the programmer, providing only a minimum amount of pre–supplied human knowledge, analysis and information.

Yet, while allowing to explore and evaluate general configurations, this approach can lead to a severely ill–conditioned synthesis problem. A suitable stabilization is therefore obtained by imposing problem–specific requirements, in our case the periodicity of the the surface elements, which is directly related to the resonant frequency, and the physical parameters, which are not so relevant to the problem. We therefore let then the Genetic Programming strategy evaluate every possible shape in the solution space we delimited.

Representation is a key issue in genetic programming (Koza, 1992) because the representation scheme can severely limit the window through which the system "observes" its world.

Since GP manipulates programs by applying genetic operators, a programming language such as LISP was chosen, since it allows each individual, i.e. computer program, to be manipulated as data. Any LISP S–expression can be depicted as a rooted point-labelled tree with ordered branches, as shown in Fig.2. Therefore in GP each individual is a computer program, described through the set of instructions needed to "build" it, typically implemented in S–expressions.

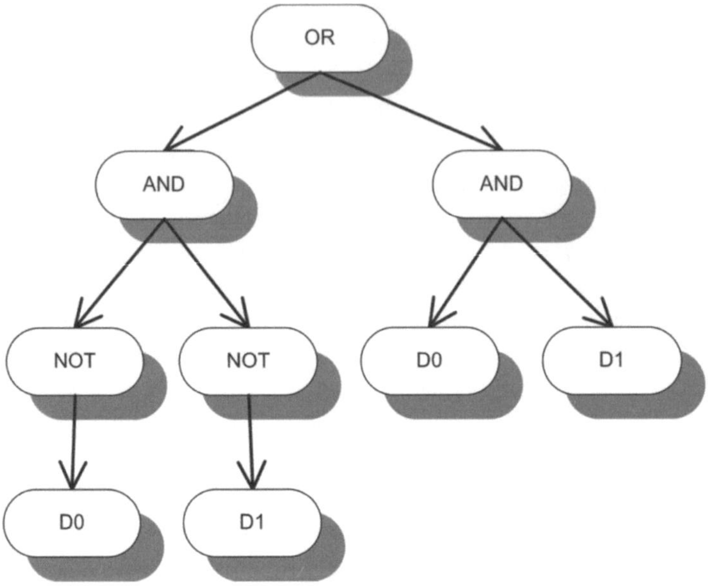

Fig. 2. S-expression: tree architecture.

While the evolutionary "strategy" is almost standard for a variety of problems (we used ECJ, a general purpose Java-based Evolutionary Computation research system and developed at George Mason University ECLab), three elements must be defined to evolve a design, i.e. "individual":

- REPRESENTATION SCHEME:definition of the design space
- VARIATION OPERATORS: code that takes one or more design representations as input and outputs a design derived from them
  - CROSSOVER:creates new S–expressions by exchanging sub S–expressions between two S–expressions. The sub S-expressions exchanged are selected randomly.
  - MUTATION: creates a new S–expression by replacing an existing argument symbol with the other possible symbol. The argument symbol replaced is selected randomly.
- FITNESS FUNCTION: a function that evaluates the individuals

In other words, representation refers to a form of data structure. Variation operations are applied to existing solutions to create new solutions. Usually, variation is based on random perturbation.

The starting point is an initial population of randomly generated computer programs composed of functions and terminals appropriate to the specified problem domain. The evolutionary strategy works in order to find the best individual, in terms of their closeness to the constraints set in the design and evaluated as a "fitness" function. This strategy let to interesting and promising results in the synthesis of EBGs, even when considering more complex structures and requirements (Deias et al., 2009a;b; 2010).

2. EBG design usign genetic programming

In the design of frequency selective surfaces, the choice of the proper element may be of utmost importance. In fact some elements are intrinsically more broad–banded or more narrow–banded than others, while some can be more easily varied by design. In literature we can find a large variety of element types, the more common are illustrated in Fig. 3. It can be now devised the great potential of Genetic Programming, since we don't really want any possible geometry to be taken into account but only feasible geometries, just like a human being would design them on a piece of paper, yet faster and with much more "fantasy".

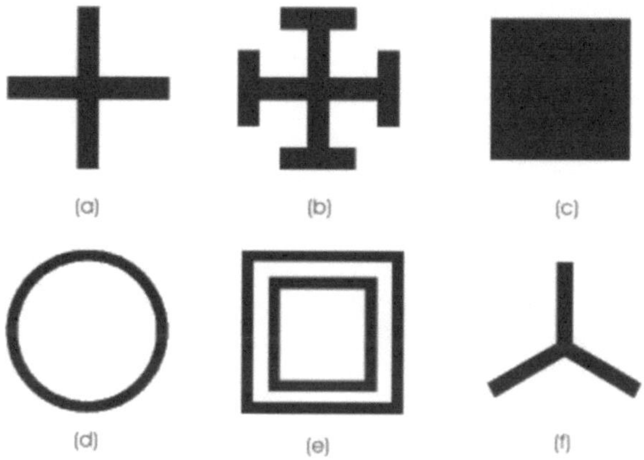

Fig. 3. Some typical FSS unit cell geometries: (a) cross dipole (b) Jerusalem cross (c) square patch (d) circular loop (e) square loops (f) tripole.

Therefore the goal of the design process is to obtain the unit cell aperture geometry of the periodic surface which fulfills the desired requirements on the resonant frequency of the periodic structure, analyzed with a full–wave technique. The GP approach has been implemented in Java, while the full–wave analysis of the periodic structure for each individual has been implemented in Fortran.

The S–expression of an individual of the population can take for example the following form:

```
 Tree 0:
 (Branch (Rettangolo 0.16589776 0.2873323
  (Ruota 90.0 (Ruota -90.0 END))) (Ruota 90.0
    (Rettangolo 0.6837649 0.2757175 (Rettangolo 0.7254247 0.23175128
     (Ruota 90.0 (Ruota 90.0 (Ruota -90.0 (Rettangolo 0.19471756 0.19593427
      END))))))))
Tree 1:
 (Branch (Branch END END) (Ruota 90.0 END))
Tree 2:
 (Branch END END)
Tree 3:
 (Ruota 90.0 (Rettangolo 0.51570845 0.25133058
   (Ruota 90.0 (Branch (Ruota -90.0 END) (Rettangolo 0.75231904 0.2833914
    END)))))
```

The corresponding input to the Fortran executable is the following, i.e. a collection of rectangular patches that form the geometry of the aperture:

```
Patch 48
-0.0 , -0.0 , 0.4572 , 0.3048
-0.4572 , -0.0 , 0.4572 , 0.3048
-0.0 , -0.3048 , 0.4572 , 0.3048
-0.4572 , -0.3048 , 0.4572 , 0.3048
-0.3048 , -0.0 , 0.3048 , 0.4572
-0.0 , -0.0 , 0.3048 , 0.4572
-0.3048 , -0.4572 , 0.3048 , 0.4572
-0.0 , -0.4572 , 0.3048 , 0.4572
0.3048 , -0.0 , 0.15239999 , 0.762
-0.4572 , -0.0 , 0.15239999 , 0.762
...
-1.3716 , -0.0 , 0.762 , 0.3048
0.6096 , -0.0 , 0.762 , 0.3048
-0.0 , -0.0 , 0.1524 , 0.3048
-0.1524 , -0.0 , 0.1524 , 0.3048
-0.0 , -0.3048 , 0.1524 , 0.3048
-0.1524 , -0.3048 , 0.1524 , 0.3048
-0.3048 , -0.0 , 0.3048 , 0.1524
-0.0 , -0.0 , 0.3048 , 0.1524
-0.3048 , -0.1524 , 0.3048 , 0.1524
-0.0 , -0.1524 , 0.3048 , 0.1524
```

And the corresponding geometry of the aperture is shown in Fig.4.

Fig. 4. Aperture Geometry.

In our specific problem, we thus obtain a set of rectangular patches which describe the aperture geometry of the EBG unit cell.

The fitness criteria measure the quality of any given solution. The selection method uses the score obtained for each solution to determine which to save and which to eliminate from the population at each generation. Those solutions that survive are the "parents" of the next generation. The initialization of an evolutionary algorithm can be completely at random, or can incorporate human or other expertise about solutions that may work better than others.

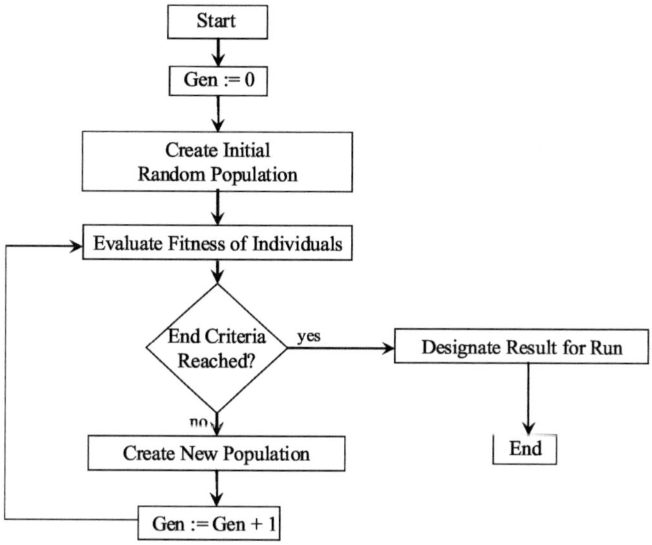

Fig. 5. Flowchart.

The fitness function employed by our GP optimization is:

$$FF = \omega_1 * \omega_2 * e^{\frac{Phase(f_C)}{20}} \quad (1)$$

where ω_1 and ω_2 are penalty coefficients developed in order to avoid geometries with high number of discretization elements within the unit cell and allowing, when using a three frequency points evaluation, individuals with larger bandwidth to prevail.

A full–wave simulation using Method of Moments is carried out on a single design, i.e. unit cell of the infinite periodic surface and individual of a set population. The starting point is a random initial population of individuals. The external parameters of the structure (i.e. substrate thickness and dielectric constant) are fixed, altogether with the periodicity of the planar EBG. The geometry of the aperture is subject to the GP optimization, while the unit square cell within which each individual/design can evolve is fixed together with its discretization step.

The set of admissible solution is then composed by every geometry/aperture that can be designed in this square, built as series of segments that can evolve in every direction, with no limit on the number of segments, on their width and length and number of subsequent ramifications. We have fixed the discretization step within the unit cell in order to pose a limit not only to the computational burden but also to the geometrical spatial bandwidth of the possible solutions. In such a way we obtain a stabilizing effect on the problem, as much as shown in (Collin, 1985) for a different problem, namely wire antennas.

The phase of the reflection coefficient, computed at one or more frequencies depending on the goal, is then incorporated into the GP strategy for the evaluation procedure, the fitness function determining the environment within which the solutions "live" and the best individual of a generation is selected. The evolutionary operators (reproduction, crossover and mutation) are then applied to the best individual leading to the subsequent evolution of the population and the next generation. In this way the geometrical properties are optimized to evolve the best solution (zero phase at the desired resonant frequency).

3. Single and multi–layer EBG full–wave analysis

Various techniques have been proposed in literature in order to analyze frequency selective surfaces (FSS) (Bardi et al., 2002; Bozzi & Perregrini, 1999; Harms et al., 1994; Mittra et al., 1988; Wu, 1995). The standard method of moments (MoM) approach is based on the induced electric currents on the FSS but it can be prohibitively costly from a computational point of view, if not even impractical in some cases, and this could explain why MoM is so unpopular.

Frequency Selective Surfaces (FSS) consist of two-dimensional periodic arrays of metallic patches patterned on a dielectric substrate or apertures etched on a metal screen, either entities can be isolated or connected in a rectangular grid. Hence, for any FSS, we can consider either the periodicity of the metallic patches or that of the apertures. By taking properly into account the continuity of electric or magnetic currents along adjacent cells we can conveniently adopt one of the two approaches, considering as unknowns of the problem either the electric or magnetic currents.

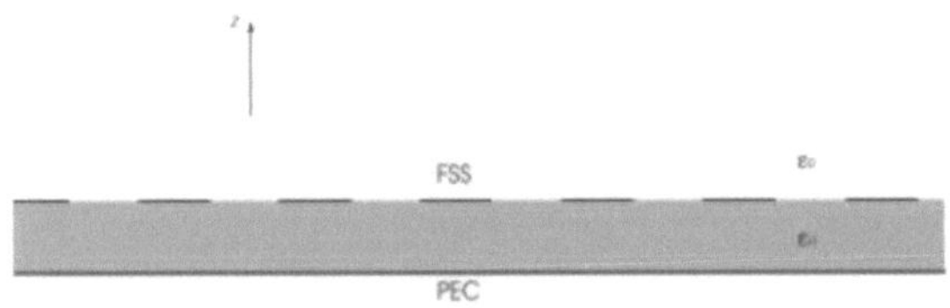

Fig. 6. Cross section view of a single layer FSS with dielectric substrate on ground plane.

The integral equation for the electric field (EFIE) applied to the periodic array of metal patches and the integral equation for the magnetic field (MFIE) applied to the periodic array of apertures thus represent two alternative formulations of the same problem, yielding to the same solution. The corresponding equivalent circuit for each approach is shown in Fig.8

Using a MoM approach based on the apertures, a very effective procedure can be devised (Deias & Mazzarella, 2006).

The aperture oriented approach (i.e. MFIE formulation) starts by applying the equivalence theorem (Balanis, 1996). The aperture is closed by a conductive sheet, and two unknown magnetic current densities $\mathbf{M}_A$ and $-\mathbf{M}_A$ on the opposite sides of the conductive sheet, as shown in Fig.7, are defined so that the continuity of the tangential electric field is guaranteed.

The MoM matrix is then obtained by imposing the continuity of the tangential component of the magnetic field across the aperture. The magnetic field is computed using a Green function in the spectral domain, too. But the relevant equivalent circuit, shown in Fig.8(b), allows to decouple the two regions below and above the metallization. This is a key feature not available in the conventional method based on the EFIE: in Fig.8(a) we can see that the equivalent circuit in this case bounds together the different regions.

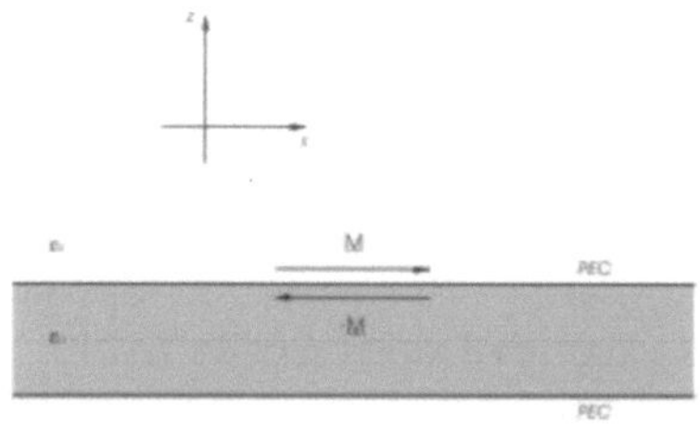

Fig. 7. Unit cell geometry of a Frequency Selective Surface with a dielectric substrate on ground plane, and magnetic currents equivalent to an aperture. The air–dielectric interface is a PEC.

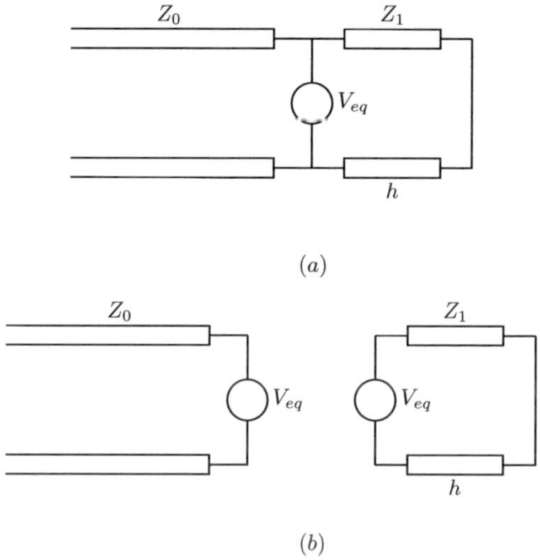

Fig. 8. Equivalent circuits resulting from the equivalence applied to a FSS (in the spectral domain): (a) metallic patches; (b) apertures.

As a consequence of our aperture approach, the MoM matrix can be decoupled in the sum of "localized" admittance matrices, each one relevant to a single region. For multi–layered FSS the advantages of the aperture approach are far more greater (Asole et al., 2007).

Let's consider the 3–layer AMC structure as shown in Fig.9, since it contains all the features of the method. The generalization to N-layer FSS is really straightforward.

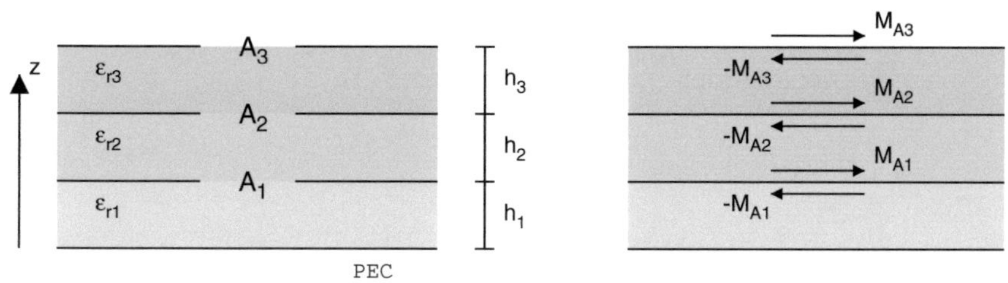

Fig. 9. Geometry of a 3–layer structure and magnetic currents equivalent to the apertures (central cell).

The dielectric layers can be chosen with the desired physical parameter and a superstrate can also be used to cover the structure. The unit cell of the periodic printed surfaces on the different layers may be of different shape but they must have the same periodicity.

According to the equivalence theorem, each aperture is closed by a PEC, and two unknown magnetic current densities $\mathbf{M}_{A_{nm}}$ and $-\mathbf{M}_{A_{nm}}$ on the opposite sides are defined in order to guarantee the continuity of the tangential electric field. By the Floquet theorem, the equivalent currents on the $(n, m)th$ cell of the periodic surface are connected to the central one by:

$$\mathbf{M}_{A,nm} = e^{-j\beta n d_x x} e^{-j\beta m d_y y} \mathbf{M}_A \tag{2}$$

where $\mathbf{M}_A$ is the current in the central cell.

As a consequence, only the current in the central cell is the unknown of the problem, and need to be discretized as a linear combination of N_b basis functions:

$$\mathbf{M}_A = \sum_{n=1}^{N_b} m_n^A \, \mathbf{f}_n \tag{3}$$

Then we obtain the MoM linear equation system by imposing the continuity of the tangential component of the magnetic field across each aperture. Again, by the Floquet theorem, this continuity needs to be forced only on the central cell (see Fig.**??**). It can be easily seen that, in the multilayered structure, we can distinguish three cases, namely an aperture lying on the first layer, on an intermediate layer or on the last layer.

For the first layer, i.e. for the discontinuity A_1, the continuity of the transverse magnetic field can be written as:

$$\mathbf{H}_{1,t}^{A_1A_1} = \mathbf{H}_{2,t}^{A_1A_1} + \mathbf{H}_{2,t}^{A_2A_1} \tag{4}$$

which is then enforced in a *weak* form. A set of testing functions is selected and the boundary condition on the magnetic field is multiplied by each testing function and the result integrated over aperture. Choosing the Galerkin method, i.e. the testing functions used are the same basis functions used to express the unknown magnetic currents, we obtain for layer A_1:

$$\int_{A_1} \mathbf{H}_{1,t}^{A_1A_1} \cdot \mathbf{f}_m = \int_{A_1} \mathbf{H}_{2,t}^{A_1A_1} \cdot \mathbf{f}_m + \int_{A_1} \mathbf{H}_{2,t}^{A_2A_1} \cdot \mathbf{f}_m, \qquad m = 1, \ldots, N_b \tag{5}$$

which, introducing the Green Functions can be written as:

$$\int_{A_1} \langle \hat{\mathbf{G}}_1^{A_1A_1}, -\mathbf{M}_{A_1} \rangle \cdot \mathbf{f}_m = \int_{A_1} \langle \hat{\mathbf{G}}_2^{A_1A_1}, \mathbf{M}_{A_1} \rangle \cdot \mathbf{f}_m + \\ \int_{A_1} \langle \hat{\mathbf{G}}_2^{A_2A_1}, \mathbf{M}_{A_2} \rangle \cdot \mathbf{f}_m, + \int_{A_1} \langle \hat{\mathbf{G}}_2^{A_2A_1}, \mathbf{M}_{A_2} \rangle \cdot \mathbf{f}_m, \qquad m = 1, \ldots, N_b \tag{6}$$

Expressing the unknown magnetic currents in the form (3):

$$\sum_{n=1}^{N_b} -m_n^{A_1} \left(\int_{A_1} \left(\int_{A_1} \hat{\mathbf{G}}_1^{A_1A_1} \cdot \hat{\mathbf{f}}_n^{A_1} \right) \cdot \mathbf{f}_m^{A_1} + \int_{A_1} \left(\int_{A_1} \hat{\mathbf{G}}_2^{A_1A_1} \cdot \hat{\mathbf{f}}_n^{A_1} \right) \cdot \mathbf{f}_m^{A_1} \right) \\ = \sum_{n=1}^{N_b} m_n^{A_2} \int_{A_1} \left(\int_{A_1} \hat{\mathbf{G}}_2^{A_2A_1} \cdot \hat{\mathbf{f}}_n^{A_2} \right) \cdot \mathbf{f}_m^{A_1}, \qquad m = 1, \ldots, N_b \tag{7}$$

where the l.h.s. term represents the self–term, i.e. the effect of the magnetic current $\mathbf{M}_{A_1}$ on the aperture A_1 , and the r.h.s. term is the coupling term, i.e. the effect of the magnetic current $\mathbf{M}_{A_2}$ on the aperture A_1.

For the second discontinuity A_2, as for any intermediate aperture, enforcing the boundary condition we obtain:

$$\mathbf{H}_{2,t}^{A_2A_2} + \mathbf{H}_{2,t}^{A_1A_2} = \mathbf{H}_{3,t}^{A_2A_2} + \mathbf{H}_{3,t}^{A_3A_2} \tag{8}$$

which leads to:

$$\sum_{n=1}^{N_b} -m_n^{A_2} \left(\int_{A_2} \left(\int_{A_2} \hat{\mathbf{G}}_2^{A_2A_2} \cdot \hat{\mathbf{f}}_n^{A_2} \right) \cdot \mathbf{f}_m^{A_2} + \int_{A_2} \left(\int_{A_2} \hat{\mathbf{G}}_3^{A_2A_2} \cdot \hat{\mathbf{f}}_n^{A_2} \right) \cdot \mathbf{f}_m^{A_2} \right) =$$
$$\sum_{n=1}^{N_b} m_n^{A_3} \int_{A_2} \left(\int_{A_2} \hat{\mathbf{G}}_3^{A_3A_2} \cdot \hat{\mathbf{f}}_n^{A_3} \right) \cdot \mathbf{f}_m^{A_2} - \sum_{n=1}^{N_b} m_n^{A_1} \int_{A_2} \left(\int_{A_2} \hat{\mathbf{G}}_2^{A_1A_2} \cdot \hat{\mathbf{f}}_n^{A_1} \right) \cdot \mathbf{f}_m^{A_2} \tag{9}$$

where $m = 1, \ldots, N_b$ and the l.h.s. term represents the self–term and the r.h.s. term the coupling terms, i.e. the effect of the magnetic currents $\mathbf{M}_{A_3}$ and $\mathbf{M}_{A_1}$ on the aperture A_2.

For the last (third) discontinuity A_3, the boundary condition leads to:

$$\mathbf{H}_{3,t}^{A_3A_3} + \mathbf{H}_{3,t}^{A_2A_3} = \mathbf{H}_{4,t}^{A_3A_3} + \mathbf{H}_{inc,t} \tag{10}$$

and we obtain:

$$\sum_{n=1}^{N_b} -m_n^{A_3} \left(\int_{A_3} \left(\int_{A_3} \hat{\mathbf{G}}_3^{A_3A_3} \cdot \hat{\mathbf{f}}_n^{A_3} \right) \cdot \mathbf{f}_m^{A_3} + \int_{A_3} \left(\int_{A_3} \hat{\mathbf{G}}_4^{A_3A_3} \cdot \hat{\mathbf{f}}_n^{A_3} \right) \cdot \mathbf{f}_m^{A_3} \right)$$
$$= - \sum_{n=1}^{N_b} m_n^{A_2} \int_{A_3} \left(\int_{A_3} \hat{\mathbf{G}}_3^{A_2A_3} \cdot \hat{\mathbf{f}}_n^{A_2} \right) \cdot \mathbf{f}_m^{A_3} - \int_S \mathbf{H}_{inc} \cdot \mathbf{f}_m^{A_3}(x,y)\, dS \tag{11}$$

where $m = 1, \ldots, N_b$ and, on the l.h.s we have the self–term and the first r.h.s. term is the coupling term, i.e. the effect of the magnetic current $\mathbf{M}_{A_2}$ on the aperture A_3. The second term of the r.h.s represent the known terms of the MoM system.

The presence of the coupling terms is the critical point in the multi–layer formulation. The key feature of the aperture approach, as previously illustrated, relies on the decoupling of the different regions. If we have a more general structure made up of N layers, we know that when dealing with the aperture A_1, which lies in the bottom layer, we only have to consider the effect of the magnetic current on the aperture A_2. When considering the aperture A_N, the last one, we only have to consider the effect of the magnetic current on the aperture A_{N-1}, and when considering any intermediate aperture A_i, for $i = 2, ..., N-1$ we have to take into account only the effect of the apertures immediately below and above, i.e. A_{i-1} and A_{i+1}. Equations (7),(9),(11) can be written in matrix form:

$$\mathcal{Y} \cdot \mathcal{M} = \mathcal{T} \tag{12}$$

where $\mathcal{Y}$ is the coefficient matrix, $\mathcal{M}$ is the vector of the unknown coefficients, i.e. the magnetic current on the apertures, and $\mathcal{T}$ is the r.h.s, which represent the incident magnetic field on the last aperture A_3.

The solution of the above equation (12) yields the unknown induced magnetic current on the apertures. The reflection coefficient is then easily calculated from the magnetic current on the upper layer.

It is important to stress out that inserting more layers will result in a larger matrix, affecting thus the overall computational time, yet for each additional layer only two blocks have to be computed.

The explicit evaluation of MoM system matrix $\mathcal{Y}$ requires the computation of the magnetic fields we previously introduced in (6),(8),(10). These are total magnetic fields,i.e. fields due to the magnetic currents of all the periodicity cells. Since the problem is linear, we start computing the magnetic field due to the magnetic current of a single cell. This field can be expressed in terms of a spectral domain (dyadic) Green function $\hat{\mathbf{G}}(u,v)$ connecting the transverse magnetic current density on the aperture to the transverse magnetic field:

$$\mathbf{H}_t^{unit\ cell}(x,y) = \int_{-\infty}^{+\infty} \int_{-\infty}^{+\infty} \hat{\mathbf{G}}(u,v) \cdot \hat{\mathbf{M}}(u,v)\, e^{-j(ux+vy)}\, du\, dv \tag{13}$$

where $\mathbf{M}$ is the magnetic current on the aperture of the central cell.

We can then write the overall magnetic transverse field of the infinite planar array by summing (13) on the infinite cells of the AMC:

$$\begin{aligned} \mathbf{H}_t(x,y) &= \sum_r \sum_s e^{-jk_0 r d_x \sin\theta\cos\phi}\, e^{-jk_0 s d_y \sin\theta\sin\phi} \,\cdot \\ &\cdot \int_{-\infty}^{+\infty} \int_{-\infty}^{+\infty} \hat{\mathbf{G}}(u,v) \cdot \hat{\mathbf{M}}(u,v)\, e^{-j(ux+vy)}\, du\, dv \end{aligned} \tag{14}$$

where d_x and d_y are the periodicity along the x–axis and the y–axis respectively, (θ, ϕ) is the direction of the incident field and k_0 is the free–space propagation constant.

Equation (14) can be simplyfied taking into account the periodicity of the structure. The fields on either side of the AMC can be expanded in terms of the Floquet space harmonics (Munk, 2000). Therefore the cell summation in (14) can be rearranged (Pozar & Schaubert, 1984) using the Poisson formula. As a result, we obtain the following compact expression:

$$\mathbf{H}_t(x,y) = \frac{4\pi^2}{d_x d_y} \sum_r \sum_s \hat{\mathbf{G}}(u_r, v_s) \cdot \hat{\mathbf{M}}(u_r, v_s)\, e^{-j(u_r x + v_s y)} \tag{15}$$

where the summation is now on the Floquet modes.

Equation (15) can be used for both fields (above and below the metallization) by using the pertinent Green function, which can be derived from the equivalent circuit shown in Fig.8*(b)* and setting $\mathbf{M} = \pm\mathbf{M}_A$ (current on the pertinent layer of the central cell).

4. Single–layer EBG design results

Significative and promising results have been found when this approach was applied to a single–layer EBG structure (Deias et al., 2009a).

One of the most interesting structures that can be found in literature is the Itoh's UC–EBG structure (Yang et al., 1999), shown in Fig.10(a) consisting of a Jerusalem cross aperture.

The reference structure has periodicity $d_x = d_y = 3.048mm$, $h = 0.635mm$ and $\varepsilon = 10.2$, displaying a resonant frequency at 14.25GHz. The reflection coefficient is shown in Fig.10(b).

We decided to use these parameters as constraints for our test, in order to compare the results obtained using the GP approach with Itoh's UC–EBG.

We used for the full–wave analysis of the periodic structure our previously developed Fortran code, implementing the aperture approach described in the previous section. Furthermore

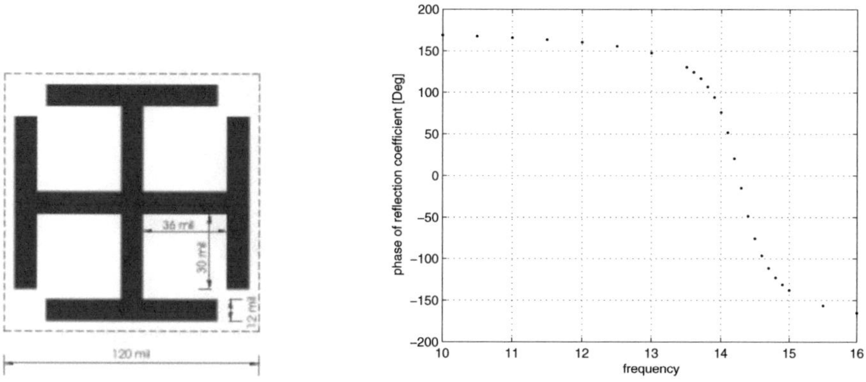

(a) Geometry (b) Reflection Coefficient

Fig. 10. UC–EBG

this approach is irrespective of the particular shape of the metallization, which in the GP optimization stage cannot be known in advance.

The GP optimization was launched to find the best structure, keeping as fixed parameters the periodicity, substrate dielectric constant and thickness, as shown in the following table.

Fixed Parameters	
Center Frequency	14.2 GHz
Periodicity (unit square cell)	$d_x = 3.048$ mm; $d_y = 3.048$ mm
Substrate dielectric constant	$\epsilon_r = 10.2$
Substrate thickness	0.635 mm
Discretization step	0.1016 mm $(30 \times 30$ *grid*$)$

Table 1. Data

In Fig.11 and Fig.12 we can see two different runs of our optimization. And in Tab.2 it is shown how the goal is achieved, in comparison also to the reference structure.

	Phase of Refl. Coeff.		
	14 GHz	**14.2 GHz**	**14.4 GHz**
Itoh (lit.)	76.05°	20.52°	-48.56°
Case A	57.8°	2.2°	-63.16°
Case B	39.28°	0.83°	-35.16°
Case C	13.06°	-6.13°	-24.93°

Table 2. Results

In Case A we used for the unit cell a discretization step of 0.1016mm, making the playground for the GP optimization a 30x30 grid. The penalty coefficients, both for the number of unknowns (i.e. borders of the discretized geometry) and for the bandwidth were initially not too strict, making the evolution process quickly evolve towards quite bulky structures. The resonant frequency was reached after few generations (for case A at gen. 13 the phase of the reflection coefficient is 2.2°).

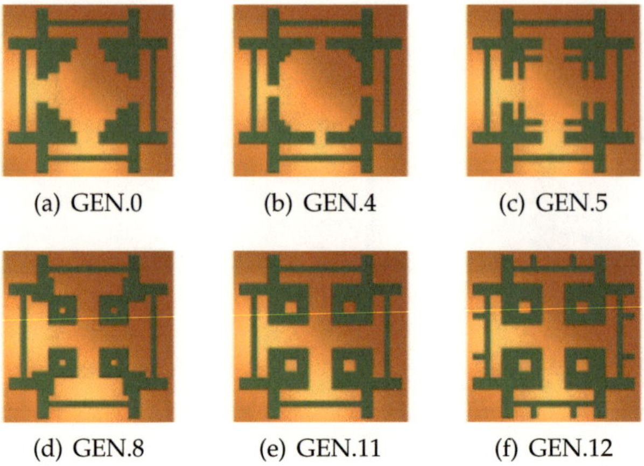

(a) GEN.0 (b) GEN.4 (c) GEN.5

(d) GEN.8 (e) GEN.11 (f) GEN.12

Fig. 11. Case A.

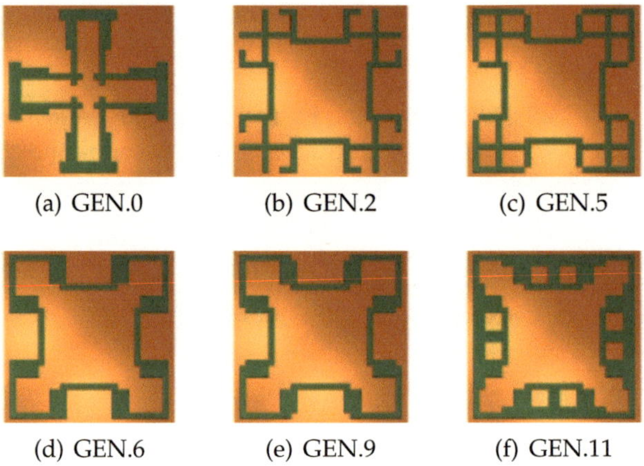

(a) GEN.0 (b) GEN.2 (c) GEN.5

(d) GEN.6 (e) GEN.9 (f) GEN.11

Fig. 12. Case B.

In Case B we then acted more stringently on the penalty coefficients. The structure remains spatially more limited, as the number of unknowns is kept reasonably low, and the bandwidth is largely improved. The number of generations in order to achieve a good result is still low.

As expected, the geometries resemble the reference one and other well–known configurations, since the only stringent requirement in this case is the resonant frequency. The fitness in this case is relatively simple, aimed almost uniquely to direct the evolution process towards a structure that resonates at the desired frequency. We then can insert an additional penalty coefficient in order to maximize the bandwidth, a secondary objective overlapped to the main goal.

Finally, as we can see in Fig.14(c), we allowed the geometry to touch the borders of the unit cell, thus considering a continuous aperture (while the geometry of the metallic patch will therefore be finite), and we found out a further improvement in the bandwidth. In Fig.15 the entire surface is shown.

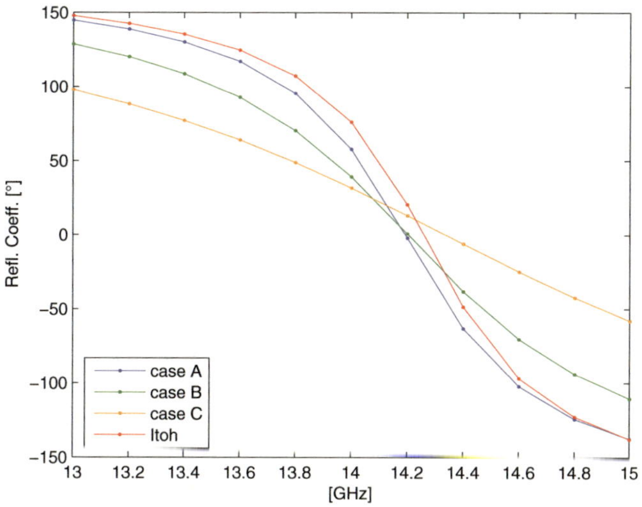

Fig. 13. Reflection Coefficient.

(a) Case A

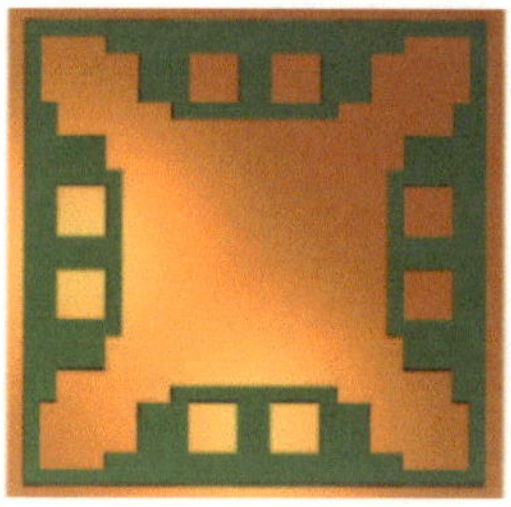

(b) Case B

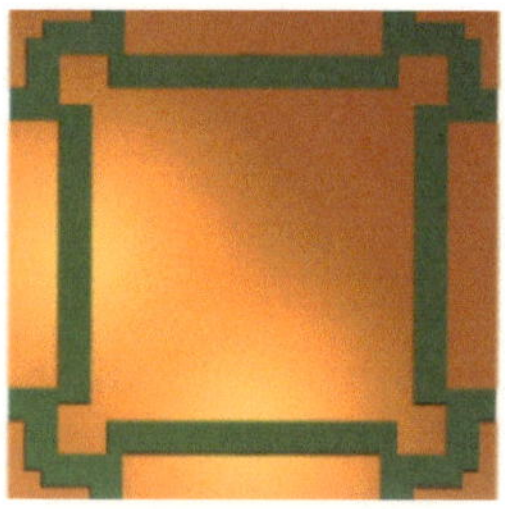

(c) Case C

Fig. 14. Best individual

We can exploit this approach using a more complex fitness function in order to obtain, for example, larger bandwidth (Deias et al., 2009a),as shown in case B, or at more interesting frequencies (Deias et al., 2010).

It is well known that there is a growing interest in antennas integrated with EBG surfaces for communication system applications, covering the 2.45 GHz and the 5 GHz wireless networking bands (Hung-Hsuan et al., 2007; Zhu & Langley, 2009). As proof of the effectiveness of our approach a simple EBG surface at this working frequencies was found with a yet limited computational effort.

Fixed parameters in this case are the following:

In Fig.16(b) we can see the reflection coefficient for the best individual shown Fig.16(a), as obtained after nine generations. The performance of our optimization is quite good leading to a simple and effective solution with a yet limited computational effort.

These results prove that the main advantage of the genetic approach is to be identified in the generalization of the "solution space", evolution being able to breed geometries that we could

Fig. 15. EBG Surface (Case C)

Fixed Parameters	
Center Frequency	2.45 GHz
Periodicity (unit square cell)	$d_x = 40$ mm; $d_y = 40$ mm
Substrate dielectric constant	$\epsilon_r = 1.38$
Substrate thickness	2.2 mm
Discretization step	0.8 mm $(50 \times 50\ grid)$

Table 3. Data

never think of. Evolutionary computation on the other hand let us scope through geometries as though they were human–described.

5. Multi–layer EBG design results

The periodic surfaces in the single–layer configuration shown so far exhibit perfect reflection or transmission only at resonance. However, many applications, exploiting the filtering properties of frequency selective surfaces, seek for a resonant curve with a flat top and faster roll off. This goal can be achieved by using two or more periodic surfaces cascaded with dielectric slabs sandwiched in between . By using multiple layers of frequency selective surfaces as part of the substrate we operate in a manner similar to that used for designing broad–band microwave filters. With such a configuration typically we can obtain a bandwidth that is considerably larger than that of a single structure.

Genetic Programming strategy in conjunction with the flexible aperture approach previously described proved to be effective in the design and optimization of EBG surfaces in the case of a two–layer structure with the same geometry in both layers. When considering more layers or different geometries we only have to take into consideration an increased computational effort.

In the case of a two–layer configuration with the following fixed parameters:

We obtained the following geometry after two generations, each generation with a population of 150 individuals:

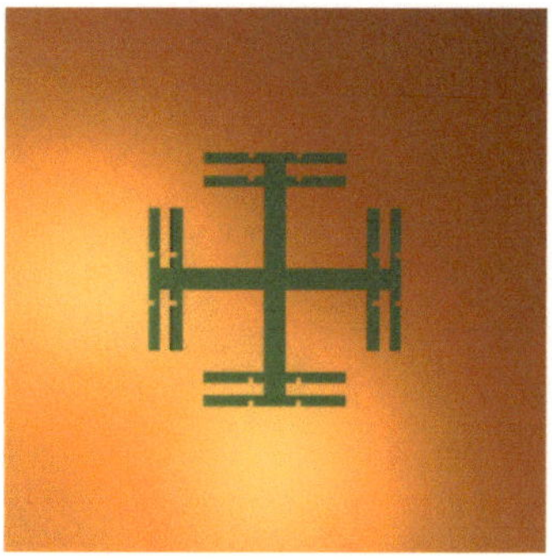

(a) Geometry

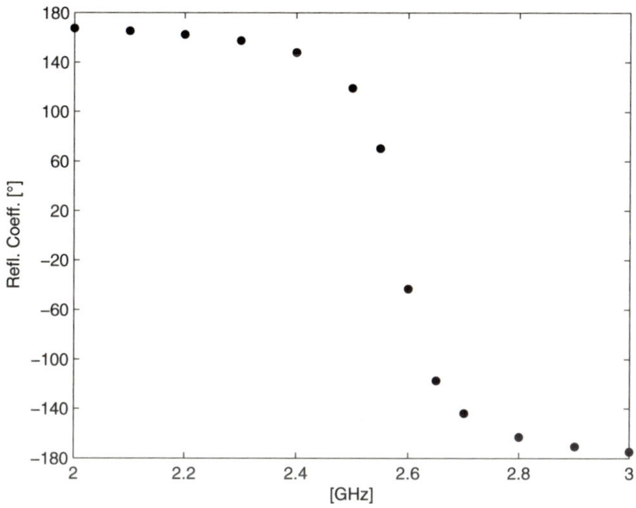

(b) Reflection Coefficient

Fig. 16

Fixed Parameters	
Center Frequency	2.45 GHz
Periodicity (unit square cell)	$d_x = 3.048$ mm; $d_y = 3.048$ mm
Substrate dielectric constant	$\epsilon_r = 10.2$
Substrate thickness	3.354772 mm
Discretization step	0.1524 mm $(20 \times 20$ *grid*$)$

Table 4. Data

This quite simple result proves that genetic programming can be efficiently adopted for this kind of design synthesis and optimization even in the case of complex structures consisting of multiple layers of EBG surfaces. Moreover the fitness function can be modified in order to take into consideration other requirements.

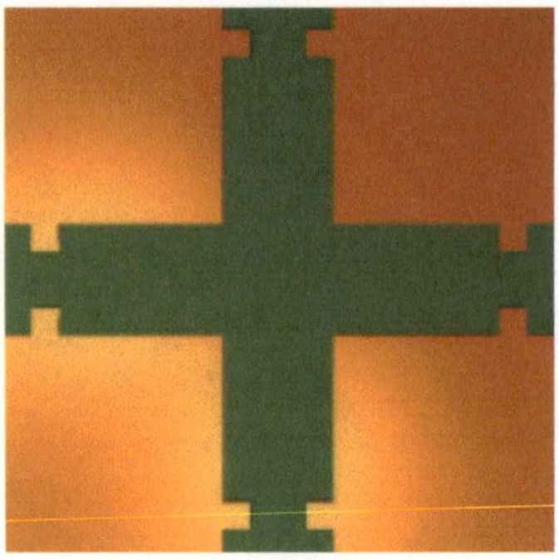

(a) Geometry

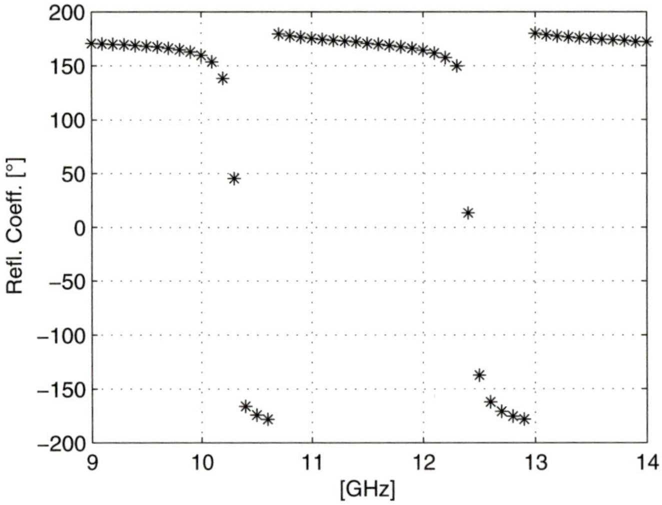

(b) Reflection Coefficient

Fig. 17

6. Acknowledgements

Work partially supported by Regione Autonoma della Sardegna, under contract CRP1_511: "Valutazione e utilizzo della Genetic Programming nel progetto di strutture a radiofrequenza e microonde".

7. References

Asole, F., Deias, L. & Mazzarella, G. (2007). A flexible full–wave analysis of multilayered amc using an aperture oriented approach, *J. Electrom. Waves Appl.* 21(No. 14): 2059–2072.

Balanis, C. A. (1996). *Antenna Theory – Analysis and Design*, Wiley, New York.

Bardi, I., Remski, R., Perry, D. & Cendes, Z. (2002). Plane wave scattering from frequency–selective surfaces by the finite element method, *IEEE Trans. Magn.* 38(No 2): 641–644.

Bozzi, M. & Perregrini, L. (1999). Efficient analysis of thin conductive screens perforated periodically with arbitrarily shaped apertures, *IEE Electronics Letters* 35(No 13): 1085–1087.

Bray, M., Bayraktar, Z. & Werner, D. (2006). Ga optimized ultra–thin tunable ebg amc surfaces, *IEEE Antennas and Propagation Society International Symposium* pp. 410 – 413.

Collin, R. E. (1985). *Antennas and Radiowave Propagation*, McGraw–Hill.

Deias, L. & Mazzarella, G. (2006). Aperture oriented approach to the analysis of artificial magnetic conductors and first–order circuit model validation, *12th International Symposium on Antenna Technology and Applied Electromagnetics (ANTEM) and Canadian Radio Sciences (URSI/CNC)*.

Deias, L., Mazzarella, G. & Sirena, N. (2009a). Bandwidth optimization of ebg surfaces using genetic programming, *LAPC 2009. Antennas and Propagation Conference.* pp. 593–596.

Deias, L., Mazzarella, G. & Sirena, N. (2009b). Synthesis of ebg surfaces using evolutionary optimization algorithms, *EuCAP 2009. 3rd European Conference on Antennas and Propagation.* pp. 99 –102.

Deias, L., Mazzarella, G. & Sirena, N. (2010). Ebg substrate synthesis for 2.45 ghz applications using genetic programming, *IEEE Antennas and Propagation Society International Symposium (APSURSI)* pp. 1–4.

Feresidis, A. P., Goussetis, G., Wang, S., & Vardaxoglou, J. C. (2005). Artificial magnetic conductor surfaces and their application to low–profile high–gain planar antennas, *IEEE Trans. Antennas Propagat.* Vol. 53(No. 11): 209–215.

Fogel, D. (2006). *Evolutionary Computation: Toward a New Philosophy of Machine Intelligence*, IEEE Press, New York.

Ge, Y., Esselle, K. & Hao, Y. (2007). Design of low–profile high-gain ebg resonator antennas using a genetic algorithm, *IEEE Antennas and Wireless Propagation Letters* 6: 480 – 483.

Gonzalo, R., de Maagt, P. & Sorolla, M. (1999). Enhanced patch-antenna performance by suppressing surface waves using photonic-bandgap substrates, *IEEE Trans. Microw. Theory Tech.* Vol. 47(No. 11): 2131–2138.

Harms, P., Mittra, R. & Wae, K. (1994). Implementation of the periodic boundary condition in the finite–difference time–domain algorithm for fss structures, *IEEE Trans. Antennas Propagat.* 42: 1317–1324.

Hosseini, M., Pirhadi, A. & Hakkak, M. (2006). Design of a novel amc with little sensitivity to the angle of incidence and very compact size, *IEEE Antennas and Propagation Society International Symposium*.

Hung-Hsuan, L., Chun-Yih, w. & Shih-Huang, Y. (2007). Metamaterial enhanced high gain antenna for wimax application, *TENCON*.

Jones, E. A. & Joines, W. T. (2000). Evolutionary optimization of yagi–uda antennas, *Antennas and Propagation Magazine* 42(No. 3).

Kovács, P., Raida, Z. & Lukes, Z. (2010). Design and optimization of periodic structures for simultaneous ebg and amc operation, *COMITE 15th International Conference on Microwave Technique* pp. 195 – 198.

Koza, J. R. (1992). *Genetic Programming: On the Programming of Computers by Means of Natural Selection*, Mit Press.

Lohn, J. D., Kraus, W. F., Linden, D. S. & Colombano, S. (2001). Evolutionary optimization of yagi–uda antennas, *Proceedings of the 4th International Conference on Evolvable Systems: From Biology to Hardware* pp. 236–243.

Michalewicz, Z. (1992). *Genetic Algorithms + Data Structures = Evolution Programs*, Springer-Verlag.

Mittra, R., Chan, C. & Cwik, Y. (1988). Techniques for analyzing frequency selective surfaces – a review, *Proceedings IEEE*, pp. 1593–1615.

Munk, B. (2000). *Frequency Selective Surfaces: Theory and Design*, Wiley Interscience, New York.

Pozar, D. M. & Schaubert, D. H. (1984). Scan blindness in infinite phased arrays of printed dipoles, *IEEE Trans. Antennas Propagat.* 32(No. 6): 602–610.

Sievenpiper, D., Zhang, L., Broas, R., Alexopolou, N. & Yablonovitch, E. (1999). High-impedance electromagnetic surfaces with a forbidden frequency band, *IEEE Trans. Microw. Theory Tech.* Vol. 47(No. 11): 2059–2074.

Tavallaee, A. & Rahmat-Samii, Y. (2007). A novel strategy for broadband and miniaturized ebg designs: hybrid mtl theory and pso algorithm, *IEEE Antennas and Propagation Society International Symposium* pp. 161 – 164.

Wu, T. K. (1995). *Frequency Selective Surface and Grid Array*, Wiley, New York.

Yang, F. R., Ma, K. P., Quian, Y. & Itoh, T. (1999). A novel tem waveguide using uniplanar compact photonic-bandgap (uc-pbg)structure, *IEEE Trans. Microw. Theory Tech.* Vol. 47(No. 11): 2092–2098.

Yeo, J., Mittra, R. & Chakravarty, S. (2002). A ga–based design of electromagnetic bandgap (ebg) structures utilizing frequency selective surfaces for bandwidth enhancement of microstrip antennas, *IEEE Antennas and Propagation Society International Symposium* 2: 400 – 403.

Zhu, S. & Langley, R. (2009). Dual-band wearable textile antenna on an ebg substrate, *IEEE Trans. Antennas Propagat.* Vol. 57(No. 4): 926–935.

14

Magnonic Metamaterials

V.V. Kruglyak et al.*
School of Physics,
University of Exeter, Exeter,
United Kingdom

1. Introduction

A large proportion of the recent growth of the volume of electromagnetics research has been associated with the emergence of so called electromagnetic metamaterials[1] and the discovered ability to design their unusual properties[2,3] by tweaking the geometry and structure of the constituent "meta-atoms"[4]. For example, negative permittivity and negative permeability can be achieved, leading to negative refractive index metamaterials[2]. The negative permeability could be obtained via geometrical control of high frequency currents, e.g. in arrays of split ring resonators[5], or alternatively one could rely on spin resonances in natural magnetic materials[6,7], as was suggested by Veselago in Ref. 2. The age of nanotechnology therefore sets an intriguing quest for additional benefits to be gained by structuring natural magnetic materials into so called *magnonic* metamaterials, in which the frequency and strength of resonances based on spin waves (magnons)[8] are determined by the geometry and magnetization configuration of meta-atoms. Spin waves can have frequencies of up to hundreds of GHz (in the exchange dominated regime)[6-9] and have already been shown to play an important role in the high frequency magnetic response of composites[10-14]. Moreover, in view of the rapid advances in the field of magnonics[9,15,16,], which in particular promises devices employing propagating spin waves, the appropriate design of magnonic metamaterials with properties defined with respect to propagating spin waves rather than electromagnetic waves acquires an independent and significant importance.

*M. Dvornik[1], R.V. Mikhaylovskiy[1], O. Dmytriiev[1], G. Gubbiotti[2,3], S. Tacchi[3], M. Madami[3], G. Carlotti[3], F. Montoncello[4], L. Giovannini[4], R. Zivieri[4], J.W. Klos[5], M.L. Sokolovskyy[5], S. Mamica[5], M. Krawczyk[5], M. Okuda[6], J.C. Eloi[6], S. Ward Jones[6], W. Schwarzacher[6], T. Schwarze[7], F. Brandl[7], D. Grundler[7], D.V. Berkov[8], E. Semenova[8] and N. Gorn[8]
[2]*Istituto Officina dei Materiali del Consiglio Nazionale delle Ricerche (IOM-CNR), Sede di Perugia, c/o Dipartimento di Fisica, Perugia, Italy*
[3]*CNISM, Unità di Perugia and Dipartimento di Fisica, Perugia, Italy*
[4]*Dipartimento di Fisica and CNISM, Università di Ferrara, Ferrara, Italy*
[5]*Faculty of Physics, Adam Mickiewicz University, Poznan, Poland*
[6] *H. H. Wills Physics Laboratory, Bristol, United Kingdom*
[7]*Lehrstuhl fuer Physik funktionaler Schichtsysteme, Physik Department, Technische Universität München, Garching bei Muenchen, Germany*
[8]*Innovent Technology Development, Jena, Germany*

In this Chapter, we review the recent advances in studies of magnonic metamaterials and challenges that have to be overcome in order for the rich opportunities of exploitation of such metamaterials to be realised. We start from a discussion of the notion of a metamaterial and its relevance to areas beyond that of common electromagnetics. We introduce band gap and effectively continuous magnonic metamaterials and demonstrate their properties and functionalities, presenting either experimental or theoretical evidence as appropriate for the illustration.

2. What makes a metamaterial?

To answer the question in the title in the most general sense (i.e. going beyond the topic of more common electromagnetic metamaterials), one has to learn first what makes standard (i.e., "nature-made") *materials.* One talks about a material when many atoms or ions are joined together to collectively create a novel quality with properties that are not observed in the constituent atoms in isolation. For example, isolated atoms have discrete electronic energy levels and associated discrete electromagnetic spectra. In a material consisting of many of such atoms, each discrete electronic level is split into a continuous "electronic band", and accordingly the electromagnetic spectrum of the material is also continuous.

As a result of joining into a material, atoms also collectively acquire new properties that are less trivially connected with those of isolated atoms. Firstly, different classes of materials are formed, including e.g. dielectrics, semiconductors, or metals, ferro-, piezo-, or segneto-electrics, dia-, para-, antiferro- or ferro-magnets, superconductors, materials with a range of different mechanical properties, etc. Secondly, new classes of waves (excitations) and their quanta (so called quasi-particles) are observed, including plasmons, magnons, phonons, excitons etc. One can also distinguish many "hybrids" of the solid state quasi-particles among themselves as well as with photons, e.g. surface plasmon-polaritons, magneto-excitons etc.

A material does not have to have a periodic arrangement of atoms and can be amorphous or quasi-crystalline. The classification of materials and their properties often does not have strict boundaries, while the differences are often quantitative. For example, dielectrics and semiconductors are different only by the size of the band gap; the same material can be ferromagnetic below and paramagnetic above a certain temperature, etc. Some electronic energy levels can remain discrete, therefore retaining their atomistic character. Materials can also consist of rather complex building blocks, e.g. molecules, crystal unit cells (often consisting of many atoms or ions), or crystallites. The most complex nature-made (truly functional!) materials - biological tissues - consist of living cells composed of millions of atoms.

This brings us to the idea of a *metamaterial* - an artificial assembly of "man-made" building blocks with tailored properties. Living aside static properties and biological tissues, the most practical ways to "tailor" properties of the building blocks are via their geometrical shaping and compositional modulation. The former leads to confinement of solid state excitations (e.g. those listed above) and hence to formation of a discrete spectrum of the allowed modes. Brought sufficiently close together, the blocks can form a metamaterial with the discrete spectrum split into a band structure.

Some metamaterials (e.g. arrays of voids) are simpler to consider as created by geometrical modification of previously continuous materials. This however does not exclude their treatment as if they were constructed from building blocks considered above. Also, one has to note that, from the point of view of electromagnetic waves, vacuum is a continuous material in that important sense that it supports propagation of electromagnetic waves. Therefore, even arrays of entirely disconnected blocks are metamaterials for electromagnetic waves. This distinguishes electromagnetic waves from the other excitations.

Generally, various dynamical properties of metamaterials can be considered relative to excitations with wavelength either comparable to or much greater than the characteristic size of the building blocks. The former case is associated with studies of artificial "band gap crystals", e.g. photonic[17], plasmonic[18], phononic[19], and magnonic[20] crystals. The latter case allows one to study the metamaterials as some effectively continuous media. The physical object under study is nonetheless the same in both cases. For example, metamaterials with artificial periodic modulation of the refractive index with periodicity comparable to the wavelength of electromagnetic waves in the visible range are known as photonic band gap structures, using which it is possible to enable propagation of light in particular directions, to localize it in chosen channels or zones, or even to completely prohibit its propagation. In the microwave frequency range, the same structures would behave as effectively continuous materials.

In a particular frequency region of interest, the same metamaterials can behave differently with respect to excitations of different kinds, e.g. electromagnetic, sound or spin waves. Attractive opportunities arise from use of one excitation, with respect to which the sample behaves as a discrete structure (e. g. a band gap metamaterial), to design a resonance feature for another excitation, with respect to which the structure behaves as quasi-continuous metamaterial. For example, plasmonic resonances can be used to alter effective electromagnetic properties in THz to visible frequency range. Magnonic and LC resonances can be used for the same purpose in GHz-THz and GHz frequency ranges, respectively.

A thorough reader will probably find that, although term "metamaterial" emerged recently as a result of ever increasing pressure to spice up funding applications and publications in high profile journals, the underpinning research can be sometimes traced back to the middle or even beginning of the 20th century. Yet, one can also argue that it is the recent advances in nanofabrication and experimental tools that justify the boom by allowing researchers to revisit the established notions at reduced length scales and to apply results to modern technology.

3. Spin waves and effectively continuous magnonic metamaterials

The main reason for electrically neutral atoms to be able to interact with electromagnetic radiation is that they themselves possess electrical structure. Indeed, each atom consists of positively charged core and negatively charged electrons. The core and electrons are not rigidly coupled, and so, the atom can be electrically polarised by and thereby interact with an external electric field.

However, the electric charge is not the only property responsible for the interaction of atoms or materials with electromagnetic radiation. In particular, spins and associated magnetic moments of elementary particles are responsible for their interaction with an external magnetic field. Being always electrically neutral, some atoms still have a net magnetic moment and hence are called "magnetic". Generally, electromagnetic field interacts with charges more strongly than with spins, and so, the associated resonance frequencies are weaker in spin resonance experiments. However, via the quantum mechanical Pauli Exclusion Principle, the electron spin governs the order in which the electronic bands are populated by electrons, and thereby strongly affects the electromagnetic spectrum. As the other side of the same phenomenon, the Pauli Exclusion Principle and electrostatic interaction result in the exchange interaction between spins and are therefore responsible for the ordering of spins observed in e.g. ferromagnetic materials. The perturbations of the magnetic ordering are called spin waves – the central object of magnonics and magnonic metamaterials.

Spin waves exhibit most of the properties inherent to waves of other origins, including the excitation and propagation[21,22,23], reflection and refraction[24,25,26,27,28,29], interference and diffraction[30,31,32,33,34,35], focusing and self-focusing[36,37,38,39,40,41], tunnelling[42,43], Doppler effect[44,45,46], and formation of spin-wave envelope solitons[47,48,49,50]. Spin-wave quantization due to the finite size effect was discovered in thin films[51,52] and more recently in laterally confined magnetic structures[53,54,55,56,57,58], with their effect upon the high frequency permeability and the observation of a negative permeability discussed in Refs. 7 and 59, respectively.

The possibility of using spin waves for the design of high frequency permeability follows already from their first direct observation in cavity ferromagnetic resonance (FMR) experiments[60]. In the latter measurements, the precession of magnetisation[60] is detected by measuring spectra of the absorption of microwaves in the cavity containing the magnetic sample under study. The spectra are determined by the density of states of spin waves that can resonantly couple to the microwave field. The very long wavelength of microwaves, as compared to the length scale of magnetic structures of interest, limits the application of the FMR technique to studies of magnonic modes with significant Fourier amplitude at nearly zero values of the wave vector[52]. However, this also mimics potential applications in which either the electromagnetic response of a magnonic device containing nano-structured functional magnetic elements is read out by the effectively uniform electromagnetic field, or the magnonic (meta-) material[61] is supposed to absorb the incident electromagnetic radiation. Continuous magnetic materials and arrays of non-interacting magnetic elements appear preferred for such applications near the frequency of the uniform FMR. However, more sophisticated micromagnetic engineering is required to push up the frequency of operation of such materials e.g. using the exchange field[62], which originates from the strongest of the magnetic interactions, rather than the uniform anisotropy or applied magnetic field. An example of using the concept can be found e.g. in Ref. 7 and is also illustrated in Figure 1.

The FMR is conventionally used to study magnetisation dynamics at frequencies up to about 100 GHz. At higher frequencies, the mismatch between the linear momentum of free space electromagnetic radiation (photon) and that of a magnon increasingly prohibits an efficient coupling. Therefore, higher frequencies require one to use different experimental and technical concepts by which to interrogate and measure e.g. THz magnons. Here, methods, known e.g. from plasmonics, might help to couple light to magnons. For example, the attenuated total reflection technique has been successfully applied to studies of magnons in antiferromagnets[63]. However, this field of research is still at its infancy and is not reviewed here to any significant extent.

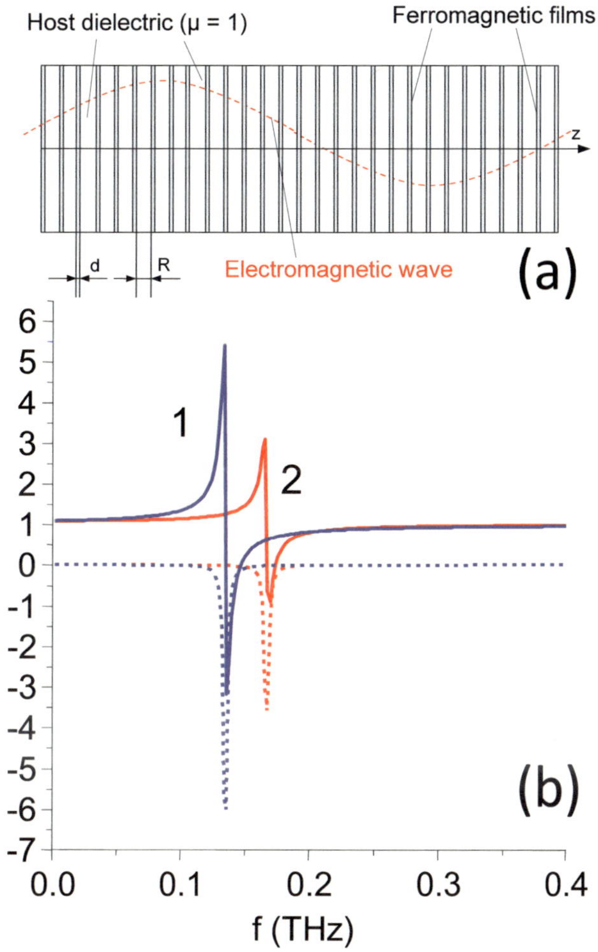

Fig. 1. (b) The effective permeability calculated for the structure shown in (a) is plotted as a function of the frequency for cases of in-plane (1) and out-of-plane (2) magnetizations. The solid and dashed lines denote the real and imaginary parts of the effective permeability, respectively. The structure under study represents an array of CoFe films with thickness of 5 nm. The filling factor of $\rho = 0.25$ is assumed. The spins are perfectly pinned at one side of each film and are free at the other. (After Ref. 7)

The VNA-FMR technique represents a relatively new twist in the FMR spectroscopy where VNA highlights the use of a broadband vector network analyser (VNA) operated in the GHz frequency regime. Microwaves applied to a waveguide locally excite spin waves that in turn induce a high-frequency voltage due to precessing magnetisation (Figure 2 (a)). The VNA-FMR technique measures spectra of both the amplitude and phase change of microwaves passing through a magnetic sample integrated with the waveguide[64,65,66]. The geometrical parameters of the waveguide determine the spatial distribution of the rf magnetic field and therefore the wavelength spectrum addressed by the microwave field. Hence, the VNA-FMR can be also referred to as a "near field" FMR. Due to the large penetration depth of microwaves, both thin film and bulk samples can be successfully investigated using this technique.

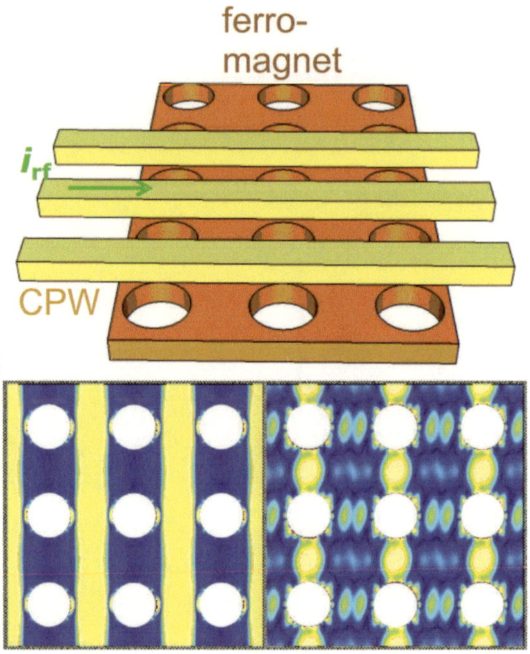

(a)

(b)

Fig. 2. (a) Sketch of a coplanar waveguide (CPW) integrated to an antidot lattice prepared from a thin Permalloy film. The CPW consists of three metallic leads (ground-signal-ground leads). Adjusting the dimensions of the CPW allows one to vary the profile of the magnetic field generated by microwave current i_{rf} supplied by the VNA. This defines the wave vector transferred to the sample. The same CPW picks up the voltage induced by precessing spins. The CPW is isolated from the ferromagnet by an insulating layer (not shown). (b) Simulated spatial distributions of spin precession amplitudes reflecting two different standing spin-wave excitations. The film is assumed to be 26 nm thick. The period (hole diameter) is 490 (240) nm. A magnetic field of a few 10 mT is applied in horizontal direction. The mode pattern shown on the right belongs to a localized mode that has frequency higher than that of the extended mode shown on the left. (After Ref. 91) Bright colours correspond to large amplitudes. The holes are shown in white.

An experimental insight into the structure of magnonic modes in nanostructured magnonic metamaterials, e.g. such those shown in Figure 2, can be achieved with state of art dynamic magnetic imaging techniques, e.g. the time-resolved scanning Kerr microscopy (TRSKM)[61]. In a TRSKM experiment, the sample is pumped so as to excite spin waves, with the pump stimulus being both repetitive and coherent, i.e. having a well-defined phase with respect to the probe beam. To probe, one uses ultrashort optical pulses and controls their arrival time relative to the pump. By changing the optical path of the probe pulse one can trace the time evolution of the excited dynamics. By scanning the position of the optical probe on the surface of the sample, one acquires images of the dynamic magnetisation with a spatial resolution of down to 250 nm in real space[67-70], and is suitable for studying both continuous and nanostructured samples, as demonstrated in Figure 3. The temporal resolution of TRSKM can be well on the sub-ps time scale, therefore offering the detection of spin waves in the THz frequency regime. The TRSKM performs a 3D vectorial analysis of the time dependent magnetization[71] and is therefore phase sensitive. Alternatively, one can combine the magneto-optical detection with a VNA-FMR setup to image spin wave modes in the frequency rather than time domain[72].

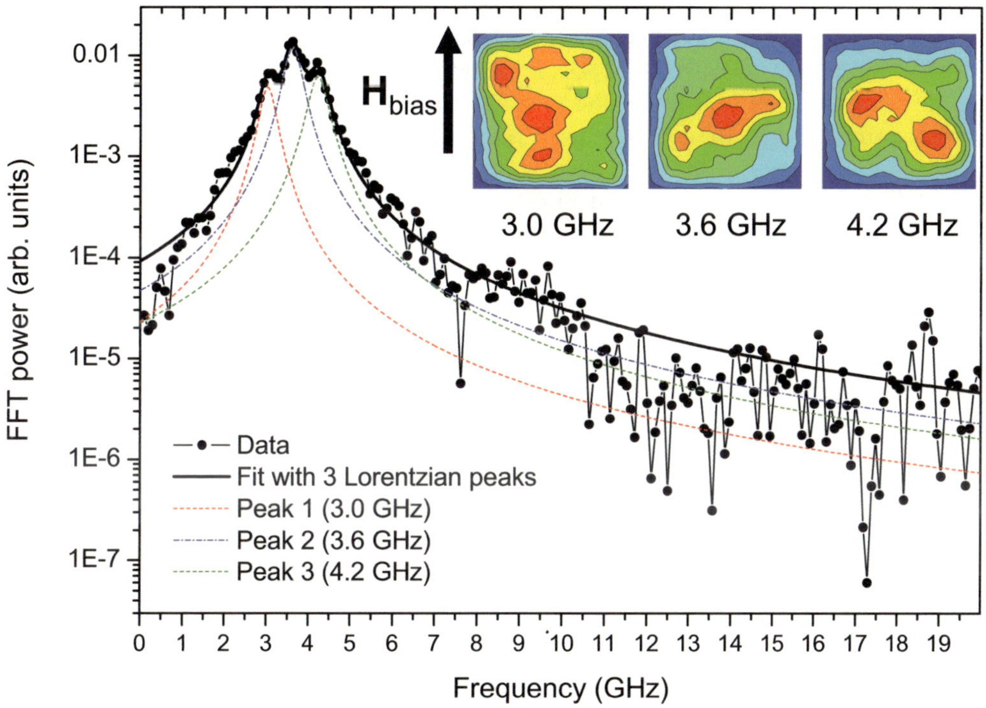

Fig. 3. The fast Fourier transform (FFT) power spectrum calculated from a time resolved Kerr signal acquired from the center of a 4 x 4 μm² array of 40 x 80 nm² stadium shaped ferromagnetic elements at a bias magnetic field of 197 Oe is shown on a logarithmic scale together with the fit to a Lorentzian 3-peak function. The inset shows images of the modes confined within the entire array and corresponding to the peak frequencies identified from the fit. The darker shades of gray correspond to greater mode amplitude. With respect to the long wavelength spin wave modes, the array acts as a continuous element made of a magnonic metamaterial. Such arrays will also act as metamaterials with respect to microwaves. After Ref. 61.

4. Band gap magnonic metamaterials

Periodically modulated magnetic materials have been shown to form magnonic crystals, i.e., a magnetic analogue of photonic crystals. Indeed, the spin wave spectrum is modified by patterning[73] and shows a tailored band structure in periodic magnetic materials[74]. The band spectrum consists of bands of allowed magnonic states and forbidden-frequency gaps ("band gaps"), in which there are no allowed magnonic states. One of the first attempts to study the propagation of spin waves in periodic magnetic structures was made by Elachi[75]. Nowadays, the number of studies on this topic has surged and continues to grow at a fast pace.

The recent advances in the studies of the band gap magnonic metamaterials are associated with advances in the Brillouin light scattering (BLS) technique, which has proved to be a very powerful tool for the investigation of magnetization dynamics in magnonic structures[76]. Thanks to the wave vector conservation in the magnon–photon interaction, one has the possibility to measure the dispersion relation (frequency versus wave vector) of spin waves. In particular, the BLS technique is suitable for measuring the magnonic band gap dispersion, provided that the periodicity of the magnonic crystal is such that the Brillouin zone (BZ) boundary lies in the accessible wave vector range (up to 2.2×10^5 rad/cm). The magnonic dispersion can be measured in different scattering geometries that differ by the relative orientation of the exchanged wave vector and direction of the applied magnetic field. So planar 1D magnonic crystals formed by arrays of closely spaced Permalloy stripes of identical[77] or alternating width[78] were studied, showing the existence of tuneable band gaps. Furthermore, alternating stripes of two different magnetic materials were studied in Ref. 79. In addition, it has recently been shown that, using a large aperture microscope objective, BLS can be used as a scanning probe technique, therefore permitting the map-out of the spatial distribution of magnonic normal modes with a lateral resolution of down to a few hundred nanometres[80,81].

The studied sample consist of long chains of Permalloy rectangular dots with rounded corners and lateral dimensions of 715 x 450 nm^2, thickness of 40 nm, edge-to-edge separation Δ = 55 nm. The magnetic material was deposited on Silicon substrate at room temperature. In our calculations, each dot was divided into cells with size $\Delta_x \times \Delta_y \times \Delta_z$ = 5 nm × 5 nm × 40 nm. The ground state was obtained using a micromagnetic code. The magnetization was assumed to be uniform in each cell and to precess around its equilibrium direction along effective field $\mathbf{H}_{\text{eff}}$. Contributions arising from the external (Zeeman), demagnetising and exchange fields were included in $\mathbf{H}_{\text{eff}}$. The following magnetic parameters obtained from the fit to the BLS frequencies of the Damon-Eshbach mode in the reference 40 nm thick continuous Permalloy film were used: saturation magnetization $4\pi M_s$ = 9 kG, $\gamma/2\pi$ = 2.94 GHz/kOe, and $A = 1.1\ 10^{-6}$ erg/cm with A being the exchange stiffness constant.

First, the collective dynamics was studied in the Voigt geometry, namely with the wave vector $\boldsymbol{q}$ perpendicular to the applied magnetic field $\mathbf{H}$ and with $\boldsymbol{q} = \boldsymbol{K}$, where $\boldsymbol{K}$ is the Bloch wave vector.

The collective modes of the array of dots were studied by using the Dynamical Matrix Method (DMM) extended to periodic magnetic systems[82]. On the basis of the number and the direction of nodal planes n = 0,1,2,.. of the dynamic magnetization inside each dot, we have classified collective modes into: the F mode with no nodal planes, Damon-Eshbach-like

(nDE) modes characterized by nodal planes parallel to the applied magnetic field **H**, Backward-like (nBA) modes with nodal planes perpendicular to **H** and nEnd-Modes (nEM) localized at the edges of each dot with nodal planes of the DE type.

Figure 4 (a) shows a scanning electron microscopy (SEM) image of the studied sample together with a reference frame and the directions of $\boldsymbol{q}$ and **H**. In Figure 4 (b) black lines denote frequencies of magnonic modes for $q = 0$, while the frequency curves corresponding to the edge of the first Brillouin zone (1BZ) at $q_{BZ} = \pi / a$ with a the periodicity are indicated by dashed red lines[83].

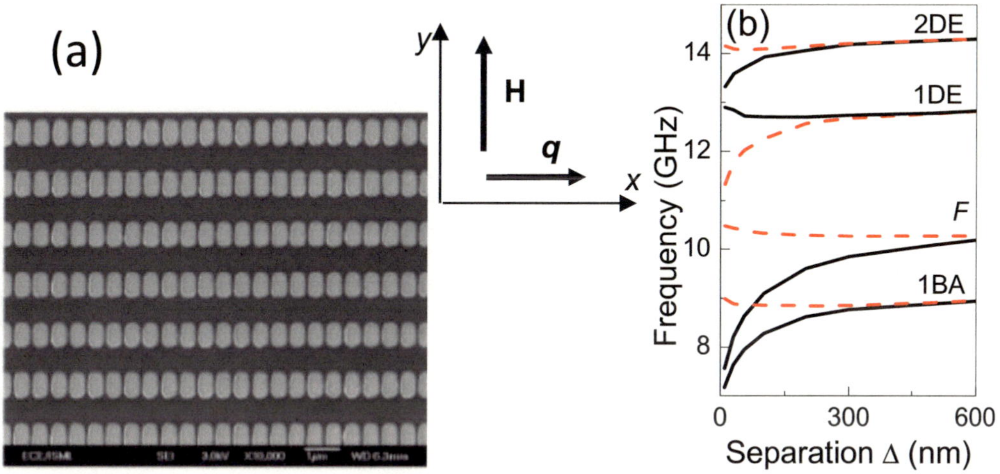

Fig. 4. (a) SEM image of the sample: Permalloy rectangular dots have lateral dimensions 715 × 450 nm^2 and interdot separation $\Delta = 55$ nm. A reference frame with the direction of **H** along the y-axis (easy axis) and of the wave vector $\boldsymbol{q}$ is also shown. (b) Calculated frequency behaviour vs. interdot separation for the sample of dots 715 x 450 nm^2 in the Voigt geometry for an applied magnetic field of magnitude $H = 1$ kOe. Full black lines: frequencies at $q = 0$. Dashed red lines: frequencies at $q = \pi / a$.

As expected, for large interdot separations, each mode is characterised by a single frequency value and frequency is independent of $\boldsymbol{q}$. On decreasing the separation, interdot coupling gives rise to the appearance of bands. Due to the effect of stray magnetic field within each band the frequency of the collective modes depends on $\boldsymbol{q}$. The largest band width is that of the F mode that has the largest stray field at any separation. Another significant feature is the narrowing of the band gaps, as $\Delta \rightarrow 0$, either at $q = 0$ or at $q_{BZ} = \pi / a$. In particular, the band gap between the F and the 1DE mode is smaller for $q_{BZ} = \pi / a$, while that between the 1DE and the 2DE mode is smaller for $q = 0$. As a matter of fact, for small separation the energetic cost required to excite the F mode at $q_{BZ} = \pi / a$, namely in the anti-phase configuration, is almost the same as that of the 1-DE mode. The difference is represented by the band gap between the two modes at the 1BZ boundary. As an example the band width for a small interdot separation $\Delta = 10$ nm of the most representative modes shown in panel (b) was estimated. Calculated band width of the F mode turned out to be about 3.8 GHz, the largest one among those of most representative modes, but note that also the other collective modes of the spectrum have an important calculated band width in this limit (larger than 0.8 GHz).

Finally, the investigation was completed by assuming the Bloch wave vector parallel to the applied magnetic field[84]. This geometry corresponds to the so called backward volume magnetostatic spin-wave (BWVMS) geometry. In Figure 5, the dependence of the spin-wave frequencies on Δ calculated at $q = 0$ is compared to the case of $q = \pi / a$, i.e. at the edge of the 1BZ. As it can be seen, for separation values of about 300 nm, the frequencies at the centre and at the edge of the 1BZ are almost degenerate and both tend to the value of the mode frequency of an isolated dot. On reducing the value of Δ, however, the dynamic dipolar magnetic coupling becomes strong enough to remove the degeneracy. This leads to the appearance of magnonics bands whose widths increase with decreasing the interdot distance. Also in this geometry there is a narrowing of band gaps, as $\Delta \rightarrow 0$, between given couples of adjacent modes, either at $q = 0$ or at $q_{BZ} = \pi / a$.

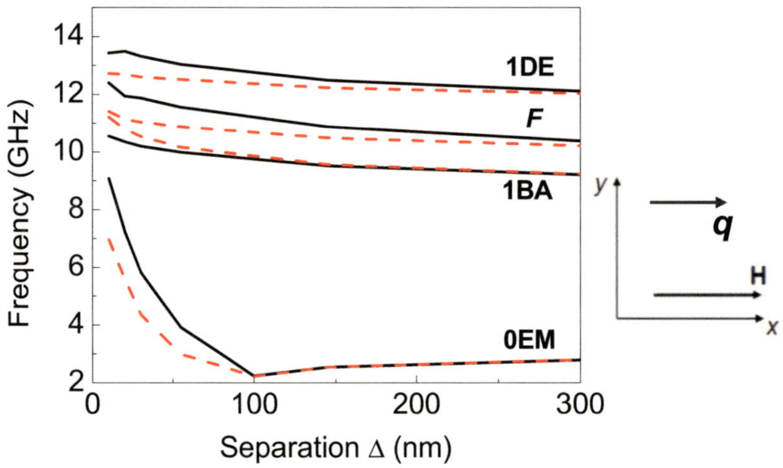

Fig. 5. Calculated mode frequencies behavior vs. interdot separation in the BWVMS geometry, i.e. for $\boldsymbol{q}$ parallel to **H**. Black lines: frequencies for $q = 0$. Dashed (red) lines: frequencies for $q_{BZ} = \pi / a$. A magnetic field of intensity $H = 1.5$ kOe was applied along the x-axis (hard axis) for the sample shown in Fig. 4 (a). The wave vector $\boldsymbol{q}$ was along the chains. The direction of $\boldsymbol{q}$ and **H** is also shown.

In particular, among the modes shown, there is a narrowing at $q_{BZ} = \pi / a$ between the 1BA and the F mode. Interestingly, for $\Delta < 100$ nm, the frequency increment of the 0EM is much more accentuated than that of collective modes in the higher part of the spectrum. This behaviour (which also concerns the others n-EM, not shown here) is related to the fact that the nEM are localized at some portions (typically one corner) of the adjacent edges of neighbouring dots.

2D magnonic crystals were proposed in Refs. 85,86, where the spectrum of dipole-exchange spin waves propagating in the plane of a magnonic crystal was discussed. The magnonic crystal consisted of periodically arranged infinitely long ferromagnetic cylinders embedded in a matrix of a different ferromagnetic material. The position and width of band gaps in the magnonic spectrum were investigated as a function of the period of the structure and the depth of modulation depth ("contrast") of the magnetic parameters. It was found that the depth of modulation of the exchange parameter has a drastic effect upon the position and width of the band gaps. Collective dynamics of lattices of magnetic vortices was studied in Refs. 87,88. FMR and time resolved scanning Kerr microscopy (TRSKM) were used to study localisation of spin waves in an array of antidots formed in a metallic ferromagnetic film in Ref. 89. VNA-FMR measurements and micromagnetic simulations were used to demonstrate a control of spin wave transmission through a similar array of antidots by an external magnetic field in Refs. 90,91 (Figure 2).

The BLS technique has been exploited to achieve a complete mapping of the spin-wave dispersion curves, along the principal symmetry directions of the first BZ, for a 2D magnonic crystal consisting of a square array of 50 nm thick NiFe disks (Figure 6 (a)). The disks have a diameter d = 600 nm and are arranged in a square matrix with the edge-to-edge interdot separation of 55 nm (period a = 655 nm). This corresponds to a square first BZ of side $2\pi/a = 2\text{x}4.8 \cdot 10^4$ rad/cm. The spin-wave frequency dispersion was studied along the principal directions of the first BZ, i.e. ΓX, ΓY, XM and YM, as shown in Figure 6 (b), for external field H = 1.0 kOe applied along the [10] direction of the disk array.

In Figure 6 (c), the dispersion curves of the most representative modes are shown along the symmetry directions of the first BZ, showing a very good agreement between experimental points and calculated curves. Since the magnetic modes maintain a symmetry character similar to those found for the isolated dot, they can be labelled in the same way. Depending on the number m (n) of nodal lines perpendicular (parallel) to the direction of the magnetization (x direction), the modes are named as backward-volume-like modes m-BA (Damon-Eshbach-like n-DE), while modes with mixed character are denoted as m-BA×n-DE. The mode without nodal lines is classified as the fundamental mode (F), while modes with dynamic magnetization localized at the ends of the particle[92] are labelled as n-EM, depending on the number of nodes n. The mode type, i.e. the two indices m and n, together with the Bloch wave vector $\boldsymbol{K}$, uniquely identifies the excitation.

Figure 6 (c) shows that several modes exhibit an appreciable dispersion and are therefore propagating modes, while other modes show a frequency that remains almost constant with the wave vector, at least within the frequency resolution of the experiment. The width of each magnonic band (allowed miniband) is proportional to the mean square dynamic magnetization inside a single dot as found by DMM approach. Hence, the dispersion is largest for the F mode, while it decreases rapidly for the higher order modes, which are characterised by an increasing number of oscillations within the single dot. Starting from the Γ point, the measured frequency of the F mode increases (decreases) along the ΓY (ΓX) direction, reaching its maximum (minimum) at the Y (X) point. This behaviour reflects the properties of the dipole-exchange spin waves in the reference continuous film (open squares in Figure 6 (c)). However, the F mode is significantly downshifted with respect to the spin waves in the continuous film, which is due to the static demagnetizing field in the array along the applied field direction (x-direction). As a consequence of this in-plane anisotropy induced by the applied field, one can see that the frequency of the DE mode in the continuous film coincides near the Y-point with that of the 1-DE mode of the array. At the same time, the backward volume spin wave in the continuous film has a frequency that is in the middle of the band gap of the array, i.e. quite far from that of the corresponding 1-BA mode at X-point. In the latter point, an overlap between the F and 1-BA bands is observed.

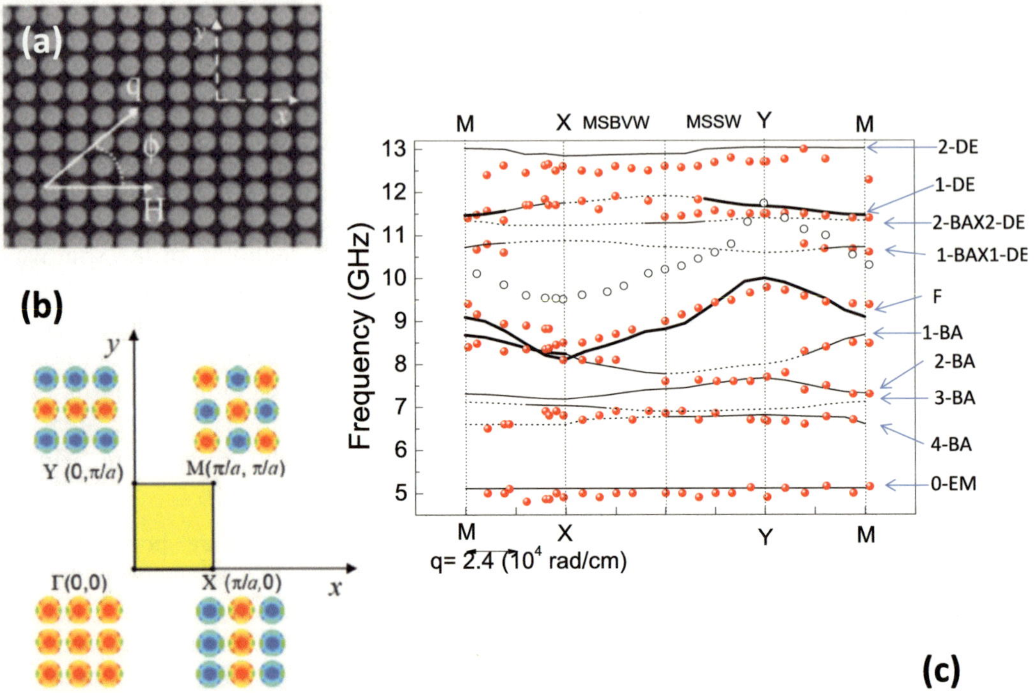

Fig. 6. (a) SEM image of the array of Permalloy disks is shown. BLS spectra were measured applying the external field H along the [10] array direction (x-axis) and changing the q_x and q_y components of the in-plane transferred wave-vector **q**. (b) Surface BZ of the 2D periodic array is shown. The behaviour of the dynamic magnetization of the fundamental mode is schematically shown in four point of the BZ for a 3x3 sub-matrix of dots. The two different colours represent out-of-plane dynamic magnetization of opposite sign. (c) Measured frequencies (dots) are shown as a function of the spin-wave wave vector along the principal direction of the first BZ, for an external magnetic field H = 1.0 kOe. The calculated dispersion curves of the most significant modes are also reported. Bold, solid or dotted lines refer to modes whose calculated cross section is comparable, smaller than 1/10 or smaller than 1/100 with respect to that of the F mode, respectively. For the sake of comparison, the experimental dispersion of the DE mode of the unpatterned film is also reported as open squares.

In Ref. [93] the dispersion of different modes was interpreted in terms of effective wave vector $\boldsymbol{k}^{\mathrm{eff}}$ introduced as an auxiliary variable that includes and replaces the band index, i.e. the mode type, and the Bloch wave vector. It can be defined in the extended zone scheme as:

$$\boldsymbol{k}^{\mathrm{eff}} = \boldsymbol{K} + P(m)\left[m + \frac{1-P(m)}{2}\right]\frac{\pi}{a}\hat{\mathbf{x}} + P(n)\left[n + \frac{1-P(n)}{2}\right]\frac{\pi}{a}\hat{\mathbf{y}} \tag{1}$$

where P(i) is the parity function (=+1,-1 for i even or odd, respectively), m-BA×n-DE is the mode type, and $\boldsymbol{K}$ is assumed to vary in the reduced BZ. The effective wave vector represents the overall oscillation of the magnetization in the array, taking into account the oscillation within the dot due to the mode character (second and third terms in Eq. (1)) and the change between adjacent dots due to the Bloch wave vector (first term). The introduction of k^{eff} helps to understand the frequency dispersion of the magnonic crystal, because, in the limit of a continuous medium, k^{eff} becomes the real wave vector of the spin excitation of the continuous film, whose dispersion curves have the following properties. The mode frequency increases (decreases) when the modulus of the wave vector increases in a direction perpendicular (parallel) to the applied field, corresponding to the MSSW (MSBVW) geometry.

To realize band gap magnonic metamaterials with a band structure of higher in-plane symmetry arrays of circular nanomagnets have been considered where an out-of-plane magnetic field stabilizes the so-called vortex state in each of the unit cells of the periodic nanodisk lattice. The dipolar interaction via nanoscale air gaps leads to allowed minibands and forbidden frequency gaps which exhibit a four-fold symmetry in in-plane directions if nanodisks are arranged in a square lattice [94]. This higher symmetry goes beyond the magnonic crystals where an in-plane field governs the symmetry of the band structure.

5. Theory of band gap magnonic metamaterials

The knowledge of the physical mechanisms which govern the dynamical behaviour of nanoscale magnetic elements is of fundamental importance for understanding the general properties of metamaterials. Because of that, the theoretical derivations of the frequency and the spatial profile of magnonic modes become necessary to gain a physical understanding of the processes observed at a macroscopic level[95-97]. Several analytical and numerical methods are used to derive the profile of normal modes. Analytical models require some preliminary assumptions (approximations) regarding the mode profiles[98,99]. Numerical tools based on micromagnetic simulations has been developed for solving the equation of motion in the time domain and to successively perform a Fourier analysis of that output signal[100,101]. The same equation can also be solved in the frequency domain. The Dynamical Matrix Method (DMM)[102] belongs to the latter approach, in which the sample is subdivided into cells and the linearized Landau-Lifshitz equation of motion is recast as a generalized eigenvalue problem, which is numerically solved by means of a finite-element method. Band structures of periodic composites can then be calculated with the help of the Bloch theorem, which reduces the number of independent variables of a periodic system to that of the corresponding unit cell[82].

The band structures of spin-waves in materials with discrete translational symmetry can be calculated also by the plane wave method (PWM). The PWM is a popular tool commonly used for studying electronic, photonic and phononic crystals because of its conceptual simplicity and applicability to any type of lattice and shape of scattering centers. The method was also adapted to the calculations of the magnonic band structures and is constantly improved, with its field of application extending to new problems[85,103]. Recently, the PWM has also been used for the calculation of spin-wave spectra of 1D magnonic crystals of finite thickness[104] and 2D thin-film magnonic crystals[105]. Only very recently, the PWM has been employed for the first time for calculating spin-wave spectra of 2D antidot arrays based on a square lattice with a good agreement with experimental results obtained[106].

The 2D magnonic crystals composed of two ferromagnetic materials in thin film geometry were studied theoretically in Refs.105,107. The plane wave method with supercell formulation was used to study the edge effect in magnonic crystals in Ref. 105. It was shown that localisation at the edges of corners of the 2D magnonic crystals with finite lateral extension is possible. The PWM was powerful to remodel the magnonic miniband formation in short-period antidot arrays as well as the tunable metamaterial properties of large-period antidot arrays prepared in $Ni_{80}Fe_{20}$ [106,108].

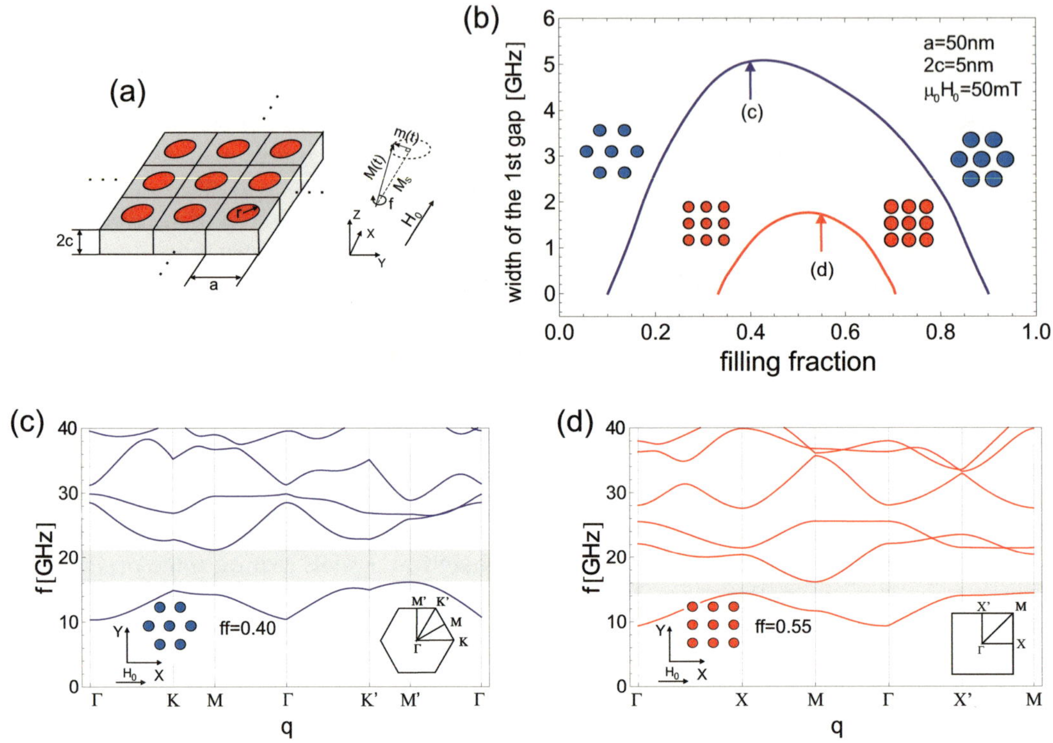

Fig. 7. The magnonic spectrum of periodic slab of finite thickness composed of the Ni inclusions embedded in Fe matrix (a) is strongly dependent on the filling fraction: the ratio of inclusion volume to the volume of unit cell, and depend on the lattice type. The maximum of the width of the first magnonic gap is reached for the intermediate values of the filling fraction (b). Note that the first absolute magnonic gap is wider for the triangular lattice (c) than for the square lattice (d).

The results of a comparative study of 2D magnonic crystals with square and triangular lattice of the Ni inclusions in Fe matrix are shown in Figure 7. The magnonic band structures for triangular and square lattices are shown in Figure 7 (c) and (d), respectively. We find that in the range of small lattice constants the triangular arrangement of cylindrical dots support opening of a magnonic gap. This gap exists for a greater range of filling fraction values and is much wider for a triangular lattice (see Figure 7 (b)). This is similar to the results of similar studies of photonic and phononic crystals. An increase of the lattice constant results in changes in mutual relation between the exchange and magnetostatic interactions. This leads to lowering frequencies of spin waves and decreasing the gap width up to closing it. In the magnetostatic regime, i.e. when the magnetostatic interactions dominate, the nonuniformity of the demagnetizing field is crucial for low frequency spin waves, especially for edge modes.

The elliptical deformation of cylindrical dots in 2D magnonic crystals was investigated in Ref. 107 by means of the plane wave method. The use of rods in the shape of elliptic cylinders as scattering centres in 2D magnonic crystals implies the introduction of two additional structural parameters: the cross-sectional ellipticity of the rods and the angle of their rotation in the plane perpendicular to the rod axis (the plane of spin-wave propagation). In contrast to the lattice constant, a change of which will strongly modify the magnonic spectrum, these new parameters allow fine tuning of the width and position of the bands and band gaps. For specific in-plane rotation angles, changing the rod ellipticity will modify the position, width, and number of bands (Figure 8). Thus, an appropriate use of rods of elliptical cross section offers additional possibilities in the design of magnonic filters with precisely adjusted passbands and stopbands.

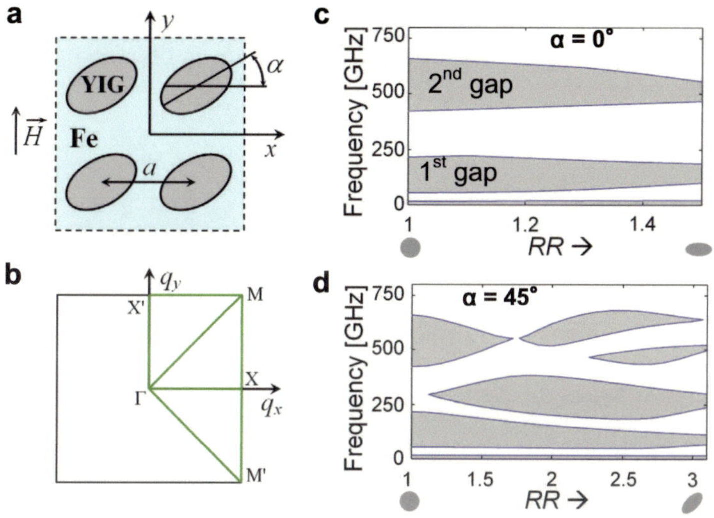

Fig. 8. Fine tuning of the magnonic band structure in the thin film 2D magnonic crystal. (a) Schematic view of the 2D magnonic crystal under study, section in the plane of periodicity. (b) High-symmetry path over the 2D Brillouin zone for ellipses arranged in a square lattice. (c)-(d) The lowest magnonic gaps (shaded) vs. rod ellipticity for 2D YIG/Fe magnonic crystals with two angles of the in-plane rotation of the rods: (c) α = 0° and (d) α = 45°. The other parameters are: lattice constant 10 nm, filling fraction 0.5, film thickness 50 nm, and external magnetic field 0.1 T.

3D band gap and effectively continuous metamaterials are the least studied objects in magnonics, due to both increased difficulty of their theoretical treatment and currently limited outlook for their fabrication and experimental investigation. Collective spin wave modes in 3D arrays of ferromagnetic particles in non-magnetic matrices were studied in Refs. 109,110. Magnonic band structure of 3D all-ferromagnetic magnonic crystals was calculated by Krawczyk and Puszkarski[111,112]. Here, again the depth of modulation of magnetic parameters is essential to generate magnonic bands and forbidden-frequency gaps of significant width.

An example of magnonic band structure for a 3D magnonic crystal resulting from the numerical solution of Landau-Lifshitz equations with the plane-wave method is shown in

Figure 9. The assumed value of the simple cubic (sc) lattice constant was 10 nm; the magnetic parameters of the matrix material were close to those of YIG, and the magnetic parameters of the ferromagnetic material of the spheres corresponded to iron. Two magnonic gaps in the resulting spectrum were observed. The first magnonic gap is delimited by the two lowest spin-wave excitations, one localized in the Fe spheres and the other in the matrix branches I and II, respectively (Figure 9 (b) and (c)). Among three cubic structures studied in Ref. 112 the magnonic crystal with face centred cubic (fcc) lattice is most suitable for gap opening.

In Ref. 113, a detailed study of all the possible combinations of 3D magnonic crystal component materials from: Co, Ni, Fe, and Py for spheres and matrix was performed to find optimal material configurations for which either absolute or partial magnonic gaps occur in the magnonic spectrum of the 3D magnonic crystal with hexagonal structure. Among the MCs considered in this study, an absolute magnonic gap is obtained in a crystal with Ni spheres embedded in Fe.

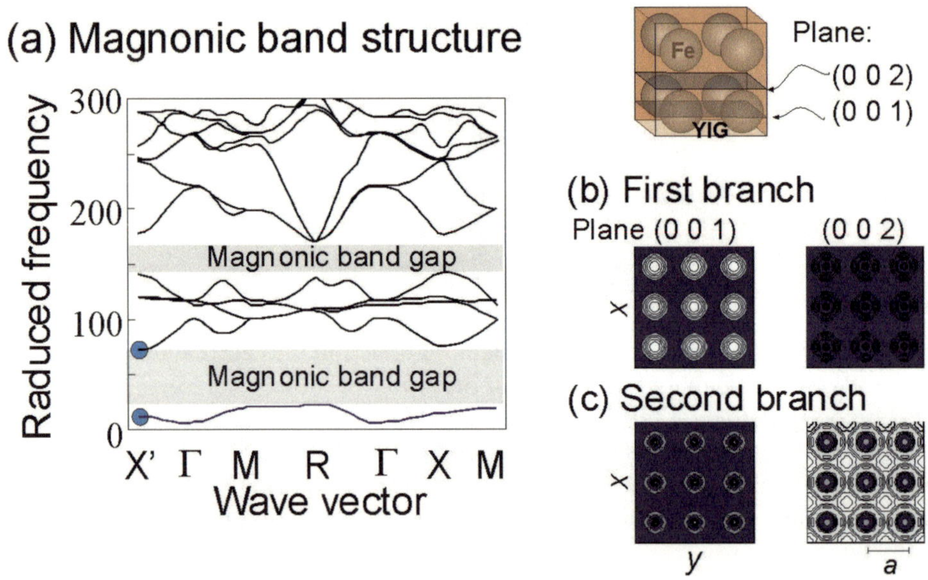

Fig. 9. (a) The magnonic band structure of a simple cubic (sc) magnonic crystal with a lattice constant of 10 nm. The magnonic crystal is composed of Fe spheres of radius 26.28 Å disposed in sites of the sc lattice and embedded in YIG. Blue circles indicate the beginnings of the first two branches at point at which profiles of spin-waves are shown in (b) and (c). (b) and (c) Profiles of squared dynamic magnetization component in two adjacent planes, (002) and (001) right and left column, respectively. The planes are shown in the inset on the top-right. White colour corresponds to maximum values of amplitudes dynamic magnetization. Reproduced from Ref. 112. Copyright 2008, American Physical Society.

Using the method from Refs. 112 and 113, we performed calculations for 3D magnonic crystals based on magnetoferritin crystals (mFT) described in the next section. In the calculations, we assumed mFT spheres (diameter 8 nm) in fcc lattice (with lattice constant 14 nm) immersed in Co. Assuming magnetic and structural parameters taken from literature for dehydrated mFT crystals and cobalt, we showed in Figure 10 (a) that a wide absolute magnonic band gap should exist in the magnonic spectrum of the structure, well above 100 GHz. It means that replacing the protein shell in magnetoferritin crystals with ferromagnetic metals should allow for opening the magnonic band gap. The gap is absolute and wide, and so, it is very promising for application of 3D magnonic crystals.

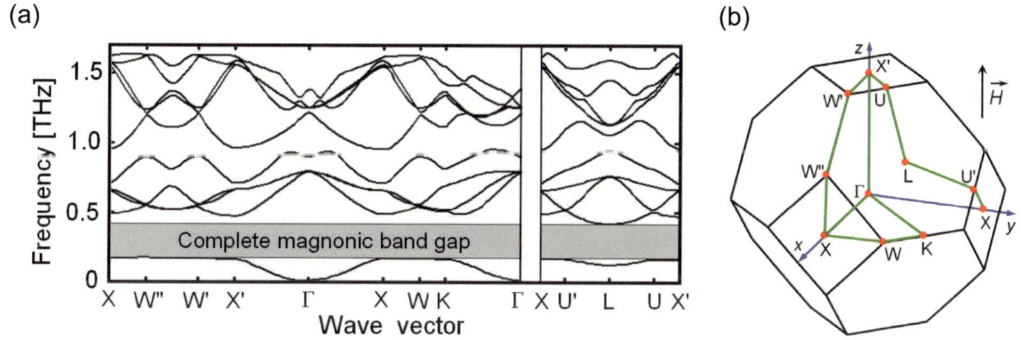

Fig. 10. (a) The magnonic band structure of fcc magnonic crystal with lattice constant of 14 nm is shown. The magnonic crystal is composed of magnetoferritin spheres of diameter 8 nm embedded in Co. The external bias magnetic field 0.1 T is applied along z-axis. (b) First Brillouin zone of the fcc lattice, with the path along which we calculate the magnonic band structure in (a).

The magnonic dispersion can also be calculated from the results of micromagnetic simulations by Fourier transforming them in both temporal and spatial dimensions into the reciprocal space. For 1D samples, the method was realised in Ref. 16, and then used in a number of further studies[114,115]. So, Figure 11 shows the magnonic dispersion of a stack of dipolarly coupled magnetic nanoelements, studied in Ref. 115. The sign of the magnonic dispersion along the stacking direction is determined by the spatial character and ellipticity of the corresponding modes of an isolated nanoelement. Moreover, there exists a critical value of the ellipticity at which the sign of the magnonic dispersion changes from negative to positive in a discontinuous way. The discovered effect suggests a novel way of tailoring the dispersion of collective spin waves in magnonic bandgap metamaterials, unique to magnonics.

Reprogrammable dynamic response has been demonstrated through different remanent states of planar arrays of nanomagnets[116]. The reconfiguration of a 1D magnonic crystal has recently been demonstrated via variation of the orientation of neighbouring ferromagnetic nanowires from a parallel to anti-parallel magnetic states (Figure 12)[117,118]. Experiments and simulations have shown that spin waves propagating perpendicular to the long axis of such coupled nanowires experience different artificial magnonic band structures in configurations (a) and (b). Magnonic dispersions have thus become reprogrammable.

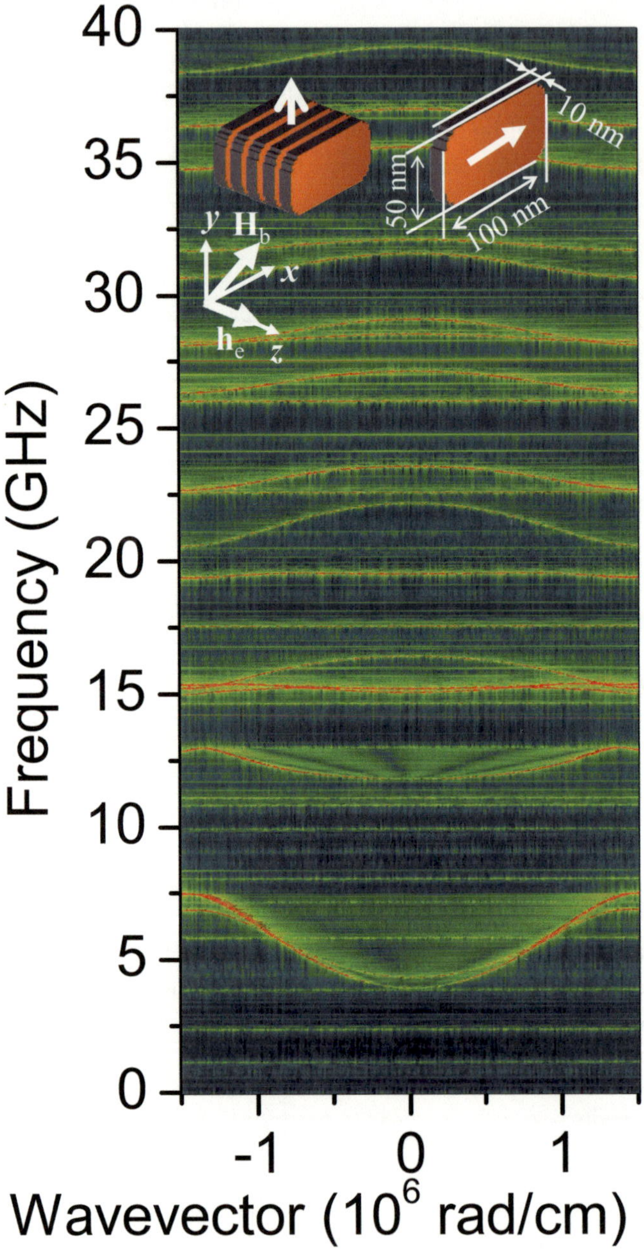

Fig. 11. The magnonic dispersion is shown for a stack of 240 stadium shaped magnetic elements with dimensions of 100x50x10 nm^3. The inset schematically shows the studied sample and geometry of the problem. After Ref. 115.

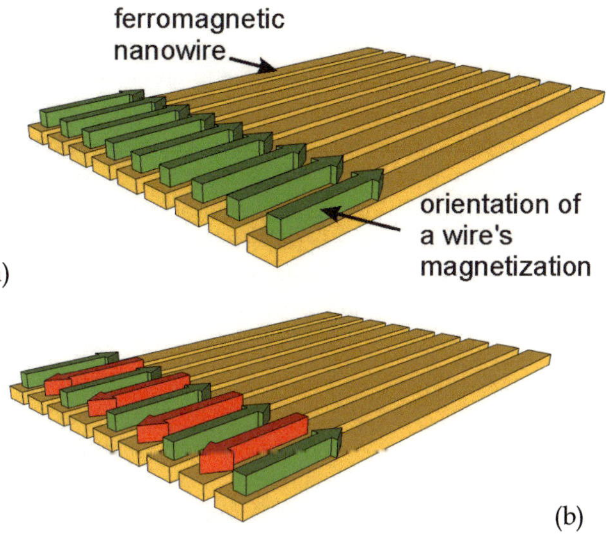

(a)

(b)

Fig. 12. Two different remanent magnetic configurations of a 1D magnonic crystal formed by interacting ferromagnetic nanowires are shown for (a) parallel and (b) anti-parallel alignment of neighbouring nanowires. In (b), the magnetic unit cell of the magnonic crystal is twice as large as the geometrical one, leading to zone folding effects of magnon dispersions in the reciprocal space [117].

6. Theory of effectively continuous magnonic metamaterias

The theory of the effective properties of magnonic metamaterials in situations when they behave as "effectively continuous" can often be derived as the long wavelength limit of the corresponding band gap theory. The latter can however be quite complex, so that it becomes more practical and useful to develop the effectively continuous theory without a reference to the band gap one. In the case of the effective permeability of magnetic composites and metamaterials, the majority of analytical models employ the so-called macrospin approximation, in which each magnetic inclusion within a non-magnetic matrix is considered as a single giant spin and is therefore characterized by a single magnetic resonance. However, it is well known that the spin wave spectrum of magnetic nano-structures and nano-elements has a complex structure, featuring series of resonances due to spatially non-uniform spin wave modes[54-56,61,64,67-69,71,92-99,115,119]. Each of the resonances is expected to contribute to the susceptibility tensor of the magnetic constituents and correspondingly to the permeability tensor of the whole metamaterial. The resonance frequencies can be controlled and reconfigured by the external magnetic[119-121] and electric[122,123] fields, and the same functionalities should therefore be inherited by the magnonic metamaterials.

A method of calculation of the effective permeability that takes full account of the complex spectrum of the metamaterial's individual magnetic constituents has been demonstrated in

Ref. 59. In this method, the susceptibility tensor of an isolated inclusion is calculated numerically and then used as an input to an analytical expression (a so-called "mixing rule") for the permeability of the whole metamaterial. Finding the susceptibility tensor of the isolated inclusion is a standard problem for micromagnetics, and can be addressed using a number of different approaches. For example, full-scale numerical micromagnetic simulations could be performed using one of the available micromagnetic packages (e.g. Nmag[124], OOMMF[125], or MicroMagus[126]). Alternatively, the dynamical matrix method, which has already been introduced above, can be used a form modified to facilitate the susceptibility calculations. The results of the application of the method to a model metamaterial representing an array of magnetic nanodisks embedded into a non-magnetic matrix (inset in Figure 13) is shown in Figure 14. Figure 13 shows the region of geometrical parameters of such a metamaterial, in which one of the components of the permeability tensor becomes negative within a certain frequency range. The method also presents a useful way by which to compare the different micromagnetic methods in order to evaluate the accuracy to be expected from micromagnetic simulations. In particular, we find that the results produced by the state-of-art micromagnetic simulations agree with each other within an error bar of about 5%, which has to be taken into account when micromagnetic calculations are used to model experimental data.

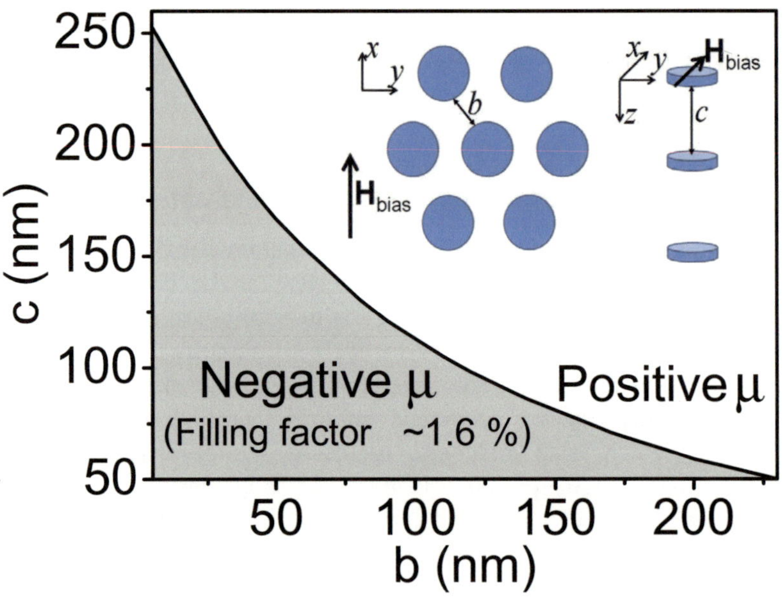

Fig. 13. The range of values of the in-plane edge-to-edge separation (b) and the distance between layers (c) in which μ_{yy} component of the permeability becomes negative near the frequencies of the dominant magnonic resonances. The inset shows the geometry of the metamaterial consisting of magnetic discs in a non-magnetic matrix. The discs are located in nodes of a hexagonal lattice. The disk diameter is d = 195 nm, the in-plane edge-to-edge separation is b = 20 nm, the distance between the layers is c = 140 nm and is much greater than disk thickness l = 5 nm.

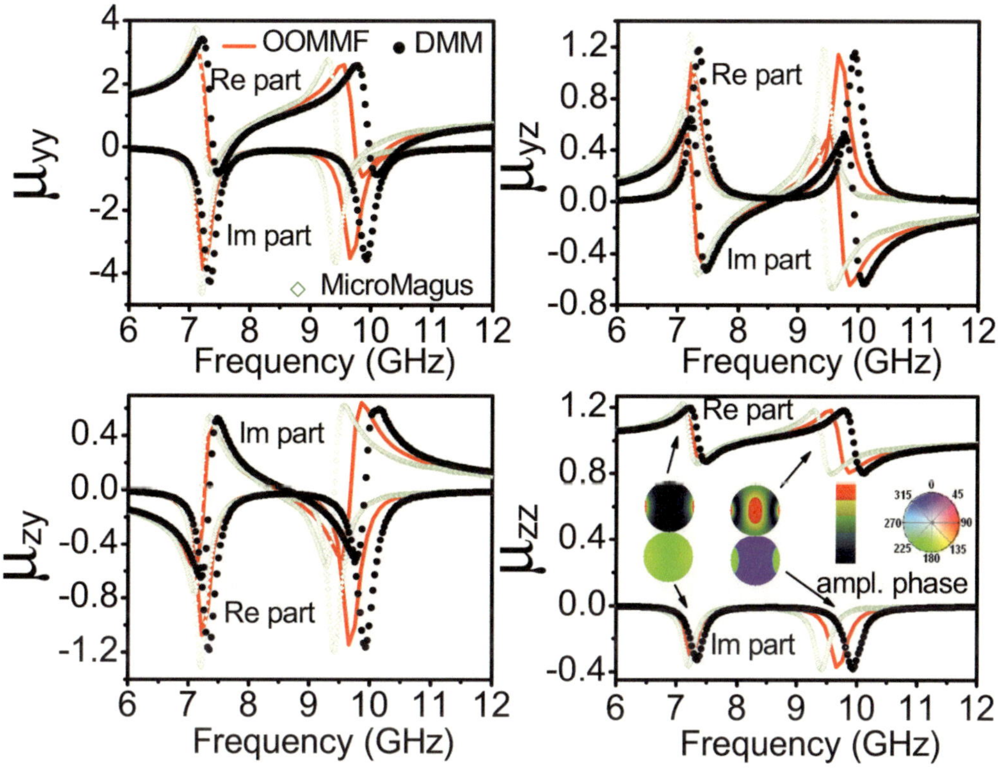

Fig. 14. The real and imaginary parts of the four components of the effective permeability tensor are shown as functions of frequency for the metamaterial depicted in Figure 13. The filling factor is 2.48% and the constant external magnetic field is H_{bias} = 933 Oe. The field is applied in the plane of the layers along the x axis. The insets show the spatial profiles of the mode amplitude (top) and phase (bottom) for the two dominant modes.

7. Bottom-up technologies for metamaterials

In the context of nanostructured magnonic metamaterials, the main nanomanufacturing challenge is to fabricate large-scale periodic structures consisting of or containing magnetic materials precisely and controllably tailored at the nanometre scale. Being at the limit of current lithographic tools, the challenge requires bottom-up technologies be exploited instead. For example, protein based colloidal crystallisation techniques can be used to produce macroscopic 3D ordered magnetic arrays[127,128].

There are a number of possible ways to achieve a periodic 3-dimensional magnetic structure. Apart from top-down lithography, which is generally limited to structures with a height much less than their in-plane dimensions, self-assembly offers a number of promising routes. For example, a number of researchers have used colloidal crystallization to generate periodic 3-dimensional arrays of magnetic nanoparticles[129]. A variation of this approach, which has the advantage that it separates nanoparticle functionality and array formation, exploits the ability of certain proteins both to act as templates for nanoparticle formation and to crystallize. The first demonstration of this method used the ubiquitous iron storage protein ferritin as a template for the growth of ferrimagnetic magnetite-maghemite nanoparticles[127,130]. Ferritin consists of a spherical protein shell with an outer diameter of 12 nm and an inner diameter of 8 nm, and the *in vivo* incorporation of Fe into apoferritin (ferritin without its mineral core) is achieved by the oxidation of Fe^{2+} ions transported through its ion channels.

Following magnetite nanoparticle synthesis, the magnetoferritin (this is the name given to ferritin containing a synthetic ferrimagnetic core) is purified by ion-exchange chromatography, and then passed through a stainless-steel-packed column in a uniform 0.7-0.8 T magnetic field to separate any protein including non- or poorly-magnetic nanoparticles from the magnetoferritin. In a further purification stage size-exclusion chromatography is used to separate magnetoferritin monomers from dimers and oligomers before crystallization, which used the sitting-drop vapour diffusion technique and Cd^{2+} as a crystallization agent. A schematic of this process is shown in Figure 15.

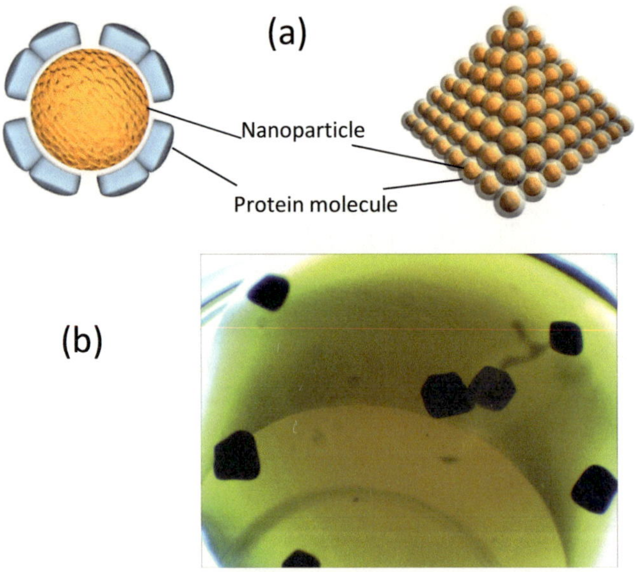

Fig. 15. (a) Schematic showing how crystallizing a protein used as a template for nanoparticle growth gives a periodic 3D array of nanoparticles. (b) Optical image showing magnetoferritin crystals. Each crystal is a periodic 3-dimensional array of magnetite nanoparticles.

To study the metamaterials properties of several magnetoferritin crystals in parallel, fifty of the as-prepared crystals were mounted on a coplanar waveguide (CPW) in order to perform all electrical spin-wave spectroscopy[131,132]. In contrast to earlier studies on nanoparticle arrays where a microwave cavity at a fixed frequency was used[133,134], the CPW-based technique allows one to measure over a broad frequency range. The CPW with the magnetoferritin crystals on top of it is shown in Figure 16. Using a VNA, which is connected to the CPW via microwave probe tips, we apply a microwave current to the CPW. This provokes a microwave magnetic field $\mathbf{h}_{rf}$ around the inner conductor of the CPW. Considering the frequencies ranging from 10 MHz up to 26.5 GHz, we address spin excitations in the magnetoferritin crystals when exploring the metamaterials properties. Using a 20 μm wide inner conductor, we excite spin waves with a distribution of wave vectors **k** given by the current distribution through the CPW[135]. Experiments are carried out in a cryogenic setup allowing us to perform measurements over a wide temperature range with an external field of up to 2.5 T. The field is applied in a direction perpendicular to the plane of the CPW. This direction is chosen so that torque $\tau \sim \mathbf{M} \times \mathbf{h}_{rf}$ is maximised when $\mathbf{M}$ ($\mathbf{h}_{rf}$) is out-of-plane (in-plane). Torque τ excites the spin waves.

We start our discussion from room temperature data taken at 290 K (Figure 16). We measure the dynamic response as a function of applied external field. We observe pronounced absorption with a linear dependency indicated by the white dashed line. Interestingly, the resonance starts to appear at fields larger than 0.1 T, whereas at 0 T there is no resonance observed. We attribute the observed behaviour to the paramagnetic response of the protein crystals, in which the magnetization vectors of the nanoparticles are randomly aligned at zero field. An anisotropy term does not seem to be present to provoke a non-zero resonance frequency at small field. The individual nanoparticles show superparamagnetic behaviour, which is expected for particles of this size.

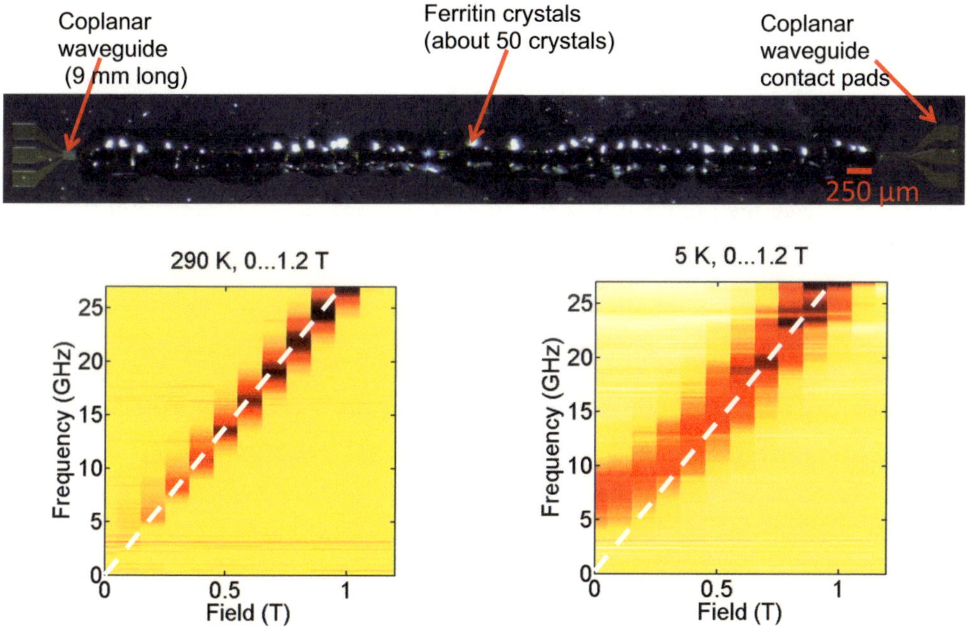

Fig. 16. (Top) Microscopy image of the 9 mm long coplanar wave guide containing 50 magnetoferritin crystals as shown in Figure 15. (Bottom) Spectroscopy data taken at 290 K (left) and 5 K (right) as a function of perpendicular magnetic field. Dark colour indicates spin-wave excitation. The broken lines are guides for the eyes reflecting both the field dependency at 290 K to facilitate comparison.

The data at 5 K show a different behaviour, as we observe a clear absorption peak even at zero field. Here, the resonance is measured to be at 6 GHz. For fields larger than about 0.5 T, we observe a linear dependence of the resonance frequency with the applied field as well, but it is slightly shifted towards higher frequencies if compared to the room temperature measurement (white dashed line). The resonance frequency at zero field implies an anisotropy term aligning spins even without external field.

In order to use 3D nanoparticle assemblies as all-magnetic metamaterials, it would be relevant that researchers combine the protein based nano-manufacturing with advanced 3D material deposition techniques such as atomic layer deposition (ALD)[136] and electrodeposition tailored for use with multiple magnetic materials. From this all-ferromagnetic 3D magnonic metamaterials might result. ALD film growth is self-limited, thereby achieving atomic scale control of the deposition. Recently, ALD was used to deposit ferromagnetic thin-films (such as Ni, Co, Fe_3O_4) into deep-etched trenches and membranes[137,138,139]. The complementary topology is also possible by, e.g., conformal coating of templates consisting of tailored nanowires[140]. Such possibilities make ALD a promising tool by which to fabricate 3D magnonic devices. Electrodeposition is also very well suited for deposition into complex templates[141]. It is fast and thereby suitable for scaling-up to produce large numbers of devices. For example, arrays of cylindrical magnetic nanowires deposited electrochemically within porous membranes[142,143,144] have attracted much attention due to their potential for use as microwave[145] and THz[13] devices.

8. Conclusions

The modern research on fundamental properties of materials is increasingly driven by their anticipated potential for technological applications. In this Chapter, we have reviewed research that has been conducted within the MAGNONICS project funded by the European Commission to reveal the potential of magnonic metamaterials nanoscale building blocks of which are made of magnetic materials. A particular attention is devoted to the use of spin wave resonances tailored in magnonic crystals for design of novel features in the electromagnetic properties in the GHz-THz frequency range. These entirely new electromagnetic metamaterials could be designed to tune the transmission, absorption and reflection of electromagnetic radiation in the GHz-THz frequency range. Such magnonic metamaterials could also find their use within magnonic filters and logic gates.

The field of magnonics and magnonic metamaterials is very young. However, the growing community of magnonics researchers has already demonstrated that they are up to the challenges existing in the magnonics science and technology. In particular, this Chapter demonstrates important advances recently achieved in the direction of the development of effectively continuous and band gap magnonic metamaterials.

9. Acknowledgments

We thank S. Neusser for providing simulation results, P. Berberich, G. Duerr, R. Huber, J. Topp, and T. Rapp for experimental support. The authors gratefully acknowledge funding received from the European Community's Seventh Framework Programme (FP7/2007-2013) under Grant Agreements no 233552 (DYNAMAG) and 228673 (MAGNONICS), from the Engineering and Physical Sciences Research Council (EPSRC) of United Kingdom (project EP/E055087/1), from the German excellence cluster "Nanosystems Initiative Munich (NIM)".

10. References

[1] A. De Baas (Editor), "Nanostructured metmaterials" (The European Commission, Publications Office, 2010).
[2] V. G. Veselago, Sov. Phys. Uspekhi 10, 509 (1968).
[3] J. B. Pendry, D. Schurig, and D. R. Smith, Science 312, 1780 (2006).
[4] J. B. Pendry, Phys. Rev. Lett. 85, 3966 (2000).
[5] J. B. Pendry, A. J. Holden, D. J. Robbins, and W. J. Stewart, IEEE Trans. Microwave Theory Tech. 47, 2075 (1999).
[6] O. Acher, J. Magn. Magn. Mater. 321, 2093 (2009).
[7] R. V. Mikhaylovskiy, E. Hendry, and V. V. Kruglyak, Phys. Rev. B 82, 195446 (2010).
[8] A. G. Gurevich and G. A. Melkov, Magnetization oscillations and waves (CRC Press, 1996).
[9] V. V. Kruglyak, S. O. Demokritov, and D. Grundler, J. Phys. D: Appl. Phys. 43, 264001 (2010).
[10] A. Aharoni, J. Appl. Phys. 69, 7762 (1991); *ibid.* 81, 830 (1997).
[11] D. Mercier, J. - C. S. Levy, G. Viau, F. Fievet-Vincent, F. Fievet, P. Toneguzzo, and O. Acher, Phys. Rev. B 62, 532 (2000).
[12] J. Ramprecht and D. Sjöberg, J. Phys. D - Appl. Phys. 41, 135005 (2008).
[13] G. S. Makeeva, M. Pardavi-Horvath, and O. A. Golovanov, IEEE Trans. Magn. 45, 4074 (2009).
[14] V. Boucher, L. - P. Carignan, T. Kodera, C. Caloz, A. Yelon, and D. Ménard, Phys. Rev. B 80, 224402 (2009).
[15] V. V. Kruglyak and A. N. Kuchko, J. Magn. Magn. Mater. 272-276, 302 (2004).
[16] V. V. Kruglyak and R. J. Hicken, J. Magn. Magn. Mater. 306, 191 (2006).
[17] E. Yablonovitch, Phys. Rev. Lett., 58, 2059 (1987).
[18] W. L. Barnes, A. Dereux, and T. W. Ebbesen, Nature (London) 424, 824 (2003).
[19] L. M. Brekhovskikh and O. A. Godin, *Acoustics Of Layered Media I: Plane And Quasi-Plane Waves* (Springer-Verlag, Berlin, 1990).
[20] S. A. Nikitov, P. Tailhades, and C. S. Tsai, J. Magn. Magn. Mater. 236, 320 (2001).
[21] S. Tamaru, J. A. Bain, R. J. M. van de Veerdonk, T. M. Crawford, M. Covington, and M. H. Kryder, Phys. Rev. B 70, 104416 (2004).
[22] V. E. Demidov, B. Hillebrands, S. O. Demokritov, M. Laufenberg, and P. P. Freitas, J. Appl. Phys. 97, 10A717 (2005).
[23] Z. Liu, F. Giesen, X. Zhu, R. D. Sydora, and M. R. Freeman, Phys. Rev. Lett. 98, 087201 (2007).
[24] J. Gouzerh, A. A. Stashkevich, N. G. Kovshikov, V. V. Matyushev, and J. M. Desvignes, J. Magn. Magn. Mater. 101, 189 (1991).
[25] Y. I. Gorobets and S. A. Reshetnyak, Techn. Phys. 43, 188 (1998).
[26] A. M. Kuchko, Metallofiz. Noveish. Tekhn. 27, 511 (2005).
[27] A. V. Vashkovsky and E. H. Lock, Phys. Uspekhi 49, 389 (2006).
[28] S. K. Kim, S. Choi, K. S. Lee, D. S. Han, D. E. Jung, and Y. S. Choi, Appl. Phys. Lett. 92, 212501 (2008).
[29] V. E. Demidov, S. O. Demokritov, D. Birt, B. O'Gorman, M. Tsoi, and X. Li, Phys. Rev. B 80, 014429 (2009).
[30] A. V. Vashkovskii, K. V. Grechushkin, A. V. Stalmakhov, and V. A. Tyulukin, Radiotekhn. Elektronika 32, 2295 (1987).

[31] V. K. Dugaev, P. Bruno, B. Canals, and C. Lacroix, Phys. Rev. B 72, 024456 (2005).
[32] S. Yang, Z. Song, and C. P. Sun, Europ. Phys. J. B 52, 377 (2006).
[33] S. K. Choi, K. S. Lee, and S. K. Kim, Appl. Phys. Lett. 89, 062501 (2006).
[34] K. Perzlmaier, G. Woltersdorf, and C. H. Back, Phys. Rev. B 77, 054425 (2008).
[35] D. R. Birt, B. O'Gorman, M. Tsoi, X. Li, V.E. Demidov, and S.O. Demokritov, Appl. Phys. Lett. 95, 122510 (2009).
[36] F. Morgenthaler, IEEE Trans. Magn. MAG8, 550 (1972).
[37] V. S. L'vov, A. M. Rubenchik, V. V. Sobolev, and V. S. Synakh, Fiz. Tverd. Tela 15, 793 (1973).
[38] O. Büttner, M. Bauer, S. O. Demokritov, B. Hillebrands, Y. S. Kivshar, V. Grimalsky, Y. Rapoport, M. P. Kostylev, B. A. Kalinikos, and A. N. Slavin, J. Appl. Phys. 87, 5088 (2000).
[39] R. Khomeriki, Europ. Phys. J. B 41, 219 (2004).
[40] V. Verrakumar and R. E. Camley, IEEE Trans. Magn. 42, 3318 (2006).
[41] V. E. Demidov, J. Jersch, S. O. Demokritov, K. Rott, P. Krzysteczko, and G. Reiss, Phys. Rev. Lett. 102, 177207 (2009).
[42] S. O. Demokritov, A. A. Serga, A. Andre, V. E. Demidov, M. P. Kostylev, B. Hillebrands, and A. N. Slavin, Phys. Rev. Lett. 93, 047201 (2004).
[43] A. Kozhanov, D. Ouellette, M. Rodwell, S. J. Allen, A. P. Jacob, D. W. Lee, and S. X. Wang, J. Appl. Phys. 105, 07D311 (2009).
[44] E. A. Vilkov, Phys. Solid State 48, 1754 (2006).
[45] D. D. Stancil, B. E. Henty, A. G. Cepni, and J. P. Van't Hof, Phys. Rev. B 74, 060404 (2006).
[46] V. Vlaminck and M. Baileul, Science 322, 410 (2008).
[47] A. M. Kosevich, B. A. Ivanov, and A. S. Kovalev, Phys. Rep. 194, 117 (1990).
[48] A. N. Slavin, S. O. Demokritov, and B. Hillebrands, Top. Appl. Phys. 83, 35 (2002).
[49] Y. K. Fetisov, C. E. Patton, and V. T. Synogach, JETP Lett. 83, 488 (2006).
[50] M. Z. Wu, P. Krivosik, B. A. Kalinikos, and C. E. Patton, Phys. Rev. Lett. 96, 227202 (2006).
[51] M. H. Seavey and P. E. Tannenwald, Phys. Rev. Lett. 1, 168 (1958).
[52] C. Kittel, Phys. Rev. 110, 1295 (1958).
[53] C. Mathieu, J. Jorzick, A. Frank, S. O. Demokritov, A. N. Slavin, B. Hillebrands, B. Bartenlian, C. Chappert, D. Decanini, F. Rousseaux, and E. Cambril Phys. Rev. Lett. 81, 3968 (1998).
[54] A. Barman, V. V. Kruglyak, R. J. Hicken, J. M. Rowe, A. Kundrotaite, J. Scott, and M. Rahman, Phys. Rev. B 69, 174426 (2004).
[55] M. Bailleul, R. Höllinger, K. Perzlmaier, and C. Fermon, Phys. Rev. B 76, 224401 (2007).
[56] I. Neudecker, F. Hoffmann, G. Woltersdorf, and C. H. Back, J. Phys. D – Appl. Phys. 41, 164010 (2008).
[57] K.-D. Lee, J.-W. Kim, J.-W. Jeong, S.-C. Shin, J. Appl. Phys. 106, 113904 (2009).
[58] M. Krawczyk, J. Magn. Magn. Mater. 322, 562 (2010).
[59] O. Dmytriiev et al (unpublished).
[60] J. H. E. Griffiths, Nature (London) 158, 670 (1946).
[61] V. V. Kruglyak, P. S. Keatley, A. Neudert, R. J. Hicken, J. R. Childress, and J. A. Katine, Phys. Rev. Lett. 104, 027201 (2010).
[62] H. Al-Wahsh, Eur. Phys. J. B 73, 527-537 (2010).

[63] R. E. Camley, M. R. F. Jensen, S. A. Feiven, and T. J. Parker, J. Appl. Phys. 83, 6280 (1998).
[64] F. Giesen, J. Podbielski, B. Botters, and D. Grundler, Phys. Rev. B 75, 184428 (2007).
[65] S. S. Kalarickal, P. Krivosik, M. Wu, C. E. Patton, M. L. Schneider, P. Kabos, T. J. Silva, and J. P. Nibarger, J. Appl. Phys. 99, 093909 (2006).
[66] M. Belmeguenai, T. Martin, G. Woltersdorf, M. Maier, and G. Bayreuther, Phys. Rev. B 76, 104414 (2007).
[67] W. K. Hiebert, A. Stankiewicz, and M. R. Freeman, Phys. Rev. Lett. 79, 1134 (1997).
[68] R. J. Hicken, A. Barman, V. V. Kruglyak, and S. Ladak, J. Phys. D: Appl. Phys. 36, 2183 (2003).
[69] C. H. Back, D. Pescia, and M. Buess, Top. Appl. Phys. 101, 137 (2006), and references therein.
[70] J. Li, M.-S. Lee, W. He, B. Redeker, A. Remhof, E. Amaladass, C. Hassel, and T. Eimueller, Rev. Sci. Instr. 80, 073703 (2009).
[71] P. S. Keatley, V. V. Kruglyak, A. Neudert, M. Delchini, R. J. Hicken, J. R. Childress, and J. A. Katine, J. Appl. Phys. 105, 07D308 (2009).
[72] M. L. Schneider, J. M. Shaw, A. B. Kos, T. Gerrits, T. J. Silva, and R. D. McMichael, J. Appl. Phys. 102, 103909 (2007).
[73] B. Hillebrands and A. Thiaville (Eds.), *Spin dynamics in Confined Structures I-III* (Springer, 2001-2006).
[74] G. Gubbiotti, S. Tacchi, G. Carlotti, N. Singh, S. Goolaup, A. O. Adeyeye, and M. Kostylev, Appl. Phys. Lett. 90, 092503 (2007).
[75] C. Elachi, IEEE MAG-11, 36 (1975).
[76] G. Gubbiotti, S. Tacchi, M. Madami, G. Carlotti, A. O. Adeyeye, and M. Kostylev, J. Phys. D: Appl. Phys. 43, 264003 (2010).
[77] G. Gubbiotti, S. Tacchi, G. Carlotti, N. Singh , S. Goolaup , A. O. Adeyeye, and M. Kostylev, Appl. Phys. Lett. 90, 092503 (2007).
[78] S. Tacchi, M. Madami, G. Gubbiotti, G. Carlotti, S. Goolaup, A. O. Adeyeye, N. Singh and M. P. Kostylev, Phys. Rev. B 82, 184408 (2010).
[79] Z. K. Wang, V. L. Zhang, H. S. Lim, S. C. Ng, M. H. Kuok, S. Jain, and A. O. Adeyeye, Appl. Phys. Lett. 94, 083112 (2009).
[80] G. Gubbiotti, G. Carlotti, M. Madami, S. Tacchi, P. Vavassori, and G. Socino, J. Appl. Phys. 105, 07D521 (2009).
[81] M. Madami, S. Bonetti, S. Tacchi, G. Carlotti, G. Gubbiotti, G. Consolo, F. Mancoff, M.A. Yar, and J. Åkerman, Nat. Nanotech. 6, 635 (2011) DOI: 10.1038/NNANO.2011.140.
[82] L. Giovannini, F. Montoncello, and F. Nizzoli, Phys. Rev. B 75, 024416 (2007).
[83] R. Zivieri, F. Montoncello, L. Giovannini, F. Nizzoli, S. Tacchi, M. Madami, G. Gubbiotti, G. Carlotti, and A. O. Adeyeye, Phys. Rev. B 83, 054431 (2011).
[84] R. Zivieri, F. Montoncello, L. Giovannini, F. Nizzoli, S. Tacchi, M. Madami, G. Gubbiotti, G. Carlotti, and A. O. Adeyeye, IEEE Trans. Magn. 47, 1563 (2011).
[85] J. O. Vasseur, L. Dobrzynski, B. Djafari-Rouhani, and H. Puszkarski, Phys. Rev. B 54, 1043 (1996).
[86] H. Puszkarski and M. Krawczyk, Solid State Phenom. 94, 125 (2003).
[87] A. Y. Galkin, B. A. Ivanov, and C. E. Zaspel, Phys. Rev. B 74, 144419 (2006).
[88] R. Antos, Y. Otani, and J. Shibata, J. Phys. Soc. Jap. 77, 031004 (2008), and references therein.

[89] M. J. Pechan, C. T. Yu, R. L. Compton, J. P. Park, and P. A. Crowell, J. Appl. Phys. 97, 10J903 (2005).
[90] S. Neusser, B. Botters, and D. Grundler, Phys. Rev. B 78, 054406 (2008).
[91] S. Neusser, B. Botters, M. Becherer, D. Schmitt-Landsiedel, and D. Grundler, Appl. Phys. Lett. 93, 122501 (2008).
[92] V. V. Kruglyak, A Barman, R. J. Hicken, J. R. Childress, and J. A. Katine, Phys. Rev. B 71, 220409 (2005).
[93] S. Tacchi, F. Montoncello, M. Madami, G. Gubbiotti, G. Carlotti, L. Giovannini, R. Zivieri, F. Nizzoli, S. Jain, A. O. Adeyeye, and N. Singh, Phys. Rev. Lett. 107, 127204 (2011).
[94] R. Huber and D. Grundler, Proc. SPIE 8100, 81000D (2011).
[95] R. D. McMichael and M. D. Stiles, J. Appl. Phys. 97, 10J901 (2005).
[96] F. Montoncello, L. Giovannini, F. Nizzoli, P. Vavassori, M. Grimsditch, T. Ono, G. Gubbiotti, S. Tacchi, and G. Carlotti, Phys. Rev. B 76, 024426 (2007).
[97] F. Montoncello, L. Giovannini, F. Nizzoli, R. Zivieri, G. Consolo, and G. Gubbiotti, J. Magn. Magn. Mater. 322, 2330 (2010).
[98] K. Y. Guslienko and A. N. Slavin, J. Appl. Phys. 87, 6337 (2000).
[99] R. Zivieri and R. L. Stamps, Phys. Rev. B 73, 144422 (2006).
[100] M. Donahue and D. Porter, OOMMF: Object Oriented MicroMagnetic Framework, Interagency report NISTIR 6376, National Institute of Standards and Technology, Gaithersburg, MD. http://math.nist.gov/oommf.
[101] M. Slodička and L. Baňas, Appl. Math. Comp. 158, 79 (2004).
[102] M. Grimsditch, L. Giovannini, F. Montoncello, F. Nizzoli, G. K. Leaf, and H. G. Kaper, Phys. Rev. B 70, 054409 (2004).
[103] R. P. Tiwari and D. Stroud, Phys. Rev. B 81, 220403(R) (2010).
[104] M. L. Sokolovskyy and M. Krawczyk, Journal of Nanoparticle Research 13, 6085 (2011), DOI: 10.1007/s11051-011-0303-5.
[105] J. W. Klos, M. Krawczyk, and M. L. Sokolvskyy, J. Appl. Phys. 109, 07D311 (2011).
[106] S. Neusser, G. Duerr, S. Tacchi, M. Sokolovskyy, M. Madami, G. Gubbiotti, M. Krawczyk, and D. Grundler, Phys. Rev. B 84, 094454 (2011).
[107] S. Mamica and M. Krawczyk, *Tuning of the spin-wave band structure in 2D magnetic composites*, Proceedings of ICCM-18, August 21-26, 2011, Jeju Island, Korea (in press).
[108] S. Neusser, H.G. Bauer, G. Duerr, R. Huber, S. Mamica, G. Woltersdorf, M. Krawczyk, C.H. Back, and D. Grundler. Phys. Rev. B 84, 184411 (2011).
[109] P. Chu, D. L. Mills, and R. Arias, Phys. Rev. B 73, 094405 (2006).
[110] E. Tartakovskaya, W. Kreuzpaintner, and A. Schreyer, J. Appl. Phys. 103, 023913 (2008).
[111] M. Krawczyk and H. Puszkarski, Cryst. Res. Techn. 41, 547 (2006).
[112] M. Krawczyk and H. Puszkarski, Phys. Rev. B 777, 054437 (2008).
[113] M. Krawczyk, J. Klos, M. L. Sokolovskyy, and S. Mamica, J. Appl. Phys. 108, 093909 (2010).
[114] S. - K. Kim, J. Phys. D: Appl. Phys. 43, 264004 (2010), and references therein.
[115] M. Dvornik and V. V. Kruglyak, Phys. Rev. B 84, 140405 (2011).
[116] F. Giesen, J. Podbielski, T. Korn, M. Steiner, A. van Staa, and D. Grundler, Appl. Phys. Lett. 86, 112510 (2005).

[117] J. Topp, D. Heitmann, M. Kostylev, and D. Grundler, Phys. Rev. Lett. 104, 207205 (2010).
[118] J. Topp, G. Duerr, K. Thurner, and D. Grundler, Pure Appl. Chem. 83, 1989 (2011).
[119] V. V. Kruglyak, P. S. Keatley, A. Neudert, M. Delchini, R. J. Hicken, J. R. Childress, and J. A. Katine, Phys. Rev. B 77, 172407 (2008).
[120] E. J. Kim, J. L. R. Watts, B. Harteneck, A. Scholl, A. Young, A. Doran, and Y. Suzuki, J. Appl. Phys. 109, 07D712 (2011).
[121] V. L. Zhang, Z. K. Wang, H. S. Lim, S. C. Ng, M. H. Kuok, S. Jain, and A. O. Adeyeye, J. Nanosci. Nanotechn. 11, 2657 (2011).
[122] P. Rovillain, R. de Sousa, Y. Gallais, A. Sacuto, M. A. Measson, D. Colson, A. Forget, M. Bibes, M. Barthelemy, and M. Cazayous, Nature Mater. 9, 975 (2010).
[123] A. Kumar, J. F. Scott, and R. S. Katiyar, Appl. Phys. Lett. 99, 062504 (2011).
[124] T. Fischbacher, M. Franchin, G. Bordignon, and H. Fangohr. IEEE Trans. Magn. 43, 2896 (2007), http://nmag.soton.ac.uk.
[125] M. Donahue, and D.G. Porter, OOMMF User's guide, Version 1.0, Interagency Report NISTIR 6376, NIST, 1999, http://math.nist.gov/oommf.
[126] D. V. Berkov, and N. L. Gorn, MicroMagus - package for micromagnetic simulations, http://www.micromagus.de.
[127] O. Kasyutich, A. Sarua, and W. Schwarzacher, J. Phys, D - Appl. Phys. 41, 134022 (2008).
[128] O. Kasyutich, D. Tatchev, A. Hoell, F. Ogrin, C. Dewhurst, and W. Schwarzacher, J. Appl. Phys. 105, 07B528 (2009).
[129] D. F. Farrell, Y. Ijiri, C. V. Kelly, J. A. Borchers, J. J. Rhyne, Y. Ding, and S. A. Majetich, J. Magn. Magn. Mater. 303, 318 (2006).
[130] F. C. Meldrum, V. J. Wade, D. L. Nimmo, B. R. Heywood, and S. Mann, Nature 349, 684 (1991).
[131] J. Podbielski, F. Giesen, M. Berginski, N. Hoyer, and D. Grundler, Superlattices and Microstructures 37, 341 (2005).
[132] S. Neusser, G. Duerr, H. G. Bauer, S. Tacchi, M. Madami, G. Woltersdorf, G. Gubbiotti, C. H. Back, and D. Grundler, Phys. Rev. Lett. 105, 067208 (2010).
[133] H. Li, M. T. Klem, K. B. Sebby, D. J. Singel, M. Young, T. Douglas, and Y. U. Idzerda, J. Magn. Magn. Mater. 321, 175.
[134] N. Guskos, E.A. Anagnostakis, and A. Guskos, JAMME 24, 26 (2007).
[135] K. J. Kennewell, M. Kostylev, and R. L. Stamps, J. Appl. Phys. 101, 09D107 (2007).
[136] T. Suntola and J. Antson, "Method for producing compound thin films", US Patent 4058430 (1974).
[137] B. S. Lim, A. Rahtu, and R. G. Gordon, Nature Mater. 2, 749 (2003).
[138] M. Daub, M. Knez, U. U. Gösele, and K. Nielsch, J. Appl. Phys. 101, 09J111 (2007).
[139] J. Bachmann, J. Jing, M. Knez, S. Barth, H. Shen, S. Mathur, U. Gösele, and K. Nielsch, J. Am. Chem. Soc. 129, 9554 (2007).
[140] A. F. I. Morral, D. Spirkoska, J. Arbiol, M. Heigoldt, J. R. Morante, and G. Abstreiter, Small 4, 899 (2008).
[141] I. Kazeminezhad and W. Schwarzacher, Electrochem. Solid-State. Lett. 11, K24-6 (2008).
[142] P. Aranda and J. M. García, J. Magn. Magn. Mater. 249, 214 (2002), and references therein.
[143] P. R. Evans, G. Yi, and W. Schwarzacher, Appl. Phys. Lett. 76, 481 (2000).

[144] A. Robinson and W. Schwarzacher, J. Appl. Phys. 93, 7250 (2003).
[145] A. Saib, D. Vanhoenacker-Janvier, I. Huynen, A. Encinas, L. Piraux, E. Ferain, and R. Legras, Appl. Phys. Lett. 83, 2378 (2003).

Section 3

The Applications of Metamaterials

15

Investigation of Dipole Antenna Loaded with DPS and DNG Materials

Amir Jafargholi and Manouchehr Kamyab
K. N. Toosi University of Technology
Iran

1. Introduction

The increasing demands on compact multifunctional devices have necessitated the development of multi-frequency printed dipoles which can be integrated into familiar devices such as laptop computers and mobile phones. The typical difficulties encountered in designing compact antennas include narrow bandwidth, and low radiation efficiency. In order to achieve a good efficiency, considerable effort must be expended on the matching network. Other researchers have found that the bandwidth of the dipole antenna can be enhanced by loading the antenna with parallel lumped element circuits (Rogers et al., 2003). Over the last decade, increasing demands for low profile multifunctional antennas have resulted in considerable interest by the electromagnetic research community in Metamaterials (MTMs). Due to unique electromagnetic properties, MTMs have been widely considered in monopole and dipole antennas to improve their performance (Erentok et al., 2005; Erentok et al., 2008; Liu et al., 2009; Jafargholi et al., 2010). The applications of Composite Right/Left Handed (CRLH) structures to load the printed dipole have been investigated both numerically (Iizuka et al., 2006; Iizuka et al., 2007; Borja et al., 2007) and analytically (Rafaei et al., 2010). However, main drawbacks of this method are low gain and low efficiency. The use of transmission-line based MTMs to realize a tri-band monopole antenna has been recently investigated in (Zhu et al., 2010). However, the cross polarization levels of the proposed antenna in (Zhu et al., 2010) are very high. It is also known that the antenna properties can be improved by covering the metal radiating parts or filling the antenna volume. For instance, the bandwidth of the microstrip patch antenna can be significantly improved by replacing the dielectric substrate with the magneto-dielectric one (Mosallaei et al., 2007). Recently, (Erentok et al., 2008) have considered the use of Double Negative (DNG) cover to match an electrically small electric dipole antenna to free space. The effect of complex material coverings on the bandwidth of the antennas has been also investigated in (Tretyakov et al., 2004).

In this section, first, the influence of the material inclusions on the input impedance of the loaded dipoles excited by a delta function is analytically investigated. Novel and accurate analytical expressions for the input impedance of the loaded dipoles are proposed based on the mode matching technique. The boundary conditions are also enforced to obtain several simultaneous equations for the discrete modal coefficients inside the radiating region. Study of the input impedance of the whole multilayered structure is accomplished by the cascade connection of mediums as characterized by their constitutive parameters. New and accurate analytical formulas for the loaded dipole antenna are derived and successfully validated through a proper comparison with the results obtained with the commercial software CST Microwave Studio.

Moreover, a compact multiband printed dipole antenna loaded with reactive elements is proposed. The reactive loading of the dipole is inspired by the Epsilon-Negative (ENG) and DNG- MTM inclusions, which enable the loaded dipole to operate in multiband. The reactive loads are realized by two rake-shaped split ring resonators (SRRs) facing each other. Investigations reveal that the loaded dipole radiates at two or three separated bands depending on symmetrical or asymmetrical loading and load locations. The new resonance frequencies are lower than the natural resonance frequency of the conventional half wavelength dipole. In this range of frequencies, the radiation efficiency of the composite antenna is high. In order to validate the simulation results, a prototype of the proposed printed dipole is fabricated and tested. The agreement between the simulated and measured results is quite good.

2. Full-wave analysis of loaded dipole antennas using mode-matching theory

In recent years, introducing MTMs opened the way for many researcher groups to enhance the antenna performances. Due to unique electromagnetic properties, MTMs have been widely considered in monopole and dipole antennas to improve their performance (Jafargholi et al., 2010). The problem of dielectric loaded wire antenna is heretofore analyzed using numerical methods, e.g., Method of Moment (MoM) (Shams et al., 2007), Finite Difference Time Domain (FDTD) (Beggs et al., 1993), and simulations based on commercial software (Kennedy et al., 2006). However, the analytical analysis of the dielectric loaded dipole antennas has not been reported in the literature.

The novelty of this section is to introduce a mode-matching analysis of a dipole antenna loaded with material inclusions. In this section, a theoretical formulation for a multiply dielectric loaded slotted spherical antenna is proposed based on the mode-matching method, to predict the behavior of the loaded dipole. It is worth noting that the radiation pattern of a finite length small angle biconical antenna differs only slightly from the pattern of a dipole (Kraus et al., 2002). Here, since the biconical antenna can be exactly analyzed and it also reduces, in the limiting case, to a cylindrical dipole antenna (Collin et al., 1969), this structure is considered for the analytical investigations. The obtained analytical formulas confirm the general conclusions recently presented in (Shams et al., 2007; Beggs et al., 1993), regarding the effect of material inclusions on the dipole antenna performance. It is demonstrated that the inclusion influence on the input impedance of a dipole is significant only for DNG-MTM inclusions.

2.1 Field analysis

Fig. 1(a) illustrates a slotted dielectric loaded hollow conducting sphere of radius *a*, containing a Hertzian dipole $\overline{J}=\hat{z}J\delta(\overline{r}-\overline{r}')$, placed at the center $(r=r',\theta=0,\phi)$, Here (r,θ,ϕ) are the spherical coordinates and δ is a delta function. The time convention of is $e^{-j\omega t}$ suppressed throughout. Due to azimuthally symmetry, the fields depend on (r,θ) and the fields are then TM waves, which can be expressed in terms of magnetic vector potentials. The total magnetic vector potential for the un-slotted sphere (First region, I) is a sum of the primary and secondary magnetic vector potentials, (Ock et al., 2009).

$$A^{i}(r,\theta)=\hat{z}A_{z}^{p}(r,\theta)+\hat{r}A_{r}^{s}(r,\theta) \quad (1)$$

While, the primary magnetic vector potential is a free-space Green's function as

$$A_{z}^{p}(r,\theta)=\frac{\mu_1 J}{4\pi}\frac{e^{ik_1R}}{R} \quad (2)$$

where $\hat{z}$ and $\hat{r}$ are unit vectors and $R=\sqrt{r^2+r'^2-2rr'\cos\theta}$. And the secondary magnetic vector potential is

$$A_{r}^{s}(r,\theta)=\sum_{n=0}^{\infty}a_n\hat{J}_n(k_1r)P_n(\cos\theta) \quad (3)$$

where $\hat{J}_n(.)$ is the spherical Bessel function and $P_n(.)$ is the Legendre function. The coefficient is (Ock et al., 2009)

$$\begin{aligned} a_n &= \frac{\mu_1 aJ}{8\pi k_1\hat{J}_n'(k_1a)}\frac{2n+1}{n(n+1)}\int_0^{\pi}\Omega\frac{\partial P_n(\cos\theta)}{\partial\theta}\sin^2\theta d\theta \\ \Omega &= \left\{\left(a^2-2r'^2+ar'\cos\theta\right)\left(ik_1\tilde{R}-1\right)+k_1^2\tilde{R}^2\left(a^2-ar'\cos\theta\right)\right\}\frac{e^{ik_1\tilde{R}}}{\tilde{R}^5} \end{aligned} \quad (4)$$

Now consider a slotted conducting sphere, as shown in Fig. 1(a). The total magnetic vector potential in region (I) consists of the incident A^{i} and scattered A_{r}^{I} potentials as

$$A_{r}^{I}(r,\theta)=\sum_{n=0}^{\infty}C_n\hat{J}_n(k_1r)P_n(\cos\theta) \quad (5)$$

Here, C_n is an unknown modal coefficient. The *r*-component of the magnetic vector potential in region (II, III, IV, and V) of the *l*-th slot is

$$A_{r}^{\gamma}(r,\theta)=\sum_{v=0}^{\infty}R_{v}^{l,\gamma}(\cos\theta)\left[D_{v}^{l,\gamma}\hat{J}_{\xi_v^l}(k_\gamma r)+E_{v}^{l,\gamma}\hat{N}_{\xi_v^l}(k_\gamma r)\right], \qquad \gamma=\mathrm{II,III,IV,V} \quad (6)$$

Where

$$R_{v}^{l,\gamma}(\cos\theta)=\begin{cases}\begin{cases}Q_{\xi_v^l}(\cos\theta) & v=0 \\ Q_{\xi_v^l}(\cos\alpha_2^l)P_{\xi_v^l}(\cos\theta)-P_{\xi_v^l}(\cos\alpha_2^l)Q_{\xi_v^l}(\cos\theta) & v\ge 1\end{cases} & \mathrm{II,V} \\ P_{\xi_v^l}(\cos\theta)+G_{\xi_v^l}^{\gamma}Q_{\xi_v^l}(\cos\theta) & \mathrm{III,IV}\end{cases} \quad (7)$$

The *r*-component of the magnetic vector potential in region (VI) is

$$A_{r}^{VI}(r,\theta)=\sum_{v=0}^{\infty}F_n\hat{H}_n^{(2)}(k_{VI}r)P_n(\cos\theta) \quad (8)$$

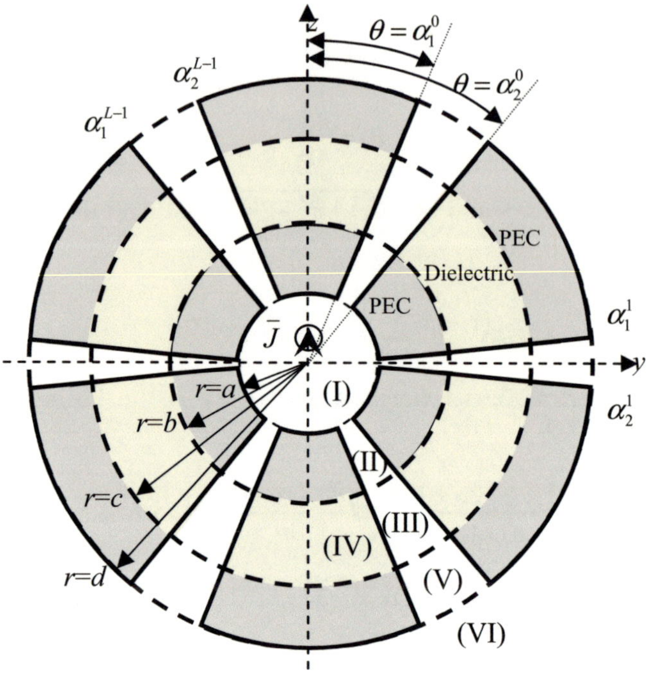
z
θ = α_1^0
θ = α_2^0
α_2^{L-1}
α_1^{L-1}
PEC
Dielectric
PEC
α_1^1
α_2^1
J
y
r=a
r=b
r=c
r=d
(I)
(II)
(III)
(IV)
(V)
(VI)

(a)

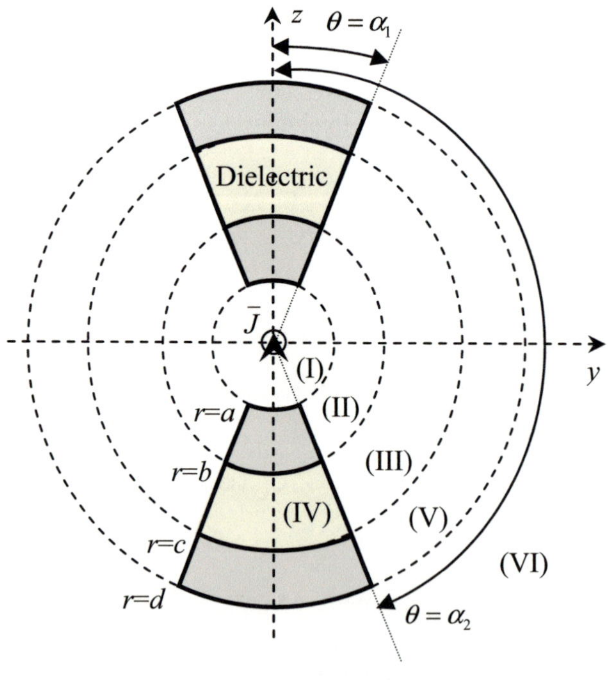
z
θ = α_1
Dielectric
J
y
(I)
(II)
(III)
(IV)
(V)
(VI)
r=a
r=b
r=c
r=d
θ = α_2

(b)

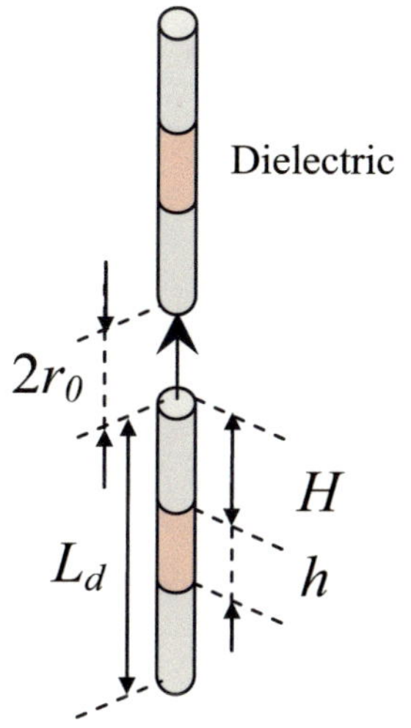

(c)

Fig. 1. (a) Multiply- (b) single slotted dielectric loaded conducting hollow sphere, and (c) dielectric loaded dipole antenna: cross-sectional view, a=0.1mm, b=2.5mm, $h=|c\text{-}b|$=0.5mm, d=5mm, r_0=0.1mm, H=2.4mm, L_d=4.9mm, and the dipole radius, r_d, is equal to 0.1mm. From (Jafargholi et. al., 2012), copyright © 2012 by the Electromagnetics, Taylor & Francis Group, LLC.

where F_n is an unknown modal coefficient and $\hat{H}_n^{(2)}(.)$ is the spherical Hankel function of the second kind. To determine the modal coefficients, we enforce the field continuities as Table 1.

Boundary	layers	Electric Fields	Magnetic Fields	Limits
1	I , II	$E_\theta^{II} = \begin{cases} E_\theta^{I} & \alpha_1^q < \theta < \alpha_2^q \\ 0 & \text{otherwise} \end{cases}$	$H_\phi^{i} + H_\phi^{I} = H_\phi^{II}$	$r=a$, $\alpha_1^q < \theta < \alpha_2^q$
2	II , III	$E_\theta^{III} = \begin{cases} E_\theta^{II} & \alpha_1^q < \theta < \alpha_2^q \\ 0 & \text{otherwise} \end{cases}$	$H_\phi^{III} = H_\phi^{II}$	$r=b$, $\alpha_1^q < \theta < \alpha_2^q$
3	III , V	$E_\theta^{III} = \begin{cases} E_\theta^{V} & \alpha_1^q < \theta < \alpha_2^q \\ 0 & \text{otherwise} \end{cases}$	$H_\phi^{III} = H_\phi^{V}$	$r=c$, $\alpha_1^q < \theta < \alpha_2^q$
4	V, VI	$E_\theta^{VI} = \begin{cases} E_\theta^{V} & \alpha_1^q < \theta < \alpha_2^q \\ 0 & \text{otherwise} \end{cases}$	$H_\phi^{VI} = H_\phi^{V}$	$r=d$, $\alpha_1^q < \theta < \alpha_2^q$
5	III , IV	$E_r^{IV} = E_r^{III}$	$H_\phi^{VI} = H_\phi^{III}$	$b<r<c$, $\theta = \alpha_{1,2}^q$

Table 1. Boundary conditions

Applying orthogonal integrals and mathematical manipulation some can write the equations as follow.

$$C_n = -\sqrt{\frac{\mu_I \varepsilon_I}{\mu_{II} \varepsilon_{II}}} \frac{2n+1}{2n(n+1)} \frac{1}{\hat{J}'_n(k_I a)} \sum_{l=0}^{L-1} \sum_{v=0}^{\infty} \left[D_v^{l,II} \hat{J}'_{\xi_v^l}(k_{II} a) + E_v^{l,II} \hat{N}'_{\xi_v^l}(k_{II} a) \right] I_{vn}^{l,II} \tag{9}$$

$$\sum_{l=0}^{L-1} \sum_{v=0}^{\infty} \left\{ D_v^{l,II} \left[\hat{J}'_{\xi_v^l}(k_{II} a) X_{uv}^{ql} - \hat{J}_{\xi_v^l}(k_{II} a) K_v^{l,II} \delta_{ql} \delta_{uv} \right] + E_v^{l,II} \left[\hat{N}'_{\xi_v^l}(k_{II} a) X_{uv}^{ql} - \hat{N}_{\xi_v^l}(k_{II} a) K_v^{l,II} \delta_{ql} \delta_{uv} \right] \right\}$$
$$= -\frac{\mu_{II} J a^2}{4\pi} L_u^q + \frac{\mu_{II}}{\mu_I} \sum_{n=0}^{\infty} a_n \hat{J}_n(k_I a) I_{un}^{q,II} \tag{10}$$

$$\sum_{l=0}^{L-1} \sum_{v=0}^{\infty} \left[D_v^{l,\gamma} \hat{J}'_{\xi_v^l}(k_\gamma r) + E_v^{l,\gamma} \hat{N}'_{\xi_v^l}(k_\gamma r) \right] K_v^{l,\gamma}$$
$$= \sqrt{\frac{\mu_\gamma \varepsilon_\gamma}{\mu_{\gamma'} \varepsilon_{\gamma'}}} \sum_{l=0}^{L-1} \sum_{v'=0}^{\infty} \left[D_{v'}^{l,\gamma'} \hat{J}'_{\xi_{v'}^l}(k_\gamma r) + E_{v'}^{l,\gamma'} \hat{N}'_{\xi_{v'}^l}(k_\gamma r) \right] K_{v'}^{l,\gamma\gamma'} \qquad \gamma = II, V \;,\; \gamma' = III \;\; r = b, c \tag{11}$$

$$\sum_{l=0}^{L-1} \sum_{v=0}^{\infty} \left[D_v^{l,\gamma} \hat{J}_{\xi_v^l}(k_\gamma r) + E_v^{l,\gamma} \hat{N}_{\xi_v^l}(k_\gamma r) \right] K_v^{l,\gamma}$$
$$= \frac{\mu_\gamma}{\mu_{\gamma'}} \sum_{l=0}^{L-1} \sum_{v'=0}^{\infty} \left[D_{v'}^{l,\gamma'} \hat{J}_{\xi_{v'}^l}(k_\gamma r) + E_{v'}^{l,\gamma'} \hat{N}_{\xi_{v'}^l}(k_\gamma r) \right] K_{v'}^{l,\gamma\gamma'} \qquad \gamma = II, V \;,\; \gamma' = III \;\; r = b, c \tag{12}$$

$$\sum_{v=0}^{\infty} \xi_v^l (\xi_v^l + 1) \left[D_v^{l,III} U_v + E_v^{l,III} U_{vw} \right] R_v^{l,III}(\cos\theta_0) =$$
$$\frac{\mu_{III} \varepsilon_{III}}{\mu_{IV} \varepsilon_{IV}} \sum_{v'=0}^{\infty} \xi_{v'}^l (\xi_{v'}^l + 1) \left[D_{v'}^{l,IV} U_{v'v} + E_{v'}^{l,IV} U_{v'w} \right] R_{v'}^{l,IV}(\cos\theta_0), \qquad \theta_0 = \alpha_1^l, \alpha_2^l \tag{13}$$

$$\sum_{l=0}^{L-1} \sum_{v=0}^{\infty} \left\{ D_v^{l,V} \left[\hat{J}'_{\xi_v^l}(k_V d) \Psi_{uv}^{ql} - \hat{J}_{\xi_v^l}(k_V d) K_v^{l,V} \delta_{ql} \delta_{uv} \right] + E_v^{l,V} \left[\hat{N}'_{\xi_v^l}(k_V d) \Psi_{uv}^{ql} - \hat{N}_{\xi_v^l}(k_V d) K_v^{l,V} \delta_{ql} \delta_{uv} \right] \right\} = 0 \tag{14}$$

$$F_n = -\sqrt{\frac{\mu_{VI} \varepsilon_{VI}}{\mu_V \varepsilon_V}} \frac{2n+1}{2n(n+1)} \frac{1}{\hat{H}_n'^{(1)}(k_{VI} d)} \sum_{l=0}^{L-1} \sum_{v=0}^{\infty} \left[D_v^{l,V} \hat{J}'_{\xi_v^l}(k_V d) + E_v^{l,V} \hat{N}'_{\xi_v^l}(k_V d) \right] I_{vn}^{l,V} \tag{15}$$

The required definitions are illustrated in the appendix. For a single slot configuration (biconical antenna loaded with a dielectric, Fig. 1(b), due to the magnetic field boundary condition between region III and IV, $R_v^{l,\gamma}(\cos\theta)$ has been simplified as

$$R_v^{l,\gamma}(\cos\theta) = \begin{cases} \begin{cases} Q_v(\cos\theta) & v = 0 \\ Q_v(\cos\alpha_2) P_v(\cos\theta) - P_v(\cos\alpha_2) Q_v(\cos\theta) & v \geq 1 \end{cases} & II, V \\ P_v(\cos\theta) + \left(\frac{\mu_{III}}{\mu_{IV}} - 1 \right) \left[\frac{P'_v(\cos\alpha_1) - P'_v(\cos\alpha_2)}{Q'_v(\cos\alpha_1) - Q'_v(\cos\alpha_2)} \right] Q_v(\cos\theta) & III \\ P_v(\cos\theta) & IV \end{cases} \tag{16}$$

Finally, the unknown coefficients are

$$C_n,\ D_v^{II},\ E_v^{II},\ D_v^{III},\ E_v^{III},\ D_v^{IV},\ E_v^{IV},\ D_v^{V},\ E_v^{V},\ F_n$$

2.2 Numerical analysis

From the formulas presented in the previous section it is straightforward to write short programs that illustrate the difference between the different types of material inclusions. To this aim, the cone angle of the biconical antenna is selected to be as small as possible, e.g., $2\alpha_1$=2.5 degree. To clear this selection, it is should be noted that, based on (Collin et al., 1969), it is well known that the input impedance of a biconical antenna changes significantly by changing cone angle. Hence the input impedance of a biconical antenna is investigated with regards to its cone angle. The inverse radiation impedance Z_v of biconical antennas, for the small feed gap condition ($k_1 a << 1$) is given by (Ock et al., 2009; Saoudy et al., 1990)

$$Z_v = \frac{j\eta_2 \sin\alpha_1 \ln\left(\cot\frac{\alpha_1}{2}\right)\sum_{v=0}^{\infty}\left[D_v^{II}\hat{J}_{\xi_v}(k_{II}b)+E_v^{II}\hat{N}_{\xi_v}(k_{II}b)\right]\left.\frac{\partial R_{\xi_v}(\cos\theta)}{\partial\theta}\right|_{\theta=\alpha_1}}{\pi\left[D_0^{II}\hat{J}_0'(k_{II}b)+E_0^{II}\hat{N}_0'(k_{II}b)\right]} \tag{17}$$

The analytic simulations have been compared with CST simulation results of an equivalent dipole antenna (radius, r_d). The results have been presented in Fig. 2. According to these results for the antenna radius r_d <0.01λ (≈biconical antenna $2\alpha_1$≤3.4 degree, with regards to f=25GHz as main frequency) the loaded dipole may be considered as a limit case of a loaded biconical antenna (the approximation meet numerical simulations with good agreement). The simulation parameters have been considered as: a=0.1mm, b=2.5mm, $h=|c-b|$=0.5mm, d=5mm, r_0=0.1mm, H=2.4mm, L_d=4.9mm, and the dipole antenna filled with DPS material inclusions, (ε_r=2.2 and μ_r=1). It is should be noted that for the radius $0.01\lambda<r_d<0.02\lambda$, the antenna input impedance has been extracted approximately; and larger values cause significant errors in impedance computations.

a. Dielectric-Covered Biconical Antennas

To validate the proposed method, it is useful to consider a conventional covered biconical antenna (Fig. 3) as first limiting case. The input impedance of a thin biconical antenna embedded in dielectric material has been derived by (Tai et al., 1958). A slightly more general expression applicable to a biconical antenna embedded in a lossless material of arbitrary permeability and permittivity has been given by (Polk et al., 1959). Assuming $L=1$, the region III fills by PEC, and $k_1 a << 1$; the slotted conducting sphere becomes a biconical antenna, as shown in Fig. 3. In Fig. 4, the effects of the numbers of modes in computation convergence have been depicted. It is clear that good convergence has been achieved.

In Fig. 5, the analytic results for the impedance of a biconical antenna have been compared with CST simulation results. As it is stated before, the antenna cone angle has been chosen as $2\alpha_1$=2.5 degree. According to this figure, a good agreement has been achieved between analytic and numeric simulations.

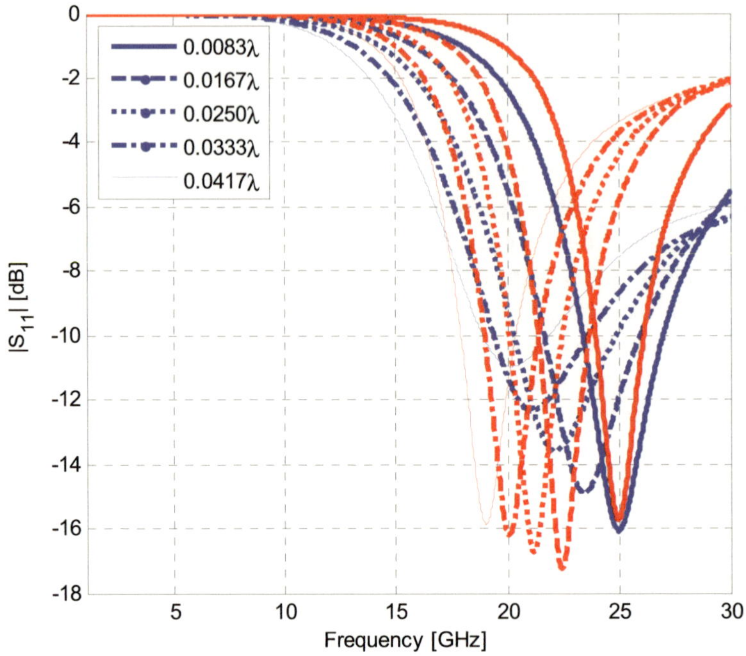

Fig. 2. The $|S_{11}|$ [dB] for a DPS-loaded dipole antenna: analytical (Blue) against numerical results (Red). Analytical results are obtained using proposed analytical expressions; while the numerical results are extracted using CST software. From (Jafargholi et. al., 2012), copyright © 2012 by the Electromagnetics, Taylor & Francis Group, LLC.

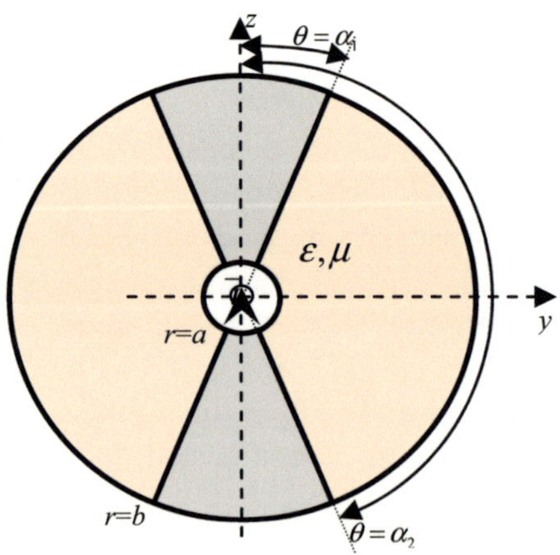

Fig. 3. Dielectric-covered biconical antenna: cross-sectional view, a=0.1mm, b=5mm, $\varepsilon_r = \mu_r = 1$. From (Jafargholi et. al., 2012), copyright © 2012 by the Electromagnetics, Taylor & Francis Group, LLC.

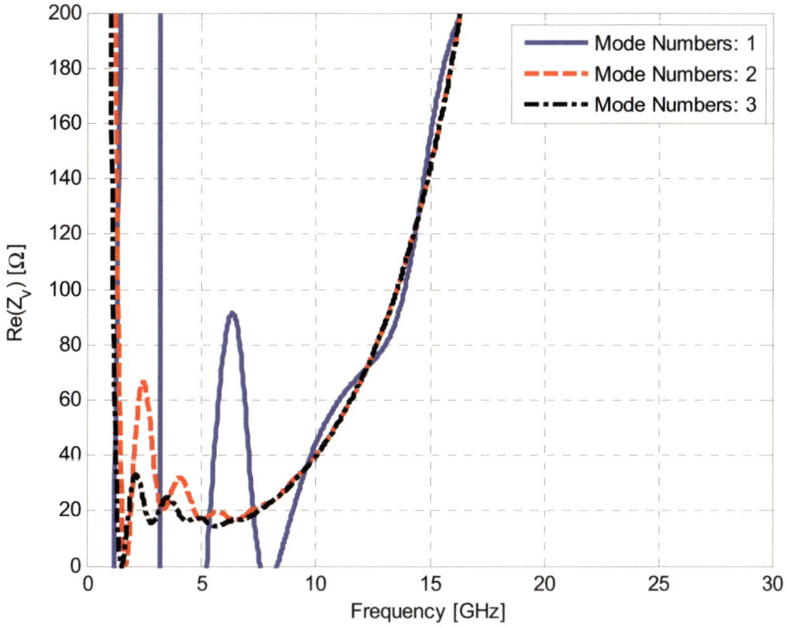

(a)

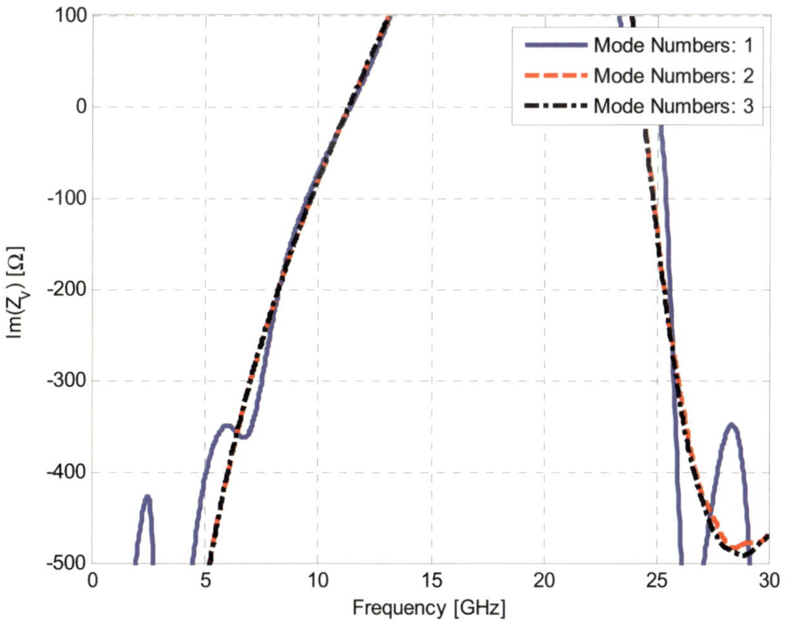

(b)

Fig. 4. Convergence analysis of the dielectric- covered biconical antenna, input impedance (a) real, and (b) imaginary parts. From (Jafargholi et. al., 2012), copyright © 2012 by the Electromagnetics, Taylor & Francis Group, LLC.

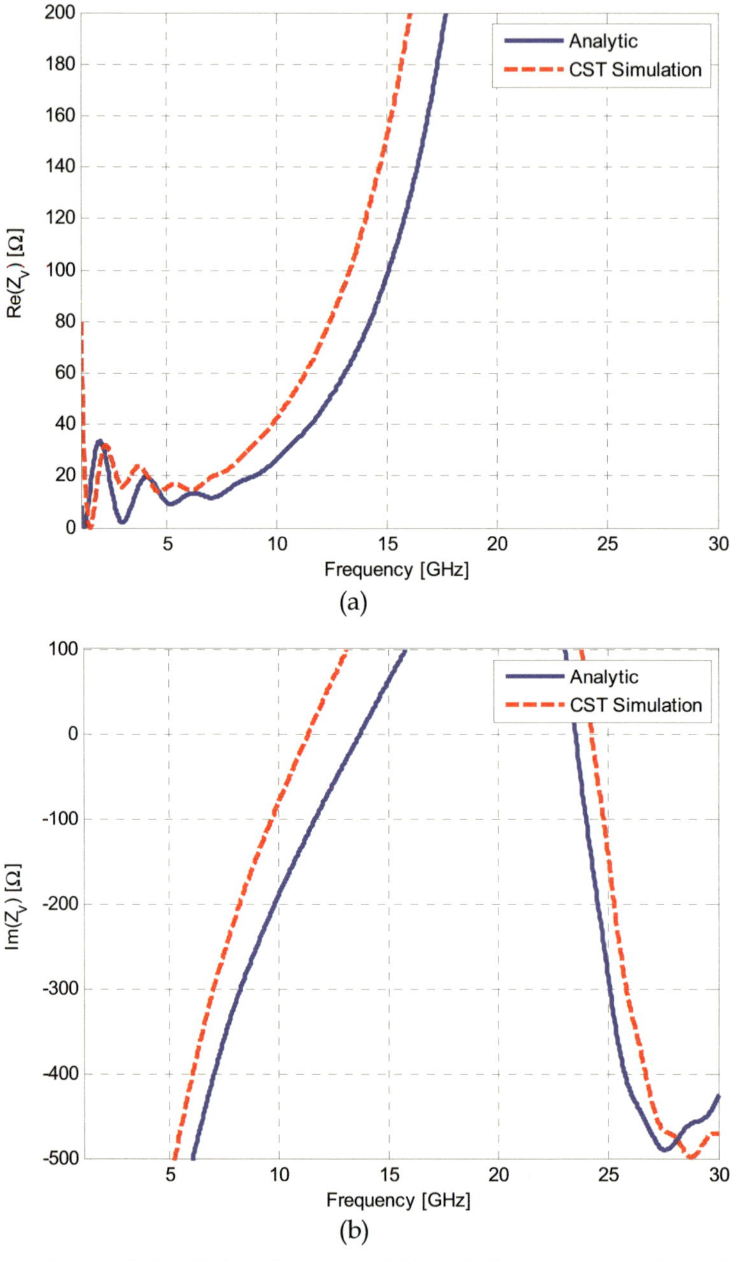

Fig. 5. Input impedance of the dielectric-covered biconical antenna: analytical against numerical results. Analytical results are obtained using proposed analytical expressions; numerical results are computed by CST software; (a) real, and (b) imaginary parts. From (Jafargholi et. al., 2012), copyright © 2012 by the Electromagnetics, Taylor & Francis Group, LLC.

In Fig. 6, the analytic result for impedance of dielectric covered biconical antenna versus dielectric material has been illustrated. The antennas parameters are b=5mm, $2a_1$=2.5 degree,

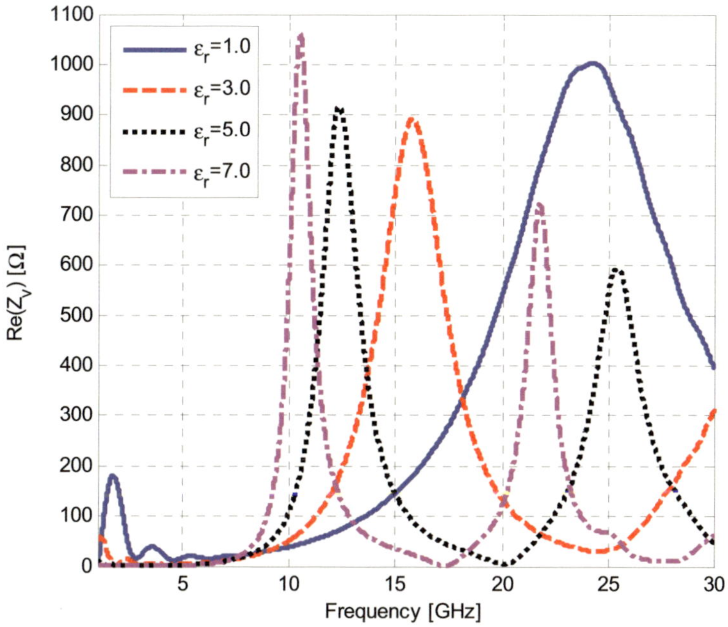

(a)

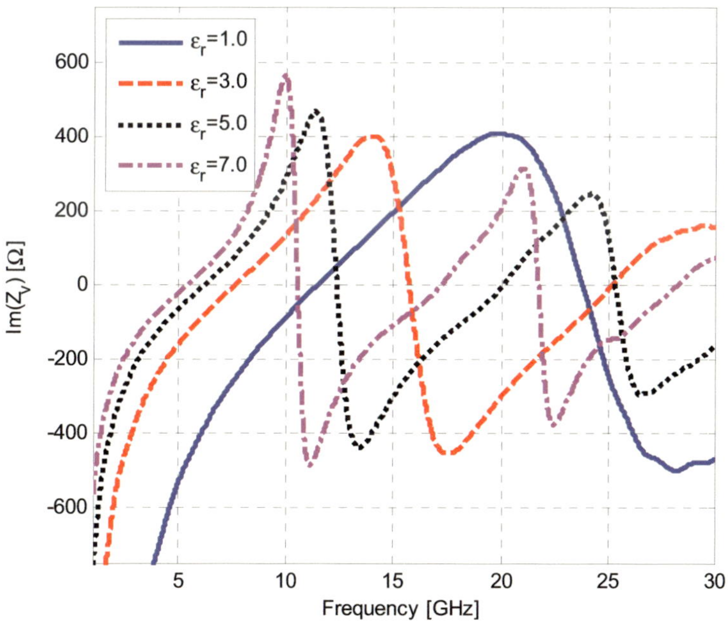

(b)

Fig. 6. Input impedance of the dielectric-loaded biconical antenna: analytical against numerical results. Analytical results are obtained using proposed analytical expressions; numerical results are computed by CST software; (a) real, and (b) imaginary parts. From (Jafargholi et. al., 2012), copyright © 2012 by the Electromagnetics, Taylor & Francis Group, LLC.

a=0.1mm. It seems clearly that the antenna input impedance affects significantly with dielectric material. The insertion of dielectric material into the biconical antenna structure causes increased frequency dependence of the antenna. The larger material permittivity, the more rapid is the variation of the input impedance. Similar results have been obtained by (Saoudy et al., 1990; Tai et al., 1958; Polk et al., 1959).

b. Dielectric Loaded Biconical Antenna

In order to demonstrate the capability of the MTM loading to realize a miniaturized antenna, two examples are studied here. The first one is a dipole antenna filled with Double Positive (DPS) material inclusions, (ε_r=2.2 and μ_r=1). A DNG-loaded dipole antenna, whose parameters are labeled in Fig. 1(c), is also studied. Here, the Drude model (Jafargholi et al., 2011) is used to simulate the MTM inclusions, since it can yield a negative real part of the permittivity/permeability over a wide frequency range. For the DNG inclusions, both μ and ε obey the Drude model (with plasma frequency ω_p=15×10^{10} rad/s and collision frequency f_c =0.01GHz) as bellow

$$\xi_r(\omega) = \xi_\infty - \frac{\omega_p^2}{\omega(\omega - iv_c)} \qquad \xi \in \{\varepsilon, \mu\} \tag{18}$$

In Fig. 7, the effects of the numbers of modes in computation convergence have been presented. Again, it is clear that good convergence has been obtained. The computation time of the analytic model is about 5 minutes for all frequency points compared to several hours using CST over a frequency range of 0 to 30GHz, while this time increased in CST for higher permittivity and permeability materials. (For a 3.2GHz dual core CPU with 2GByte RAM).

The analytical results for the input impedance (both real and imaginary parts) of the DPS- and DNG-loaded dipole antennas are presented in Fig. 8. As a reference, the simulated input impedance of equivalent DPS- and DNG- loaded dipoles are also plotted in this figure. As can be seen in this figure, the analytical results for the input impedance of the loaded dipoles are in good agreement with the CST simulation results. Simulations show that for the dipole antenna loaded with DNG-inclusions, an additional resonance frequency is introduced at the frequencies lower than the antenna resonant frequency where the antenna radiates an omnidirectional radiation pattern. In contrast, for the dipoles loaded with DPS-inclusions, changing DPS locations on the antenna arms causes no resonances at frequencies lower than the main resonant frequency.

3. A compact multi-band printed dipole antenna loaded with single-cell MTM

Now, the effect of material inclusions embedded in a simple dipole antenna has been investigated. The numerical investigations result in some general conclusions regarding the effect of material inclusions on the dipole antenna performance. It is demonstrated that in contrast to the DPS and Mu-Negative (MNG) MTMs, ENG- and DNG-MTM inclusions can provide multi-band performance. To practically realize this method, a compact multiband printed dipole antenna is designed using reactive loading, which is inspired by ENG-MTM inclusions. To this aim, a novel printed MTM element is proposed and successfully tested. The proposed MTM cell shows ENG behavior at around the antenna operating frequency.

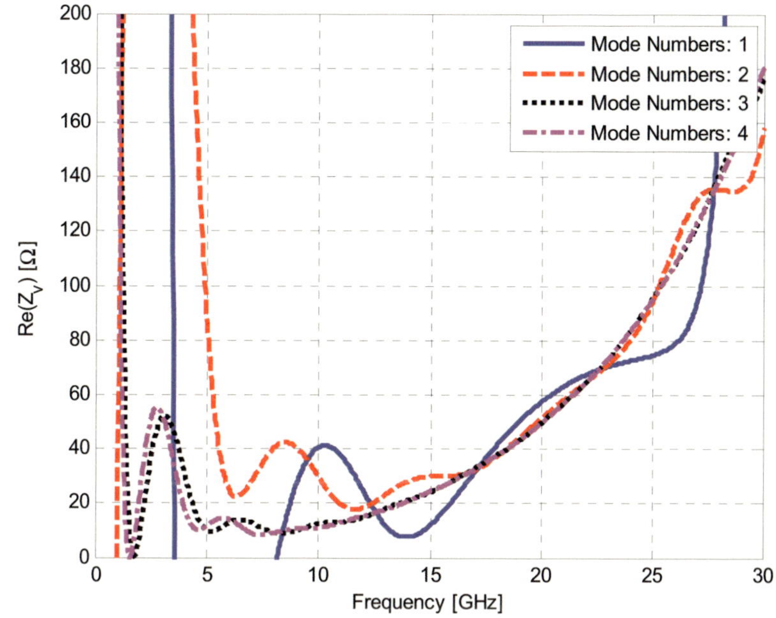

(a)

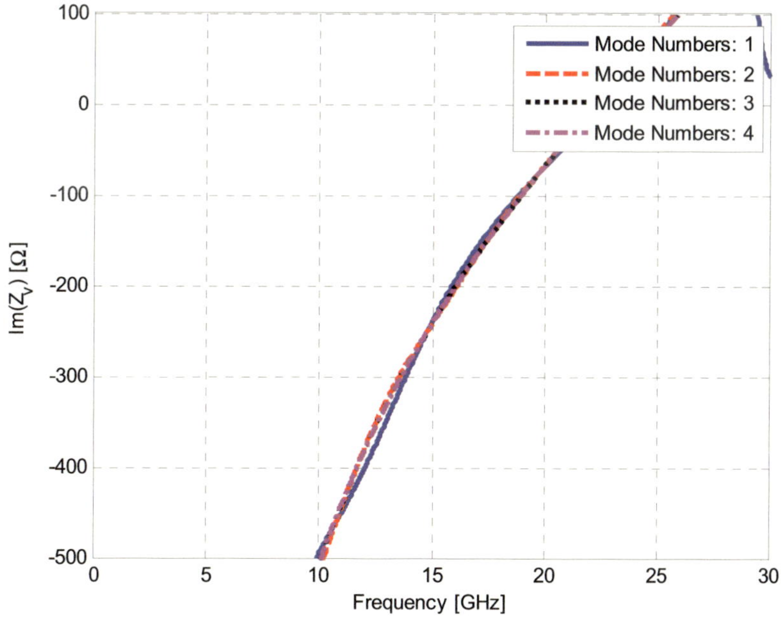

(b)

Fig. 7. Convergence analysis of the dielectric- loaded biconical antenna, input impedance (a) real, and (b) imaginary parts. From (Jafargholi et. al., 2012), copyright © 2012 by the Electromagnetics, Taylor & Francis Group, LLC.

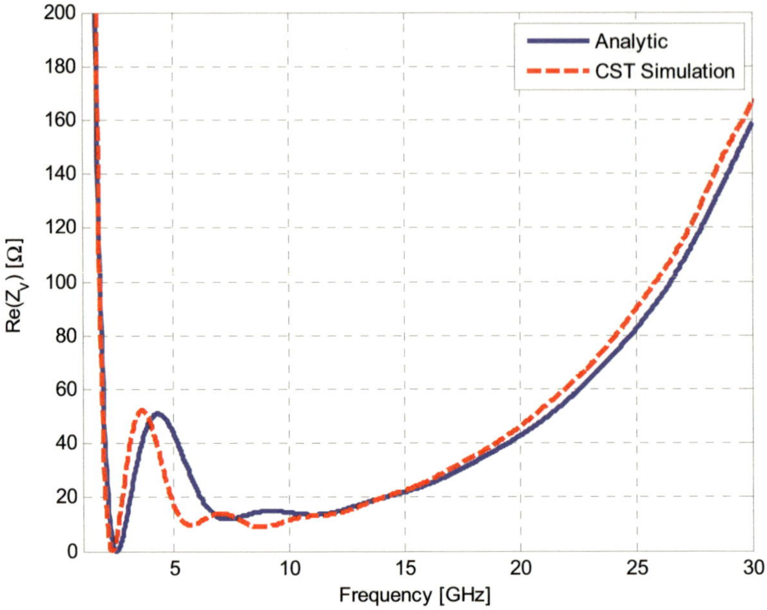
Analytic
CST Simulation
Re(Z_V) [Ω]
Frequency [GHz]
200
180
160
140
120
100
80
60
40
20
0
5
10
15
20
25
30

(a)

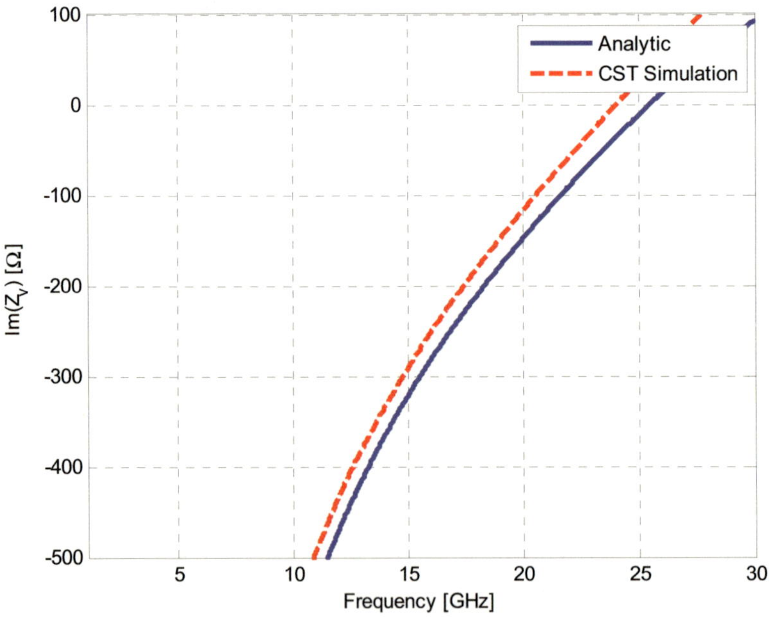
Analytic
CST Simulation
Im(Z_V) [Ω]
Frequency [GHz]
100
0
-100
-200
-300
-400
-500
5
10
15
20
25
30

(b)

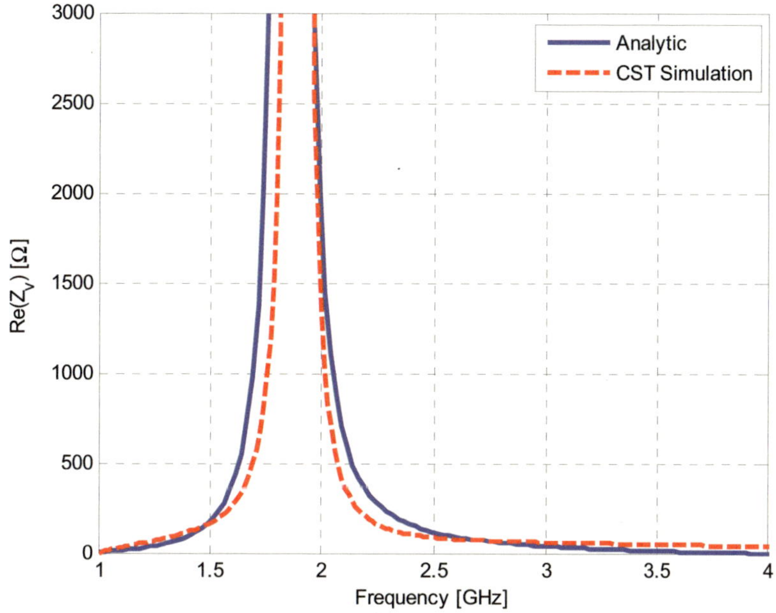

(c)

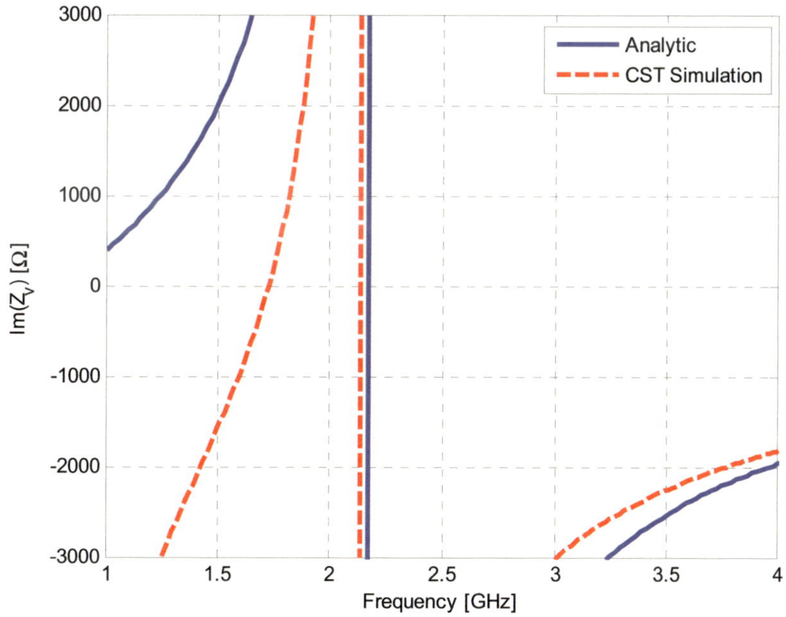

(d)

Fig. 8. Input impedance of the loaded biconical antenna: analytical against numerical results. Analytical results are obtained using proposed analytical expressions; numerical results are computed by CST software; (a) real, and (b) imaginary parts for dielectric materials and (c) real, and (d) imaginary parts for DNG metamaterials. From (Jafargholi et. al., 2012), copyright © 2012 by the Electromagnetics, Taylor & Francis Group, LLC.

The dimensions of the proposed MTM cell is optimized to meet the specifications of the mobile bands (890.2MHz-914.8MHz, and 1710MHz-1784MHz) while maintaining its compact size. The antenna radiation efficiency at the first resonance frequency is significantly higher than those reported for other miniaturized printed dipoles in the literature (Iizuka et al., 2006; Iizuka et al., 2007; Borja et al., 2007; Rafaei et al., 2010). It is worthwhile to point out here that the subject of single-cell MTM loading is not new and has been studied by other authors (Zhu et al., 2010).

3.1 A dipole antenna loaded with MTM inclusion

It is known that the resonance frequencies of an original monopole/dipole are harmonics of the main resonant frequency ω_1. However, ominidirectional radiation pattern distortion and low directivity are two major disadvantages associated with monopole/dipole antenna resonating at higher order harmonics ($\omega_m > \omega_1$) (Balanis, 1989; Jafargholi et al., 2010).

In this section, a simple and intuitive rule for determining the beneficial filling material type for dipole antennas has been introduced. A dipole antenna loaded with cylindrical dispersive MTM inclusions is shown in Fig. 9. It is assumed that the MTM inclusions are embedded in the both arms of the dipole. Here, the Drude model (Engheta et al., 2006) is used to simulate the MTM inclusions, since it can yield a negative real part of the permittivity/permeability over a wide frequency range. Depending on the MTM type either μ or ε (or both) obey the Drude model (with plasma frequency ω_p=1.8×10^{10} rad/s and collision frequency f_c =0.2GHz) and are equal to one otherwise. The distance from the location of the MTM inclusions to the feed point is denoted as d_{MIF}.

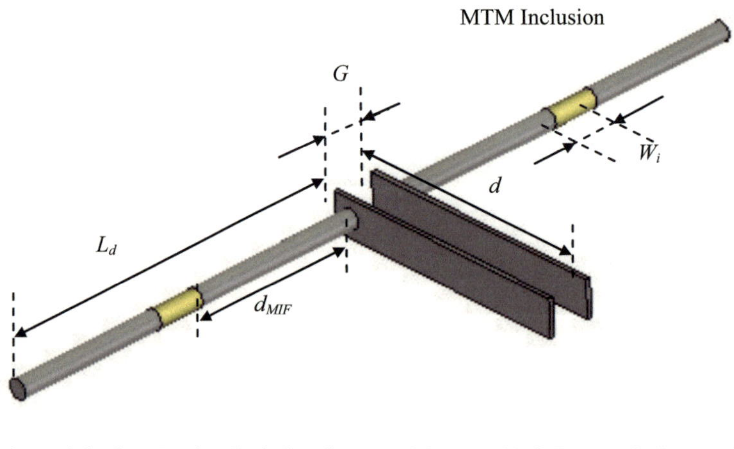

Fig. 9. An ideal model of MTM loaded dipole: L_d=120mm, W_i=2.5mm, G=5mm, d=27mm. From (Rafaei et al., 2011), copyright © 2011 by the IET Microwaves, Antennas & Propagation.

The behaviors of the loaded dipole as a function of the MTM type and the distance of the MTM inclusions from the antenna feed point, d_{MIF}, have been studied. Fig. 10 shows the antenna reflection coefficient for the dipoles loaded with DPS-, MNG-, DNG-, and ENG-inclusions, with d_{MIF} as a parameter. As the ENG- or DNG- inclusions are added, the antenna resonant behavior changes.

It can be concluded from Fig. 10 that for the dipole antenna loaded with DNG- or ENG-inclusions, an additional resonance frequency is introduced at the frequencies lower than the antenna resonant frequency where the antenna radiates an omnidirectional radiation pattern. In contrast, for the dipoles loaded with DPS- or MNG-inclusions, changing DPS/MNG locations on the antenna arms causes no resonances at frequencies lower than the main resonant frequency, as shown in Fig. 10(a,b).

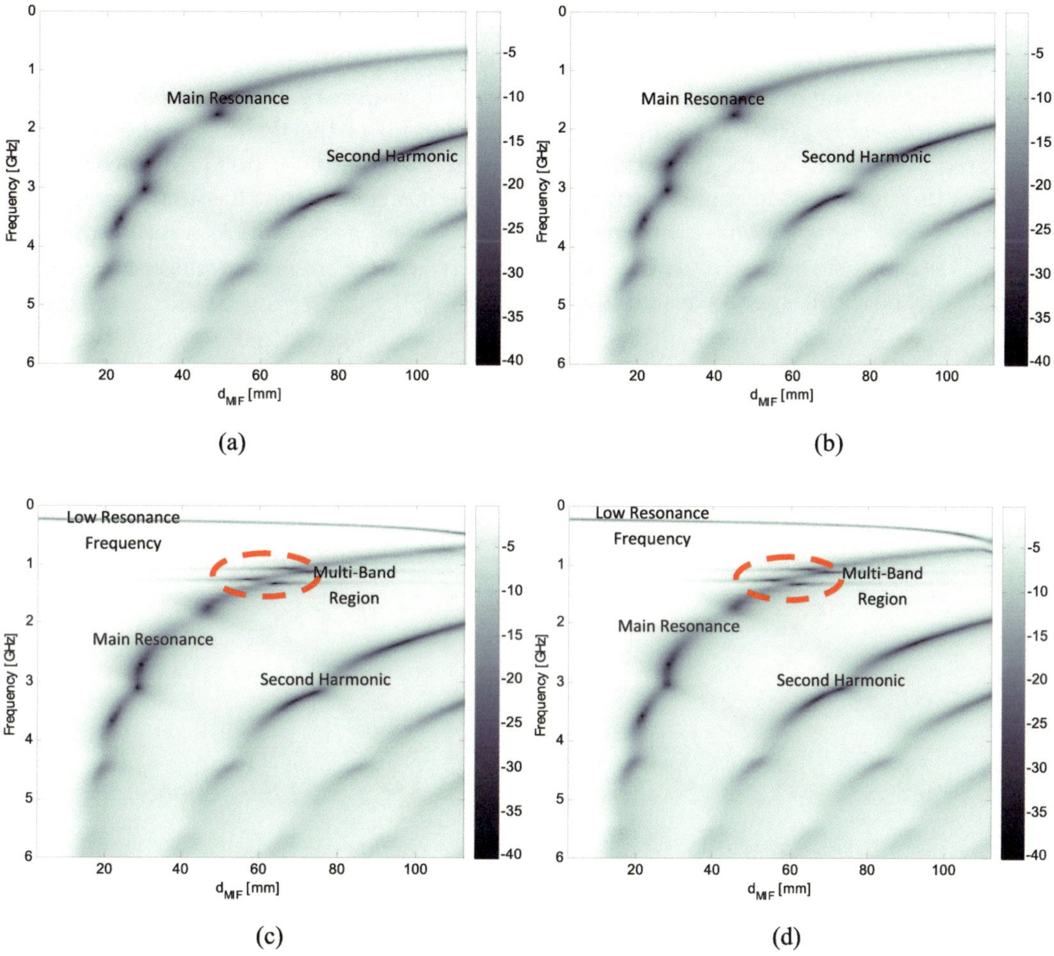

Fig. 10. CST simulation results for $|S_{11}|$ [dB], versus location of (a) DPS-, (b) MNG-, (c) DNG- and (d) ENG-inclusions. From (Rafaei et al., 2011), copyright © 2011 by the IET Microwaves, Antennas & Propagation.

As the distance between the ENG-/DNG-inclusions and the feed point is increased, the main resonant frequency decreases while the low resonant frequency is almost unchanged. This feature provides the ability to choose the second resonance frequency arbitrarily based on provision dictated by application. And thus the frequency ratio between these two frequencies can be readily controlled by adjusting the inclusion locations.

In addition, for the case of the dipoles loaded with DNG-/ENG-blocks and 50mm<d_{MIF}<75mm, more than one resonance is introduced at around the antenna main resonant frequency where the antenna radiates omnidirectional radiation patterns, as shown in Fig. 10(c,d).

To make the concept more clear, three DNG loaded dipoles are designed and simulated. The reflection coefficient results for the dipole antennas loaded with different DNG blocks and different d_{MIF} are shown in Fig. 11. For comparison purposes, the reflection coefficient of an unloaded dipole antenna is also presented in Fig. 11. As can be seen, all the antennas have multi-resonance behavior. The first frequency bands of the proposed loaded dipoles are narrow. This narrow frequency bands are the direct consequence of the resonant nature of the MTM inclusions. The gain, efficiency, and bandwidth of the three loaded dipoles are compared in Table 2. For the first design, the antenna bandwidth at first resonance is quite good but its gain is low. In contrast, for the second design, the antenna has a high gain at the first resonance frequency but at the expense of a narrower bandwidth. As a result, the type of the DNG-inclusion is a result of a trade-off between the antenna radiation efficiency (gain) and bandwidth, such as design III.

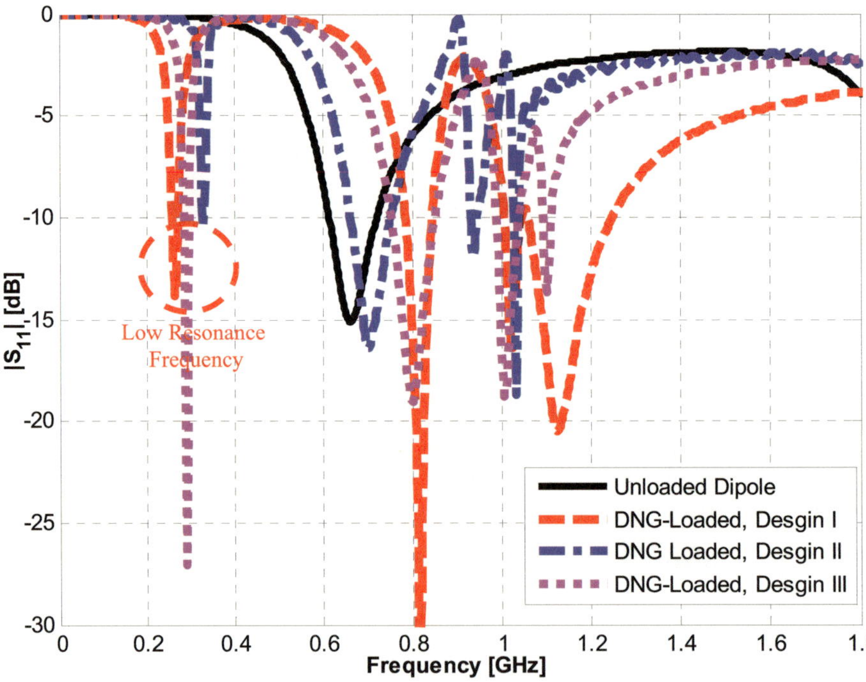

Fig. 11. Reflection coefficient results for dipole antennas loaded with different DNG blocks; Design I: d_{MIF}=72mm and Drude model with ω_p=1.8×10^{10}rad/s, and fc =0.2GHz, Design II: d_{MIF}=100mm, ω_p=1.8×10^{10}rad/s and f_c =0.01GHz, and Design III: d_{MIF}=85mm, ω_p=1.8×10^{10}rad/s and f_c =0.1GHz. As a reference, an unloaded dipole antenna is also simulated. From (Rafaei et al., 2011), copyright © 2011 by the IET Microwaves, Antennas & Propagation.

Design Type	Design I				Design II				Design III			
	f_{r1}	f_{r2}	f_{r3}	f_{r4}	f_{r1}	f_{r2}	f_{r3}	f_{r4}	f_{r1}	f_{r2}	f_{r3}	f_{r4}
Center Frequency (GHz)	0.26	0.8	1.0	1.1	0.33	0.7	0.93	1.02	0.29	0.8	1.0	1.1
Gain (dBi)	-7.2	1.25	0.0	2.1	1.3	2.1	1.9	3.4	-3.5	2.0	0.5	2.2
Efficiency (%)	12.2	78	65	91	84	99	100	100	25	91	80	95
Bandwidth (%)	5.6	7.5	5.4	21.8	1.1	13.1	2.4	1.5	4.5	12.1	5	1.8

Table 2. Gain, efficiency, and bandwidth characteristics of the dipole antenna loaded with different DNG inclusions

The behaviors of the loaded dipole as a function of the plasma frequency for d_{MIF}=72mm, have been also studied. Fig. 12 shows the antenna reflection coefficient for the dipoles loaded with ENG-, DNG-, and MNG-inclusions. It can be concluded from Fig. 12 that for the dipole antennas loaded with DNG- or ENG-inclusions, an additional resonance frequency is introduced at the frequencies lower than the antenna main resonant frequency. In addition, as can be seen from Fig. 12(a,b), in the DNG-loaded case, an additional resonance has been appeared, especially for higher values of plasma frequency, as compared to the ENG-loaded dipole. In contrast, for the dipoles loaded with MNG-inclusions, changing plasma frequency causes no resonances at frequencies lower than the main resonant frequency, as shown in Fig. 12(c).

In Fig. 13, the effect of permittivity and permeability of the DPS inclusions on the resonance frequency of the dipole antenna has been studied. As can be seen, increasing the permittivity of the loaded DPS results in a dual band operation in which the frequency separation ratio increases as the permittivity increases. However, for the dipoles loaded with magnetic inclusions, changing permeability causes no resonance frequency change.

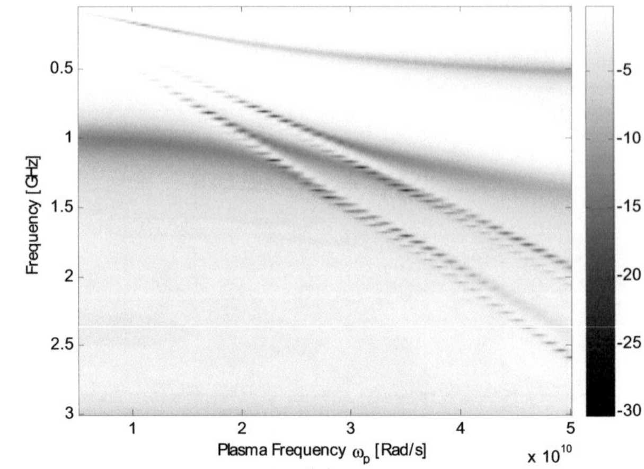

(a)

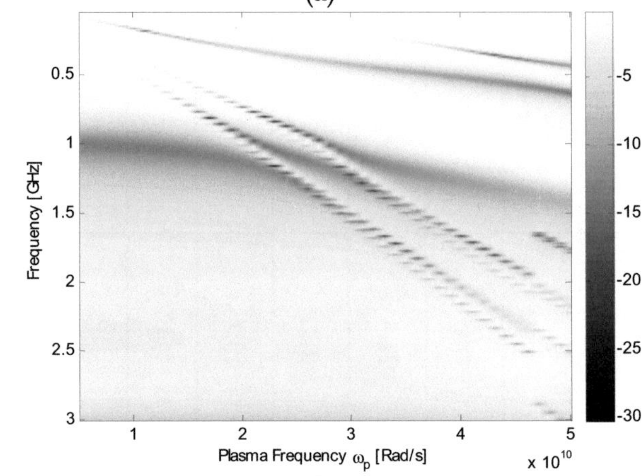

(b)

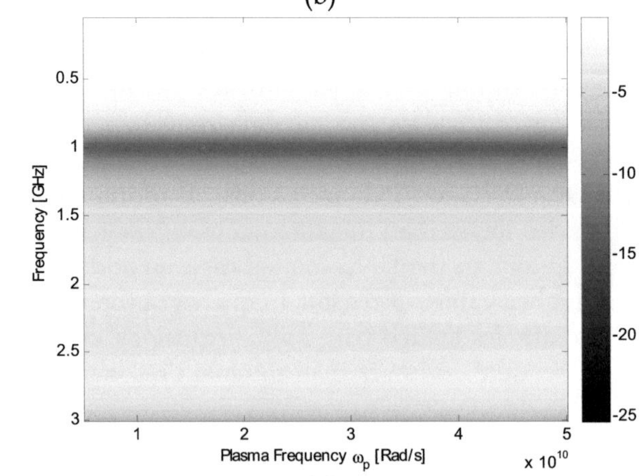

(c)

Fig. 12. CST simulation results for $|S_{11}|$ [dB] versus plasma frequency, d_{MIF}=72mm, (a) ENG-, (b) DNG-, (c) MNG-inclusions. From (Jafargholi et. al., 2012), copyright © 2012 by the ACES Journal.

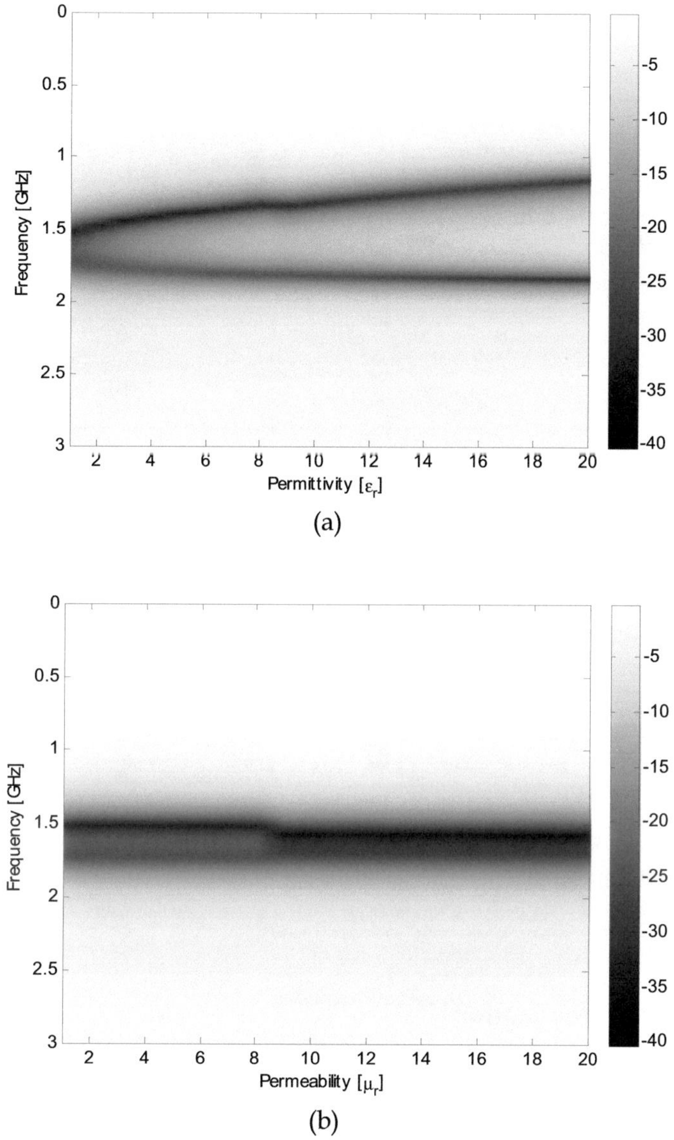

Fig. 13. CST simulation results for $|S_{11}|$ [dB], d_{MIF}=45mm, versus material (a) permittivity, μ_r=1, and (b) permeability, ε_r=1. From (Jafargholi et. al., 2012), copyright © 2012 by the ACES Journal.

The behaviors of the loaded dipole as a function of the inclusion width for d_{MIF}=45mm, have been also studied. Fig. 14 shows the antenna reflection coefficient for the dipoles loaded with DPS-, ENG-, DNG-, and MNG-inclusions. It can be concluded from Fig. 14(c,d) that, for the dipole antennas loaded with DNG-/ENG- inclusions, the additional resonance frequency is significantly affected by the width of the inclusions. In contrast, for the dipoles loaded with DPS- and MNG-inclusions, changing inclusion widths causes no resonances at frequencies lower than the main resonant frequency, as shown in Fig. 14(a,b).

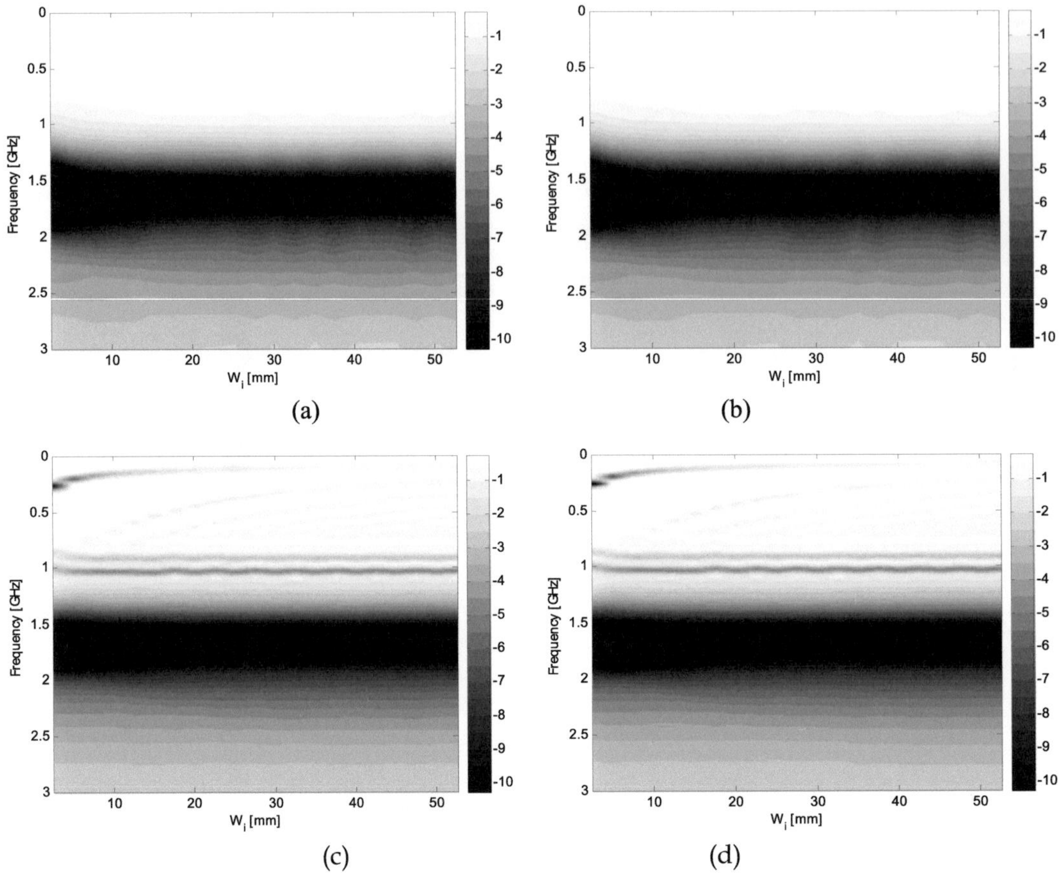

Fig. 14. CST simulation results for $|S_{11}|$ [dB], d_{MIF}=45mm, versus width of (a) DPS-, (b) MNG-, (c) DNG- and (d) ENG-inclusions. From (Jafargholi et. al., 2012), copyright © 2012 by the ACES Journal.

4. Simulations and realization

In the previous section, it was revealed that the use of the ENG- and DNG-inclusions has led to a multi-resonance behavior. In this section, a new printed MTM cell is introduced to realize the ENG-inclusions. Fig. 15 shows a schematic of the proposed MTM cell along with its design parameters. The proposed MTM cell is printed on a FR4 substrate with a thickness of 0.8mm and a dielectric constant of 4.4. An important feature of the proposed MTM is that it offers more degrees of freedom than conventional MTM cells (Engheta et al., 2006).

In order to retrieve the constitutive parameters of the proposed metamaterial, a unit cell positioned between two perfect electric conductors (PEC) in x direction and two perfect magnetic conductors (PMC) in z direction is simulated, and used to model an infinite periodic structure (Veysi et al., 2010). The resultant scattering parameters obtained from CST microwave studio are exerted to the Chen's algorithm (Veysi et al., 2010). The normalized impedance (z) and refractive index (n) of the under-study medium can be calculated as following:

$$z = \pm\sqrt{\frac{(1+s_{11})^2 - s_{21}^2}{(1-s_{11})^2 - s_{21}^2}}, \mathrm{real}(z) \geq 0 \quad (19)$$

$$n = \frac{1}{k_0 d}\left\{\left[\left[\ln\left(e^{ink_0 d}\right)\right]'' + 2m\pi\right] - i\left[\ln\left(e^{ink_0 d}\right)\right]'\right\} \quad (20)$$

Where

$$\mathrm{Im}\,(n) \geq 0,\ e^{ink_0 d} = \frac{S_{21}}{1 - S_{11}\frac{z-1}{z+1}} \quad (21)$$

$$\mathrm{Re}(n(\omega)) = 1 + \frac{2}{\pi}\mathrm{P.V.}\left[\int_0^{\infty}\frac{\omega'\,\mathrm{Im}(n(\omega'))}{\omega'^2 - \omega^2}d\omega'\right] \quad (22)$$

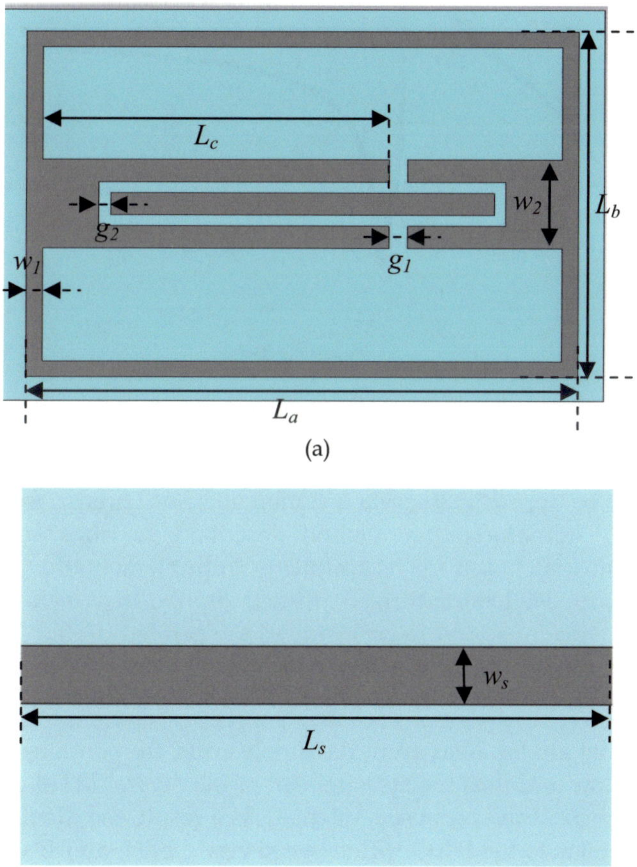

Fig. 15. Schematic of the proposed MTM unit cell and its design parameters, (a) front view, (b) back view: L_a=23.54mm, L_b=15.55mm, L_c=14.78mm, w_1=0.7mm, g_1=0.8mm, w_2=4mm, g_2= 0.5mm, w_s=2.5mm, and L_s=26.75mm. From (Rafaei et al., 2011), copyright © 2011 by the IET Microwaves, Antennas & Propagation.

The ambiguity of the value of m in (20) is resolved by using Kramers-Kronig (KK) relating the real and imaginary parts of the index of refraction (Lucarini et al., 2004). Where, *P.V.* denotes the principal value of the integral. The effective permittivity (ε) and permeability (μ) of the medium can be expressed as: $\varepsilon=n/z$, $\mu=nz$. Fig. 16 shows the retrieved effective parameters of the proposed metamaterial cell. As can be seen, the proposed MTM cell has the permittivity that exhibits Drude behaviour at frequencies lower than 1.1GHz and Lorentz behaviour (Engheta et al., 2006) at frequencies higher than 1.1GHz.

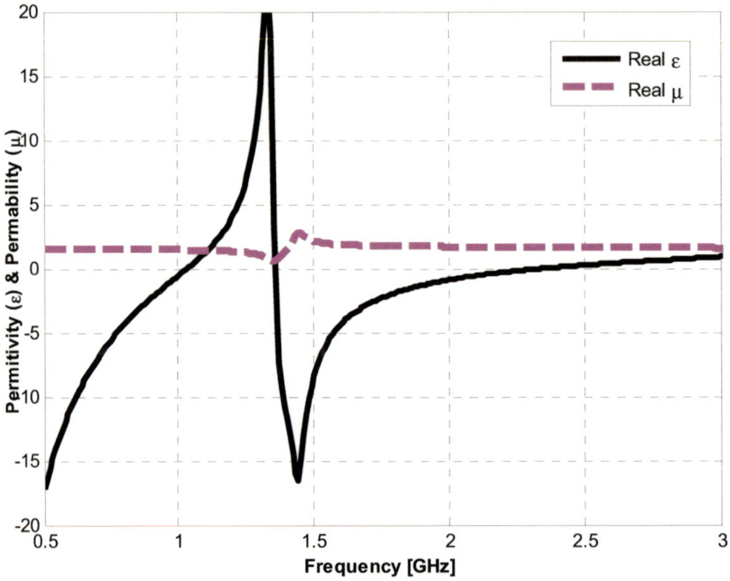

Fig. 16. Retrieved effective parameters of the Proposed MTM cell. From (Rafaei et al., 2011), copyright © 2011 by the IET Microwaves, Antennas & Propagation.

Thus, this MTM can be approximated via a combination of Lorentz and Drude models. In order to realize the miniaturization method described previous section, double-sided printed dipole antenna is chosen for its simplicity in implementation and its low profile. Fig. 17 shows the proposed miniaturized printed dipole, in which a pair of proposed MTM cells is symmetrically added to each side of the printed dipole. The proposed MTM cells and dipole are printed on a FR4 substrate with a thickness of 0.8mm and a dielectric constant of 4.4 to reduce the cost of the antenna and to make it more rigid in construction.

For the MTM cells that are far away from the dipole arms, the coupling levels of them with the dipole arms are low and thus the arrangement of the several MTM cells has no effect on the frequency behaviour of the proposed antenna. As a result, the dipole is just loaded with single cell MTM. Similar to the DNG- (Ziolkowski et al., 2003) and ENG- (Alu et al., 2003) MTMs, the proposed MTM cell can be modelled as a parallel resonant LC circuit. Thus, the proposed metamaterial cell is modelled as a resonant LC circuit parallel to the dipole, and the radiation into the free space is modelled as a resistor (Sievenpiper et al., 2006). A prototype of the proposed miniaturized dual-band printed dipole is fabricated to confirm the simulation results. Fig. 18 shows a photograph of the fabricated antenna.

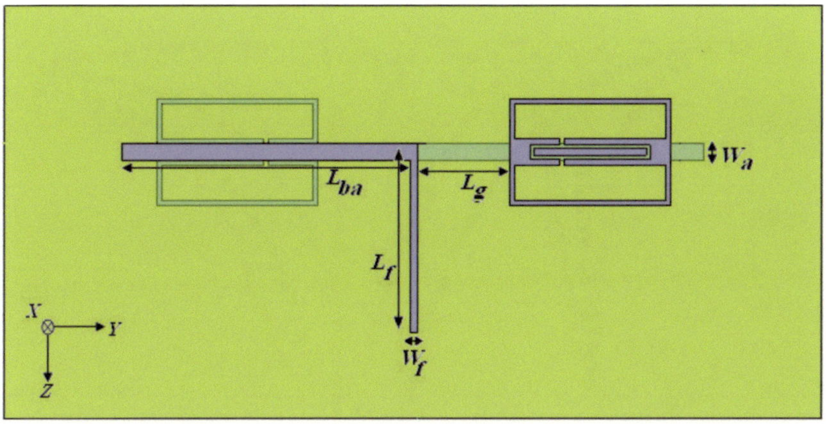

Fig. 17. Printed dipole symmetrically loaded with single cell MTM: L_{ba}=42.05mm, L_f=27.5mm, L_g=12.52mm, W_a= 2.5mm, W_f=0.8mm. From (Rafaei et al., 2011), copyright © 2011 by the IET Microwaves, Antennas & Propagation.

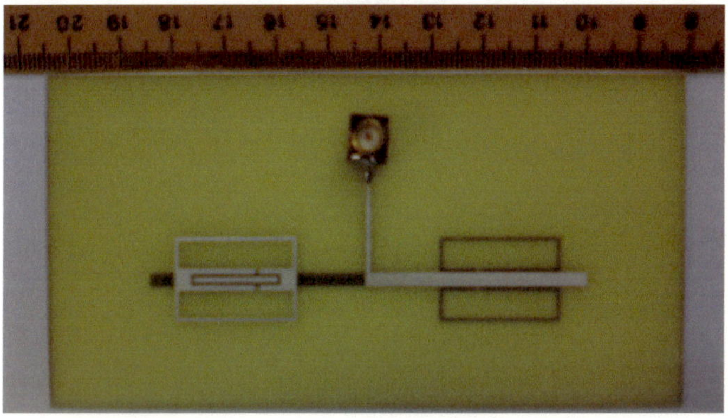

Fig. 18. Prototype of proposed miniaturized printed dipole antenna loaded with single cell MTM. From (Rafaei et al., 2011), copyright © 2011 by the IET Microwaves, Antennas & Propagation.

Fig. 19 shows the reflection coefficient of the proposed symmetrically loaded dipole with the gap length, g_1, of 0.8mm as well as the unloaded dipole antenna. As can be seen, the dipole antenna along with the loading elements provides good matching at both resonance frequencies. For comparison purposes, a simple dipole antenna loaded with lossy ENG inclusions, with the same retrieved effective parameters of the proposed MTM cell (See Fig. 16), is also simulated. As can be seen from Fig. 19, the reflection coefficient of the dipole loaded with ENG inclusions correlates nicely to that obtained for the single cell MTM loaded dipole. The co-polarized and cross-polarized radiation patterns of the proposed loaded dipole are measured at the resonant frequencies of 940MHz and 1.7GHz.

The measured and simulated radiation patterns at first and second resonant frequencies are shown in Fig. 20. As expected, the radiation patterns at both resonant frequencies are similar to that of the conventional unloaded dipole antenna. The gain of the proposed antenna at low resonant frequency is high compared to that of the other miniaturized MTM loaded dipoles (Iizuka et al., 2006; Iizuka et al., 2007; Borja et al., 2007; Rafaei et al., 2010).

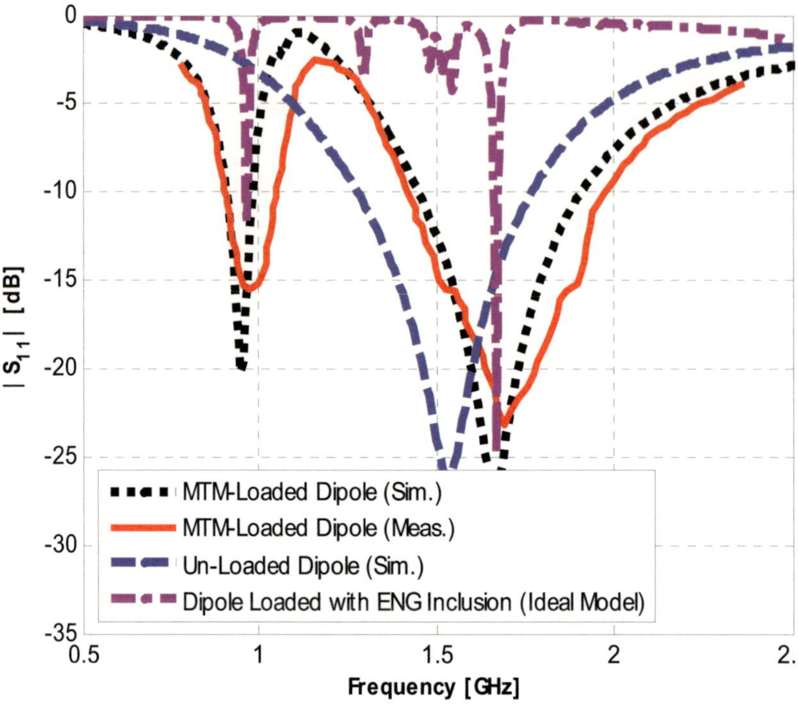

Fig. 19. Reflection coefficient of the proposed miniaturized printed dipole antenna loaded with single cell MTM. As a reference, an unloaded dipole and an ideal model of the ENG-Loaded dipole are also simulated. From (Rafaei et al., 2011), copyright © 2011 by the IET Microwaves, Antennas & Propagation.

The antenna gains at first and second resonant frequencies are -2.679dBi and 1dBi, respectively. The proposed antenna has a broad bandwidth of 15.96% at 940MHz (which is significantly wider than the bandwidth of other miniaturized MTM loaded dipoles (Iizuka et al., 2006; Iizuka et al., 2007; Borja et al., 2007; Rafaei et al., 2010)) and 32.35% at 1.7GHz. An important advantage of the proposed antenna is that the dipole length does not need to be increased to lower the resonant frequency. Consequently, a compact antenna is obtained.

Finally, the effect of the MTM location is investigated to obtain some engineering guidelines for loaded dipole designs. Thus, the loading elements move along the antenna arms and the antenna reflection coefficient is plotted in Fig. 21 for each stage. The gain, bandwidth and efficiency of the loaded dipoles with different MTM locations are also compared in Table 3.

As can be seen, the first resonant frequency remains approximately unchanged while the second one reduces as the MTM cells move away from the antenna feed point. Thus, when the MTM elements move closer to the dipole ends, the separation of the two resonances decreases. In addition, when the MTM cells are placed close to the antenna feed point, the proposed antenna cannot match very well to a 50Ω transmission line. Moreover, as can be seen from Figs. 10, 11 and 21 the single cell MTM loaded printed dipole follows closely the frequency behavior of the dipole antenna loaded with cylindrical dispersive ENG-inclusions, as d_{MIF} or L_g increases.

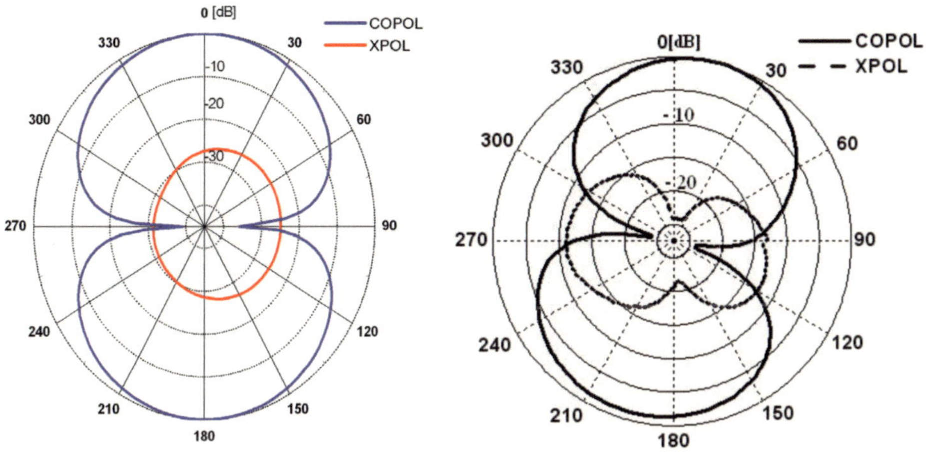

(a)

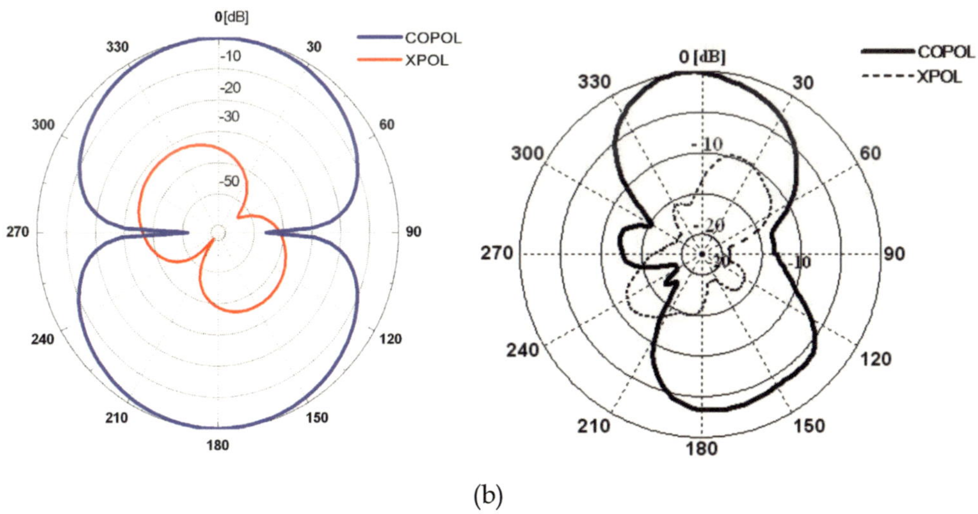

(b)

Fig. 20. Radiation patterns of the proposed printed dipole antenna at (a) 940MHz and (b) 1.7GHz. (Right hand figures are measurements). From (Rafaei et al., 2011), copyright © 2011 by the IET Microwaves, Antennas & Propagation.

In the previous design, a printed dipole was symmetrically loaded with single cell MTMs to realize a dual band operation. Here, the design parameters of the proposed dual band antenna are changed to provide a tri-band dipole antenna. As a result, two different tri-band printed dipoles are provided in the following subsections. Since changing the locations and dimensions of the MTM cells does not have any significant effect on the antenna radiation patterns, the proposed antennas radiate omnidirectional radiation patterns at all resonant frequencies. However, these are not plotted here for the sake of brevity.

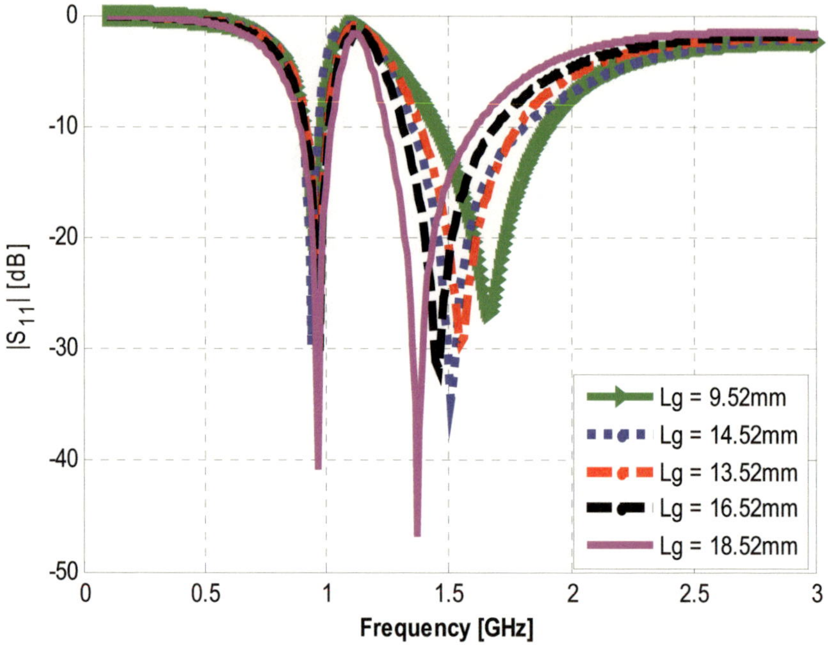

Fig. 21. The effect of MTM location on the reflection coefficient of the printed dipole antenna. From (Rafaei et al., 2011), copyright © 2011 by the IET Microwaves, Antennas & Propagation.

5. Acknowledgment

The authors would like to thank Iran Telecommunication Research Centre (ITRC) for its financial supports.

6. References

Alu A. , and Engheta N.; "Pairing an Epsilon-Negative Slab With a Mu-Negative Slab: Resonance, Tunneling and Transparency" *IEEE Trans. Antennas Propagat.*, vol. 51, pp. 2558–2571, Oct. 2003.

Balanis C. A., *Advanced Engineering Electromagnetics*. New York: Wiley, 1989.

Beggs R., Luebbers J., and Chamberlin K., "Finite difference time-domain calculation of transients in antennas with nonlinear loads," *IEEE Trans. Antennas Propag.*, vol. 41, no.5, pp.566-, May 1993.

Borja A. L., Hall P. S., Liu Q. and Iizuka H., "Omnidirectional left-handed loop antenna," *IEEE Antennas and Wireless Propagation Lett.*, Vol. 6, 495-498, 2007.

Collin R. E., and Zucker F. J., *Antenna Theory*, McGraw-Hill, 1969.

Engheta N. and Ziolkowski R., *Metamaterials: Physics and Engineering Explorations,* John Wiley & Sons Inc., 2006.

Erentok A., and Ziolkowski R. W., "Metamaterial-Inspired Efficient Electrically Small Antennas" *IEEE Trans. Antennas Propagat.*, vol. 56, pp. 691–707, March 2008.

Erentok A., Luljak P., and Ziolkowski R. W., "Antenna performance near a volumetric metamaterial realization of an artificial magnetic conductor," *IEEE Trans. Antennas Propagat.*, vol. 53, pp. 160–172, Jan. 2005.

Iizuka H. and Hall P. S., "Left-handed dipole antennas and their implementations," *IEEE Trans. Antennas Propag.*, vol. 55, no. 5, 1246–1253, May 2007.

Iizuka H., Hall P. S., and Borja A. L., "Dipole antenna with left-handed loading," *IEEE Antennas Wireless Propag. Lett.*, vol. 5, pp. 483–485, 2006.

Jafargholi A., and Kamyab M., Full-Wave Analysis of Double Positive/Double Negative Loaded Dipole Antennas, Electromagnetics, 32:2, 103-116, (2012).

Jafargholi A., and Kamyab M., *Metamaterials in Antenna Engineering, Theory and Applications*, LAP Lambert Academic Publishing, Germany, 2011.

Jafargholi A., Kamyab M., and Veysi M., "Artificial magnetic conductor loaded monopole antenna" *IEEE Antennas Wireless Propag. Lett.*, vol. 9, pp. 211-214, 2010.

Jafargholi A., Kamyab M., Rafaei M., Veysi M., "A compact dual-band printed dipole antenna loaded with CLL-based metamaterials", *International Review of Electrical Engineering*, vol. 5, no. 6, 2710-2714, 2010.

Kennedy T. F., Fasenfest K. D., Long S. A. and Williams J. T., "Modification and Control of Currents on Electrically Large Wire Structures Using Composite Dielectric Bead Elements, "IEEE Trans. Antennas Propag., vol. 54, no. 12, 3608–3613, 2006.

Kraus J. D., and Marhefka R. J., *Antennas for All Applications*, McGraw-Hill, 2002.

Liu Q., Hall P. S., and Borja A. L.," Efficiency of Electrically Small Dipole Antennas Loaded With Left-Handed Transmission Lines," *IEEE Trans. Antennas Propagat.*, vol. 57, no. 10, pp. 3009–3017, Oct. 2009.

Lucarini V., Saarinen J. J., Peiponen K. E., and Vartiainen E. M., "Kramers-Kronig Relation in Optical Materials Research", *Springer Series in optical sciences*, 2004.

Mosallaei H., and Sarabandi K., "Design and Modeling of Patch Antenna Printed on Magneto-Dielectric Embedded-Circuit Metasubstrate" *IEEE Trans. Antennas Propag.*, vol. 55, no. 1, 1031–1038, Jan 2007.

Ock J. S., and Eom H. J., "Radiation of a Hertzian Dipole in a Slotted Conducting Sphere," *IEEE Trans. Antennas Propagat.*, vol. 57, nov 12, pp. 3847–3851, Dec. 2009.

Polk C., "Resonance and Supergain Effects in Small Ferromagnetically or Dielectrically Loaded Biconical Antennas, "*IRE Trans. Antennas Propag.*, pp. 414-423; December, 1959.

Rafaei M., Kamyab M., Jafargholi A., and Mousavi S. M., "Analytical modeling of the printed dipole antenna loaded with CRLH structures," *Progress In Electromagnetics Research B*, vol. 20, 167-186, 2010.

Rogers S. D., Butler C. M., and Martin A. Q., "Design and Realization of GA-Optimized Wire Monopole and Matching Network With 20:1 Bandwidth" *IEEE Trans. Antennas Propag.*, vol. 51, no. 3, pp. 493–502, March. 2003.

Saoudy S. A. and Hamid M., "Input admittance of a biconical antenna with wide feed gap," *IEEE Trans. Antennas Propag.*, vol. 38, no. 11, pp. 1784–1790, Nov. 1990

Shams K. M. Z. and Ali M., "*Analyses of a Dipole Antenna Loaded by a Cylindrical Shell of Double Negative* (DNG) Metamaterial," International Journal of Antennas and Propagation, vol. 2007, Article ID 97481, 10 pages.

Sievenpiper D., "Chapter11: Review of theory, fabrication, and applications of high impedance ground planes," in *Metamaterials: Physics and Engineering Explorations*, edited by Erentok A., and Ziolkowski R. W.

16

Antenna Designs with Electromagnetic Band Gap Structures

Dalia M.N. Elsheakh, Hala A. Elsadek and Esmat A. Abdallah
Electronics Research Institute, Giza, Egypt

1. Introduction

The word "meta", in Greek language, means beyond. It implies that the electromagnetic response of metamaterials (MTMs) is unachievable or unavailable in conventional materials. Many efforts have been done to search for an adequate definition for MTMs. In 2002, J.B. Pendry wrote in a conference paper: "meta-materials, materials whose permeability and permittivity derive from their structure". Later, in 2006, C. Caloz and T. Itoh wrote: "Electromagnetic metamaterials are broadly defined as artificial effectively homogeneous electromagnetic structures with unusual properties not readily available in nature" [1]. Perhaps, a serious obstacle on the road to a universal definition for the term MTMs is the fact that researchers working with these objects do not commonly agree on their most essential characteristics. In [2] and [3], some of the problematic aspects of the non-naturality definition were raised, like the difficulty in separating classical composites from the new class of metamaterials. Another argument against the "not found in nature" property is that it unnecessary excludes impressive examples of natural media that could be called metamaterials par excellence, such as structural colors [4].

MTMs cover an extremely large scientific domain which ranges from optics to nanoscience and from material science to antenna engineering. In this chapter, we focus primarily on the subject of MTMs in the electromagnetic field. Personally, We prefer the definition given by D.R. Smith: Electromagnetic metamaterials are artificially structured materials that are designed to interact with and control electromagnetic waves [5]. The term "artificial" refers to the fact that the electromagnetic response of these materials is dominated by scattering from periodically or amorphously placed inclusions (e.g., metallic or dielectric spheres, wires, and loops) [6].

In the family of MTMs, "left-handed" (LH) media drew an enormous amount of interest. This concept was first put forward by a Russian physicist, Victor Veselago, in 1968, for whom the medium is characterized by a simultaneously negative electric permittivity and negative magnetic permeability [7]. Veselago argued that such media are allowed by Maxwell's equations and that electromagnetic plane waves can propagate inside them, but the phase velocity of such a plane wave is in the opposite direction of the Poynting vector. Hence, some researchers use the term "backward wave media" (BWM) to describe these LH materials [8]. When such media are interfaced with conventional dielectrics, Snell's Law is reversed, leading to the negative refraction of an incident plane wave as shown in figure 1.

Nevertheless, Veselago's conjecture was essentially ignored for thirty years due to the absence of naturally occurring materials or compounds that possess simultaneously negative permittivity and permeability.

In 2000, a metamaterial, based on conducting wires [9] and split-ring resonators (SRRs) [10], was demonstrated to have a negative refractive index over a certain range of microwave frequencies [10-13]. Wires, either continuous or with periodic breaks, can provide a positive or a negative effective permittivity. Planar SRRs or wound coils (also known as Swiss Rolls) can provide a positive or a negative effective permeability. Harnessing the phenomenon of negative refraction, these metamaterials offer a good potential for all kinds of applications, such as "perfect" lens [14], imaging [15], resonators [16], and cloaking [17].

Metamaterials possessing these properties are also frequently named "Negative Refractive Index (NRI)" and "Double Negative (DNG) material". In addition to the materials with simultaneously negative permittivity and negative permeability, the single negative metamaterials have also drawn a great interest. Applications are found for these materials either with a negative permittivity "Epsilon Negative (ENG)" [17] or a negative permeability "Mu Negative (MNG)" [18]. Besides, materials with the properties of "Epsilon near Zero (ENZ)" [19] and "Mu Near Zero (MNZ)", known as "nihility" materials have also been studied. A simple synopsis of these metamaterials can be found in figure 2, where the angular frequencies ωpe and ωpm represent, respectively the electric and magnetic plasma frequency [20]. Up to now, we talked about metamaterials who exhibit their great performances by artificially tailoring the permittivity or permeability. Besides, the term "metamaterial" has also been used by some authors to describe other periodic structures such as electromagnetic bandgap (EBG) structures or photonic crystals, when the period is much smaller in physical size than the wavelength of the impinging electromagnetic wave. The electromagnetic response of such structures is dominated by Bragg-type scattering and involves higher order spatial harmonics (Bloch-Floquet modes) [20]. In this chapter, we focus on such a kind of metamaterial, the so-called "electromagnetic band gap" (EBG). Electromagnetic has received great attention among researchers all over the world because of its immense civilian and defense applications. During the Second World War, the use of radar and thereafter the wide use of microwave communication systems facilitated the transformation from radio to microwave frequency. This dramatic change demanded more advanced materials for high frequency performance and opened up new dimensions in the field of electromagnetic materials. Nano-composites and electromagnetic band-gap structures are examples of metamaterials under right hand rules. Electromagnetic band-gap (EBG) structures have attracted increasing interest in the electromagnetic community. Because of their desirable electromagnetic properties [21], they have been widely studied for potential applications in antenna engineering. Hundreds of EBG papers have been published in various journals and conferences in the last 5 years. EBG are periodic arrangements of dielectric or metallic elements in one, two or three dimensional manners. EBG inhibits the passage of electromagnetic wave at certain angles of incidence at some frequencies. These frequencies are called partial band-gap. At a specific frequency band, EBG does not allow the propagation of wave in all directions and this frequency region is called the complete band-gap or global band-gap [22, 23]. Physicists put the original idea of EBG forward and some recent studies revealed the interesting fact that EBG exists in living organisms. The well known examples are the butterfly wing scales and eyes of some insects. In this case, a metallic like reflection effect is obtained by using refractive index differences.

A multilayer thin film with different refractive indices in animals is a good example for this. Recently, these ideas were undergone a preliminary study for its commercialization such as paints for certain applications.

The concept of electromagnetic band-gap (EBG) structures originates from the solid-state physics and optic domain, where photonic crystals with forbidden band-gap for light emissions were proposed in [27–28] and then widely investigated in the [29–33]. Thus, the terminology, photonic band-gap (PBG) structures, was popularly used in the early days. Since then, a profusion of scientific creativity has been witnessed as new forms of electromagnetic structures are invented for radio frequency and microwaves. EBG can be realized in one, two and three dimensional forms. The dimensionality depends on the periodicity directions. Three dimensional EBG are more appropriate for getting a complete band-gap because they can inhibit waves for all incident angles. The band-gap in EBG is analogous to a forbidden energy gap in electronic crystals. Hence EBG are also termed as photonic crystals (PCs). The first attempts towards three-dimensional structures were realized in the form of face centered cubic (fcc) lattice structures [35]. At the initial stages of EBG research, due to the lack of theoretical predictions, a 'cut and try' approach was adopted in experimentally predicting the band-gap. At the beginning, the investigations of EBG were mainly on wave interactions of these structures at optical frequencies and hence PBG emerged with the name of photonic band-gap structures. Now, vast extensions of EBG at microwave [36], millimetre [37] and sub-millimetre wave frequencies [38] are electromagnetic band-gap (EBG) structures. A periodic structure can give rise to multiple band-gaps. However, it should be noted that the band-gap in EBG is not only due to the periodicity of the structure but also due to the individual resonance of one element. A study revealed the mechanisms to form a band-gap in an EBG [21]. The band-gap formation in EBG is due to the interplay between macroscopic and microscopic resonances of a periodic structure. The periodicity governs the macroscopic resonance or the Bragg resonance. It is also called the lattice resonance. Microscopic resonance is due to the element characteristics and it is called the Mie resonance [20]. When the two resonances coincide, the structure possesses a band-gap having maximum width. Depending on the structural characteristics and polarization of the wave, one resonance mechanism (i.e. either the multiple scattering resonance or the single element scattered resonances) can dominate over the other. The characteristic property of stop bands at certain frequencies enables many applications using EBG. At this stop band, all electromagnetic wave will be reflected back and the structure will act like a mirror. At other frequencies, it will act as transparent medium. This concept is illustrated in figure 3.

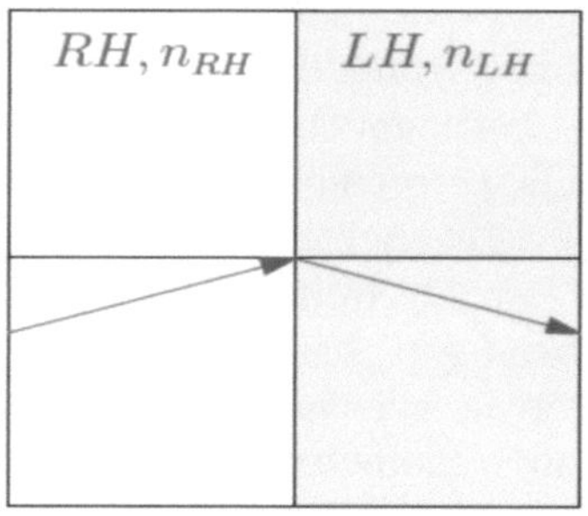

Fig. 1. A negative reflection.

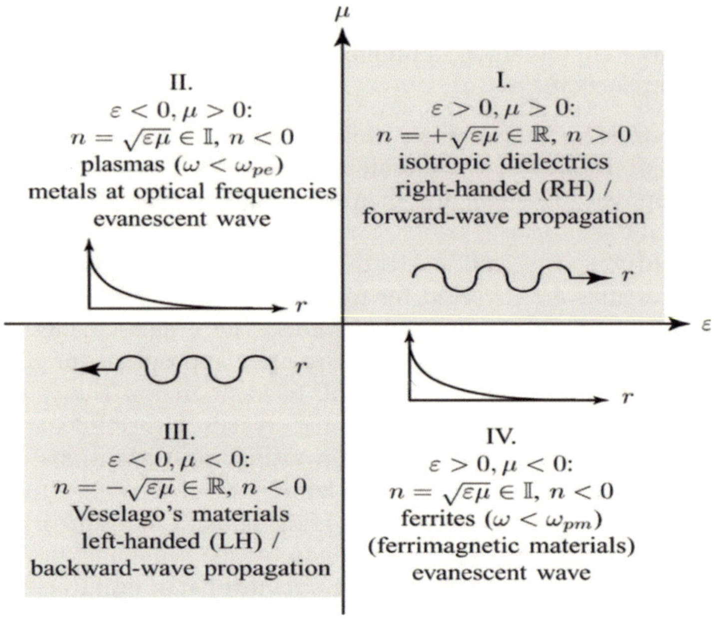

Fig. 2. Permittivity, permeability and refractive index diagram [20].

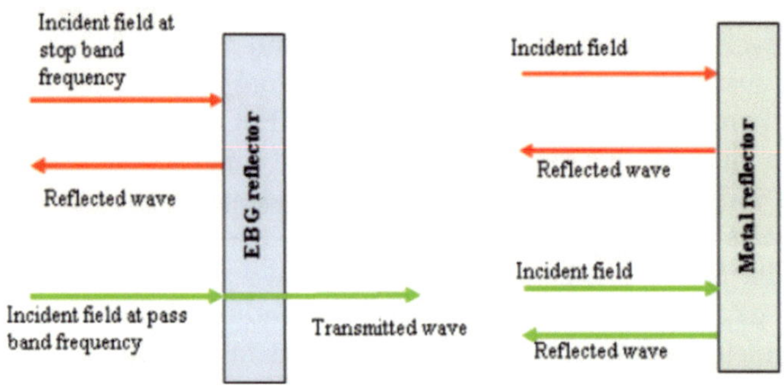

Fig. 3. Diagram illustrating the application of EBG as a mirror and its comparison with a metal reflector [21].

Microstrip antennas mounted on a substrate can radiate only a small amount of its power into free space because of the power leak through the dielectric substrate [32]. In order to increase the efficiency of the antenna, the propagation through the substrate must be prohibited. In this case, the antenna can radiate more towards the main beam direction and hence increase its efficiency. Recently, there has been much interest in the field of Metallo-Dielectric Electromagnetic Band-Gap (MDEBG) structures because of the promising future applications and the important role these artificially engineered periodic materials may play in the field of antennas. The name "Photonic (or Electromagnetic, which is more appropriate for the frequency band of applications) band-gap" has its origin in the fact these structures effectively prevent the propagation of electromagnetic waves within a specific frequency range (the band gap). Two examples of the qualitative geometry of such structures are given in figure 4 [33]. As shown in figure 4, a MDEBG structure is essentially a surface comprising a plurality of elements. Each of the elements is interconnected with each other to form an array of metallic parts embedded in a slab of dielectric. In other words, they are periodical structures of densely packed planar conducting patches separated from a solid metal plane by a dielectric layer. Sometimes metallic pins (or via) are introduced to prevent electromagnetic waves from traveling in the waveguide between the array and the ground. Each unit cell, which is periodically repeated to form the array, essentially behaves as a microwave resonant circuit. The plurality of the resonant elements is parameterized to substantially block surface wave's propagation in the device within a predetermined frequency band gap [35].

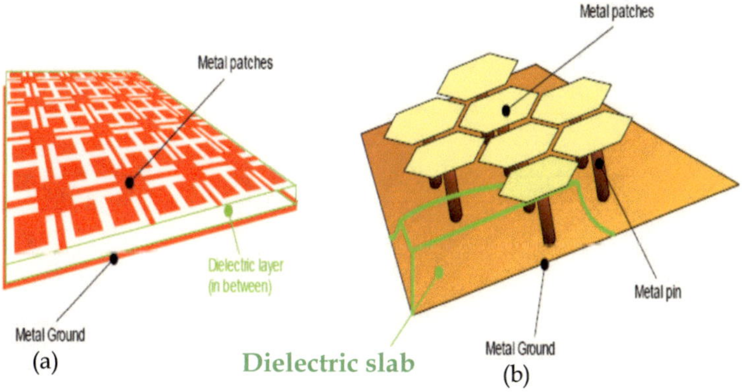

Fig. 4. Simple examples of Metallo-Dielectric EBG structures [32].

The objective of this chapter is to investigate the EM properties of the microstrip antennas based on metallo-dielectric electromagnetic band-gap structures. It is important to emphasize the fact that simple embedded electromagnetic band-gap (EEBG) structures presented in this work are targeting operating frequency band at 0.5 - 20 GHz range. Since simple EEBG structures have impractical geometrical sizes in the 500 MHz to 2 GHz frequency range, more complex EEBG structures need to be employed. In addition, from wireless communication application, miniaturization of electronic systems requires the availability of miniaturized EEBG structures with appropriate patch sizes. The concept introduced in this chapter can be generally applied regardless of the size of the EEBG structures. For wideband radiation, reduction from hundreds of MHz to few GHz either a combination of different methods or use of advanced EEBG structures is the best solution as shown in figure 5 [40]. Figure 5 shows the efficacy range of EEBG structures covered in [40].

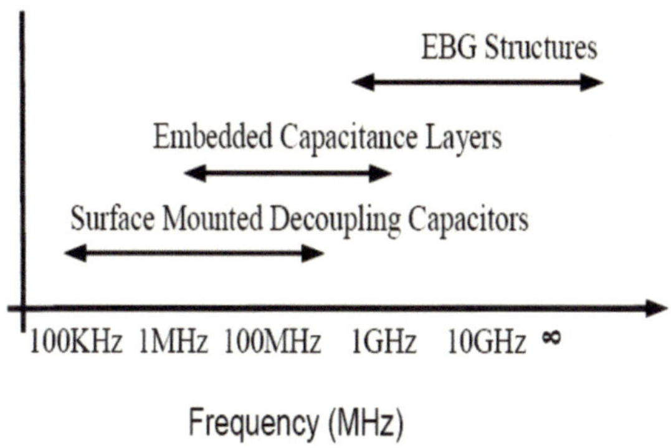

Fig. 5. Efficacy range of different reduction methods.

2. Disadvantages of metal ground planes and the MDEBG solution

MDEBG structures are useful where the presence of classic electric conductors as antenna ground planes adversely affects the performance of the entire electromagnetic device. As it is known, classic conductive surfaces are extensively used as antenna reflectors: they redirect one half of the radiation into the opposite direction potentially improving the antenna gain by 3 dBi.

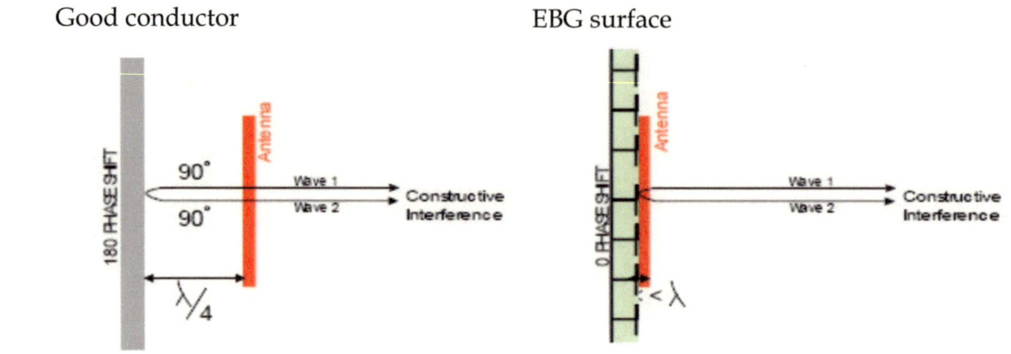

Fig. 6. An antenna separated by ¼ λ from the ground plane (on the left) and the alternative MDEBG layout (on the right).

However, they do have two main disadvantages: first, they reverse the phase of the reflected wave and second, they support propagating surface waves, which can have unwanted effects on the antenna performance. The fact that they reverse the phase 180 degrees is due to the most obvious constraint that the tangent electric field on a classic conductive surface must be zero, so the electromagnetic waves experience a 180 degrees phase shift on reflection. Because of the phase reversal, the image currents cancel the antenna currents, resulting in poor radiation efficiency when the antenna is too close to the conductive surface. This problem is often solved by including a quarter wavelength between the radiating element and the ground plane (see figure 6), but the disadvantage of this solution is the fact that the structure requires a minimum thickness of $\lambda/4$. It will be shown that, by using the novel MDEBG structures as ground planes, the antenna can be almost be attached to the ground plane, resulting in a useful reduction of volume. It will be proved that, at the frequency where the MDEBG structure does not give any reflection phase shift, a design of MDEBG structures even 8 times thinner than the classic ones (which implement a $\lambda/4$ spacing between antenna and ground plane) is possible. As stated before, another issue is the propagation of surface waves when normal ground planes are used: these are propagating electromagnetic waves bound to the interface between metal and free space and they will radiate if scattered by bends, discontinuities or surface textures. The unwanted result is a kind of multipath interference, which can be seen as ripples in the radiation pattern (see figure 7) [41].

Again, by using MDEBG structures, it will be shown that surface waves can be suppressed. It follows that, when multiple antennas share the same normal conductive ground plane, like it happens in phased arrays, the above mentioned surface waves may cause undesired mutual coupling between the antennas (see figure 8). Once again, by using MDEBG surfaces structures, it is possible to alter the surfaces properties of the ground plane and avoid this mutual coupling. Therefore, it can be easily understood that the Metallo-Dielectric EBG structure is a useful alternative to antenna classic metallic reflectors.

From above discussion, to realized high performance antenna:

- A reflector which lacks edge currents that radiate power into the back hemisphere of the antenna is needed;
- Surface waves on a ground plane associated with an antenna have to be suppressed to provide more efficient antennas, and reduce coupling.

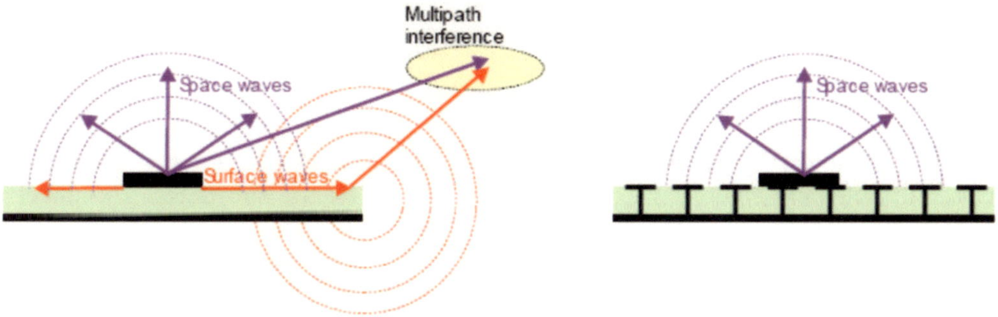

Fig. 7. Multipath interference due to surface waves on a normal ground plane (on the left) and the alternative MDEBG layout (on the right).

Nowadays, one of the most commonly used solution to prevent the propagation of unwanted surface waves is the so called "choke ring" which provides excellent electrical performance for GPS antennas. They are usually very large and heavy. While size and weight are not issues for most base station applications, for a GPS surveyor carrying one around in the field, size and weight are important factors. Moreover, choke rings are also expensive, typically costing thousands of dollars.

3. Metallo EBG structures as novel ground planes for antenna applications

The modern trends in communication systems require wide bandwidth, small size and low profile antennas. A planar microstrip antenna (PMA) is the good candidate for use in UWB wireless technology because they can reach wide impedance bandwidth and nearly omnidirectional azimuthally radiation pattern by different methods. One simple but powerful technique is to replace the cylindrical wires with the plate elements, such as rectangular (square), elliptical (circular), triangular shapes, and others. Another way to increase the impedance bandwidth of the monopole antennas can be achieved by modifying the ground plane. Different shapes of modified ground plane as semi circular is used to increase bandwidth. With the intention to overcome this handicap, a thick, high permittivity substrate is used, and potential surface waves are suppressed applying so called Electromagnetic Band-gap (EBG) [43]. The surface wave propagation is a serious problem of MPA. Surface waves reduce antenna efficiency and gain, limit bandwidth, increase end-fire radiation and cross-polarization levels. To avoid this, the substrate is periodically loaded so that the surface waves cannot propagate along the substrate. Also, other surface wave coupling effects like mutual coupling between array elements and interference with board systems can be suppressed [44].

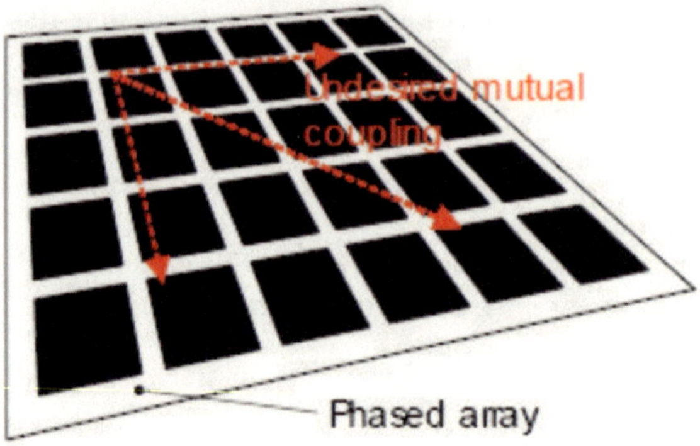

Fig. 8. Mutual coupling for multiple antennas due to the presence of surface waves.

To avoid using costly hybrid technology, innovative design must be developed to successfully integrate antenna with circuitry on high dielectric substrates. The novel RF system architecture reduces size, weight, losses and it is also suitable for integration with micro-electromechanical systems to realize reconfigurable circuits and antennas. Two technologies have been mainly pursued so far to achieve MPA on high-dielectric substrate with optimum performance. One is based on micromachining technology hile the other makes use of the concept of photonic band-gap substrates

A defected ground structure (DGS) has gained significant interests. It rejects certain frequency bands, and hence it is called electromagnetic band-gap (EBG) structures. Liu et al. has presented a novel DGS based meander microstrip line providing a broad stop-band [46]. A novel defected ground structure with islands (DGSI) is proposed by T. Itoh. in [48]. The DGS is realized on the bottom plane with two islands placed at both sides of the microstrip line on the upper plane. Due to their excellent pass and rejection frequency band characteristics, DGS circuits are widely used in various active and passive microwave and millimeter-wave devices such as filters, dividers, couplers, amplifiers, resonators and antennas.

4. Applications

One of the main purposes of this chapter is to deeply go into all the promising possible applications of MDEBG structures by satisfy their challenging design requirements of the antennas. Medical application as image scan is very important nowadays. A very promising way to eradicate the problems created by surface waves in this application (e.g. scan blindness), while at the same time improving performance, is to use electromagnetic band-gap structures instead of standard dielectric antenna substrates. Within the MDEBG, in fact, the unwanted effect of surface waves will be efficiently suppressed.

Another microwave application is high precision GPS. High precision GPS is going to be used in many situations. By very accurately determining the phase of the signal, it is hoped to reach a position accuracy of a few millimeters. However, in order to avoid errors in measurements (due to multi-path), the backward radiated field has to be at least a few orders of magnitude below the frontward field. Such a requirement cannot be obtained if the antenna excites surface waves and electromagnetic band-gap materials could again be used as design solutions. Another application is mobile and wireless communications. As the world goes, wireless, data and voice transmissions are bound to become even more common. A lot of attention is now focused on Bluetooth band hence its implementation in different wireless systems increases in everyday life. Moreover, for other applications like cell-phones, more attention is being paid to the shielding offered by the antenna and the potential health hazard.

Other than that, the EBG structure also can be used as a band reject especially for ultra wide band applications which operate at very wide frequency ranges. The simulation and measurement results for ultra wide band with and without band rejection are investigated in this chapter.

EBG structures are used to prevent some operating modes and make harmonic control. These techniques can increase the usability of antenna systems. The design and simulation of the antennas with EBG structures have recently received more attention. Initial concepts only have been proven with limited fabricated devices. A lot of research efforts is still expected in this field. This chapter is a step to investigate and develop how the EBG structure/ground plane can be used to optimize the antenna performance to satisfy several applications. The investigated antennas are miniaturized with broadband characteristics to be suitable for multi-band/ multi-function operation in wireless communications and medical applications.

5. Problems led to the use of EBG

Antenna designs have experienced enormous advances in the past several decades and they are still under research and development. Many new technologies have emerged in the modern antenna design arena and one exciting breakthrough is the discovery/ development of EBG structures. The applications of EBG structures in antenna designs have become interesting topic for antenna scientists and engineers. The recent explosion in antenna developments has been fueled by the increasing popularity of wireless communication systems and devices. From the traditional radio and TV broadcast systems to the advanced satellite system and wireless local area networks, wireless communications have evolved into an indispensable part of people's daily lives. Antennas play a paramount role in the development of modern wireless communication devices, ranging from cell phones to portable GPS navigators, and from the network cards of laptops to the receivers of satellite TV. A series of design requirements, such as low profile, compact size, broad bandwidth, and multiple functionalities, keep on challenging antenna researchers and propelling the development of new antennas. Progress in computational electromagnetic, as another important driving force, has substantially contributed to the rapid development of novel antenna designs. It has greatly expanded the antenna researchers' capabilities in improving and optimizing their designs efficiently.

Various numerical techniques, such as the method of moments (MOM), finite element method (FEM), and the finite difference time domain (FDTD) method, have been well developed over the years. As a consequence, numerous commercial software packages have emerged. Nowadays with powerful personal computers and advanced numerical techniques or commercial software, antenna researchers are able to exploit complex electromagnetic materials in antenna designs, resulting in many novel and efficient antenna structures with most of required characteristics.

For these reasons, EBG structures and their applications in antennas have become a new research direction in the antenna researches community. It was first proposed to respond to some antenna challenges in wireless communications [40]. For example,

- How to suppress surface waves in the antenna ground plane?
- How to design an efficient low profile antenna near a ground plane?
- How to increase the gain of an antenna?

In the novel RF system architecture of reducing the size, weight, losses and suitability for integration with micro-electro-machine-system lead to achieve reconfigurable antenna with required features. If a conventional substrate is used, then most of the antenna radiation is emitted from the substrate (since it has a higher dielectric constant than air). A big part of this radiation is trapped inside the substrate because of total internal reflection. As a consequence of this, more than 50% of the radiated energy is lost. Also heat dissipation and temperature effects arise in the substrate.

Two technologies have been used to achieve microstrip antenna on high dielectric substrate with optimum performance:

- One is based on micromachining technology.
- Second is photonic band-gap (PBG) substrate.

A photonic crystal essentially behaves much like a band-stop filter, rejecting the propagation of energy over a fixed band of frequencies. An appropriate EBG substrate is selected, then all of the energy can be directed towards the radiating direction (total reflection by the EBG structure), thus improving the antenna directivity and eliminating the substrate heat dissipation [41].

Due to the complexity of the EBG structures, it is usually difficult to characterize them through analytical methods. Instead, full wave simulators that are based on advanced numerical methods have been popularly used in EBG analysis. Dispersion diagram, surface impedance, and reflection phase features are explored for different famous EBG structures. The interaction of antennas and EBG structures are extensively investigated. In summary, the EBG research has flourished since the beginning of this new millennium.

6. Electromagnetic Band-Gap (EBG) structure

6.1 The parameters of EBG (figure 9)

- Permittivity of the dielectric materials used (ε_r)
- Dimensions of the mushroom patches (a)
- Periodicity (P)
- Incident angle of electromagnetic waves (θ_i)

A periodic structure is characterized by the following parameters:

1. $\lambda_r = 2.a$ (1)
2. Shape of individual patches.
3. Filling factor ratio between size of the patches and the periodicity of unit cell (a/P).

For best performance [40].

- For 3D-EBG $0.9 < a/P < 0.95$
- For 2D-EBG $0.65 < a/P < 0.75$

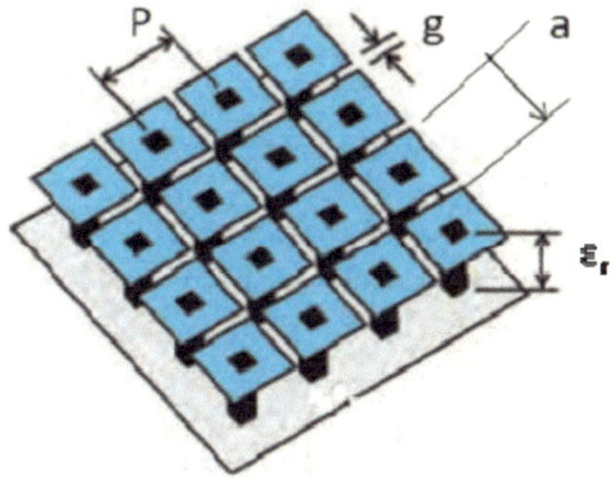

Fig. 9. The parameters of EBG.

6.2 The features of EBG

The main feature of EBG structures is their capability to affect the radiative dynamics within the structure so that there are no electromagnetic modes available within the dielectric. This feature is analogous to periodically arranged atomic lattice of a semiconductor which gives rise to the allowed values of energy that an electron can have at the valence band and at the conduction band, with an energy band-gap separating the two. The optical analogy to this situation is a periodic dielectric structure with alternating high and low values of permittivity, which gives rise to a photonic band-gap [50].

6.3 Applications of EBG

In optical domain new highly efficient opto-electronic devices are considered such as very efficient laser diodes [44, 45], microscale light circuits, multiplexers or demultiplexers based on inhibition of spontaneous emission, photoluminescence, wave-guiding, and superprism phenomenon [46-49]. Studies of frequencies occurring for metal photonic crystals have also shown that the frequency can be controlled and could appear in the microwave region [50]. In the microwave domain, many developments concern the direct control of the electromagnetic energy and its transmission: mirrors, electromagnetic windows, and radiation pattern control. We find also the high impedance material of Sievenpiper et al [51]. They proposed their structure as perfect magnetic wall to reduce the leaky waves in antenna array. The material developed allows the realization of antennas, low loss coplanar lines and compact integrated filter [52-54].

Other applications include duplexers and controllable PBG materials. Due to a certain easiness of fabrication in this frequency domain, the challenge of the electronically controlled photonic crystals has a significant interest. Industrial applications of these crystals are under development, concern mainly aerospace, and telecom domains [43].

6.3.1 Antenna substrates for surface wave suppressions

Surface waves are by-products in many antenna designs. It directs electromagnetic wave propagation along the ground plane instead of radiation into free space, consequently reduce the antenna efficiency and gain. The diffraction of surface waves increases the back lobe radiations, which may deteriorate the signal to noise ratio in wireless communication systems such as GPS receivers. In addition, surface waves raise the mutual coupling levels in array designs as shown in figure 10, resulting in the blind scanning angles in phased array systems. The band-gap feature of EBG structures has found useful applications in suppressing the surface waves in various antenna designs. For example, an EBG structure is used to surround a microstrip antenna to increase the antenna gain and reduce the back lobe as shown in figure 7. In addition, it is used to replace the quarter-wavelength choke rings in GPS antenna designs. Many array antennas also integrate EBG structures to reduce the mutual coupling level more over to increase antenna gain used super-state EBG as shown in figure 10 [40].

Fig. 10. EBG substrate for surface wave suppression with low mutual coupling microstrip array.

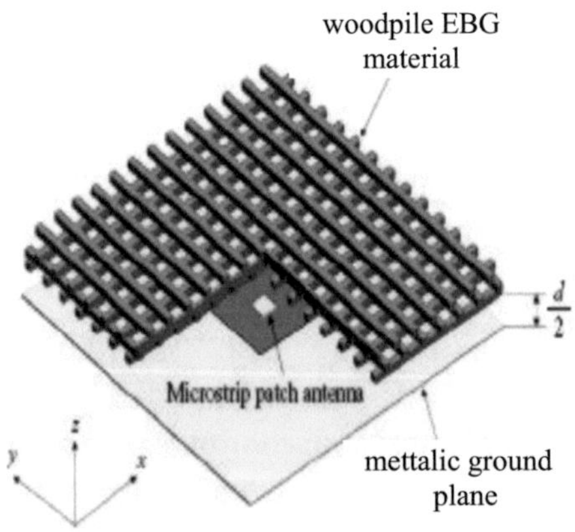

Fig. 11. A high gain resonator antenna design using a woodpile EBG structure.

6.3.2 Antenna substrates for efficient low profile antenna design

Another favorable application of EBG is to design low profile wire antennas with good radiation efficiency, as shown in figure 11, which is desired in modern wireless communication systems. To illustrate the fundamental principle, Table 1 compares the EBG with the traditional PEC ground plane in antenna designs [43].

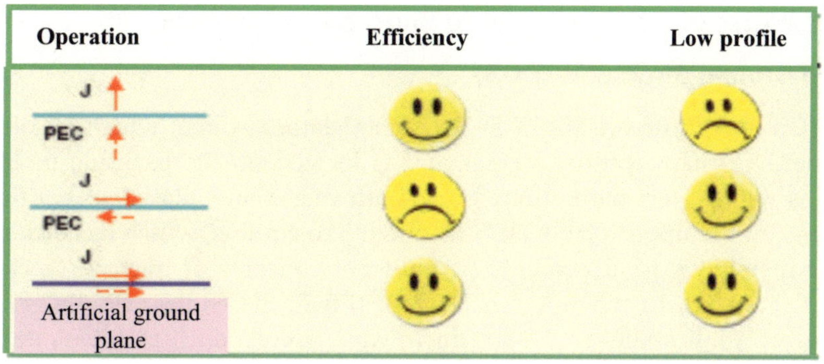

Table 1. Comparison of conventional PEC and artificial ground planes in antenna designs [43].

When an electric current is vertical to a PEC ground plane, the image current has the same direction and reinforces the radiation from the original current. Thus, this antenna has good radiation efficiency, but suffers from relative large antenna height due to the vertical placement of the current. To realize a low profile configuration, one may position an antenna horizontally close to the ground plane. However, the problem is the poor radiation efficiency because the opposite image current cancels the radiation from the original current. In contrast, the EBG surface is capable of providing a constructive image current within a certain frequency band, resulting in good radiation efficiency. In summary, the EBG surface exhibits a great potential for low profile efficient antenna applications as shown in figure 12. Based on this concept, various antennas have been constructed on the EBG ground plane [59-62]. Typical configurations include dipole antenna, monopole antenna, and spiral antenna. EBG surfaces have also been optimized to realize better performance such as multi-band and wideband designs.

Fig. 12. EBG substrate with low profile antenna.

6.4 Advantages of EBG

Utilizing electromagnetic band-gap crystals in a patch antenna with an air gap appears to perform five key functions:

4. Increase operation bandwidth
5. Reduce side-lobe levels
6. Increase front to back (F/B) ratios
7. Increase directivity and consequently gain improvement
8. Harmonic control (suppression of resonance at the harmonic frequencies of the antenna)

7. Classifications of electromagnetic band-gap structures

7.1 Defected Ground Structures (DGS)

A Defected Ground Structure (DGS) is an etched lattice shape, which is located on the ground plane. DGS has arbitrary shapes and is located on the backside metallic ground plane. DGS is realized on the bottom plane with one island placed at both sides of the microstrip line on the upper plane. DGS for the microtrip line, which has etched defects in the backside metallic ground plane, is one hotspot concepts of microwave circuit design nowadays. Compared to photonic band-gap (PBG), DGS has simple structure and potentially great applicability to design microwave circuits such as filters, amplifiers and oscillators. DGSs have gained significant interests. It rejects certain frequency bands, and hence it is called electromagnetic band-gap (EBG) structures as shown in figure 13 [57]. The DGS cell has a simple geometrical shape, such as rectangle. Novel fractal DGS cell is proposed. Its band-gap and slow-wave characteristics are better than the conventional ground plane. DGSs have gained quite significance in filter design [58] showing optimal pass-band and stop-band responses plus sharp selectivity and ripple rejection. Application of CPW-based spiral-shaped DGS to MMIC for reduced phase noise oscillator [57], active devices (BJT and FET) can also be mounted using DGS technique. High amount of isolation is achieved in microstrip diplexer and harmonic control can also be achieved on microstrip antenna structures using DGS. Figure 14 gives the schematic of such a DGS with its approximate surface area. A novel DGS based meander microstrip line providing a broad stop band is presented in [59]. Novel Defected Ground Structures with Islands (DGSI) is proposed in [58]. Careful selection of the line width guarantees 50Ω characteristic impedance (Z_0).

The EM simulation results of the DGSI are compared with circuit simulation results using extracted parameters; showing excellent agreement between the two in wide band. Examination of stop band characteristics is studied using concentric circular rings in different configurations. Metallic backing significantly reduces interference effects, harmonics and phase noise. Several novel 1D DGS are presented for microwave integrated circuits (MIC), monolithic MIC (MMIC), low temperature cored ceramic (LTCC) including RF front-end applications. Significant change in the characteristics including slow-wave factor (SWF) of periodic structures like transmission lines is achieved using quite a few unconventional DGS like spiral-shaped and vertically periodic DGS. Vertically periodic DGSs (VPDGS) have been used in reducing the size of MIC and amplifiers, thus increasing SWF significantly. Harmonic control can also be achieved on microstrip antenna structures using 1-D DGS [59].

Table 2 presents the difference between defected ground plane structure and band-gap structure. The characteristics of the defected ground structure are:

- Disturbs shielding fields on the ground plane.
- Increases effective permittivity.
- Increases effective capacitance and inductance of transmission line.
- Has one-pole LPF characteristics (3dB cutoff and resonance frequency).
- Size reduction for the component.

Comparisons	Electromagnetic Band-gap Structure (EBG)	Defected Ground Structure DGS
Geometry	Periodic etched structure	One or few etched structure
Microwave Circuit properties	Similar	Similar
Equivalent Circuit extraction	Very difficult	Relatively simple

Table 2. Comparison between defected ground structure and EBG..

7.2 Photonic Band-Gap (PBG) structures

7.2.1 Historical background of PBG

Photonic band-gap (PBG) structures are periodic structures that manipulate electromagnetic radiation in a manner similar to semiconductor devices manipulating electrons. Semiconductor material exhibits an electronic band-gap where there are electrons cannot exist. Similarly, a photonic crystal that contains a photonic band-gap does not allow the propagation of electromagnetic radiation with specific frequencies within the band-gap [60]. This phenomenon results from the destructive Bragg diffraction interference due to the periodic boundary conditions of PBG structures. This property has a significant importance in many microwave and optical applications to improve their efficiency. The photonic band-gap structures were first investigated in [61] by Yablonovitch. He introduced vector-spherical-wave expansion method or the vector Koringa-Kohn-Rostker (KKR) method to calculate the dispersion relation and the transmittance for the regular array of dielectric spheres. This was the first self-consistent treatment of the electromagnetic eigen modes in 3-D dielectric systems with large periodic modulation of the dielectric constant. A remarkable step was made by Yablonovitch who pointed out the possibility of the realization of photonic band-gap, localized defect modes, and their applications to various optoelectronic devices. His ideas stimulated many researchers, and energetic research activities including his own studies were initiated. The process to realize the photonic band-gap is described in [62]. Lee et al [49] discussed the strong localization of electromagnetic waves in disordered PBG structures.

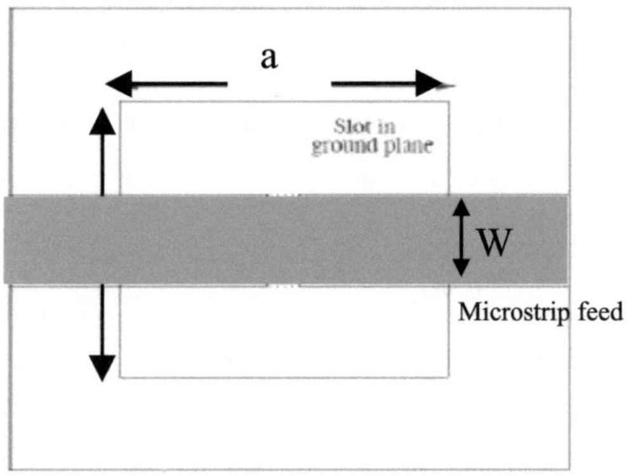

Fig. 13. Schematic diagram of a unit DGS cell.

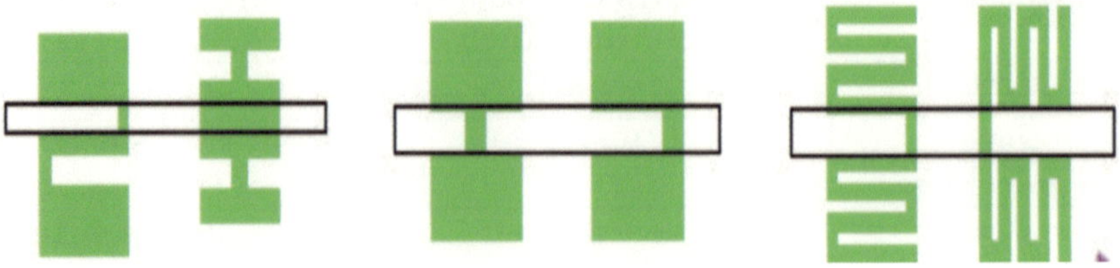

Fig. 14. Different shapes of DGS structures.

Since then, many researchers in various fields such as physics, electronics, waves, optics, fabrication, and chemistry have been engaged in the realization of photonic band-gap, localized defect modes, and other microwave and optical properties peculiar to the PBG structures [63]. They have also been collaborating to make new structures and measure their properties. PBG structures were initially applied to optical applications like high-quality optical mirrors and resonators. One of the most exciting projected applications of PBG structures is the production of optical circuits by manipulating light with optical waveguides. Currently, fiber-optic signals must be converted to electronic signals in order to be analyzed by computers or other devices connected to the optical line. This slows the signal considerably and the electronic circuitry is fairly inefficient as well. A device that runs completely on light would revolutionize technology in several areas, leading eventually to an all optical network (AON).

The integration of photonic components such as lasers, detectors, couplers, and waveguides is still at a very primitive stage due to difficulties with implementing integrated optical components smaller than certain sizes. For instance, small bends or curves of waveguides lead to the leakages of optical signals, so the bends have to be bigger than certain length. PBG structures are proposed as a possible solution to this problem. Adding a defect in a PBG structure opens a path along which electro-magnetic radiation can propagate. Because the PBG structure can be tailored to completely reflect certain frequencies, it is possible to turn corners with light at a distance of the light's wavelength order. Another application of PBG structures is the use of these crystals to greatly improve the efficiency of lasers and light-emitting diodes. Due to the scalability of PBG structures, several applications at microwave and millimeter-wave frequencies have been developed [64]. Other applications such as resonant cavities can be designed by using PBG structures are also presented. Applications based on the guiding and localized mode properties of the PBG structures are given. Some of these common applications in both microwaves and optics are power splitters, switches, directional couplers, high quality filters, and channel drop filters.

7.2.2 Basic PBG structures

Figure 15, 16 shows different configurations of PBG structures composed of two different materials. These configurations include 1D, 2D and 3D periodicities. The two different materials of a PBG structure can be two different dielectric materials or a metal and a dielectric material. Metallic photonic band-gap systems have received far less attention than dielectric PBG structures. However, it has been suggested that periodic metallic structures have important applications, such as cavities, waveguides, and antennas [65]. The main advantage of dielectric PBG structures is that they do not include metallic loss which is usually the dominant loss at high frequencies. Dielectric materials that can be used to construct PBG structures are widely available for almost all frequency bands from ultraviolet to microwave range which facilitate the design and fabrication. Many basic characteristics are common in both ordinary and photonic crystals. These characteristics are utilized to build the fundamental theories of photonic crystals. However, a major difference between ordinary and photonic crystals is the scale of the lattice constant. In the case of ordinary crystals, the lattice constant is in the order of angstroms. On the other hand, lattice constant is constructed with millimeter dimensions for microwave range. Figure 16 shows examples of different PBG structures used in the design of different applications. Various technologies have been developed and applied for manufacturing PBG structures in the last ten years such as stacking slabs of 2D materials for 3D structures implementations. For practical implementation of an infinite 2D PBG structure, a 2D PBG slab of finite thickness is surrounded by two perfect electric conductor (PEC) plates. This modification is implemented to guide the waves in the normal direction to the periodic cell. In optical applications, confinement of electromagnetic waves inside a 2D PBG slab can be obtained by surrounding the slab by a different material of another dielectric constant [66].

(a) (b) (c)

Fig. 15. Examples of different PBG, (a) 1D, (b) 2D and (c) 3D configurations.

Fig. 16. Typical examples of three-dimensional photonic crystals.

i. The Unit Cell

As stated in the American Heritage Dictionary of the English Language, definition of a crystal (unit cell) is "a homogenous solid formed by a repeating, three-dimensional pattern of atoms, molecules, or shapes, having fixed distances between constituent parts". This replicating pattern is referred to as the unit cell of a crystal.

Figure 17 illustrates the unit cell for the circular crystal lattice. The unit cell contains all the pertinent information of the crystal such as the crystal geometry (shape, thickness, etc.), material properties (dielectric or magnetic), and the lattice spacing (shown as the dimension "p". It is this replicating unit cell that provides the periodicity in the crystal, and controls the location and extent of the band-gap.

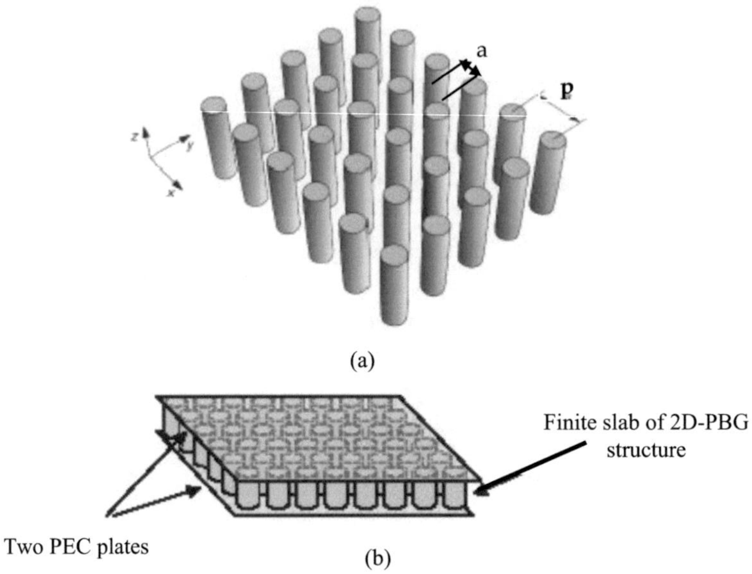

Fig. 17. Practical implementation of 2D PBG structure, (a) Finite slab of 2D PBG structure surrounded by air, (b) Finite slab of 2D PBG structure surrounded by two PEC plates.

ii. The Band-gap

As a first order approximation, a band-gap is obtainable in a high dielectric material with integrated photonic crystals when an incident electromagnetic field propagates with a guide-wavelength, approximately equal to the lattice spacing of the crystal.

$$\lambda_r = \frac{c}{f_r\sqrt{\varepsilon_r}} = 2a \tag{2}$$

This rough approximation locates the center of the band-gap, which can extend higher than ±10% of the center frequency for high dielectric constant materials [67] by using a typical transmission coefficient (S_{21}) plot for a 2-port network, such as a microstrip transmission line. This curve, figure19, illustrates that as port 1 of the transmission line is excited, over 15dB of attenuation is experienced as the energy propagates from port 1 to port 2. Thus, the photonic crystal introduces a stop-band filter response. Indeed, the formation of the band-gap is heavily dependent on (1) the periodicity of the crystal, (2) the refractive index (dielectric constant) ratios between the base material (the substrate as a whole) and the impurities that form the crystal. Typically, the refractive index ratio must be at least 2:1 (substrate-to-impurity) ratio for the band-gap to exist. For the 2-D triangular structure, the broadest band-gap is obtainable when the impurities (the cylindrical post) are of air (ε_r =1), while the base material is a high dielectric constant (for example, ε_r =10). A 10:1 dielectric ($\sqrt{10}$ = 3.16:1 refractive index) ratio would satisfy the index requirement and form a broad band-gap, with proper crystal spacing. This explains the need for a high dielectric substrate for a patch antenna designed on a photonic crystal substrate [68].

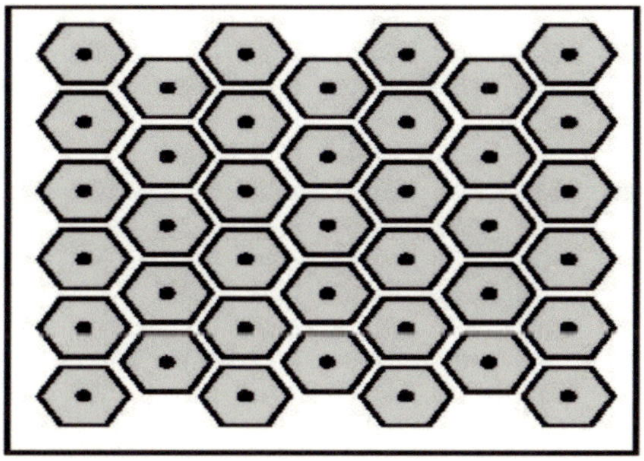

Fig. 18. Top view of the high-impedance surface.

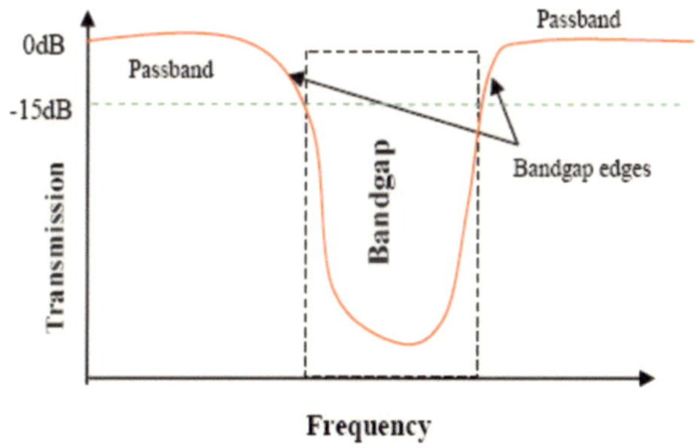

Fig. 19. Transmission loss plot illustrates band gap in a microstrip transmission line at microwave frequencies.

iii. Defects in Periodicity

A photonic crystal essentially behaves much like a band-stop filter, rejecting the propagation of energy over a fixed band of frequencies. However, once a defect is introduced such that it disrupts the periodicity in the crystal, an area to localize or "trap" electromagnetic energy is established. In this region, a pass-band response is created. This ability to confine and guide electromagnetic energy has several practical applications at microwave frequencies as filters, couplers, and especially antennas. This rather simple concept of placing defects in a photonic crystal structure introduces a new methodology in the design of microstrip (patch) antennas. The idea is to design a patch antenna on a 2D photonic crystal substrate, where the patch becomes the "defect" in the crystal structure.

In this case, crystal arrays of cylindrical air holes are patterned into the dielectric substrate of the patch antenna. By not patterning the area under the patch, a defect is established in the photonic crystal, localizing the EM fields. Surface waves along the XY plane of the patch are forbidden from forming due to the periodicity of the photonic crystal in that plane. This prevention of surface waves improves operational bandwidth and directivity, while reducing side-lobes and coupling, which are common concerns in microstrip antenna designs [68]. Using these concepts, a photonic crystal patch antenna was developed.

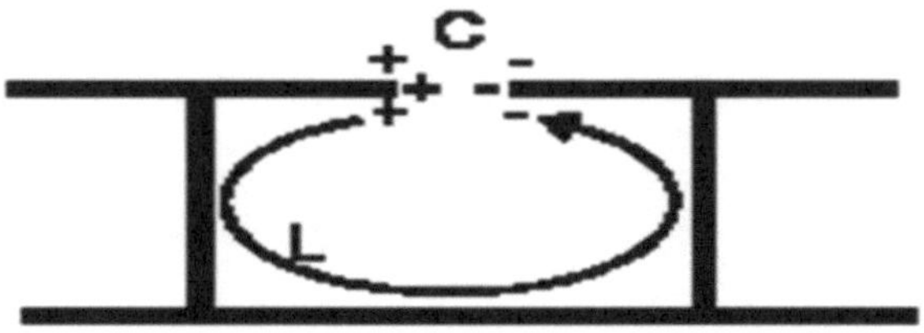

Fig. 20. Origin of the equivalent circuit elements.

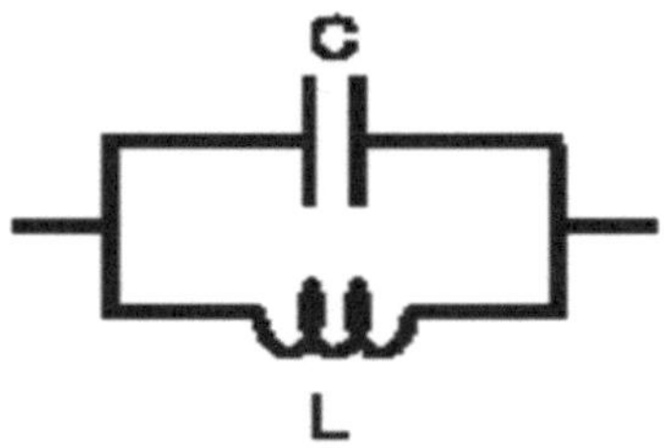

Fig. 21. Equivalent circuit model for the high-impedance surface

7.3 High Impedance Electromagnetic Surface (HIES)

7.3.1 Introduction and background of HIES

A new type of metallic electromagnetic structure has been developed. It is characterized by high surface impedance. Although it is made of continuous metal and conducting DC currents, it does not conduct AC currents within a forbidden frequency band. Unlike normal conductors, this new surface does not support propagating surface waves, and it reflects electromagnetic waves with no phase reversal. The geometry consists of a metal sheet, textured with a two-dimensional lattice of resonant elements, which act as a two-dimensional filter to prevent the propagation of electric currents. The surface can be described using a lumped parameter circuit model, which accurately predicts many of its electromagnetic properties. This unique material is applicable to a variety of electromagnetic problems, including new kinds of low-profile antennas. By incorporating a special texture on a conducting surface, it is possible to alter its radio-frequency electromagnetic properties [69].

In the limit where the period of the surface texture is much smaller than the wavelength, the structure can be described using an effective medium model, and its qualities can be summarized into a single parameter: the surface impedance. A high-impedance surface, shown in figure 18, consists of an array of metal, protrusions on a flat metal sheet. They are arranged in a two-dimensional lattice, and are usually formed as metal plates, connected to the continuous lower conductor by vertical posts as shown in figure 20. They can be visualized as mushrooms or thumbtacks or other shapes protruding from the surface. High Impedance Surfaces as two dimensional EBG structures can be used as microstrip antenna substrate to eliminate the surface wave [70].

7.3.2 High-impedance surfaces properties

The properties of the new high-impedance surface are similar to those of the corrugated slab. The quarter-wavelength slots have simply been folded up into lumped elements, capacitors and inductors that are distributed in two dimensions. The two-dimensional array of resonant elements can be explained using a simple circuit model. The capacitance is due to the proximity of the top metal patches, while the inductance originates from current loops within the structure, as shown in figure 20. The electromagnetic properties of the surface can be predicted by using an equivalent LC circuit, shown in figure 21. The impedance of a parallel resonant LC circuit, given in Eq. 3, is qualitatively similar to the tangent function that describes the impedance of the corrugated surface.

$$Z_s = \frac{j\omega L}{1 - \omega^2 LC} \tag{3}$$

It is inductive at low frequencies, and thus supports TM surface waves. It is capacitive at high frequencies, and supports TE surface waves. In a narrow band around the LC resonance, the impedance is very high. In this frequency range, currents on the surface radiate very efficiently, and the structure suppresses the propagation of both types of surface waves. Having high surface impedance, it also reflects external electromagnetic waves without the phase reversal that occurs on a flat conductor. By using lumped elements, we retain the reflection phase and surface wave properties of the quarter-wave corrugated slab, while reducing the overall thickness to a small fraction of a wavelength [72-74].

7.3.3 Improved surface wave current

Surface waves are excited on microstrip antenna when the substrate $\epsilon_r > 1$. Besides end fire radiation, surface waves give rise to coupling between various elements of an array. Surface waves are launched into the substrate at an elevation angle θ lying between π / 2 and $\sin^{-1}(1/\sqrt{\varepsilon_r})$. These waves are incident on the ground plane at this angle, get the reflected from there, then meet the dielectric-air interface, which also reflect them. Following this zig-zag path, they finally reach the boundaries of the microstrip structure where they are reflected back and diffracted by the edges giving rise to end-fire radiation [94]. On other way in the boundary, if there is any other antenna in proximity, the surface wave can become coupled into it. Surface waves will decay as $1/\sqrt{\varepsilon_r}$ so that coupling also decreases away from the point of excitation. Surface wave are TM and TE modes of the substrate. These modes are characterized by waves attenuating in the transverse direction (normal to the antenna plane) and having a real propagation constant above the cut-off frequency. The phase velocity of the surface waves is strongly dependent on the substrate parameters h and ϵ_r. Figure 22 shows the propagation of the surface wave in microstrip antenna [70].

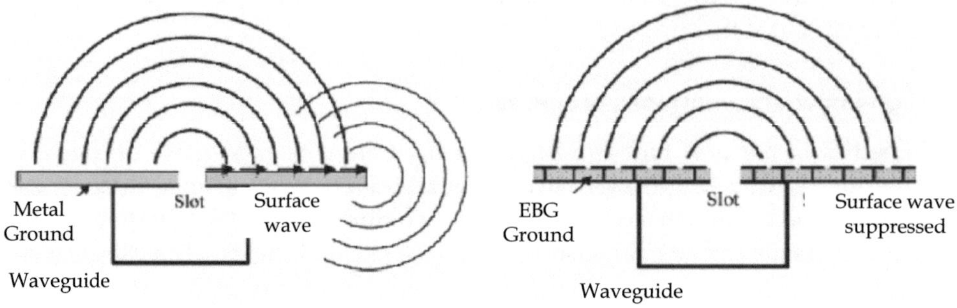

(a) (b)

Fig. 22. (a) The substrate without EBG structure, (b) with EBG structure.

Surface wave propagation is a serious problem in microstrip antennas. It reduces antenna efficiency and gain, limits bandwidth, increases end-fire radiation, increases cross-polarization levels, and limits the applicable frequency range of microstrip antennas. Two solutions to the surface wave problem are available now. One of the approaches is based on the micromachining technology in which part of the substrate beneath the radiating element is removed to realize a low dielectric constant environment for the antenna. In this case the power loss through surface wave excitation is reduced and coupling of power to the space wave is enhanced. The second technique relies on electromagnetic band-gap structure (EBG) engineering. In this case, the substrate is periodically loaded so that the surface wave dispersion diagram presents a forbidden frequency range (stop-band or band-gap) about the antenna operating frequency. Because the surface waves cannot propagate along the substrate, an increase amount of radiating power couples to the space waves. Also, other surface wave coupling effects like mutual coupling between array elements and interference with onboard systems are now absent [43].

7.3.4 Artificial Magnetic Conductors (AMC)

Background of AMC

Artificial magnetic conductors (AMC), also known as high-impedance surfaces [76] as shown in figure 23(a) and (b), have received considerable attention in recent years. An AMC is a type of electromagnetic band-gap (EBG) material or artificially engineered material with a magnetic conductor surface for a specified frequency band. AMC structures are typically realized based on periodic dielectric substrates and various metallization patterns. Several types of AMC ground planes have already been extensively studied. AMC surfaces have two important and interesting properties that do not occur in nature and have led to a wide range of microwave circuit applications.

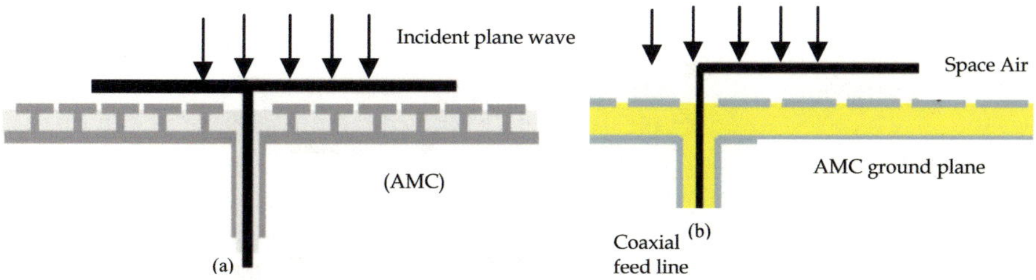

Fig. 23. Horizontal patch antenna on Artificial Magnetic Conductors ground plane.

In the artificial surfaces such as artificial magnetic conductors, the high-impedance surface has proven useful as an antenna ground plane. AMC takes advantage of both the suppression of surface waves and the unusual reflection phase. As a result of the suppression of surface waves, an antenna on Artificial Magnetic Conductors produces a smoother radiation profile than a similar antenna on a conventional metal ground plane, with less power wasted in the backward direction. This can be applied to a variety of antenna designs, including patch antennas, which often suffer from the effects of surface waves. For phased arrays, the suppression of surface waves can reduce inter-element coupling, and help to eliminate blind angles. AMC is particularly applicable to the field of portable hand-held communications, in which the interaction between the antenna and the user can have a significant impact on antenna performance. Using this new ground plane shown in figure 23 as a shield between the antenna and the user in portable communications equipment can lead to higher antenna efficiency, longer battery life, and lower weight.

Parameter	Dimension Change	Center Frequency	Band edge Effect	Bandwidth
Effect of h_1	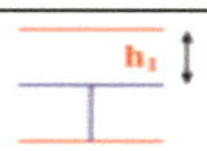	Negative Correlation	Lower Side Band	Positive Correlation
Effect of ε_r	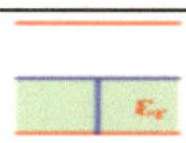	Negative Correlation	Upper Side Band	Negative Correlation
Effect of radius	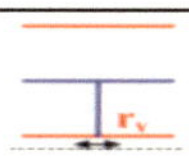	Negative Correlation	Upper Side Band	Negative Correlation
Effect of gap	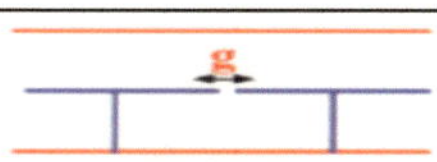	Positive Correlation	Upper Side Band	Positive Correlation
Via position	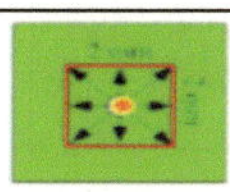	Negative correlation	Upper side band	Negative Correlation
Via type	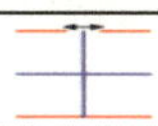	No Effect	No Effect	No Effect

Table 3. Parameters Analysis of EBG Design.

AMC surfaces have very high surface impedance within a specific limited frequency range, where the tangential magnetic field is small, even with a large electric field along the surface [73]. Therefore, an AMC surface can have a reflection coefficient of +1 (in-phase reflection). Generally, the reflection phase is defined as the phase of the reflected electric field which is normalized to the phase of the incident electric field at the reflecting surface. It can be called in-phase (or out-of-phase) reflection, if the reflection phase is 0 (or not). In practice, the reflection phase of an AMC surface varies continuously from +180 to -180 relative to the frequency, and crosses zero at just one frequency (for one resonant mode). The useful bandwidth of an AMC is generally defined as 135 to 45 on either side of the central frequency. Thus, due to this unusual boundary condition, in contrast to the case of a conventional metal plane, an AMC surface can function as a new type of ground plane for low-profile wire antennas, which is desirable in many wireless communication systems.

Difference	EEBG	UP-EBG
Substrate structure	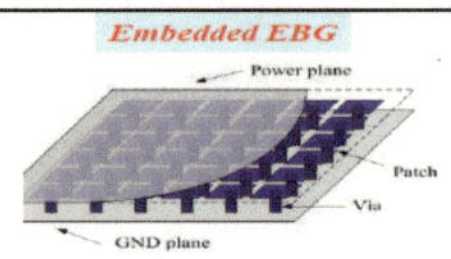	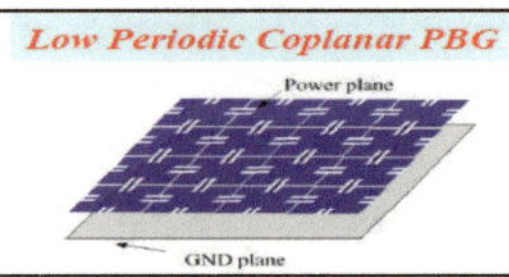
Impedance	Low impedance Surface (LIS)	*High impedance Surface (HIS)*
Defacement shapes		
Equivalent circuit	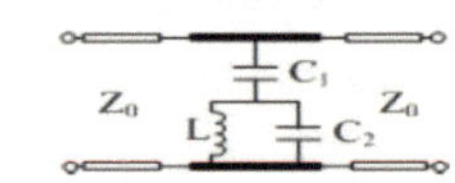	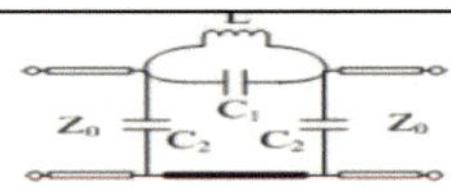
Resonance frequency	$f_o = \dfrac{1}{2\pi\sqrt{L(C_1 + C_2)}}$	$f_o = \dfrac{1}{2\pi\sqrt{LC_1}}$
Bandwidth	$\pm\Delta\omega = \omega_o C_1 \dfrac{(1 - L_1 L_2 \omega^2{}_o) Z_0}{4C}$	$\Delta\omega / \omega_o = \dfrac{Z_o}{\eta}$

Table 4. Comparison between different characteristics of electrical equivalent circuit for both embedded EBG and UP- EBG.

8. Design of single patch antennas with DGS

8.1 Miniaturized design consideration

A novel design method using the first iteration of fractal carpet gasket is utilized as DGS cell to study the size miniaturization. Figure 24(a) shows schematics of the proposed DGS cell. The DGS is etched on the back side of the metallic ground plane as shown in figure 24(b). The substrate is RT/Duroid with 0.813mm thickness and dielectric constant ε_r=3.38. The feed line width *w =1.85mm* is chosen for the characteristics impendence of 50 Ω microstrip line at frequency 5.25GHz. The defect unit is with length L=12mm and width W=9mm, L_g=1mm, L_s=1mm, W_g=1mm and W_s=3mm. The effect of one as well as two unit cells of rectangular DGS (RDGS) on inset feed MPA performance is studied by etching the defect from certain reference point as a start at distance **X**=0 mm as shown in figure 24(b) and gradually increase **X** to change the position of DGS with step 2mm until DGS is far away from the image projection under the radiating surface. For further reduction two unit cells of the proposed RDGS are used, which are etched in face to face with separation 3mm. it is noted that the maximum size reduction is 35% in case of one unit cell at defected displacement **X**=10mm from antenna edge. In case of two unit cells the maximum reduction ratio reached 53% at same defect position [82].

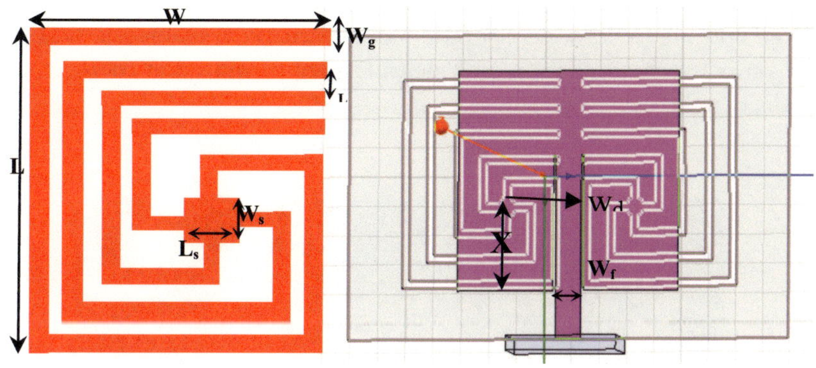

Fig. 24. (a) The geometry of one unit cell of DGS, (b) the proposed antenna prototype

8.2 Multiband design consideration

In many applications it is also important to design single feed antenna for multiband resonant frequencies. Spiral defected ground structure DGS is used to provide both multi resonant frequencies and compact size. Figure 24(a) shows schematic of the proposed one cell spiral DGS. Spiral structures, however, are known to produce large cross polarization. Therefore to completely eliminate the cross polarization, a four-arm spiral is explored, as shown in figure 24(b). Four spiral branches, each with a 0.01 $\lambda_{5.2GHz}$ width, split from the center and rotate outwards are used in the design.

As it may be seen, this geometry is symmetric not only in +/-x or +/-y directions, but also in x and y directions. A design using one unit cell of the spiral DGS is simulated and it is clear that the effect of increasing distance ***X*** on antenna performance of one cell as well as two unit cells. As distance **X** increases the resonant frequencies are reduced until certain distance and after that the trend is reversed. Maximum reduction in the fundamental resonant frequency was achieved at distance **X**=9mm (0.16$\lambda_{5.2GHz}$) and DGS shifted from width center $\boldsymbol{W_d}$ by 3mm (0.083$\lambda_{5.2GHz}$). More reduction in resonant frequency was achieved by using two unit cells of spiral DGS which are placed face to face and at a separation distance from center to center $\boldsymbol{2W_d}$ equal to 10mm (0.18λ_0) with separation $\mathbf{W_g}$ equal to 3mm (0.054$\lambda_{5.2GHz}$). We use same previous substrate and antenna dimensions are 15 X16 mm^2 and DGS is etched in the bottom of the metallic ground plane as in figure 24 (b). The dimension of the largest arms length L=13.25mm and largest width W=*11.25mm*, air-gap *g=0.5mm and* L_g= *2mm* with spacing patch L_s=W_s=*1mm* with 0.5mm inner square is used in the design. The line width W_f =*1.85mm* is chosen for the characteristics impendence of *50 Ω* microstrip line at frequency *5.25GHz*. The fundamental MPA resonance frequency is reduced by using one unit cell spiral DGS which is shifted from length center by 0.011λ_0. In case of using two unit cells, maximum reduction in resonant frequency occurs when the cells are face to face and centered under radiating plate but far from width center by 0.05λ_0.

8.3 Simulation and measurement of miniaturized multi-band antenna

The final two proposed antenna configurations were fabricated as shown in figure 25 by using photolithographic techniques. The reduction in antenna resonant frequency is due to increase in both electrical and magnetic coupling from ground to radiating antenna plate. The measured and simulated results are shown in figure 26. Second proposed antenna was fabricated also and the multi-band operation and reducing antenna size was achieved and the comparison between simulation and measurement of reflection coefficient is shown in figure 27 [81].

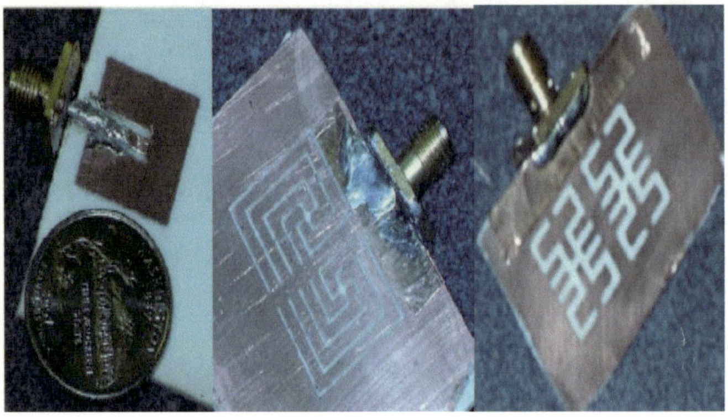

Fig. 25. Photo of the fabricated two antennas.

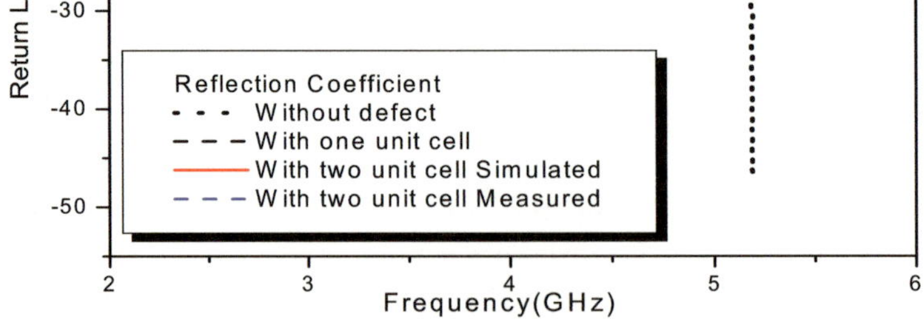

Fig. 26. Comparison between measured and simulated reflection coefficient of antenna without and with one and two gasket unit cell

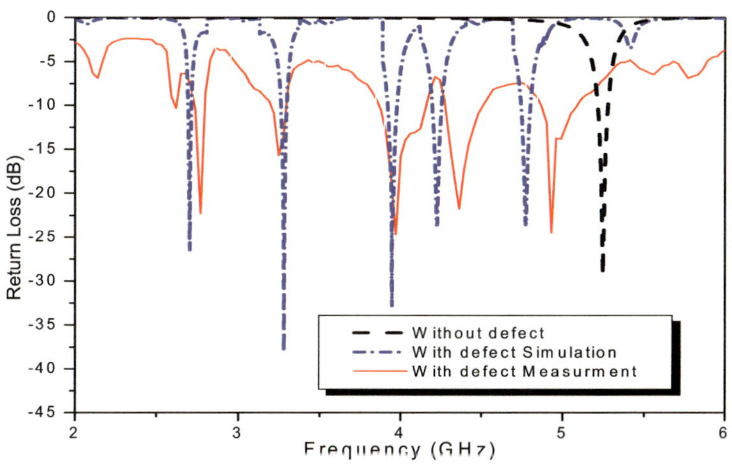

Fig. 27. Comparison between measured and simulated reflection coefficient of antenna without and with two unit cell of spiral DGS simulated and measured.

9. Design of single patch antennas with EBG

The conventional half-wavelength size is relatively large in modern portable communication devices. Various approaches have been proposed, such as using shorting pins, cutting slots, and designing meandering microstrip lines. Increasing the dielectric constant of the substrate is also a simple and effective way in reducing the antenna size. Applications of MPAs on high dielectric constant substrate are of growing interest due to their compact size and conformability with monolithic microwave integrated circuits (MMIC). However, there are several drawbacks with the use of high dielectric constant substrate, namely, narrow bandwidth, low radiation efficiency, and poor radiation patterns, which result from strong surface waves excited in the substrate. The narrow bandwidth can be expanded by increasing the substrate thickness, which, however, will launch stronger surface waves. As a result, the radiation efficiency and patterns of the antenna are further degraded. To quantify this phenomenon, a comparative study of MPAs on substrates with different dielectric constants and different thicknesses is performed in this section. Table 5 illustrates the four samples under study. Two of them with low dielectric constant substrate (ε_r=2.2) and the other two are built on the high dielectric constant substrate (ε_r=10.2).

Example	Patch size mm^2	Dilectric constant (ε_r)	Height (mm)
1	18 X 10	2.2	1
2	16X 13	2.2	2
3	9 X 6	10.2	1
4	8 X 6	10.2	2

Table 5. The antennas parameters.

Figure 28 shows the simulated S_{11} of these four structures. By tuning the patch size and the feeding probe location, all the antennas match well to 50 Ω around 5.1 GHz. It is noticed that the patch sizes on high dielectric constant substrate are remarkably smaller than those on low dielectric constant substrate as shown in table 5, which is the main advantage of using high dielectric constant substrate. However, the antenna bandwidth (S_{11} < 10 dB) on 1 mm substrate height is decreased from 1.38% to 0.61% when the ε_r is increased from 2.2 to 10.2. Similar phenomenon is observed for the 2mm height, the bandwidth is decreased from 2.40% to 1.71%. For the same dielectric constant substrates, the antenna bandwidth is enhanced when the thickness is doubled. For example, the antenna bandwidth on the high dielectric constant substrate is increased from 0.61% to 1.71% when the substrate thickness is increased from 1 mm to 2 mm. It's important to point out that the bandwidth of example (4) is even larger than that of example (1), which means that the bandwidth of MPAs on high permittivity substrate can be recovered by increasing the substrate thickness.

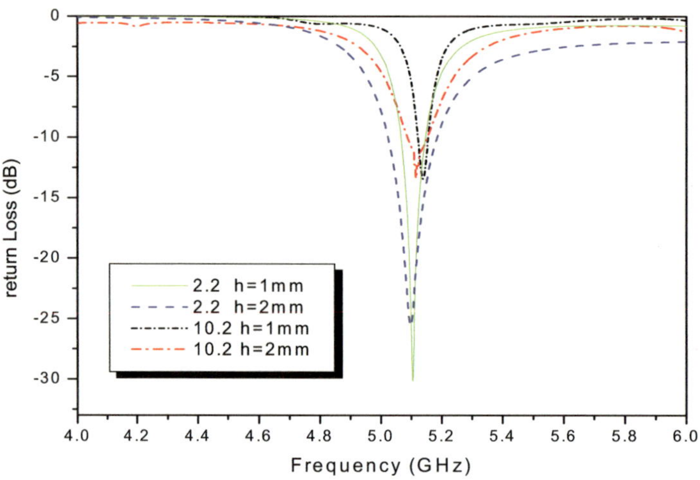

Fig. 28. Return loss comparison of patch antennas with different dielectric constants and substrate height.

Figure 29 compares the H-plane radiation patterns of these four antennas. A finite ground plane of $\lambda \times \lambda$ size is used in the simulations, where λ is the free space wavelength at 5.1 GHz. The antennas on the high dielectric constant substrates exhibit lower directivities and higher back radiation lobes than those on the low dielectric substrates. For antennas on the same dielectric constant substrate, when the thickness increases, the antenna directivity decreases, especially for those on high dielectric constant substrates. Similar observations are also found in the E-plane patterns.

These phenomena can be explained from the excitation of surface waves in the substrate. When a high dielectric constant and thick substrate is used, strong surface waves are excited. This causes reduction of the radiation efficiency and directivity. In addition, when the surface waves diffract at the edges of the ground plane, the back radiation is typically increases.

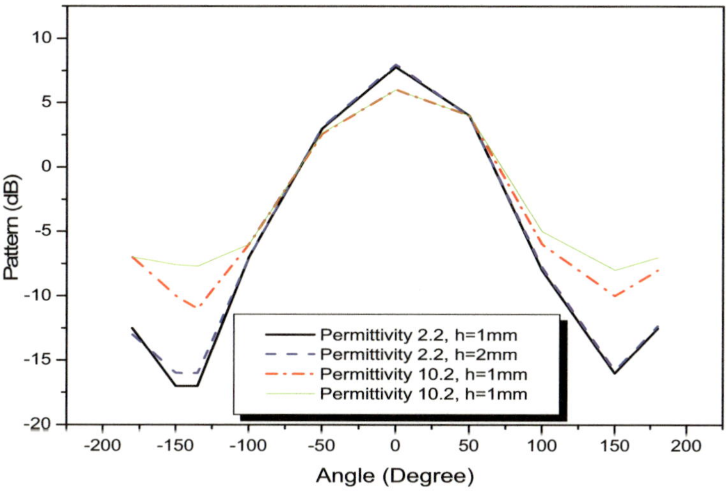

Fig. 29. H-Plane radiation pattern of patch antennas with different dielectric constants and substrate heights.

i. Gain Enhancement of a Single Patch Antenna

To overcome the drawbacks of using the thick and high dielectric constant substrate, several methods have been proposed to manipulate the antenna substrate. One approach suggested is to lower the effective dielectric constant of the substrate under the patch using micromachining techniques [94]. A shortcoming of this approach is the larger patch size than that on an unperturbed substrate. Another approach is to surround the patch with a complete band gap structure or synthesized low dielectric constant substrate so that the surface wave's impact can be reduced. A MPA design is proposed that does not excite surface waves. In this section, EBG structure is applied in patch antenna design to overcome the undesirable features of the high dielectric constant substrates while maintaining the desirable features of utilizing small antenna size.

ii. Patch Antenna Surrounded by EBG Structures

Figure 30 sketches the geometry of a MPA surrounded by a mushroom-like electromagnetic band gap (EBG) structure. The EBG is designed so that its surface wave band gap covers the antenna resonant frequency. As a result, the surface waves excited by the patch antenna are inhibited from propagation by the EBG structure. To effectively suppress the surface waves, four rows of EBG cells are used in the design. It is worthwhile to point out that the EBG cell is very compact because of the high dielectric constant and the thick substrate employed. Therefore, the ground plane size can remain small, such as $1\lambda \times 1\lambda$. For comparison, MPA designed on a step-like substrate is investigated, as shown in figure 31. The idea is to use a thick substrate under the patch to keep the antenna bandwidth and use a thin substrate around the patch which reduces the surface waves. The distance between the patch and the step needs to be carefully chosen. If the distance is too small, the resonant feature of the patch will change and the bandwidth will decrease. However, when the distance is too large, it cannot reduce the surface waves effectively. To validate the above design concepts, four antennas were simulated on RT/Duroid 6010 (ε_r = 10.2) substrate with a finite ground plane of 52 × 52 mm². Two of them are normal patch antennas built on 1.27 mm and 2.54 mm thick substrates as references. The step-like structure stacks two 1.27 mm thick substrates under the patch and the distance from the patch edge to the step is 10 mm. The EBG structure is built on 2.54 mm height substrate and the EBG patch size is 2.5 × 2.5 mm² with 0.5 mm separation. Figure 32 compares the measured S_{11} results of these four antennas. All the four patches are tuned to resonate at the same frequency 5.1 GHz. It is noticed that the patch on the thin substrate has the narrowest bandwidth of only 1% while the other three have similar bandwidths of about 3–4%. Thus, the thickness of the substrate under the patch is the main factor determining the impedance bandwidth of the antenna. The step substrate and the EBG structure, which are located away from the patch antenna, have less effect on the antenna bandwidth.

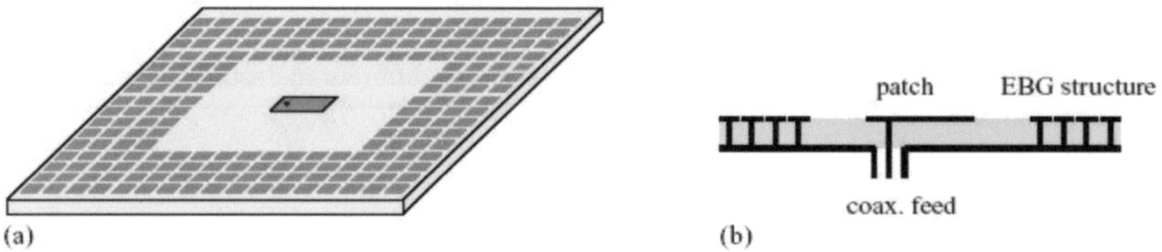

Fig. 30. Patch antenna surrounded by a mushroom-like EBG structure: (a) geometry and (b) cross section.

The antenna on the height 2.5mm has the lowest front radiation while its back radiation is the largest. When the substrate thickness is reduced, the surface waves become weaker and the radiation pattern improves. The step-like structure exhibits similar radiation performance as the antenna on the thin substrate. The best radiation performance is achieved by the EBG antenna structure. Due to successful suppression of surface waves, its front radiation is the highest, which is about 3.2 dB higher than the thick case. Since the surface wave diffraction at the edges of the ground plane is suppressed, the EBG antenna has a very low back lobe, which is more than 10 dB lower than other cases. Table 6 lists the simulated results of these antennas. Note that the radiation patterns are normalized to the maximum value of the EBG antenna.

Antenna	Bandwidth %	Front Radiation (dB)	Back Radiation (dB)
Thin	1	-2.3	-15.5
Thick	4	-3.2	-12
Step stair	4.7	-2	-14
EBG	3	0	-25

Table 6. Simulated performance of four different MPA designs on the high dielectric constant substrate.

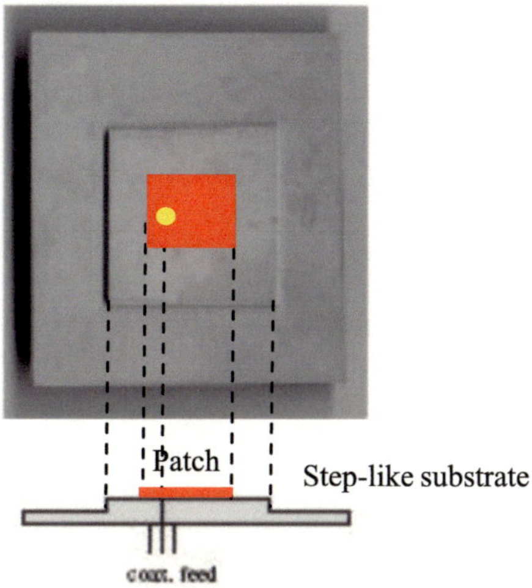

Fig. 31. Patch antenna on a step-like substrate.

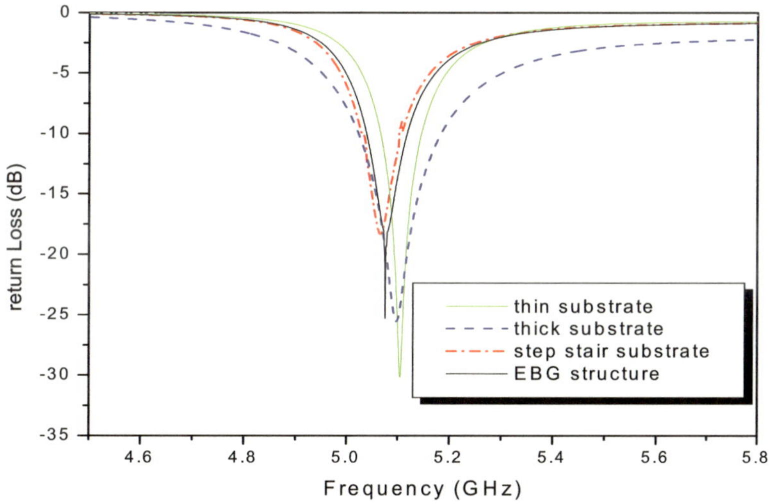

Fig. 32. Comparison of the measured return loss of the four MPA structures.

It is also interesting to notice that in the E-plane the beam-width of the EBG case is much narrower than the other three cases whereas in the H-plane it is similar to other designs. The reason is that the surface waves are mainly propagating along the E-plane as shown in figure 33. Once the EBG structure stops the surface wave propagation, the beam becomes much narrower in the E-plane. From above comparisons it is clear that the EBG structure improves the radiation performances of the patch antenna while maintaining its compact size and adequate bandwidth.

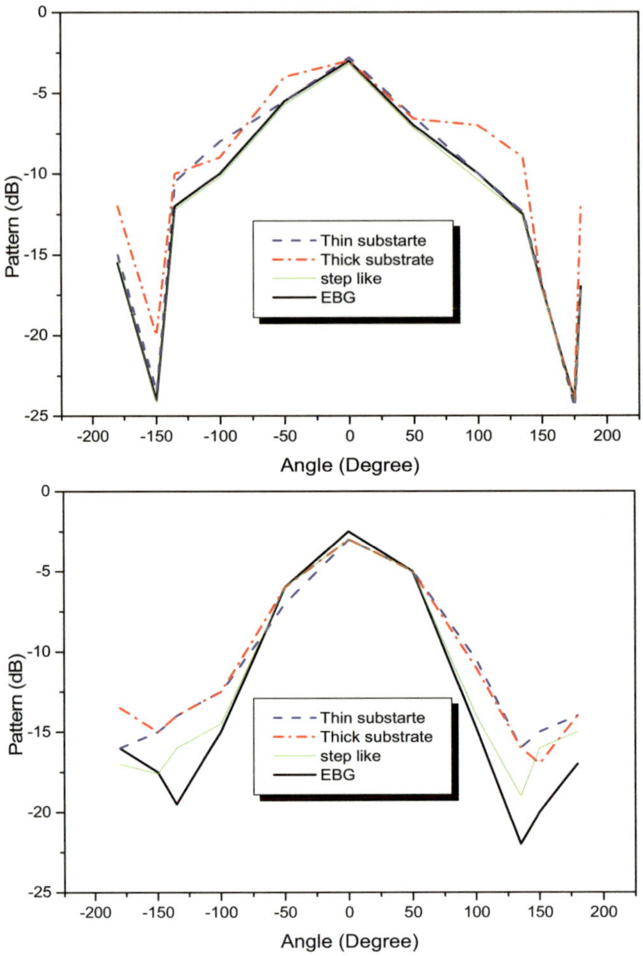

Fig. 33. Simulated radiation patterns of different patch antennas: (a) E-plane and (b) H-plane pattern.

10. Antenna design for ultra-wideand wireless communications applications

10.1 Enhancement of microstrip monopole antenna bandwidth by using EBG structures

The low profile, light weight, and low cost of manufacturing of monopole microstrip patch antennas have made them attractive candidates in many applications ranging from very high-data rate and short-range wireless communication systems, to modern radar systems. The limited bandwidth of the MPA, however, needs to be further improved to facilitate applications in UWB systems. Spiral arms shaped metallo EBG was used to increase the bandwidth of planar monopole antennas. This approach resulted in size reduction and cceptable performance from 1 to 35GHz. In this section, we present a new approach based on using a variety of shapes and sizes of embedded EBG structures. The designs of the EBG structures as well as the improvement steps in the antenna performance are described in the following sections [89].

i.Antenna Geometry and Design A low profile microstrip monopole patch antenna is proposed as shown in figure 34. Rectangular and circular geometries were simulated to optimize the performance.

Starting with a rectangular patch antenna on a rectangular ground plane and ended up with an umbrella semi-circular shape, for both the radiator and the ground plane. A semi-circular ground plan with radius R_g =15mm was used as shown in figure 34. As a last step, circular shapes for both the radiator and the ground plane with radii of R_r =12mm for the antenna and R_g=15mm for the ground plane are used as shown in figure 34. The microstrip feed line length is L_f =16mm, and width W_f =1.9mm. The antenna is printed on FR4 substrate with ε_r=4.7 and thickness 3.2mm (0.034λ_o). After optimizing the basic antenna design with circular radiator and ground plane, various designs of EBG structures are embedded to improve the bandwidth, enhance gain, and reduce the antenna size beyond that was previously reported. The proposed metallic patch antenna layout was first surrounded by an EBG lattice with circular holes with dimension computed as described in [51], hence is given by $d = \lambda_g$ where **d** is the diameter of drilled hole and $\boldsymbol{\lambda_g}$ is the guided wavelength. Then embedded circular and square patches in a periodical electromagnetic band-gap structure were used as shown in figure 34, and the side view is shown in figure 34. The periodic patches have radius=3mm for the circular patch and, half-square side =3mm for the square patch. The periodicity in the two cases is P=7mm, and the vias radius equals 0.25mm. These dimensions were used based on optimized simulation results. The optimization process includes embedded circular and square patches dimensions as well as antenna parameters such as gain, bandwidth, and radiation efficiency. The fabricated antenna is shown in figure 34.

ii. Simulation and Experimental Results

The simulation results are shown in figure 35 (a). Then partial rectangular ground plane is used with dimensions 30x15 mm^2. Results for this case are also shown in figure 35 (a). The partial rectangular ground plane is then converted to half circular plane to increase the bandwidth by creating adjacent staggered resonance modes. Both the radiator and the ground plane shapes are then converted to half circular shapes with radii of 12mm and 15mm, respectively. The final design for the basic antenna shows ultra wideband characteristics shown in figure 35 (b) but with bandwidth discontinuity from 7 to 10GHz and 12.5 to 17.5GHz. To remove these discontinuities, EBG structures are embedded. The effect of EBG on the antenna performance is examined first by drilling cylindrical holes each with a radius equals to 3mm and periodicity P=7mm. Then circular and square EBG patches are embedded in the substrate. Simulation results to illustrate the performance of each of the EBG structures (circular or square) as well as the case of the drilled holes are shown in figure 37. These results were obtained by calculating the transmission coefficient of a 50 Ω microstrip transmission line placed on the top of a substrate with either square or circular EBG structure or circular holes. From figure 37, it is seen that each of these embedded structures has different effect on the bandwidth of the microstrip line. The pass-band and stop-band for the embedded square are larger and deeper than that of circular embedded EBG. Embedded circular patch of radius 3mm and square patch of side length d=6mm EBG structures with periodicity P=7mm are combined with the half circular-monopole antenna as illustrated in figure 45(a) and 45(b), respectively. The comparison between simulation and measurement of reflection coefficient S_{11} of both embedded EBG are shown in figures 37 and 38, respectively. From these figures it is noted that not only improvement in the bandwidth was achieved but also the antenna size was reduced to about 34% from that of the basic half circular monopole antenna without EBG. This gives more than 60% size reduction higher than that published in [92] by 10%. The final antenna with square embedded patches has average antenna gain of about 6.5dBi. The gain value changes throughout the operating band as shown in figure 38. At lower frequencies, the gain is about 3dBi and started to increase with frequency until it reaches its maximum value of 8.5dBi at 12.5GHz. After this, it starts to decrease again with frequency. However, the antenna gain with EBG is larger than without EBG throughout the entire frequency band by an average value of about 2dBi. The antenna radiation efficiency has the highest value from 1 to 7.5GHz and then it stays between 60-70% up to 35GHz. The final antenna design has an average efficiency of 73.5% as compared to only 56.5% for the basic antenna without embedding EBG structures.

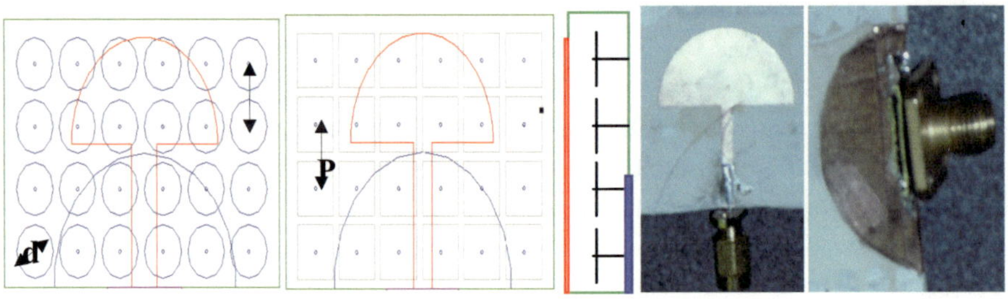

a b c d e

Fig. 34. The proposed antenna with embedded EBG (a) Circular EBG, (b) Square EBG (c) side view (d) fabricated radiator monopole antenna, and (e) ground plane on back of the substrate.

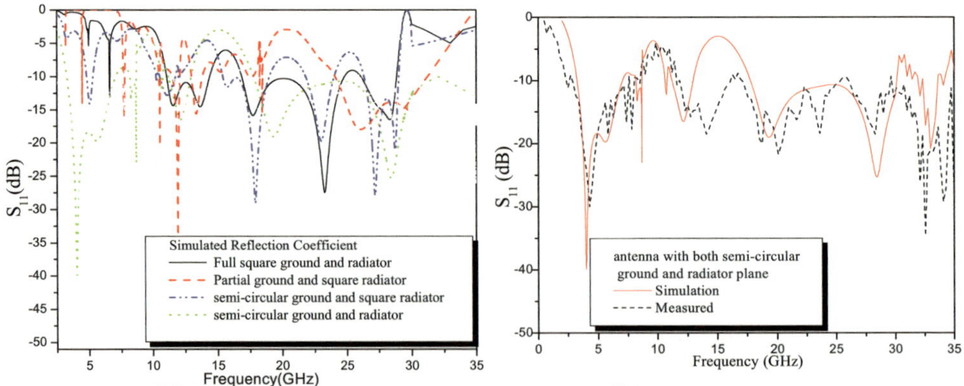

(a) (b)

Fig. 35. (a) Simulated S_{11} vs. frequency for various design parameters and (b) comparison between simulation and measurement for the proposed.

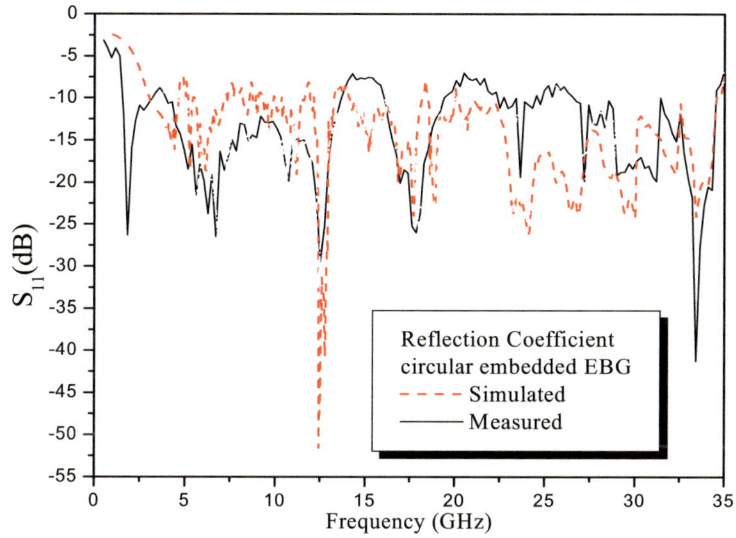

Fig. 36. Comparison between measured and simulated reflection coefficient for umbrella monopole antenna with embedded circular EBG patches.

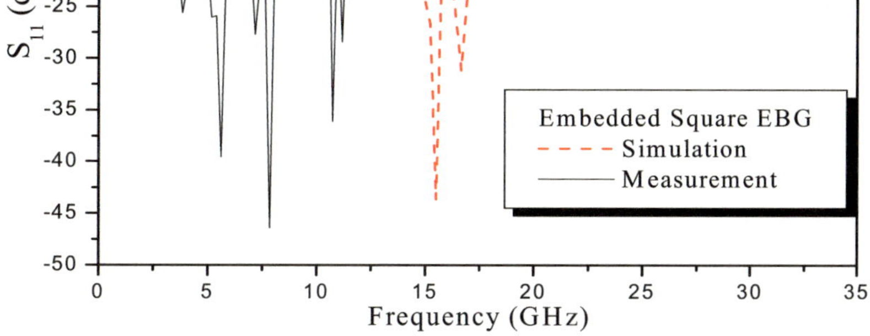

Fig. 37. Comparison between measured and simulated reflection coefficients of umbrella monopole antenna with embedded square EBG patches.

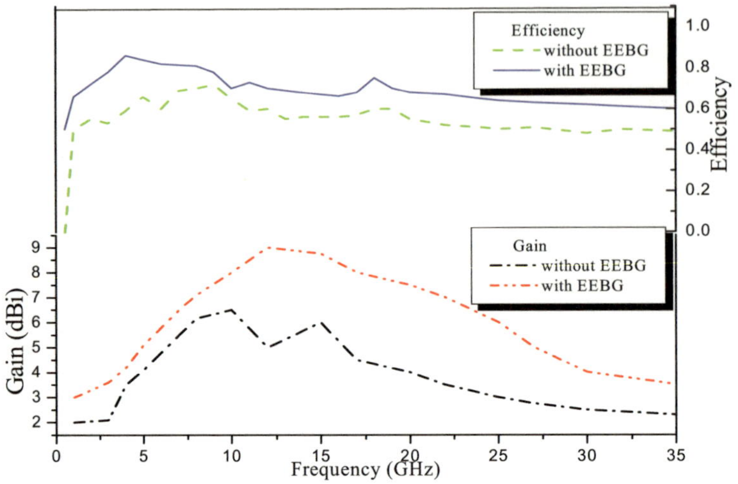

Fig. 38. Antenna gain and radiation efficiency vs. frequency for antennas with and without square EEBG. EBG structure

10.2 Ultra-wide bandwidth umbrella shaped MMPA using Spiral Artificial Magnetic Conductor (SAMC)

i. Proposed Antenna Geometry and Design

The proposed antenna has semi-circular shape. The thickness of the substrate used is approximately $0.034\lambda_{3.3GHz}$ = (3.2 mm). The semi-circular patch of radius R_r mm, is placed on one side of dielectric substrate with relative permittivity 4.7 and tan δ =0.02. The dimension for the substrate is $L_s \times W_s$ mm^2. The antenna is fed by 50Ω microstrip feed-line of width 'W_f'. The semi-circular radiator is placed 'L_f' distance from one edge of the substrate. The design of the proposed antenna started from conventional shape of printed rectangular microstrip monopole. To improve the bandwidth, the radiator was modified to be semi-circular patch with radius 12mm. The antenna size is thus reduced by 22% from the original size. The dimension of FR4 substrate is 40x40 mm^2 but in this case discontinuities in bandwidth were found between frequency band from 1GHz to 5GHz and from 6GHz to 10GHz as shown in following sections.

ii. Spiral AMC Design

In this section, two typical printed spiral geometries are investigated as shown in figure 39. The operation principle of the AMC surface can be simply explained by an equivalent LC circuit theory. To increase the value of the inductance, a single spiral is placed on top of the grounded substrate to replace the conventional ground plane. The parameters of the substrate remain the same as the reference conventional monopole. The width of the spiral is 1mm=$0.011\lambda_{3.3GHz}$ with gap=1mm. Two shapes of spiral AMC are used in the proposed design, one arm spiral as in figure 40(a) and four arms spiral as in figure 40(b). The two spiral shapes are applied on the antenna ground plane as shown in figure 40 (a) and 40(b) to improve the performance such as decreasing the antenna size, reducing the bandwidth discontinuities and increasing the antenna gain. After achieving the best possible results from this approach, holes were drilled to further improve the bandwidth and enhance antenna gain as shown in figure 40(c). The design of the drilled holes EBG structure is straight forward, hence, d=λ_g where *d* is the diameter of drilled holes and λ_g is the guided wavelength. d=1mm and a periodicity =3mm were chosen and the fabricated antenna is shown in figure 41(a) and 41(b). The diameter of the drilled holes *d* is chosen to be sufficiently applied at the desired discontinuities frequencies from 6 to 10GHz. This makes the presence of the holes too small to affect the dielectric constant of the substrate.

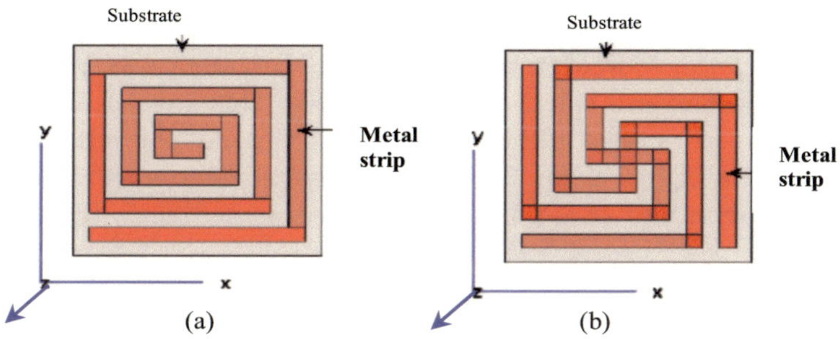

Fig. 39. The spiral shape as AMC; (a) one arm, (b) four arms.

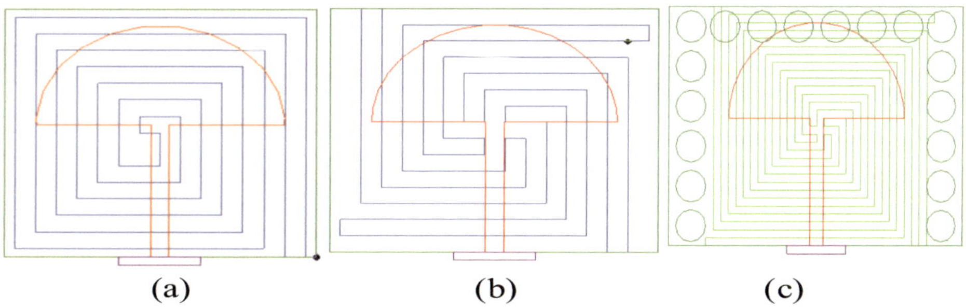

Fig. 40. Umbrella shape with spiral ground plane (a) one arm (b) four arms and (c) four arms with surrounded holes.

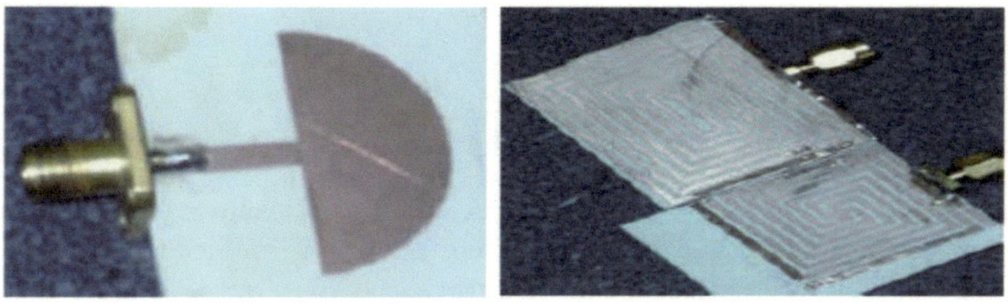

(a) Antenna radiator plane (b) Spiral ground plane

Fig. 41. Fabricated umbrella shape antenna with spiral ground plane.

iii. Simulation and Measurement results

In order to provide design criteria for the proposed antenna, the effects of each developing geometrical shape are studied. We started with rectangular plates for both the radiator and the ground plane. The radiator dimensions are (24x12) mm^2 and the ground plane has the dimensions of (40x40) mm^2. The simulation results are shown in figure 42. Second attempt involved converting the rectangular radiator to a semi-circular plate with radius =12mm as shown in figure 42 (same steps as in the previous sections). The electromagnetic AMC band-gap structure was used to enhance the bandwidth, increase the antenna gain and reduce the antenna size. Both the transmission line approach and reflection phase methods were used to study and redesign the performance of spiral shapes. First, the reflection coefficient and transmission coefficients S_{11} and S_{21} were calculated as shown in figure 43 for the two types of spiral. Figure 43 indicates that the performance of the four arms spiral ground plane is better than one arm spiral. Moreover, a large cross polarization was observed in many frequency bands in the one arm spiral one performance. This cross polarization resulted from the asymmetric geometry of the one arm spiral. Second, the reflection angle was calculated as shown in figure 44 for two types of spiral. Figure 44 indicates that the performance of spiral with four arms ground plane gives larger bandwidth and near from zero reflection phase.

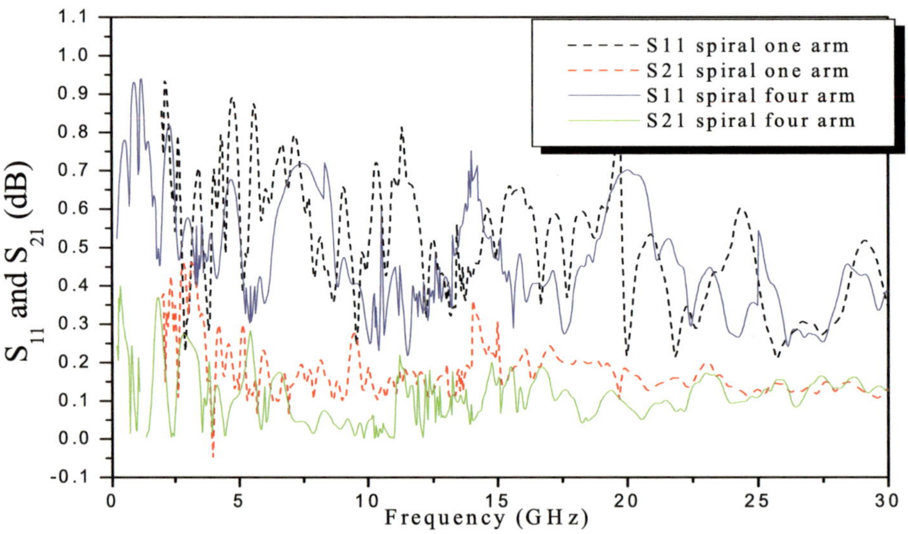

Fig. 42. Simulated results of S_{11} and S_{21} of spiral ground plane with one and four arms.

Then each of the two types of spirals was integrated as an antenna ground plane. The comparison between simulated and measured reflection coefficient for the antenna integrated with SAMC ground plane with one arm is shown in figure 45. A large frequency discontinuity was also observed in this case in the frequency band from 1 GHz up to 40 GHz. Although, the results as shown in figure 46 seem to meet the desired requirement in terms of surface compactness, the resulting high cross polarization makes it unacceptable in many applications. Due to the significant cross polarization, the single arm spiral geometry was not a good candidate for applications that require low cross polarization. Therefore, an alternative design that consists of four arms spiral is used [91].

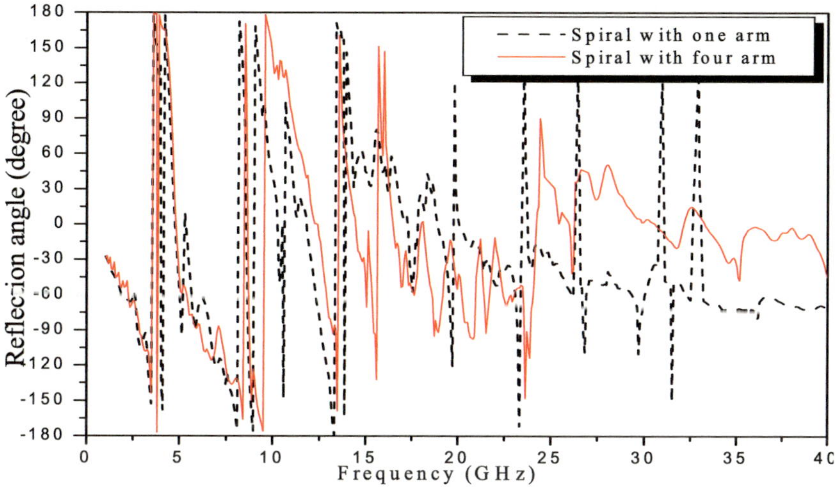

Fig. 43. The S_{11} phase of spiral AMC with one and four arms.

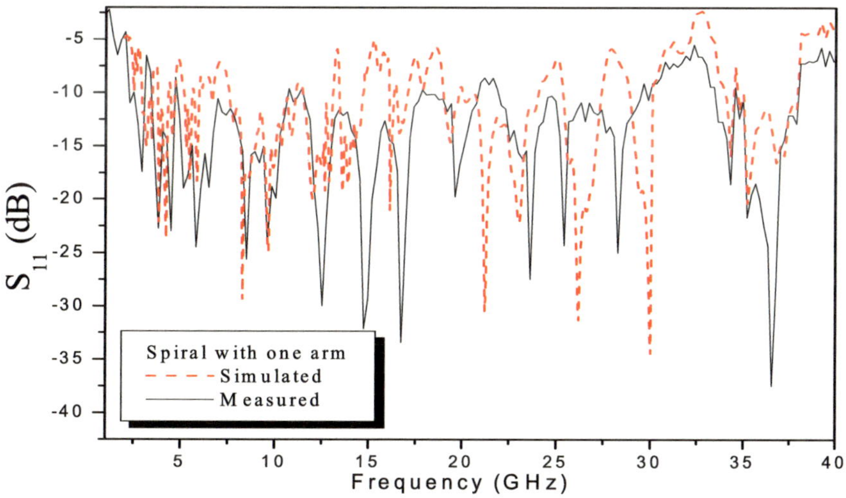

Fig. 44. Comparison between simulated and measured S_{11} of proposed antenna with one arm spiral ground plane.

The four arms spiral is then applied on an antenna ground plane. Four spiral branches, each with a 1mm $=0.011\lambda_{3.3GHz}$ width, split from the center and rotate in clockwise direction. The antenna reflection coefficient is shown in figure 46. Simulation results show that there is small discontinuity in operated antenna bandwidth between frequencies from to 0.1GHz to

1GHz, from 5.5GHz to 6.5 GHz, from 7GHz to 7.5GHz, 13.5GHz to14.5GHz In this case, the resonant frequency decreases as the number of spiral arms increases. The first resonant frequency is 49.45% lower than the reference geometry. This significant reduction in size for a single element leads to an attractive design feature for many wireless communication applications. Finally, both electromagnetic band-gap (EBG) drilled holes and artificial magnetic conductor (AMC) were merged to optimize the antenna performance as is shown in figure 42(c).

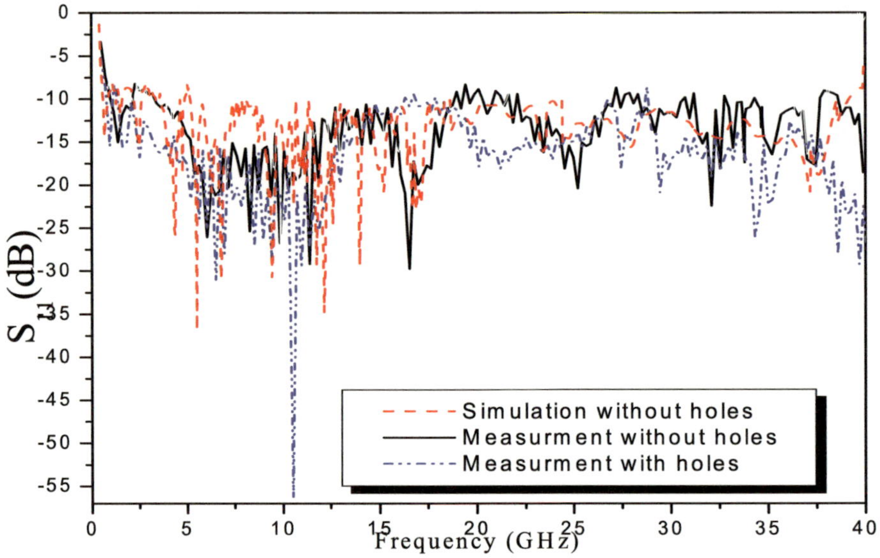

Fig. 45. The comparison between simulated and measured S_{11} of the proposed antenna with four arms spiral ground plane.

10.3 Design MMPA by using printed unequal arms V-shaped slot

i. Design Considerations

The printed monopole antennas give very large impedance bandwidth with reasonably good radiation pattern in azimuth plane, which can be explained in two ways. To estimate the lower band-edge frequency of printed monopole antennas, the standard formulation given for monopole antenna can be used with suitable modification. The equation was worked out for the planar monopole antennas. If h is the height of substrate in cm, which is taken the same as that of an equivalent monopole and L_L in cm is the highest effective length of the V- unequal arms monopole antenna and L_f is the length of 50Ω feed line in cm, then the lower band-edge frequency is given as shown in figure 47:

$$f_L = c/\lambda = \frac{7.2}{(h + L_L + L_f)\sqrt{\varepsilon_r}} \text{GHz} \tag{4}$$

Where all dimensions h, L_L and L_f in cm.

ii. Antenna Design Methodology and Geometry

A V-shaped slot monopole with two unequal arms is proposed to achieve compact size and ultra wideband design. For the lower band (0.75GHz), two different-length arms of a V-shaped patch were used to excite two closely staggered resonant modes [84-86] as shown in figure 47. The two patches (triangular shaped patch and the electromagnetically coupled V shape are excited in the TM_{01} mode. The design achieves ultra bandwidth in lower band with antenna thickness less than $0.01\lambda_o$. Multi frequency operation is achieved by etching V slot with different spacing between feeding plate and triangular patch. It is placed on one side of dielectric substrate RT/D 6010 with relative permittivity 10.2 and tan δ =0.002. The dimension for the substrate is $L_s \times W_s$ =50x50mm². The lowest frequency of operation is f_L=0.75GHz. Monopole antenna passes by many steps to reach the final novel proposed shapes. The design of the proposed antenna started from conventional shape of microstrip monopole antenna with square substrate equal to 50 mm using rectangular ground plane and triangular plate for antenna radiator. The dimensions are optimized by choosing a 50 Ω feeding of W_f =2.5mm and L_f=16.5mm. The main objective of optimization is producing broadband antenna. The second step to improve the antenna bandwidth is modifying the ground plane to partial rectangular ground plane, the ground plane dimensions are 17.5x50 mm². Third step is using V-shaped slot with unequal arms for radiator plate as shown in figure 47. Fourth step is modifying the ground plane and optimization is used to enhance the antenna bandwidth and reduce the electrical antenna size. The triangular radiator patch with base and height W_p=17mm and L_p=15mm with arms width W_L=12mm, W_s=4mm and arms length L_L=36mm, L_s=26mm and air gap separation between triangular patch and V shaped arms g_L=2mm, g_s=3.5mm with feeding length 17.7mm. The ground plane with L_t=12mm and radius 14.4mm.

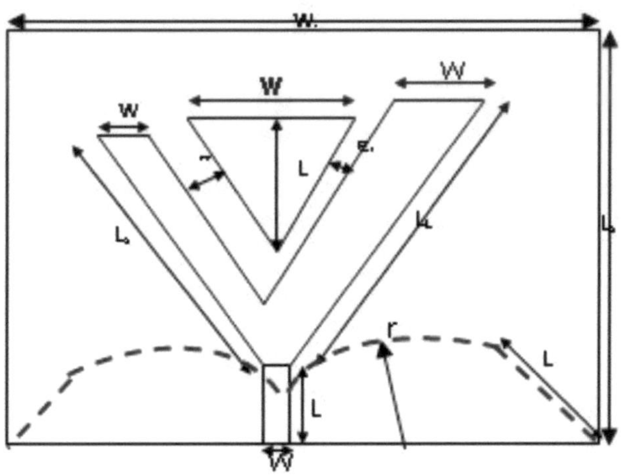

Fig. 46. The proposed antenna geometry.

$$2a = \lambda_{guide} \quad (5)$$

$$a / p = 0.8 \sim 0.9 \quad (6)$$

where a, P and λ_{guide} are side length, periodicity, and guided wavelength, respectively.

Finally, the proposed antenna is printed on metamaterial surface, first shape is using artificial magnetic conductor (AMC). As embedded spiral AMC with four arms is used to improve the impedance matching and reduce the antenna size as shown in figure 48(a) with arm width g = air gap = 2mm at height h=1.25mm. Secondly, using embedded uni-planar square EBG without via and with square side length 5mm for further improvement in antenna gain and bandwidth discontinuity as shown in figure 48(b) with side length a=5mm and periodicity P=5.5mm at height h=1.25mm. Finally, using electromagnetic band-gap structure (EBG) as embedded square EBG with square side length 4mm, periodicity 5mm and via radius =0.25mm at the same previous height as shown in figure 48(c) to enhance the antenna efficiency and gain. These dimensions were selected by recomputation in a conventional way. The EBG substrate does not interfere with the near field of the antenna, and it suppresses the surface waves, which are not included in the patch antenna design [84-86]. The design of an EBG antenna has been straightforward as shown in Eqs. 5 and 6.

iii. EBG Methodologies and Geometry

The application of EBG in printed antenna design has received significant attention recently. They have mainly been used to achieve microstrip antennas on thick, high-dielectric constant substrate with optimum performance. The approach is based on using artificial substrates made of periodic metalo-dielectric resonant implants in order to have a complete forbidden band gap around the desired antenna operative frequency. Because surface waves cannot propagate along the substrate, an increased amount of radiated power couples to space waves reducing antenna losses while increasing its gain and bandwidth. In this section, a compact AMC design that minimizes the cross polarization effect of printed spiral geometry will be presented. As the number of arms or turns order increases, the equivalent inductance increases, resulting in a lower resonant frequency. Several typical printed spiral geometries are investigated in this section and their reflection phase characteristics are reported. Due to the significant cross polarization, the single and double spiral geometries are not good candidates for applications requiring low cross polarization.

A four-arm spiral is explored to eliminate the cross polarization, as shown in figure 48(a). Each arm is rotated 90°. Therefore, this symmetrical condition guarantees the same scattering response to the x- and y-polarized incident waves. As a result, no cross polarization is observed from this structure. Second, using two dimension embedded EBG without via simplifies the fabrication process and is compatible with microwave and millimeter wave circuits. The embedded EBG surface and the AMC surface each has his own advantages. Third, a mettalo 3D-EBG with via is used: Advantages of the 3D-EBG surface is obtaining: a lower frequency and a wider bandwidth. At a given frequency, its size is smaller than the AMC design. Advantages of the uni-planar surface are obtaining a lower frequency and wider bandwidth. In addition, it is less sensitive to the incident angle and polarization.

iv. Results and Discussion

The antenna performance was investigated by both simulation and measurement as shown in figure 49. In order to provide design criteria for the proposed antenna, the effects of each developing geometrical shape are analyzed.

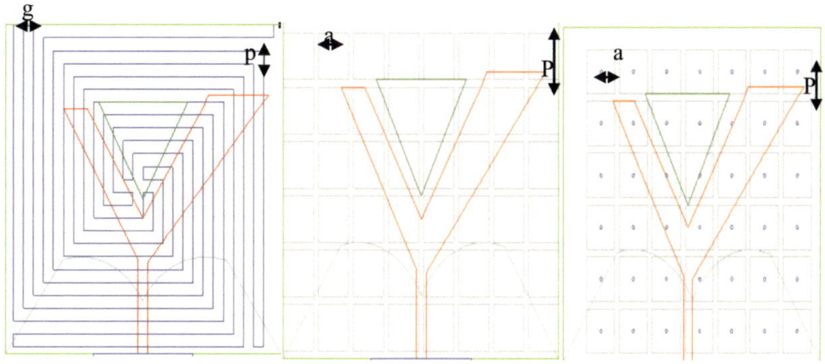

(a) (b) (c)

Fig. 47. The configuration of three EBG techniques.

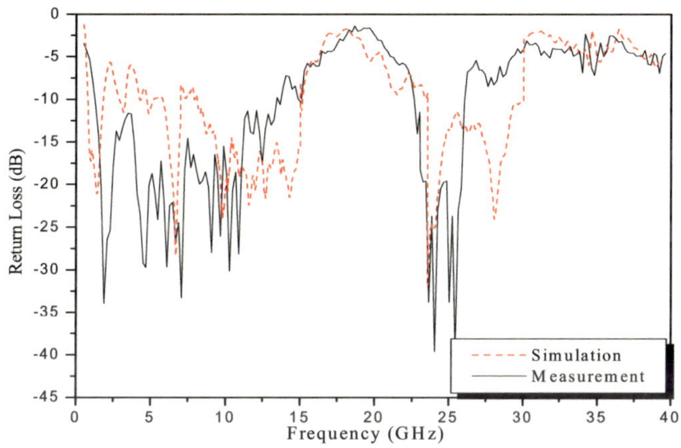

Fig. 48. Comparison between simulated and measured reflection coefficient.

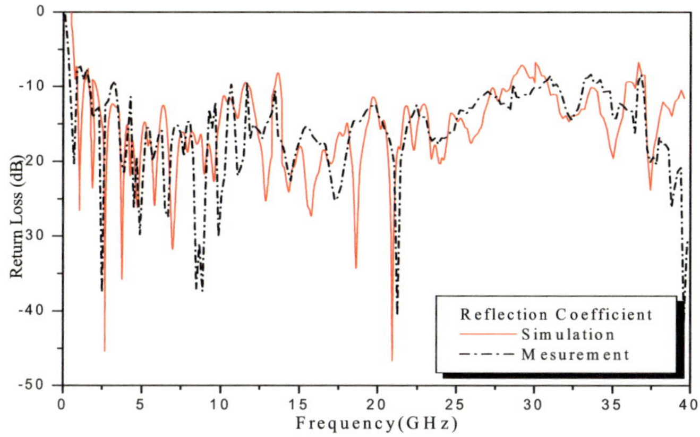

Fig. 49. Comparison between simulated and measured reflection coefficient for the proposed antenna with embedded spiral AMC

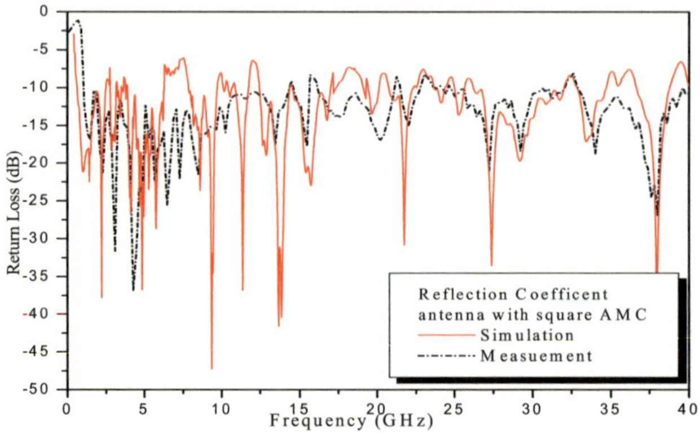

Fig. 50. Comparison between simulated and measured reflection coefficient of the proposed antenna with embedded 2D-EBG.

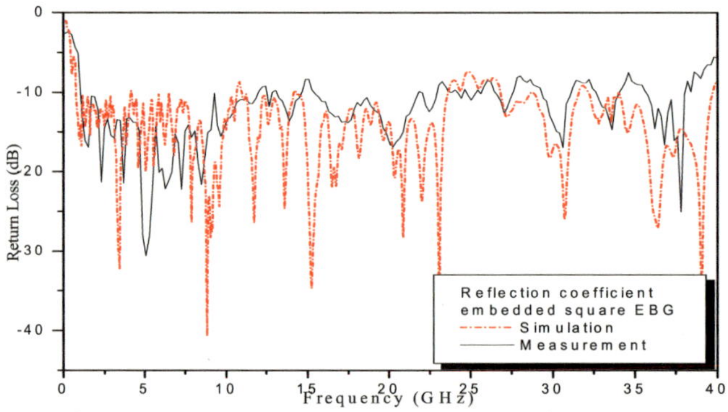

Fig. 51. Comparison between simulated and measured reflection coefficient of antenna with embedded 3D-EBG.

Metamaterials structures are used to enhance the antenna performance as bandwidth, gain and improve the reflection phase. Starting with embedded spiral AMC, the comparison between simulation and measurement of reflection coefficient is shown in figure 50, and there are bandwidth discontinuity found at frequencies from 2 to 2.5GHz, 12 to 12.5GHz, 14 to 14.5GHz and 27 to 32GHz. However, the electrical antenna size reduced to 0.6GHz with reduction 65% from original size of the proposed antenna. Secondly, embedded square uni-planar 2D-EBG were used. The comparison between simulation and measurement is shown in figure 51 and the bandwidth discontinuities occur at frequencies from 5 to 8 GHz, from 12 to 13, 17 to 18.5GHz and from 32 to 34.5GHz. Thirdly, embedded 3D-EBG is used for further improvement in antenna gain and bandwidth. The bandwidth discontinuities occur at frequencies from 11 to 12GHz and from 26 to 26.5GHz as shown in figure 52. It may be noted that not only improvement in the bandwidth is achieved but also the antenna size was reduced to about 65% from that of the proposed monopole antenna without EBG. From these figures, one can notice that there are small discrepancies between the simulated and measured results. This may be attributed due to the same reasons stated before. Antenna gain for three antenna structures are shown in figure 53. This figure shows that, the antenna with embedded 3D-EBG has the antenna gain response along the operating band with average gain of about of 13dBi followed by 2D-EBG with average gain 11dBi and spiral AMC with 9dBi. The fabricated antenna is shown in figure 54.

10.4 Co-planar boat MPA with modified ground plane by using EBG

i. Antenna Geometry

The geometry of the proposed antenna is shown in figure 55, where an equi-lateral triangular patch with L_r=70mm is placed co-planar to a finite ground plane that has a trapezoidal shape with size of $\mathbf{W_{g1}}$ =30mm, $\mathbf{W_{g2}}$=63mm, hieght $\mathbf{L_g}$=21mm. The dielectric substrate used is FR4 with dielectric constant ε_r=4.7 and dimension 100x100mm^2 with thickness **h** = 3.2mm. The patch is proximity fed by a 50Ω microstrip line at the fundamental frequency 3.3GHz with line length and width of $\mathbf{L_f}$ = 63mm and $\mathbf{W_f}$ = 5.5 mm, respectively. The top and side views of the proposed antenna are shown in figure 55. To obtain a good impedance matching, the end of the feed line has to extend beyond the centre of the patch. Initially, different dimensions of the MPA were used in order to minimize the size of the patch antenna with maximizing the bandwidth at the same time. The antenna geometry of the whole structure should be optimized (the ground plane dimension, separation between the patch and the ground and feed line position) to obtain the best possible impedance bandwidth [87].

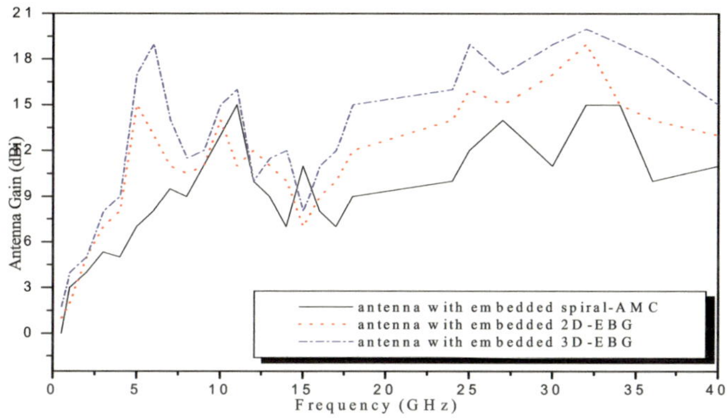

Fig. 52. Comparison of gain for the studied EBG structures.

Second part of this section, is using 2D-EBG etched in the feeding line to improve impedance matching with head square dimension a= 2mm, slot length $\mathbf{L_d}$=1.25mm with width g=0.5mm and periodicity **P**=4mm. An etched 2D-EBG in 50Ω feed line disturbs the shield current distribution in the feed line. This disturbance can change the characteristics of the transmission line since it increases the effective capacitance and inductance of a transmission line, respectively.

Finally, using the four arms spiral AMC to reduce the antenna size by adding inductance component. A larger equivalent inductance may be realized with a larger number of spiral turns. However, as revealed in this chapter, if the unit geometry is not symmetric with respect to the polarizations of the incident waves, the AMC surface generates a high level of cross polarization. Thus, the behavior of reflection phase may not be applicable in the designated frequency band of operation. Using four arms spiral shape, and each arm rotates 90° can exactly recover itself. Therefore, this symmetrical condition guarantees the same scattering response to the x- and y-polarized incident waves. As a result, no cross polarization was observed from this structure. Therefore, with this design the compactness of geometry is achieved without generating the cross polarization level. The dimension of the spiral arm width $\mathbf{W_d}$ is equal to separation between arms =5mm (0.05λ_g) and the largest spiral length $\mathbf{L_s}$= 80mm as shown in figure 55(b).

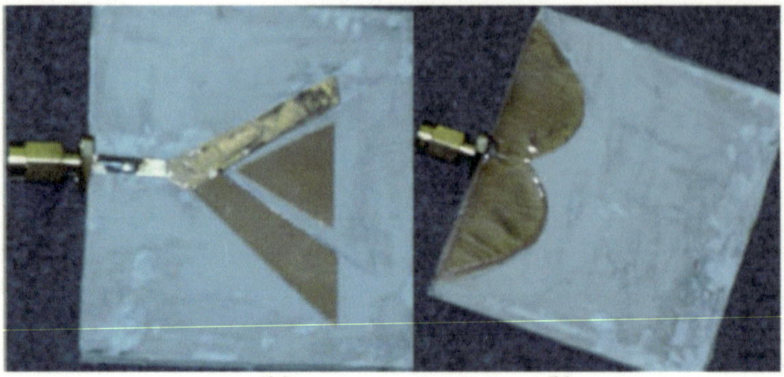

Fig. 53. The fabricated antenna (a) radiator and (b) ground plane

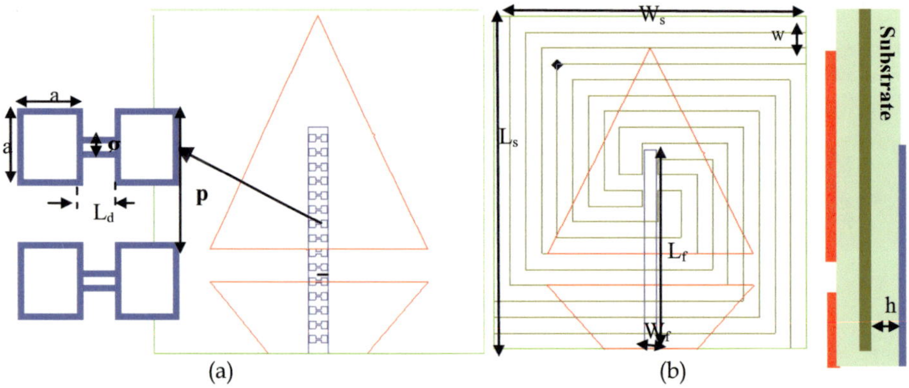

Fig. 54. Top and side view of the proposed antenna., (a) the proposed antenna with EBG elevation and (bc) side view.

ii. Simulated and Measured Results

The antenna performance was investigated by using simulations and verified experimentally by fabricating the antenna using photolithographic techniques. In order to provide design criteria for the proposed antenna, the effects of each developing geometrical dimensions were analyzed. The effect of feeding line length was studied. There is an optimum value of the feed length which is 63mm. The simulated return loss with various ground plane width, $\mathbf{W_{g1}}$ studied and it is found that larger width gives a broader bandwidth as well as a lower return loss magnitude. The proposed antenna is sensitive to $\mathbf{L_g}$ and in fact broadband performance is obtained for $\mathbf{L_g}$= 21 mm. It is known that in proximity fed patch antennas the position of the feed line under patch is important. The simulated return loss with various separations between the triangular radiator and the trapezoidal ground $\mathbf{W_{gap}}$ plane is also studied as shown in figure 56 where it can be seen that larger gap width gives a lower bandwidth as well as a lower return loss magnitude. However, there is an optimum separation $\mathbf{W_{gap}}$ which gives good reflection coefficient and bandwidth as shown in figure 57. Figure 58 shows a comparison between simulated and measured reflection coefficient of the proposed antenna with optimum dimensions. Figure 58 shows comparison between simulated and measured reflection coefficient. There is an improvement in bandwidth from 1GHz to about 40GHz with discontinuity band in the operating region. An embedded four arms spiral AMC was added at height h= 1.6mm from the ground plane to decrease the bandwidth discontinuity and cross polarization as well as increasing the antenna gain.

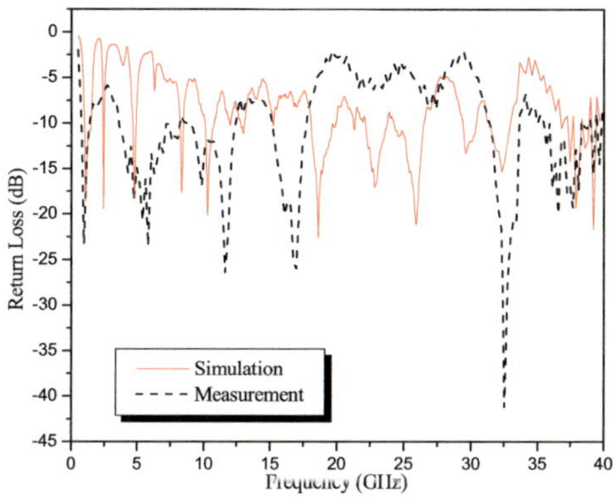

Fig. 55. Comparison between simulated and measured reflection coefficient of the antenna without EBG

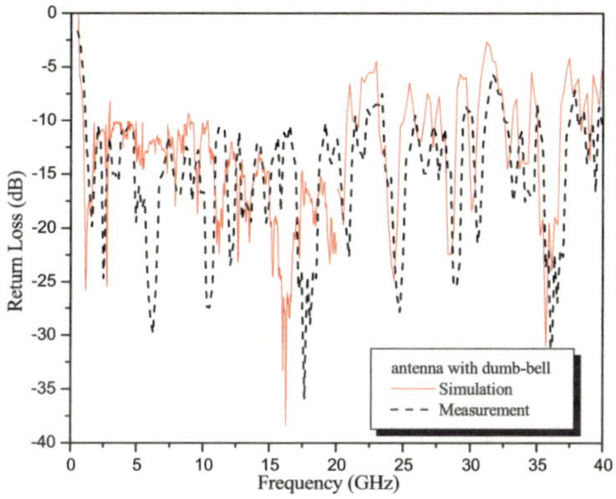

Fig. 56. Comparison between simulated and measured return loss for the antenna with dumb-bell

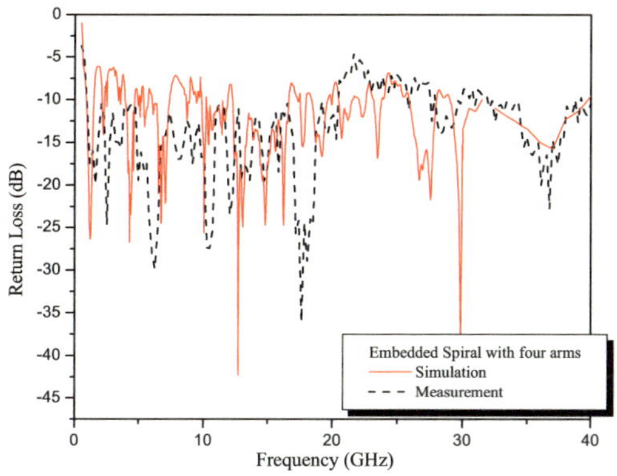

Fig. 57. Comparison between simulated and measured return loss for the proposed antenna with embedded SAMC under the antenna radiator.

11. Design of microstrip patch antenna arrays with EBG

After discussing EBG applications in single microstrip patch antennas, we now present how EBG can help to improve the performance of MPAA. The mutual coupling between array elements as an important parameter in array design. Strong mutual coupling could reduce the array efficiency and cause the scan blindness in phased array systems. [93] Therefore, the electromagnetic band gap (EBG) structures are used to reduce the coupling between array elements.

There are diverse forms of EBG structures [97], and novel designs such as EBG structures integrated with active device and multilayer EBG structures have been proposed recently. This section focuses on different types of EBG structures. Its band-gap features are revealed for two reasons, suppression of surface-wave propagation, and in-phase reflection coefficient. The feature of surface-wave suppression helps to improve antenna's performance such as increasing the antenna gain and reducing back radiation [92]. Meanwhile, the in-phase reflection feature leads to low profile antenna designs [94]. This section concentrates on the surface-wave suppression effect of the EBG structure and its application to reduce the mutual coupling of MPAA. To explore the surface-wave suppression effect, the propagating fields of an infinitesimal dipole source with and without the EBG structure are simulated and a frequency stop-band for the field propagation is identified. Furthermore, the propagating near fields at frequencies inside and outside the band gap is graphically presented for a clear understanding of the physics of the EBG structure. It is worthwhile to point out that this band-gap study is closely associated with specific antenna applications such as MPAA. Applications of MPAA on high dielectric constant substrates are of special interest due to their compact size and conformability with the monolithic microwave integrated circuit (MMIC). However, the utilization of a high dielectric constant substrate has some drawbacks. Among these are narrower bandwidths and pronounced surface waves. The bandwidth can be recovered using a thick substrate, yet this excites severe surface waves. The generation of surface waves decreases the antenna efficiency and degrades the antenna pattern. Furthermore, it increases the mutual coupling of the antenna array which causes the blind angle of a scanning array. Several methods have been proposed to reduce the effects of surface waves. One suggested approach is the synthesized substrate that lowers the effective dielectric constant of the substrate either under or around the patch. Another approach is to use a reduced surface wave patch antenna. The EBG structures are also used to improve the antenna performance. However, most researchers only study the EBG effects on a MPA element, and to the best of our knowledge there are no comprehensive results reported for antenna arrays. The mutual coupling of MPAA is parametrically investigated, including the E- and H-coupling directions, different substrate thickness, and various dielectric constants. In both coupling directions, increasing the substrate thickness will increase the mutual coupling. However, the effect of the dielectric constant on mutual coupling is different at various coupling directions. It is found that for the E-plane coupling is stronger on a high permittivity substrate than that on a low permittivity substrate. In contrast, for the H-plane coupled cases the mutual coupling is weaker on a high permittivity substrate than that on a low permittivity substrate. This difference is due to surface waves propagating along the E-plane direction. To reduce the strong mutual coupling of the E-plane coupled MPAA on a thick and high permittivity substrate, the EBG structure is inserted between antenna elements. When the EBG parameters are properly designed, the pronounced surface waves are suppressed, resulting in a low mutual coupling. This method is compared with previous methods such as cavity backed patch antennas. The EBG structure exhibits a better capability in lowering the mutual coupling than other approaches.

11.1 Different types of EBG for mutual coupling reduction

Utilization of electromagnetic band-gap (EBG) structures is becoming attractive in the electromagnetic and antenna community. In this chapter we describe three ways to improve the performance of microstrip antenna arrays by using 3D-EBG, 2D-EBG and defected ground structure DGS. At the end of this chapter as conclusion, we make comparison between the effects of these methods on the array characteristics.

11.2 Mutual coupling reduction by using the 2D-EBG structure

Surface waves are undesired because when a patch antenna radiates, a portion of total available radiated power becomes trapped along the surface of the substrate. It can extract total available power for radiation to space wave as well as there is harmonic frequency created [96]. For arrays, surface waves have a significant impact on the mutual coupling between array elements. One solution to reduce surface waves is using electromagnetic band-gap (EBG) or photonic band-gap structure (PBG). Many shapes of EBG slot have been studied for single element microstrip antenna such as circles, dumb-bells and squares. However, not many have realized in antenna arrays. It has been demonstrated that the EBG structure will lead to a reduction in the side-lobe levels and improvements in the front to back ratio and overall antenna efficiency for the radiation pattern. However, the antenna in the above mentioned references has only one patch. The unique capability of the EBG structure to reduce the mutual coupling between elements of an antenna array was demonstrated. The side lobe of the antenna with one patch is due to surface-wave diffraction at the edges of the antenna substrate. For antenna array, the side lobe is related to the pattern of the individual antenna, location of antenna in the array and the relative amplitudes of excitation. In addition, the mutual coupling between radiators affects the current distribution on the antenna and resulted in increased side lobes.

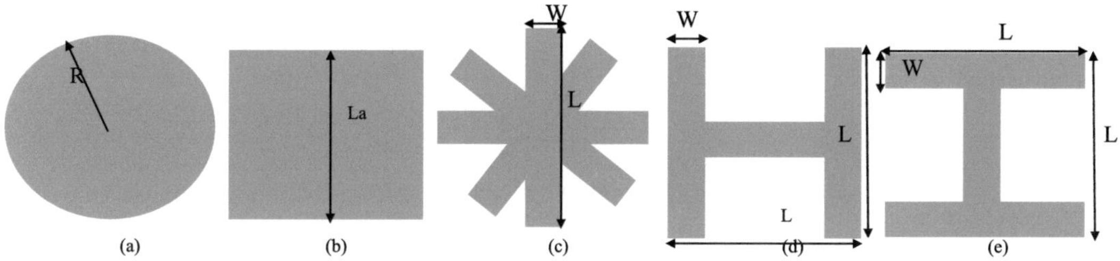

Fig. 58. The different shapes of one unit cell of 2D-EBG: (a) conventional circle, (b) conventional square, (c) star, (d) H-shape and (e) I-shape.

In this section, MPA array with three different shapes of 2D-EBG as star, H shape and I-shape slot etched on the ground plane are designed, simulated and measured. In this study, harmonic suppression and reduction of the mutual coupling effect are investigated by proposing these new shapes of 2D-EBG. The obtained results demonstrate that the 2D-EBG not only reduces the mutual coupling between the patches of antenna array, but also suppresses the second harmonic, reduces the side lobe level and gives results better than conventional 2D-EBG shapes as circle and square. It is also shown that the novel shapes of 2D-EBG on the ground plane increases the gain of the antenna array.

i. Configurations of 2D-EBG Shapes

Three different shapes of 2D-EBG are presented, as shown in figure 59; the three shapes are compared with familiar conventional shapes as circular and square shapes 2D-EBG by using transmission line approach. The proposed EBG units are composed of several rectangular-shape slots (of length L and width W). This EBG cell can provide a cutoff frequency and attenuation pole. It is well known that an attenuation pole can be generated by a combination of the inductance and capacitance elements, which presents circuit model for the cell for all 2D-EBG structures. Here, the capacitance is provided by the transverse slot and the inductance by different shapes slots. For star shape there are four rectangular slots with L = 5 mm, W = 1 mm at angles = 0°, 45°, 90° and 135°. The second shape, which is the H shaped slot, consists of three rectangular slots with the same previous dimensions. The third shape is the I-shaped slot, which is obtained by rotating H-shape by 90°. The substrate with a dielectric constant of 10.2, loss tangent of 0.0019 and thickness of 2.5 mm is considered here. The microstrip feeding line on top plane has a width W_f = 2.3mm, corresponding to 50Ω characteristic impedance. 2D-EBG cells are etched on the ground plane with periodicity P= 7mm and ratio L/p ≅ 0.7. Then the reflection and transmission coefficients (S_{11} and S_{21}) are calculated using the high frequency structure simulator (HFSS).

ii. Antenna Array Design

Consider an ordinary antenna array with two elements, at 5.2 GHz, the dimensions of the patches are patch width W_p= 8mm, patch length L_p=7.5mm and microstrip feed line with length L_f, L_s =20mm, 13mm, respectively, and the distance between the patches is L_B= 19 mm ($0.44\lambda_{5.2GHz}$.). It can be seen that the antenna will radiate energy at a harmonic frequency of 7.5GHz. In order to suppress such harmonics, the band-stop characteristic of the EBG structure may be used. In this section, which is simply 2D-EBG cells are etched on the ground metal sheet. The two separate array elements are also studied to measure the mutual coupling between the two patches in MIMO arrays.

iii. Results and Discussion

The response that is shown in figure 60 presents both conventional shapes as circular and square 2D-EBG as well as the three new shapes. The second step, is applying these shapes to the ground plane of two element array antenna. The results indicate that the harmonic at 7.5 GHz is indeed suppressed as well as reducing array antenna size by about 7.5% for H-shape, 8% for star-shape and for I-shape reduction in size reaches 15%. According to the characteristics of EBG, the surface wave can also be suppressed.

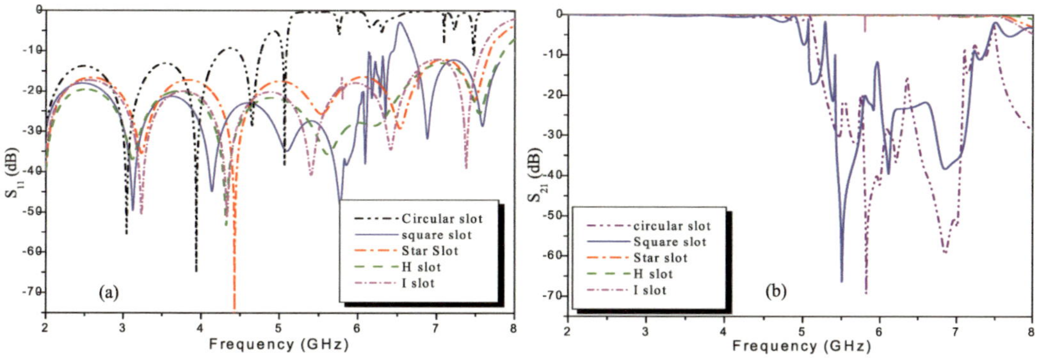

Fig. 59. (a) The reflection and (b) transmission coefficients for different 2D-EBG shapes.

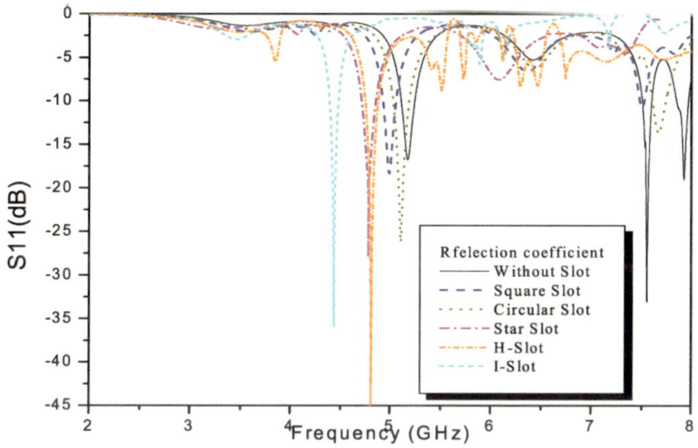

Fig. 60. The reflection coefficient for different MPAA with 2D -EBG.

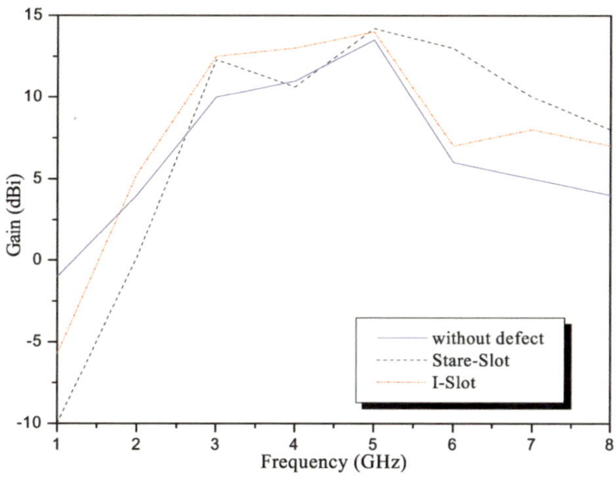

Fig. 61. The Gain response for different shapes of MPAA without and with 2D-EBG shapes.

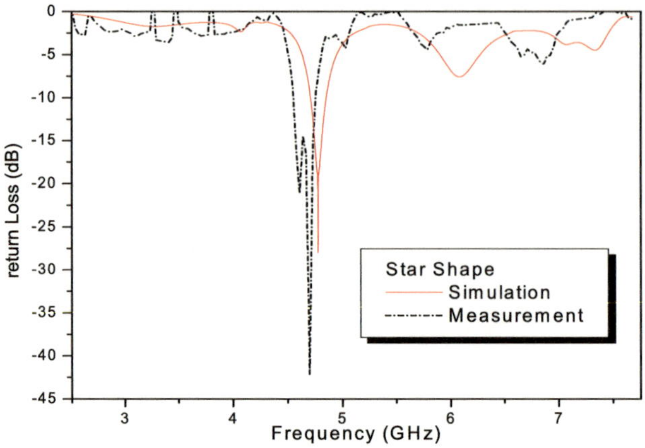

Fig. 62. Comparison between measured and simulated reflection coefficients for star shaped slot 2D-EBG.

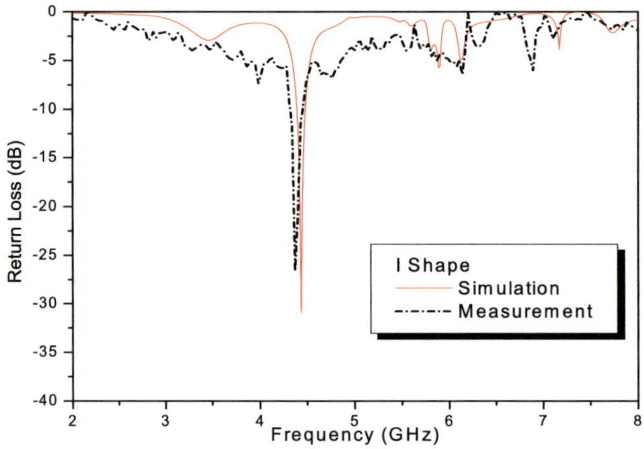

Fig. 63. Comparison between measured and simulated reflection coefficients for I shaped slot 2D-EBG.

For effective suppression of the harmonics, and for effective suppression of the surface waves, a periodic structure surrounding the patches, in addition to underneath the patches, are necessary [98]. The array performance of the conventional and the three new shapes of 2D-EBG are given in table 7. From table 7 and figure 61, one can notice that; I shape gives maximum reduction in resonant array frequency than other shapes so reduce the electrical array size, star shape gives maximum average antenna gain and minimum mutual coupling while H shape gives larger antenna bandwidth than others. By using conventional shapes as circular and square EBG, the concentration of surface current decreases but not eliminated. However, the gain of the two-element array antenna is also studied for different array antennas with and without 2D-EBG as shown in figure 62. The comparisons between measured and simulated reflection coefficients are shown in figures 63 and 64 for the star and I shape, respectively. From table 7 it notes that, at higher frequencies the antenna gain with 2D-EBG is better than that without EBG by about 9dBi maximum difference and 3dBi in average over the entire antenna band which verifies the harmonic suppression behavior.

In addition the average efficiency of the array is also studied over the operating band. The average array efficiency with conventional 2D-EBG is lower than that without by about 15% while with three new shapes of 2D-EBG is lower by about 10%.

Shape of 2D-EBG	Antenna Gain dB @5.2GHz	Mutual Coupling $\|S_{21}\|$	Harmonic level dB	Reflection Coefficient	F_o GHz	BW %	Defect Geometry, Periodicity X direction /Y direction
Without	10	-16dB	-35	-17dB	5.1	3	
Square	13.5	-18.5dB	-10	-17.5dB	5	4	Side dimension 4mm, 6mm, 6mm
Circular	13	-19dB	-15	-20dB	4.8	5	Radius 2mm, 2mm, 2mm
Star	13.75	-40dB	-9	-30dB	4.75	5	Side length 1x $4mm^2$, 6mm, 6mm
H	12.75	-20dB	-10	-40dB	5	5.1	Side length $1x5mm^2$, 6mm, 6mm
I	11	-30dB	-7	-45dB	4	5	Side length $1x5mm^2$, 6mm, 6mm

Table 7. The effect of different 2D-EBG shapes on the antenna performance.

11.3 Novel shapes of low mutual coupling 2X2 MPAA by using DGS

In this section, we propose new shapes of DGS structures integrated with microstrip array elements that suppress surface wave and lead to a high isolation between array elements, hence reduce the mutual coupling. Many published papers have made use of relatively complex periodic electromagnetic band-gap (EBG) structures to reduce the mutual coupling. In this section we propose improvement in E and H plane coupling by about 4dB than dumb-bell shapes presented in [95].

i. DGS Configurations and Response

Figure 65 displays the different shapes of the DGS etched on the metallic ground plane. The different cell shapes of DGS are dumb-bell, H, E, H with inverted H and back-to-back E as shown in figure 65 (a)-(e), respectively.

The responses of these shapes are compared in figure 66 by using the conventional 50Ω microstrip transmission line approach. The transmission and reflection coefficients of the conventional dumbbell and new shapes of DGS are presented in figure 66(a) and 66(b), respectively. It is demonstrated that when an aperture etched on the ground plane, the effective inductance of the microstrip increases and the width of the connecting gap determines the shunt capacitance. The inductance and capacitance level may be independently controlled by the equivalent aperture area and gap width, respectively.

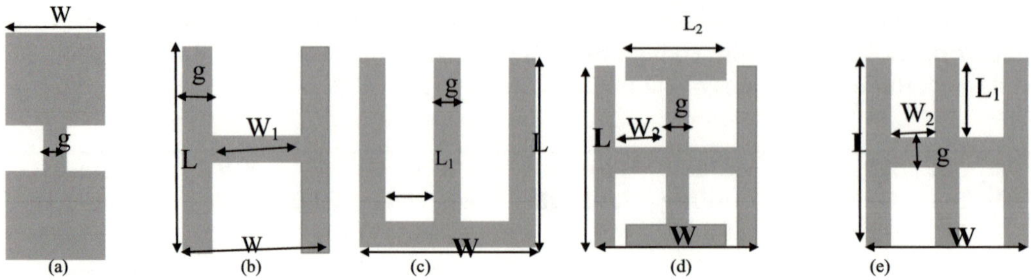

Fig. 64. Different DGS shapes (a) dumbbell, (b) H shape, (c)E, (d) H with inverted H and (e) back to back E shape.

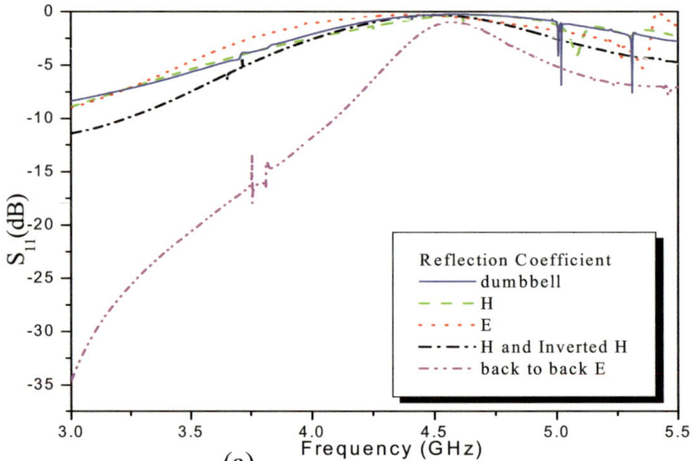

(a)

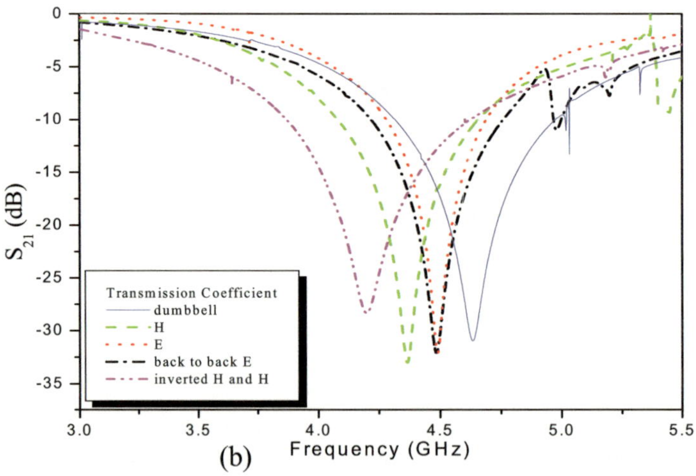

(b)

Fig. 65. (a) Reflection coefficient, and (b) Transmission coefficient for different DGS.

On the other hand, it should be noted that etching holes in the ground plane of DGS structures degrades the isolation characteristic of the ground plane thus obtaining large backward radiation pattern. However, by using these new shapes of the DGS structures, the effective apertures area are reduced thus there are improvements on the array E- and H-plane radiation patters. Figure 66 indicates that all new shapes (E, H, back-to-back E and H with inverted H) give wider stop-band filter response than conventional dumbbell shapes, hence improve the antenna characteristics.

11.3.1 Study of mutual coupling reduction

Due to high excitation of surface waves in both E *and H*-plane coupling between microstrip array patches an investigation was carried out by simulation to study the effect of the newly DGS shapes on the mutual coupling. Two substrates with thickness of 1.6 mm and 2.5mm and permittivity of 4.7 and 10.2, respectively, were used in the simulations. This difference is due to surface waves propagating along the E-plane direction, which can be easily viewed from the provided near field plots. For microstrip feed array as shown in figure 67 (a) the array element is with dimensions W_p x L_p = 16mm x 12mm and L_a=15mm. While the edge-to-edge separation is $L_B>\lambda_0/2$=30mm at a designed resonant frequency 5.25GHz and printed on FR4 substrate with dielectric constant 4.7, height 1.6mm and substrate dimensions W_s x L_s = 53mm × 53mm $(\lambda_0\times\lambda_0)$. Figure 67(b) illustrates the layout of a 4-element coaxial feed microstrip array on the defected ground plane with substrate dimensions 63mm × 63mm= $(\lambda_0\times\lambda_0)$ at a designed frequency of 4.75 GHz. The array patch is with dimensions Wp x Lp = 10.5mm x 8.5mm with edge to edge separation distance $d>\lambda_0/2$ =35mm and printed on RT/D6010 substrate with dielectric constant 10.2 and height 2.5mm. In coaxial feed, only element 2 is excited while other elements 1, 3, 4 are 50Ω terminated. The DGS dimensions for a band-gap at the resonant frequency of the antenna are optimized shapes. It consists of rectangular strips, each with dimensions L X g = 7.5mm x 1.5mm and width W=7.5mm, W=3.5mm and W=1.5mm with length L= 3mm and L= 3.5mm for E-shape, H with inverted H shape and back to back E shaped, respectively. Figure 68 shows the reflection coefficient of the microstrip line feed 2x2 MPAA at low dielectric constant FR4 substrate without and with different shapes of DGS [92]. It is observed that DGS antenna resonant frequency shifts towards lower values with respect to the conventional antenna. This small frequency shift is due to wave slowing effects of DGS. Both back-to-back E and H with inverted H give better response than other shapes. Table 8 summarizes all the MPAA characteristics without and with DGS. Figure 69 shows the comparison between simulated and measured reflection coefficient of the MPAA with back-to-back E and H with inverted H DGS. Figure 70 is a plot of all responses of the coaxial fed 2x2 MPAA at high dielectric constant RT/D6010 substrate with different shapes of DGS together with the conventional array. S_{11}, figure 70 (a), shows that DGS structure shifts down the antenna resonant frequency as compared to the conventional MPAA. The *E*-plane mutual coupling S_{21} are shown in figure 70 (b). The conventional antenna shows a very strong coupling of −9.63 dB due to surface waves pronounced in thick, high permittivity substrate. Since the resonant frequency 4.75GHz of the antenna falls inside the DGS band-gap, surface waves are suppressed and simulations show that mutual coupling drops to −20dB with H and inverted H-shape that is lower than the conventional by 11.28 dB. The *H*-plane mutual coupling results are shown in figure 70 (c) as S_{31}. Again, it is reduced by about 6dB compared to the conventional when using DGS shape. S_{41} which represents, orthogonal coupling, is reduced by 8dB than the conventional case as shown in figure 70 (d). Table 9, summarized all the obtained results.

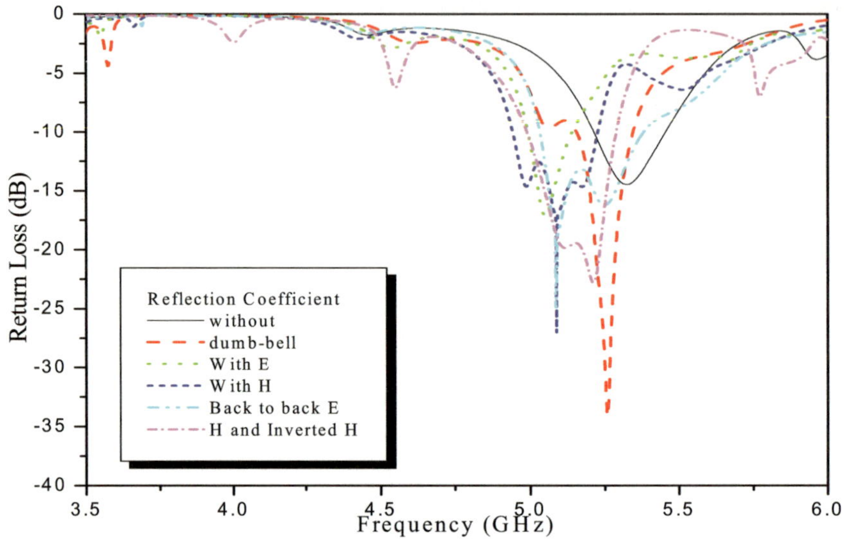

Fig. 66. The reflection coefficient comparison between antenna without and with DGS shapes for line feed array.

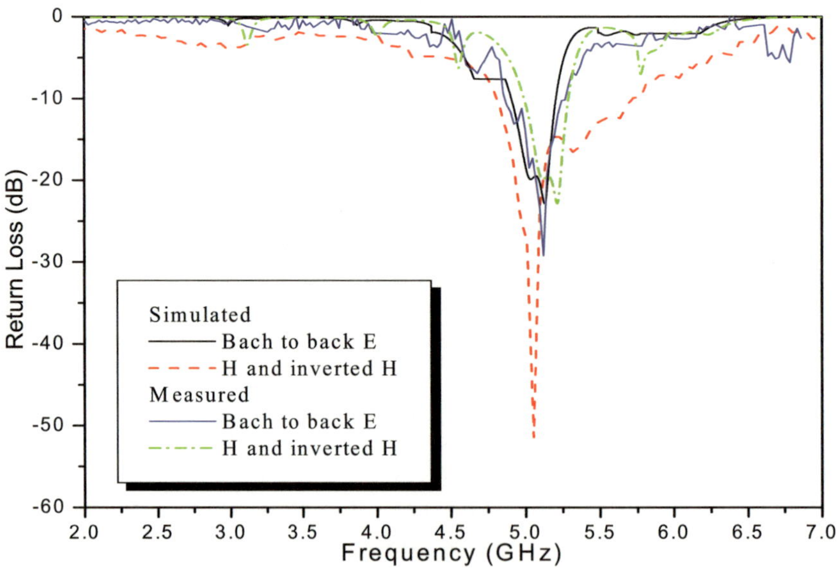

Fig. 67. Comparison between simulated and measured reflection coefficient of MPAA with defect back-to-back E and H with inverted H.

Antenna Shape	Resonant Frequency (GHz)	S_{11} (dB)	BW(MHz)	Antenna Gain (dBi)
Without Defect	5.3	-15	200	8.7
With dumb-bell	5.25	-18	250	9
With H	5.15	-17	175	9.5
With E	5.17	-30	270	9.2
Back to Back E	5.15	-28	350	10.2
H and Inverted H	5.23	-20	200	9.7

Table 8. The characteristics of the microstrip line feed array.

Antenna Shape	F_o (GHz)	Frequency response S_{11}, S_{21}, S_{31}, S_{41} (dB)	BW (MHz)	Antenna Gain (dB)
Without Defect	4.75	-11, -13, -9.63, -12	150	4
With Dimple	4.7	-14, -15, -22, -25	200	4.2
With H	4.65	-15, -20, -20, -25	225	4.2
With E Shape	4.4	-14, -20, -21, -25	200	4.3
Back to Back E	4.63	-20, -20, -20, -25	250	5
H and Inverted H	4.25	-17, -20, -20, -25	200	6.5

Table 9. The characteristics of MPAA with coaxial feed

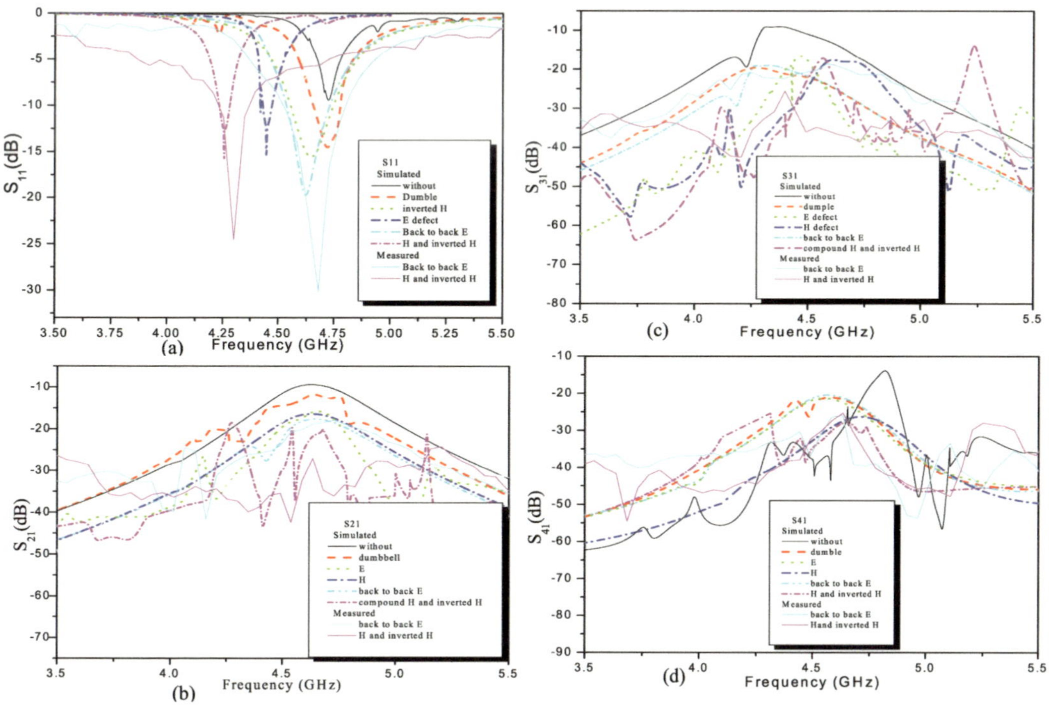

Fig. 68. The 2X2 coaxial MPAA response, (a) S_{11}, (b) S_{21}, (c) S_{31} and (d) S_{41}.

11.3.2 Influence of the different shapes of DGS on the MPAA performance

From above discussion and results in last sections, it can be concluded that:

- 1- Etching DGS on the ground plane in the middle way between the array patches improves both E-and H-plane coupling. This improvement is due to the surface waves suppression because they do not propagate along the E-Plane direction. So, the mutual coupling of MPAA is determined by both the directions of surface waves and antenna size.
- 2- The new shapes of DGS as back to back E and H with inverted H give better performance than conventional DGS shapes as dumb-bell shape.

The MPAA with back-to-back E and H with inverted H DGS shapes are fabricated using photolithographic techniques as shown in figure 71. The measured results show good agreement with simulated ones.

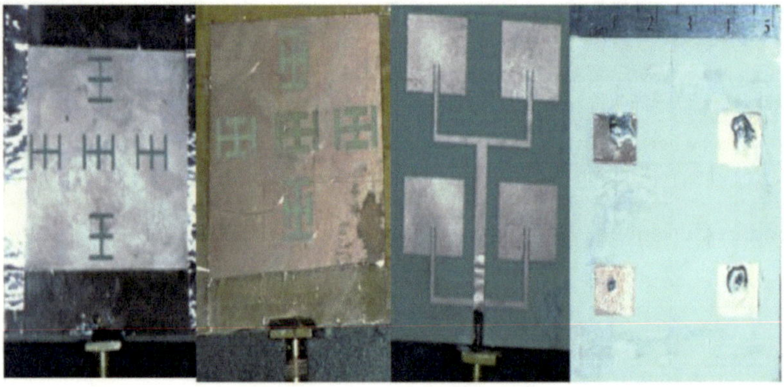

(a) (b) (c) (d)

Fig. 69. (a) The back-to-back E, (b) H and inverted H, (c) inset line feed array and (d) coaxial feed array.

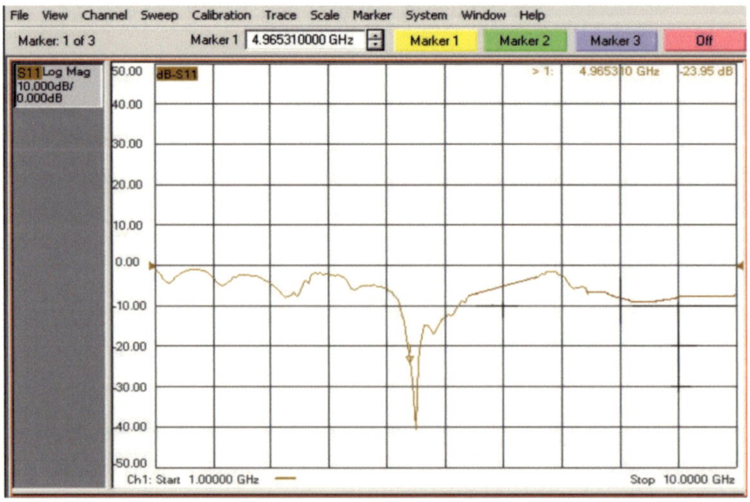

Fig. 70. The measured reflection coefficient of line feed MPAA.

11.3.3 Influence of the DGS on other antenna array characteristics

1. Harmonic Control

Harmonic radiation is a drawback of active integrated MPAA. DGS structures are suggested to reduce the higher-order harmonics in the MPAA. The DGS antenna strongly eliminates the harmonic resonances as shown in figure 72.

2. PIFA Array Design

For further reduction in the array size, planar inverted F antenna (PIFA) are used for array's by using patch length less than $\lambda_0/4$. The same dimensions length and width of the coaxial array patch as previous design are kept the same but shorting wall at the top end edge of each element is added. Figure 73 shows the frequency response of the 2x2 PIFA array with same excitation conditions. From the results, the newly DGS shapes improved the array by more than 5dB in S-parameters as compared with the conventional PIFA array. The reduction in array size is around 55% as compared to conventional array.

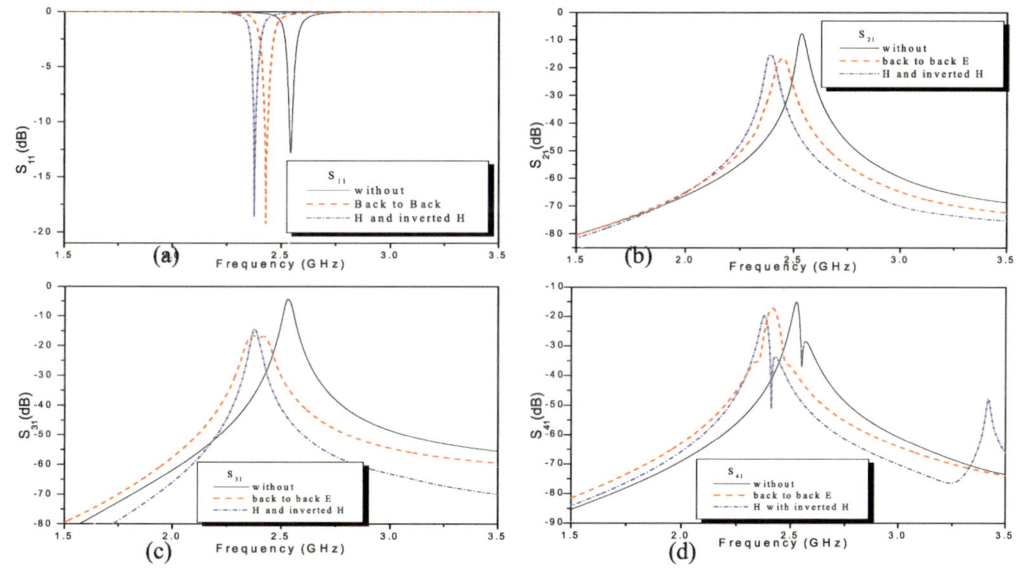

Fig. 71. The 2x2 microstrip PIFA array responses without and with back-to-back E and H with I DGS.

11.4 Ultra-wide bandwidth 2x2 MPAA by using EBG

Various types of EBG structures have been studied. In one of the first applications, a planar antenna mounted onto an EBG substrate was considered to increase the overall radiation efficiency of the device. Increasing antenna directivity was studied using an EBG structure. A compact spiral EBG structure was studied for microstrip antenna arrays. There are diverse forms of EBG structures, and novel designs such as EBG structures integrated with active device and multilayer EBG structures have been proposed recently [98].

Figure 74 shows four types of the spiral EBG structures used in this study. Comparing to other EBG structures such as dielectric rods and holes, the proposed spiral structure has a unique feature of compactness, which is important in wireless communication applications. Specifically, in this study, we investigate the use of four shapes of spiral EBG to help increase the bandwidth of a 2x2 MPAA. We choose the shape of a four-arm spiral AMC, figure 74 (b), for complete cancellation of the antenna cross polarization. This four-arm AMC is then embedded with other EBG structures to further improve the performance of the 2x2 MPAA. First it is embedded with a large four-arms spiral (LSAMC) as shown in figure 74(c), then with small spiral SSAMC patch cells with periodicity P as shown in figure 74(d), and finally with a mushroom-type EBG with spiral patches as shown in figure 74 (e). Obtained results show that the LSAMC design improves the antenna bandwidth and reduces size, while the SSAMC improves the antenna reflection phase as well as reduces the antenna array size. The embedded spiral electromagnetic band gap structure (ESEBG) in figure 74 (e), was found to improve the antenna array bandwidth and gain. Details of the specific design dimensions and the obtained simulation and experimental results are described in the following sections [98]. The concept of spiral ground plane like spiral antenna as given before in the section before,

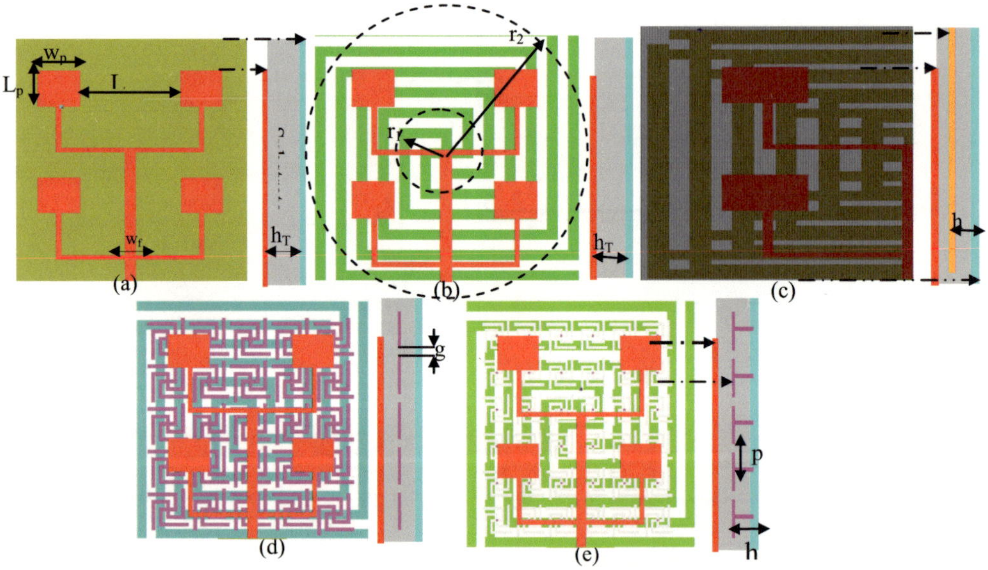

Fig. 72. Different prototype shapes of 2x2 MPAA with different EBG configurations.

11.4.1 Electromagnetic band-gap structure techniques

A typical 2x2 patch antenna array is shown in figure 74(a), where single patch length L_p =7mm, width w_p =9mm, with patch separation L_B=18mm=0.35$\lambda_{5.2GHz}$ and substrate thickness h_T = 2.5mm with conventional rectangular ground plane of dimension 50x50mm^2. The material of the substrate is RT/D6010 of dielectric constant ε_r =10.2. Figure 74 (b) represents the first type of EBG with four arms spiral AMC ground plane. The dimensions of the ground spiral arms, according to optimization of the transmission coefficient response of the spiral, is arm width equal to separation between arms =2mm and it rotates outward counter clockwise.

Figure 74 (c) presents prototype array antenna with second type of EBG as embedded large spiral AMC in the middle of the substrate h=1.25mm with previous ground. The dimensions of the embedded spiral is kept the same of the ground plane but less number of spiral turns and centered under the 2x2 array antenna. To add further improvement in antenna response, small cells of spiral patches with patch size dimensions 6x6 mm^2 with spiral arm width equal to gap separation=0.5mm and periodicity **P**=7.5mm are added at the same substrate height **h**=1.25mm with same pervious ground as shown in figure 74(d). The transmission coefficient response for this spiral structure is also shown in figure 75. From figure 75, embedded EBG gives best transmission performance followed by embedded small cells spiral with AMC then embedded large spiral AMC and the worst response is for spiral AMC in ground plane. Finally, embedded small four arms spiral patches are used as electromagnetic band-gap structure (EBG) with same pervious dimensions and with vias in the dielectric layer with radius 0.25mm.

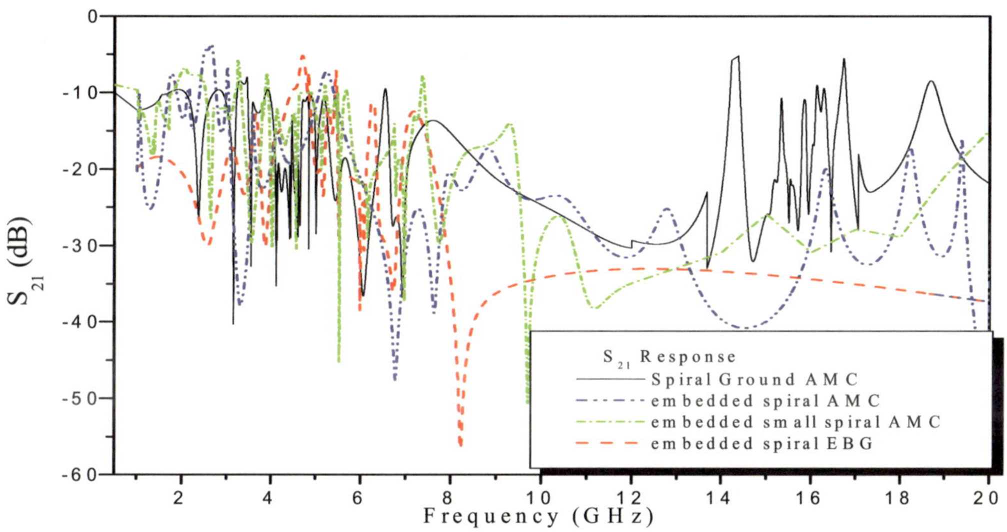

Fig. 73. The transmission response for spiral AMC ground, embedded spiral AMC, embedded small spiral AMC and embedded spiral EBG.

11.4.2 Antennas results and discussion

We started with conventional 2x2 MPAA with resonant frequency 5.2GHz. The array resonates at 5.2GHz with average gain 6.5dBi and the first harmonic appears at 7.8GHz then the conventional ground plane is reduced by four arms spiral AMC ground plane. The antenna size is reduced by 50% and the bandwidth extended from 2.5 to 19GHz with bandwidth discontinuities in the operating antenna sub-bands as shown in figure 76. Figure 77 shows the comparison between measured and simulated reflection coefficient of array antenna with spiral ground. The antenna bandwidth extended from 3GHz to 19GHz with discontinuities in bandwidth, average antenna gain is 7.8dBi. For further improvement in antenna performance as bandwidth and antenna size reduction, another embedded spiral with less number of arm turns are added at height 1.25mm from the spiral ground plane. The antenna bandwidth extended from 1.25 to 19GHz with decreased number of sub-band discontinuities as shown in figure 77 with average antenna gain 8.8dBi. Thirdly, small spiral patches are added with periodicity **P**=7.5mm to improve the bandwidth discontinuities and reflection phase especially at antenna operating frequency at 5.2GHz. To achieve optimum performance, $2a = \lambda_g$ where **a** is the side of embedded patch and λ_g is the guided wavelength. The bandwidth extended from 0.75 to 20GHz as shown in figure 77 with average antenna gain increased to 9dBi, so it is increased from original gain by about 4dBi.

Embedded electromagnetic band gap structure with four arm spiral patches are added at the same height to improve the antenna gain to 15.5dBi and good extended bandwidth without any discontinuities from 0.5 to 3.5 and from 4 to 19GHz as shown in figure 78. Figure 79 shows antenna array gain versus frequency. From this figure, all configurations give antenna gain better than the conventional array used over the entire operating band and the embedded spiral EBG gives the best performance followed by small spiral embedded AMC then the large embedded spiral AMC. To Table 10, summarizes all results of antennas array characteristics. The four different configurations were fabricated as shown in figure 80 by using the photolithographic techniques.

Antenna Ch/cs	**Conventional Antenna**	**Antenna with SAMC Ground**	**Antenna with E LSAMC**	**Antenna with E SSAMC**	**Antenna ESEBG**
Res. Freq.	5.25GHz	2.5GHz	2.25GHz	1.25GHz	0.75z
-10 dB BW (GHz)	Fundamental resonant 5.1-5.3	Extended from 1 to 3, from 3.25 to 5, from 5.2 to 9 and from 9.5 to 19GHz	Extended from 1.25 to 3, from 4 to 5, from 6 to 12 and from 12.5 to 16GHz	Extended from 1.25 to 2.5, from 2.5 to 3.5 from 4 to 5 from 5 to 7 and from10 to17 GHz	Extended from 0.5 to 3.5, and from 4 to 19.5GHz
Average Rad. Effic.	0.96	0.75	0.8	0.8	0.85
Average Antenna gain	6.5dBi	7.8dBi	8.8dBi	9dBi	10.5dBi
Fabrication	Easy	Easy	Moderate	Moderate	Hard
Reflection Phase	Bad	Medium	Good	Good	Very Good

Table 10. Characteristics of various antennas.

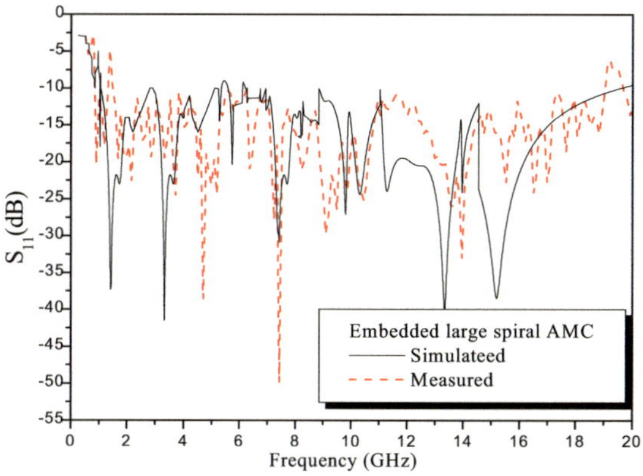

Fig. 74. The comparison between measured and simulated reflection coefficient of embedded large SAMC.

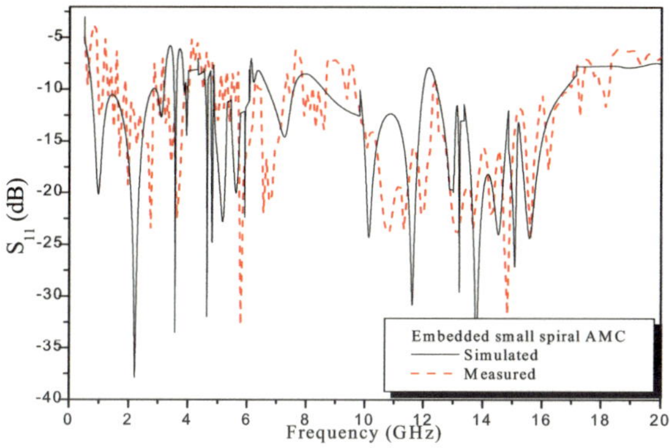

Fig. 75. Comparison between measured and simulated reflection coefficient of embedded small spiral AMC.

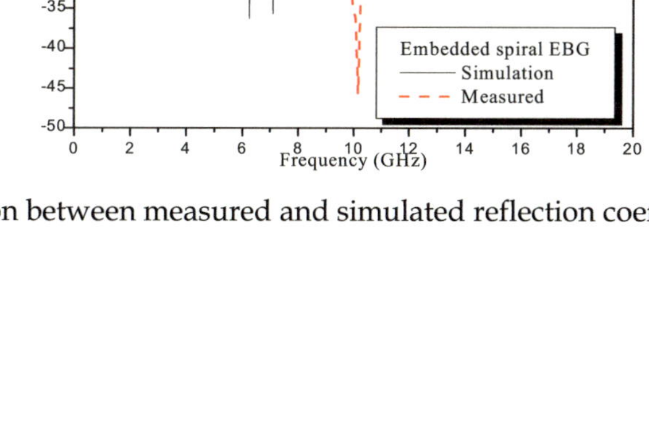

Fig. 76. Comparison between measured and simulated reflection coefficient of embedded small spiral EBG.

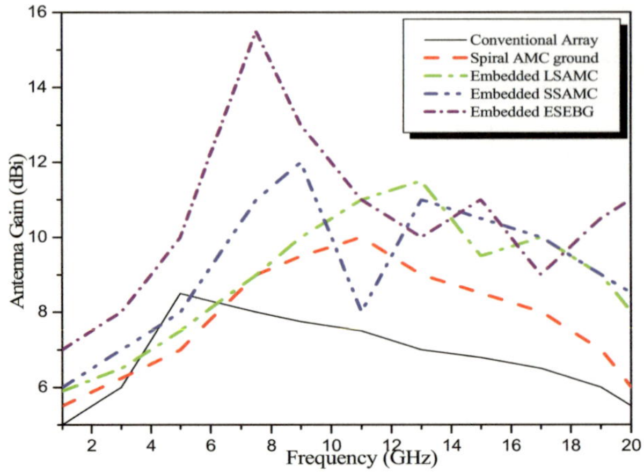

Fig. 77. Array antenna gain -vs-frequency for different configurations.

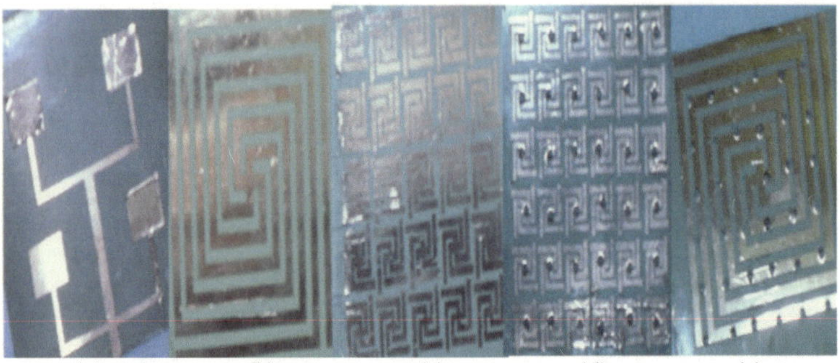

(a) (b) (c) (d) (e)

Fig. 78. Fabricated 2x2 MPAR(a) Conventional radiator patches, (b) embedded large SAMC, (c) embedded small SAMC, (d) embedded SEBG and (e) ground plane.

12. References

[1] C. Caloz, and T. Itoh, Electromagnetic Metamaterials: Transmission Line Theory and Microwave Applications, A John Wiley & son, Inc., Canada, 2006.

[2] A. Sihvola, "Electromagnetic emergence in metamaterials. Deconstruction of terminology

[3] of complex media", *Advances in Electromagnetics of Complex Media and Metamaterials, NATO Science Series II: Mathematics, Physics and Chemistry*, Kluwer Academic Publishers, vol. 89, pp. 1 - 17, 2003

[4] A. Sihvola, "Metamaterials in electromagnetics", *Metamaterials*, vol. 1, no. 1, pp. 2 - 11.

[5] A. Sihvola, "Metamaterials: A Personal View," *Radioengineering*, vol. 18, no. 2, pp. 90 - 94, 2009.

[6] D. R. Smith, "What are Electromagnetic Metamaterials?," *Novel Electromagnetic Materials*, Retrieved Auguest, 2010.

[7] http://people.ee.duke.edu/~drsmith/about_metamaterials.html.

[8] J. G. Webster, *Metamaterial*, Wiley Encyclopedia of Electrical and Electronics Engineering, Jan. 2010.

[9] V. G. Veselago, "The Eletrodynamics of Substances with Simultaneously Negative Values of ε and μ," *Sov. Phys. Usp*, vol. 10, pp. 509 - 514, 1968.
[10] I. V. Lindell, S. A. Tretyakov, K. I. Nikoskinen, and S. Ilvonen, "BW Media–Media with Negative parameters, Capable of Supporting Backward Waves," *Microwave Opt. Tech. Lett.*, vol. 31, no.2, pp. 129-133, 2001.
[11] J. B. Pendry, A. J. Holden,W. J. Stewart, and I. Youngs, "Extremely Low Frequency Plasmons in Metallic Mesostructures," *Phys. Rev. Lett.*, vol. 76, pp. 4773 - 4776, 1996.
[12] J. B. Pendary, A. J. Holden, D. J. Robbins, and W. J. Stewart, "Magnetism from Conductors and Enhanced Nonlinear Phenomena," *IEEE Trans. Microwave Theory Tech.*, vol. 47, pp. 2075 - 2084, 1999.
[13] D. R. Smith, W. Padilla, D. C. Viers, S. C. Nemat-Nasser, and S. Schultz, "A Composite Medium with Simultaneously Negative Permeability and Permittivity,"*Phys. Rev. Lett.*, vol. 84, pp. 4184 - 4187, 2000.
[14] D. V. Smith, and J. B. Pendry, "Homogenization of Metamaterials by Field Averaging," *J. Opt. Soc, Qm.*, vol. 23, no. 3, pp. 391 - 403, 2006.
[15] J. B. Pendry, "Negative Refraction Makes a Perfect Lens," *Phys. Rev. Lett.*, vol. 85, pp. 3966 - 3969, 2000.
[16] M. C. K. Wiltshire, J. V. Hajnal, J. B. Pendry, D. J. Edwards, and C. J. Stevens, "Metamaterial Endoscope for Magnetic Field Transfer: Near Field Imaging with Magnetic Wires," *Optics Express*, vol. 11, no. 7, pp. 709 - 715, 2003.
[17] N. Katsarakis, T. Koschny, M. Kafesaki, E. N. Economou, and C. M. Soukoulis, "Electric Coupling to the Magnetic Resonance of Split Rring Resonators", *Applied Phys. lett.*, vol. 84, no. 15, pp. 2943 - 2945, 2004.
[18] D. Schurig, J. J. Mock, B. J. Justice, S. A. Cummer, J. B. Pendry, A. F. Starr, and D. R. Smith, "Metamaterial Electromagnetic Cloak at Microwave Frequencies," *Science*, vol. 314, no. 5801, pp. 977 - 980, 2006.
[19] A. Alu, and N. Engheta, "Polarizabilities and Effective Parameters for Collections of Spherical Nanoparticles Formed by Pairs of Concentric Double-Negative, Single-Negative, and/or Double-Positive Metamaterial Layers," *Journal of Applied Physics*, vol. 97, no. 9, pp. 094310 1 - 12, 2005.
[20] S. F. Mahmoud, "A New Miniaturized Annular Ring Patch Resonator Partially Loaded by A Metamaterial Ring with Negative Permeability and Permittivity," *IEEE Antennas and Wireless Propagation Letters*, vol. 3, pp. 19 - 22, 2004.
[21] A. Alu, M. G. Silveirinha, A. Salandrino, and N. Engheta, "Epsilon-Near-Zero Metamaterials and Electromagnetic Sources: Tailoring the Radiation Phase Pattern," *Physical Review B*, vol. 75, no. 15, 2007.
[22] F. Yang and Y. Rahmat-Samii, "Applications of Electromagnetic Band-Gap (EBG) Structures in Microwave Antenna Designs," Proc. of 3rd International Conference on Microwave and Millimeter Wave Technology, pp.528–31, 2002.
[23] R. Marqués and F. Medina, "An Introductory overview on right-handed metamaterials", Proc. the 27th ESA Antenna Workshop on Innovative Periodic Antennas, Spain, pp. 35-41, 2004.
[24] C. R. Simovski, P. Maagt, and I. V. Melchakova, "High impedance surfaces having resonance with respect to polarization and incident angle," IEEE Trans. Antennas Propagat., Vol. 53, no. 3, pp.908–14, 2005.

[25] G. Gampala, Analysis and design of artificial magnetic conductors for X-band antenna applications, M.Sc. thesis at The University of Mississippi, 2007.
[26] M. A. Jensen, Time-Domain Finite-Difference Methods in Electromagnetics: Application to Personal Communication, Ph.D. dissertation at University of California, Los Angeles, 1994.
[27] F. Yang and Y. Rahmat-Samii, "Bent monopole antennas on EBG ground plane with reconfigurable radiation patterns," 2004 IEEE APS Int. Symp. Dig., Vol. 2, pp. 1819–1822, Monterey, CA, June 20–26, 2004.
[28] E. Yablonovitch, "Inhibited Spontaneous Emission in Solid-State Physics and Electronics," *Physical Review Letters*, vol. 58, no. 20, pp. 2059-2062, 1987.
[29] W. Barnes, T. Priest, S. Kitson, J. Sambles, "Photonic Surfaces for Surface-Plasmon Polaritons", *Phys. Rev. B*, vol. 54, pp. 6227, 1996.
[30] S. Kitson, W. Barnes, J. Sambles, "Full Photonic Band Gap for Surface Modes in the Visible", *Phys. Rev. Lett.* , vol. 77, pp. 2670, 1996.
[31] L. Brillouin, *Wave Propagation in Periodic Structures; Electric Filters and Crystal Lattices*, 2nd ed., Dover Publications, New York (1953)
[32] F. Yang, and Y. Rahmat-Samii, "Reflection Phase Characterizations of the EBG Ground Plane for Low Profile Wire Antenna Applications," *IEEE Trans. on Antennas Propag.*, vol. 51, no. 10, Oct. 2003.
[33] J. R. Sohn, H. S. Tae, J. G. Lee, and J. H. Lee, "Comparative Analysis of Four Types of High Impedance Surfaces for Low Profile Antenna Applications," *Ant. and Propagat. Society International Symposium*, vol. 1A, pp. 758 - 761, 2005.
[34] B. Wu, B. Li, T. Su, and C.-H. Liang, "Equivalent-circuit analysis and lowpass filter design of split-ring resonator DGS," Journal of Electromagnetic Waves and Applications, Vol. 20, No. 14, 1943–1953, 2006.
[35] J. Chen, Z.-B.Weng, Y.-C. Jiao, and F.-S. Zhang, "Lowpass filter design of Hilbert curve ring defected ground structure,", PIER 70, pp. 269–280, 2007.
[36] R. Sharma, T. Chakravarty, and S. Bhooshan, "Design of a Novel 3 dB Microstrip Backward Wave Coupler Using Defected Ground Structure," Progress In Electromagnetic Research, PIER 65, pp.261–273, 2006.
[37] V. Radisic, Y. Qian, R. Coccioli, and T. Itoh, "Novel 2-D photonic bandgap structure for microstrip lines," IEEE Microw. Guided Wave Lett., Vol. 8, No. 2, 69–71, Feb. 1998.
[38] F. Falcone, T. Lopetegi, and M. Sorolla, "1-D and 2-D photonic bandgap microstrip structures," Microw. Opt. Technol. Lett., Vol. 22, No. 6, 411–412, Sep. 1999.
[39] S. Zouhdi, Ari Sihvola and A. P. Vinogradov, "Applications of EBG in low profile antenna designs: what have we learned" Springer Netherlands, 2009.
[40] D. Cabric, M.S.W. Chen, D.A. Sobel, J. Yang, R.W. Brodersen, "Future wireless systems: UWB, 60GHz, and integrated Circuits Conference," pp. 793–796 cognitive radios," IEEE Proceedings of the Custom Sept. 2005.
[41] D. Elsheakh, "Electromagnetic Band-Gap (EBG) Structure for Microstrip Antenna Systems (Analysis and Design)", PhD thesis, Ain Shams University 2010.
[42] F. R. Yang, Novel Periodic Structures for Applications to Microwave Circuits, Ph.D. Dissertation, Electrical Engineering Dept., University of California, Los Angeles, 1999.

[43] M. Bradley "One-dimensional photonic band-gap structures and the analogy between optical and quantum mechanical tunneling," Eur. J. Phys, pp.108-112, 1997.
[44] N. Engheta and R. Ziolkowski, "Metamaterials: Physics and Engineering Explorations", John Wiley & Sons Inc., 2006.
[45] L. Yang, M. Fan, F. Chen, J. She and Z. Feng, "A novel compact electromagnetic-bandgap (EBG) structure and its application for microwave circuits", IEEE Trans. Microwave Theory Tech., Vol. 53, No. 1, Jan. 2005.
[46] J. Rarity, and C. Weisbuch, "In micro cavities and photonic bandgaps," Physics and Applications, Kluwer Academic Publishers, Dordrecht, 1996.
[47] Park, J. S., Kim, C. S., Kang, H. T. et al., "A novel resonant microstrip RF phase shifter using defected ground structure", 30th European Microwave Conf., pp.72-75, France, 2000.
[48] K. B. Chung, Hong, S. W. "Wavelength demultiplexers based on the superprism phenomena in photonic crystals" IEEE Trans. Antennas and Propag., vol. 81, pp. 1549-1551, Augst 2002.
[49] C. C. Chang,Y. Qian, and T. Itoh, "Analysis and applications of uniplanar compact photonic bandgap structures," PIER, vol. 41, pp. 211–235, 2003.
[50] Dahrele, S. John and Lee, "Strong Localization of Photons in Certain Disordered Dielectric Super Lattices", Physical Review Letters, vol. 58, pp. 2486 - 2489, June 1985.
[51] H. Kosaka, T. Kawashima, A. Tomita, M. Notomi, T. Tamamura, T. Sato, and S. Kawakami, "Photonic Crystals for Micro Light Wave Circuits Using Wavelength-Dependent Angular Beam Steering," Appl. Phys. Lett., vol. 74, pp. 1370–1372, Sept. 1999.
[52] D. Sievenpiper, "Chapter 11: Review of Theory, Fabrication, and Applications of High Impedance Ground Planes," in Metamaterials: Physics and Engineering Explorations, edited by N. Engheta and R. Ziolkowski, John Wiley & Sons Inc., 2006.
[53] G.-H. Li,X.-H. Jiang, and X.-M. Zhong, "A novel defected ground structure and its application to a low pass filter," Microwave and Optical Technology Letters, vol. 48, pp. 453– 456, Sep. 2006.
[54] C. S. Kim, J. S. Lim, S. Nam, K. Y. Kang, and D. Ahn, "Equivalent circuit modeling of spiral defected ground structure for microstrip line," Electron. Lett., vol. 38, pp. 1109–1120, 2002.
[55] M. Martinez-Vazquez and R. Baggen, "Characterization of printed EBG surfaces for GPS applications," IEEE Int. Workshop on Antenna Technology Small Antennas and Novel Metamaterials, pp. 5–8, March 2006.
[56] I. Garcia, "Electromagnetic band-gap (EBG) structure in antenna design for mobile communication," Project National Science Foundation, 2002-2003.
[57] Y. Ning, C. Zhining, W. Yunyi and C. M. Y. W, " A novel two-layer compact electromagnetic bandgap (EBG) structure and its applications in microwave circuits", Vol. 46 No. 4 Science in China (Series E) August 2003.
[58] X. Q. Chen, X. W. Shi, Y. C. Guo, and C. M. Xiao, " A novel dual band transmitter using microstrip defected ground structure", Progress In Electromagnetics Research, PIER 83, pp. 1–11, 2008.

[59] C. S. Kim, J. S. Park, D. Ahn, and J. B. Lim, "A novel 1-D periodic defected ground structure for planar circuits," IEEE Microwave Guided Wave Lett., vol. 10, pp. 131–133, Apr. 2000.
[60] M. Fallah-Rad and L. Shafai, "Enhanced performance of a microstrip patch antenna using high impedance EBG structure," IEEE APS Int. Symp. Dig., vol. 3, pp. 982–5, June 2003.
[61] C. C. Chiau, X. Chen, and C. G. Parini, "A microstrip patch antenna on the embedded multi-period EBG structure," Proceeding of the 6th Int. Symp. Antennas, Propagation and EM Theory, pp. 96–106, 2003.
[62] E. Yablonovitch, "Photonic band-gap structures", Journal of the Optical Society of America B, vol. 10, No. 2, pp. 283-295, Feb. 1993.
[63] H. Mosallaei and Y. Rahmat-Samii, "Periodic bandgap and effective dielectric materials in electromagnetics: characterization and applications in nanocavities and waveguides," IEEE Trans. Antennas and Propag., vol. 51, pp. 549–63, April 2003.
[64] M. Rahman and M. Stuchly, "Wide-band microstrip patch antenna with planar PBG structure," in Proc. IEEE APS Dig., vol. 2, pp.486–489, 2001,.
[65] G. Guida, A. de Lustrac, and A. Priou, "An introduction to photonic band gap (PBG) materials", PIER, vol. 41, pp.1-20, 2003.
[66] L. Qing-chun, Z. Fang-ming, and H. E Sai-ling, "A New Photonic Band-Gap Cover for a Patch Antenna with a Photonic Band-Gap Substrate", Journal of Zhejiang University Science, vol. 5, No. 3, pp. 269-273, Mar. 2004.
[67] M. A. I. El-Dahshory "Design and analysis of photonic band gap structures", Master Thesis, Cairo University, 2008.
[68] T. Sundström," Analysis of photonic crystal waveguides by the use of FDTD with regularization", Report presented to Royal Institute of Technology, 2004.
[69] A. Aminian, F. Yang, and Y. Rahmat-Samii, "Bandwidth determination for soft and hard ground planes by spectral FDTD: a unified approach in visible and surface wave regions," IEEE Trans. Antennas Propag., vol. 53, pp.18–28, January 2005.
[70] A. S. Barlevy and Y. Rahmat-Samii, "Characterization of electromagnetic band-gaps composed of multiple periodic tripods with interconnecting vias Concept, analysis, and design," IEEE Trans. Antennas Propag., vol. 49, pp. 343–353, Jun. 2001.
[71] C. C. Chiau, X. Chen, and C. G. Parini, "A multi-period EBG structure for microstrip antennas," Proceedings of 2003 ICAP, vol. 2, pp. 727–730, 2003.
[72] G. Goussetis, A. P. Feresidis, and J. C. Vardaxoglou, "FSS printed on grounded dielectric substrates resonance phenomena, AMC and EBG characteristics," IEEE APS Int. Symp. Dig., vol. 1B, pp. 644–647, July 2005.
[73] G. Goussetis, A. P. Feresidis, and J. C. Vardaxoglou, "Tailoring the AMC and EBG characteristics of periodic metallic arrays printed on grounded dielectric substrate," IEEE Trans. Antennas Propagat., vol. 54, pp. 82–9, Jun. 2006.
[74] J. R. Sohn, K. Y. Kim, and H.-S. Tae, "Comparative study on various artificial magnetic conductor for low profile", PIER 61, pp.27–37, 2006
[75] M. G. Bray and D. H. Werner, "A novel design approach for an independently tunable dual-band EBG AMC surface," IEEE APS Int. Symp. Dig., Vol. 1, pp. 289–292, June 2004.
[76] A.R. Butz "Alternative algorithm for Hilbert's space filling curve", IEEE Trans. On Computers, No. 20, pp.424-442, April 1971.

[77] P. Feresidis, A. Chauraya, G. Goussetis, J. C. Vardaxoglou and P. de Maagt, "Multiband artificial magnetic conductor surfaces", Proc. IEE Seminar on Metamaterials, for Microwave and (Sub) Millimetre Wave Applications, 24 pp. 1-4, Nov. 2003, London, UK.
[78] H. Sagan, "Space-Filling Curves," Springer-Verlag, New York, 1994.
[79] V K.J. Vinoy, K.A. Jose, V.K. Varadan, and V.V. Varadan, "Hilbert Curve Fractal Antenna: A Small Resonant Antenna for VHF/UHF Applications," Microwave & Optical Technology Letters, Vol. 29, pp. 215-219, March 2001.
[80] X. Wang Y. Hao Hall, P.S, "Dual-Band Resonances of a Patch Antenna on UC-EBG substrate," Microwave Conference Proceedings, APMC 2005.
[81] D. Nashaat, H. A. Elsadek, E. Abdallah, H. Elhenawy and M. F. Iskander "Electromagnetic Analyses and an Equivalent Circuit Model of Microstrip Patch Antenna with Rectangular Defected Ground Plane" Proceedings of IEEE international symposium on antenna and propagation AP-S, June 2009.
[82] D. Nashaat, H. A. Elsadek, E. Abdallah, H. Elhenawy and M. F. Iskander "Multiband and Miniaturized Inset Feed Microstrip Patch Antenna Using Multiple Spiral-Shaped Defect Ground Structure (DGS)" Proceedings of IEEE international symposium on antenna and propagation AP-S, June 2009.
[83] D. Nashaat, H. A. Elsadek, E. Abdallah, H. Elhenawy and M. F. Iskander "Miniaturized and Multiband Operations of Inset feed Microstrip Patch Antenna by Using Novel Shape of Defect Ground Structure (DGS) in Wireless Applications" PIERS 2009 in Moscow Progress in Electromagnetics Research Symposium, August, 2009, Moscow, RUSSIA.
[84] D. Nashaat, H. A. Elsadek, E. Abdallah, H. Elhenawy and M. F. Iskander, "Ultra-Wideband and Miniaturization of the Microstrip Monopole Patch antenna (MMPA) with Modified Ground Plane for Wireless Applications," PIERL journal, Vol. 10, pp.171-184, 2009.
[85] D. Nashaat, H. A. Elsadek, E. Abdallah, H. Elhenawy and M. F. Iskander "Ultra-Wideband Microstrip Monopole Antenna by Using Unequal Arms V- Shaped Slot printed on Metamaterial Surface" Proceedings 3rd International Congress on Advanced Electromagnetic Materials in Microwaves and Optics, London, UK, Aug., 2009.
[86] Roy, S., Foerster, J.R., Somayazulu, V.S., and D.G. Leeper. 2004. "Ultrawideband Radio Design: The Promise of High-Speed, Short-Range Wireless Connectivity." Proceedings of the IEEE. Vol. 92. pp. 295-311. Feb 2004.
[87] D. Nashaat, H. A. Elsadek, E. Abdallah, H. Elhenawy and M. F. Iskander, "Enhancement of Ultra-Wide Band Microstrip Monopole Antenna by Using Unequal Arms V-Shaped Slot Printed on Metamaterial Surface", Microwave and Optical Technology letters, Vol. 52, No. 10, pp:2203-2208, October 2010.
[88] D. Nashaat, H. A. Elsadek, E. Abdallah, H. Elhenawy and M. F. Iskander, "Ultra-Wideband Co-planar Boat Microstrip Patch with Modified Ground Plane by Using Electromagnetic Band-Gap Structure (EBG) for Wireless Communication" published in Microwave antenna Optical Technology letters, Vol 52, issue 5, pp.1159-1164, 2010.
[89] A. Yu and X. Zhang, "A low profile monopole antenna using a dumbbell EBG structure," IEEE APS Int. Symp. Dig., Vol. 2, pp. 1155–8, 20–25 June 2004.

[90] D. Nashaat, H. A. Elsadek, E.t Abdallah, H. Elhenawy and M. F. Iskander, "Enhancement of Microstrip Monopole Antenna Bandwidth by Using EBG Structures published in IEEE Antennas and Wireless Propag., Letters, vol. 8, pp.959-963, 2009.

[91] D. Nashaat, H. A. Elsadek, E. Abdallah, H. Elhenawy and M. F. Iskander, "Ultra-Wide Bandwidth Umbrella Shaped Microstrip Monopole Antenna Using Spiral Artificial Magnetic Conductor (SAMC)" IEEE Antennas and Wireless Propag., Letters, vol.8, pp.1225-1229, 2009.

[92] D. Nashaat, H. A. Elsadek, E. Abdallah, H. Elhenawy and M. F. Iskander, "Investigated New Embedded Shapes of Electromagnetic Bandgap Structures and Via Effect for Improvement Microstrip Patch Antenna Performance " PIERB journal, vol. 12, pp.90-107, 2010.

[93] D. Nashaat, H. A. Elsadek, E. Abdallah, H. Elhenawy and M. F. Iskander, "Reconfigurable Single and Multiband Inset feed Microstrip Patch Antenna for Wireless Communication Devices" PIERC journal, Vol. 12, pp.191-201, 2010.

[94] F. Yang and Y. Rahmat-Samii, Mutual coupling reduction of microstrip antennas using electromagnetic band-gap structure, Proc IEEE AP-S Dig. 2 pp. 478–481, 2001.

[95] W. Yun and Y. J Yoon, "A Wideband aperture-coupled microstrip array antenna using inverted feeding structures," IEEE Trans. Antenna and Propg., Vol. 53, No. 2, PP: 861-862, Feb. 2005.

[96] R. Gonzalo Garcia, P. de Maagt and M. Sorolla, "Enhanced patch-antenna performance by suppressing surface waves using photonic-band-gap substrate," IEEE Trans. Microwave Theory Tech., Vol. 47, No. 11, PP: 213 1-2138, November 1999.

[97] D. Nashaat, H. A. Elsadek, E. Abdallah, H. Elhenawy and M. F. Iskander, "Microstrip Array Antenna with New 2D-Electromagnetic Band Gap Structure Shapes to Reduce Harmonics and Mutual Coupling", published in PIERC journal, Vol. 20, pp.203-213, 2010.

[98] M. Salehi and A. Tavakoli, A novel low coupling microstrip antenna array design using defected ground structure, Int J Electron Commun 60, pp. 718–723, 2006.

[99] D. Nashaat, H. A. Elsadek, E. Abdallah, H. Elhenawy and M. F. Iskander,, "Low Mutual Coupling 2X2 Microstrip Patch Array Antenna by Using Novel Shapes of Defect Ground Structure (DGS)" published in Microwave antenna Optical Technology letters, Vol 52, issue 5, pp.1208-1215, 2010.

[100] D. Nashaat, H. A. Elsadek, E. Abdallah, H. Elhenawy and M. F. Iskander, "Ultra-Wide Bandwidth 2x2 Microstrip Patch Array Antenna by Using Electromagnetic Band-gap Structure (EBG)", IEEE Trans., Antenna and Propagation, vol.5, pp., 2011.

[101] R. Elliot, "On the Theory of Corrugated Plane Surfaces", *IRE Trans. Ant. Prop.*, vol. 2, pp. 71 - 81, 1954.

[102] W. Rotman, "A Study of Single-Surface Corrugated Guides", *Proc. IRE*, vol. 39, pp. 952 - 959, 1951.

[103] P.-S. Kildal, "Artificially Soft and Hard Surfaces in Electromagnetics", *IEEE Trans. Ant. Prop.*, vol. 38, pp. 1537 - 1544, 1990.

17

Designs of True-Time-Delay Lines and Phase Shifters Based on CRLH TL Unit Cells

J. Zhang, S.W. Cheung and T.I. Yuk
Department of Electrical and Electronic Engineering,
The University of Hong Kong, Hong Kong,
China

1. Introduction

1.1 Brief description on definition of electromagnetic metamaterials

Electromagnetic (EM) Metamaterials have specific EM properties that cannot be found in nature (Caloz & Itoh, 2006). These specific EM properties are obtained by the artificial structures, rather than the composition of the metamaterials, and affect the propagations of EM waves when the average structure sizes are much smaller than the guided wavelengths λ_g, i.e., at least smaller than $\lambda_g/4$. Since the propagation of an EM wave is related to its electric and magnetic fields, the EM properties of Electromagnetic (EM) Metamaterials can be described by using its permittivity ε and permeability μ.

Fig. 1 shows the four possible combinations of permittivity and permeability of materials. In quadrant I, we can find materials such as isotropic dielectrics which have $\varepsilon > 0$ and $\mu > 0$. In quadrant II, where $\varepsilon < 0$ and $\mu > 0$, we can find materials like plasmas to have such properties. While in quadrant IV, we also can find ferromagnetic materials to have $\varepsilon > 0$ and $\mu < 0$. However, in quadrant III where both permittivity and permeability are negative, i.e. $\varepsilon < 0$ and $\mu < 0$, there is no natural material having such properties. However, one can construct artificial structures that have negative permittivity and permeability, and these artificial materials are called left-handed (LH) materials as will be explained in the following sections.

1.2 Left-handed (LH) materials

When an EM wave propagates in a conventional material, the electric field **E** and magnetic field **H** are orthogonal to each other and also orthogonal to the wave-vector **k** which has the same direction as the power flow density (also known as Poynting vector **S**). These three vectors, **E**, **H** and **k**, form a right-hand (RH) triplet, so conventional materials can also be called right-handed (RH) materials. In the 1960s, Russian physicist Viktor Veselago theoretically investigated the existence of materials with negative permittivity ε and also negative permeability μ (Veselago, 1968). He speculated that such materials would also satisfy Maxwell's equations but allow the electric field **E**, magnetic field **H** and wave-vector **k** of an EM wave to form a left-handed (LH) triplet. For this reason, the term 'left-handed' was used to describe these theoretical materials. Based on the fact that, in these materials, **E**,

H and **k** formed a LH triplet and **E**, **H** and **S** formed a RH triplet, Veselago showed that, for a uniform plane wave propagating in such materials, the direction of the wave vector **k** would be anti-parallel to the direction of the Poynting vector **S**. The phenomenon is in contrary to the case of plane wave propagation in conventional or RH materials.

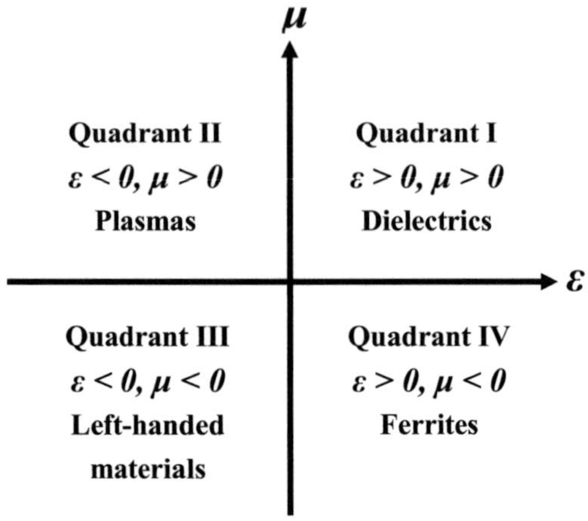

Fig. 1. Permittivity-permeability diagram

1.3 Wave propagation in LH materials by Maxwell's equations

The general form of time-varying Maxwell's equations can be written as (Ulaby, 2004):

$$\nabla \times \mathrm{E} = -\frac{\partial \mathrm{B}}{\partial t} \tag{1a}$$

$$\nabla \times \mathrm{H} = \mathrm{J} + \frac{\partial \mathrm{D}}{\partial t} \tag{1b}$$

$$\nabla \bullet \mathrm{D} = \rho \tag{1c}$$

$$\nabla \bullet \mathrm{B} = 0 \tag{1d}$$

where ∇ : vector operator $\frac{\partial}{\partial x}\hat{\mathrm{x}} + \frac{\partial}{\partial y}\hat{\mathrm{y}} + \frac{\partial}{\partial z}\hat{\mathbf{z}}$,

- $\bullet$: vector dot product,
- **E:** electric field intensity, in V/m,
- **H:** magnetic field intensity, in A/m,
- **D:** electric flux density, in C/m^2,
- **B:** magnetic flux density, in W/m^2,
- **J:** electric current density, in A/m^2, and
- ρ : electric charge density, in C/m^3.

The relationships between the field intensities **E** and **B**, flux densities **D** and **H** and current density **J** are (Ulaby, 2004):

$$D = \varepsilon E, \quad B = \mu H, \quad J = \sigma E \tag{2}$$

where *ε*, *μ* and σ are the electric permittivity (also called the dielectric constant), permeability and conductivity, respectively, of the material under consideration.

Consider a wave having a single frequency *ω*. Then by introducing the time factor $e^{+j\omega t}$ to (1), the time derivatives in Maxwell's equations of (1) can be replaced by *jω* and Maxwell's equations can be re-written as:

$$\nabla \times E = -j\omega\mu H \tag{3a}$$

$$\nabla \times H = (\sigma + j\omega\varepsilon)E \tag{3b}$$

$$\nabla \bullet E = \rho / \varepsilon \tag{3c}$$

$$\nabla \bullet H = 0 \tag{3d}$$

In free space which is lossless, *J*, σ and *ρ* are all zero and (3) becomes:

$$\nabla \times E = -j\omega\mu H \tag{4a}$$

$$\nabla \times H = j\omega\varepsilon E \tag{4b}$$

$$\nabla \bullet \mathbf{E} = 0\rho / \varepsilon \tag{4c}$$

$$\nabla \bullet H = 0 \tag{4d}$$

It can be readily shown from (4) that

$$\nabla^2 E + k^2 E = 0 \tag{5a}$$

$$\nabla^2 H + k^2 H = 0 \tag{5a}$$

where $k = 2\pi/\lambda$ is the wave number (which is real if ε and *μ* are real) with *λ* being the wavelength.

(5) is known as the wave equation which has many solutions. A plane wave is a wave having a constant phase over a set of planes, while a uniform-plane wave is a wave having both magnitude and phase constant. An EM wave in free space is a uniform-plane wave having the electric field **E** and magnetic field **B** mutually perpendicular to each other and also to the direction of propagation, i.e., a transverse electromagnetic (TEM) wave. For convenience and without lost of generality, assume that the EM wave considered is propagating in the +z direction in free space. Under these conditions, the solution for (5) is (Ulaby, 2004):

$$E(z) = E_0 e^{-j\mathbf{k}\bullet\mathbf{z}} \tag{6a}$$

$$H(z) = \mathbf{e_k} \times \frac{1}{\eta_0} E(z) = \frac{\mathbf{e_k} \times E_0(z)}{\eta_0} e^{-j\mathbf{k}\bullet\mathbf{z}} \tag{6b}$$

where $E_0 = E_m e^{j\phi}\mathbf{e_x}$ (with $E_m = |\mathbf{E_0}|$ and $\mathbf{e_x}$ being the unit vector along +x direction)
- η_0 : free space impendence 377 Ω
- **k**: wave vector having magnitude of k at **+z** direction .

Substituting (6a) and (6b) into (4a) and (4b), respectively, yields

$$\mathrm{k} \times \mathrm{E} = \omega\mu \mathrm{H} \tag{7a}$$

$$\mathrm{k} \times \mathrm{H} = -\omega\varepsilon \mathrm{E} \tag{7b}$$

in which, for positive values of ε and μ, the triplet (**E**, **H**, **k**) can be used to produce an orthogonality diagram shown in Fig. 2(a) using our right hand (RH). However, if ε and μ are both negative, (7) becomes:

$$\mathrm{k} \times \mathrm{E} = -\omega|\mu|\mathrm{H} \tag{8a}$$

$$\mathrm{k} \times \mathrm{H} = \omega|\varepsilon|\mathrm{E} \tag{8b}$$

in which now the triplet (**E**, **H**, **k**) forms an orthogonality diagram shown in Fig. 2(b) using our left hand (LH). This result verifies Velago's speculation and this is the primary reason for materials with negative ε and μ to be called LH materials.

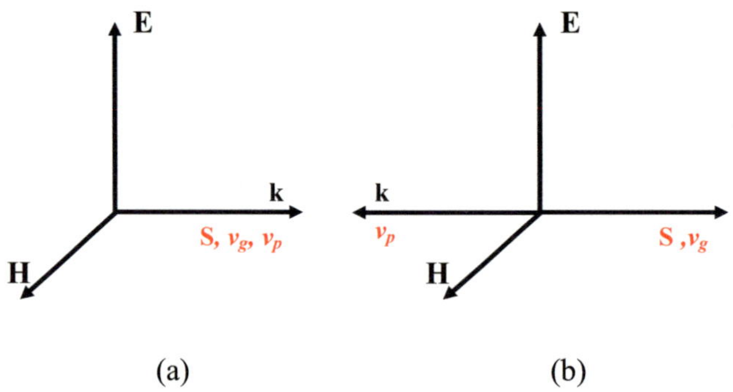

Fig. 2. Orthogonality diagram of E, H, k for uniform plane wave in (a) RH material and (b) LH material

The power flow density of an EM wave, also known as the Poynting vector, is defined as (Ulaby, 2004):

$$\mathrm{S} = \mathrm{E} \times \mathrm{H} \tag{9}$$

(9) indicates that the power flow density is only determined by **E** and **H** but not the signs of ε and μ. Thus in both RH and LH materials, the triplets **(E, H, S)** have the same form of orthogonality, i.e., following our right hand, as shown in Figs. 2(a) and 2(b), and the directions of energy flow are the same.

The group velocity v_g of an EM wave is given by (Ulaby, 2004):

$$v_g = \frac{\partial \omega}{\partial k} \tag{10}$$

which expresses the velocity of power flow and so has the same direction as the Poynting vector **S**, as shown in Fig. 2.

The phase velocity v_p is the velocity of wave-front given by (Ulaby, 2004):

$$v_p = \frac{\omega}{k} \tag{11}$$

and so has the same sign as k, i.e., $v_p > 0$ for $k > 0$ in RH materials and $v_p < 0$ for $k < 0$ in LH materials. As a result, the v_p in LH materials and in RH materials are anti-parallel as indicated in Fig. 2.

Table 1 summarizes the aforementioned analysis for a plane uniform EM wave in the RH and LH materials.

	triplet **(E,H,k)**	k	v_p	triplet **(E,H,S)**	v_g
RH material	RH orthogonality	> 0	> 0	RH orthogonality	> 0
LH material	LH orthogonality	< 0	< 0	RH orthogonality	> 0

Table 1. Characteristics of RH and LH material

2. Transmission line (TL) approach of metamaterials

2.1 Left-handed transmission lines (LH TLs)

A microstrip transmission line fabricated on a substrate is shown in Fig. 3(a). The transmission line is also called a right-handed transmission line (RH TL) because when an EM wave travelling through it, the triplet (**E, H, k**) forms a RH orthogonality, as will be shown later. The equivalent circuit of the transmission line is shown in Fig. 3(b), where L_R models the RH series inductance along the transmission line and C_R models the RH shunt capacitance between the transmission line and the ground on the other side of the substrate (Pozar, 2004).

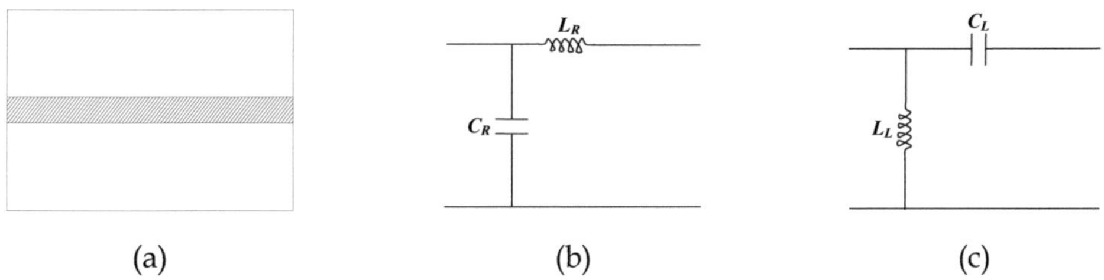

Fig. 3. (a) Structure of RH TL (b) equivalent circuit of RH TL and (c) equivalent circuit of LH TL

To simplify our analysis, we assume that the transmission line is lossless. The complex propagation constant γ, the phase constant β, the phase velocity v_p and the group velocity v_g of the RH TL are, respectively, given by (Pozar, 2004):

$$\gamma = j\beta = \sqrt{ZY} = j\omega\sqrt{L_R C_R} \tag{12a}$$

$$\beta = \omega\sqrt{L_R C_R} > 0 \tag{12b}$$

$$v_p = \frac{\omega}{\beta} = \frac{1}{\sqrt{L_R C_R}} > 0 \tag{12c}$$

$$v_g = \frac{\partial\omega}{\partial\beta} = \frac{1}{\sqrt{L_R C_R}} > 0 \tag{12d}$$

By making a duality of the equivalent circuit in Fig. 3(b), i.e. replacing the series inductance L_R with a series capacitance C_L and the shunt capacitance C_R with a shunt inductance L_L, we can have the equivalent circuit of a left-handed transmission line (LH TL) (Lai et al., 2004) shown in Fig. 3(c). It is called a LH TL simply because when an EM wave travelling through it, the triplet (**E, H, k**) forms a LH orthogonality, as will be shown later. Now the complex propagation constant γ, the phase constant β, the phase velocity v_p and the group velocity v_g of the LH TL can be found as

$$\gamma = j\beta = \sqrt{ZY} = -j\omega\frac{1}{\sqrt{L_L C_L}} \tag{13a}$$

$$\beta = -\frac{1}{\omega\sqrt{L_L C_L}} < 0 \tag{13b}$$

$$v_p = \frac{\omega}{\beta} = -\omega^2\sqrt{L_L C_L} < 0 \tag{13c}$$

$$v_g = \frac{\partial\omega}{\partial\beta} = \omega^2\sqrt{L_L C_L} > 0 \tag{13d}$$

Here the phase constant β (= $2\pi/\lambda$) in (13b) (equivalent to the wave number k used in wave propagation previously) is negative and so the phase velocity v_p in (13c) associated with the direction of phase propagation is negative. However, the group velocity v_g in (13d) indicating the direction of power flow (Poynting vector S) remains positive. This characteristic agrees with that of LH materials shown in Table 1, so the LC circuit shown in Fig. 3(c) can be used to realize LH materials.

2.2 Composite right/left-handed transmission line (CRLH TL)

The transmission line structure shown in Fig. 4(a) was proposed in (Lai et al., 2004) to realize the series capacitance C_L and shunt inductance L_L in Fig. 3(c) for the LH materials (Lai et al., 2004). The structure consists of a series inter-digital capacitors to realize the series capacitance C_L and a via shunted to ground on the other of the substrate to realize the shunt inductance L_L. However, when an EM wave travels along the structure, the current flowing along the upper metal trace induces a magnetic field, creating an inductive effect. This effect is modeled by the series inductance L_R in Fig. 4(b). Moreover, the potential difference generated between the upper metal trace and the ground plane on the other side produces an electric field, creating a capacitive effect. This effect is modeled by the shunt capacitance C_R in Fig. 4(b). Since the inductive and capacitive effects caused by the series inductance L_R and shunt capacitance C_R, respectively, cannot be avoided in practical implementation of LH TLs, the term "composite right/left-handed transmission line" (CRLH TL) is used to describe such a structure.

(a)

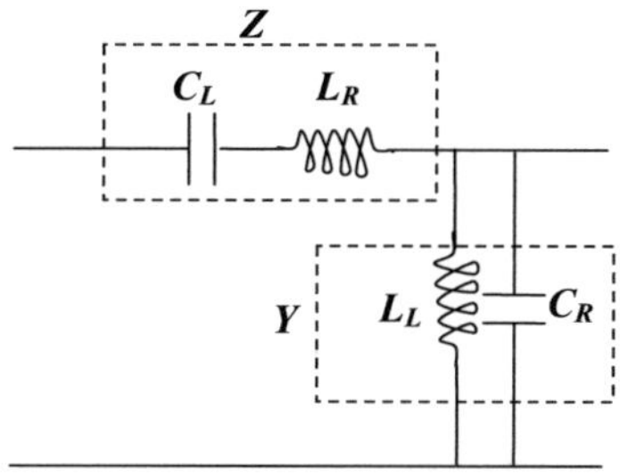

(b)

Fig. 4. (a) CRLH TL structure and (b) equivalent circuit

2.3 Dispersion diagrams

Here, we show that the structure of Fig. 4(a) indeed has LH and RH properties shown in Table 1. The complex propagation constant of the CRLH TL structure in Fig. 4(b) can be written as (Caloz & Itoh, 2006; Lai et al., 2004)

$$\gamma = \alpha + j\beta = \sqrt{ZY} \tag{14}$$

where α and are β are the attenuation and phase constants, respectively. Assume the structure, as represented by Fig. 4(a), is lossless,and so has no attenuation, i.e. $\alpha = 0$ in (14). The propagation constant is an imaginary number:

$$\gamma = j\beta = \sqrt{ZY} = \sqrt{(j\omega L_R + \frac{1}{j\omega C_L})(j\omega C_R + \frac{1}{j\omega L_L})} \tag{15a}$$

$$= \sqrt{(-1)}\sqrt{\omega^2 L_R C_R + \frac{1}{\omega^2 L_L C_L} - (\frac{L_R}{L_L} + \frac{C_R}{C_L})} \tag{15b}$$

$$= js(\omega)\sqrt{\omega^2 L_R C_R + \frac{1}{\omega^2 L_L C_L} - (\frac{L_R}{L_L} + \frac{C_R}{C_L})} \tag{15c}$$

where

$$s(\omega) = \pm 1 \tag{16}$$

In the equivalent circuit of Fig. 4(b), the components C_L and L_R form a series-tune circuit with a resonant frequency at $1/\sqrt{L_R C_L}$. At frequencies larger than $1/\sqrt{L_R C_L}$, the circuit is inductive. While the components C_R and L_L form a parallel-tune circuit resonating at $1/\sqrt{L_L C_R}$. At frequencies larger than $1/\sqrt{L_L C_R}$, the circuit is capacitive. Thus, at frequencies $\omega > \max(1/\sqrt{L_R C_L}, 1/\sqrt{L_L C_R})$, the series-tune circuit is inductive and the parallel-tune circuit is capacitive. The CRLH TL structure will have an equivalent-circuit model similar to one shown in Fig. 3(b) and so behaves like a RH TL. Under this condition, from (12b), the propagation constant is a positively imaginary number, i.e., $\gamma = j\beta$, and so $s(\omega) = +1$ in (16). At frequencies less than $1/\sqrt{L_R C_L}$, the series-tune circuit formed by C_L and L_R is capacitive. At frequencies less than $1/\sqrt{L_L C_R}$, the parallel-tune circuit formed by C_R and L_L is inductive. Thus at frequencies $\omega < \min(1/\sqrt{L_R C_L}, 1/\sqrt{L_L C_R})$, the series-tune circuit is capacitive and the parallel-tune circuit is inductive. Now the CRLH TL structure has an equivalent-circuit model similar to one shown in Fig. 3(c) and so behaves like a LH TL. Under this condition, from (13b), the propagation constant is a negatively imaginary number i.e., $\gamma = -j\beta$ and so $s(\omega) = -1$ in (16). Between these two limits, i.e., $\min(1/\sqrt{L_R C_L}, 1/\sqrt{L_L C_R}) < \omega < \max(1/\sqrt{L_R C_L}, 1/\sqrt{L_L C_R})$, the radicand in (15c) is purely imaginary and so the propagation constant is purely real at $\gamma = \beta$. There is only attenuation in the CRLH TL structure which behaves as a stop-band filter. This stop band is a unique characteristic of the CRLH TL.

Therefore, the propagation constant of a CRLH TL structure can be re-written from (15) as:

$$\gamma(\omega) = \begin{cases} j\beta(\omega) = j\sqrt{\omega^2 L_R C_R + \dfrac{1}{\omega^2 L_L C_L} - \left(\dfrac{L_R}{L_L} + \dfrac{C_R}{C_L}\right)} \\ \quad \text{If } \omega > \max\left(\dfrac{1}{\sqrt{L_R C_L}}, \dfrac{1}{\sqrt{L_L C_R}}\right) \\ j\beta(\omega) = -j\sqrt{\omega^2 L_R C_R + \dfrac{1}{\omega^2 L_L C_L} - \left(\dfrac{L_R}{L_L} + \dfrac{C_R}{C_L}\right)} \\ \quad \text{If } \omega < \min\left(\dfrac{1}{\sqrt{L_R C_L}}, \dfrac{1}{\sqrt{L_L C_R}}\right) \\ \beta(\omega) = \sqrt{-\omega^2 L_R C_R - \dfrac{1}{\omega^2 L_L C_L} + \left(\dfrac{L_R}{L_L} + \dfrac{C_R}{C_L}\right)} \\ \quad \text{If } \min\left(\dfrac{1}{\sqrt{L_R C_L}}, \dfrac{1}{\sqrt{L_L C_R}}\right) < \omega < \max\left(\dfrac{1}{\sqrt{L_R C_L}}, \dfrac{1}{\sqrt{L_L C_R}}\right) \end{cases} \tag{17}$$

The dispersion diagrams of the RH TL, LH TL, and CRLH TL plotted using (12b), (13b) and (17), respectively, are shown in Fig. 5 (Caloz & Itoh, 2006; Lai et al., 2004). For the RH TL, Fig. 5(a) shows that the group velocity (i.e., $v_g = d\omega/d\beta$) is positive and has values only for $\beta > 0$. Since phase velocity is defined as $v_p = \omega/\beta$, Fig. 5(a) shows that the RH TL has $v_p > 0$. Fig. 5(b) shows that the LH TL also has a positive group velocity ($v_g = d\omega/d\beta$) but only for $\beta < 0$ which leads to $v_p < 0$ (because $v_p = \omega/\beta$). These characteristics agree with those of the RH and LH materials shown in Table 1. Thus the circuit models in Figs. 3(b) and 3(c) realized using transmission lines have the EM properties of the RH and LH materials, respectively. In Fig. 5(c), the regions for $\beta > 0$ and $\beta < 0$ are known here as the RH and LH regions, respectively.

Now, consider the dispersion diagram in Fig. 5(c). It can be seen that the CRLH TL behaves like a LH TL for $\beta < 0$ ($v_p < 0$, $v_g > 0$) and like a RH TL for $\beta > 0$ ($v_p > 0$, $v_g > 0$). Thus the circuit of Fig. 4(b) indeed has both the LH and RH properties. Fig. 5(c) shows that the CRLH TL has quite small group velocities in the LH region and much large group velocities in the RH region. The LH region with low group velocities can be used to implement TTDLs with high time-delay efficiencies as described later. Moreover, the CRLH TL has a stopband in the frequency range: $\min(1/\sqrt{L_R C_L}, 1/\sqrt{L_L C_R}) < \omega < \max(1/\sqrt{L_R C_L}, 1/\sqrt{L_L C_R})$, where $\beta = 0$, leading to a zero group velocity v_g and meaning zero power flow.

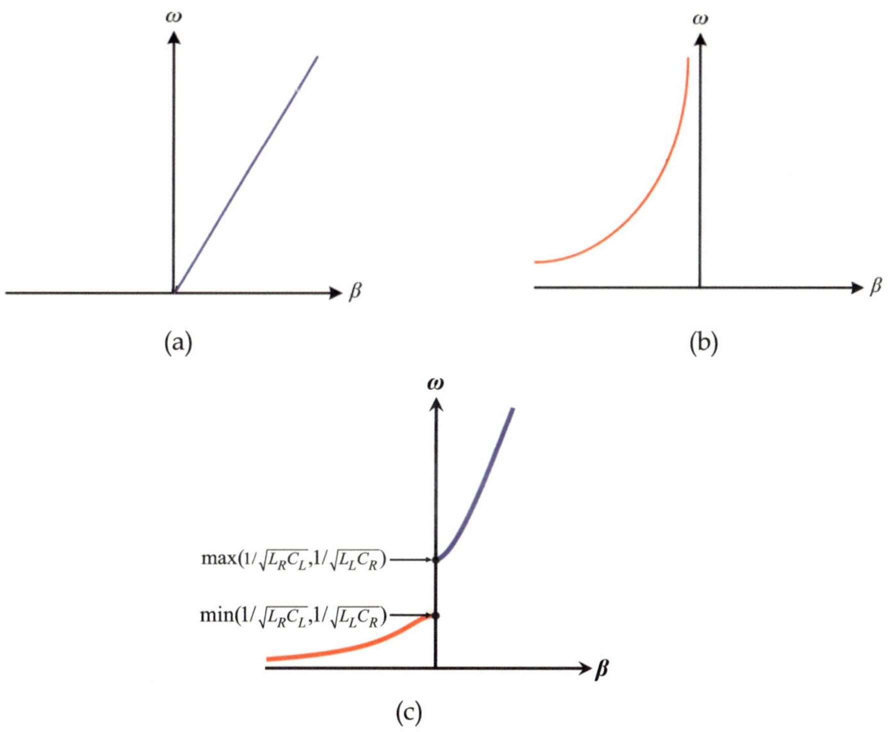

Fig. 5. Dispersion diagrams of (a) RH TL, (b) LH TL and (c) CRLH TL

3. True-time-delay lines (TTDLs) using CRLH TL unit cells

3.1 Introduction

True-time-delay lines (TTDLs) are widely employed in various microwave devices and subsystems. They find applications in phased arrays, feed-forward amplifiers, delay-lock loops, phase noise measurement systems and oscillators (Lee, 2004). There are different approaches to implement the TTDLs. For example, in magnetostatic wave (MSW) TTDLs (Fetisov & Kabos, 1998) and surface-acoustic wave (SAW) TTDLs (Smith & Gerard, 1969), the group velocities were slowed down by transferring the microwave signals into the magnetostatic waves and surface acoustic waves, respectively. These designs have quite high time-delay efficiencies, but the bulky and complicated transducers required are not conducive for planar microwave circuits. Moreover, SAW TTDLs have extremely narrow bandwidths of only several MHz and MSW TTDLs have very large insertion losses. Optical TTDLs (Xu et al., 2004) have very small insertion losses, so we can use very long optical fiber cables to achieve very large TDs. However, optical TTDLs also need complicated transducers to transfer the microwave signals into the optical waves.

Microstrip lines, with the advantages of simplicity in feeding and compatibility with planar circuits, are widely used in communications systems, particularly for small communications devices. A simple RH TL can also be used to implement TTDL. However, due to the low time-delay efficiency, it is difficult to achieve a long TD using RH TL. To increase the time-delay efficiency of RH TL, various slow-wave structures have been proposed. Some employ periodic discontinuities such as the Electromagnetic Bandgap Structures (EBG) (Kim & Drayton, 2007) and defected ground structure (DGS) (Woo et al., 2008). Others use periodic equivalent LC networks (Zhang & Yang, 2008; Zhang et al., 2011). All of these designs adapt the same basic concept,of increasing the series inductance and shunt capacitance per unit length, hence the effective dielectric, to reduce the guided wavelength of the EM waves.

As described previously, the CRLH TL has a unique dispersion characteristic, i.e., having small group velocities in the LH region. Thus a CRLH TL operating in the LH region can be used to realize a TTDL with a high time-delay efficiency. In this section, a TTDL using four symmetrical CRLH TL unit cells is studied.

3.2 TTDL realized by transmission line (TL)

The true-time delay (TTD) of a transmission line (TL) is the time it takes for an EM wave to travel through it, so a TL can be used to design a TTDL. For a TL with a length of *L*, the TTD is:

$$\tau = L / v_g \tag{18}$$

where v_g is the group velocity given by

$$v_g = d\omega / d\beta \tag{19}$$

with β and ω being the propagation constant and frequency in rad/s, respectively. (18) and (19), show that the TTD is inversely proportional to the group velocity v_g. For a given length of TL, the smaller is the v_g, the longer will be the TD.

3.3 Symmetrical CRLH TL unit cell

To design a TTDL with a large TTD, we propose to cascade a number of CRLH TL structures together, each having a small group velocity. For matching purpose, the CRLH TL structure has to be designed so that S_{11} = S_{22} over the operating frequency band. However, this is not easy to do using the CRLH TL structure shown in Fig. 4(a) because adjusting any of the structural parameters will change both S_{11} and S_{22}. To overcome such difficulty, we propose a new structure as shown in Fig. 6 which has a symmetrical structure and is here called a symmetrical CRLH TL unit cell (Zhang et al., 2009). Compared with the CRLH TL structure as shown in 4(a), our proposed CRLH TL unit cell has two stubs, instead of one stub, having a grounded via, making the whole structure centrosymmetrical. To deign such unit cell for use in our TTDL, we only need to match the input impedance to a 50-Ω coaxial cable, which is relatively easy to do. Once this matching is designed, due to its symmetrical structure, the two Z-parameters, Z_{11} and Z_{22}, will be the same and equal to 50 Ω. The symmetry of the unit cell leads to an equivalent-symmetrical π-model shown in Fig. 6(c) in which L_R models the RH series inductance along all the horizontal fingers, C_L models the LH coupling capacitance between the fingers, $2L_L$ models the LH shunt inductance of each stub having a via at the end to the ground, and $C_R/2$ models the RH shunt capacitance between the fingers and the ground on each side of the unit cell.

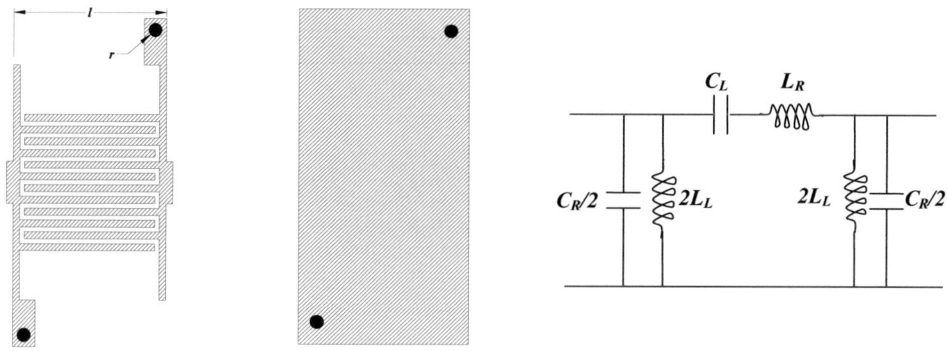

(a) (b) (c)

Fig. 6. Proposed CRLH TL unit cell. (a) Top view, (b) bottom, and (c) equivalent circuit.

Here, we will show that the CRLH TL unit cell in Fig. 6(a) has the same dispersion diagram shown in Fig. 5 (c), so that we can have the LH region to operate the unit cell. The expressions for L_R, C_R, L_L and C_L in Fig. 6(c) are, respectively, given by (Bahl, 2001; Marc & Robert, 1991; Pozar, 2004):

$$L_R = \frac{Z_0\sqrt{\varepsilon_{re}}}{Nc}l \tag{20a}$$

$$C_R = \frac{N\sqrt{\varepsilon_{re}}}{Z_0 c}l \tag{20b}$$

$$C_L = (\varepsilon_r + 1)l\left\{4.409(N-3)\tanh\left[0.55\left(\frac{h}{W}\right)^{0.45}\right] + 9.92\tanh\left[0.52\left(\frac{h}{W}\right)^{0.5}\right]\right\}\times 10^{-12} \tag{20c}$$

$$L_L = \frac{\mu_0}{2\pi}\left[h\cdot\ln\left(\frac{h+\sqrt{r^2+h^2}}{r}\right)+\frac{3}{2}(r-\sqrt{r^2+h^2})\right]+\frac{Z_0\sqrt{\varepsilon_{re}}}{c}l' \tag{20d}$$

where h is the thickness of the substrate, r is the radius of the ground via, ε_r is the relative dielectric constant, ε_{re} is the effective dielectric constant, l' is the distance from the ground via to the port, l is the length of the finger, N is the number of fingers, W is the width of all the fingers together and Z_0 is the characteristic impedance of each of the fingers. To derive the complex propagation constant of this symmetrical CRLH TL unit cell, we separate the L_R and C_L in Fig. 6(c) into two capacitances $2C_L$ and two inductances $L_R/2$, respectively. This results in two sub-circuits as shown in Fig. 7, each having the same propagation constant γ'. The total propagation constant γ is the sum of the propagation constants for these two sub-circuits, i.e., $\gamma = 2\gamma'$. In Fig. 7, the propagation constant of the symmetrical CRLH TL unit cell is:

$$\gamma = j\beta = 2\gamma' = 2\sqrt{ZY} = 2\sqrt{(j\omega\frac{L_R}{2} + \frac{1}{j\omega 2C_L})(j\omega\frac{C_R}{2} + \frac{1}{j\omega 2L_L})}$$

$$= \sqrt{(-1)}\sqrt{\omega^2 L_R C_R + \frac{1}{\omega^2 L_L C_L} - (\frac{L_R}{L_L} + \frac{C_R}{C_L})}$$

$$= js(\omega)\sqrt{\omega^2 L_R C_R + \frac{1}{\omega^2 L_L C_L} - (\frac{L_R}{L_L} + \frac{C_R}{C_L})} \quad (21)$$

$$\text{where } s(\omega) = \pm 1$$

(21) shows that the propagation constant is exactly the same as (15c). Thus the symmetrical CRLH TL unit cell has the same dispersion diagram shown in Fig. 5(c).

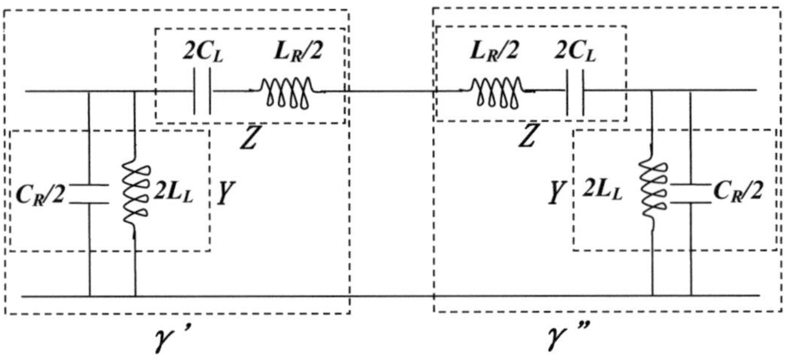

Fig. 7. Equivalent circuit of Fig. 6(c)

3.4 TTDL realization using symmetrical CRLH TL unit cells

From (18) and (19), the TD of a TL is given by

$$\tau = \frac{L}{v_g} = \frac{L}{d\omega / d\beta} = \frac{Ld\beta}{d\omega} = \frac{Ang(s_{21})}{2\pi df} \quad (22)$$

where $Ang(S_{21})$ is the phase of S_{21}. Figure 5(c) shows that the CRLH TL has quite small group velocities v_g in the LH region and much higher group velocities in the RH region. Thus the LH region with low group velocities can be used to implement TTDLs with longer TTDs and high time-delay efficiencies as indicated in (22). It can be see that (22) also indicate that, for a given length L, the larger is the slope of phase response, i.e., $Ang(S_{21})/df$, the longer will be the TD or the higher will be the time-delay efficiency. Moreover, by adjusting

L_R, C_R, L_L and C_L using thc structural parameters through (20), the CRLH TL unit cell can also be designed to operate at different frequencies and TDs.

3.5 Simulation and measurement results

3.5.1 Single CRLH TL unit cell

The proposed symmetrical CRLH TL unit cell shown in Fig. 6 has been designed with a center frequency at around 3 GHz on a Rogers substrate, RO4350, with a thickness of 0.762 and a permittivity of 3.48 using computer simulation. It has a total a length of 6.8 mm. The design is optimized for the criteria for wide impedance bandwidth, small insertion and large phase response. The simulated results on return loss, -10log$|S_{11}|$, and insertion loss, -10log$|S_{21}|$, of the CRLH TL unit cell are shown in Fig. 8(a), while the phase response, (Ang(S_{21})), shown in Fig. 8(b). For comparison, the results of a RH TL with the same length of 6.8 mm are also shown in the same figure. Figure 8(a) shows that the CRLH unit cell has an operating bandwidth of 2.2-4.0 GHz, a return loss of large than 15 dB and insertion loss of less than 1 dB within the operating bandwidth. The slope of phase response, as shown in Fig. 8(b), is 41.3 degree/GHz, about 3 times larger that of the RH TL at 13.3 degree/GHz. Thus we can expect that the CRLH TL unit cell can achieve a TTD 3.1 times longer than that of a RH TL for the same length and so has the time delay efficiency 3.1 times higher than that of the RH TL.

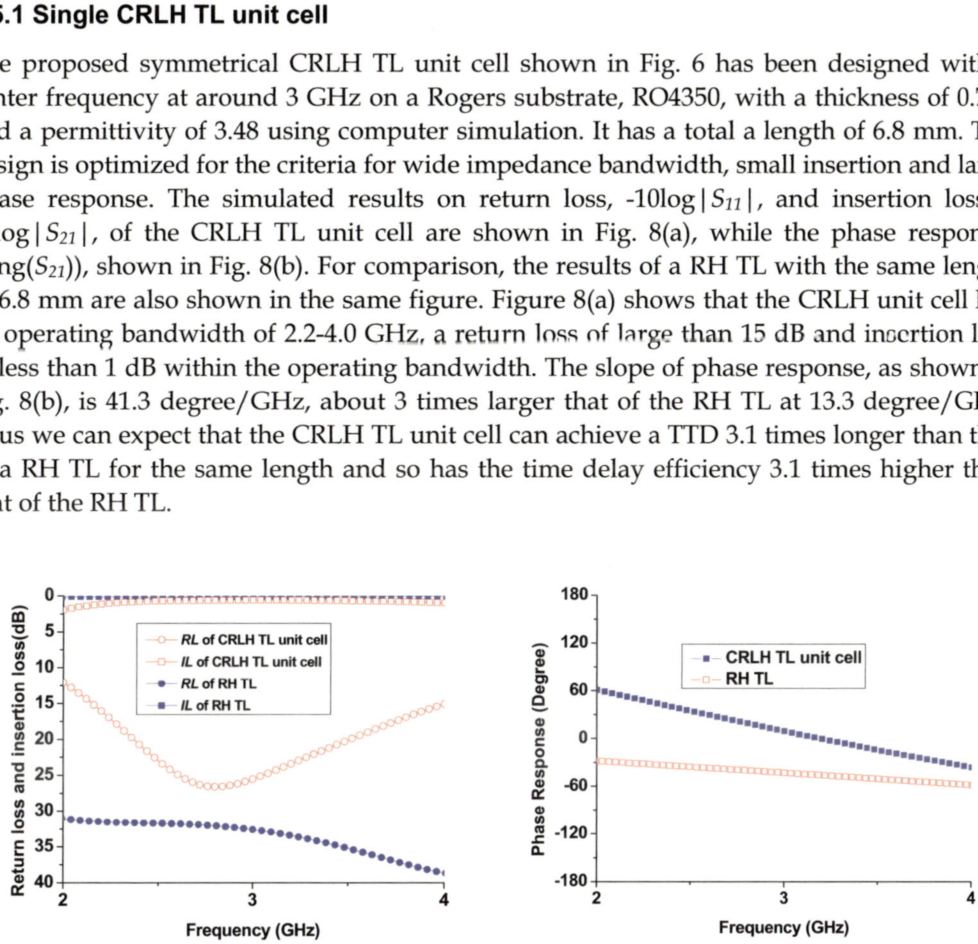

(a) (b)

Fig. 8. Simulated (a) return loss (RL) and insertion loss (IL) and (b) phase response of CRLH TL unit cell and RH TL

3.5.2 Multi-CRLH TL unit cells

A TTDL constructed by cascading four symmetrical CRLH unit cells, with a total length of 30 mm, is shown in Fig. 9(a). The TTDL has been designed, studied and optimized using computer simulation. The final design has also been implemented on a Rogers substrate, RO4350, with a thickness of 0.762 and a permittivity of 3.48 and measured for verification of simulation results. For comparison, a TTDL implemented using a RH TL with the same length of 30 mm, as shown in Fig. 9(b), has also been designed and simulated using the same substrate. The prototype-modules of the two TTDLs with the same dimension of 30 mm×15 mm×2 mm are shown in Fig. 10.

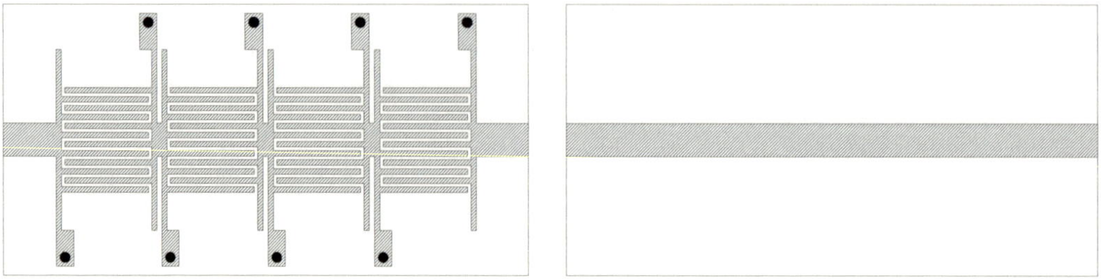

(a) (b)

Fig. 9. TTDLs using (a) four CRLH TL unit cells and (b) RH TL

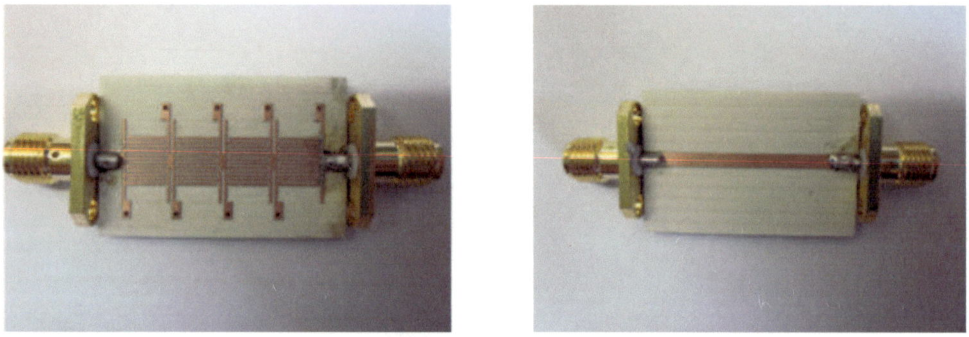

(a) (b)

Fig. 10. Prototypes of TTDLs using (a) four CRLH TL unit cells and (b) RH TL

The simulated and measured return losses, insertion losses and TTDs of the two TTDLs are shown in Fig. 11. It can be seen that the simulated and measured results show good agreements. Figure 11(a) show that the measured return losses and the insertion losses of the TTDLs are more than 15 dB and less than 1 dB, respectively, across the frequency band from 2.2 - 3.7 GHz. The measurement results in Fig. 11(b) show that the TTDL using CRLH TL unit cells achieves a TTD of 510 ps, about 3.2 times larger than the RH TL having a TTD of 160 ps. This result is consistent with the result obtained for a single CRLH TL unit cell. The measurement results in Fig. 11(b) show that the maximum TTD error for the TTDL using the CRLH TL unit cells is about -21 ps or -4.1% at the frequency of 2.4 GHz. For the TTDL using RH TL, although the maximum TD error is about -7.8 ps at the frequency of 2.8 GHz, the percentage is higher at -4.8%.

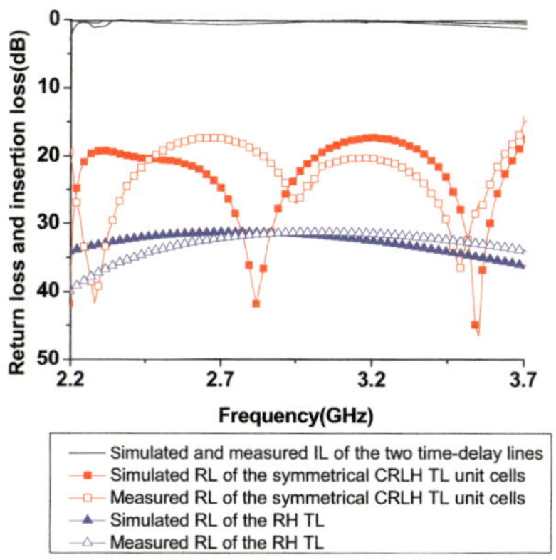

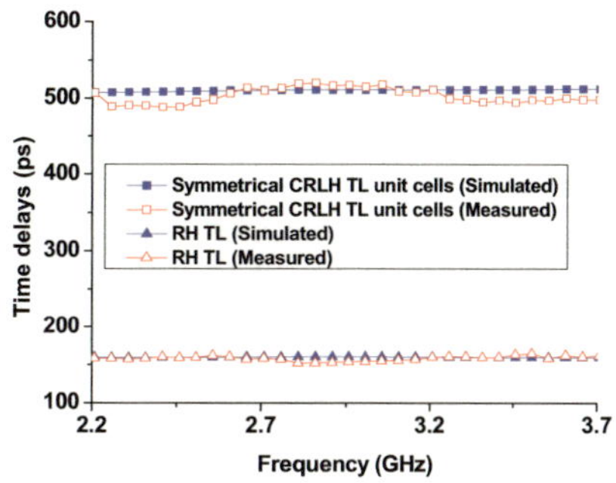

(a) (b)

Fig. 11. Simulated and measured (a) return loss and insertion loss and (b) TTDs of TTDLs using CRLH TL unit cells and RH TL

4. Digital-phase shifters using CRLH unit cells

4.1 Introduction

Phase shifters are essential components in radar and phased array systems. In the designs of phase shifters, insertion loss, size and power-handling capability are major factors for considerations. Phase shifters can be classified into passive and active. Passive phase shifters can be implemented using ferrite technology (Adam et al., 2002) to achieve higher power capabilities, but they have large sizes and heavy weights. Active phase shifters implemented using solid-state devices such as FET and CMOS technologies have smaller sizes. However, their low power-handling capabilities and nonreciprocal characteristics limit their applications.

In digital-phase shifters, phase shift is usually obtained by switching between two transmission lines of different lengths or between lumped-element low-pass and high-pass filters (Keul & Bhat, 1991). Usually, the switches are implemented using solid-state devices such as PIN diodes which have typical power-handling capabilities of just a few Watts and this limits the power-handling capabilities of the phase shifters. Moreover, if an n-bit phase shifter is constructed by cascading several phase shifters together to provide the required phase shifts, the physical size and insertion loss will be undoubtedly increased.

Recently, different design approaches of phase shifters based on using metamaterials have been proposed and studied (Antoniades & Eleftheriades, 2003; Damm et al., 2006; Kim et al., 2005; Kholodnyak et al., 2006; Lapine et al., 2006; Vendik et al., 2009;). These designs share one of the major drawbacks, i.e., the power-handling capability is limited by the power-handling capabilities of the switches, tunable diodes and tunable capacitors used in the designs. A phase shifter based on CRLH TL employing MEMS has also been proposed and studied (Monti et al., 2009), but the reliability of the design needs more studies (Rebeiz et al., 2002).

In this section, we present an approach for the design of n-bit phase shifters using the CRLH TL unit cells. The phase shifters designed using this approach have the advantages of compact size, high power-handling capability, low insertion loss, arbitrary phase-shift range and arbitrary step size. PIN diodes mounted on the fingers of the CRLH TL unit cell are used as switches to control the phase shift. Different phase shifts are achieved by using different states of these switches and the controlling bits are used to select one of those switch states for the required phase shift.

4.2 Description of CRLH TL unit cell using ABCD-parameters

The symmetrical CRLH TL unit cell used for the designs of digital-phase shifters is shown in Fig. 12(a) with the equivalent π-model circuit shown in Fig. 12(b). The symmetrical CRLH TL unit cell and the equivalent π-model circuit are similar to those shown in Fig. 6(a), thus the expressions for L_R, C_R, L_L and C_L are same as given in (20a)-(20d).

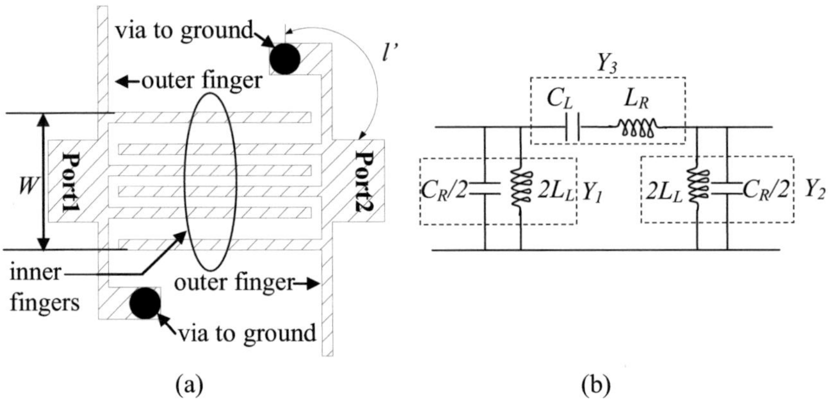

Fig. 12. Structure and equivalent circuit of symmetrical CRLH TL unit cell.

The transmission ABCD matrix of the π-model circuit in Fig. 12 (b) is (M.D. Pozar, 2004):

$$\begin{bmatrix} A & B \\ C & D \end{bmatrix} = \begin{bmatrix} 1+Y_2/Y_3 & 1/Y_3 \\ Y_1+Y_2+Y_1Y_2/Y_3 & 1+Y_1/Y_3 \end{bmatrix} \tag{23}$$

where $Y_1 = Y_2 = j\omega C_R/2 + 1/2j\omega L_L$ and $Y_3 = \dfrac{1}{j\omega L_R + 1/j\omega C_L}$.

Using (23), the expressions for S_{11}, S_{21} and phase incursion Ang(S_{21}) can be easily obtained, respectively, as:

$$S_{11} = \frac{(\frac{b}{Z_0})^2 - Z_0^2 a^2 [2-ab]^2}{4[1-ab]^2 + [Z_0 a^2 b - 2Z_0 a - (\frac{b}{Z_0})]^2} + j\frac{2[1-ab][2Z_0 a - \frac{b}{Z_0} - a^2 b Z_0]}{4[1-ab]^2 + [Z_0 a^2 b - 2Z_0 a - \frac{b}{Z_0}]^2} \tag{24a}$$

$$S_{21} = \frac{4[1-ab]}{4[1-ab]^2 + [Z_0a^2b - 2Z_0a - \frac{b}{Z_0}]^2} - 2j\frac{[Z_0a^2b - 2Z_0a - \frac{b}{Z_0}]}{4[1-ab]^2 + [Z_0a^2b - 2Z_0a - \frac{b}{Z_0}]^2} \tag{24b}$$

and

$$Ang(S_{21}) = \tan^{-1}[-\frac{Z_0a^2b - 2Z_0a - \frac{b}{Z_0}}{2(1-ab)}] \tag{24c}$$

$$\text{where } a = \frac{1-\omega^2 L_L C_R}{2\omega L_L} \text{ and } b = \frac{1-\omega^2 L_R C_L}{\omega C_L}$$

have been used in (24a)–(24c). As expressed in (20a)–(20d), the structural parameters of the CRLH TL unit cell can be used to determine the values of L_R, C_R, L_L and C_L and in turn to determine the phase incursion through (24c) and operation frequency through (24a) and (24b).

4.3 Designs of digital-phase shifters using CRLH unit cells

4.3.1 Basic ideas

The symmetrical CRLH TL unit cell in Fig. 12 (a) is used as the basic cell to design digital-phase shifters here. The advantage of using symmetrical CRLH TL unit cells is that when more unit cells are cascaded together to provide a more states, there is no need to perform any matching between adjacent unit cells.

Here, we use Fig.13 to illustrate our proposed idea of using CRLH TL unit cells to design digital-phase shifters. In the figure, we mount four switches on four different fingers of a unit cell. For an n-bit phase shifter, there will be a total of 2^n states, from 0 to 2^n-1, determined by "closed" or "open" states of the switches on the unit cell. The n-controlling bits are used to select a particular state from these 2^n-1 states and hence to provide a particular phase shift. The "closed" or "open" states in the switches determine the values L_R, C_R, L_L and C_L through (20a)-(20d) and, in turn, determine the phase incursion $Ang(S_{21})$ in (24c), so each switch state can be used to provide a particular phase shift, i.e.,

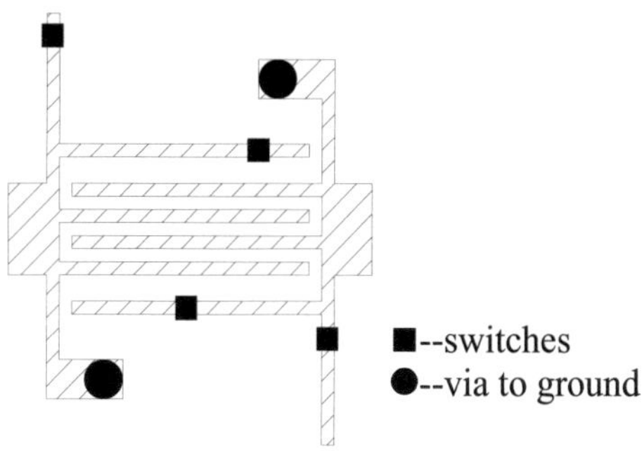

Fig. 13. Symmetrical CRLH TL unit cell mounted with four switches

$$\Delta\theta_m = Ang(S_{21})_m - Ang(S_{21})_0 \tag{25}$$

where m is the index (from 0 to 2^n-1) for the switch states, and $(S_{21})_m$ and $(S_{21})_0$ are the values of S_{21} in the mth-switch and zeroth-switch states, respectively. For convenience and without lost of generality, the zeroth-switch state is taken as the state with all switches closed.

For high power applications such as radars, power-handling capability is one of the important concerns in the design of phase shifters. In our design, the power-handling capacity of the switches (i.e., PIN diodes in our case) used in the CRLH TL unit cell determines the power-handling capacity of the phase shifter. Here, surface-current density distribution is used to study the power-handling capacity of the switches. Computer simulation results on the surface-current density distribution of the CRLH TL unit cell on a Rogers substrate, RO5880, with eight fingers and four switches, are shown in Fig. 14. Figs. 14(a) and 14(b) show the surface-current density distributions with all four switches "opened" or "closed", respectively, at 9.5 GHz. The arrowheads indicate the positions of the switches on the fingers of the CRLH TL unit cell. With all switches "opened", Fig. 14 (a) shows that the largest surface-current density flowing through the switches is about -28 dB below (or 1/25.1 of) those flowing through the input and output ports. While with all switches "closed", Fig. 14 (b) shows that the largest surface-current density flowing through the switches is about -16 dB below (or 1/6.3 of) those flowing through the input and output ports. The width of the finger is only 1/7 of the port, so the largest surface-current through the switches is 16.43 dB below (or 1/44 of) those flowing through the ports. Thus the power-handling capability of the phase shifter is about 44 times higher than the power-handling capability of a switch (PIN diode).

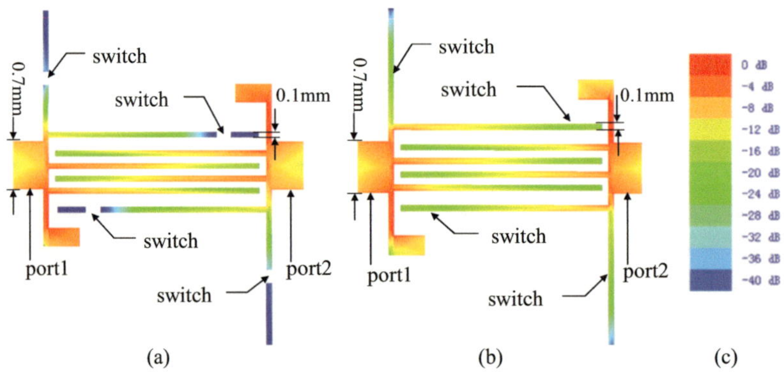

Fig. 14. Simulated surface-current density distribution on symmetrical CRLH TL unit cell with (a) all switches "opened", (b) all switches "closed", and (c) color scale

4.3.2 2-bit phase shifter

A 2-bit phase shifter using a single CRLH TL unit cell is shown in Fig. 15(a), where two switches are mounted on two different fingers (Zhang et al., 2010). The positions of the switches in the "open" and "close" states determine the values of L_R, C_R, L_L and C_L of the unit cell through (20a) - (20d) and hence the phase incursion. The state with all switches closed is taken as the zeroth-switch state. Simulation studies are used to determine the positions of the switches on the top and bottom fingers of the unit cell in order to achieve the phase shifts of 45^0 and 22.5^0, respectively. The phase shifts of 67.5^0 and 0^0 are provided by "opening" and "closing", respectively, both two switches. Table 2 shows the phase shift for different input logic patterns. It should be noted that, although the phase shifts of 0^0, 22.5^0, 45^0 and 67.5^0 with equal step size of 22.5^0 are used in our design, other phase shift values and step sizes are just possible and easily be achieved. The layout of the 2-bit phase shifter with the DC bias circuits for the PINs (the switches) is shown in Fig. 15(b), where RF_{in} and RF_{out} are the input and output RF signals, respectively, and the input bits have the DC voltages of 0 and 5 V representing the two corresponding logic levels "1" and "0".

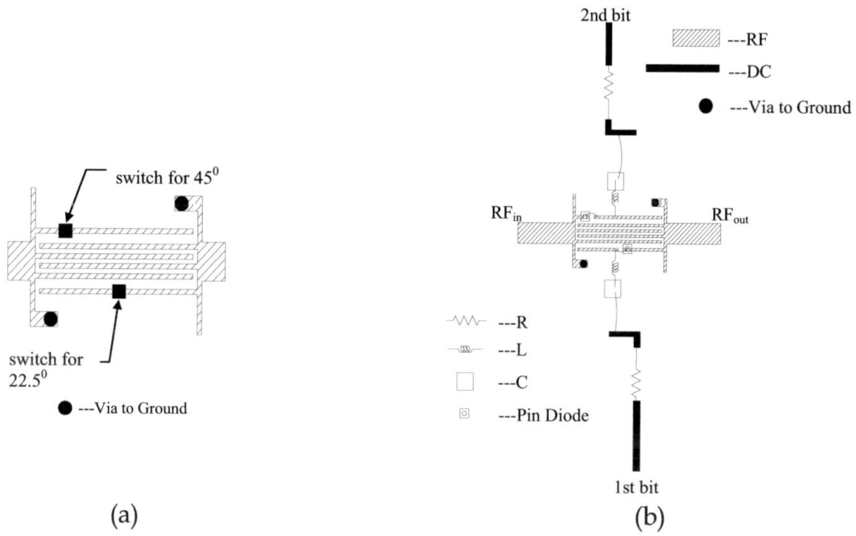

Fig. 15. (a) 2-bit phase shifter using CRLH TL unit cell and (b) layout of 2-bit phase shifter with DC bias circuits

States	Phase shifts
00	0^0
01	22.5^0
10	45^0
11	90^0

Table 2. Phase shifts for different states of 2-bit phase shifter

4.3.3 3-bit phase shifter

Fig. 16(a) shows a 3-bit phase shifter using a single CRLH TL unit cell where four switches are used to achieve eight different switch states and hence eight phase shifts of 0^0, 22.5^0, 45^0, 67.5^0, 90^0, 112.5^0, 135^0 and 157.5^0 (Zhang et al., 2010). Note that the number of controlling bits is less than the number of switches on the unit cell because two switches may be used to provide a phase shift. The state with all switches closed is taken as the zeroth-switch state. Again, computer simulation is used to determine the positions of these switches on the fingers in order to achieve these phase shifts. In Fig. 16(a), the two switches on the inner fingers are used to provide the phase shifts of 22.5^0 and 45^0. The two switches on the outer fingers serving in pair are used to provide a total phase shift of 90^0. Different switch states are used to provide other phase shifts such as 67.5^0, 112.5^0, 135^0 and 157.5^0. When all switches are closed, the phase shift is 0^0. Table 3 shows the phase shift for different input logic pattern. The layout of the 3-bit phase shifter with DC bias circuit is shown in Fig. 16(b).

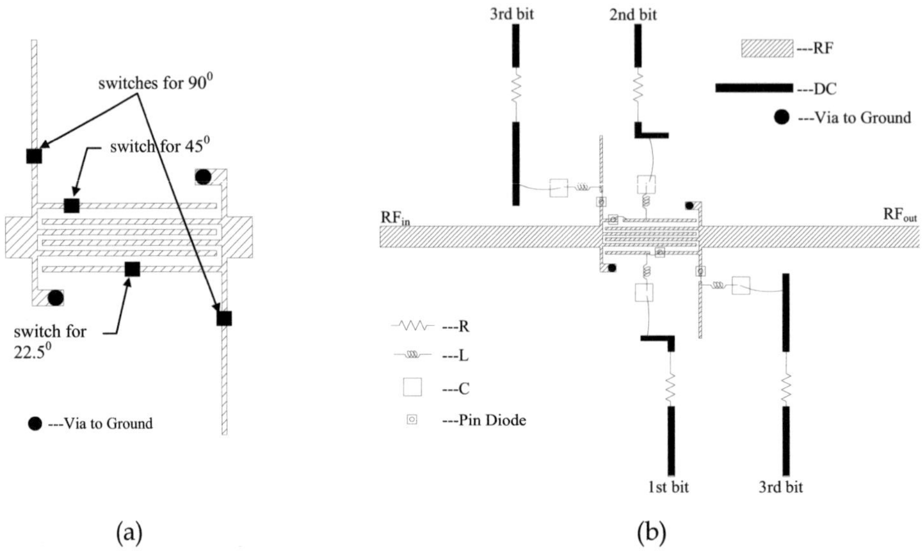

Fig. 16. (a) 3-bit phase shifter using one CRLH TL unit cell and (b) layout of 3-bit phase shifter with DC bias circuits

States	Phase shifts
000	0^0
001	22.5^0
010	45^0
011	67.5^0
100	90^0
101	112.5^0
110	135^0
111	157.5^0

Table 3. Phase shifts for different states of 3-bit phase shifter

4.3.4 6-bit phase shifter

To design a 6-bit phase shifter, we employ two symmetrical CRLH TL unit cells, as shown in Figs. 17(a) and 17(b), in cascade to provide the required phase shifts (Zhang et al., 2010). The state with all switches closed is taken as the zeroth-switch state. In the unit cell of Fig. 17(a), we use four switches to achieve sixteen different switch states and hence sixteen different phase shifts. The two switches on the inner fingers are used to provide the phase shifts of 5.625^0 and 11.25^0. The two switches on the outer fingers are used to provide the phase shifts of 22.5^0 and 45^0. Different switch states are used to provide other phase shifts such as 16.875^0, 28.125^0, 34.75^0, 39.375^0, 50.625^0, 56.25^0, 61.875^0, 67.5^0, 74.125^0,

78.75^0 and 84.375^0. Thus these switch states can be used to provide the phase shifts from 0^0 up to 84.375^0 at a step size of 5.625^0. In the unit cell of Fig. 17(b), we use two switches in a pair to provide the large phase shifts of 90^0 or 180^0. The pair of switches on the inner fingers provides a phase shift of 90^0, while the pair of switches on the outer fingers provides a phase shift of 180^0. The phase shift of 270^0 is achieved by "opening" all four switches. This phase shifter therefore can provide the phase shifts from 0^0 up to 270^0 at a step size of 90^0. By cascading these two unit cells, any phase shifts at a multiple number of 5.625^0 can be achieved by using the 6 controlling bits according to (25), i.e. the phase shifter can provide the phase shifts from 0^0 up to 354.375^0 at a step size of 5.625^0. Table 4 shows the phase shift for different input logic pattern. The layout of the 6-bit phase-shifter with DC bias circuits is shown in Fig. 17(c).

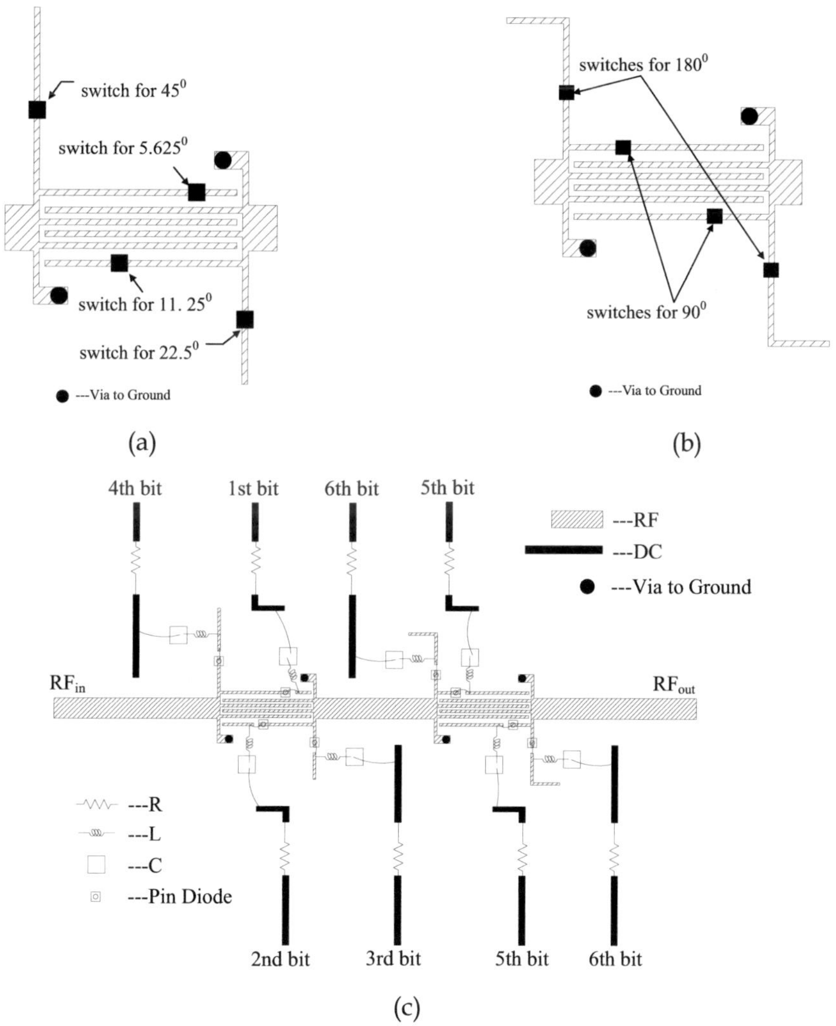

Fig. 17. 6-bit phase shifter using two CRLH TL unit cells. (a) CRLH unit cell in 1st stage, (b) CRLH unit cell in 2nd stage, and (c) layout of 6-bit phase shifter with DC bias circuits

States	Phase shifts	States	Phase shifts	States	Phase shifts	States	Phase shifts
000000	0^0	010000	90^0	100000	180^0	110000	270^0
000001	5.625^0	010001	95.625^0	100001	185.625^0	110001	275.625^0
000010	11.25^0	010010	101.25^0	100010	191.25^0	110010	281.25^0
000011	16.875^0	010011	106.875^0	100011	196.875^0	110011	286.875^0
000100	22.5^0	010100	112.5^0	100100	202.5^0	110100	292.5^0
000101	28.125^0	010101	118.125^0	100101	208.125^0	110101	298.125^0
000110	33.75^0	010110	123.75^0	100110	213.75^0	110110	303.75^0
000111	39.375^0	010111	129.375^0	100111	219.375^0	110111	309.375^0
001000	45^0	011000	135^0	101000	225^0	111000	315^0
001001	50.625^0	011001	140.625^0	101001	230.625^0	111001	320.625^0
001010	56.25^0	011010	146.25^0	101010	236.25^0	111010	326.25^0
001011	61.875^0	011011	151.875^0	101011	241.875^0	111011	331.875^0
001100	67.5^0	011100	157.5^0	101100	247.5^0	111100	337.5^0
001101	73.125^0	011101	163.125^0	101101	253.125^0	111101	343.125^0
001110	78.75^0	011110	168.75^0	101110	258.75^0	111110	348.75^0
001111	84.375^0	011111	174.375^0	101111	264.375^0	111111	354.375^0

Table 4. Phase shifts for different states of 6-bit phase shifter

4.4 Simulation and measurement results

The 2-bit, 3-bit and 6-bit phase shifters in Figs. 15(b), 16(b) and 17(c), respectively, have been designed to operate in an operating frequency band of 9-10 GHz using computer simulation. The substrates used in our designs were Rogers, RO5880, with a thickness of 0.254 mm and a permittivity of 2.2. The switches used were PIN diodes from SKYWORKS Co. Ltd, having a dimension of 0.35 mm×0.35 mm×0.15 mm, operating frequency range of 100 MHz-18 GHz and instantaneous power-handling capability of 2.5 W (34 dBm). Surface-mount-technology components such as resistors, capacitors and inductors were used to construct the bias circuits for the PIN diodes and also the circuits to isolate the RF signal from DC. For verification, all the final designs have been fabricated using Rogers substrate, RO5880, the same substrate used in our simulations. The prototyped modules of the 2-bit, 3-bit and 6-bit phase-shifters are shown in Fig. 18, having the dimensions of 35 mm×40 mm×2 mm, 50 mm×40 mm×2 mm and 60 mm×40 mm×2 mm, respectively.

The return losses ($-10\log|S_{11}|$), insertion losses ($-10\log|S_{21}|$) and phase shifts of these prototyped phase shifters have been measured using a network analyzer to verify the simulation results. The simulation and measurement results for the 2-bit, 3-bit and 6-bit phase shifters are shown in Figs. 19, 20 and 21, respectively. It can be seen that the simulated and measured results show good agreements.

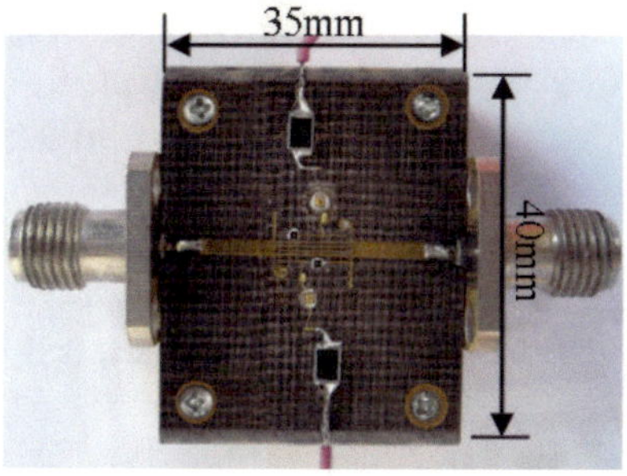

(a)

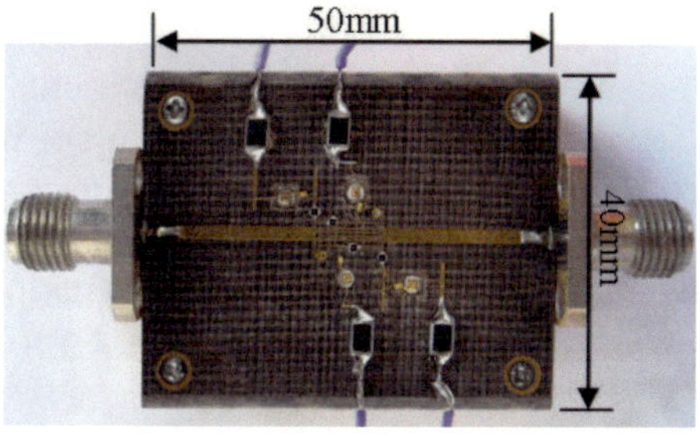

(b)

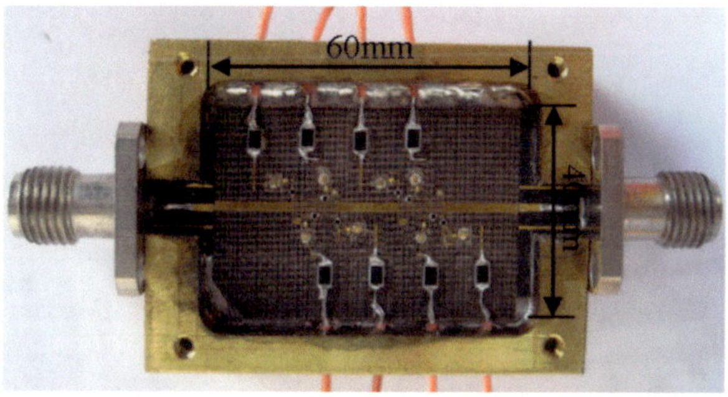

(c)

Fig. 18. Prototyped (a) 2-bit phase shifter, (b) 3-bit phase shifter and (c) 6-bit phase shifter

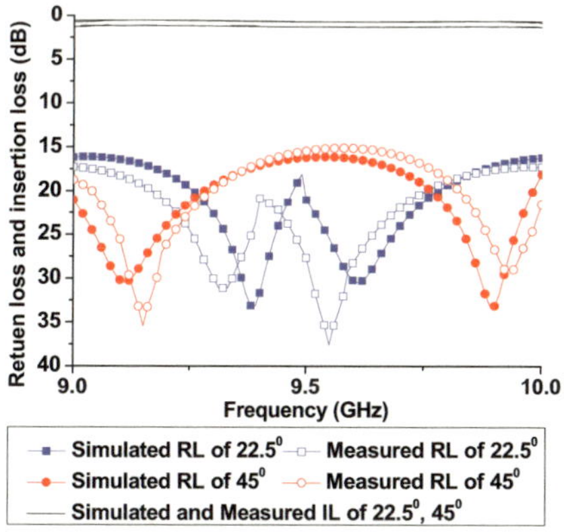

(a)

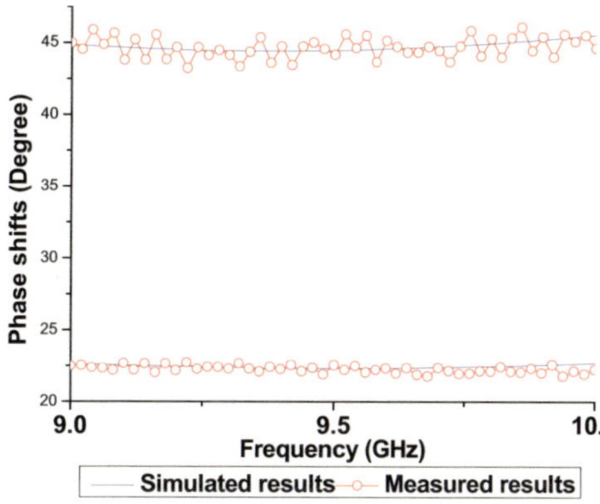

(b)

Fig. 19. Simulated and measured results of 2-bit phase shifter; (a) return loss and insertion loss and (b) phase shift

The insertion losses of the three phase shifters are all less than 1.3 dB in the frequency band of 9-10 GHz. Figs. 19(a) and 20(a) show that the return losses of the 2-bit and 3-bit phase shifters are more than 15 dB in the frequency band of 9-10 GHz. For the 6-bit phase shifter, Fig. 21(a) shows the return loss is more than 15 dB for the frequency band from 9.2 to 9.8 GHz. Regarding accuracy, the measurement results in Fig. 19(b) show that the maximum error for the 2-bit phase shifter is about -2^0 at the phase shift of 67.5^0 and frequency of 9.73 GHz. For the 3-bit phase shifter, Fig. 20(b) shows that the maximum errors is $+4.5^0$ at the phase shift of 157.5^0 and frequency of 9.10 GHz. While for the 6-bit phase shifter, it is -6.3^0 at the phase shift of 180^0 and frequency of 9.66 GHz.

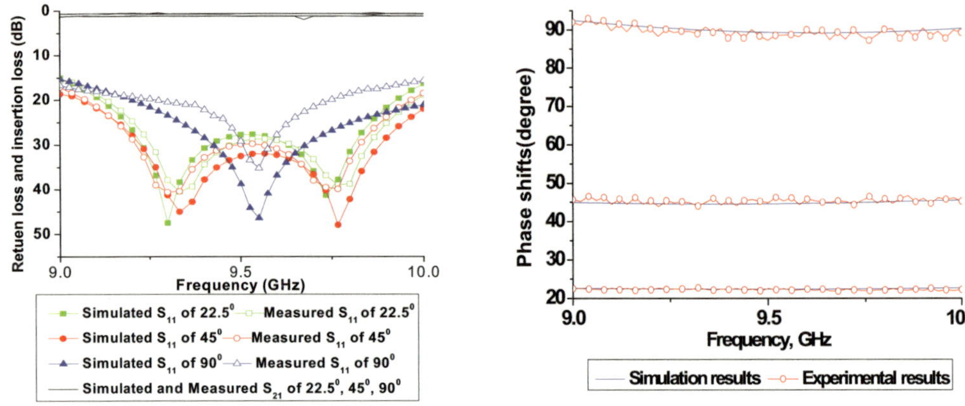

(a) (b)

Fig. 20. Simulated and measured results of 3-bit phase shifter; (a) return loss and insertion loss and (b) phase shift

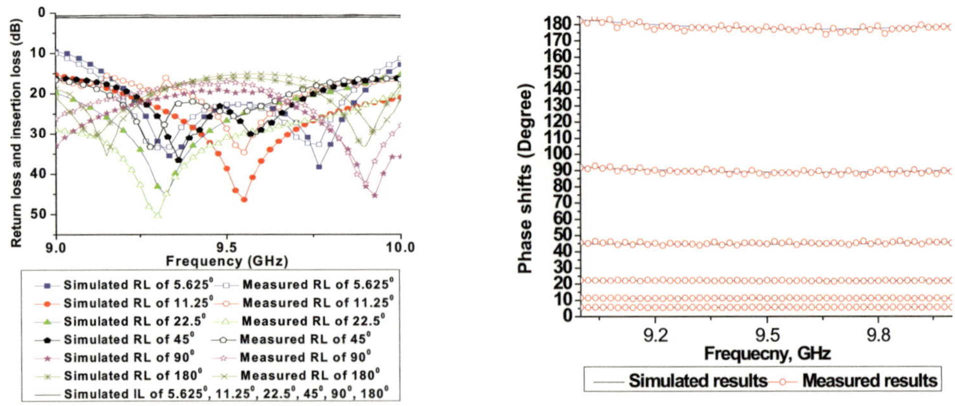

(a) (b)

Fig. 21. Simulated and measured results of 6-bit phase shifter; (a) return loss and insertion loss and (b) phase shift

The power-handling capability of the phase shifters has been studied using a setup with the block diagram shown in Fig. 22. The solid-state transmitter is basically an AM modulator which takes in the signal from the IF signal source, uses it to modulate the carrier signal at a much higher frequency and amplifies the modulated signal to a high-power level. The amplified signal is fed to the phase shifter under tested via an isolator which prevents any high-power signal from reflecting back and damaging the solid-state transmitter. A 40-dB coupler is used to couple a small portion of the high-power signal from the output of the phase shifter and feed it to the frequency spectrograph for measuring the power capability of the phase shifter. Majority of the high-power signal from the phase shifter is fed to a high-power load for power dissipation.

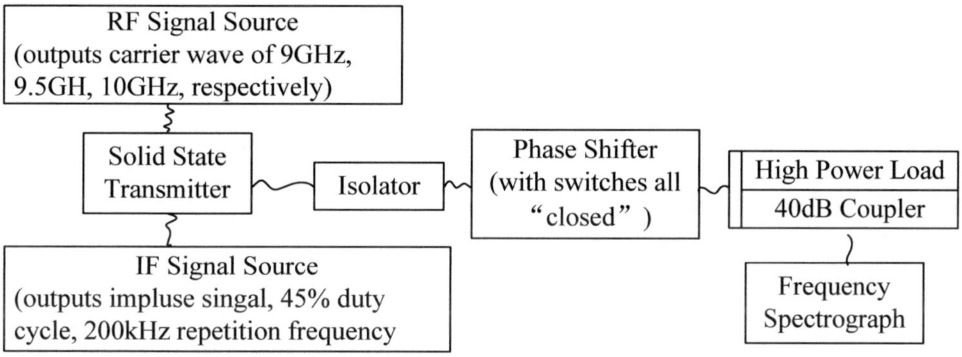

Fig. 22. Equipment setup for measuring power-handling capability of phase shifter

Frequency (GHz)		9	9.5	10
Peak Power (dBm)	2-bit	51.2	51.3	51.1
	3-bit	51.1	51.2	51.4
	6-bit	51.3	50.9	51.1

Table 5. Power-Handling Capability of Phase Shifters

In this study, the maximum transmitted peak power causing a diode to burn off is taken as the power-handling capability of the phase shifter and the results are shown in Table 5. Since the PIN diodes used in our design have an instantaneous power-handling capability of 34 dBm, the expected power-handling capabilities for the 2-bit, 3-bit and 6-bit phase shifters are about 50.43 dBm, i.e. 44 times higher as described previously. Table 4.4 shows that the power-handling capabilities of the three phase shifters at 9, 9.5 and 10 GHz are over 51.1, 51.1 and 50.9 dBm, respectively, corresponding to about 51.5, 51.5 and 49.2 times of the power-handling capability of the PIN diodes.

5. Conclusions

In this chapter, the EM metamatreials and the realization of EM metamaterials using the transmission line approach have been briefly described. A CRLH TL unit cell with a symmetrical structure has been proposed for the designs of TTDLs and digital-phase shifters. TTDLs using a single unit cell and four unit cells in cascaded have been designed and studied. Simulation and measurement results have shown that the TTDLs have the return losses of more than 15 dB and insertion losses of less than 1 dB. For the same length of 30 mm, the TTDL constructed using four CRLH TL unit cells in cascade can achieve a much larger TD, about 3.2 times larger, than that using the RH TL. Digital-phase shifters constructed using the CRLH TL unit cells have the advantages of small sizes, arbitrary phase-shift ranges and arbitrary step sizes. Three digital-phase shifters, 2-bit, 3-bit and 6-bit, have been designed and studied. Simulation and experimental results have shown that the digital-phase shifters have the low insertion losses of about 1.3 dB and return losses of larger than 15 dB across the operation bandwidths. Moreover, they have the much higher power-handling capabilities, about 50 times higher, than that of the PIN diodes used as switches in our designs for the digital-phase shifters.

6. References

Adam, J.D.; Davis, L.E.; Dionne, G.F.; Schloemann E.F. & Stitzer, S.N. (2002). Ferrite devices and materials. *IEEE Transactions on Microwave Theory and Techniques*, Vol. 50, 2002, pp. 721-737

Antoniades, M.A. & Eleftheriades, G.V. (2003). Compact linear lead/lag metamaterial phase shifters for broadband applications. *IEEE Antennas and Wireless Propagation Letters*, Vol. 2, 2003, pp.103-106

Bahl, I. (2011). *Lumped Elements for RF and Microwave Circuits*, Artech House, Boston& London

Caloz, C. & Itoh, T. (2006). *Electromagnetic Metamaterials: Transmission Line Theory and Microwave Applications: the Engineering Approach*, John Wiley & Sons, New Jersey

Damm, C.; Schüßler, M.; Freese J. & Jakoby, R. (2006). Artificial line phase shifter with separately tunable phase and line impedance, *36th European Microwave Conference*, 2006

Fetisov, Y.K. & Kabos, P. (1998). Active Magnetostatic Wave Delay Line. *IEEE Transactions on Magnetics*, Vol. 34, 1998, pp. 259 – 271

Keul, S. & Bhat, B. (1991). *Microwave and millimeter wave phase shifters*, Artech house, MA

Kim, H.; Kozyiev, A.B.; Karbassi, A. & van der Weide, D.W. (2005). Linear Tunable Phase Shifter Using a Left-handed Transmission-line. *IEEE Microwave and Wireless Components Letters*, Vol. 15, 2005, pp. 366-368

Kholodnyak, D.V.; Serebryakova, E.V.; Vendik, I.B. & Vendik, O.G. (2006). Broadband Digital Phase Shifter Based On Switchable Right- and Left-handed Transmission Line Sections. *IEEE Microwave and Wireless Components Letters*, Vol. 1, 2006, pp. 258-260

Kim, H. & Drayton, R.F. (2007). Size Reduction Method of Coplanar Waveguide (CPW) Electromagnetic Bandgap (EBG) Structures Using Slow Wave Design, *Topical Meeting on Silicon Monolithic Integrated Circuits in RF Systems*, 2007

Lai, A.; Itoh, T. & Caloz, C. (2004). Composite right/left-handed transmission line metamaterials. *IEEE Microwave Magazine*, Vol. 5, Sep 2004, pp. 34 - 50

Lee, T.H. (2004). *Planar Microwave Engineering*, Cambridge University Press, UK

Lapine, M.; Nefedov, I.S.; Saeily, J. & Tretyakov, S.A. (2006). Artificial lines with exotic dispersion for phase shifters and delay lines, *36th European Microwave Conference*, 2006

Marc, E.G. & Robert, A.P. (1991). Modelling Via Hole Grounds in Microstrip. *IEEE Microwave and Guided Wave Letters*, Vol. 1, 1991, pp. 135-137

Monti, G.; DePaolis, R. & Tarricone, L. (2009). Design of a 3-state reconfigurable CRLH transmission line based on MEMS switches, *Progress In Electromagnetics Research*, PIER 95, 2009, pp. 283-297

Pozar, M.D. (2004). *Microwave Engineering*, John Wiley & Sons, New Jersey

Rebeiz, G.M.; Tan, G.L. & Hayden, J.S. (2002). RF MEMS phase shifters: design and applications. *IEEE Microwave Magazine*, Vol. 2, 2002, pp. 72-81

Smith, W.R. & Gerard, H.M. (1969). Design of Surface Wave Delay Lines with Interdigital Transducers. *IEEE Transactions on Microwave Theory and Techniques*, Vol. 17, 1969, pp. 865 – 873

Ulaby, F.T. (2004). *Fundamentals of Applied Electromagnetics*, Pearson/Prentice Hall, New Jersey

Veselago, V.G. (1968). The electrodynamics of substances with simultaneously negative values of ε and μ, *Soviet Physics Uspekhi*, January-Feburary 1968

Vendik, I.B.; Kholodnyak, D.V. & Kapitanova, P.V. (2009). Microwave phase shifters and filters based on a combination of left-handed and right-handed transmission lines, In: *Metamaterials Handbook Vol. II. Applications of Metamaterials*, pp. 1-21, CRC Press

Woo, D.J.; Lee, J.W. & Lee, T.K. (2008). Multi-band Rejection DGS with Improved Slow-wave Effect, *38th European Microwave Conference*, 2008

Xu, J.; Lu, Z. & Zhang, X.C. (2004). Compact Involute Optical Delay Line. *Electronic Letters*, Vol. 40, 2004, pp. 1218 – 1219

Zhang, Y.Y. & Yang, H.Y.D. (2008). Ultra slow-wave periodic transmission line using 3D substrate metallization, *Microwave Symposium Digest 2008 IEEE MTT-S International*, 2008

Zhang, J.; Zhu, Q.; Jiang, Q. & Xu, S.J. (2009). Design of time delay lines with periodic microstrip line and composite right/left-handed transmission line. *Microwave and Optical Technology Letters*, Vol. 51, 2009, pp. 1679-1682

Zhang, J.; Cheung, S.W. & Yuk, T.I. (2010). Design of n-bit Phase Shifters with High Power Handling Capability Inspired by Composite Right/Left-handed Transmission Line Unit Cells. *IET Microwaves, Antennas & Propagation*, Vol. 4, August 2010, pp. 991-999

Zhang, J.; Cheung, S.W. & Yuk, T.I. (2010). A 3-bit Phase Shifter with High-power Capacity Based on Composite Right/Left-handed Transmission Line. *Microwave and Optical Technology Letters*, Vol. 52, August 2010, pp.1778-1782

Zhang, J.; Cheung, S.W. & Yuk, T.I. (2011). UWB True-time-delay Lines Inspired by CRLH TL Unit Cells. *Microwave and Optical Technology Letters*, Vol. 53, September 2011, pp.1955-1961

18

Perfect Metamaterial Absorbers in Microwave and Terahertz Bands

Qi-Ye Wen*, Huai-Wu Zhang, Qing-Hui Yang, Zhi Chen,
Bi-Hui Zhao, Yang Long and Yu-Lan Jing
State Key Laboratory of Electronic Films and Integrated Devices,
University of Electronic Science and Technology of China, Chengdu,
China

1. Introduction

Recently, resonant metamaterial absorbers (MAs) at microwave and terahertz (THz) bands have attracted much attention due to the advantages such as high absorption, low density, and thin thickness [1–6]. The MA generally composed of a metamaterial layer and a metal plate layer separated by a dielectric spacer. With this kind of novel device, unity absorptivity can be realized by matching the impedance of MA to free space. Besides that, wide-angle, polarization insensitive and even multi-bands/wide-band absorption can be achieved through properly device designing [7-11]. Furthermore, our previous investigations on MA show that the absorber traps the incident electronmagnetic (EM) wave into some specific spots of the devices, and then converts it into heat [9, 12]. All these features make MAs very useful in areas such as EM detector/imager, anti-electromagnetic interference, stealth technology, phase imaging, spectroscopy and thermal emission.

Tunable devices, which allow one to real-time control and manipulate of EM radiation, are emerging as an interesting issue in metamaterials fields [13]. Combined with pin diodes, switchable microwave MAs were developed and an electronic control strategy was demonstrated [14, 15]. However, the device fabrication process is complex, and the structure with pin diodes is hard to be scaled down to higher frequency such as THz and visible regimes. In this chapter, we proposed a VO_2 based switchable MA in microwave band. VO_2 is known to exhibit a transition from an insulating phase to a metallic state (IMT) when it is thermally, electrically or optically triggered. By this unique property, VO_2 films have already been used to tune the resonance characteristics of metamaterial in near IR [16] and THz regimes [17, 18]. The switchable microwave MA presented in this work is realized by placing VO_2 thin film between the electronic split ring resonator (eSRR) and the dielectric layer. It is found that by triggering the IMT of the VO_2, the absorption amplitude of the device can be significantly switched between absorber and reflector with high speed. This VO_2 based MA has the advantages such as simple fabrication, strong tunability and easy to scale to terahertz and optical bands. This tunable MA also has the potential to be a self-

*Corresponding Author

resetting "smart" EM absorber, since the absorbed wave would transfer into heat and that can also trigger the IMT of the VO_2 film.

2. Basic structure and working principle of metamaterial absorber

The first metamaterial based absorber was proposed by N. L. Landy et al in the microwave band [1]. It is called "perfect metamaterials absorber" because nearly 100% absorption can be achieved theoretically. A single unit cell of the absorber consisted of three layers as shown in Fig.1. The top layer is the electric split-ring resonator (eSRR), the middle layer is isolation layer (such as polyimide), and the bottom layer is rectangular metal strip. Due to the lithography alignment and multi-step lithography process, the preparation process of the first MA is complex. Experimentally, the maximum absorptivity only reaches to 70% at 1.3THz due to the fabrication tolerance. An improved MA was proposed by H. Tao et al, with the bottom metal strips replaced by a continuous metal film [2]. This improved absorber operates quite well for both TE and TM radiation over a large range of incident angles (0-50°), and the measured absorbance was further improved to 97%. Therefore, the MA with a continuous metal ground plane become the most commonly structure in the researches.

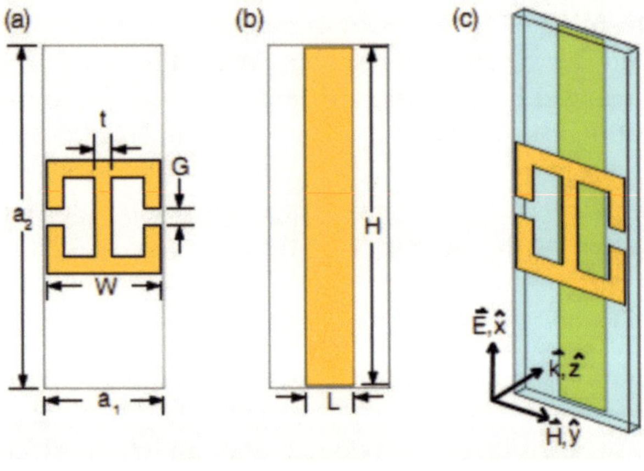

Fig. 1. Basic structure of three layers metamaterial absorber (a)the layer of metamaterial;(b) the bottom metallic layer;(c)the unit structure include the dielectric layer. Figure adapted from N. L. Landy et al. [1]

The absorption mechanism of the MA is as follows: First, by changing the geometry of the SRR and the thickness of the spacer, the impedance Z (ω) of the absorber can be designed to match the impedance of free space at a specific frequency (center frequency) resulting in zero reflection; Second, electromagnetic waves can not pass through the metallic ground plane, giving rise to zero transmission too. Thus, electromagnetic waves will be completely restricted in the device and finally be consumed. In principal, the metamaterial absorber can absorb 100% of the narrow-band electromagnetic waves. It can be used in microwave, terahertz (THz) and even light wave band by adjusting the feature size of the unit cell.

Though more and more attentions have been paid to MA, the mechanism of the near-unity absorption is still under studying. It has been suggested the matching between the effective permittivity and permeability may be able to interpret the perfect absorption. However, the effective medium theory has some problems in describing MA because the three-layer structured device doesn't exactly satisfy the homogeneous-effective limit, according to Caloz [20]. A typical case is that the strong asymmetric absorption phenomenon cannot be fully explained by effective medium model [21]. The simulation results show dramatically different behaviors when the electromagnetic waves incident from the two opposite directions. For example, when light is incident from the front to the resonators the device acts as a perfect absorber, while when light is incident from the back to the ground plane the device behaves like a perfect mirror. Furthermore, the MA consists of only two metallic layers, thus are strongly inhomogeneous in the wave propagating direction, which is obviously in contrast to the effective medium model.

Q.Y. Wen et al have proposed a transmission line (TL) mode based on the equivalent RLC model [12]. In the TL model, it is assumed that the transverse electromagnetic (TEM) wave propagates through free space and the substrate with intrinsic impedances *Zi* and *Zo* respectively. There are two assumptions for constructing the TL model. One is that coupling capacitor or coupling inductor between the eSRR layer and wires layer should be ignorable, so that these two layers can be individually modeled, as demonstrated in Fig. 1. Another is that the THz wave normally incidents on the absorber plane with the electrical field parallel to the split gap of the eSRR. The TL model of eSRR proposed by A. K. Azad [22] is used to describe the eSRR layer, in which the LC resonance and dipole resonance each is represented by one group of L, C and R respectively, and the coupling between these two resonances is specified by the parameter M. The wires layer part is mimicked by the TL model developed by L. Fu [23], with the only resonance expressed by one group of L, C and R. The function of isolation layer is modeled by a transmission line which contains all EM related properties of the isolation layer such as ε, μ and thickness. It connects the eSRR part and wire part. All the parameters are needed to be optimized until the S-parameters calculated by the TL model fit the simulation results.

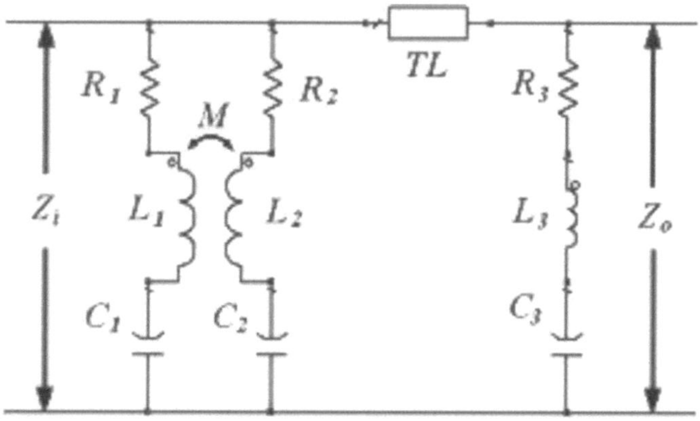

Fig. 2. Transmission line model of metamaterial absorber

By this TL model, the asymmetric phenomenon of THz absorption is unambiguously demonstrated and explained. The strong absorption is found to be mainly related to the LC resonance of the eSRR structure. The isolation layer in the absorber, however, is actually an impedance transformer and plays key role in producing the perfect absorption. The studies by TL model also show that the electromagnetic wave is concentrated on some specific location in the absorber. It indicates that the trapped electromagnetic wave in the absorber can be converted into thermal energy, electric energy or any kinds of other energy depending on the functions of the spacer materials. This feature as electromagnetic wave trapper has many potential applications such as radiation detecting bolometers and thermal emitter.

3. Progress in metamaterial absorber

The perfect absorber first proposed by N. L. Landy is an anisotropic absorber. They further proposed a polarization independent absorber, as shown in Fig.3 [5]. One of the main features of the structure is that the eSRR has fourfold-rotational symmetry about the propagation axis and was therefore polarization insensitive. The measured absorptivity of this absorber is 65% at 1.145 THz. Experimental results confirmed that the polarization-independent metamaterial absorber can be realized by chosen a fourfold rotational symmetry SRR structure.

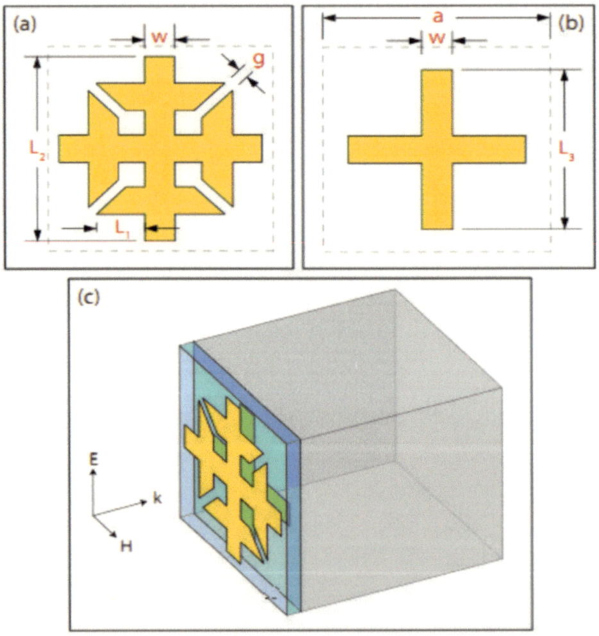

Fig. 3. Polarization-independent metamaterial absorber. Figure adapted from N. I. Landy et al. [5]

H. Tao et.al designed a wide-angle terahertz metamaterial absorber [13]. Most importantly, the device was fabricated on a highly flexible polyimide substrate with a total thickness of 16 um. This novel design enables its use in nonplanar applications as it can be easily wrapped around objects as small as 6 mm in diameter. They demonstrated, through simulation and experiment, that this metamaterial absorber operates over a very wide range of angles of incidence for both transverse electric (TE) and transverse magnetic (TM) configuration.

Apart from single-frequency absorber, dual-band and multi-band absorbers also draw attention from researchers. A dual-band metamaterial absorber was demonstrated in terahertz band, as shown in Fig 4 [9]. The special feature of this absorber is that its eSRR unit possesses two kinds of split gaps and therefore exhibits two well-separated LC resonances. Theoretical calculation shows that there are two distinct absorptive peaks located around 0.50 and 0.94THz, each with absorptions over 99.99%. The measured absorption is 81% for low frequency absorption and 63.4% for high frequency absorption. The experimental result suggests that the design of an eSRR with multiple LC resonances is a key step toward a multi-frequency absorber. A similar result was also reported by H. Tao et al [10].

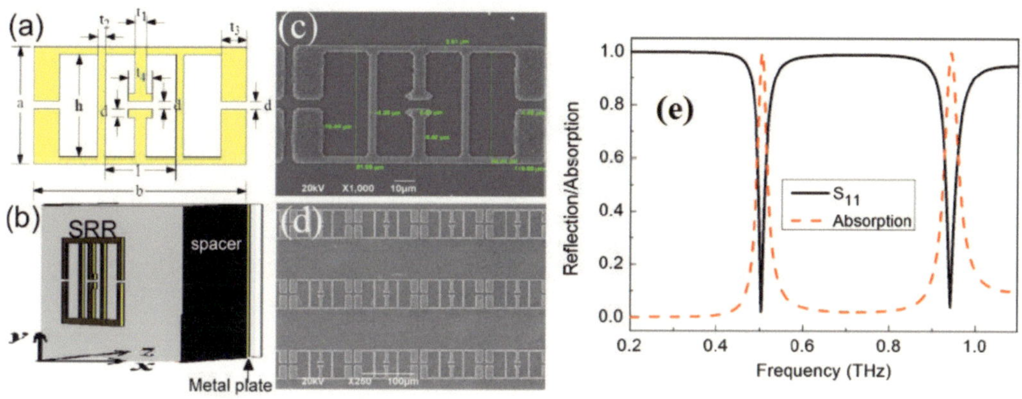

Fig. 4. Dual-band THz metamaterial absorber. (a) Designed electric split ring resonator, (b) Perspective view of the designed absorber. (c) A unit cell of the experimentally realized absorber. (d) Photograph of a portion of the fabricated absorber. (e) The simulated reflection (solid line) and absorption (dotted line) curve of the absorber.

X. P. Shen et al. have developed a wide angle triple-band absorber structure very recently, as show in Fig.5 [11]. The top layer consists of an array of three nested copper closed ring resonator arrays, which is primarily responsible for the electric response to the incident field. The bottom layer is a copper plane, which is used to zero the transmission and is responsible for the magnetic response. The experimental results show three absorption peaks at frequencies 4.06GHz, 6.73GHz and 9.22GHz with absorptivity of 99%, 93%, and 95%, respectively, which agrees well with the simulation results.

In certain applications, broadening the absorption bandwidth is also of importance. Fig.6 shows an omnidirectional polarization-insensitive absorber with a broadband feature in the terahertz regime is proposed by Q.Y. Ye et al. [8]. They demonstrated that the bandwidth of the absorption can be effectively improved by using a multilayer structure, while the wide-angle feature remains. A simple cross-shaped pattern was used as the resonator. The experiment results show that with the increasing of metamaterial layers, the absorption peak is gradually broadened. A bandwidth of nearly 1000 GHz with perfect absorption (more than 97%) was achieved in the three layer absorber.

A frequency tunable metamaterial absorber is proposed by incorporated a pin-diode between two resonators [15]. Simulation and measurement results show that by forward or reverse biasing the diodes so as to change the coupling between the resonators, the absorber can be dynamically switched to operate in two adjacent frequency bands with nearly perfect peak absorption. It is also shown that by tuning the loading position of the diodes, it is able to adjust the frequency difference between the two switchable absorbing bands.

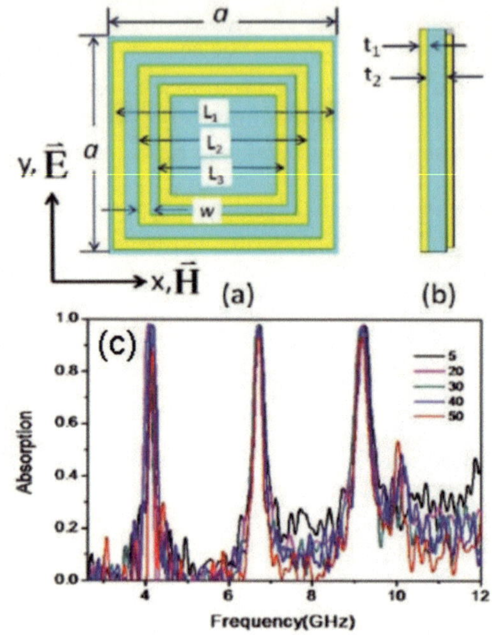

Fig. 5. (a) The front and (b) side view of the unit cell of the triple band microwave absorber, and (c) simulation and experimental results of the triple-band absorber at various angles of incidence. Figure adapted from X. P. Shen ea al [11].

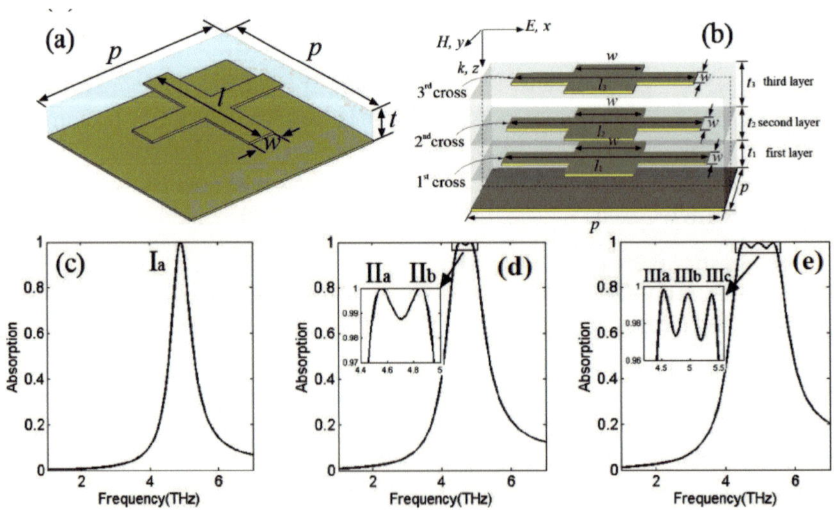

Fig. 6. (a) Unit cell of the absorbers, (b) Schematic diagram of a 3-layer cross structure, and bsorption spectra for (c) 1-layer cross structure; (d) 2-layer cross structure; (e) 3-layer cross structure.

4. Switchable metamaterials absorber based on vanadium dioxides

Though tunable metamaterial absorber has been demonstrated in microwave band, the device fabrication process is complex, and the structure with pin diodes is hard to be scaled down to higher frequency such as THz and visible regimes. In this chapter, we proposed a novel switchable MA in microwave band, which is simple and can be tuned thermally, electrically and optically. The MA under studied has a similar structure to our previously designed dual band THz absorber [9] except that a thin VO_2 film is incorporated between the eSRR and the substrate, as shown in Fig.7. The eSRR unit cell is composed of two kinds of SRR with one (the inner resonator) invaginated in the other (the outer resonator pairs). It should be noted that, for the purpose of comparison, only the inner resonator has an underlying VO_2 patch and the outer resonators contact to the substrate directly. As shown in Fig.7 (b), the area of the VO_2 patch is equal to the inner resonator. Utilizing the commercial software CST Microwave Studio ™ 2009, the size parameters of the eSRR marked in Fig.7 (a) was optimized to obtain two strong absorptions. In the simulation, the microwave MA is built on C-cut sapphire substrate with one side covered with metallic ground plane and another side with eSRR arrays. Both of the metallic eSRR and ground plane are modeled as copper sheet with conductivity of 5.8×10^7 S/m. A 200nm thick VO_2 film was applied in the simulation. The permittivity and the permeability of VO_2 were set to be 3 and 1, respectively. The conductivity (σ) was swept from 0.02S/m to 2000S/m to mimic the phase transition process. The incident microwave wave is normal to the devices plane with the electronic field perpendicular to the split gaps, e.g. along y axis as indicated in Fig.7 (b). The optimized parameters are: a=2mm, b=4mm, d=0.1mm, h=1.8mm, l=1mm, t1=t2=0.2mm, t3=0.3mm, t4=0.5mm and the unit cell is 3mm × 5mm. The copper thickness of eSRR layer is 0.2μm and that for ground plane is 0.8μm.

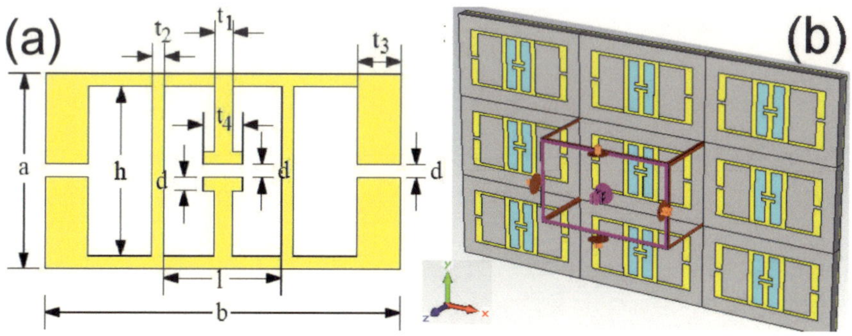

Fig. 7. Dual-band microwave metamaterial absorber. (a) Schematic of the designed eSRR unit cell. (b) Perspective view of the absorber sheet with VO_2 patterns. Gray represents the sapphire substrate, yellow the copper and dark green the VO_2 film.

Fig.8 shows the simulated S_{11} parameters of the MA with different conductivity of the VO_2 patch. When σ is 0.02S/m, e.g, the insulating state of the VO_2 film, two distinct reflection peaks are observed around 9.03 GHz and 17.6 GHz. Since the transmission of MA is zero due to the metallic bottom plane, the absorptivity can be calculated using $A=1-|S_{11}|^2$. High absorptivity of 92.7% and 99.4% was obtained at 9.03 GHz and 17.6 GHz, respectively. The surface current distributions on the top eSRR and bottom plane layer at resonance frequencies were monitored, as shown in Fig.9. The results confirmed that both resonances come from circulating currents in the eSRR resonators [9, 10]. Furthermore, it shows that the low frequency response is determined mainly by the outer ring pairs (Fig 9 (a) & (c)), while the high frequency response is induced mainly by inner resonator (Fig.9 (b) & (d)).

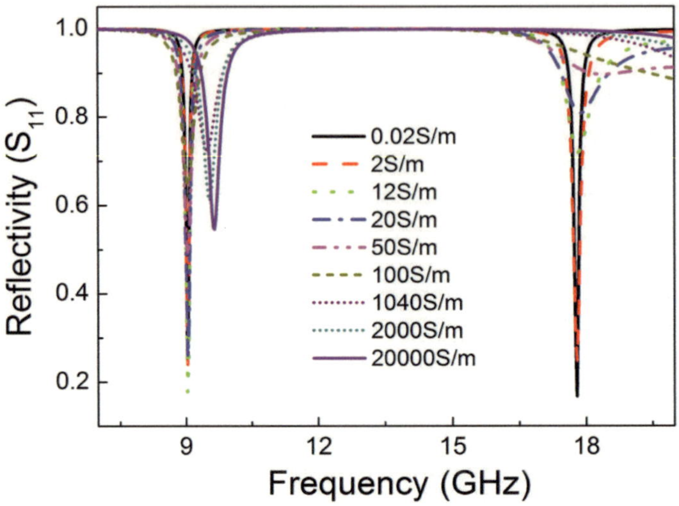

Fig. 8. Simulated reflection curve for metamaterial absorber with different conductivity of VO_2

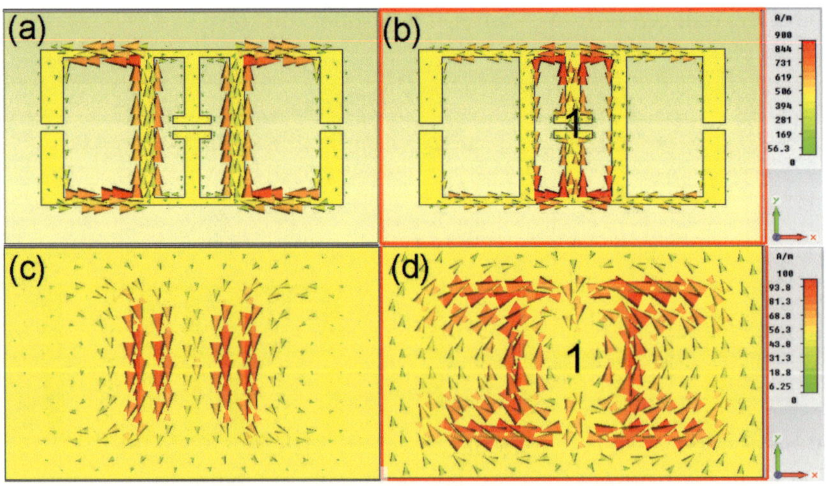

Fig. 9. Surface current distributions in the top and bottom metallic layers. (a) and (c) for the low frequency absorption, and (b) and (d) for high frequency absorption

When the value of σ increases to 2S/m, the high frequency reflectivity begins to increase. With the increase of σ from 2S/m to 50S/m, the peak reflectivity around 17.6GHz increases abruptly from 0.25 to 0.90, corresponding to a decrease of absorptivity from 93.75% to 19%. Further increase of σ gives rise to a slight increase of the reflectivity. As to the low frequency absorption, however, the situation is different. It seems that the variation of σ from 0.02S/m to 20S/m have little effect on absorption. Further increase σ from 20 S/m to 20000 S/m leads to a moderate increase of peak reflectivity from 0.26 to 0.54. Interestingly, for both absorptions a blue-shift about 0.6GHz of the peak frequency is clearly observed. All these results indicate that the IMT of VO_2 does affect the absorption of the MA, and the mechanism will be discussed later in detail.

With the optimized size parameters the microwave MA device was fabricated. Firstly, a VO_2 layer of almost 200 nm in thickness was deposited on 0.5mm-thick C-type sapphire substrates by using the reactive magnetron sputtering technique. The phase transition temperature (TP) of the films was measured to be around 340K and the conductivity exhibits nearly three-orders decrease [18]. The VO_2 film was then etched using CF_4/O_2 plasma to 1.2mm × 2mm patches in a period of 3mm × 5mm, which is the identical period of the eSRR array. After that copper film with thickness of 0.2μm was sputtering deposited on top side of the VO_2/Substrate, following by a conventional lithography process to form the designed eSRR structure with its inner resonator overlapping on the VO_2 patch. Finally another 0.8μm copper film was deposited on the backside of the substrate.

Based on the principle of Arc Tracking Test method, a vector network analyzer (Agilent 8720ES) with two horn antennas was used to transmit EM waves onto the sample sheet and receive the reflected signals. The incident and receive angle is less than 5° from normal in the experiment. Fig.10 shows the measured S_{11} curves of the VO_2 based MA under different temperature. Two strong reflection peaks are clearly appeared and they are distinct from each other, which agrees with the simulation well. At room temperature (RT), two reflective minimums were appeared at 9.36GHz and 18.6GHz, respectively. By $A=1-|S_{11}|2$ the peak absorptivity at low and high frequency were calculated to be 84.8% and 92.1%. For both absorptions, the measured peak frequencies and absorption amplitudes have slight derivations from the theoretical results, which are probably induced by the fabrication tolerance or the difference of the material parameter between simulation and experiment. Variations of the absorption characteristic of the MA were observed when the device temperature increased from RT to 345K. For the low-frequency absorption, the reflectivity minimum shows a small increase from 0.39 to 0.57, corresponding to an absorption decrease from 84.8% to 67.5%. Significant change of the absorption was happened in the high frequency case. With the increase of temperature from RT to 345K, the reflectivity minimum increases notably from 0.28 to 0.83, accompanying with a blue-shift of the peak frequency form 18.6 GHz to 19.12 GHz. Therefore, a deep amplitude modulation about 60% to the microwave absorption and a moderate frequency shift of about 0.5 GHz was realized. All these results agree with the simulated results very well, confirming that the VO_2 based microwave MA is thermally tunable.

In order to further clarify the tunability of MA, the temperature dependence of the reflectivity amplitude and corresponding peak frequency for high-frequency absorption was summarized and plotted in Fig.10 (b). It can be seen that the violent variation of the absorption characteristics, including the amplitude and peak frequency, occurs during a very narrow temperature range from 337K to 345K, which is the phase-transition temperature range of the VO_2 films under studied. Combining the theoretical and experimental results, we can conclude that it is the thermally triggered IMT of VO_2 that induces the tunability of the MA. To understand how the IMT of VO_2 affects the properties of MA, the distribution of the absorption at low and high frequency was examined at RT and compared in Fig.11. In the low-frequency case, the absorption mainly takes place in the vicinity of the two outer gaps, with a small portion occurring at the four corners and inner split gap. In the high-frequency case, almost all the absorption occurs at inner gap. Therefore, the working principles of our tunable microwave MA can be explained as follows: At RT, the VO_2 is in its insulating state and allows dual band absorptions. When the temperature increase beyond T_P, the VO_2 film changes into the metallic state and electronically shorts the inner split gap of the eSRR. The impedance match for the high frequency absorption is seriously broken thus the absorption is strongly attenuated. For low frequency absorption, though the absorption from the inner resonator is also attenuated as temperature increasing, the absorption from the outer resonator pairs is still strong since there is no underlying VO_2 film. That's why the low frequency absorption is not seriously crippled. The resonance frequency of eSRR can be described as $\omega=(LC)^{-1/2}$, where L and C represents the inductance and capacitance of the resonator [19]. Therefore, the blue-shift of the peak frequencies can be ascribed to the decrease of inductance as a result of the metallization of VO_2 film. These results confirmed that by triggering the IMT of the VO_2 films both the amplitude and frequency of the absorption can be tuned.

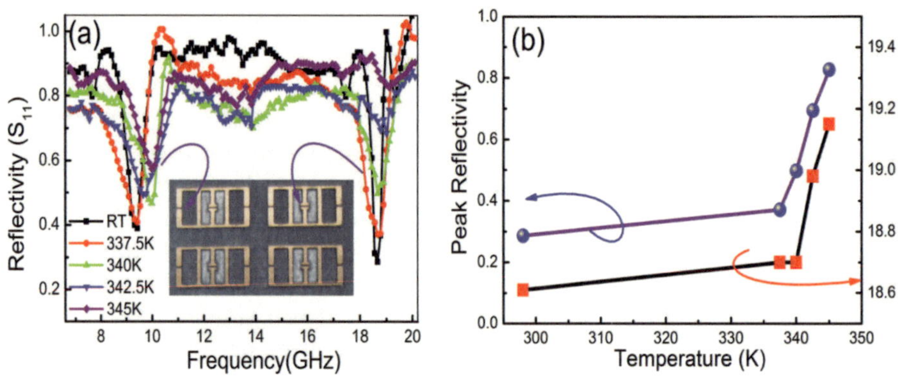

Fig. 10. (a) Measured reflectivity curves of the VO_2 based MA with respect to the devices temperature. Inset is the image of the fabricated device; (b) the temperature dependence of the reflection amplitude and corresponding peak frequency for high frequency absorption.

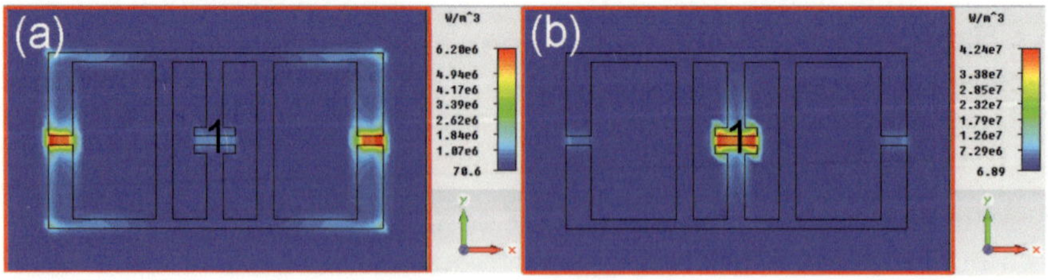

Fig. 11. Distributions of absorption densities for (a) low frequency absorption and (b) high frequency absorption.

5. Summary

Metamaterial absorbers with strong absorbance in microwave and terahertz band are very useful in many regimes such as detecting THz, creating thermal radiation, or cloaking. The metamaterial absorber could be designed to be polarization independent, broadband and even tunable with a small volume, small thermal mass, and high absorption coefficient. In this chapter, a tunable microwave absorber was demonstrated theoretically and experimentally. It is found that by thermally triggering the IMT of the VO_2, the absorption amplitude of the device can be significantly tuned. This tunable MA also has the potential to be a self-resetting "smart" EM absorber, since the absorbed wave would transfer into heat and that can also trigger the IMT of the VO_2 film.

6. Acknowledgement

This work is supported by NSFC under Grant No. 61131005 and 61021061, National Basic Research Program of China (973) under Grant No. 2007CB310407, the "New Century Excellent Talent Foundation" under Grant No. NCET-11-0068, Sichuan Youth S & T foundation under No. 2011JQ0001, and Sichuan International S&T Cooperation Program under Grant No. 2010HH0026. This work is partly supported by "the Fundamental Research Funds for the Central Universities" under Grant No. ZYGX2010J034.

7. References

[1] N. I. Landy, S. Sajuyigbe, J. J. Mock, D. R. Smith, and W. J. Padilla, "Perfect Metamaterial Absorber," Phys. Rev. Lett. 100, 207402 (2008).

[2] H. Tao, C. M. Bingham, A. C. Strikwerda, D. Pilon, D. Shrekenhamer, N. I. Landy, K. Fan, X. Zhang, W. J. Padilla and R. D. Averitt, "Highly flexible wide angle of incidence terahertz metamaterial absorber: Design, fabrication, and characterization," Phys. Rev. B 78, 241103R (2008)

[3] H. Tao, N. I. Landy, C. M. Bingham, X. Zhang, R. D. Averitt, and W. J. Padilla, "A metamaterial absorber for the terahertz regime: Design, fabrication and characterization," Opt. Express 16, 7181-7188 (2008).

[4] Y. Avitzour, Y. A. Urzhumov, and G. Shvets, "Wide-angle infrared absorber based on a negative-index plasmonic metamaterial," Phys. Rev. B 79, 045131 (2009).

[5] N. I. Landy, C. M. Bingham, T. Tyler, N. Jokerst, D. R. Smith and W. J. Padilla, "Design, theory, and measurement of a polarization-insensitive absorber for terahertz imaging," Phys. Rev. B 79, 125104 (2009).

[6] R. Huang, Z. W. Li, L. B. Kong, L. Liu, and S. Matitsine, "Analysis and design of an ultra-thin metamaterial absorber," Progress In Electromagnetic Research B 14, 407-429 (2009).

[7] B. Wang, T. Koschny, and C. M. Soukoulis, "Wide-angle and polarization-independent chiral metamaterial absorber," Phys. Rev. B, 80, 033108 (2009).

[8] Y. Q. Ye, Y. Jin, and S. L. He, "Omnidirectional, polarization-insensitive and broadband thin absorber in the terahertz regime," J. Opt. Soc. Am. B 27, 498-504 (2010)

[9] Q. Y. Wen, H. W. Zhang, Y. S. Xie, Q. H. Yang, and Y. L. Liu, "Dual Band Terahertz Metamaterial Absorber: Design, fabrication, and characterization," Appl. Phys. Lett. 95, 241111 (2009)

[10] H Tao, C M Bingham, D Pilon, Kebin Fan, A C Strikwerda, D Shrekenhamer, W J Padilla, Xin Zhang and R D Averitt, "A dual band terahertz metamaterial absorber," J. Phys. D: Appl. Phys. 43, 225102 (2010)

[11] X. P. Shen, T. J. Cui, J. M. Zhao, H. F. Ma, W. X. Jiang, and H. Li, "Polarization-independent wide-angle triple-band metamaterial absorber," Opt. Express 19, 9401-9407, (2011)

[12] Q. Y. Wen, Y. S. Xie, H. W. Zhang, Q. H. Yang, Y. X. Li, and Y. L. Liu, "Transmission line model and fields analysis of metamaterial absorber in the terahertz band," Opt. Express 17, 20256–20265 (2009)

[13] H. T. Chen, W. J. Padilla, J. M. O. Zide, A. C. Gossard, A. J. Taylor, and R. D. Averitt, "Active terahertz metamaterials devices," Nature, 444, 597 (2006)

[14] A. Tennant. and B. Chambers, "A single-layer tunable microwave absorber using an active FSS," IEEE Microw. Wirel. Compon. Lett., 14, 46-47 (2004)

[15] B. Zhu, C. Huang, Y. Feng, J. Zhao, and T. Jiang, "Dual band switchable metamaterial electromagnetic absorber," Progress In Electromagnetics Research B, 24, 121-129 (2010)

[16] M. J. Dicken, K. Aydin, I. M. Pryce, L. A. Sweatlock, E. M. Boyd, S. Walavalkar, J. Ma, and H. A. Atwater, "Frequency tunable near-infrared metamaterials based on VO_2 phase transition," Opt. Express 17, 18330-18339 (2009)

[17] T. Driscoll, S. Palit, M. M. Qazilbash, M. Brehm, F. Keilmann, B. G. Chae, S. J. Yun, H. T. Kim, S. Y. Cho, N. M. Jokerst, D. R. Smith, and D. N. Basov, "Dynamic tuning of an infrared hybrid-metamaterial resonance using vanadium dioxide," Appl. Phys. Lett. 93, 024101 (2009)

[18] Q. Y. Wen, H. W. Zhang, Q. H. Yang, Y. S. Xie, K Chen, and Y. L. Liu, "Terahertz Metamaterials with VO_2 Cut-wires for Thermal Tunability," Appl. Phys. Lett. 97, 021111 (2010)

[19] J. Q. Gu, J. G. Han, X. C. Lu, R. Singh, Z. Tian, Q. R. Xing, and W. L. Zhang, "A close-ring pair terahertz metamaterial resonating at normal incidence," Opt. Express 17, 20307-20312 (2009).

[20] A. C. Caloz, and T. Itoh. "Electromagnetic Metamaterial: Transmission Line Theory and Microwave Applications", John wiley & Sons, 2005.

[21] Y. X Li, Y. S. Xie, H. W. Zhang, Y. L. Liu, Q. Y. Wen, W. W. Lin, " The strong non-reciprocity of metamaterial absorber: characteristic, interpretation and modelling", J Phys. D: Appl. Phys. 42 095408(2009)

[22] A. K. Azad, A. J. Taylor, E. Smirnova, J. F. O'Hara, " Characterization and analysis of terahertz metamaterials based on rectangular split-ring resonators", Appl. Phys. Lett. 92, 011119 (2008)

[23] L. Fu, H. Schweizer, H. Guo, N. Liu, H. Giessen, " Synthesis of transmission line models for metamaterial slabs at optical frequencies", Phys. Rev. B 78, 115110 (2008)

19

Metamaterial-Based Compact Filter Design

Merih Palandöken
Berlin Institute of Technology
Germany

1. Introduction

Emerging requirements of increasingly complex wireless systems necessitate novel design methods of wireless components to be developed for the fulfillment of many performance criteria simultaneously. One of these wireless components to be enhanced for high data rate transmission systems and matched with the new challenges in new generation communication systems is the microwave filter. This chapter is mainly dealing with the novel microwave filter design methods based on artificial materials. How to design electrically small resonators and to couple each of these resonators with the successive resonators in a periodic/aperiodic manner in addition to the feeding line for the bandpass and bandstop filter designs are highlighted throughout this chapter.

In this chapter, because the current trend in filter design is the filter miniaturization for more compact wireless systems, basic approaches in electrically small filter design based on artificial metamaterials are explained. The compact resonators result the signal suppression level of the incoming signal in bandstop filter designs and signal selectivity of the transmitted signal in bandpass filter designs to be higher as a performance enhancement. It is due to the target geometrical compactness, resulting into the possibility of cascading more electrically small resonator cells in a restricted area. Because these performance parameters are highly related with Q factor of each individual resonator, the eigenmodes of periodically loaded negative permittivity and negative permeability resonators are analytically calculated in Section 2. The passband frequencies and Q factor of the filter to be designed can be engineered with the geometrical and topological parameters of the resonators, which is the fundamental idea of artificial material based filter designs. This analytical calculation points out the effect of electromagnetic material parameters of negative permittivity and permeability materials on filter performance. In Section 3 and Section 4, two compact filter designs are proposed and explained along with the numerical results as metamaterial based filter examples. One of these designs is the fractal resonator based bandstop filter. The other design is the compact bandpass filter based on thin wire loaded spiral resonator cells. These are quite good examples of compact filter designs with negative permeability and left-handed metamaterial unit cells, respectively. These numerically calculated filters points out the compactness and performance enhancement of novel designs in comparison to the conventional filter designs in a restricted volume.

2. Theoretical analysis

The motivating principle of theoretical research on LHM is first introduced by Veselago with his theoretical paper in 1968 [Veselago, 1968]. He considered electromagnetic wave propagation through a homogenous isotropic electromagnetic material in which both permittivity and permeability were assumed to have negative real values. Because the direction of the Poynting vector of a monochromatic plane wave is opposite to that of its phase velocity in such a material, he referred to this medium as left-handed medium. This material property makes in turn these materials to support backward-wave propagation due to its negative refractive index. The natural inexistence of such exotic materials had unfortunately led the motivating ideas of Veselago on negative refraction, its various electromagnetic and optical consequences to receive little attention in the scientific community [Engheta & Ziolkowski, 2006; Caloz & Itoh, 2005; Eleftheriades & Balmain, 2005]. However, the work of Pendry on the electromagnetic engineering of magnetic permeability and electric permittivity of the materials with electrically small metallic inclusions has made Veselago's ideas realisable [Pendry et al.,1999; Pendry et al., 1996]. In 2000, Smith inspired by the work of Pendry and constructed a composite LH medium in the microwave regime by arranging the periodic arrays of small metallic wires and SRRs [Smith & Kroll, 2000]. He demonstrated the anomalous refraction at the interface of this LH medium with the air, which is the result of negative refraction in this artificial material. The effective electromagnetic parameters were also retrieved experimentally and numerically from the transmission and reflection data to prove the negative refractive index [Smith et al., 2005 ; Alexopoulos et al., 2007; Chen et al., 2004 ; Smith et al., 2000]. There have now been several theoretical and experimental studies that have been reported confirming negative refractive index. There are some engineering applications derived from this concept such as phase compensation and electrically small resonators [Engheta, 2002], negative angles of refraction [Kong et al., 2002; Kolinko & Smith, 2003; Ziolkowski, 2003], sub-wavelength waveguides with lateral dimensions below diffraction limits [Alu & Engheta, 2003,2004], enhanced focusing [Grbic & Eleftheriades,2004], backward wave antennas [Grbic & Eleftheriades,2002; Caloz & Itoh, 2005], enhanced electrically small antennas [Ziolkowski & Kipple, 2003] and compact microwave filters [Marques et al., 2008].

The main principle of negative refractive index in LHMs can be deduced quite easily by the calculation of TE and TM wave impedances of waveguide modes. The TE_{mn} and TM_{mn} wave impedances are calculated in a general form as

$$\begin{aligned} Z_{TEmn} &= \frac{\omega\mu}{k_{zmn}} = -\frac{\omega|\mu_{LHM}|}{k_{zmn}} \\ Z_{TMmn} &= \frac{k_{zmn}}{\omega\varepsilon} = -\frac{k_{zmn}}{\omega|\varepsilon_{LHM}|} \end{aligned} \tag{1}$$

with the longitudinal wave number, k_{zmn} at the operation frequency of ω with the magnetic permeability, μ and electric permittivity, ε of the waveguide filling material. Because the wave impedance of a passive microwave component is always positive, the phase constant has to be correspondingly negative in the impedance formulation due to the negative permittivity and permeability. As a result, the effective refractive index is also negative. This is an alternative explanation to understand the underlying reasoning why the refractive index has to be negative in the artificial materials with negative permittivity and permeability. Due to the exploitation of electrically small resonators in the resonator design, compact resonators can be cascaded to increase the frequency selectivity and stop-band rejection level in a restricted volume. The accompanying degradation of Q factor due to the periodic arrangement of intercoupled cells can be compensated by the implementation of complementary resonators, in which virtual magnetic currents are the excited resonant sources instead of electrical currents flowing through the lossy metal [Marques et al., 2008].

In this chapter, the subwavelength resonance feature of periodically arranged negative permittivity and permeability materials in a rectangular waveguide is explained by the calculation of eigenmode equation for TE modes. The eigenmode calculation can be similarly done for TM modes by replacing TE wave impedance with TM wave impedance in ABCD matrix formulation. The periodically loaded waveguide model is shown in Fig. 1.

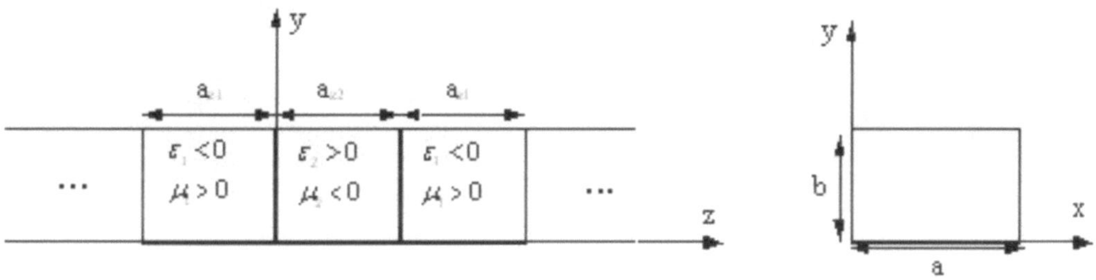

Fig. 1. Waveguide model of periodically arranged negative permeability and permittivity materials

ABCD matrix of negative permittivity material of length a_{z1} is formulated with TE-wave impedance, Z_{H1} and phase constant, k_{z1} as

$$M_{ABCD\_negeps} = \begin{bmatrix} \cos(k_{z1}a_{z1}) & jZ_{H1}\sin(k_{z1}a_{z1}) \\ j\dfrac{\sin(k_{z1}a_{z1})}{Z_{H1}} & \cos(k_{z1}a_{z1}) \end{bmatrix}$$

$$k_{z1} = \sqrt{\omega^2\mu_1\varepsilon_1 - (\frac{m\pi}{a})^2 - (\frac{n\pi}{b})^2} \qquad (2.a)$$

$$Z_{H1} = \frac{\omega\mu_1}{k_{z1}}$$

where a and b are the waveguide side lengths in x and y directions, respectively.

In the same manner, ABCD matrix of negative permeability material of length a_{z2} is formulated with TE-wave impedance, Z_{H2} and phase constant, k_{z2} as

$$M_{ABCD\_negmu} = \begin{bmatrix} \cos(k_{z2}a_{z2}) & jZ_{H2}\sin(k_{z2}a_{z2}) \\ j\dfrac{\sin(k_{z2}a_{z2})}{Z_{H2}} & \cos(k_{z2}a_{z2}) \end{bmatrix}$$

$$k_{z2} = \sqrt{\omega^2 \mu_2 \varepsilon_2 - (\frac{m\pi}{a})^2 - (\frac{n\pi}{b})^2}$$

$$Z_{H2} = \frac{\omega\mu_2}{k_{z2}} \tag{2.b}$$

Thus, ABCD matrix of one unit cell of length, $a_{z1}+a_{z2}$, consisting of the negative permeability and permittivity materials is calculated by cascading ABCD matrices of the respective materials as

$$M_{ABCD\_cell} = \begin{bmatrix} \cos(k_{z2}\frac{a_{z2}}{2}) & jZ_{H2}\sin(k_{z2}\frac{a_{z2}}{2}) \\ j\frac{\sin(k_{z2}\frac{a_{z2}}{2})}{Z_{H2}} & \cos(k_{z2}\frac{a_{z2}}{2}) \end{bmatrix} \begin{bmatrix} \cos(k_{z1}a_{z1}) & jZ_{H1}\sin(k_{z1}a_{z1}) \\ j\frac{\sin(k_{z1}a_{z1})}{Z_{H1}} & \cos(k_{z1}a_{z1}) \end{bmatrix}$$
$$\begin{bmatrix} \cos(k_{z2}\frac{a_{z2}}{2}) & jZ_{H2}\sin(k_{z2}\frac{a_{z2}}{2}) \\ j\frac{\sin(k_{z2}\frac{a_{z2}}{2})}{Z_{H2}} & \cos(k_{z2}\frac{a_{z2}}{2}) \end{bmatrix}$$

$$M_{ABCD\_cell} = \begin{bmatrix} A & B \\ C & D \end{bmatrix} \tag{2.c}$$

where the matrix elements are calculated as in (2.d).

$$\begin{aligned}
A &= \cos(k_{z1}a_{z1})\cos(k_{z2}a_{z2}) - \frac{1}{2}\left(\frac{Z_{H2}}{Z_{H1}} + \frac{Z_{H1}}{Z_{H2}}\right)\sin(k_{z1}a_{z1})\sin(k_{z2}a_{z2}) \\
B &= jZ_{H2}\cos(k_{z1}a_{z1})\sin(k_{z2}a_{z2}) + j\frac{\sin(k_{z1}a_{z1})}{2Z_{H1}}\left(\left(Z_{H1}^2 - Z_{H2}^2\right) + \left(Z_{H1}^2 + Z_{H2}^2\right)\cos(k_{z2}a_{z2})\right) \\
C &= j\frac{1}{Z_{H2}}\cos(k_{z1}a_{z1})\sin(k_{z2}a_{z2}) + j\frac{\sin(k_{z1}a_{z1})}{2Z_{H1}Z_{H2}^2}\left(\left(-Z_{H1}^2 + Z_{H2}^2\right) + \left(Z_{H2}^2 + Z_{H1}^2\right)\cos(k_{z2}a_{z2})\right) \\
D &= \cos(k_{z1}a_{z1})\cos(k_{z2}a_{z2}) - \frac{1}{2}\left(\frac{Z_{H2}}{Z_{H1}} + \frac{Z_{H1}}{Z_{H2}}\right)\sin(k_{z1}a_{z1})\sin(k_{z2}a_{z2})
\end{aligned} \tag{2.d}$$

Thus, the dispersion relation of the periodic array of negative permeability and permittivity materials is calculated from ABCD parameters to determine the complex propagation constant, γ_{zeff} as

$$\cosh(\gamma_{zeff}(a_{z1} + a_{z2})) = \cos(k_{z1}a_{z1})\cos(k_{z2}a_{z2}) - \frac{1}{2}\left(\frac{Z_{H1}}{Z_{H2}} + \frac{Z_{H2}}{Z_{H1}}\right)\sin(k_{z1}a_{z1})\sin(k_{z2}a_{z2}) \tag{2.e}$$

This dispersion relation can be formulated in an alternative form for the case of ideal lossless material parameters as

$$\cosh(\gamma_{zeff}(a_{z1}+a_{z2}))=\cosh(\alpha_{z1}a_{z1})\cosh(\alpha_{z2}a_{z2})-\frac{1}{2}\left(\frac{Z_1}{Z_2}+\frac{Z_2}{Z_1}\right)\sinh(\alpha_{z1}a_{z1})\sinh(\alpha_{z2}a_{z2}) \quad (3.a)$$

with the wave reactances, $Z_{1,2}$ and attenuation constants, $\alpha_{z1,2}$ of the negative permittivity and permeability media, respectively. The wave reactances and attenuation constants are calculated from (2.a) and (2.b) in the form of

$$Z_{H1}=\frac{\omega\mu_1}{-j\sqrt{\omega^2\mu_1|\varepsilon_1|+(\frac{m\pi}{a})^2+(\frac{n\pi}{b})^2}}=jZ_1$$
$$Z_{H2}=\frac{\omega|\mu_2|}{j\sqrt{\omega^2|\mu_2|\varepsilon_2+(\frac{m\pi}{a})^2+(\frac{n\pi}{b})^2}}=-jZ_2 \quad (3.b)$$

$$\alpha_{z1}=\sqrt{\omega^2|\varepsilon_1|\mu_1+(\frac{m\pi}{a})^2+(\frac{n\pi}{b})^2}$$
$$\alpha_{z2}=\sqrt{\omega^2\varepsilon_2|\mu_2|+(\frac{m\pi}{a})^2+(\frac{n\pi}{b})^2} \quad (3.c)$$

As it is deduced from (3.b), the wave impedances of negative permittivity and permeability materials are inductive and capacitive, respectively. Thus, the wave propagation can be in principle obtained at any frequency by engineering the material parameters and adjusting the slab thicknesses correspondingly to satisfy the resonance condition. Because of the parametric dependence of eigenmode equation on monotonically increasing functions, only one resonance frequency is calculated for a certain phase shift per cell. This results only one frequency band to be obtained in the dispersion diagram for each transversal wave number in the case of lossless material parameters with low material dispersion. This property makes the novel cavity resonators with one resonance frequency for each transversal wave number to be designed by loading the negative permittivity material with the negative permeability material. However, this is not possible in the resonator designs by longitudinal pairing of two RHMs or one RHM with LHM [Engheta & Ziolkowski, 2006].

Another potential application is the design of monomode waveguides with arbitrary thick lateral dimension by loading the waveguide walls with the negative permeability and permittivity materials laterally instead of longitudinally [Alu & Engheta, 2003]. Another important issue is the design of a cavity resonator with negative material parameters at a predetermined resonance frequency, which can be derived from the eigenmode equation. Eigenmode equation indicates that the resonance frequency of a cavity resonator is only dependent on the ratio of each slab thickness rather than the total slab thickness unlike in the conventional resonators [Engheta, 2002]. The same result can also be concluded for the transmission line and resonator designs, which are based on pairing of LH and RH materials in the same unit cell but not of two RH materials. This property leads the compact subwavelength transmission media and resonators to be designed by pairing any slab thicknesses of negative permittivity and permability materials.This is also the main principle in the design of subwavelength guided wave structures with the lateral dimensions below the diffraction limits [Grbic & Eleftheriades,2004].

As a case study, one negative permittivity material of Drude type electric response and one negative permeability material of Lorentzian type magnetic response are periodically arranged inside a rectangular waveguide. The slab lengths of negative permittivity and permeability materials are 1mm and 2mm, respectively. The magnetic resonance, magnetic plasma frequency and loss parameter are 2.10 GHz, 2.21 GHz and 100 Hz for the negative permeability material. The electric plasma frequency and loss parameter are 10 GHz and 100 Hz for the negative permittivity material. As deduced from the material parameters, the material losses are taken into account in the model, however, kept quite small to have low transmission loss and observe the transmission band broadening. The dispersion diagram and Bloch impedance are analytically calculated and shown in Fig. 2 between the magnetic resonance and plasma frequencies of the negative permeability medium.

As it is deduced from Fig. 2a, the propagation constant is negative between 2.106 GHz and 2.146 GHz with 40 MHz bandwidth. It results approximately into Q factor of 53 in LH passband. However, in addition to this LH band, RH band is also obtained between 2.173 GHz and 2.21GHz with 37 MHz bandwidth. This RH passband has an approximate Q factor of 60 with the center frequency of 2.19GHz. There is 27 MHz bandgap between LH and RH bands, extending from 2.146 GHz to 2.173 GHz with the highest signal rejection level at the stop band frequency of 2.16 GHz. The emergence of these two bands is mainly related with the included material dispersion and loss. As shown in Fig. 2b, Bloch impedance is high ohmic at the lower edge and low ohmic at the higher edge with no reactive part. Thus, the composite material can be modeled as a combination of parallel and series resonant circuits with the resonance frequencies of 2.146 GHz and 2.173 GHz, respectively. In order to confirm this issue and the emergence of RH band at the higher frequencies, the equivalent circuit model of negative permittivity and permeability materials are illustrated in Fig. 3 [Engheta & Ziolkowski, 2006].

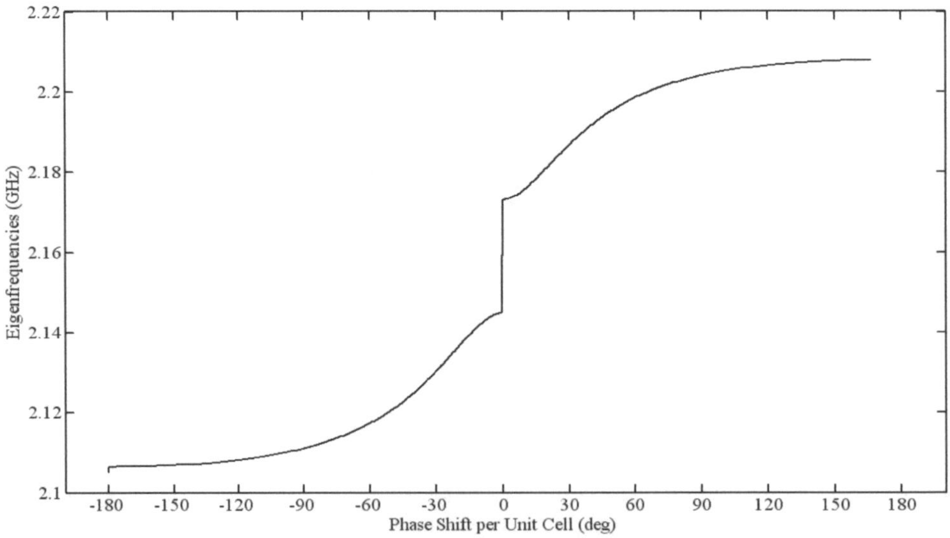

Fig. 2a. 1D dispersion diagram of periodically arranged negative permeability and permittivity materials

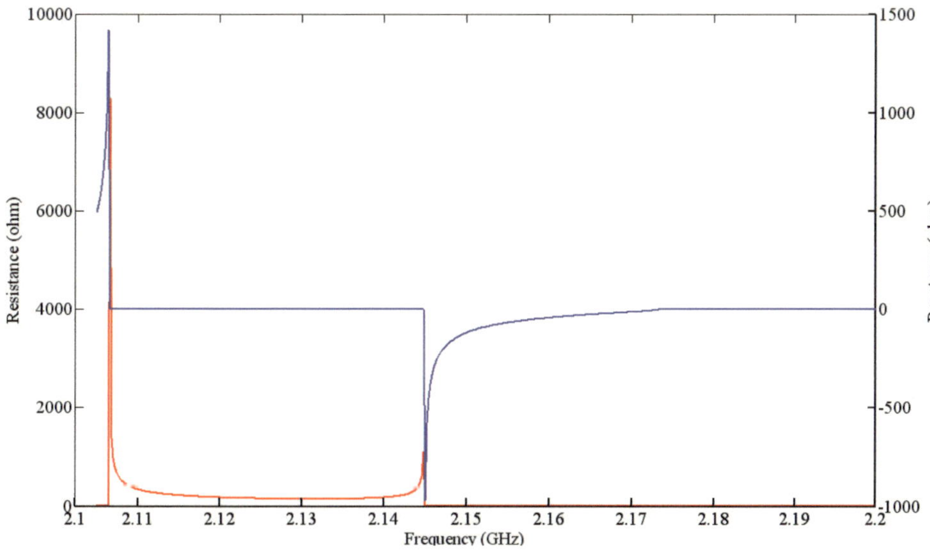

Fig. 2b. Resistance (red) and reactance (blue) of periodically arranged negative permeability and permittivity materials

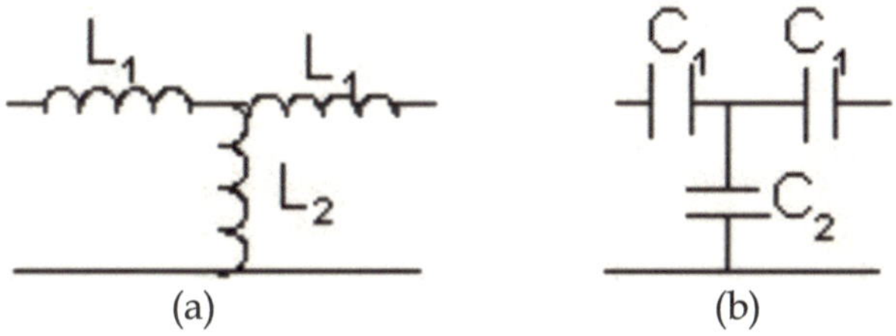

Fig. 3. Equivalent circuit model of one unit cell of (a) negative permittivity and (b) negative permeability materials

The periodic pairing of negative permittivity material with the negative permeability material results one unit cell to be modeled as

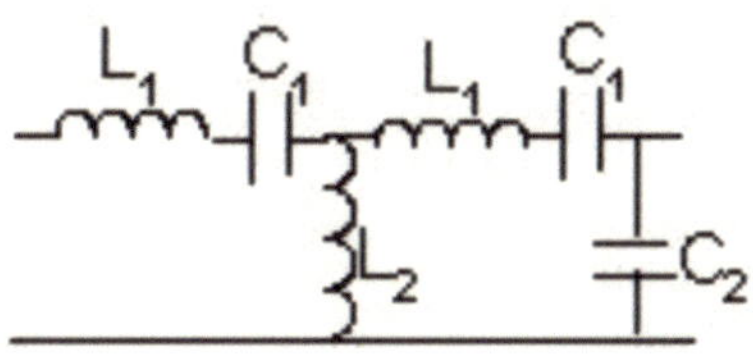

Fig. 4. Equivalent circuit model of one unit cell of periodically loaded negative permittivity and permeability materials

As deduced from Fig.4, for the frequencies larger than the series resonance frequency, the impedance of series branches is inductive. In addition, the resulting impedance of parallel branches are capacitive for the frequencies larger than the parallel resonance frequency. Thus, this circuit model has an equivalent form of series inductor loaded with the shunt capacitor as conventional RH transmission lines at the frequencies larger than the parallel and series resonance frequencies. This is the main reason why this composite material has an additional RH band at the frequencies larger than the LH band. In a similar manner, the equivalent circuit model of LH materials can be derived by the investigation of the same circuit model for the frequencies smaller than the parallel and series resonance frequencies. For this case, the circuit model has an equivalent form of series capacitor loaded with the shunt inductor. It is the dual of RH circuit model. One important conclusion from these circuit models is that the negative permittivity and permeability materials can be alternatively designed without relying on high lossy resonance phenomenon. In other words, rather than embedding the electrically small resonant metallic inclusions into the host material, the left handed feature can also be realized with low loss by periodic loading of conventional microstrip transmission lines with series capacitors and shunt inductors in planar microwave technology [Caloz & Itoh, 2005; Eleftheriades & Balmain, 2005]. Many microwave circuits have been implemented by using this strategy such as compact broadband couplers [Nguyen & Caloz 2007], broadband phase shifters [Eleftheriades & Balmain, 2005], compact wideband filters[Gil et al., 2007], compact resonant antennas [Lee et al., 2005, 2006; Schüßler et al., 2004; Sanada et al. 2004].

3. Spiral fractal resonator based compact band-stop filter

In this section, a compact, low insertion loss, high selective band-stop filter is explained. The filter is composed of two unit cells of electrically small artificial magnetic metamaterials, which have the topological form of fractal spiral resonators in [Palandoken & Henke, 2009]. The geometry of fractal spiral resonators is formed with the direct connection of two concentric Hilbert fractal curves of different dimensions as in a spiral form to excite the magnetic resonance. The operation principle of band-stop filter is based on the excitation of two electrically coupled fractal spiral resonators through direct connection with the feeding line.

The geometry of one unit cell of periodic artificial magnetic material is described shortly in Section 3.1 along with the reflection/transmission parameters and the resonant field pattern at the magnetic resonance frequency. In Section 3.2, the proposed filter topology is depicted with the geometrical and material parameters. The logical approach of the filter design is explained. In Section 3.3, the transmission and reflection parameters are illustrated to verify the desired filter performance and the operation principle is clarified in the light of simulated current distribution at the stopband frequency.

3.1 Structural design of fractal spiral resonator

The geometry of the artificial magnetic material is shown in Fig. 5. Each of the outer and inner rings are the mirrored image of the first order Hilbert fractal to form the ring shape. They are then connected at one end to obtain the spiral form from these two concentric Hilbert fractal curves. The marked inner section is the extension of the inner Hilbert curve so as to increase the resonant length due to the increased inductive and capacitive coupling between the different sections. The substrate material is standard 0.5 mm thick FR4 with dielectric constant 4.4 and tan(δ) 0.02. The metallization is copper. The copper line width and minimum distance between any two lines are 0.2 mm. The other geometrical parameters are L1= 2.2mm, L2= 0.8mm and L3= 1mm.

The unit cell size is ax = 5mm, ay =2mm, az = 5mm. Only one side of the substrate is structured with the prescribed fractal geometry while leaving the other side without any metal layer. Because two unit cells of this spiral fractal geometry is excited through the direct connection with the feeding line in the proposed band-stop filter, it is important to verify the magnetic resonance frequency as this frequency is the stopband frequency at which no field transmission is allowed.

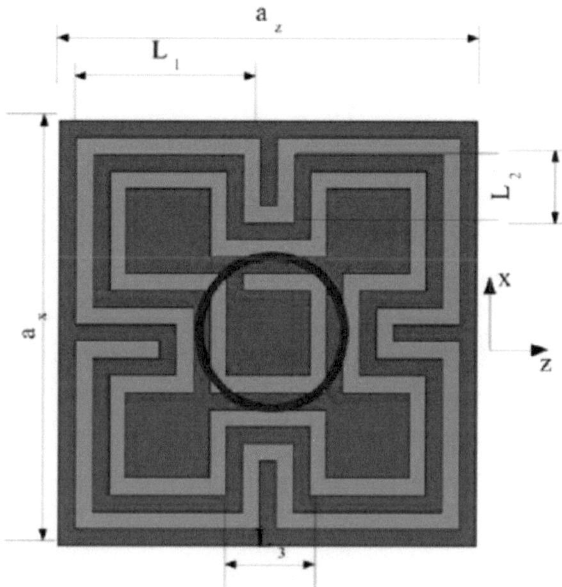

Fig. 5. Fractal spiral resonator geometry

In order to induce the magnetic resonance for the determination of stopband frequency, the structure has to be excited with out-off- plane directed magnetic field. Thus, in the numerical model, the structure is excited by z-direction propagating, x-direction polarized plane wave. Perfect Electric Conductor at two x planes and Perfect Magnetic Conductor at two y planes are assigned as the boundary conditions.

The resonance frequency and surface current distribution at the resonance frequency are numerically calculated with FEM based commercial software HFSS. The simulated S-parameters and surface current distribution are shown in Fig. 6 and Fig. 7, respectively.

Due to the spiralling form of surface current and the resulting out-off-plane directed magnetic field, the magnetic resonance is quite effective. Therefore, this electrically small structure can be regarded as a resonant magnetic dipole at 1.52 GHz. The transmission deep through one unit cell-thick artificial material in Fig.6 is effectively due to the depolarization effect of this magnetic dipole for the incoming field. This is the reason why this artificial magnetic material is regarded as a negative permeability material in a certain frequency band and the frequency of transmission deep is regarded as the magnetic resonance frequency of the electrically small structure. This frequency band is the desired band at which the proposed band-stop filter is designed to operate and the incoming field is exponentially attenuated in the propagation direction.

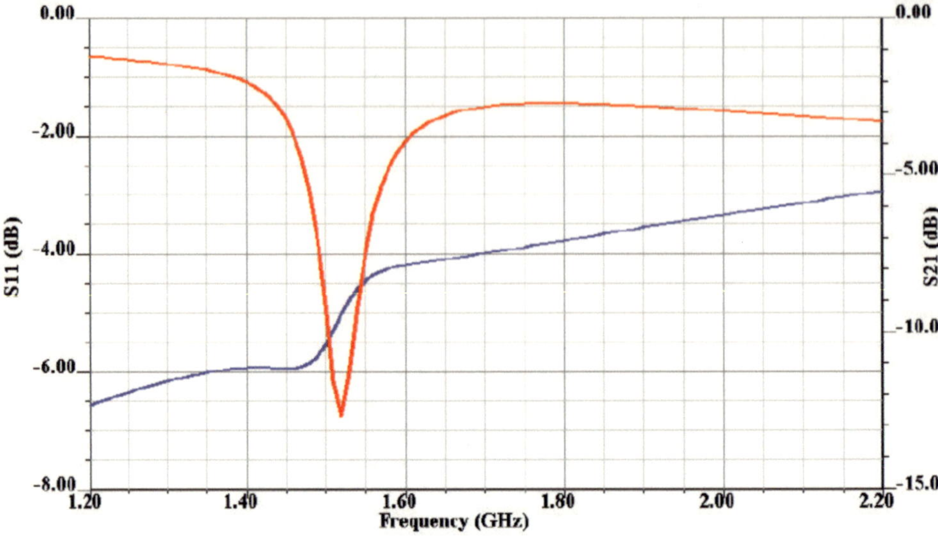

Fig. 6. Transmission (red) and reflection (blue) parameters of fractal spiral resonator

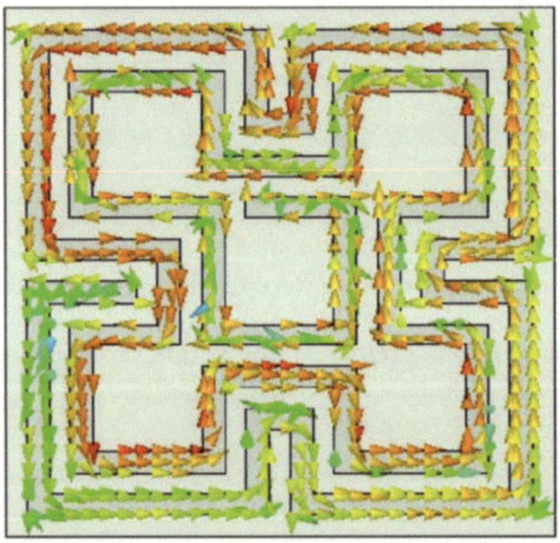

Fig. 7. Surface current distribution of fractal spiral resonator at 1.52 GHz

3.2 Band-stop filter design

The proposed band-stop filter is shown in Fig. 8. As it could be deduced from the filter topology, two fractal resonators are connected antisymmetrically along the x axis through the feeding line to have symmetrical return loss. In the filter model, the separation distance between x-direction oriented cells is 0.1 mm. Each unit cell is connected directly with the feeding line to increase the field coupling as shown with the red circles in Fig. 8 as a feeding method [Palandoken & Henke, 2010]. One conventional feeding method is to have resonator shaped slots in the ground plane for high field rejection [Marques et al., 2008]. However, this design method is not selected in order to suppress the resulting back radiation due to the slotted ground plane for lower insertion loss. In addition, this type of resonator feeding improves the filter selectivity in comparison to the proximity coupled feeding method. The separation distance between the unit cells and feeding line sections is 0.2 mm. Rather than using high lossy FR4, in the filter model low loss Rogers 4003 material with relative permittivity 3.38 and loss tangent 0.0027 is used. The main reason to use low-loss substrate instead of high-loss FR4 substrate is not only to decrease the insertion loss with low loss substrate but also with low permittivity material to increase the field coupling from the feeding line to the fractal spiral resonators.

The filter width (Wf) and length (Lf) are 10.1 mm and 5.4 mm, respectively. The width of each metallic line is 0.2 mm except the width of feeding line, which is 1.1 mm to excite both resonators effectively and couple each of the resonators with 50 Ω line impedance at both ports. The length of feeding line sections at each port is 5 mm. The total size of band-stop filter is 15.4 mm. The design principle of the proposed filter is to feed the fractal resonators directly through the feeding line with their resonant field distributions. This field distribution is the resonant field excited at the band edge of the bandgap in the Brilliouin diagram, which results into no transmission of the incoming wave because of its standing wave nature. As it could be noticed in Section 3.1, the resonance frequency of fractal spiral resonators is numerically calculated with the electrical coupling between the in-plane oriented cells and magnetic coupling between the out-plane oriented cells in the transverse plane. The feeding method, which is currently exploited in this design, results these fractal spiral resonators to couple electrically as in the numerical model in Section 3.1. As a next step, the return and insertion loss of the band-stop filter is numerically calculated in addition to the resonant surface current distribution.

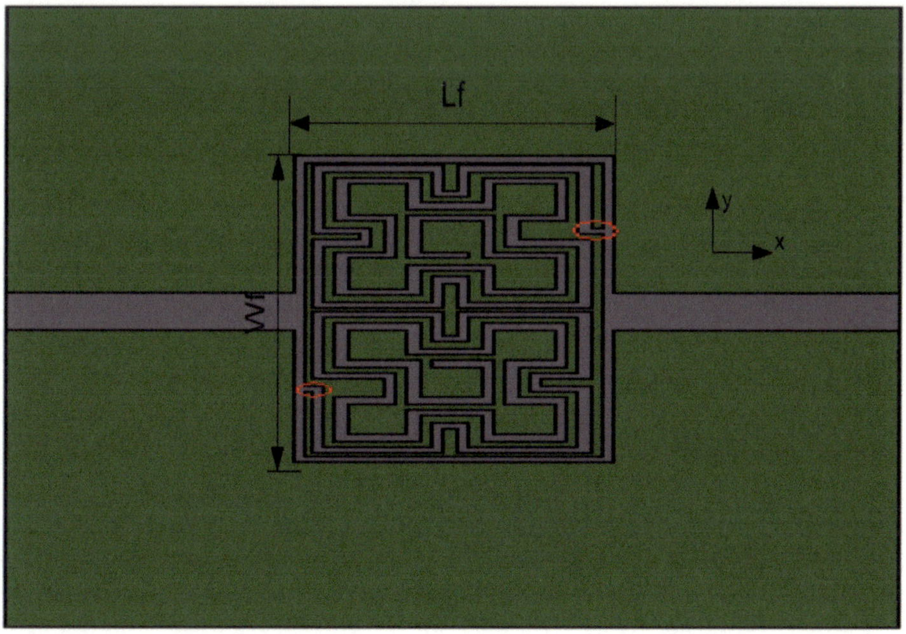

Fig. 8. Fractal spiral resonator based band-stop filter

3.3 Numerical results of fractal spiral resonator based band-stop filter

To validate the design concept, the performance of the band-stop filter is numerically calculated by using FEM based commercial software HFSS. The simulation results of the insertion and return losses are shown in Fig. 9. As shown in Fig. 9, the return loss is larger than 10 dB in two frequency bands of which is smaller than 0.75 GHz and which is larger than 1.87 GHz in the frequency span of 0.5-2.5 GHz. The insertion loss in the passband is better than 1 dB. The frequency rejection level is larger than 20 dB in the frequency band of 1.36-1.40 GHz and 27 dB at the center frequency, 1.38 GHz. The selectivity of the proposed band-stop filter is quite promising, which is 100 dB/GHz frequency selectivity with 3 dB reference insertion loss.

The physical size of the main filtering section is λo/40.18 x λo/21.48 at the center frequency, which is quite compact in comparison to the conventional stepped impedance or coupled line filters [Pozar, 2004]. On the other hand, there is no matching network, which is quite advantageous in the filter design to reduce the filter physical size. The total size of the filter even with the transmission line sections at the input and output ports is also compact, which is λo/10.85 x λo/14.46. The surface current distribution at the center frequency of the stopband, 1.38 GHz is shown in Fig. 10 to verify the design principle.

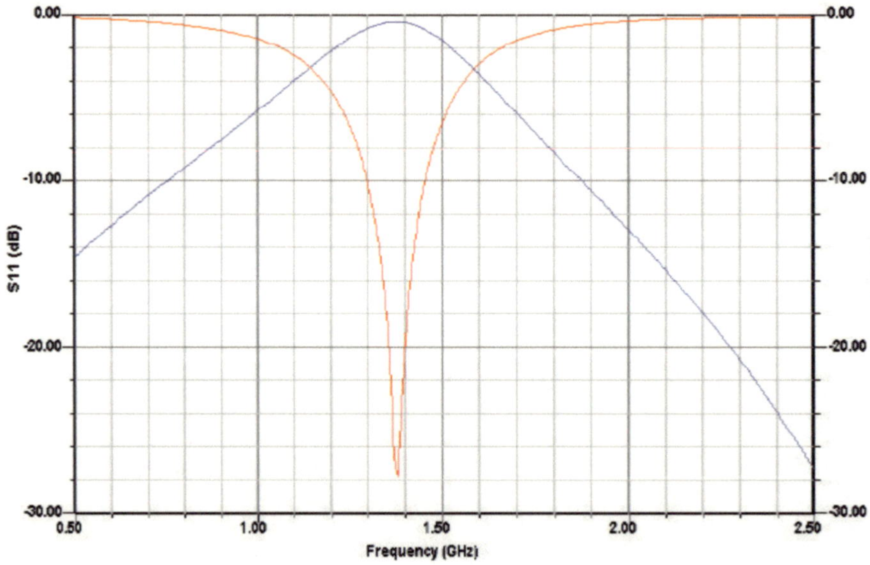

Fig. 9. Reflection(blue) and transmission (red) parameters of band-stop filter

As shown in Fig. 10, two fractal resonators are excited through the direct connection with the feeding line in their resonant current distribution. This type of resonator feeding leads these electrically small cells to have same direction directed magnetic dipole moments, which is actually the same field distribution for two electrically coupled spiral resonators. However, because of the additional electrical length due to the feeding line section for both of the spiral resonators, the center frequency of the stopband is lower than the resonance frequency of the fractal resonators.

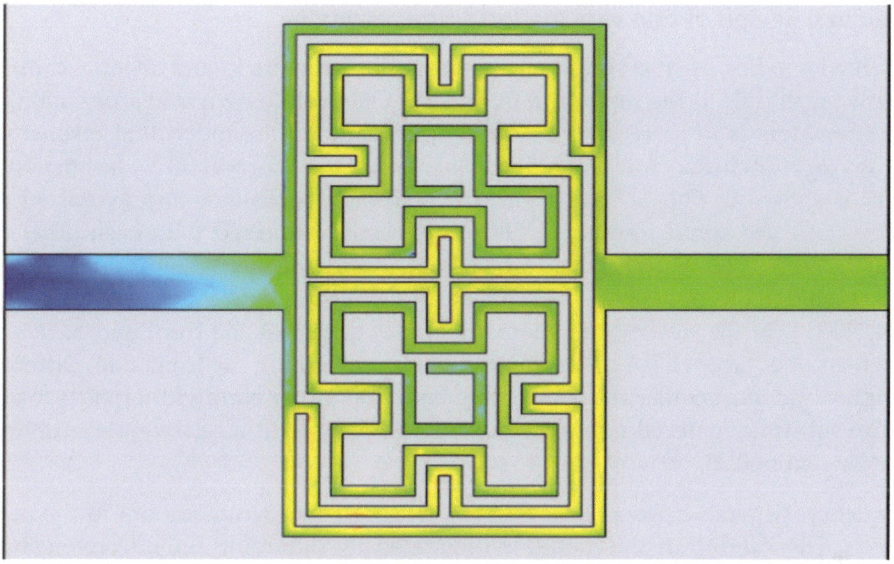

Fig. 10. Surface current distribution of fractal spiral resonator based BSF at 1.38GHz

The operation principle of the proposed filter is thus based on the suppression of the incoming field through the direct coupling of the feeding line to the spiral resonators at the resonant frequency. The resonant field, which is confined at the spiral resonators do not lead the magnetic dipole-like resonators to couple effectively to the electric dipole-like transmission line section at the second port. This reduces the transmission of the incoming field though the input feeding line and increases the reflection of the input signal at the first port.

As a result, in this section, the design of a compact, high selective, low insertion loss band-stop filter with two unit cells of novel magnetic metamaterial geometry is introduced. The physical dimensions of the designed band-stop filter are $\lambda o/10.85$ x $\lambda o/14.46$ with the additional transmission line sections at the center frequency of the stopband . No matching network is required in filter design, which is quite important to reduce the total filter size effectively. The insertion and return losses are numerically calculated and the current distribution at the center frequency of the stopband is illustrated. The proposed filter has satisfactory insertion loss, which is better than 1 dB. The selectivity of the filter is 100 dB/GHz, which is quite suitable to be used as band-stop filter in modern communication systems.

4. Thin wire loaded spiral resonator based compact band-pass filter

In this section, the design of a compact, LHM-based band-pass filter (BPF) is explained. The band-pass filter is composed of two unit cells of electrically small LHMs, which have the same geometry of thin wire loaded spiral resonators in [palandoken et al., 2009]. The LHM geometry is based on the direct connection of thin wire with the spiral resonator in order to enhance the magnetic field coupling inbetween in addition to the increased electrical length for reduced resonance frequency. The operation principle of BPF is based on the excitation of two coupled LHM resonators at two eigenfrequencies through direct connection with the feeding line.

4.1 Structural design of thin wire loaded spiral resonator

LHM behavior relies on the simultaneous excitation of electric and magnetic dipole-like electrically small cells at the resonance frequency. One well-known miniaturization method of such resonators is to increase the field coupling among the individual resonators. This design strategy is chosen for the LHM cell exploited in current BPF design. The LHM geometry is shown in Fig. 11 along with the boundary conditions and excitation sources. The wire strips and spiral resonators (SR) are directly connected with each other on both sides of the substrate. Further, instead of SRRs as in the original proposal of artificial magnetic material, SRs are used, which have half the resonance frequency of SRRs [Baena et al., 2004, 2005]. In the design, the geometrical parameters of the front and back side unit cells are the same, except a 0.6 mm shorter wire strip length on the front side. Different strip wire lengths lead to a smaller resonance frequency and larger bandwidth [palandoken et al., 2009]. The substrate material is nonmagnetic FR4-Epoxy with a relative permittivity of 4.4 and loss tangent of 0.02.

The frequency dispersive property of FR4 has not been taken into account in the numerical calculation. The validity of the model is illustrated by retrieving the effective constitutive parameters from S parameters and by the opposite direction of group and phase velocity in [palandoken et al., 2009].

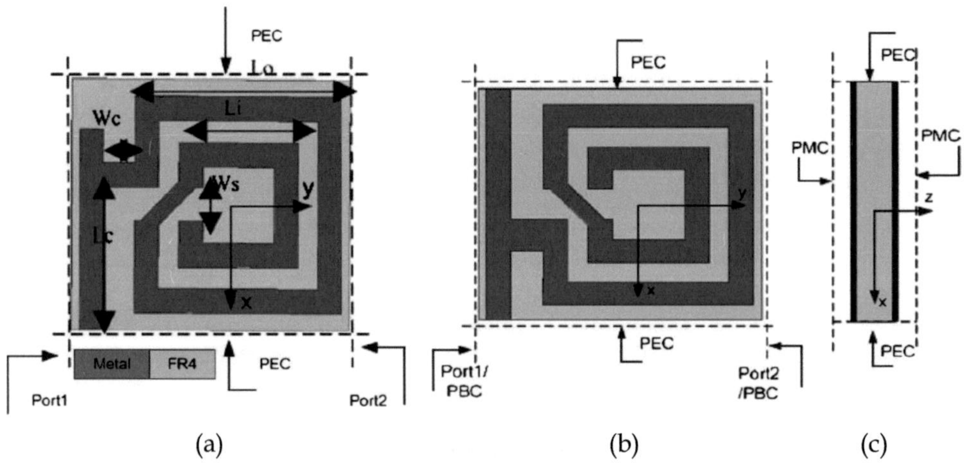

Fig. 11. LHM unit cell geometry. (a) Front and (b) back side of one LHM unit cell with indicated geometrical parameters in [palandoken et al., 2009].

4.2 Band-pass filter design

The band-pass filter is shown in Fig. 12. As it could be implied from the filter topology, two LHM resonators are connected antisymmetrically along the y axis through the feeding line in order to have symmetrical reflection parameter. In the filter model, the separation distance between x-direction oriented cells is 0.2 mm. Each unit cell is connected directly with the feeding line to excite each of LHM resonators effectively and couple the excited LHM resonator electrically and magnetically with the another resonator depending on the mode of excitation in addition to the extended feeding line sections as an effective feeding method. LHM resonator shaped slots in the ground plane are not exploited in the current band-pass filter due to the resulting back radiation from the slotted ground plane. The separation distance between one unit cell and feeding line sections in y-direction is 0.8 mm. One LHM resonator is shifted from the another resonator along y-direction with the distance of 0.4 mm to obtain optimum field intercoupling. Instead of using high lossy FR4, in the filter model low loss Rogers 5880 material with the relative permittivity 2.2 and loss tangent 0.0009 is used. The main reason to use low-loss substrate instead of high-loss FR4 substrate is not only to decrease the insertion loss with low loss substrate but also with low permittivity material to increase the field coupling from the first resonator into the second resonator in addition to the field coupling from the feeding line to each of LHM resonators.

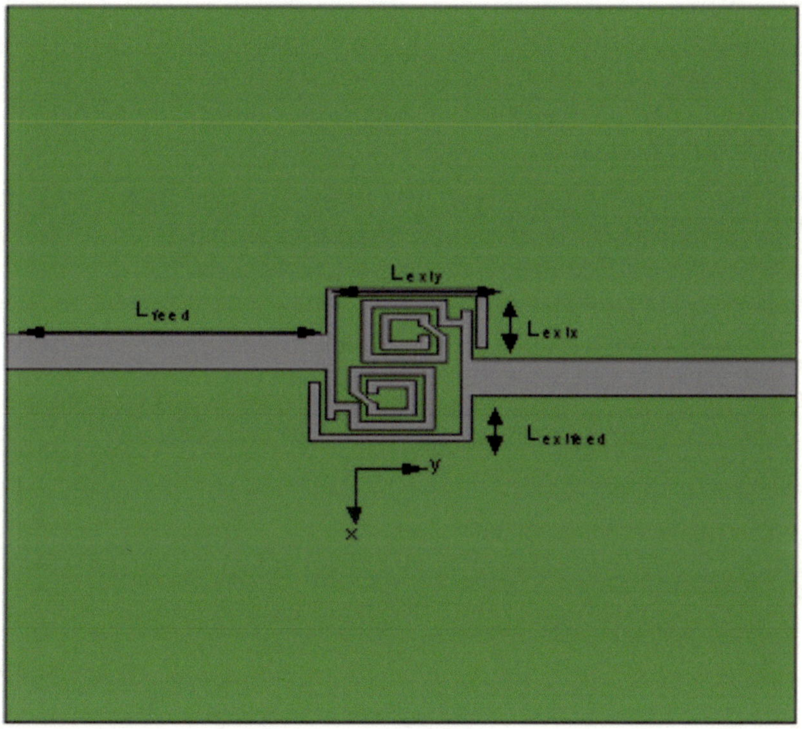

Fig. 12. Thin wire loaded spiral resonator based BPF

The lengths of extended feeding line in x- (L_{extx}) and y-directions (L_{exty}) are 2.5mm and 4.9mm, respectively. The width and length (L_{feed}) of feeding line are 1.5mm and 9.8mm, respectively. The direct connection length of the feeding line, ($L_{extfeed}$) in the extended sections is 1.9mm. The width and length of Rogers substrate are 30 mm and 24.2mm, respectively. The total size of band-pass filter is 6.4mm x 5.4 mm.

The design principle of the proposed filter is to feed one of the LHM resonators directly through the feeding line with its resonant field distribution. The highly concentrated field distribution results the directly excited first resonator to couple to the second LHM resonator electrically and magnetically depending on the excited 0 and 180 modes. The excited field in the first LHM resonator is enhanced by the magnetic field of the extended section of the feeding line. The feeding method, which is currently exploited in this design, results these LHM resonators to couple feeding line in an optimal manner. As a next step, the return and insertion loss of the band-pass filter is numerically calculated in addition to the resonant surface current distribution at two eigenfrequencies.

4.3 Numerical results of thin wire loaded spiral resonator based band-pass filter

To validate the filter design concept, the reflection and transmission parameters of band-pass filter are numerically calculated by using FEM based commercial software HFSS. The numerical results are shown in Fig. 13. As deduced from Fig. 13, the return loss is larger than 10 dB with the insertion loss smaller than 1.3 dB in the frequency band from 4.49GHz upto 4.92 GHz. The lowest insertion loss is 0.4 dB at 4.83GHz. The filter selectivities , roll-off factors, are approximately 400 dB/GHz and 62 dB/GHz at the lower and higher edge of the passband, respectively. The filter response is quite similar to the transfer function of Chebychev filter due to existing passband ripples and steep roll-off from the passband edges to the stop-band frequencies.

The physical size of the filtering section without the feeding line sections is λo/10 x λo/12 at the center frequency of 4.7GHz. It is quite compact in comparison to the conventional stepped impedance or coupled line filter designs. One important advantage of the current filter design is to have no additional matching network, which reduces the filter physical size significantly. The surface current distributions at two resonance frequencies, 4.52 GHz and 4.83 GHz, are shown in Fig. 14 to verify the design principle.

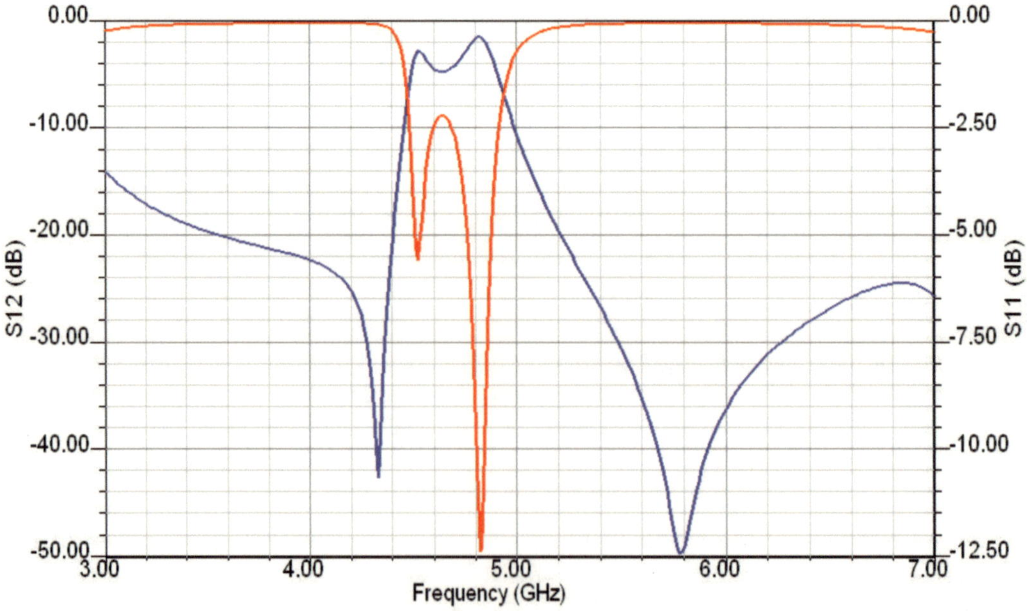

Fig. 13. Reflection(blue) and transmission (red) parameters of band-pass filter

(a) (b)

Fig. 14. Surface current distribution of thin wire loaded spiral resonator based BPF at (a) 4.52 GHz and (b) 4.83GHz

As shown in Fig. 14, one of LHM resonators is directly fed through the feeding line without any metallic connection from the excited resonator to the another LHM resonator. The resonant field distribution of LHM resonator and extended feeding line leads the incoming field to be coupled from the input port to the output port in 0 mode at 4.52GHz and 180 (π) mode at 4.83 GHz. The electric coupling at the lower resonance frequency results both LHM resonators and extended feeding line sections to radiate in comparison to the magnetic coupling at the higher resonance frequency. This is the main reason why the insertion loss is higher at the lower resonance frequency than that at the higher resonance frequency. The transmission principle of incoming field in the passband is based on the excitation of output LHM resonator by the electric and magnetic coupling of the resonant field excited in the directly fed resonator at the lower and higher resonance frequency, respectively.

As a result, in the last section of this chapter, the design of a compact band-pass filter with two unit cells of LHM geometry is explained. The physical dimensions of band-pass filter are $\lambda o/10$ x $\lambda o/12$ at the center frequency of 4.7GHz. There is no need of additional matching network in filter design, which shrinks the total filter size effectively. The insertion and return losses are numerically calculated. The current distributions at two resonance frequencies are illustrated. The proposed filter has satisfactory insertion loss, which is smaller than 1.2 dB in the frequency band between 4.49GHz and 4.92 GHz with the bandwidth of 430MHz. The filter selectivities calculated with reference to 3dB insertion loss value are 400 dB/GHz and 62 dB/GHz at the lower and higher edge of the passband, respectively.

5. References

N. G. Alexopoulos, C. A. Kyriazidou, and H. F. Contopanagos, "Effective parameters for metamorphic materials and metamaterials through a resonant inverse scattering approach," *IEEE Trans. Microw. Theory Tech.*, vol. 55, no. 2, pp. 254–267, Feb. 2007

A. Alù and N. Engheta, "Pairing an epsilon-negative slab with a mu-negative slab: Resonance, tunneling and transparency," IEEE Trans. Antennas Propag., vol. 51, no. 10, pp. 2558–2571, Oct. 2003.

A. Alù and N. Engheta, "Guided modes in a waveguide filled with a pair of single-negative(SNG), double-negative (DNG), and/or double-positive (DPS) layers," IEEE Trans. Microw. Theory Tech., vol. 52, no. 1, pp. 199–210, Jan. 2004.

J. Baena, R. Marqués & F. Medina, "Artificial magnetic metamaterial design by using spiral resonators," Phy. Rev. B, vol. 69, pp. 0144021–0144025, 2004

C.Caloz and T. Itoh, *Electromagnetic Metamaterials: Transmission Line Theory and Microwave Applications*, Piscataway, NJ: Wiley- IEEE, 2005

X. Chen, T. M. Grzegorczyk, B.-I.Wu, J. Pacheco, Jr., and J. A. Kong, "Robust method to retrieve the constitutive effective parameters of metamaterials," *Phys. Rev. Lett. E*, vol. 70, pp. 0166081–0166087, 2004.

G. V. Eleftheriades and K. G. Balmain, *Negative Refraction Metamaterials: Fundamental Principles and Applications*, New York:Wiley Interscience, 2005

N. Engheta, "An idea for thin subwavelength cavity resonators using metamaterials with negative permittivity and permeability," IEEE Antennas Wireless Propag. Lett., vol. 1, no. 1, pp. 10–13, 2002

N. Engheta and R. W. Ziolkowski, *Metamaterials Physics and Engineering Explorations*, Eds. New York: Wiley/IEEE, 2006.

M. Gil, J. Bonache, J. Garcia-Garcia, J. Martel & F. Martin, "Composite right/left-handed metamaterial transmission lines based on complementary split-rings resonators and their applications to very wideband and compact filter design," *IEEE Trans. Microw. Theory Tech.*, vol. 55, no. 6, pp. 1296–1304, Jun. 2007.

A. Grbic and G. V. Eleftheriades, "Overcoming the diffraction limit with a planar left- handed transmission lines," Phys. Rev. Lett., vol. 92, no. 11, p. 117 403, Mar. 2004.

A. Grbic and G. V. Eleftheriades, "Experimental verification of backward-wave radiation from a negative refractive index metamaterial," J.Appl. Phys., vol. 92, pp. 5930–5935, Nov. 2002.

P. Kolinko and D. R. Smith, "Numerical study of electromagnetic waves interacting with negative index materials," Opt. Express, vol. 11, pp. 640–648, Apr. 2003

J. A. Kong, B.-I. Wu, and Y. Zhang, "A unique lateral displacement of a Gaussian beam transmitted through a slab with negative permittivity and permeability," Microwave Opt. Technol. Lett., vol. 33, pp. 136–139, Mar. 2002.

C.-J. Lee, K. M. K. H. Leong, and T. Itoh, "Composite right/left-handed transmission line based compact resonant antennas for RF module integration," *IEEE Trans. Antennas Propag.*, vol. 54, no. 8, pp. 2283–2291, Aug. 2006.

C.-J. Lee, K. M. K. H. Leong, and T. Itoh, "Design of resonant small antenna using composite right/left handed transmission line," in *Antennas Propag. Soc. Int. Symp.*, 2005, vol. 2B

Ricardo Marqués, Ferran Martín, Mario Sorolla, *Metamaterials with negative parameter:theory, design, and microwave applications,* John Wiley & Sons, 2008

H. V. Nguyen and C. Caloz, "Generalized coupled-mode approach of metamaterial coupled line couplers: Coupling theory, phenomenological explanation and experimental demonstration," *IEEE Trans. Microw.Theory Tech.*, vol. 55, no. 5, pp. 1029–1039, May 2007

M. Palandöken, A. Grede & H. Henke, "Broadband microstrip antenna with Left-handedMetamaterials, "IEEE Trans. Antennas Propag., vol. 57, no.2, pp. 331-338, Feb. 2009

M. Palandöken, H. Henke, "Fractal spiral resonator as magnetic metamaterial, "Applied Electromagnetics Conference, pp. 1-4, 2009

M. Palandöken, H. Henke, "Compact LHM-based band-stop filter", Mediterranean Microwave Symposium, pp. 229–231, Aug. 2010

J. B. Pendry, A. J. Holden, D. J. Robbins & W. J. Stewart, "Magnetism from conductors and enhanced nonlinear phenomena," IEEE Trans. Microw. Theory Tech., vol. 47, no. 11, pp. 2075–2081, Nov. 1999

J. B. Pendry, A. J. Holden, D. J. Robbins & W. J. Stewart, "Low-frequency plasmons in thin wire structures," J. Phys., Condens.Matter, vol. 10, pp. 4785–4809, 1998

David Pozar, *Microwave Engineering* , 2004, Wiley

A. Sanada, M. Kimura, I. Awai, C. Caloz, and T. Itoh, "A planar zeroth- order resonator antenna using a left handed transmission line," in *34th Eur. Microw. Conf.*, Amsterdam, The Netherlands, 2004, pp. 1341–1344.

M. Schüßler, J. Freese, and R. Jakoby, "Design of compact planar antennas using LH-transmission lines," in *Proc. IEEE MTT-S Int. Microw. Symp.*, 2004, vol. 1, pp. 209–212.

D. R. Smith and N. Kroll, "Negative refractive index in left-handed materials," Phys. Rev. Lett., vol. 85, pp. 2933–2936, Oct. 2000.

D. R. Smith, D. C. Vier, N. Kroll, and S. Schultz, "Direct calculation of permeability and permittivity for a left- handed metamaterial," *App.Phys. Lett.*, vol. 77, no. 14, pp. 2246–2248, Oct. 2000

D. R. Smith, D. C. Vier, T. Koschny, and C. M. Soukoulis, "Electromagnetic parameter retrieval from inhomogeneous metamaterials," *Phys. Rev. E*, vol. 71, pp. 0366171–03661711, 2005

R.W. Ziolkowski and A. Kipple, "Application of double negative metamaterials to increase the power radiated by electrically small antennas," IEEE Trans. Antennas Propag., vol. 51, no. 10, pp. 2626–2640, Oct. 2003

R. W. Ziolkowski, "Pulsed and CW Gaussian beam interactions with double negative metamaterial slabs," Opt. Express, vol. 11, pp. 662–681, Apr. 2003

V. G. Veselago, "The electrodynamics of substances with simultaneously negative values of ε and μ," Sov. Phys. – Usp., vol. 47, pp. 509–514, Jan.–Feb. 1968.

20

Metasurfaces for High Directivity Antenna Applications

Shah Nawaz Burokur, Abdelwaheb Ourir,
André de Lustrac and Riad Yahiaoui
Institut d'Electronique Fondamentale,
Univ. Paris-Sud, CNRS UMR 8622,
France

1. Introduction

There has been a lot of study published in literature on the improvement of the performances of microstrip patch antennas. Most of the solutions proposed in the past were to use an array of several antennas. The particular disadvantage of this method comes from the feeding of each antenna and also from the coupling between each element. Other interesting solutions have then been suggested: the first one (Jackson & Alexópoulos, 1985) was to make use of a superstrate of either high permittivity or permeability above the patch antenna and the second one proposed (Nakano et al., 2004), is to sandwich the antenna by dielectric layers of the same permittivity. A Left-Handed Medium (LHM) superstrate where both permittivity and permeability are simultaneously negative has also been suggested (Burokur et al., 2005). The numerical study of a patch antenna where a Left-Handed Medium (LHM) is placed above has been done and in this case a gain enhancement of about 3 dB has been observed. However, these solutions are all based on non-planar designs which are bulky for novel telecommunication systems requiring compact low-profile and environment friendly directive antennas.

To overcome the major problem of complex feeding systems in antenna arrays, the design of compact directive electromagnetic sources based on a single feeding point has become an important and interesting research field. Different interesting solutions based on this concept have been proposed. At first, resonant cavities in one-dimensional (1-D) dielectric photonic crystals have been used (Cheype et al., 2002). Afterwards, three dimensional (3-D) structures have been used, leading to better performances (Temelkuran et al., 2000). Another interesting solution proposed by Enoch *et al.* was to use the refractive properties of a low optical index material interface in order to achieve a directive emission (Enoch et al., 2002). The authors have shown how a simple stack of metallic grids can lead to ultra-refraction. Because the resulting metamaterial structure has an index of refraction, *n*, which is positive, but near zero, all of the rays emanating from a point source within such a slab of zero index material would refract, by Snell's Law, almost parallel to the normal of every radiating aperture. We shall note that these solutions are all also based on the use of a bulky 3-D material.

Otherwise, the most common method to reach directive emission is obviously based on the Fabry-Pérot reflex-cavity mechanism (Trentini, 1956). Such cavities have first been considered quite bulky too since a thickness of half of the working wavelength is required (Akalin et al., 2002). But recently, the introduction of composite metasurfaces has shown that the half wavelength thickness restriction in a Fabry-Pérot cavity can be judiciously avoided. For example, Feresidis *et al.* showed that a quarter wavelength thick Fabry-Pérot cavity can be designed by using Artificial Magnetic Conductor (AMC) surfaces introducing a zero degree reflection phase shift to incident waves (Feresidis et al., 2005). Assuming no losses and exactly 0° reflection phase, the surface is referred to as a Perfect Magnetic Conductor (PMC), which is the complementary of a Perfect Electric Conductor (PEC). The latter AMC surfaces have been first proposed in order to act as the so called High Impedance Surface (HIS) (Sievenpiper et al., 1999). This HIS is composed of metallic patches periodically organized on a dielectric substrate and shorted to the metallic ground plane with vias, appearing as "mushroom" structures. In a particular frequency band where reflection phase is comprised between -90° and +90°, this surface creates image currents and reflections in-phase with the emitting source instead of out of phase reflections as the case of conventional metallic ground plane. The HIS allows also the suppression of surface waves which travel on conventional ground plane. However, the HIS of Sievenpiper needs a non-planar fabrication process, which is not suitable for implementation in lots of microwave and millimetric circuits.

The reflex-cavity antenna proposed by Feresidis was composed of two planar AMC surfaces and a microstrip patch antenna acting as the primary (feeding) source. The first AMC surface was used as the feeding source's ground plane so as to replace the PEC surface and hence, to achieve a 0° reflection phase. The second one acted as a Partially Reflective Surface (PRS) with a reflection phase equal to 180°. This idea has then been pushed further by Zhou *et al.* (Zhou et al., 2005). By taking advantage of the dispersive characteristics of metamaterials, the authors designed a subwavelength cavity with a thickness smaller than a 10th of the wavelength. Compared to Feresidis, Zhou made use of a non-planar mushroom structure with a dipole acting as the feeding source.

In this chapter, using a novel composite metamaterial, made of both capacitive and inductive grids, we review our recent works in the fields of low-profile and high-gain metamaterial-based reflex-cavity type antennas. First, we will show how our group has lately further reduced the cavity thickness by $\lambda/30$ for applications to ultra-thin directive antennas by using a PEC surface as the source's ground plane and one subwavelength metamaterial-based composite surface as the PRS. We will also present how an optimization of the cavity has also been undertaken in order to reduce the thickness to $\lambda/60$ by using an AMC surface instead of the PEC ground plane and a metasurface as PRS. We will then present the modeling and characterization of resonant cavities for enhancing the directivity. Finally, a phase controlled metasurface will be proposed for applications to beam steerable and frequency reconfigurable cavity antennas. Numerical analyses using Finite Element Method (FEM) based software *HFSS* and CST's Transmission Line Modeling (TLM) solver *MICROSTRIPES* together with discussions on the fabrication process and the experimental results will be presented for the different cavities mentioned above.

2. Operating principle of the Fabry-Pérot reflex-cavity

A cavity antenna is formed by a feeding source placed between two reflecting surfaces as shown in Fig. 1. In this paper, different cavities based on the schematic model presented in Fig. 1 will be discussed and used. The cavity is composed of a PEC surface acting as a conventional ground plane for the feeding source and a metamaterial-based surface (metasurface) playing the role of a transmitting window known as a PRS. Following the earlier work of Trentini, a simple optical ray model can be used to describe the resonant cavity modes (Trentini, 1956). This model is used to theoretically predict the operating mode of a low-profile high-directivity metamaterial-based subwavelength reflex-cavity antenna. Let us consider the cavity presented in Fig. 1(a). It is formed by a feeding antenna placed between two reflectors separated by a distance *h*. Phase shifts are introduced by these two reflectors and also by the path length of the wave travelling inside the cavity. With the multiple reflections of the wave emitted by the antenna, a resonance is achieved when the reflected waves are in phase after one cavity roundtrip. The resonance condition, for waves propagating vertically, can then be written as:

$$h + t\sqrt{\varepsilon_r} = (\phi_{PRS} + \phi_r)\frac{\lambda}{4\pi} \pm N\frac{\lambda}{2} \tag{1}$$

where ϕ_{PRS} is the reflection phase of the PRS reflector, ϕ_r is the reflection phase of the feeding source's ground plane, ε_r is the relative permittivity of the substrate supporting the primary source and *t* is its thickness. N is an integer qualifying the electromagnetic mode of the cavity. If the cavity and the substrate thicknesses *t* and *h* are fixed, the resonant wavelength is determined by the sum of the reflection phases $\phi_{PRS} + \phi_r$ for a fixed N. Conversely, for a given wavelength, the thickness *h* can be minimized by reducing the total phase shift $\phi_{PRS} + \phi_r$. The use of metasurfaces answers this purpose since they can exhibit an LC resonance. This resonance helps to have a reflection phase response varying from 180° to -180°, passing through 0° at the resonance frequency. By choosing a desired operating cavity frequency above the metasurface resonance where the reflection phase is negative, the sum $\phi_{PRS} + \phi_r$ can be very small leading to a very low cavity thickness. Since the reflector near the feeding antenna in Fig. 1(a) is composed of a PEC surface, then ϕ_r will be very close to 180°. On the other side, an AMC ground plane is used in Fig. 1(b) and in such case ϕ_r will show frequency dependent phase characteristics.

Therefore, taking advantage of the phase dispersive characteristics of metasurfaces, we will present several models of reflex-cavity antennas, each designed for a specific task. We will first present a λ/30 (1 mm @ 10 GHz) thick cavity antenna by using a PRS reflection phase value around -120°. This cavity antenna has a narrow beam profile in both E- and H-planes, producing a directivity of 160 (22 dBi). To further reduce the cavity thickness, we will emphasize on the use of two metasurfaces as illustrated in Fig. 1 (b), one as a PRS reflector and the other one as AMC ground plane of the primary source. The combination of these two metasurfaces, particularly the low phase values above their resonance, allows to design very low profile cavity antennas. For e.g., a λ/60 thickness has been achieved and the latter cavity presents a directivity of 78 (19 dBi).

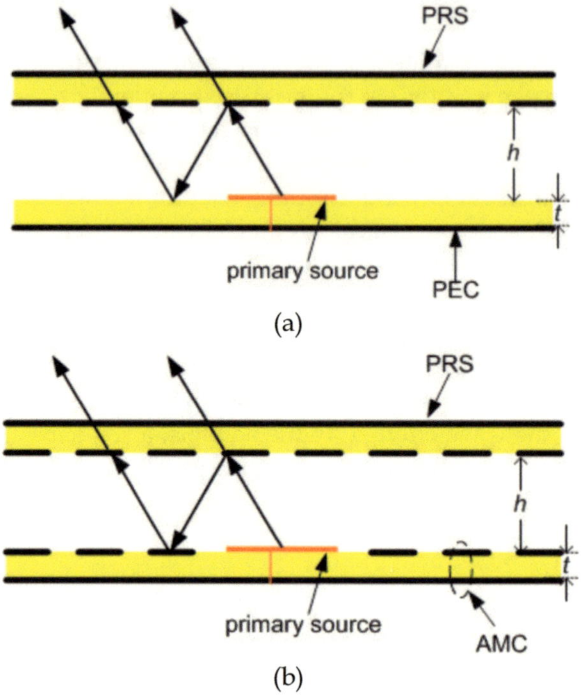

Fig. 1. Resonant cavity formed by a PEC ground plane and a metamaterial-based PRS (a) and, an AMC ground plane and a metamaterial-based PRS (b).

Since directivity depends strongly on the radiating aperture which is defined by the field distribution illuminating the PRS, we will present two ways on how we can manipulate the directivity of such reflex-cavity antennas. First, we will present the use of lateral PEC walls in the cavity antenna to form what we will refer to as metallic cavity. This method allows to enhance the directivity by 3 dB compared to the case where the cavity is open on the lateral sides. Also, the metallic cavity presents lower backward radiations due to the confinement of electromagnetic radiation, therefore increasing the front-to-back (FBR) ratio. Secondly, in order to optimize the field distribution illuminating the PRS, we will study the use of several primary sources inside the cavity. We will show how judiciously placing the different sources in the cavity helps to increase the directivity to more than 6 dB compared to single source fed cavity.

Finally, we will present beam steerable and frequency reconfigurable cavity antennas. For the beam steering, we will in a first step study a cavity where the PRS presents a locally variable phase. The latter PRS then acts as a phased array of micro-antennas, thus allowing to achieve beam steering. This concept has been pushed further by designing an electronically tunable metasurface via the incorporation of lumped elements (varactor diodes). This active metasurface can be used as PRS for two different tasks. Firstly, by applying different bias voltage along the PRS, a locally variable phase is obtained and is fully compatible for beam steering. On the other side, if we change the bias voltage of all the lumped elements similarly, then we can tune the operation frequency of the PRS so as to achieve a frequency reconfigurable reflex-cavity antenna.

3. Analysis of the planar metasurfaces

The cavity presented in Fig. 1 requires the application of a metamaterial-based surface. So in this section, we will design planar metamaterial-based surfaces for operation near 10 GHz.

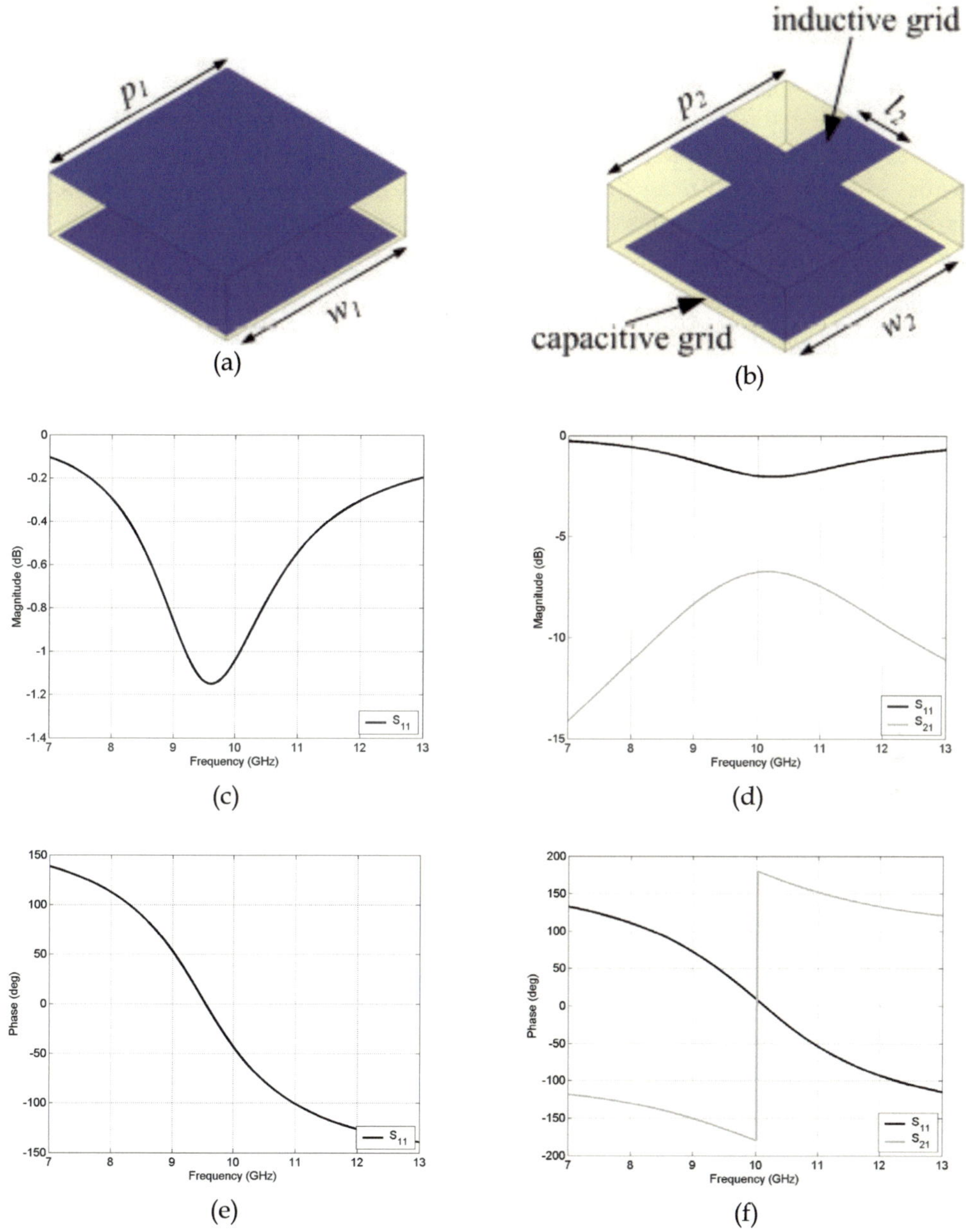

Fig. 2. Unit cell of AMC ground plane (a) and metamaterial-based PRS (b). Calculated reflection magnitude and phase of the AMC ground plane (c), (e) and reflection and transmission magnitudes and phases of the metamaterial-based PRS (d), (f).

The surface used by our group in order to achieve the AMC ground plane is made of a metamaterial composed of 2-D periodically subwavelength metallic square patches organized on one face of a dielectric substrate as illustrated in Fig. 2(a). The different dimensions of the patches are as follows: period p_1 = 4 mm and width w_1 = 3.8 mm. Another surface which we are going to use for the PRS of the cavity is made of a composite metamaterial consisting of simultaneously a capacitive and an inductive grid on the two faces of a dielectric substrate. The capacitive grid is also formed by 2-D periodic metallic patches (period p_2 = 4 mm and width w_2 = 3.6 mm) whereas the inductive grid is formed by a 2-D periodic mesh (line width l_2 = 1.2 mm) as shown in Fig. 2(b). Concerning the substrate, we have used the double copper cladded epoxy substrate of relative permittivity ε_r = 3.9, of tangential loss tanδ = 0.0197 and having a thickness of 1.2 mm. The size of the different patterns has been chosen in order to minimize the phase of the reflection coefficient near 10 GHz while providing a sufficiently high reflectance (~90%).

The metasurfaces are analyzed numerically using the finite element software *HFSS* so as to present its characteristics in terms of reflection and transmission. Simulations are performed on a unit cell together with appropriate periodic boundary conditions. The results are presented in Fig. 2(c) and Fig. 2(d). As shown, the calculated resonance frequency of the AMC surface and PRS reflector is respectively 10.4 GHz and 9.7 GHz. At resonance, phase crosses 0° as illustrated in Fig. 2(e) and Fig. 2(f).

The composite metamaterial acts as a resonant filter which presents a reflection phase varying from 180° to –180°, depending on the frequency. This variation helps to be more flexible in designing thin cavities by choosing reflection phase values below 0°.

4. Metamaterial-based low-profile highly directive cavity antenna

In this section, we discuss about the design, implementation and characterization of low profile and highly directive cavity antennas. Two different models are presented; an AMC-PRS cavity and a PEC-PRS cavity.

4.1 AMC-PRS cavity antenna

The AMC-PRS cavity antenna is formed by the AMC reflector and the metasurface reflector used as PRS together with a patch antenna designed to operate near 10 GHz (Ourir et al., 2006a). The patch antenna of dimensions 6.8 × 7 mm^2 is placed on the AMC in the cavity as shown in Fig. 1(b). The reflectors used are those presented in Fig. 2. The different phases (simulated and measured) are used to estimate the thickness h of the AMC-PRS cavity as given by Eq. (1). Fig. 3(a) shows that h first decreases with increasing frequency of the first resonant mode (N = 0) to the point where a cavity zero thickness is reached at around 10.2 GHz. Then a jump in the mode occurs leading to an abrupt variation of h and the value decreases again for N = 1. A cavity thickness h = 0.5 mm is chosen for the cavity. The thickness h of the Fabry-Perot cavity formed by the two reflectors is adjusted mechanically. The lateral dimensions of the reflector plates are 17 × 17 cm^2. This thickness leads to a good matching of the cavity at 10.1 GHz (Fig. 3(b)) corresponding to the design of a $\lambda/60$ cavity. This frequency is in good agreement with the resonance frequency calculated from the optical ray model. The directive emission of the subwavelength cavity antenna at 10.1 GHz is illustrated from the calculated and measured E-plane ($\phi = 90°$) and H-plane ($\phi = 0°$) radiation patterns in Fig. 3(c) and 3(d).

Using the formulation proposed in (Temelkuran et al., 2000), the directivity of the cavity antenna is written as:

$$D = \frac{4\pi}{\theta_1 \theta_2} \tag{2}$$

where θ_1 and θ_2 are respectively the half-power widths for the E-plane and H-plane radiation patterns. The antenna directivity is then found to be equal to 78 (19 dB) for $\theta_1 = 22°$ and $\theta_2 = 24°$.

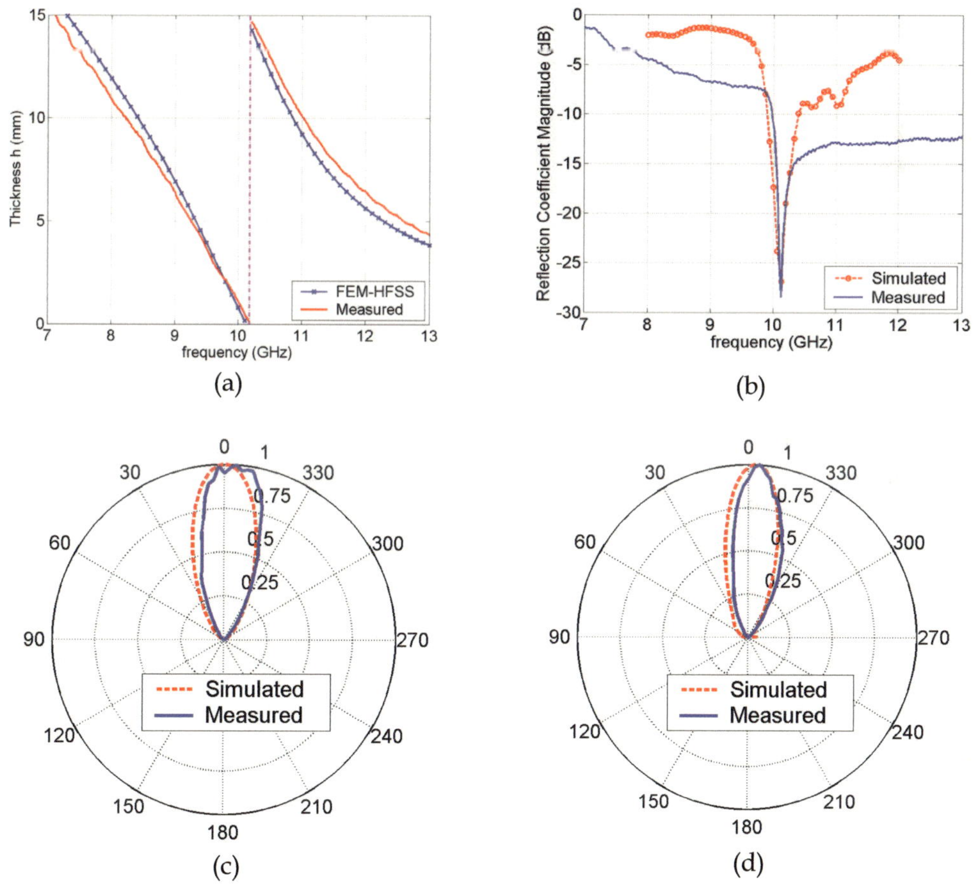

Fig. 3. (a) Evolution of the cavity thickness h versus frequency, this evolution being estimated from Eq. (1) by the calculated and measured reflection phases of the two reflectors used in the AMC-PRS cavity. (b) Calculated and measured matching of the cavity antenna. (c) E-plane ($\phi = 90°$) radiation pattern at 10.1 GHz. (d) H-plane ($\phi = 0°$) radiation pattern at 10.1 GHz.

4.2 PEC-PRS cavity antenna

In order to simplify the fabrication of the cavity antenna, another one using only one metamaterial-based surface reflector acting as the PRS and a PEC reflector (similar to the cavity shown in Fig. 1(a)) is designed (Ourir et al., 2006b). As we have seen from the reflection coefficients in Fig. 2(c) and 2(d), losses are maximum at the resonance frequency of the metamaterial-based surfaces. Thus using only one reflector has also the advantage of presenting lower losses. The PRS composed of simultaneously a capacitive and an inductive grid on the two faces of a dielectric substrate as presented in Fig. 2(b) has been designed for this purpose. Concerning the metallic patches of the capacitive grid, a period p_2 = 5 mm and a width w_2 = 4.8 mm are used. A line width l_2 = 2.2 mm is considered for the mesh of the inductive grid. This PRS having a resonance frequency of about 8 GHz presents a reflection phase close to -150° for frequencies higher than 10 GHz. The use of such a reflector in conjunction with a PEC leads also to a subwavelength cavity since the sum (ϕ_{PRS} + ϕ_r) is very close to zero between 9 GHz and 11 GHz.

A 1 mm (λ/30) thick cavity is designed with lateral dimensions of 10 × 10 cm² where the resonance is achieved at around 9.7 GHz. The antenna gain patterns in the E- and H-planes obtained from simulation and measurements are presented in Fig. 4(a) and 4(b).

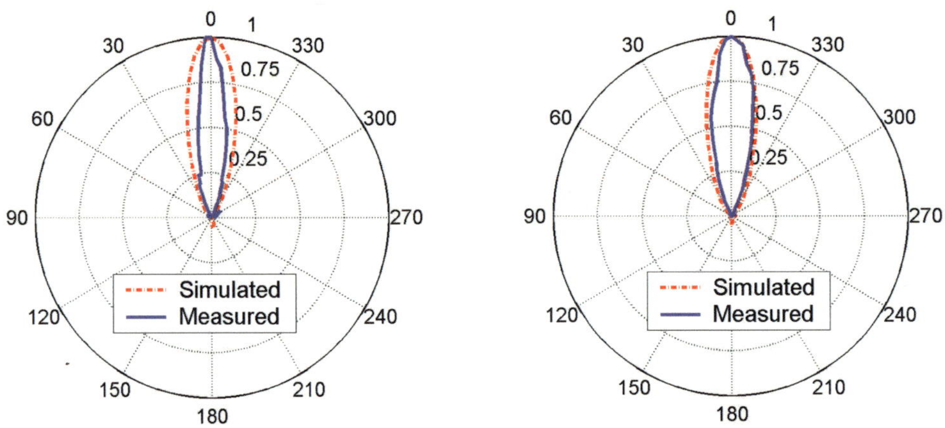

(a) (b)

Fig. 4. (a) Calculated and measured reflection phases of the PRS reflector used in the PEC-PRS cavity. (b) Calculated and measured matching of the cavity antenna. (c) E-plane (ϕ = 90°) radiation pattern at 9.7 GHz. (d) H-plane (ϕ = 0°) radiation pattern at 9.7 GHz.

In this case, despite the use of only one metamaterial-based surface as the PRS and the use of smaller lateral dimensions than the two metamaterial-based cavity, the antenna directivity is found to be twice and equal to 160 (22 dB).

5. Directivity enhancement in Fabry-Pérot cavity antennas

This section deals with the enhancement of directivity in Fabry-Pérot cavity antennas. Two different approaches are presented to achieve higher performances in terms of directivity and beamwidths. In order to reach a higher directivity, a larger surface of the

PRS must be illuminated. Therefore, a better distribution and confinement of the electromagnetic energy must be produced in the cavity. For this purpose, two innovative solutions can be considered. The first one is to shield the cavity by four metallic walls and the second one is to feed the cavity by multiple primary sources. The two methods are detailed below.

5.1 Metallic cavity antenna

The cavity antenna proposed in this section was designed at 2.46 GHz for point to point radio communication links. The metallic cavity is composed of the feeding antenna's PEC ground plane and a metamaterial-based PRS as reflectors. Furthermore, four metallic walls are also fixed on the lateral sides so as to enhance the directivity of the cavity antenna while keeping low lateral dimensions (Burokur et al., 2009a).

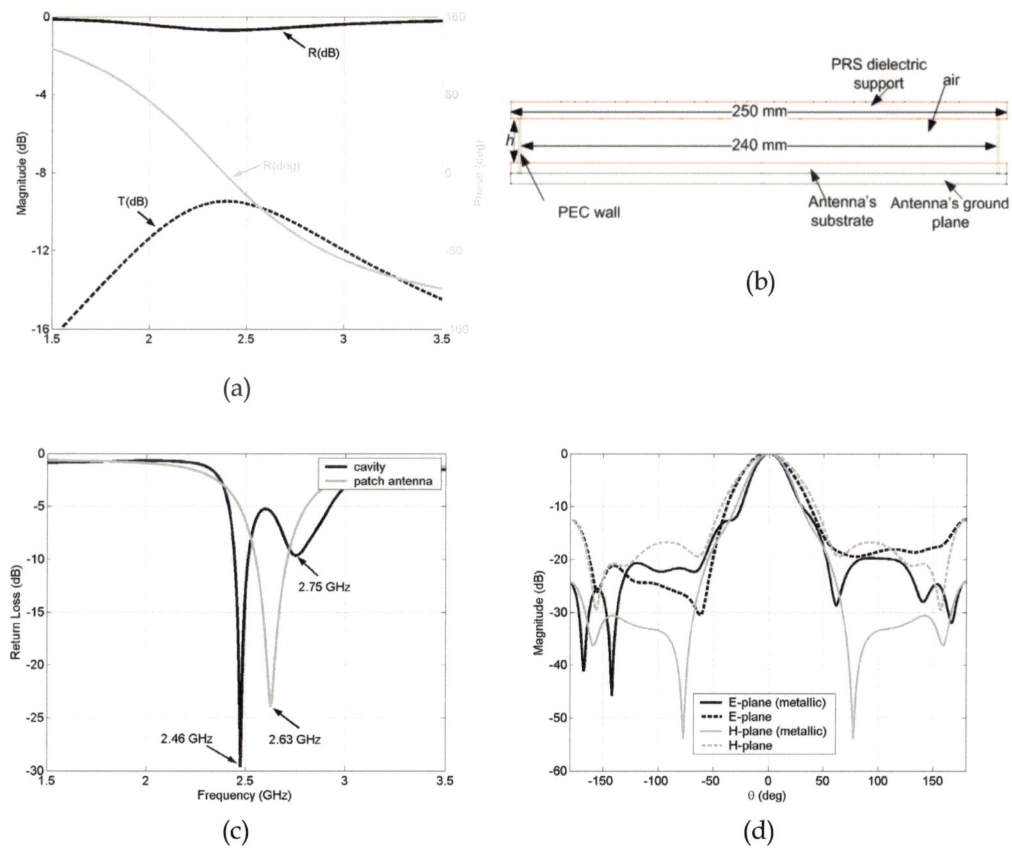

Fig. 5. (a) Calculated reflection phase (solid grey), reflection (solid dark) and transmission (dashed dark) magnitudes for the PRS reflector. (b) Schematic view of the metallic cavity antenna with $h = 21.5$ mm. (c) Return losses of the cavity antenna and the feeding patch antenna. (d) E- ($\phi = 90°$) and H-plane ($\phi = 0°$) radiation patterns at 2.46 GHz for the metallic and conventional cavity antennas.

The inductive and capacitive grids of the metasurface are printed on the faces of an 8 mm thick foam dielectric substrate (ε_r = 1.45, tanδ = 0.0058). This thickness is sufficient enough to provide a relatively smooth slope of the phase response, hence rendering the metamaterial less sensitive to fabrication tolerances. The capacitive grid is formed by 2-D periodic metallic patches lattice (period p_2 = 20 mm and width w_2 = 18.8 mm) whereas the inductive grid is formed by a 2-D periodic mesh (line width l_2 = 6 mm). The size of the different patterns has been chosen in order to have the phase of the reflection coefficient below 0° near 2.46 GHz while providing a sufficiently high reflectance (~90%). The numerical results presented in Fig. 5(a) show firstly a resonance frequency of 2.38 GHz, i.e. where the phase crosses 0°. Secondly, we can also note a pass-band behavior where the transmission level is relatively low (about -9.5 dB). Finally this figure shows a reflection phase of –15° at 2.46 GHz.

The microstrip patch feeding source having dimensions 43 mm x 43 mm is designed on a similar foam dielectric substrate of thickness 5 mm. The surface of the inductive and capacitive grids forming the PRS has dimensions 200 mm × 200 mm, while the lateral dimensions of the dielectric board supporting the grids as well as that of the cavity have been increased to 250 mm × 250 mm. However the lateral metallic walls are separated by a distance of 240 mm, as illustrated by the side view of the cavity antenna in Fig. 5(b). So with a ϕ_{PRS} = –15°, the thickness of the cavity is found to be h = 21.5 mm (< $\lambda/5$).The simulated metallic cavity presents a return loss of 22.8 dB at 2.46 GHz [Fig. 5(c)]. A second resonance is observed at 2.75 GHz corresponding to the resonance of the feeding antenna. These two resonances are situated at each side of that of the feeding patch alone due to the coupling between the patch antenna and the FP cavity.

The calculated results [Fig. 5(d)] for the E- and H-plane radiation patterns show a directivity of 15.21 dB. Compared to a similar cavity without metallic walls, an enhancement of about 3 dB and lower secondary lobes are achieved. To reach this same directivity without metallic walls, we should have used a cavity with lateral dimensions close to 400 mm × 400 mm. Also, the metallic cavity presents very low backward radiations (–24.3 dB) due to the energy confinement by the lateral walls.

A prototype of the proposed cavity has been fabricated and measured (Fig. 6). However, the responses measured with h = 21.5 mm have not shown a resonance as expected at 2.46 GHz but at 2.49 GHz. This is due to the matching of the fabricated feeding patch antenna which does not occur at 2.63 GHz as in simulation. Moreover, the responses of the PRS may also present a shift in frequency which can be attributed to the manufacturing tolerances. A modification on the thickness of the cavity has then been undertaken in order to achieve as close as possible the calculated resonance frequency. Three other different thicknesses (h = 25 mm, h = 27.9 mm and h = 28.5 mm) have shown remarkable performances. The different results are summarized in Table 1.

As the thickness increases, the resonance of the cavity antenna tends to lower frequencies. For h = 25 mm, the measurements show a return loss of 11 dB and a directivity of 12.79 dB with secondary lobes reaching a level of –26.5 dB. For h = 28.5 mm, the return loss is enhanced to 21.5 dB at 2.405 GHz but the directivity falls to 12.4 dB. The best directivity (13.4 dB) is observed at 2.405 GHz for h = 27.9 mm with secondary lobes level of -22.7 dB.

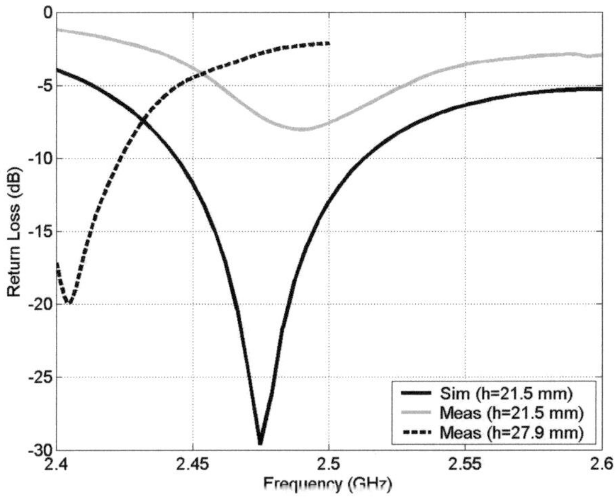

(a)

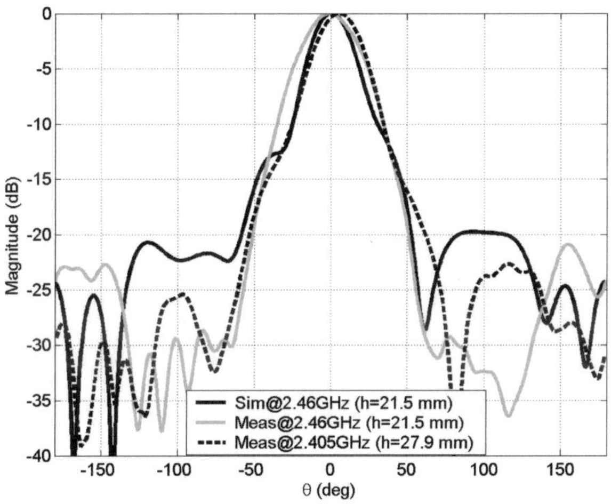

(b)

Fig. 6. (a) Measured return loss of the metallic cavity antenna. (b) Comparison between simulated and measured E-plane radiation patterns.

h (mm)	Resonant frequency (GHz)	Return loss (dB)	Directivity (dB)	Secondary lobes level (dB)
21.5 (sim)	2.46	29.5	15.3	-19.7
21.5 (meas)	2.49	8	12.36	-28.7
25 (meas)	2.46	11	12.79	-26.5
27.9 (meas)	2.405	20	13.4	-22.7
28.5 (meas)	2.4	21.5	12.4	-24.4

Table 1. Performances of the metallic cavity antenna.

5.2 Multisource-fed cavity antenna

As stated earlier, the second method to reach higher directivity is based on the use of multiple primary sources in the cavity. Therefore in this section, the cavities operating near 10 GHz are fed with a 2 x 2 microstrip patch array (Yahiaoui et al. 2009, Burokur et al., 2009b). The four patches with dimensions $W_p = L_p = 7.5$ mm are fed simultaneously via microstrip transmission lines acting as $\lambda/4$ impedance transformers and excited by a 50Ω SMA connector as shown in Fig. 7(a). The inter-element spacing a of the microstrip patch array feed plays an important role in the directivity of the cavity antenna. For this reason, the influence of this latter parameter is studied for a fixed cavity thickness $h = 1.5$ mm. The inter-element spacing a is varied from 0.5λ to 3λ. The return losses of the cavities are plotted in Fig. 7(b). We can note a very good matching (< –10 dB) around 9.25 GHz for the four different cases.

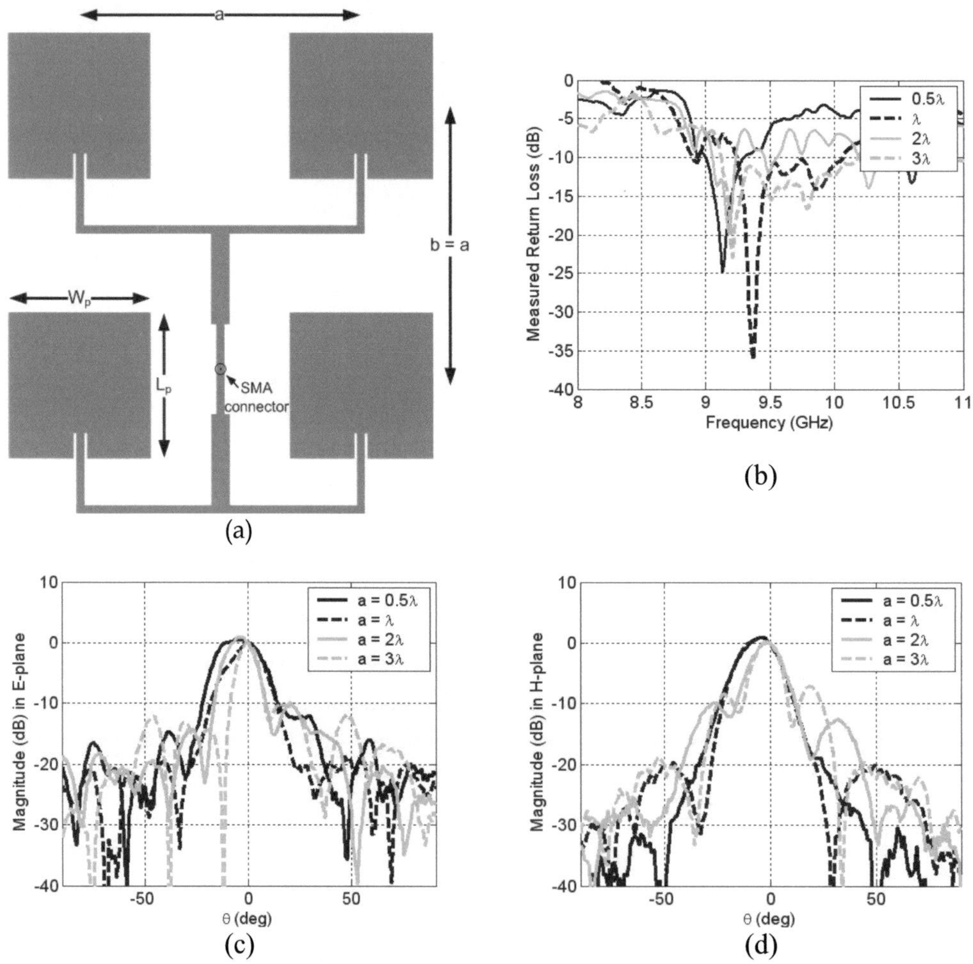

Fig. 7. (a) 2 × 2 patch array used as a multi-source. (b) Measured return losses of the cavities. (c)-(d) Measured E-plane and H-plane radiation patterns with a = 0.5λ, λ, 2λ and 3λ for a cavity thickness h = 1.5 mm.

The measured E- and H-plane radiation patterns of the cavity antennas are presented in Fig. 7(c) and 7(d). For $a = 0.5\lambda$, a measured directivity of 19 dB is obtained at 8.93 GHz. This value is very close to that of a cavity fed by a single source (see for *e.g.* Ourir et al., 2006a, 2006b). So, it is worth to note that conversely to classical antenna arrays, the directivity is not doubled each time that the number of sources is doubled. For $a = \lambda$, a measured directivity of 20.9 dB is noted at 9.07 GHz, showing clearly an enhancement of 1.9 dB with regard to the case $a = 0.5\lambda$. It is also very important to note that the sidelobes level of the patch array is considerably reduced when embedded in the cavity. This effect is highlighted in Table 2 where the performances of cavities for the different inter-element spacing are presented. 23.21 dB and 25.35 dB is respectively deduced from the measured planes for $a = 2\lambda$ and $a = 3\lambda$. When the case $a = 3\lambda$ is compared to $a = 0.5\lambda$, an increase of 6.35 dB is obtained for the directivity, which is comparable to an increase from a single patch element to a 2 × 2 patch array. The measured sidelobes level are higher (~ -8dB in the H-plane) for the case $a = 3\lambda$. However, this sidelobes level is still low compared to the sidelobes level of the source alone. It is well known that an inter-element spacing of an array higher than λ leads to high sidelobes level and also to the apparition of grating lobes.

The directivity D of the cavity antennas can be calculated using $D = 41253/(\theta_1 \times \theta_2)$ where θ_1 and θ_2 are respectively the half-power widths (in degrees) for the H-plane and E-plane patterns. The directivity values are given in Table 2 where we can observe that an increase in the inter-element spacing a in the cavity antenna gives rise to a higher directivity. This is because the radiation area at the surface of the source is bigger when a increases and therefore, a larger surface of the PRS is illuminated by the radiation source. This phenomenon is illustrated in Fig. 8 where the E-field distribution is plotted in a horizontal plane at two different locations z in the cavity antenna. $z = 0$ and $z = 1.5$ corresponds respectively to the plane of the radiating patch array source and to the thickness $h = 1.5$ mm at the inner surface of the PRS (location of the capacitive grid). This figure shows that the radiation area at the surface of the feed source in the case $a = 3\lambda$ is bigger than in the case $a = 0.5\lambda$ and therefore, a larger surface of the PRS is illuminated leading to a higher directivity. On the counter part, the side lobes level also increases.

a (mm)	Resonance frequency (GHz)	Maximum directivity (dB)	Secondary lobes level (dB)
0.5λ	9.13	19 @ 8.93 GHz	-12
λ	9.37	20.9 @ 9.07 GHz	-19
2λ	9.18	23.21 @ 8.94 GHz	-10
3λ	9.21	25.35 @ 8.96 GHz	-8

Table 2. Performances of the cavity antennas with $a = 0.5\lambda$, λ, 2λ and 3λ for a cavity thickness $h = 1.5$ mm.

6. Beam steering in Fabry-Pérot cavity antennas

In this section, we present the modeling and characterization of optimized resonant cavities for beam steering applications. Firstly, the design principle is presented for a passive cavity. The idea is then pushed further to achieve controllable beam steering by incorporating lumped elements in the metasurface reflector.

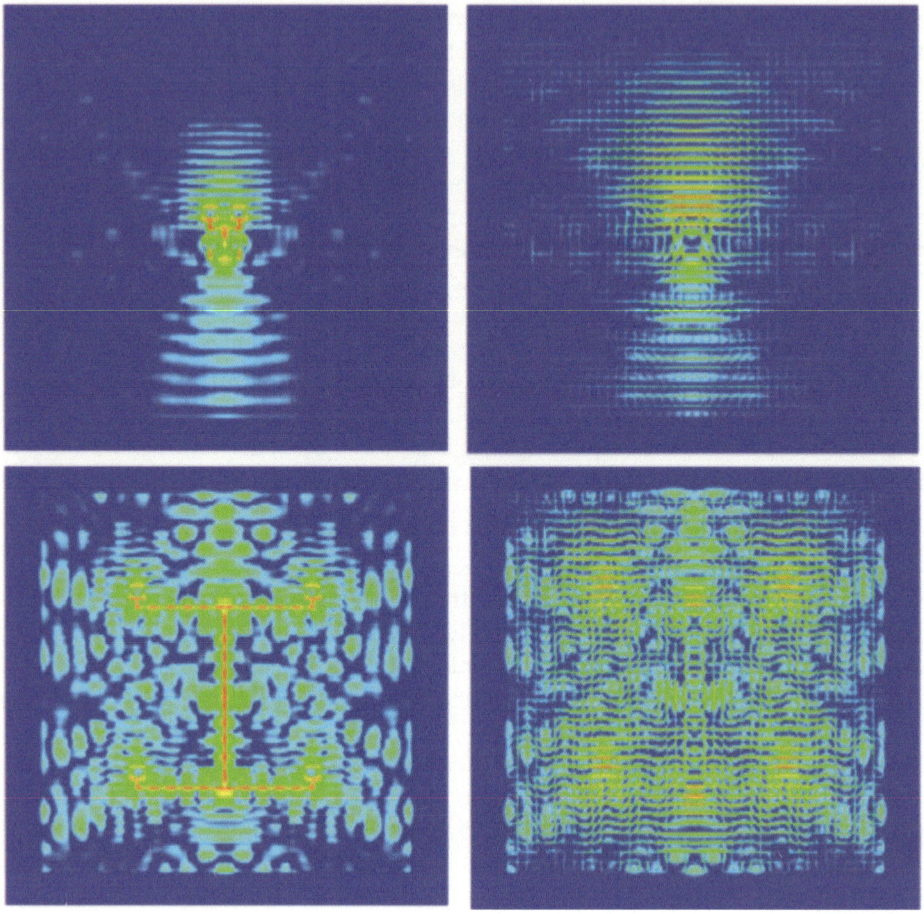

Fig. 8. E-field distribution in a horizontal plane in the cavity antenna for a = 0.5λ and a = 3λ.

6.1 Passive beam steering

Since the beam steering operation is presented in only one radiation plane, the metasurface used is composed of a 1-D array of copper strips etched on each face of a dielectric substrate as shown in Fig. 9(a).

We shall note that the gap spacing g in the capacitive grid plays a crucial role in determining the capacitance and therefore the resonance frequency of the metasurface. By changing g and keeping all the other geometric parameters unchanged, the capacitance of the metamaterial will also vary. As a consequence, the phases of the computed reflection coefficients vary. This behavior is illustrated by the numerical results shown in Fig. 9(b). We can note that the variation of g accounts for the shift of the resonance frequency. An increase in the value of g causes a decrease in the value of the capacitance created between two cells, and finally a shift of the resonance towards higher frequencies. At a particular frequency, the phase of the metasurface increases with an increase in the gap spacing. The study on the variation of g shows that it is possible to design a PRS with a continuous variation of the gap g, resulting in a local variation of the phase characteristics (Fig. 9(c)). If we consider each gap as a slot antenna, an analogy can then be made with an array of several antennas with a regular phase difference. The locally variable phase metasurface can then be applied for passive beam steering (Ourir et al. 2007a, 2009).

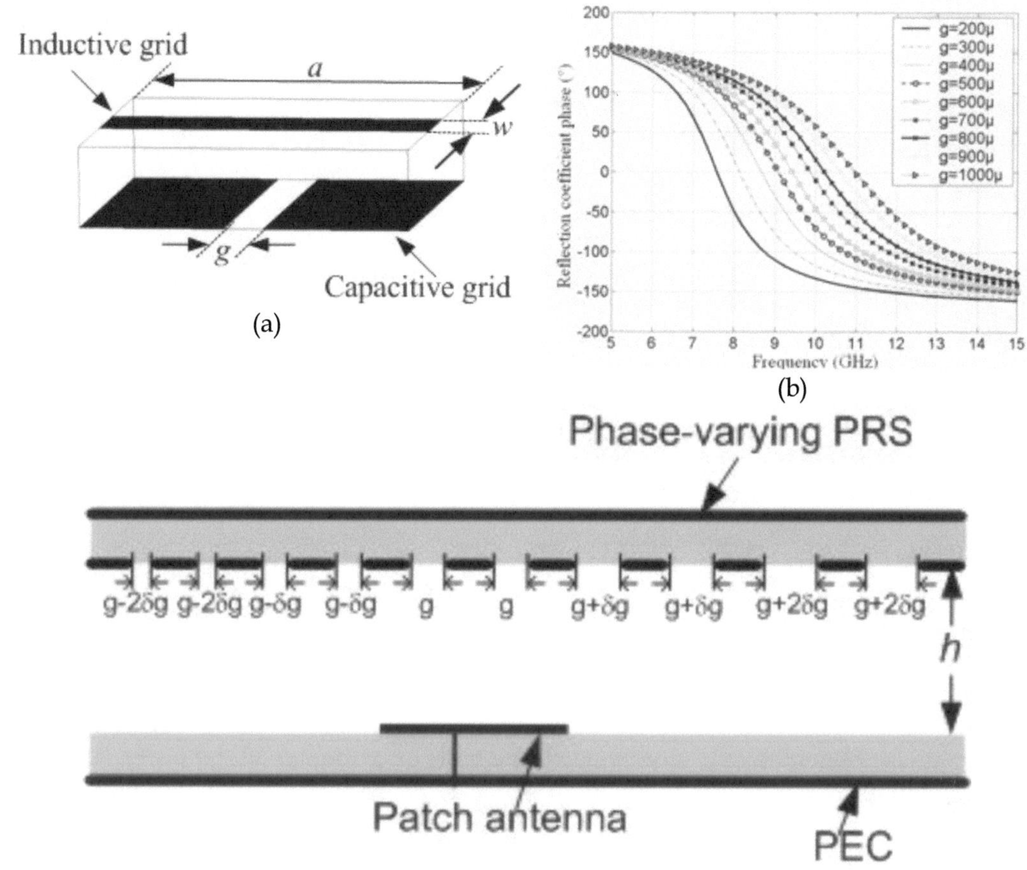

(c)

Fig. 9. (a) Elementary cell of the metamaterial composed of an inductive and capacitive grid, which is proposed for the PRS. (b) Reflection phase coefficient of the metasurface versus the gap width g. (c) Schematic view of the cavity composed of a PEC and a métasurface with a variable gap width.

To show the performances in terms of beam steering, several subwavelength cavities have been simulated and fabricated using the 1-D metasurface as PRS. The first one consists of the metamaterial PRS with the same gap spacing g = 400 µm between the metallic strips of the capacitive grid (δg = 0). This prototype will assure no deflection of the beam since it exists no phase variation of the metamaterial. The second and third ones are the prototypes incorporating respectively a variation of δg = 50 µm and δg = 100 µm along the positive x-direction. The cases where the variation δg is negative (180° rotation of the PRS around the z-axis) have also been considered. Note that here the resonance frequency of the central region of the metamaterial corresponds to that of the PRS without gap spacing variation (g = 400 μm and δg = 0), i.e. 8.7 GHz as shown in Fig. 9(b). The resonance frequency of the cavity is found to be ~10.5 GHz for the three prototypes as shown in Fig. 10(a). Best matching is observed when the metallic gap of the PRS capacitive grid increases. However, the resonance frequency remains the same for the three configurations since it depends on the gap spacing of the central region of the PRS, which is the same for the three prototypes.

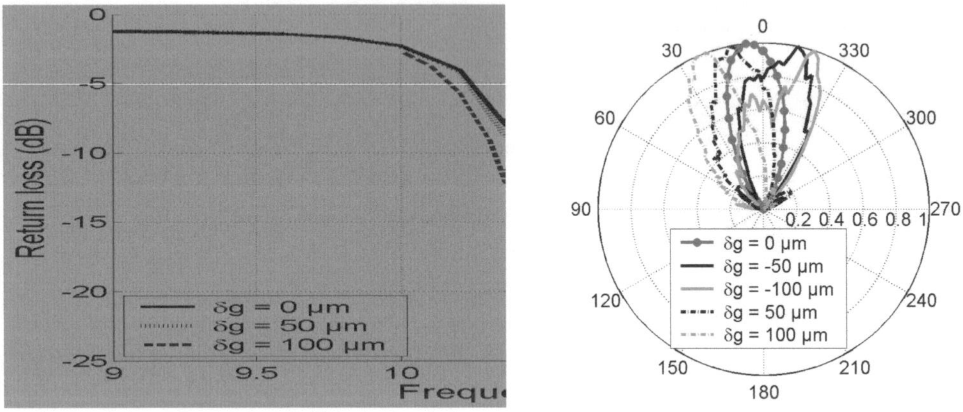

(a) (b)

Fig. 10. (a) Return loss of the antennas with different variation of gap width. (b) Measured gain patterns of the cavity antennas versus the gap width variation.

Fig. 10(b) shows the measured gain patterns of the antenna in the E (ϕ = 90°) plane at 10.5 GHz for an optimized cavity thickness h = 1 mm. For δg = 0, the beam is normal to the plane of the antenna and shows no deflection, which confirms our prediction on the constant phase metamaterial. However, in the case of a regular variation of 50 μm, a deflection of the antenna beam of about 10° can be observed either in the forward (clockwise) or backward (anti-clockwise) direction depending if δg is respectively negative or positive. Similar observations and a higher deflection of ±20° can be noted for δg = ±100 μm. The directivity of the cavity antenna can be calculated using the following expression: $D = 41253/(\theta_1 \times \theta_2)$ where θ_1 and θ_2 are respectively the half-power widths (in degrees) for the **H**-plane and **E**-plane patterns. In this case, the directivity is found to be approximately equal to 14.8 dB.

6.2 Active beam steering

The cavity antenna proposed in this section includes the use of lumped elements such as varactor diodes so as to be able to control electronically the phase of the metasurface. As a preliminary step in the design of such cavities, we present firstly the design of the active metasurface.

6.2.1 Electronically controlled metasurface

The metasurface used in this section is based on the same principle as the one illustrated in Fig. 9. But, instead of applying a linear variation of the gap spacing g in order to create a locally variable phase, we now use active components to make the phase of the metasurface shift in frequency. Varactor diodes having a capacitance value ranging from 0.5 pF to 1.0 pF are thus incorporated into the capacitive grid between two adjacent metallic strips (Fig. 11(a)) and depending on the applied bias voltage, the phase of the metasurface varies with frequency (Ourir et al. 2007b). The variable capacitive grid of the tunable phase PRS used for an operating frequency around 8 GHz consists of a lattice of metallic strips with varactor diodes connected each 6 mm (s = 6 mm) between two adjacent strips. The width of the strips and the spacing between two strips of the capacitive grid is respectively w = 1 mm and g = 2 mm.

Concerning the inductive grid, the width of the strips and the spacing between two strips are respectively w_1 = 2 mm and g_1 = 4 mm (Fig. 11(b)). Note that the inductive grid is not made tunable. RF chokes are also used in the microstrip circuit in order to prevent high frequency signals going to the DC bias system. Potentiometers are implemented in the structure to create a voltage divider circuit so as to be able to bias locally the varactors. The capacitance in each row can then be adjusted according to the bias voltage applied. This capacitance can also be varied from one row to another by the use of the voltage dividers on the prototype. By changing the bias voltage of the varactors of the PRS similarly, the capacitance of the metamaterial will also vary. As a consequence, the reflection and the transmission coefficients also vary. This behavior is illustrated by the measurement results of the reflection coefficient magnitude and phase shown in Fig. 11(c) and 11(d) respectively. These curves are obtained when the same bias voltage is applied to the different rows of varactors along the PRS. The measurements are performed in an anechoic chamber using two horn antennas working in the [2 GHz - 18 GHz] frequency band and an 8722ES network analyzer. From Fig. 11(c), we can note that the variation of the bias voltage accounts for the shift of the resonance frequency of the PRS, *i.e.* the frequency where the phase crosses 0°. An increase in the bias voltage leads to a decrease in the value of the capacitance of the metamaterial, and finally to a shift of the resonance towards higher frequencies. At a particular frequency the phase of the PRS increases with an increase in the bias voltage. This phase shift is very important since it will help to tune the resonance frequency of the cavity antenna and also to control the radiated beam direction of the antenna.

6.2.2 Active beam steering

Instead of applying a uniform variation in the periodicity of the cells composing the capacitive grid so as to create a locally variable phase as in section 6.1, we now use the electronically controlled metasurface as PRS (Ourir et al. 2009). The active components biased differently make the phase of the PRS shifts in frequency locally. As illustrated by the varactors bias system shown in Fig. 12(a), the proposed PRS is now divided into different regions, where each of them has a specific bias voltage bias. We shall note that here the resonance frequency of the cavity is imposed by the resonance frequency of the central region just above the feeding source corresponding to the bias voltage $V_4=V_1+3\delta V$. The bias voltage is thus increased uniformly with a step δV when moving from the left to the right of the metamaterial-based PRS by the use of the potentiometers. This action creates a regular variation of the phase along the PRS.

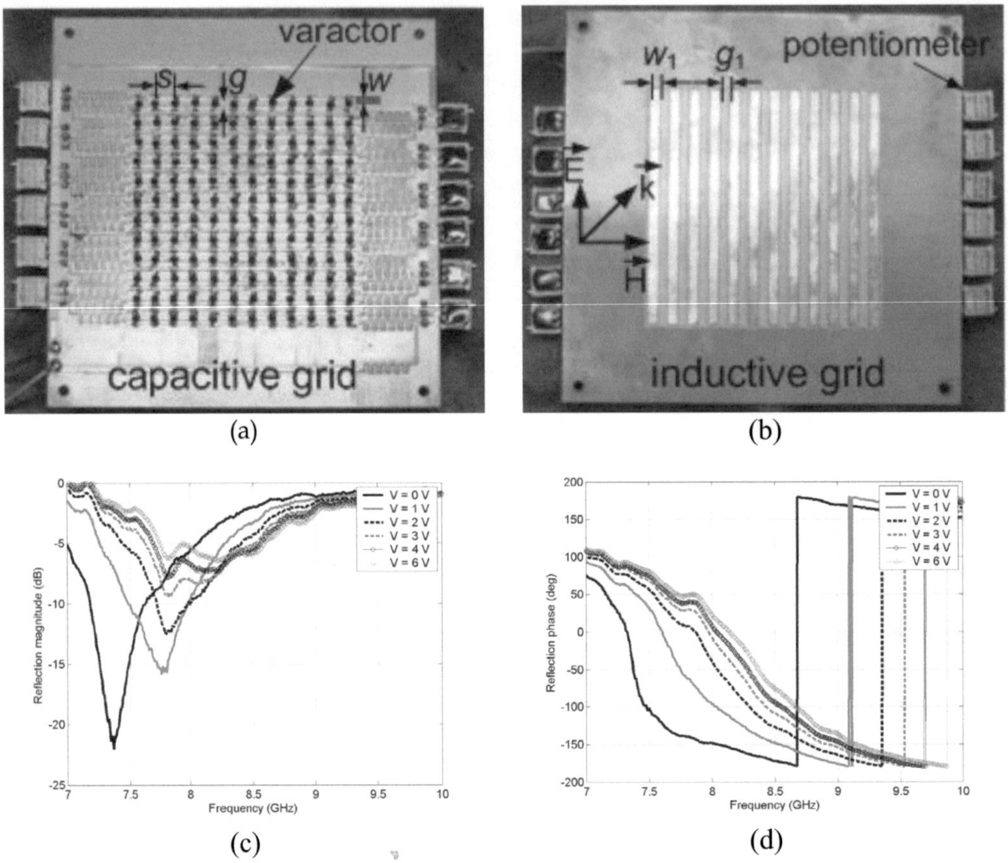

Fig. 11. (a) Electronically phase-varying metasurface. (a) Capacitive grid incorporating varactors and voltage dividers. (b) Inductive grid. (c) Measured magnitude and (d) measured phase of the reflection coefficient versus bias voltage of the varicaps.

The first configuration studied here is the antenna cavity based on the metamaterial PRS with the same null bias voltage for all the varactors. This configuration will assure no deflection of the beam since it exists no phase variation of the metamaterial. The second and third configurations are prototypes incorporating respectively a variation of δV = 0.2 V and δV = 0.3 V along the positive *x*-direction. The cases where the variation δV is negative (180° rotation of the PRS around the *z*-axis) have also been considered.

Fig. 12(b) shows the gain patterns of the antenna in the **E**-plane ($\phi = 90°$) at 7.9 GHz for the optimized cavity. For δV = 0 V, the beam is normal to the plane of the antenna and shows no deflection, which confirms our prediction on the constant phase metamaterial. However, in the case of a regular variation of δV = 0.2 V, a deflection of the antenna beam of about 7° can be observed either in the forward or backward direction depending if δV is respectively negative or positive. Similar observations and a higher deflection can be noted for respectively δV = 0.3 V and δV = -0.3 V. This figure illustrates very clearly the control of the radiation pattern of the antenna by the bias voltage of the varactors. The direction of the radiation beam depends of the direction of the variation of the bias of the varactors. If we inverse the sign of δV, the sign of the deviation changes also. This demonstration opens the door to the realization of very simple electronically beam steering ultra-compact antennas based on active metamaterials.

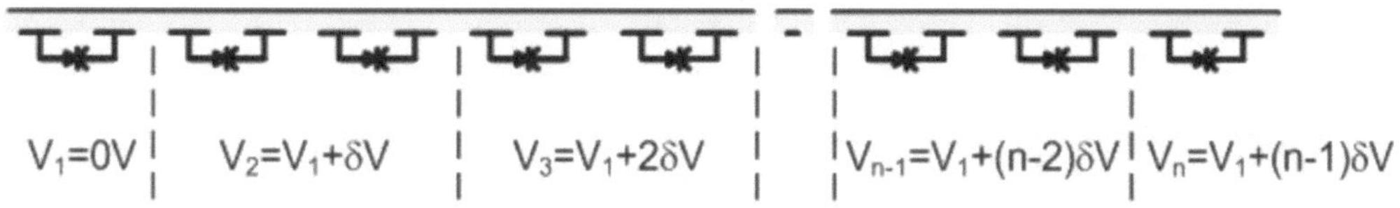

(a)

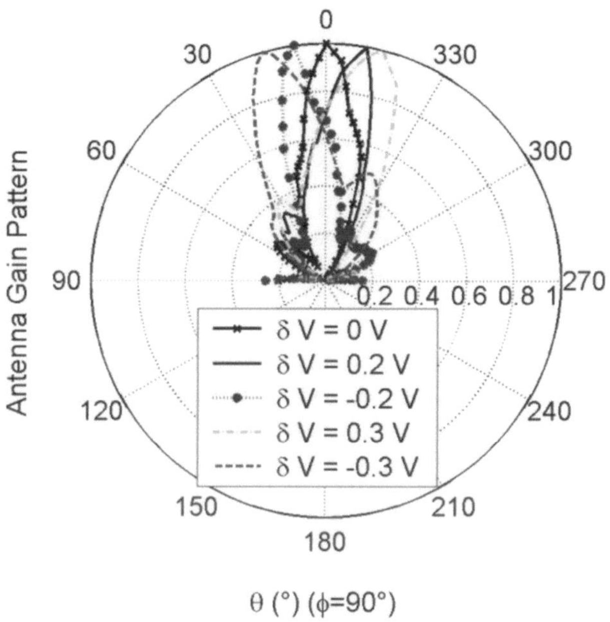

(b)

Fig. 12. (a) Variation of the bias voltage of the varactors along the phase varying PRS. (b) Measured gain patterns in the **E**-plane ($\phi = 90°$) at 7.9 GHz for δV = 0 V, δV = 0.2 V and δV = 0.3 V. The steering of the antenna's radiated beam can be clearly observed with a positive steering angle for positive bias and negative one for a negative bias.

7. Frequency agile Fabry-Pérot cavity antennas

Conversely to beam steerable cavity antennas, we do not need a locally phase-varying PRS for frequency agility applications. What we seek is the ability to change the resonance frequency of the PRS and this is possible by changing simultaneously and in the same manner the capacitance value of the varactor diodes. Here, we show that a tunable metasurface associated to an array of wideband sources in a Fabry-Pérot cavity leads to a reconfigurable directive emission on a wide frequency range (Burokur et al. 2010, 2011). A similar electronically controlled PRS as the one shown in Fig. 11(a) is designed to operate near 2 GHz in base station antennas for mobile phone communication systems. The primary source of the cavity is a wideband microstrip patch antenna designed to cover 1.8 GHz - 2.7 GHz frequency range and therefore to illuminate the PRS at any frequency within this range. This patch antenna is electromagnetically coupled to an L-probe which itself is connected to a coaxial connector. Simulations have shown a good matching (return loss < 10 dB) from 1.8 GHz to 2.7 GHz.

To demonstrate experimentally the mechanism for reconfigurable directive emissions from a metamaterial-based FP cavity, a prototype having dimensions 400*400 mm^2 (approximately 3λ*3λ) has been fabricated and tested. As it has been shown in section 5.2, the directivity is drastically enhanced when a cavity is fed by judiciously spaced multiple sources since a larger surface of the PRS is illuminated, and therefore the size of the effective radiating aperture of the cavity antenna is increased. Four elementary sources constituting a 2 x 2 wideband patch array are used as primary source; the inter-element spacing between the different sources being 200 mm. Fig. 13(a) and 13(b) shows respectively the photography of the prototype and the capacitive grid of the electronically tunable metasurface In order to experimentally estimate directivity and gain of the cavity's radiated beam, direct far field measurements are performed using a SATIMO STARLAB and the characteristics are shown in Fig. 14.

When capacitance of the metasurface reflector is changed by varying bias voltage of varactor diodes, the frequency of maximum gain is tuned as clearly shown in the different diagrams of Fig. 14. When 0 V is applied, maximum gain is observed at 1.9 GHz corresponding approximately to the simulated case with C = 6.5 pF. When DC bias voltage is increased, the capacitance value is decreased, resulting in an increase of maximum gain frequency. For 24 V, maximum gain occurs at 2.31 GHz, corresponding to lowest capacitance value. To gain more insight in the electromagnetic properties of the metamaterial-based Fabry-Pérot cavity, intensity maps of scanned far field versus frequency and elevation angle θ, in E-plane are presented. The emission frequency represented by the red spot varies from 1.9 GHz to 2.31 GHz from 0 V to 24 V as shown in Figs. 14(a), 14(c), 14(e) and 14(g). These figures demonstrate clearly the frequency reconfigurability property of the cavity. We shall also note that for each frequency the spot is situated at an elevation angle of 0°, indicating a radiated beam normal to the cavity metasurface reflector. Figs. 14(b), 14(d), 14(f) and 14(h) show radiation patterns in E- and H-planes at respectively 1.9 GHz, 2.02 GHz, 2.16 GHz and 2.31 GHz corresponding to maximum gain frequency for 0 V, 5 V, 12 V and 24 V. The tuning range of maximum gain frequency results in an effective operation bandwidth close to 20%. A wide frequency bandwidth is achieved due to the cavity thickness fixed in this particular case. With h = 15 mm, reflection phase values around 0° are needed in the 1.85 GHz - 2.25 GHz frequency band. A lower h would lead to phase values approaching -180° and the possible frequency bandwidth from the capacitance tuning range would be narrow. Actually, a high directivity (approximately 18 dBi) is obtained experimentally due to the large lateral dimensions of the fabricated cavity and also to the use of four elementary sources instead of only one where only 14 dBi is obtained.

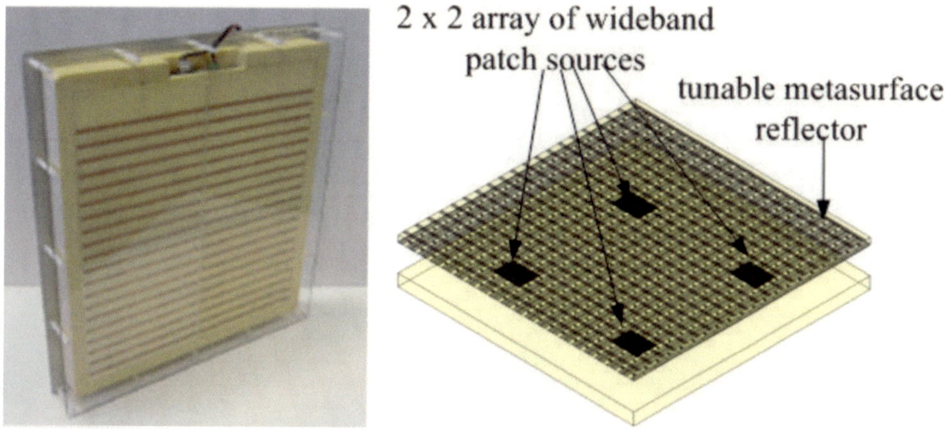

(a)

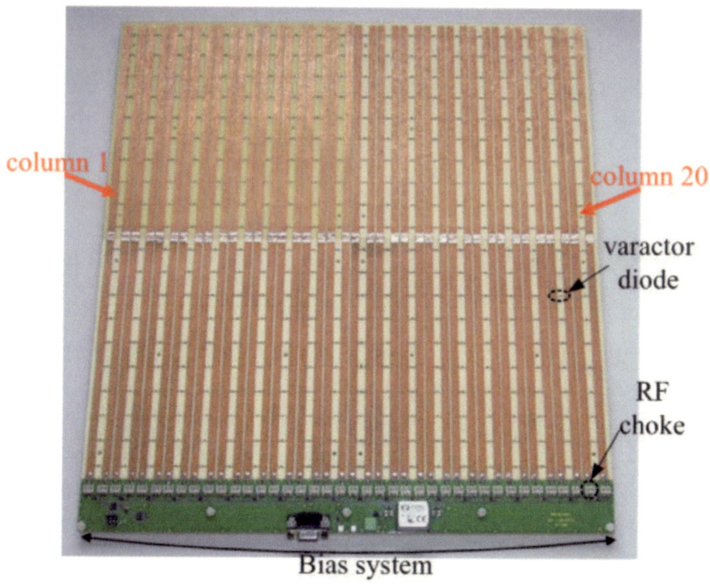

(b)

Fig. 13. (a) Photography and perspective view of the cavity antenna. (b) Electronically tunable metasurface reflector incorporating varactor diodes, RF chokes and bias system.

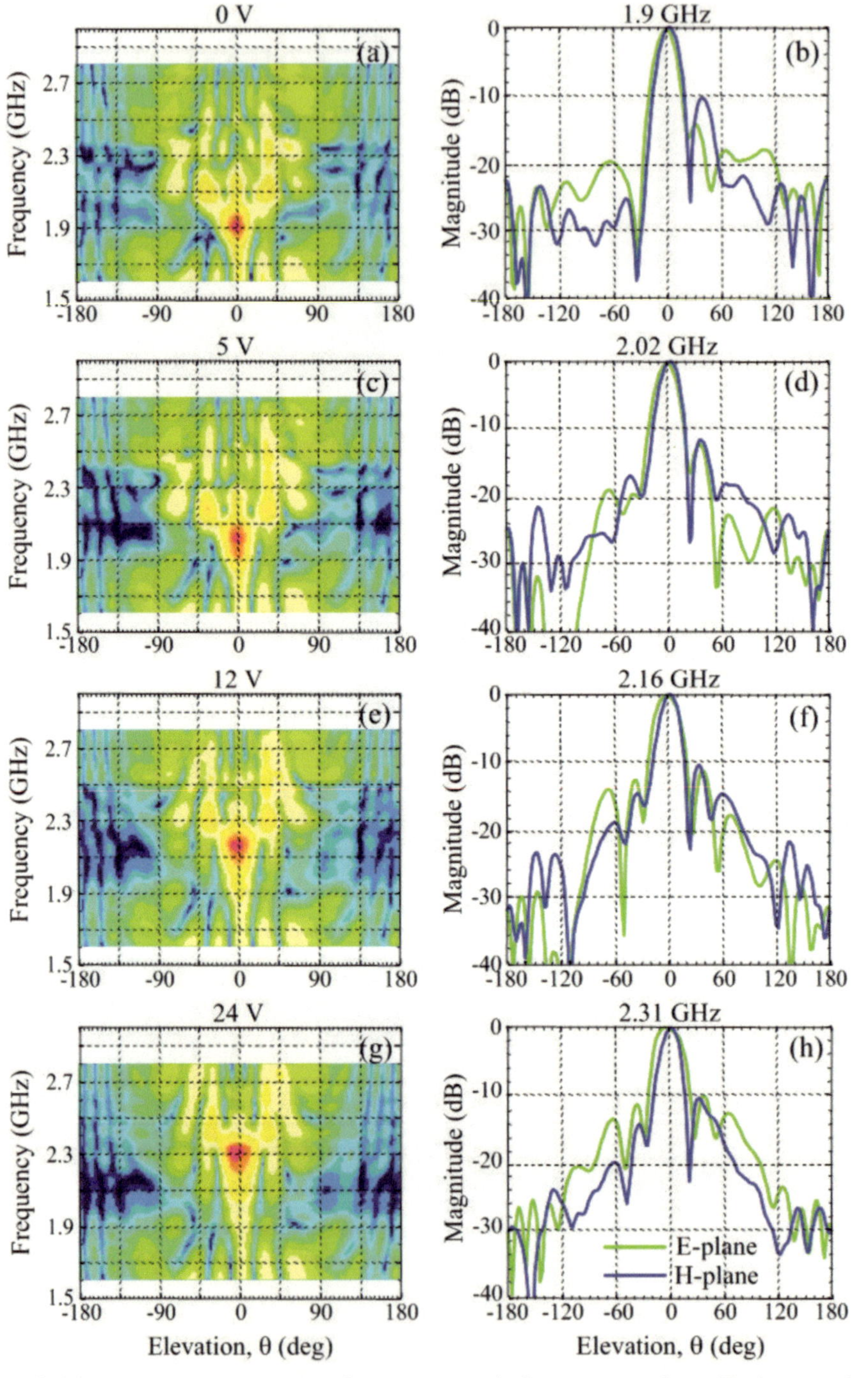

Fig. 14. Far field intensity maps versus frequency and elevation angle in E-plane and measured radiation patterns in E- and H-planes at maximum gain frequency for different bias voltage applied : (a)-(b) 0 V – 1.9 GHz, (c)-(d) 5 V – 2.02 GHz, (e)-(f) 12 V – 2.16 GHz, and (g)-(h) 24 V – 2.31 GHz.

8. Conclusion

To conclude, we have presented various aspects of reflex-cavity antennas: low-profile, high gain, beam steering and frequency agility. For each aspect, numerical calculations together with measurements have been presented. The development of these works has enabled to promote the interesting characteristics of metamaterial-based surfaces. Variable phase metasurfaces compared to conventional PEC and AMC surfaces have also shown their usefulness in reconfigurability applications. Further studies are actually performed to transpose the reflex-cavity antenna concept to industrial applications in various domains such as telecommunications, aeronautical, transport and housing.

9. Acknowledgements

The authors are very grateful to the French National Research Agency (ANR) for the financial support of the METABIP Project. These works have also been made possible by the partial financial support of the Eureka TELEMAC project. We would like also to thank our partners P. Ratajczak and J.-P. Daniel for the fabrication and characterization of antenna prototypes.

10. References

Akalin, T., Danglot, J., Vanbesien, O. & Lippens, D. (2002). A highly directive dipole antenna embedded in a Fabry–Perot-type cavity. *IEEE Microw. Wireless Component Lett.*, Vol.12, No.2, (February 2002), pp. 48-50, ISSN 1531-1309.

Burokur, S.N., Latrach, M. & Toutain, S. (2005). Theoretical investigation of a circular patch antenna in the presence of a Left-Handed Medium. *IEEE Antennas Wireless Propag. Lett.*, Vol.4, (June 2005), pp. 183-186, ISSN 1536-1225.

Burokur, S.N., Ourir, A., Daniel, J.-P., Ratajczak, P. & de Lustrac, A. (2009a). Highly directive ISM band cavity antenna using a bi-layered metasurface reflector. *Microwave Opt. Technol. Lett.*, Vol.51, No.6, (June 2009), pp. 1393-1396, ISSN 0895-2477.

Burokur, S.N., Yahiaoui, R. & de Lustrac, A. (2009b). Subwavelength metamaterial-based resonant cavities fed by multiple sources for high directivity. *Microwave Opt. Technol. Lett.*, Vol.51, No.8, (August 2009), pp. 1883-1888, ISSN 0895-2477.

Burokur, S.N., Daniel, J.-P., Ratajczak, P. & de Lustrac, A. (2010). Tunable bilayered metasurface for frequency reconfigurable directive emissions. *Appl. Phys. Lett.*, Vol.97, No.6, (August 2010), 064101, ISSN 0003-6951.

Burokur, S.N., Daniel, J.-P., Ratajczak, P. & de Lustrac, A. (2011). Low-profile frequency agile directive antenna based on an active metasurface. *Microwave Opt. Technol. Lett.*, Vol.53, No.10, (October 2011), pp. 2291-2295, ISSN 0895-2477.

Cheype, C., Serier, C., Thèvenot, M., Monédière, T., Reinex, A. & Jecko, B. (2002). An electromagnetic bandgap resonator antenna. *IEEE Trans. Antennas Propag.*, Vol.50, No.9, (September 2002), pp. 1285-1290, ISSN 0018-926X.

Enoch, S., Tayeb, G., Sabouroux, P., Guérin, N. & Vincent, P. (2002). A metamaterial for directive emission. *Phys. Rev. Lett.*, Vol.89, No.21, (November 2002), 213902, ISSN 0031-9007.

Feresidis, A.P., Goussetis, G., Wang, S. & Vardaxoglou, J.C. (2005). Artificial Magnetic Conductor Surfaces and their application to low-profile high-gain planar antennas. *IEEE Trans. Antennas Propag.*, Vol.53, No.1, (January 2005), pp. 209-215, ISSN 0018-926X.

Jackson, D.R. & Alexópoulos, N.G. (1985). Gain enhancement methods for printed circuit antennas. *IEEE Trans. Antennas Propag.*, Vol.AP-33, No.9, (September 1985), pp. 976-987, ISSN 0018-926X.

Nakano, H., Ikeda, M., Hitosugi, K. & Yamauchi, J. (2004). A spiral antenna sandwiched by dielectric layers. *IEEE Trans. Antennas Propag.*, Vol.52, No.6, (June 2004), pp. 1417-1423, ISSN 0018-926X.

Ourir, A., de Lustrac, A. & Lourtioz, J.-M. (2006a). All-metamaterial-based subwavelength cavities (λ/60) for ultrathin directive antennas. *Appl. Phys. Lett.*, Vol.88, No.8, (February 2006), 084103, ISSN 0003-6951.

Ourir, A., de Lustrac, A. & Lourtioz, J.-M. (2006b). Optimization of metamaterial based subwavelength cavities for ultracompact directive antennas. *Microwave Opt. Technol. Lett.*, Vol.48, No.12, (December 2006), pp. 2573-2577, ISSN 0895-2477.

Ourir, A., Burokur, S.N. & de Lustrac, A. (2007a). Phase-varying metamaterial for compact steerable directive antennas. *Electron. Lett.*, Vol.43, No.9, (April 2007), pp. 493-494, ISSN 0013-5194.

Ourir, A., Burokur, S.N. & de Lustrac, A. (2007b). Electronically reconfigurable metamaterial for compact directive cavity antennas. *Electron. Lett.*, Vol.43, No.13, (June 2007), pp. 698-700, ISSN 0013-5194.

Ourir, A., Burokur, S.N., Yahiaoui, R. & de Lustrac, A. (2009). Directive metamaterial-based subwavelength resonant cavity antennas – Applications for beam steering. *C. R. Physique*, Vol.10, No.5, (June 2009), pp. 414-422, ISSN 1631-0705.

Sievenpiper, D., Zhang, L., Broas, R.F.J., Alexópoulos, N.G. & Yablonovitch, E. (1999). High-Impedance Electromagnetic Surfaces with a forbidden frequency band. *IEEE Trans. Microw. Theory Tech.*, Vol.47, No.11, (November 1999), pp. 2059-2074, ISSN 0018-9480.

Temelkuran, B., Bayindir, M., Ozbay, E., Biswas, R., Sigalas, M.M., Tuttle, G. & Ho, K.M. (2000). Photonic crystal-based resonant antenna with a very high directivity. *J. Appl. Phys.*, Vol.87, No.1, (January 2000), pp. 603-605, ISSN 0021-8979.

Trentini, G.V. (1956). Partially reflecting sheet arrays. *IRE Trans. Antennas Propag.*, Vol.4, No.4, (October 1956), pp. 666-671, ISSN 0096-1973.

Yahiaoui, R., Burokur, S.N. & de Lustrac, A. (2009). Enhanced directivity of ultra-thin metamaterial-based cavity antenna fed by multisource. *Electron. Lett.*, Vol.45, No.16, (July 2009), pp. 814-816, ISSN 0013-5194.

Zhou, L., Li, H., Qin, Y., Wei, Z. & Chan, C.T. (2005). Directive emissions from subwavelength metamaterial-based cavities. *Appl. Phys. Lett.*, Vol.86, No.10, (March 2005), 101101, ISSN 0003-6951.

21

Magnetically Tunable Unidirectional Electromagnetic Devices Based on Magnetic Surface Plasmon

Shiyang Liu[1], Huajin Chen[1], Zhifang Lin[2] and S. T. Chui[3]
[1]*Institute of Information Optics, Zhejiang Normal University*
[2]*State Key Laboratory of Surface Physics (SKLSP) and Department of Physics, Fudan University*
[3]*Department of Physics and Astronomy, University of Delaware*
[1,2]*China*
[3]*USA*

1. Introduction

Plasmonic metamaterials are composite, artificial materials, consisting of metallic resonant building blocks designed with state-of-the-art configurations. Many exotic phenomena not occurring in nature such as negative refraction (Burgos et al., 2010; Dolling et al., 2006; Lezec et al., 2007; Liu et al., 2008; Pendry, 2000; Shalaev, 2007; Shelby et al., 2001; Smith et al., 2000; Valentine et al., 2008; Veselago, 1968; Zhang et al., 2005), subwavelength imaging (Fang et al., 2005; Liu et al., 2007; Pendry, 2000; Taubner et al., 2006), cloaking (Ergin et al., 2010; Leonhardt, 2006; Lai et al., 2009; Li et al., 2008; Liu et al., 2009; Pendry et al, 2006; Schurig et al., 2006), and so on can be observed in the systems made of such materials. A particular characteristic of the system is the excitation of surface plasmon polaritons (SPPs) (Barnes et al., 2003; Giannini et al., 2010; Noginov et al., 2008; Zayats et al., 2005), which originates from the coupling of the electromagnetic (EM) wave to the free electron oscillation in metallic surface. The EM waveguiding mediated by the SPPs has gained more and more attention from theorists, experimentalists, and engineers due to its promising applications in integrated optical circuit, optical storage, biosensing, and even medical therapy (Engheta, 2007; Ozbay, 2006; Zia et al., 2006). A series of plasmonic components have been proposed and realized, such as plasmonic waveguides, beam splitter, sharp bends and so on. Very recently, based on the excitation of the SPPs the plasmonic Luneburg and Eaton lens are even realized experimentally (Zentgraf, 2011).

So far, plasmonics relevant phenomena have been extensively investigated from both the theoretical and the experimental perspectives. The symmetry of the Maxwell's equations with respect to its electric and magnetic part suggests that the magnetic analogue of the surface plasmon, "magnetic surface plasmon" (MSP) (Gollub et al., 2005; Liu et al., 2008), can also be excited. It originates from the coupling of EM wave to the collective resonance of spin wave, and thus can be observed in the magnetic system. However, the relevant issues are not examined elaborately as the case for the surface plasmon in metallic materials. In addition, there appeared burgeoning activities in exploring the phenomena resulting from the time reversal symmetry (TRS) breaking for photons. Among others, the EM one-way edge modes analogous to quantum Hall edge states were discussed by Haldane and Raghu (Haldane & Raghu, 2008), and later by Wang and co-workers (Wang et al., 2008; 2009). Recently, the self-guiding unidirectional EM edge states are also realized theoretically (Ao et al., 2009) and experimentally (Poo et al., 2011). In the metamaterials designed with ferrite materials, the magnetic response is intrinsic, accordingly, termed as magnetic metamaterials (MMs), where the MSP resonance can be excited. Besides, the TRS breaking relevant phenomenon for the photons can also be observed (Liu et al., 2010; 2011; Wang et al., 2008). In particular, we can find that the MMs can mold the reflection in a dramatic manner (Liu et al., 2011).

Similar to the electric surface plasmon (ESP) resonance occurring when its permittivity $\varepsilon = -1$ in metallic rod. For a ferrite rod, the MSP resonance occurs when its effective permeability $\mu = -1$ (Liu et al., 2008). Due to the TRS breaking nature in ferrite materials, there exist unequal amounts of states with opposite angular momenta for the scattered field. Near the MSP resonance, the angular momenta contents are dominated by one sign, with that for the other sign almost completely suppressed. Accordingly, a giant circulation (clockwise or anticlockwise) of energy flow develops, so only the energy flow in one direction is supported (Chui & Lin, 2007). Consequently, in this work we consider the phenomenon resulting from the combined action of the MSP resonance and the TRS breaking (Liu et al., 2011). In addition, the working frequency can be controlled by an external magnetic field (EMF), facilitating the design of the practical EM devices.

The present chapter is organized as follows. In the second part, we give a brief introduction on the Mie theory on the ferrite rod and the multiple scattering theory used in the numerical simulations. Then, we present the reflection behavior due to the MSP resonance and its physical origin. The dependence of the behavior on the working frequency and the source-interface separation is also examined. Following this part, we show a design of a one-way EM waveguide (OEMW) based on this effect. Most importantly, the robustness of the OEMW against defect, disorder, and inhomogeneity of the EMF are examined. The manipulability of the working frequency is demonstrated as well by tuning the EMF. Some other complicate EM devices such as a sharp beam bender and a beam splitter are also designed. In addition, all the designed EM devices are shown to be operable even in the deep subwavelength scale. Our results are summarized in the conclusion part.

2. Theoretical approach

To design the MM, we use single crystal yttrium-iron-garnet (YIG) ferrite rods as building blocks. In our case, the ferrite rods are arranged periodically as a square lattice in the air with lattice constant a and the radius of the ferrite rod r_s. The rod axis are oriented along the z direction, corresponding to the direction of the EMF. When fully magnetized, the magnetic permeability can be written in the form (Pozar, 2004; Slichter, 1978)

$$\widehat{\mu} = \begin{pmatrix} \mu_r & -i\mu_\kappa & 0 \\ i\mu_\kappa & \mu_r & 0 \\ 0 & 0 & 1 \end{pmatrix}, \qquad \widehat{\mu}^{-1} = \begin{pmatrix} \mu'_r & -i\mu'_\kappa & 0 \\ i\mu'_\kappa & \mu'_r & 0 \\ 0 & 0 & 1 \end{pmatrix}, \tag{1}$$

with

$$\mu_r = 1 + \frac{\omega_m(\omega_0 - i\alpha\omega)}{(\omega_0 - i\alpha\omega)^2 - \omega^2}, \quad \mu_\kappa = \frac{\omega_m \omega}{(\omega_0 - i\alpha\omega)^2 - \omega^2}, \quad \mu'_r = \frac{\mu_r}{\mu_r^2 - \mu_\kappa^2}, \quad \mu'_r = \frac{-\mu_\kappa}{\mu_r^2 - \mu_\kappa^2},$$

where $\omega_0 = \gamma H_0$ is the resonance frequency with $\gamma = 2.8\,\text{MHz/Oe}$ the gyromagnetic ratio; H_0 the sum of the EMF applied in z direction and the shape anisotropy field (Pozar, 2004), $\omega_m = 4\pi\gamma M_s$ is the characteristic frequency with $4\pi M_s = 1750$ Oe the saturation magnetization, and α is the damping coefficient of the ferrite. In the calculation of photonic band diagram, we set $\alpha = 0$ (Wang et al., 2008), and in the simulation $\alpha = 3 \times 10^{-4}$. The relative permittivity of the ferrite rods is taken as $\varepsilon = 15 + 3 \times 10^{-3}i$. In our work, the transverse magnetic (TM) mode is considered.

To make our results reliable and convergent efficiently, we have used the Mie theory to solve the scattering properties of a single ferrite rod. For an incident EM wave with TM polarization, it can be expanded into vector cylindrical wave functions (VCWFs) (Bohren & Huffman, 1983)

$$\boldsymbol{E}_0(j) = \sum_n E_n q_n^{(j,j)} \boldsymbol{N}_n^{(1)}(k_b, \boldsymbol{r}_j), \tag{2}$$

where j suggests that the $j-$th coordinate is used, indicating that the EM fields are expanded around the $j-$th ferrite rod, $\boldsymbol{N}_n^{(1)}(k_b, \boldsymbol{r}_j)$ is the VCWF with k_b the wavenumber in the background medium, satisfying $k_b^2 = \omega^2\varepsilon_b\mu_b$, ε_b and μ_b correspond, respectively, to the permittivity and permeability of the background medium, $E_n = i^n|E_0|$ with $|E_0|$ the amplitude of the incident EM wave. For plane wave, the expansion coefficients can be easily obtained

$$q_n^{(j,j)} = e^{-in\phi_\kappa} \exp(i\boldsymbol{k}_b \cdot \boldsymbol{d}_{0j}), \tag{3}$$

where $\boldsymbol{k}_b = \kappa\cos\phi_\kappa \boldsymbol{e}_x + \kappa\sin\phi_\kappa \boldsymbol{e}_y$ is the wave vector of the incident EM wave with $\kappa = |\boldsymbol{k}_b|$, $\boldsymbol{d}_{0j} = \boldsymbol{r} - \boldsymbol{r}_j$, corresponding to the position of the $j-$th ferrite rod in the $0-$th coordinate. The scattered EM wave can be expanded into VCWFs as well in the form

$$\boldsymbol{E}_s(j) = -\sum_n E_n b_n^{(j,j)} \boldsymbol{N}_n^{(3)}(k_b, \boldsymbol{r}_j), \tag{4}$$

In Eqs. (2) and (4), the VCWFs are defined according to

$$\boldsymbol{N}_n^{(J)}(k, \boldsymbol{r}) = z_n^{(J)}(kr)e^{in\phi}\boldsymbol{e}_z, \tag{5}$$

where

$$z_n^{(J)}(kr) = \begin{cases} J_n(kr) & \text{for } J = 1, \\ H_n^{(1)}(kr) & \text{for } J = 3, \end{cases} \tag{6}$$

In Eq. (6), $J_n(kr)$ and $H_n^{(1)}(kr)$ correspond, respectively, to the first kind of Bessel function and the first kind of Hankel function.

The EM wave inside the ferrite rod can also be expanded into VCWFs, but the form is so complicated. Therefore, to save the space we do not give the corresponding result. By matching the boundary conditions, we can solve the problem and obtain the Mie coefficients, which can connect the unknown scattering coefficients with the given expansion coefficients of the incident wave in the form

$$b_n^{(j,j)} = \mathcal{S}_n^{(j)} q_n^{(j,j)}, \tag{7}$$

where $\mathcal{S}_n^{(j)}$ is the Mie coefficients of the $j-$th ferrite rod (Chen et al., 2007; Chui & Lin, 2007; Eggimann, 1960)

$$\mathcal{S}_n = \frac{\dfrac{\mu_s}{\mu_b} J_n'(x) - J_n(x)\left[\dfrac{m_s^2}{m_s'} D_n(m_s'x) + \dfrac{n\mu_k'}{x}\right]}{\dfrac{\mu_s}{\mu_b} H_n^{(1)'}(x) - H_n^{(1)}(x)\left[\dfrac{m_s^2}{m_s'} D_n(m_s'x) + \dfrac{n\mu_k'}{x}\right]}. \tag{8}$$

In Eq. (8), the superscripts "$'$" of $J_n(x)$ and $H_n(x)$ denote the derivatives with respect to the argument $x = k_b r_s$, the other parameters $k_s^2 = \omega^2 \epsilon_s \mu_s$, $m_s = k_s/k_b$, $m_s' = m_s/\sqrt{\mu_r'}$, and $D_n(m_s' x) = J_n'(m_s' x)/J_n(m_s' x)$.

By combining the Mie theory with the multiple scattering theory (Felbacq et al., 1994; Leung & Qiu, 1993; Liu & Lin, 2006; Wang et al., 1993), we can handle the scattering problem for a system consisting of multiple ferrite rods. Then, the scattering coefficients around each ferrite rod can be obtained by solving the linear equations

$$b_n^{(j)} = S_n^{(j)} \left[q_n^{(j,j)} - \sum_{l \neq j} \sum_m A_{nm} b_m^{(l)} \right]. \tag{9}$$

In Eq. (9), A_{nm} is the translational coefficient (Chew, 1995)

$$A_{nm} = i^{n-m} H_{n-m}^{(1)}(k d_{lj}) \exp\left[-i(n-m)\phi_{lj} \right], \tag{10}$$

where $\boldsymbol{d}_{lj} = \boldsymbol{r}_l - \boldsymbol{r}_j$ and (d_{lj}, ϕ_{lj}) is the polar coordinate of the position vector $\boldsymbol{d}_{lj}$. By setting $q_n^{(j,j)} = 0$ and find the corresponding eigenmodes, we can also calculate the photonic band diagrams.

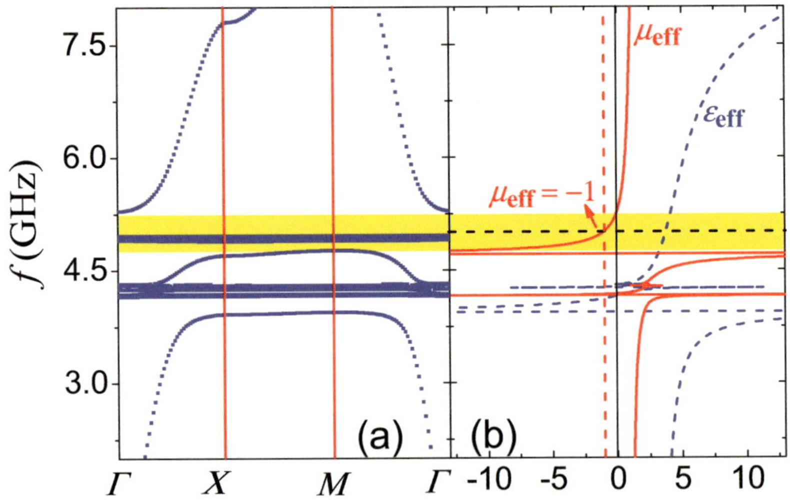

Fig. 1. (a) The photonic band diagram for an MM of square lattice with lattice constant $a = 8$ mm and under an EMF of $H_0 = 900$ Oe. (b) The retrieved effective constitutive parameters where the blue dashed line and red solid line correspond to the effective permittivity ε_{eff} and the effective magnetic permeability μ_{eff}, respectively.

3. MSP resonance in MM

The MM considered are made of single crystal YIG rods arranged periodically in a square lattice with the lattice constant $a = 8$ mm and the rod radius $r = \frac{1}{4}a = 2$ mm. Firstly, let's examine the origin of the MSP resonance. It depends on the resonance of the each single ferrite rod. The coupling of the neighboring MSP resonant states of each single ferrite rod results in the resonance of the whole system. For a ferrite rod with the permeability given by Eq. (1), the MSP resonance occurs at a frequency

$$f_s = \frac{1}{2\pi}\left(\omega_0 + \frac{1}{2}\omega_m\right), \tag{11}$$

corresponding to the case when $\mu_r + \mu_\kappa = -1$ (Liu et al., 2008), which can be regarded as the effective permeability of a single ferrite rod. In the present case, the applied EMF field is such that $H_0 = 900$ Oe, yielding a MSP resonance at $f_s = 4.97$ GHz. To gain a better understanding of the MSP resonance, we present the photonic band diagram in Fig. 1 (a). At the working frequency $f = 5$ GHz (corresponding the wavelength 60 mm) the wavelength is nearly 8 times of the lattice constant a so that the MM can be considered as an effective medium. With the effective-medium theory developed in our previous work (Jin et al., 2009), we can retrieve the effective permeability tensor,

$$\mu_e = \begin{pmatrix} \mu & -i\mu' & 0 \\ i\mu' & \mu & 0 \\ 0 & 0 & 1 \end{pmatrix}. \tag{12}$$

Then, the effective permeability μ_{eff} for TM mode is given by $(\mu^2 - \mu'^2)/\mu$, which is shown in Fig. 1(b). By comparing the photonic band diagram and the effective constitutive parameters μ_{eff} and ε_{eff}, we can find that double positive effective parameters correspond to the photonic bands, single negative effective parameter corresponds to the photonic band gap (PBG) , while the resonances correspond to the flat bands. Near the frequency corresponding to the flat bands in Fig. 1(b) (at the MSP resonance frequency $f_s = 4.97$ GHz) the effective magnetic permeability is nearly equal to -1 as is marked by the black dashed line, consistent with the analysis on the single ferrite rod. For this reason, the MSP resonance can be considered as the magnetic analogue of the surface plasmon resonance inhabited in the metallic materials. In addition, around the MSP resonance two PBGs come into existence as denoted by the yellow stripes in Fig. 1(a) where $\mu_{\text{eff}} < 0$, in the vicinity of the MSP resonance. This is also the working frequency range we select to examine the corresponding reflection behavior.

4. Molding reflection with MM slab

To explore the physical consequence resulting from the combined action of the MSP resonance and the TRS breaking, we have examined the reflection behavior of a TM wave excited by a line source working at a frequency near f_s and located near an MM slab. Typical results are demonstrated in Fig. 2, where the line source oscillating at $f = 5$ GHz is located a distance $a = 8$ mm away from a four-layer MM slab. The total electric field, the scattered electric field, together with the x-component of the Poynting vector are plotted in panels (a), (b), and (c), respectively, where a sharply asymmetric reflection (SAR) can be observed. On the left hand side (LHS) of the line source, the scattered field substantially cancels the incoming field, resulting in a darkened region near the MM surface. On the right hand side (RHS) of the line source, the scattered field significantly enhances the EM field, giving rise to a brightened region near the MM surface. Since the working frequency is selected in the PBG as can be seen from the photonic band diagram in Fig. 1 (a), the EM wave can not propagate inside the MM slab. The vanishment of the total field inside the MM slab indicates that the incident and scattered fields inside the MM have a π phase difference. For the scattered wave, on the LHS of the line source, the bright fringes both inside and outside the MM [shown in Fig. 2 (b)] remains at the same positions. This indicates that the scattered waves on the LHS are continuous in phase near the surface of MM. As the scattered field cancels the incident field

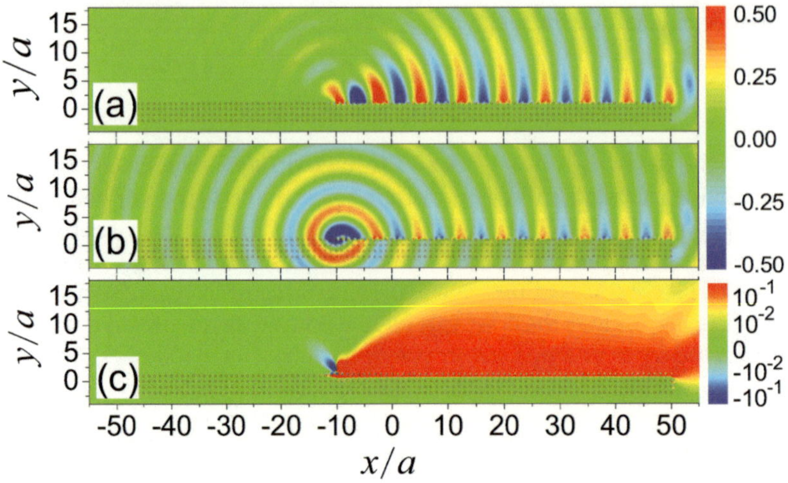

Fig. 2. The patterns of the total electric field (a), the scattered electric field (b), and the x-component of the Poynting vector (c) for a line source operating at a frequency $f = 5$ GHz close to the MSP resonance. The lattice constant of the MM slab is $a = 8$ mm and the line source is located $1.0a$ away from the surface of the MM slab.

inside the MM slab, it therefore also attenuates the incident field outside the MM slab. The situation is quite different on the RHS of the line source. As can be seen from Fig. 2 (b), the bright fringe outside the MM slab is at the same position as the dark fringe inside the MM slab near the interface. There appears a nearly half-wavelength mismatch for scattered waves inside and outside the MM slab. This suggests a phase change around π occurs inside and outside the MM slab. As a consequence, while the scattered field cancels the incident field inside the slab due to the phase mismatch, it considerably enhances the EM field near surface outside the MM slab owing to in-phase interference. In this manner, we can understand the SAR effect phenomenologically.

As can be seen from Fig. 1, the working frequency is selected near the MSP resonance, which plays a crucial role for the phenomenon occurring in the system. The excitation of the MSP resonance can lead to the appearance of the unidirectional circulation of the energy flow as schematically shown in Fig. 3. Accordingly, for a line source located near the surface of the MM slab, the leftward energy flow is inhibited while the rightward energy flow is supported

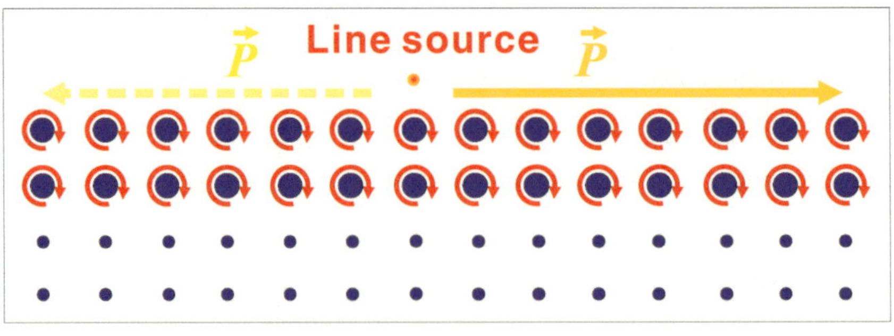

Fig. 3. The simple physical picture illustrating the excitation of the one-way circulating MSP band states, which is responsible for the occurrence of the SAR effect shown in Fig. 2.

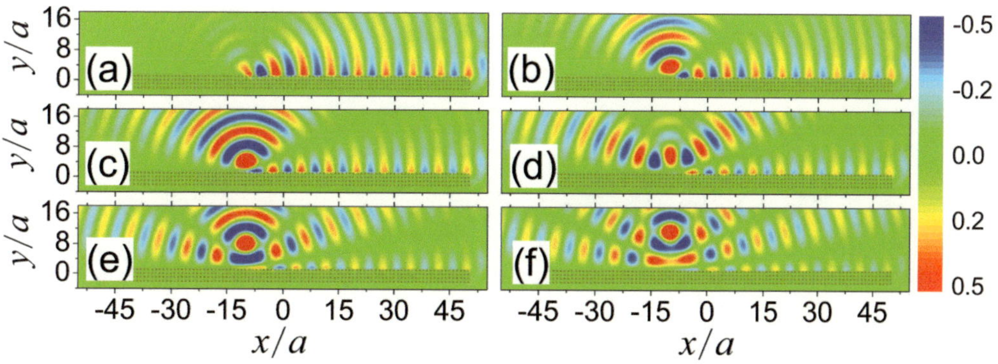

Fig. 4. The electric field patterns showing the SAR effect corresponding to different source-interface separations d. Panels (a), (b), (c), (d), (e), and (f) correspond to the cases when $d = 1a$, $2a$, $3a$, $5a$, $7a$, and $10a$, respectively. All the other parameters are the same as those in Fig. 1. The working frequency is selected at $f = 5$ GHz. The MM slab is the same as that in Fig. 2.

and reinforced. For this reason, we can observe the SAR effect shown in Fig. 2. Interestingly, by reversing the orientation of the EMF the circulation of the energy flow can also be reversed. Therefore, the EM property relevant to the SAR effect is magnetically tunable.

4.1 Separation dependence of the SAR effect

As aforementioned, the SAR effect originates from the MSP resonance. Therefore, it is necessary to examine the dependence of the SAR effect on the source-interface separation d. The corresponding results are shown in Fig. 4 where panels (a), (b), (c), (d), (e), and (f) correspond to the electric field patterns when $d = 1a$, $2a$, $3a$, $5a$, $7a$, and $10a$, respectively. The operating frequency is $f = 5$ GHz, all the other parameters of the MM slab are also the same as those in Fig. 2. When the line source is close to the interface of MM slab a dramatic SAR effect can be observed. With the increase of the separation d, the SAR effect becomes weaker and weaker, namely, more and more electric field is scattered into the outside space. When the line source is far enough from the interface ($d \approx \lambda$), the electric field near the interface becomes very weak and the electric field pattern is nearly symmetric as can be observed in Figs. 4 (e) and (f). This result can be understood from the following approximate physical picture. The incoming electric field at a position $\boldsymbol{r}$ is given by

$$E_i(\boldsymbol{r}) = \exp[i\boldsymbol{k} \cdot (\boldsymbol{r} - \boldsymbol{e}_y d)]/|\boldsymbol{r} - \boldsymbol{e}_y d|. \tag{13}$$

The scattered electric field is approximately of the spatial dependence of that from an image term

$$E_s(\boldsymbol{r}) = \mathrm{s}\exp[i\boldsymbol{k} \cdot (\boldsymbol{r} + \boldsymbol{e}_y d)]/|\boldsymbol{r} + \boldsymbol{e}_y d|, \tag{14}$$

where $\mathrm{s} = \pm 1$, depending on whether one is on the LHS or RHS of the source. For small d, there is an obvious cancellation between these two terms. As d is increased, this cancellation only remain at $y = 0$. But for other values of y, this cancellation becomes too much weaker.

4.2 Frequency dependence of the SAR effect

The MSP resonance occurs at a specified frequency. Accordingly, the SAR effect should be also dependent on the selection of the working frequency. In Fig. 5 (a), we present the photonic

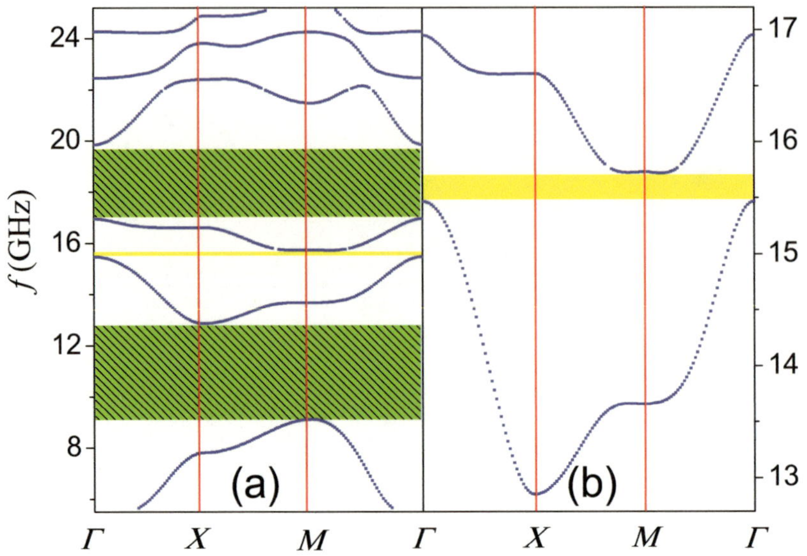

Fig. 5. (a) The photonic band diagram at the higher frequency where three PBGs arising from the bragg scattering can be observed. (b) The amplified view of the photonic band diagram around the second PBG. All the parameters involved are the same as those in Fig. 1.

band diagrams at higher frequencies, where three PBGs are identified. We first examine the reflection behavior at the frequency $f = 9.5$ GHz located at the bottom of the first PBG. The simulation result is shown in Fig. 6 (a), where the field pattern is almost symmetric, no SAR effect can be observed. Presumably, the Chern numbers of the photonic bands nearby are zero, only a tiny circulation of the energy flow occurs so that no significant asymmetry reflection is observed. The result is consistent with that in Wang and coworkers' research where they present the Chern number corresponding to the photonic bands (Wang et al., 2008). Similarly, for the upmost PBG ranging from 17 GHz to 19.5 GHz, the SAR effect can not be observed either due to the same reason. Between these two PBGs, there exist another narrow PBG as denoted by a yellow stripe in Fig. 5 (a). We select a frequency $f = 15.5$ GHz as the working frequency to examine the reflection behavior. For convenience, we have given in Fig. 5 (b) the amplified view of the photonic band diagram around this PBG. The corresponding electric field pattern is shown in Fig. 6 (b) where the electric field on the RHS is nearly vanished near the interface, while on the LHS the electric field can be supported so that the SAR effect comes into appearance. The PBG comes from the degeneracy lift resulting from the gyrotropic anisotropy of the MMs. At this frequency range, the energy circulation can be excited, which we will discuss later on.

For comparison, we have also performed the simulation to examine the reflection behavior of a line source from an MPC slab with the same parameters in Wang and coworkers' research (Wang et al., 2008). Concretely, the lattice constant is $a = 38.7$ mm, the radius of the ferrite rod is $r = 0.11a = 4.3$ mm, the corresponding matrix elements of the magnetic permeability are $\mu = 14$ and $\mu' = -12.4$, the working frequency is $f = 4.28$ GHz, and the line source is located $0.25a$ away from the interface. The electric field pattern is shown in Fig. 6 (c) where the EM wave is reflected somewhat leftwards. A weak asymmetry reflection can be observed. In this case, the Chern numbers of the photonic bands around the PBG are not zero as shown by Wang and coworkers (Wang et al., 2008). However, different from the results shown in Fig.

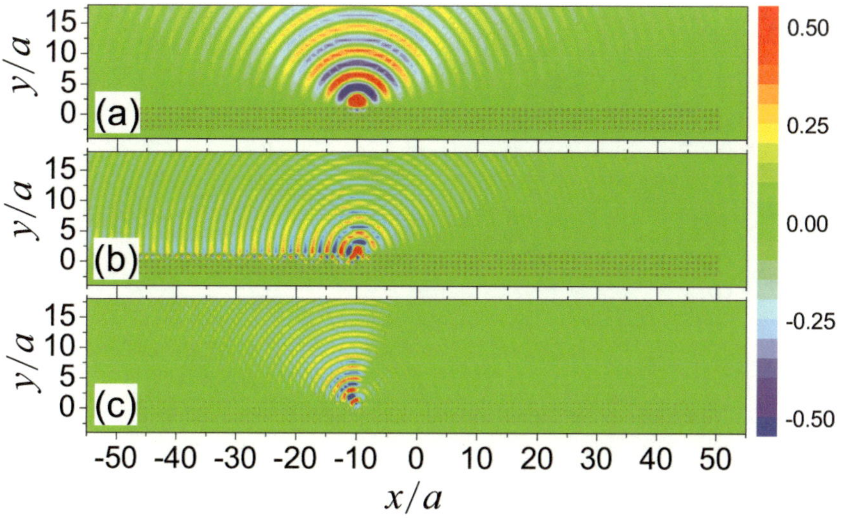

Fig. 6. The reflection behavior for the working frequency located at $f = 9.5$ GHz (a) and $f = 15.5$ GHz (b), corresponding to those in the lower and higher PBGs, respectively, due to the Bragg scattering. The reflection behavior of a line source from an MPC slab with the same parameters in (Wang et al., 2008) is also examined as shown in panel (c).

2, there appears no enhancement of EM field near the interface due to the absence of MSP resonance.

It should be noted that the direction of EM energy circulation does not depend solely on the magnetization, as can be illustrated by comparing Fig. 2 (a) to Fig. 6 (b). To determine the direction of the circulation (clockwise or anticlockwise) we need to calculate the energy circulation around the ferrite rods. For a single rod, it can be evaluated according to

$$\oint \boldsymbol{P} \cdot \mathrm{d}\boldsymbol{l} = A(\omega) \sum_n n \left| \frac{b_n}{q_n} \frac{H_n^{(1)}(x)}{J_n(x)} - 1 \right|^2 |q_n J_n(x)|^2, \tag{15}$$

where $A(\omega) = \frac{\pi |E_0|^2}{\mu_b \omega}$ is the prefactor with E_0 the amplitude of the incident wave and μ_b the magnetic permeability of the background medium, $x = k_b r_s$ is the size parameter, b_n and q_n are, respectively, the expansion coefficients of the scattered and incident wave corresponding to the angular momentum n, $J_n(x)$ is $n-$th order the Bessel function, and $H_n^{(1)}(x)$ is the $n-$th order Hankel function of the first kind. From Eq. (15), it can be found that the energy circulation depends on the difference of the scattering amplitude for the $|n|$ and the $-|n|$ terms. While the scattering amplitude corresponding to $n = 0$ is not involved. We have calculated the energy circulation around a typical ferrite rod in the first layer of the MM slab. The results show that the circulation is about $-10^{-2}A(\omega)$ for Fig. 2 (a), suggesting a clockwise energy flow, thus explaining the rightwards reflection. For Fig. 6 (a), the energy circulation is nearly zero, corresponding the nearly symmetric reflection. For Fig. 6 (b), the energy circulation is about $10^{-2}A(\omega)$, indicating that an anticlockwise energy flow is formed, explaining the leftwards reflection. While for the case in Fig. 6 (c), the energy circulation is about $10^{-4}A(\omega)$, much weaker than the case in Fig. 2 (a) and Fig. 6 (b). Accordingly, although the leftward reflection occurs, along the interface nearly no EM mode is supported.

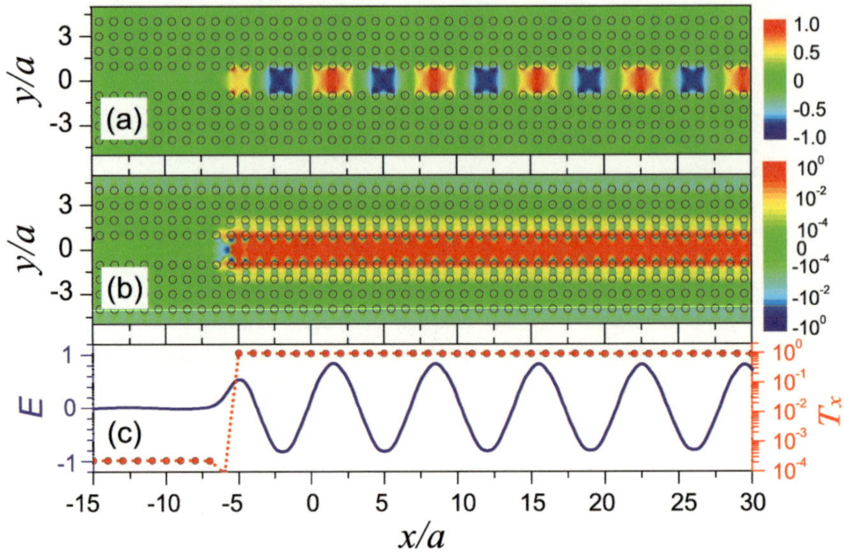

Fig. 7. The profile of the electric field (a) and the x-component of the Poynting vector (b) when a line source of $f = 5$ GHz is located at $(-5.5a, 0)$, between two MM slabs with opposite magnetization and the channel width $D = 2a$. Here $a = 8$ mm is the lattice constant of a square lattice of MM. Panel (c) displays the electric field E at $y = 0$ and the rightward transmitted power T_x versus x as denoted by blue solid line and red dotted line, respectively.

5. Design of an OEMW based on the MSP

Based on the SAR effect we can explore some possible applications. A typical example is due to the remarkable asymmetry of the Poynting vector as shown in Fig. 2 (c). Because of the existence of the PBGs around the MSP resonance, the EM wave can be confined between two MM slabs and transported only in the channel, similar to the conventional photonic crystal waveguide. If two MM slabs have the opposite magnetization, then the EM wave reflected forward from one MM slab will also be reflected forward from the other, leading to the design of an OEMW different from those proposed recently (Fu et al., 2011; 2010; He et al., 2010; Huang & Jiang, 2009; Wang et al., 2008; 2009; Yu et al., 2008; Zhu & Jiang, 2010). The performance of the OEMW is illustrated in Fig. 7 (a) where a line source is placed at $(-5.5a, 0)$, in the middle of two MM slabs with the channel width $D = 2a$. The parameters for the MM slabs and the operating frequency of the line source are the same as in Fig. 2. It can be observed that the EM wave propagates rightward as demonstrated by the x component P_x of the Poynting vector $\boldsymbol{P}$ shown in Fig. 7 (b). To further illustrate the one-way characteristic, we display in Fig. 7 (c) the electric field along $y = 0$ and the rightward transmitted power, $T_x = \int_{-3a}^{3a} P_x \mathrm{d}y$, as the functions of the position x. In addition, the EM energy exhibits an extremely low decay rate that is less than 0.1 dB/λ when taking into account of the realistic material absorption (Pozar, 2004). To search out the physical essence, in the following, we will neglect the damping for simplicity.

5.1 Channel width dependence of the OEMW

As already shown in Fig. 4, the SAR effect is dependent on the source-interface separation and it is only workable when the separation is small. This suggests that the OEMW designed

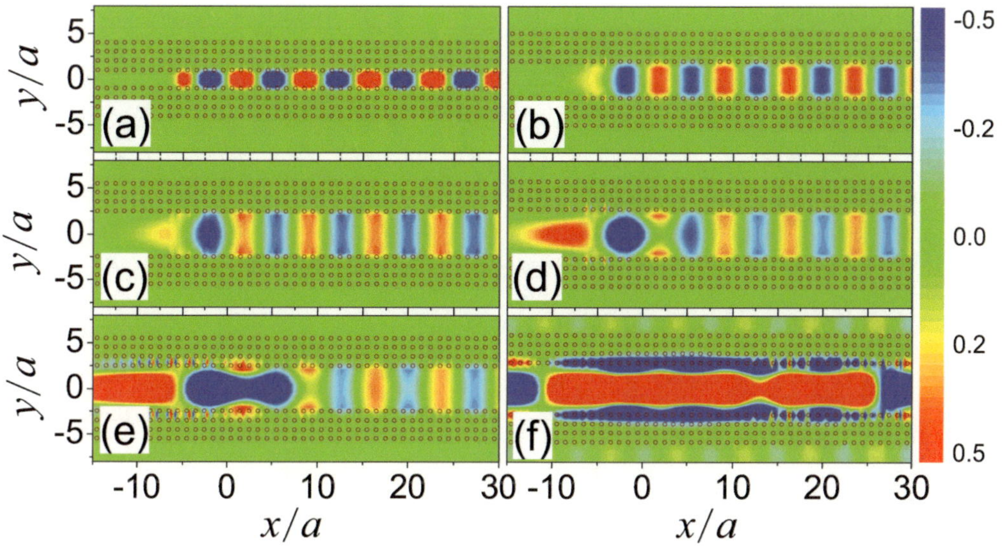

Fig. 8. The electric field patterns for the OEMW operated with different channel widths. Panels (a), (b), (c), (d), (e), and (f) correspond to the cases for the channel widths $D = 2.0a$, $D = 4.0a$, $D = 5.5a$, $D = 5.6a$, $D = 5.7a$, and $D = 6.0a$, respectively.

according to this effect should also depend on the channel width. Physically, as the channel width becomes larger than a wavelength, one part of the EM wave can propagate along the channel without experiencing any of the reflection. Therefore, the unidirectional propagation characteristic shown in Fig. 7 will be diminished. In Fig. 8, we have presented the electric field patterns for the OEMW with different channel widths. Panels (a), (b), (d), (e), and (f) correspond to channel widths $D = 2.0a$, $D = 4.0a$, $D = 5.5a$, $D = 5.6a$, $D = 5.7a$, and $D = 6.0a$, respectively. For OEMW with the channel width $D < 5.5a$, the leftward propagating EM wave is suppressed completely. The OEMW can be considered as a unidirectional device with good performance. However, with the increase of the channel width more and more EM field leaks leftwards as can be observed in Figs. 8 (d)-(f). This result is in good agreement with the result shown in Fig. 4 in that the unidirectionality becomes weaker and weaker with the increase of the source-interface separation. When the channel width is increased to $D = 6.0a$, nearly 0.8λ, no obvious difference can be observed for the leftward and rightward propagating EM waves.

5.2 Robustness against defect, disorder, and inhomogeneity of EMF

A particular important issue of current interest is the robustness of a designed EM devices against defect and disorder. We have performed the simulations to illustrate the issue, which indicates that the one-way waveguiding property based on the SAR effect appears to be immune to defect and disorders, as demonstrated typically in Figs. 9 (b-d). Figure 9 (b) simulates the electric field pattern when a finite linear array of close-packed perfect electrical conductor (PEC) rods is inserted to block the channel. The radius of the PEC rod is $r_p = \frac{1}{2}r_s$. The linear array ranges from $y = -2.5a$ to $y = 2.5a$, forming a drastic defect extending over $\frac{2}{3}\lambda$ in length. The EM wave is seen to circumvent the PEC defect, maintaining a nearly complete power transmission T_x along the channel, as shown in Fig. 9 (j). In Fig. 10, we present a

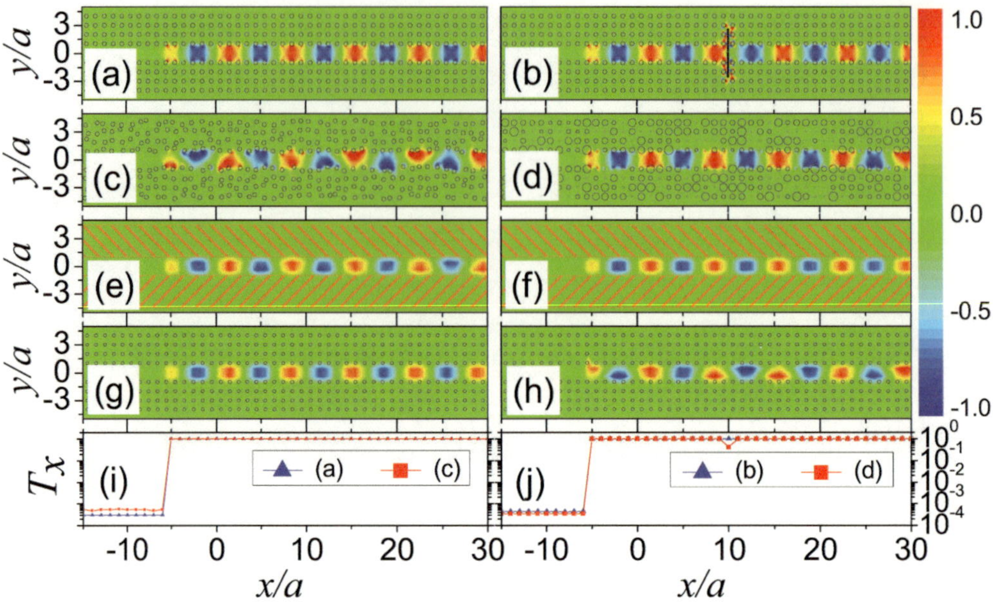

Fig. 9. The electric field patterns for the OEMW operated under the conditions (a) the same as Fig. 7 (a), given here for comparison, (b) with the introduction of a drastic PEC defect to block the channel, (c) with position disorder, and (d) with size fluctuation. The averaged electric field patterns for 20 configurations with position disorder (e) and size fluctuation (f) are also given. The inhomogeneity of the EMF is considered as well for the cases of (g) inhomogeneous and symmetric EMF, and (h) inhomogeneous and asymmetric EMF. In addition, the normalized transmitted power versus x for (a), (c) and (b), (d) are plotted and presented in panels (i) and (j), respectively.

simple schematic diagram to show how the OEMW works and how the EM wave circumvents the PEC defect, which could be helpful for the understanding of mechanism dominating the behavior. For the perfect case without any defect, the MSP resonance can induce the formation of a unidirectional energy circulation so that the channel can only support the energy flow in one direction, the OEMW is thus designed as shown in Fig. 10 (a). When a PEC defect is inserted into the channel, an equivalent waveguiding channel can be created between the PEC defect and the MM as marked by the dashed vertical arrows. Therefore, the EM wave can get around the PEC defect without experiencing any backscattering by propagating along this equivalent channel, resulting in a nearly complete energy transmission. Compared with the defect free waveguide shown in Fig. 9 (a), it can be observed that the defect changes only the phase of the rightward propagating wave, as a result of the delay due to the PEC defect.

The designed OEMW is also robust against the position disorder and the size fluctuation of the ferrite rods as illustrated in Figs. 9 (c) and 9 (d) where we present the corresponding electric field patterns when these two types of perturbations are involved. Position disorder is controlled by a series of random numbers, which introduce a maximal coordinate variation equal to $\frac{1}{4}a$. The size fluctuation of the rod radius is uniform up to 50% of the unperturbed case. It can be observed that the disorder only alters the field patterns, but not the power transmission through the channel as shown in Figs. 9 (i) and (j). Actually, for even stronger perturbation the OEMW still works well provided that the channel is not destroyed. For

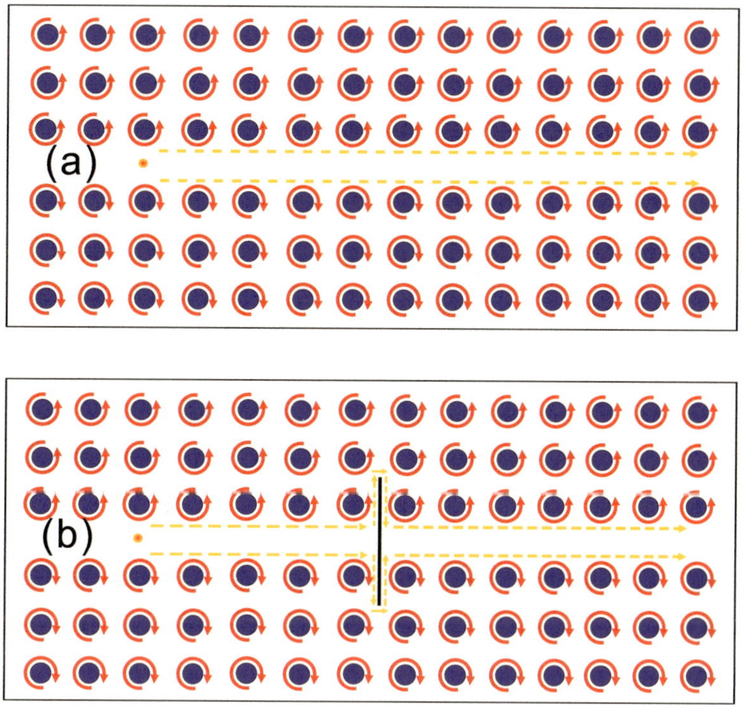

Fig. 10. The simple schematic picture to illustrate how the OEMW works (a), and how the EM wave can get around the defect without any back scattering (b) where the point denote the line source, the circular arrows denote the energy circulation supported by the ferrite rods, and the solid black line denote the PEC defect.

the edge state waveguide, (Wang et al., 2008) the working frequency lies in the Bragg type PBG, so although the system can be immune to the defect, it may suffer from disorder of the building blocks. In Figs. 9 (e) and (f), we also present the averaged electric field patterns over 20 configurations of position and radius disordered systems, respectively. It can be found that the averaged field patterns bear resemblance to that without any perturbation, owing to cancellation of the field at the irregular interface for different configurations. The cases corresponding to different amplitude of perturbation are examined as well, similar behaviors can be observed, indicating the statistical validity of the results.

In realistic situation, the EMF can not be perfectly homogeneous. For this reason, we also present the simulation results for the OEMW operated under an inhomogeneous EMF. Figure 9 (g) corresponds to the case when a symmetric inhomogeneous EMF is exerted. The EMFs at the different layers from inside to outside are, respectively, 890 Oe, 910 Oe, 930 Oe, and 950 Oe, symmetrically for the upper and lower MM slabs. The situation corresponding to the asymmetric inhomogeneous EMF is also considered as shown in Fig. 9 (h) where the EMFs at different layers from inside to outside are 870 Oe, 900 Oe, 910 Oe, 950 Oe for the upper MM slab and 920 Oe, 930 Oe, 940 Oe, 970 Oe for the lower MM slab. It can be observed that the symmetric distribution of the EMF leads to a symmetric electric field pattern and the asymmetric EMF can shape the electric field pattern into an asymmetric one. However, a good performance can still be maintained. This is also a favorable aspect for the design of the EM devices.

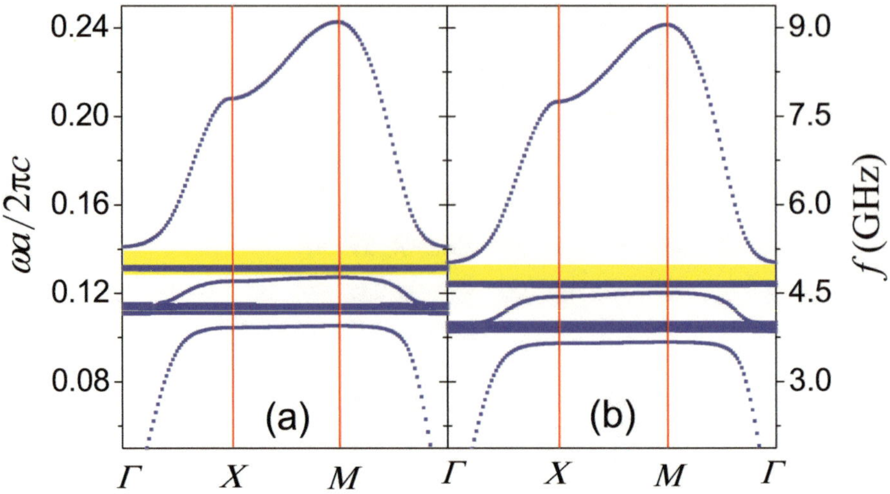

Fig. 11. The photonic band diagrams corresponding to the MM operated under two different EMFs (a) $H_0 = 900$ Oe and (b) $H_0 = 800$ Oe, respectively. The other parameters of the MM are the same as those used in Fig. 1. The working frequency ranges for the OEMW under different EMFs are denoted with the yellow stripes.

5.3 Tunability of the working frequency

A special property of the OEMW designed in our work is the tunability of the working frequency due to the dependence of the MSP resonance on the EMF. This tunability can be clearly seen by examining the photonic band diagrams of the MM under different EMFs as shown in Figs. 11 (a) and (b), corresponding to $H_0 = 900$ Oe and $H_0 = 800$ Oe, respectively. The yellow stripes mark the working frequency range of the OEMW, the flat bands there correspond to the MSP resonance. When the EMF decreases from $H_0 = 900$ Oe to $H_0 = 800$ Oe, the flat bands move downwards, so does the working frequency. In this manner, the working frequency can be controlled by an EMF. Due to the sensitivity of the MSP resonance dependent on the EMF, the working frequency of the OEMW can be manipulated easily.

To give a clear picture how the OEMW works under different EMF, we present in Fig. 12 the electric field patterns of the OEMW under $H_0 = 900$ Oe [(a), (b)] and $H_0 = 800$ Oe [(c), (d)], respectively. The other parameters of the OEMW are the same as those in Fig. 7. In panel (a), the frequency is $f = 5.2$ GHz, lying in the working frequency range. As can be observed, the EM field is confined in the channel and a good one-way propagating behavior is manifested. When the frequency is decreased to $f = 4.7$ GHz under the same EMF, it lies outside the working frequency range, the EM field leaks outside the channel as demonstrated in Fig. 12 (b). Accordingly, the EM field propagating along the channel becomes weaker and weaker so that at this frequency it can not be operated as an OEMW. Nonetheless, by tuning EMF to a lower value $H_0 = 800$ Oe, $f = 4.7$ GHz is located in the new working frequency range as can be confirmed from the photonic band diagram in Fig. 11 (b). The corresponding electric field pattern is shown in Fig. 12 (c) where we can observe that the OEMW works very well. Similar to the case shown in Fig. 12 (b), if the frequency is tuned to $f = 5.2$ GHz, lying outside the working range under $H_0 = 800$ Oe. Then, the device can not be operated as an OEMW by examining the electric field pattern shown in Fig. 12 (d) where we can see that most of the EM field leaks outside the channel. From the above analysis, we can conclude that by tuning the

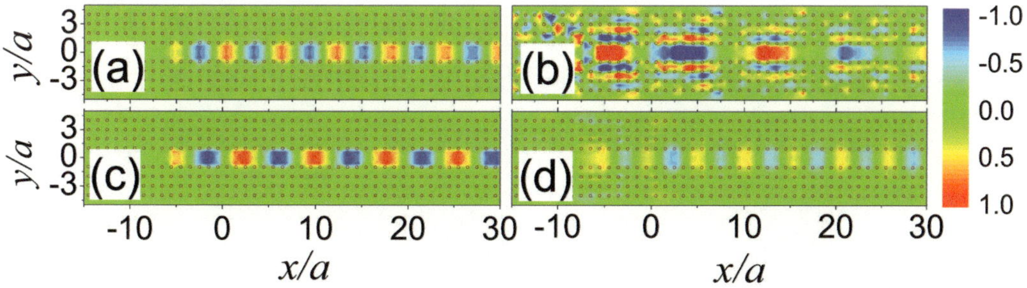

Fig. 12. The electric field patterns showing the OEMW working at frequency $f = 5.2$ GHz (a), (d) and $f = 4.7$ GHz (b), (c). The exerted EMFs are $H_0 = 900$ Oe and $H_0 = 800$ Oe for (a)-(b) and (c)-(d), respectively. All the other parameters are the same as those in Fig. 7.

EMF from $H_0 = 900$ Oe to $H_0 = 800$ Oe, the working frequency can be adjusted from 5.2 GHz to 4.7 GHz. Accordingly, by tuning the EMF we can manipulate the working frequency of the OEMW. However, for the OEMW designed based solely on the TRS breaking, the working frequency is inert to the EMF so that no manipulability can be expected.

6. Design of beam bender and splitter

Based upon the SAR effect, we can also design other EM waveguiding devices by constructing a corner configuration. Typical simulation results of such applications are presented in Fig. 13 where the configuration of the system is similar to that of the OEMW except that a cladding slab is added on top of the system. For a line source located in the vertical channel at $(0, -15a)$, the EM wave is seen to propagate upward first, operated in the same manner as an OEMW. After that, the EM waves make a 90° turn at the corner without any backward scattering so that nearly 100% power transmission is realized for the beam bender, as shown quantitatively in Fig. 13 (e) by the blue solid line. The power transmission rates are defined as T_x/T_y with T_x and T_y the EM energy propagating along the two different parts of the channel in the x (rightward) and the y (upward) directions. They are calculated numerically according to $T_x = \int_{-3a}^{3a} P_x \mathrm{d}y$ and $T_y = \int_{-3a}^{3a} P_y \mathrm{d}x$. By reversing the magnetization of the ferrite rods with the coordinates $x < 0$ in the upper cladding slab, the EM wave can be divided equally into two branches at the bifurcation point $x = 0$ as shown in Fig. 13 (b). In each wing the EM wave is transported with 50% power transmission in some frequency range, as is shown in Fig. 13 (e) by the red dashed line. In the frequency range $f > 5.04$ GHz the power transmission rate for the beam bender decreases obviously as can be observed in Fig. 13 (e). To explain this, we have presented the electric field patterns for the beam bender and splitter operated at the frequency $f = 5.1$ GHz. The results are shown in Figs. 13 (c) and (d), respectively. For the beam bender, due to the difference of the light path at the upper and lower interfaces the wave front is severely distorted. In addition, as the frequency deviates from the MSP resonance, the energy circulation around the ferrite rods becomes weaker. Therefore, some part of the EM wave leaks leftwards as shown in Fig. 13 (c), resulting in the drop of the transmission rate. Since the orientation of the magnetization can be controlled by an EMF, the function of the system can be switched between bender and splitter. This makes the device more flexible and favorable in practical applications.

The unidirectional waveguiding device designed in our work can still be operable with a good performance in a deep subwavelength scale. In Fig. 14, we present the simulation results. The

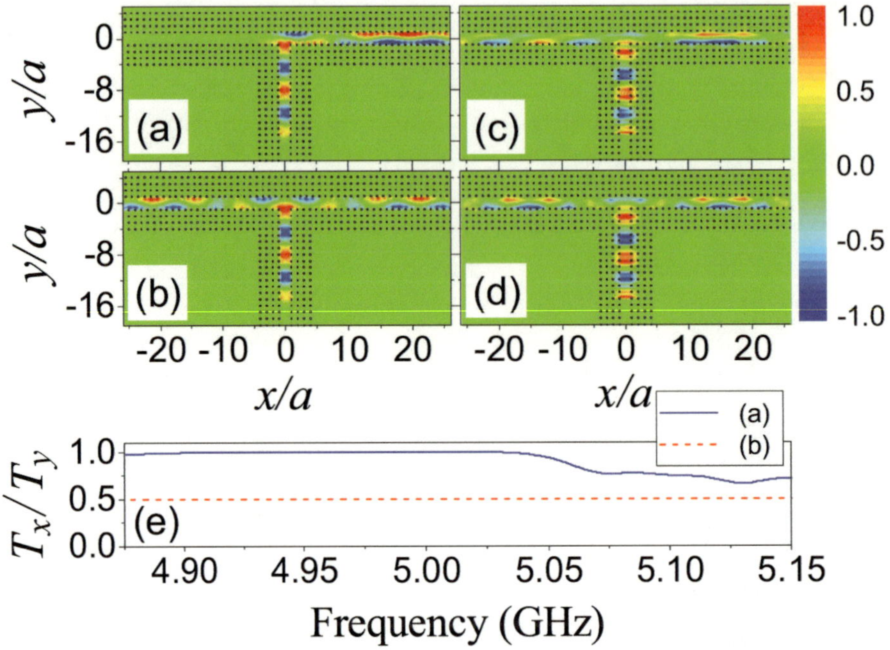

Fig. 13. The simulated electric field patterns corresponding to a 90° shaper beam bender (a) and a beam splitter (b) operating at $f = 5$ GHz under the EMF $H_0 = 900$ Oe. The electric field patterns for the beam bender and splitter working at $f = 5.1$ GHz are also presented in panels (c) and (d), respectively. The lattice constant is $a = 8$ mm, kept unchanged and the line source is located at $(0, -15a)$. The curves in (e) give the transmissivity T_x/T_y plotted as the functions of frequency.

waveguiding device considered in panels (a), (b), and (c) are designed in the same manner as those in Figs. 7 (a), 13 (a), and 13 (b), respectively, except that the MM is scaled down in size, with $a = 2$ mm and $r = \frac{1}{4}a$, while keeping the working frequency $f = 5$ GHz unchanged. In such situation, the working wavelength $\lambda = 60$ mm is nearly 30 times the lattice constant a. It can be observed from the field patterns that a superior subwavelength confining and steering is realized in a straight OEMW, a sharp beam bender, and a beam splitter. The numerical calculations indicate that the full lateral width at half maximum field intensity $w_h < 0.1\lambda$. The power transmissivity is also simulated for the beam bender and splitter as shown in Fig. 14 (d) where the red dashed line and the blue solid line correspond to panels (b) and (c), respectively. It can be seen that the waveguiding devices still exhibit a high transmission efficiency as well as a finite band width. In addition, compared with the results shown in Fig. 13 (e), it can be seen that a shift of the working frequency is demonstrated, which originates from the enhancement of the coupling strength due to the decrease of the lattice separation. Comparing with the results shown in Fig. 13, we can also find that the transmission efficiency for the beam bender is much improved. The reason lies in that the decrease of the channel width will ease up the distortion of the EM field, which is clearly demonstrated in Fig. 14 (b). Accordingly, the power leakage of the EM wave to the left wing of the horizonal channel can be decreased. Besides, just like the case shown in Fig. 13 the function of the system is still magnetically switchable between beam bender and splitter.

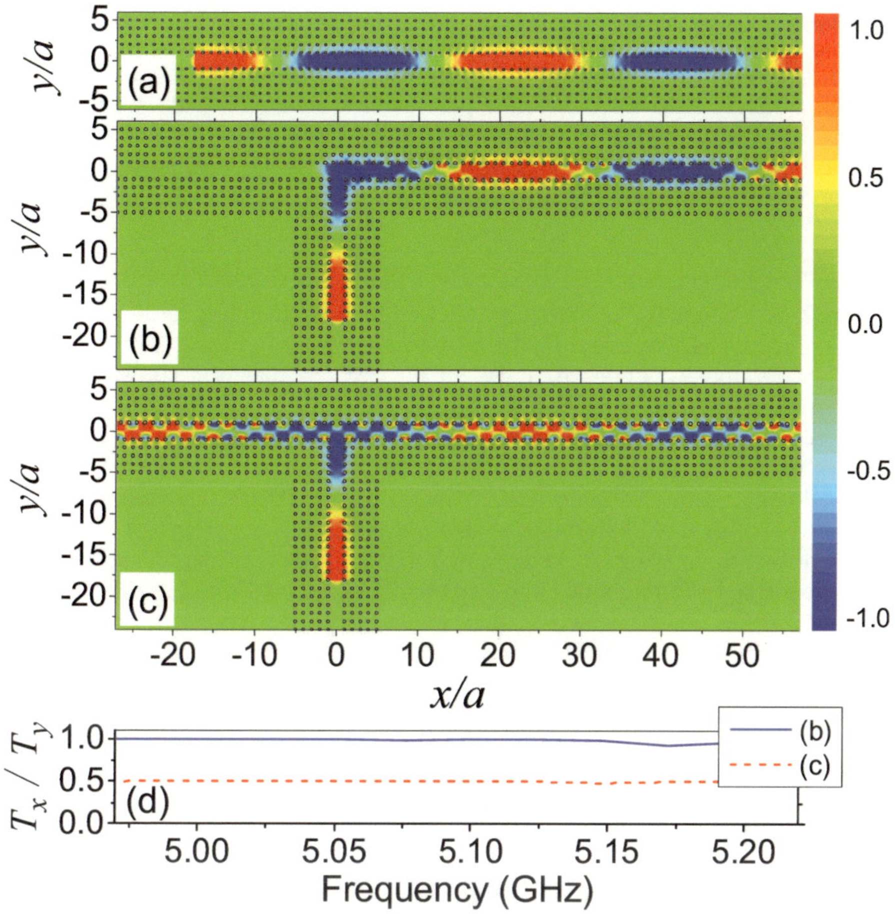

Fig. 14. The electric field patterns showing the device operable at the deep subwavelength scale. Panels (a), (b), and (c) are the same as Fig. 7(a), Fig. 13(a), and Fig. 13(b), respectively, except that the lattice constant of the MM is $a = 2$ mm and rod radius $r = \frac{1}{4}a$. The curves in (d) gives the rightward transmissivity for the beam bender and beam splitter.

7. Conclusion

In summary, we have demonstrated a very interesting SAR effect occurring at the interface of an MM slab. The Mie theory and the multiple scattering theory used in our simulation have also been introduced. Our results suggest that the SAR effect originates from combined action of MSP resonance and TRS breaking nature of MM under an EMF. We also examine the dependence of SAR effect on the frequency and the source-interface separation. Possible applications have been demonstrated by designing a straight OEMW, a sharp beam bender, and a beam splitter. An excellent performance of the device can be achieved. In particular, our design can even be operable in the deep subwavelength scale. Actually, a lot more issues can be expected with appropriate design of the MM such as the muti-channel unidirectional filter, cloaking, and also the zero index materials.

8. Acknowledgements

This work is supported by the China 973 program, NNSFC (10774028, 10904020), MOE of China (B06011), SSTC (08dj1400302), China postdoctoral science foundation (200902211), the open project of SKLSP in Fudan Univeristy (KL2011_8), Zhejiang Provincial Natural Science Foundation of China (Y12A040009) and Zhejing Normal University initiative foundation. STC is partly supported by the DOE.

9. References

Ao, X. Y.; Lin, Z. F. & Chan, C. T. (2009). One-way edge mode in a magneto-optical honeycomb photonic crystal. *Phys. Rev. B*, Vol. 80, No. 3, 033105

Barnes, W. L.; Dereux, A.; Ebbesen, T. W. (2003). Surface plasmon subwavelength optics. *Nature*, Vol. 424, No. 6950, 824-830

Bohren, C. F. & Huffman, D. R. (1983). *Absorption and Scattering of Light by Small Particles*, Wiley, New York.

Burgos, S. P.; Waele, R. de; Polman, A. & Atwater, H. A. (2010). A single-layer wide-angle negative-index metamaterial at visible frequencies. *Nat. Mater.* Vol. 9, No. 5, 407-412

Chen, P.; Wu, R. X.; Xu J.; Jiang, A. M. & Ji, X. Y. (2007). Effect of magnetic anisotropy on the stop band of ferromagnetic band gap materials. *J. Phys.: Condens. Matter*, Vol. 19, No. 10, 106205

Chew, W. C. (1995). *Waves and Fields in Inhomogenous Media*, IEEE Press, New York.

Chui, S. T. & Lin, Z. F. (2007). Probing states with macroscopic circulations in magnetic photonic crystals. *J. Phys.: Condens. Matter*, Vol. 19, No. 40, 406233

Chui, S. T.; Liu, S. Y. & Lin Z. F. (2010). Reflected wave of finite circulation from magnetic photonic crystals. *J. Phys.: Condens. Matter*, Vol. 22, No. 18, 182201

Dolling, G.; Enkrich, C.; Wegener, M.; Soukoulis, C. M. & Linden, S. (2006). Simultaneous negative phase and group velocity of light in a metamaterial. *Science*, Vol. 312, No. 5775, 892-894

Eggimann, W. H. (1960). Scattering of a plane wave on a ferrite cylinder at normal incidence. *IRE Trans. Microwave Theory and Tech.*, Vol. 8, No. 4, 440-445

Engheta, N. (2007). Circuits with light at nanoscales: Optical nanocircuits inspired by metamaterials. *Science*, Vol. 317, No. 5845, 1698-1702

Ergin, T.; Stenger, N.; Brenner, P.; Pendry, J. B. & Wegener, M. (2010). Three-dimensional invisibility cloak at optical wavelengths. *Science*, Vol. 328, No. 5976, 337-339

Fang, N.; Lee, H.; Sun, C. & Zhang, X. (2005). Sub-diffraction-limited optical imaging with a silver superlens. *Science*, Vol. 308, No. 5721, 534-537

Felbacq, D.; Tayeb, G. & Maystre, D. (1994). Scattering by a random set of parallel cylinders. *J. Opt. Soc. Am. A*, Vol. 11, No. 9, 2526-2538

Fu, J. X.; Lian, J.; Liu, R. J.; Guan, L. & Li, Z. Y. (2011). Unidirectional channel-drop filter by one-way gyromagnetic phtonic crystal waveguides. *Appl. Phys. Lett.*, Vol. 98, No. 21, 211104

Fu, J. X.; Liu, R. J.; & Li, Z. Y. (2011). Robust one-way modes in gyromagnetic photonic crystal waveguides with different interfaces. *Appl. Phys. Lett.*, Vol. 97, No. 4, 041112

Giannini,V.; Vecchi, G. & Rivas, J. G. (2010). Lighting up multipolar surface plasmon polaritons by collective resonances in arrays of nanoantennas. *Phys. Rev. Lett.*, Vol. 105, No.26, 266801

Gollub, J. R.; Smith, D. R.; Vier, D. C.; Perram, T.; Mock, J. J. (2005). Experimental characterization of magnetic surface plasmons on metamaterials with negative permeability. *Phys. Rev. B*, Vol. 71, No. 19, 195402

Haldane, F. D. M. & Raghu, S. (2008). Possible realization of directional optical waveguides in photonic crystals with broken time-reversal symmetry. *Phys. Rev. Lett.*, Vol. 100, No. 1, 013904

He, C.; Chen, X. L.; Lu, M. H.; Li, X. F.; Wan, W. W.; Qian, X. S.; Yin, R. C. & Chen, Y. F. (2010). Tunable one-way cross-waveguide splitter based on gyromagnetic phtonic crystal. *Appl. Phys. Lett.*, Vol. 96, No. 11, 111111

Huang, C. & Jiang, C. (2009). Nonreciprocal photonic crystal delay waveguide. *J. Opt. Soc. Am. B*, Vol. 26, No. 10, 1954-1958

Jin, J. J.; Liu, S. Y.; Lin, Z. F. & Chui, S. T. (2009). Effective-medium theory for anisotropic magnetic metamaterials. *Phys. Rev. B*, Vol. 80, No. 11, 115101

Leonhardt, U. (2006). Optical conformal mapping. *Science*, Vol. 312, No. 5781, 1777-1780

Lezec, H. J.; Dionne, J. A. & Atwater, H. A. (2007). Negative refraction at visible frequencies. *Science*, Vol. 316, No. 5823, 430-432

Lai, Y.; Chen, H. Y.; Zhang, Z. Q. & Chan, C. T. (2009). Complementary media invisibility cloak that cloaks objects at a distance outside the cloaking shell. *Phys. Rev. Lett.*, Vol. 102, No. 9, 093901

Leung, K. M. & Qiu, Y (1993). Multiple-scattering calculation of the two-dimensional phtonic band structure. *Phys. Rev. B*, Vol. 48, No. 11, 7767-7771

Li, J. & Pendry, J. B. (2008). Hiding under the carpet: A new strategy for cloaking. *Phys, Rev. Lett.*, Vol. 101, No. 20, 203901

Liu S. Y. & Lin, Z. F. (2006). Opening up complete photonic bandgaps in three-dimensional photonic crystals consisting of biaxial dielectric spheres. *Phys. Rev. E*, Vol. 73, No. 6, 066609

Liu, Z. W.; Lee, H.; Xiong, Y.; Sun, C. & Zhang, X. (2007). Far-field optical hyperlens magnifying sub-diffraction-limited objects. *Science*, Vol. 315, No. 5819, 1686

Liu, S. Y.; Chen, W. K.; Du, J. J.; Lin, Z. F.; Chui, S. T. & Chan, C. T. (2008). Manipulating negative-refractive behavior with a magnetic field. *Phys. Rev. Lett.*, Vol. 101, No. 15, 157407

Liu, S. Y.; Du, J. J.; Lin, Z. F.; Wu, R. X. & Chui S. T. (2008). Formation of robust and completely tunable resonant photonic band gaps. *Phys. Rev. B*, Vol. 78, No. 15, 155101

Liu, R.; Ji, P. C.; Mock, J.; Chin, J. J. Y.; Cui, T. J. & Smith, D. R. (2009). Broadband ground-plane cloak. *Science*, Vol. 323, No. 5912, 366-369

Liu, S. Y.; Lu, W. L.; Lin, Z. F. & Chui, S. T. (2010). Magnetically controllable unidirectional electromagnetic waveguiding devices designed with metamaterials. *Appl. Phys. Lett.*, Vol. 97, No. 20, 201113

Liu, S.Y.; Lu, W. L.; Lin, Z. F. & Chui, S. T. (2011). Molding reflection from metamaterials based on magnetic surface plasmons. *Phys. Rev. B*, Vol. 84, No.4, 045425

Noginov, M. A.; Zhu, G.; Mayy, M.; Ritzo, B. A.; Noginova, N. & Podolskiy, V. A. (2008). Stimulated emission of surface plasmon polaritons. *Phys. Rev. Lett.*, Vol. 101, No. 22, 226806

Ozbay, E.(2006). Plasmonics: Merging photonics and electronics at nanoscale dimensions. *Science*, Vol. 311, No. 5758, 189-193

Pendry, J. B. (2000). Negative refraction makes a perfect lens. *Phys. Rev. Lett.*, Vol. 85, No. 18, 3966-3969

Pendry, J. B.; Holden, A. J.; Stewart, W. J. & Youngs, I. (1996). Extremely low frequency plasmons in metallic mesostructures. *Phys. Rev. Lett.*, Vol. 76, No. 25, 4773-4776

Pendry, J. B.; Holden, A. J.; Holden, D. J. & Stewart, W. J. (1999). Magnetism from conductors and enhanced nonlinear phenomena. *IEEE Tran. Microwave Theory Tech.*, Vol. 47, No. 11, 2075-2084

Pendry, J. B.; Schurig, D. & Smith, D. R. (2006). Controlling electromagnetic fields. *Science*, Vol. 312, No. 5781, 1780-1782

Pozar, D. M. (2004). *Microwave Engineering*, Wiley, New York.

Poo, Y.; Wu, R. X.; Lin, Z. F.; Yang, Y. & Chan, C. T. (2011). Experimental realization of self-guiding unidirectional electromagnetic edge states. *Phys. Rev. Lett.*, Vol. 106, No. 9, 093903

Schurig, D.; Mock, J. J.; Justice, B. J.; Cummer, S. A.; Pendry, J. B.; Starr, A. F. & Smith, D. R. (2006). Metamaterial electromagnetic cloak at microwave frequencies. *Science*, Vol. 314, No. 5801, 977-980

Shalaev, V. M. (2007). Optical negative-index metamaterials. *Nature Photon.*, Vol. 1, No. 1, 41-48

Shelby, R. A.; Smith, D. R. & Schultz S. (2001). Experimental verification of a negative index of reflection. *Science*, Vol. 292, No. 5514, 77-79

Slichter, C. P. (1978) *Principle of Magnetic Resonance*, Springer, Berlin.

Smith, D. R.; Padilla, W. J.; Vier, D. C.; Nemat-Nasser, S. C. & Schultz, S. (2000). Composite medium with simultaneously negative permeability and permittivity. *Phys. Rev. Lett.*, Vol. 84, No. 18, 4184-4187

Taubner, T.; Korobkin, D.; Urzhumov, Y.; Shvets, G. & Hillenbrand, R. (2006). Near-field microscopy through a SiC superlens. *Science*, Vol. 313, No. 5793, 1595

Valentine, J.; Zhang, S.; Zentgraf, T.; Ulin-Avila, E.; Genov, D. A.; Bartal, G. & Zhang, X. (2008). Three-dimensional optical metamaterial with a negative refractive index. *Nature*, Vol. 455, No. 7211, 376-379

Veselago, V. G. (1968). The electrodynamics of substrates whith simultaneously negative values of ε and μ. *Sov. Phys. Usp.*, Vol. 10, No. 4, 509-514

Wang, X. D.; Zhang, X. G.; Yu, Q. L. & Harmon, B. N. (1993). Multiple-scattering theory for electromagnetic waves. *Phys. Rev. B*, Vol. 47, No. 8, 4161-4167

Wang, Z.; Chong, Y. D.; Joannopoulos, J. D. & Soljačić, M. (2008). Reflection-free one-way edge modes in a gyromagnetic photonic crystal. *Phys. Rev. Lett.*, Vol. 100, No. 1, 013905

Wang, Z.; Chong, Y. D.; Joannopoulos, J. D. & Soljačić, M. (2009). Observation of unidirectional backscattering-immune topological electromagnetic states. *Nature*, Vol. 461, No. 7265, 772-775

Yu, Z. F.; Veronis, G.; Wang, Z. & Fan, S. H. (2008). One-Way electromagnetic waveguide formed at the interface between a plasmonic metal under a static magnetic field and a photonic crystal. *Phys. Rev. Lett.*, Vol. 100, No. 2, 023902

Zayats, A. V.; Smolyaninov, I. I. & Maradudin, A. A. (2005). Nano-optics of surface plasmon polaritons. *Phys. Rep.*, Vol. 408, No. 3-4, 131-314

Zentgraf, T.; Liu, Y. M.; Mikkelsen, M. H.; Valentine, J. & Zhang, X. (2011). Plasmonic Luneburg and Eaton lenses. *Nat. Nanotechnology*, Vol. 6, No. 3, 151-155

Zhang, S.; Fan, W.; Panoiu, N. C.; Malloy, K. J.; Osgood, R. M. & Brueck, S. R. J. (2005). Experimental demonstration of near-infrared negative-index metamaterials. *Phys. Rev. Lett.*, Vol. 95, No. 13, 137404

Zhu, H. B. & Jiang, C. (2010). Broadband unidirectional electromagnetic mode at interface of anti-parallel magnetized media. *Opt. Express*, Vol. 18, No. 7, 6914-6921

Zia, R.; Schuller, J. A.; Chandran, A. & Brongersma, M. L. (2006). Plasmonics: the next chip-scale technology. *Materials today*, Vol. 9, No. 7-8, 20-27

22

Applications of Artificial Magnetic Conductors in Monopole and Dipole Antennas

Amir Jafargholi[1], Mahmood Rafaei Booket[2] and Mehdi Veysi[1]
[1]*K.N. Toosi University of Technology*
[2]*Tarbiat Modares University*
Iran

1. Introduction

In recent years, introducing metamaterials based on theory established by Veselago, opened the way for many researcher groups to enhance the antenna performances (Veselago, 1968; Veysi et al., 2010; Rafaei et al., 2010; Veysi et al., 2011; Rafaei et al., 2011; Jafargholi et al., 2010; Jafargholi et al., 2011). A standard procedure has been also established for the design of bulk artificial media with negative macroscopic permeability and permittivity. In the past few years, extensive research has been carried out on the metamaterial realization of the artificial magnetic conductors (AMCs) (Jafargholi et al., 2010; McVay et al., 2004). As revealed in (Erentok et al., 2005), the artificial magnetic conductor can be also realized using a volumetric metamaterial constructed from a periodic arrangement of the capacitively loaded loop (CLL) elements. Recently, it has found interesting applications in antenna engineering (Ziolkowski et al., 2003). Due to unique electromagnetic properties, metamaterials have been widely considered in monopole and dipole antennas to improve their performance (Ziolkowski et al., 2003; Liu et al., 2009; Rogers et al., 2003).

This chapter is mainly focused on two different applications of artificial magnetic conductors in the field of dipole pattern modification and dual and multiband dipole antennas. At first, the radiation patterns of the monopole and dipole antennas have been considered, especially at the second harmonic of the main resonant frequency. A closer examination on the PMC structures reveals that they can be used to modify the monopole and dipole radiation patterns. To this aim, the CLLs have been used to realize perfect magnetic conductor.

Finally, this chapter examines reactive loading technique to achieve dual band operation and further size reduction for double-sided printed dipole antennas. Here, the reactive loads are realized by two balanced CLLs placed close to the edge of the printed dipole antenna.

2. Pattern modification of monopole antenna

The linearly polarized and omnidirectional radiation pattern of a vertical monopole antenna has led to a wide range of applications in wireless communications such as wireless local area network (WLAN) and radio broadcast. It is known that the long monopole antennas ($\geq\lambda/2$) suffer from the 180 phase reversal. The fields radiated by reverse current of the long monopole antenna do not reinforce those radiated by original current, resulting in very poor radiation efficiency (Balanis, 1989). As a result, the antenna radiation pattern does not remain omnidirectional within the interested frequency range. In the conventional monopole antenna, the resonance frequencies ω_m correspond to the frequencies where the physical length L of the monopole is an odd multiple of quarter-wavelength. In other words, the antenna resonance frequencies are harmonics of the design frequency ω_1. However, omnidirectional radiation pattern distortion and low directivity are two major disadvantages associated with monopole resonating at higher order harmonics ($\omega_m > \omega_1$). In other words, conventional monopole antenna only radiates an omnidirectional radiation pattern at the design frequency ω_1. In order to have omnidirectional radiation pattern within the antenna bandwidth (ranging from f_L to f_U), the monopole length has to be less than $\lambda_U/2$, where λ_U is the free space wavelength at f_U. However, the antenna directivity decreases because of the significant reduction in the monopole length.

In this section, the use of AMCs to load a monopole antenna has been investigated. It is known that the current reversal, that occurs at frequencies much beyond the antenna natural frequency, disturbs the omnidirectional radiation pattern of the monopole antenna. The current distribution of the monopole antenna can be improved to a large extent by using PMC loading. To this aim, the CLLs are used to realize perfect magnetic conductor behavior.

2.1 PMC loaded monopole antenna

Due to the reverse current effects, the monopole radiation pattern does not remain omnidirectional at the second harmonic of the main resonant frequency (Balanis, 1989). In this section, a monopole antenna loaded with the PMC layer is proposed to increase omnidirectional radiation bandwidth. To make the concept more clear, three ideal models are simulated, all of which are partly covered by a very thin PMC shell, as shown in Fig. 1. As a reference, a conventional monopole antenna is also simulated for comparison. It has the same dimensions as the geometries in Fig. 1, except that the PMC cover is removed.

Fig. 2, shows the simulated reflection coefficient of the monopole antennas with and without the PMC cover. The resonant frequencies for case I and II are 18.5GHz and 24GHz, respectively, whereas the resonant frequency for case III remains the same as the conventional monopole antenna. For the conventional monopole antenna, distortion of the omnidirectional radiation pattern occurs at frequencies higher than 20GHz. This upper limit is indicated by dashed line in Fig. 2, and considered as an antenna length limitation.

Fig. 3, shows the radiation patterns of the monopole antennas with and without PMC cover when frequency varies from 12GHz to 30GHz in 1GHz increments. As revealed in the figure, when the monopole antenna is loaded with the PMC cover (case III), the antenna radiation pattern considerably improves as compared to that of the conventional monopole antenna, especially at the second harmonic of the main resonant frequency (29GHz).

Fig. 4, compares the simulated directivity of the monopole antennas with and without the PMC cover in the azimuth plane. As is evident from Fig. 4, the directivity curve for the case II is approximately flat while the antenna directivity for the case III significantly improves as compared to that of the conventional monopole antenna. For our discussion on the pattern modification, the results shown in Figs. 2 to 4, need to be considered simultaneously.

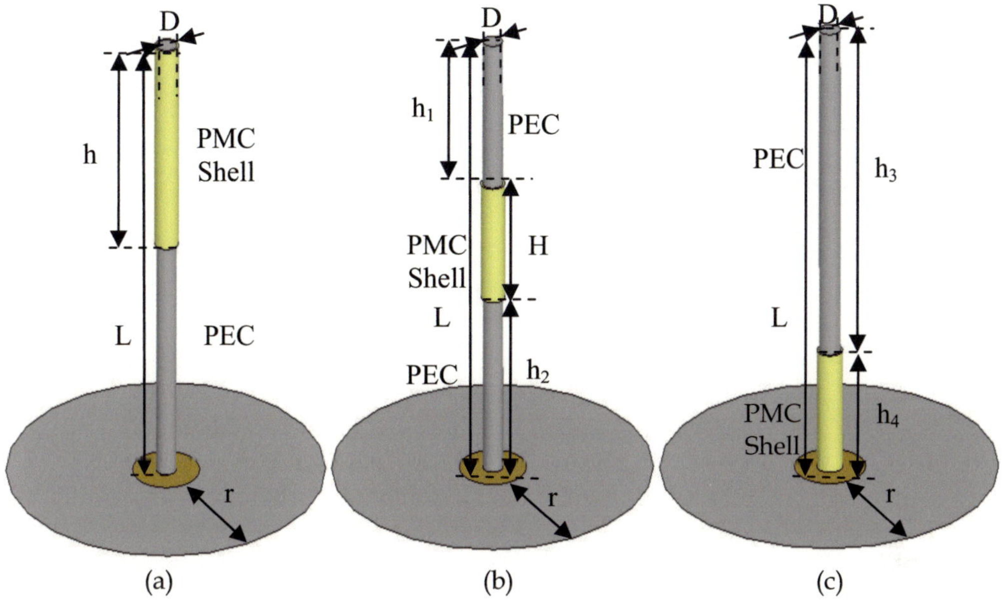

Fig. 1. Three ideal models of PMC loaded monopole antennas, (a) case I, (b) case II, and (c) case III: L=7.5mm, D=0.5mm and h=3.5mm, h_1=2.5mm, h_2=3mm, H=2mm, h_3=5.5mm, h_4=2mm, and r =4mm. From (Jafargholi et al., 2010), copyright © 2010 by the Institute of Electrical and Electronics Engineers (IEEE).

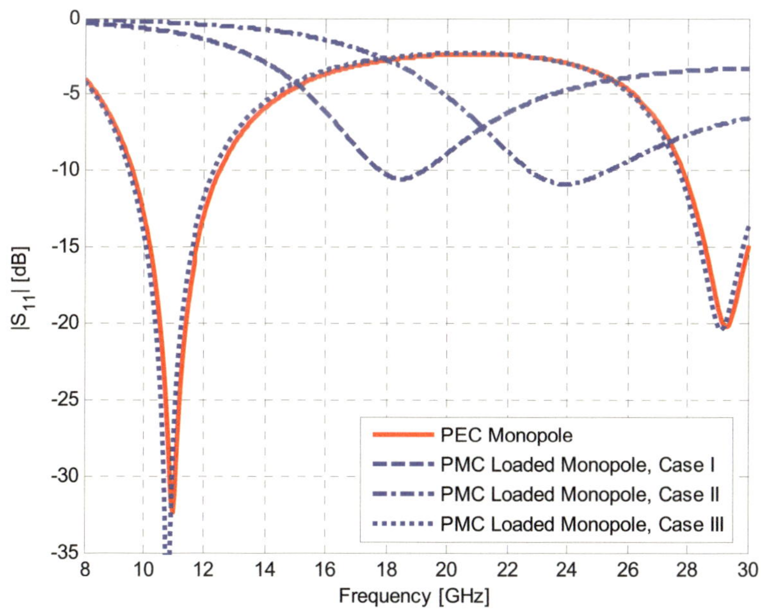

Fig. 2. Reflection coefficient of monopole antennas with and without PMC cover. From (Jafargholi et al., 2010), copyright © 2010 by the Institute of Electrical and Electronics Engineers (IEEE).

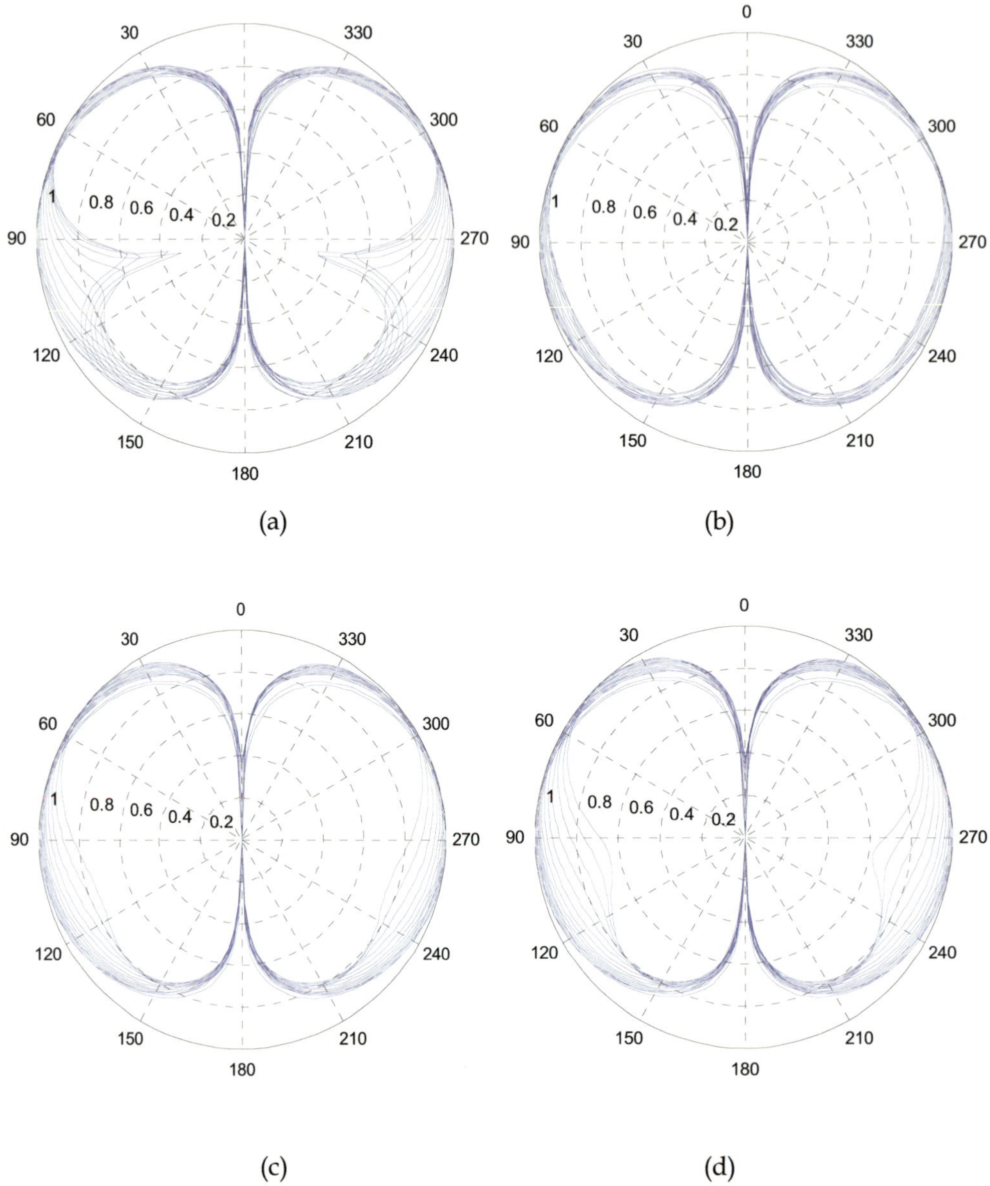

Fig. 3. Normalized Radiation Patterns v.s. frequency (a) conventional monopole, (b) PMC loaded monopole (case I), (c) PMC loaded monopole (case II) , and (d) PMC loaded monopole (case III). From (Jafargholi et al., 2010), copyright © 2010 by the Institute of Electrical and Electronics Engineers (IEEE).

Consequently, the ideal model shown in Fig. 1 (c) (case III) is considered for the practical realization. It should be pointed out that we can always use a shorter monopole antenna to improve the omnidirectional radiation pattern. However, the prices we pay are the higher resonant frequency and lower directivity due to the significant reduction in the monopole length.

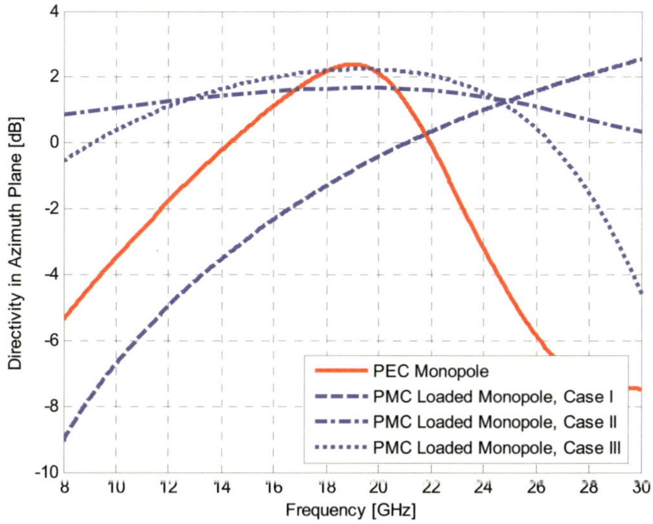

Fig. 4. Simulated directivity (maximum) of monopole antennas with and without PMC cover (on a finite ground plane) in azimuth plane. From (Jafargholi et al., 2010), copyright © 2010 by the Institute of Electrical and Electronics Engineers (IEEE).

2.2 CLL loaded monopole

In the previous section, it was revealed that the suppression of phase reversal by incorporating PMC cover has led to the improved radiation pattern, especially at the second harmonic of the main resonant frequency. In this section, the capacitively loaded loops (CLLs) are used to realize an artificial PMC (Erentok et al., 2005). Fig. 5 shows a CLL loaded monopole antenna together with the basic unit cell of the CLL structure. The finite two-CLL-deep metamaterial AMC cover was designed separately (without the ground plane) to operate at around 26GHz. The CLL dimensions were then optimized to obtain good radiation patterns. The total number of the CLL elements is 44 and the separation between the CLL elements is 0.5mm.

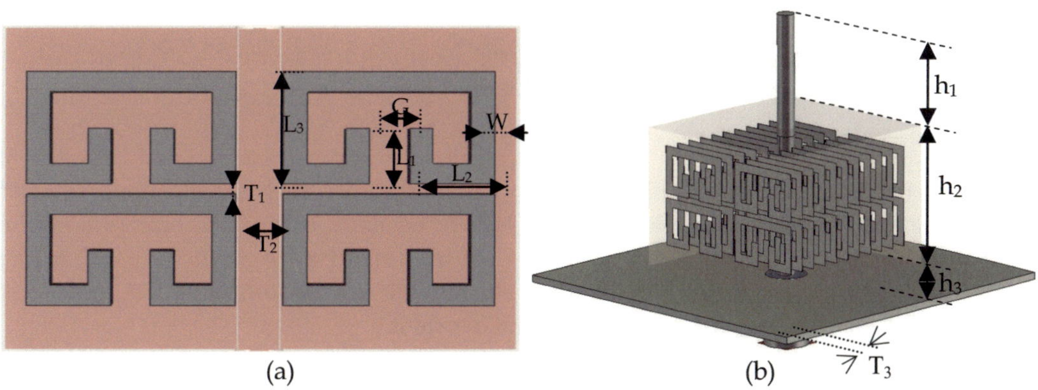

Fig. 5. A schematic view of (a) CLL unit cell: L_1=0.67mm, L_2=0.89mm, L_3=1.37mm, W= 0.247mm, G=0.411mm, T_1= 0.13mm, T_2= 0.5mm, and (b) CLL loaded monopole antenna on a finite ground plane: h_1= 3.25mm, h_2 =4mm, h_3=0.25mm, T_3=0.45mm. From (Jafargholi et al., 2010)], copyright © 2010 by the Institute of Electrical and Electronics Engineers (IEEE).

Fig. 6 shows a comparison between the reflection coefficients of the CLL loaded and unloaded monopole antennas of 7.5mm length. Furthermore, Fig. 7 shows the surface current densities on the CLL loaded and unloaded monopole antennas. It can be seen that the current in the CLL loaded region is the superposition of two currents oriented in opposite directions. However, the surface current caused by CLL cells is dominant, and thus the current in all parts of the monopole has the same phase. To more understand the operation mechanism of the CLL loaded monopole antenna, we assume that the monopole antenna is surrounded by CLL cells, where the current direction of the monopole is reversed. Fig. 8, conceptually explains the distribution of the surface current density on the CLL loaded monopole antenna.

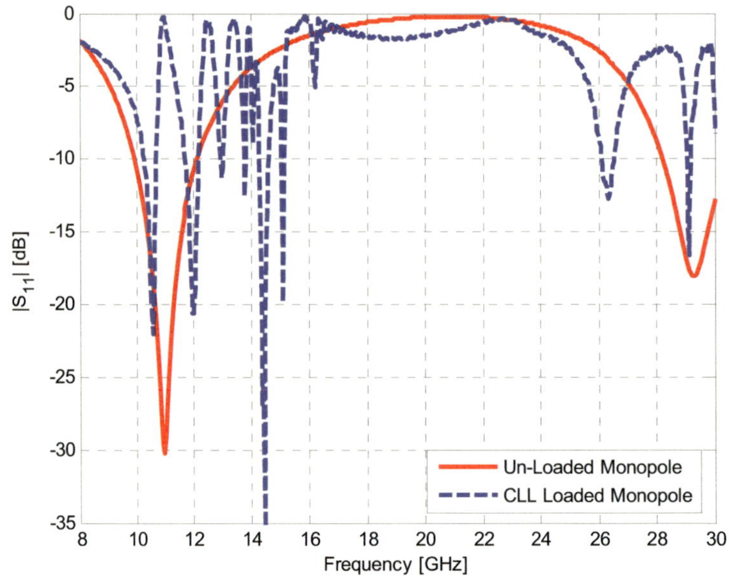

Fig. 6. Simulated reflection coefficient of the CLL loaded and unloaded monopole antennas. From (Jafargholi et al., 2010), copyright © 2010 by the Institute of Electrical and Electronics Engineers (IEEE).

A rigorous explanation must consider the complex interactions between the monopole and the CLL structure, such as the effects of the finite dimensions of the CLL structure on the current distribution. Consequently, full wave analysis methods have to be used in the antenna designs. However, to simplify the analysis, one can assume that the transverse dimensions of the CLL loaded region are infinite in extent.

Based on image theorem, when an electric current is vertical to a PMC (PEC) region, the current image has the reversed (same) direction. For the current at the bottom of the monopole flowing into the CLL loaded region, the CLL loaded region acts as a PMC cover (Erentok et al., 2005). Consequently, the direction of the image current is opposite to that of the original current, as shown in Fig. 8.

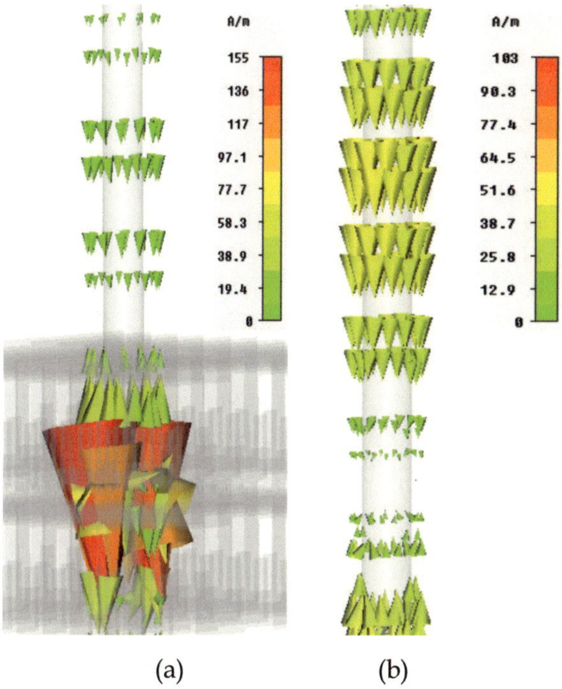

Fig. 7. Surface current densities on the (a) CLL loaded monopole and (b) unloaded monopole at f= 25GHz. From (Jafargholi et al., 2010), copyright © 2010 by the Institute of Electrical and Electronics Engineers (IEEE).

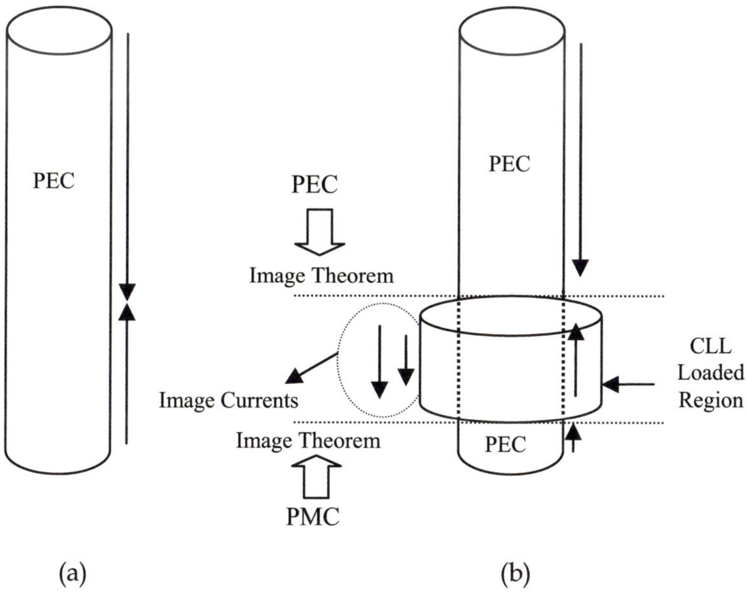

Fig. 8. Conceptual schematics of the (a) conventional and (b) PMC-loaded monopoles. From (Jafargholi et al., 2010)], copyright © 2010 by the Institute of Electrical and Electronics Engineers (IEEE).

In contrast, for the current at the top of monopole flowing into the CLL loaded region, the CLL loaded region acts as an artificial electric conductor (AEC) (Erentok et al., 2005), and thus the image current in the CLL loaded region has the same direction as the original current. The total surface current in the CLL loaded region is obtained as the sum of the two image currents and original current. Consequently, the current phase of the CLL loaded monopole antenna remains unchanged throughout the antenna, as shown in Fig. 8. The radiation patterns of the conventional and CLL loaded monopole antennas on a finite ground plane are shown in Figs. 9 and 10, respectively. As can be seen, when the monopole antenna is loaded with the CLL structure, the radiation patterns improve significantly, especially at the second harmonic (26GHz, 29GHz, and 30GHz) of the main resonant frequency where antenna is matched well.

Also, the antenna radiation efficiency is reasonably high over a wide frequency window, despite the material loss in copper and the CLL metamaterial features. Although, simulation results confirm the modification of the antenna radiation patterns at frequencies up to around 30GHz, the antenna reflection coefficient needs to be modified by impedance matching techniques, especially at frequencies far from the antenna resonant frequencies (Erentok et al., 2008). The simulated gains of the CLL loaded and unloaded monopole antennas in azimuth plane are compared in Fig. 11. As compared to the conventional monopole antenna, the gain of the CLL loaded monopole antenna significantly increases, especially at the higher frequencies. These comparisons demonstrate the unique capability of the AMCs to improve the antenna radiation bandwidth. It is worth noting that the antenna radiation patterns are not completely symmetric in the frequency band from 12GHz to 30GHz because of the asymmetric geometry.

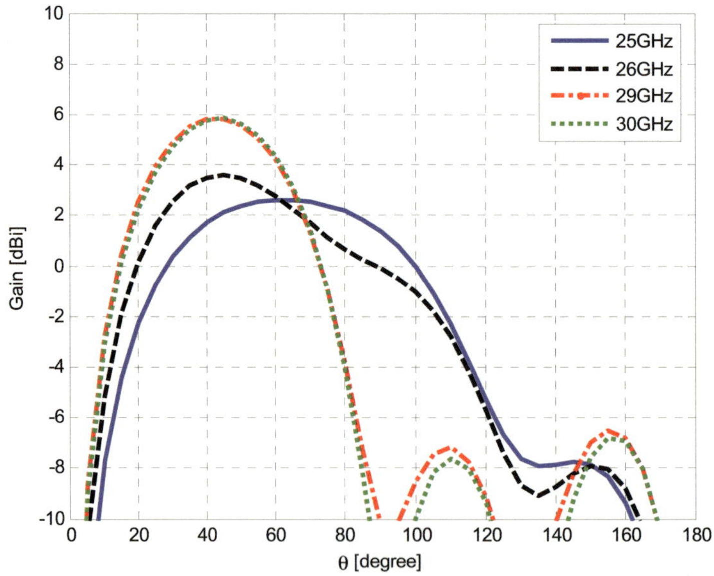

Fig. 9. Radiation patterns of the conventional monopole antenna versus frequency at φ=0°. From (Jafargholi et al., 2010), copyright © 2010 by the Institute of Electrical and Electronics Engineers (IEEE).

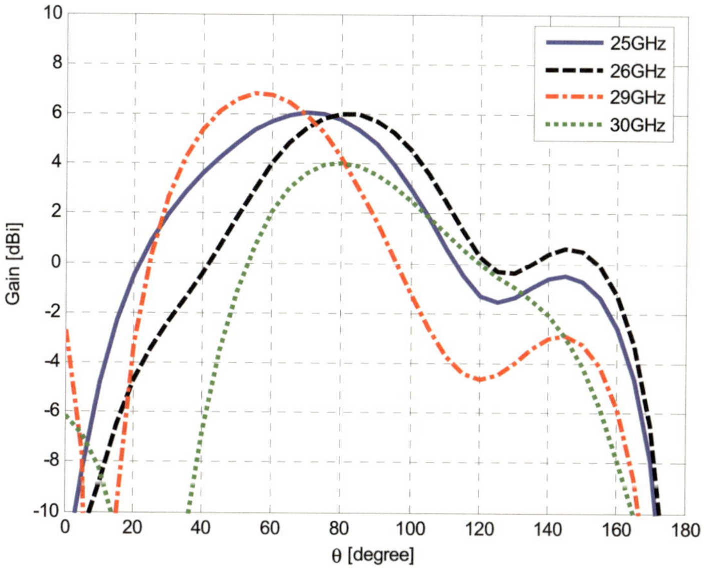

Fig. 10. Radiation patterns of the CLL loaded monopole antenna versus frequency at $\varphi=0^0$. From (Jafargholi et al., 2010), copyright © 2010 by the Institute of Electrical and Electronics Engineers (IEEE).

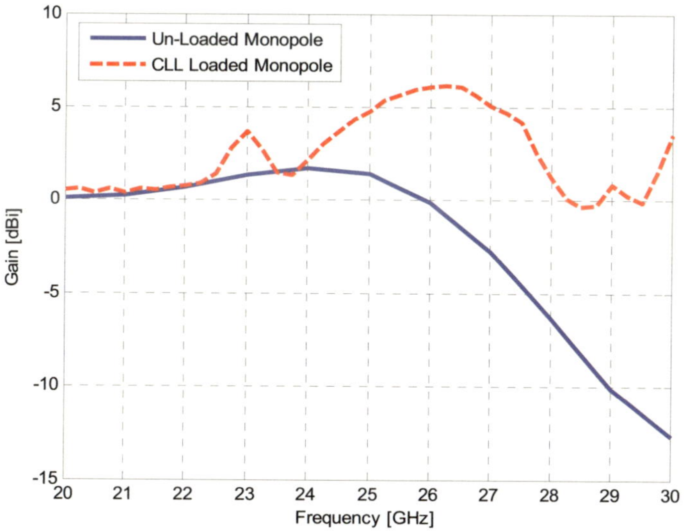

Fig. 11. Comparison of the maximum gains of the monopole antennas with and without CLL covers (on a finite ground plane) in the azimuth plane. From (Jafargholi et al., 2010), copyright © 2010 by the Institute of Electrical and Electronics Engineers (IEEE).

2.3 Pattern modification of dipole antenna

Dipole antennas are preferable in modern wireless communication systems, especially their printed types, due to their low profile, light weight and low fabrication cost as well as their compatibility with microwave and millimeter wave circuits. It was revealed in (Erentok et al., 2007; Erentok et al., 2008) that an efficient and electrically small magnetic based antenna can be realized by adding a planar interdigitated CLL element to a rectangular semi-loop antenna, which is coaxially-fed through a finite ground plane. The performance of a printed dipole antenna near a 3D-CLL block has been also examined in (Zhu et al., 2010). Recently, the use of TL-MTM to load antennas has been investigated in (Antoniades et al., 2009; Zhu et al., 2009). A miniaturized printed dipole loaded with left-handed transmission lines is also proposed in (Iizuka et al., 2006; Iizuka et al., 2007).

However, it is known that the conventional dipole antenna only radiates an omni-directional radiation pattern at the design frequency ω_1. As described in the previous section, the suppression of phase reversal by incorporating PMC cover has led to the improved radiation pattern (Jafargholi et al., 2010). This section is focused on the pattern modification of the wire dipole antenna using artificial magnetic conductors.

Fig. 12, shows a CLL loaded dipole antenna. The finite two-CLL-deep metamaterial Artificial Magnetic Conductor (AMC) cover was designed separately to operate at around 27GHz. The CLL dimensions were then optimized and placed optimally on both dipole arms to obtain good radiation patterns. The total number of the CLL elements is 88 (4×11×2) which symmetrically coupled to dipole antenna arms. The separation between the CLL elements is also fixed at 0.5mm.

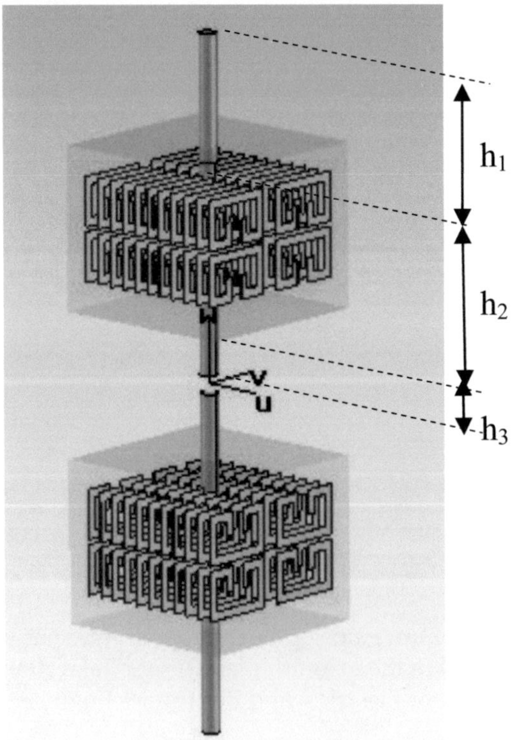

Fig. 12. A schematic view of CLL loaded dipole antenna, unit cell parameters (See Fig. 5a): L_1=0.67mm, L_2=0.89mm, L_3=1.37mm, W= 0.247mm, G=0.411mm, T_1= 0.13mm, T_2= 0.5mm, h_1= 3.25mm, h_2 =4mm, h_3 =2mm, and antenna feed gap = 0.5mm, and length of dipole antenna =19mm.

Fig. 13, shows a comparison between the reflection coefficient and input impedance of CLL loaded and unloaded wire dipole antennas of 19mm length. It was revealed in (Jafargholi et al., 2010) that ideally the dipole antenna input impedance does not change significantly by using AMC loading. However, here, the input impedance for the realized CLL loaded dipole changes due to the existence of the CLL resonance and the interaction between the CLL elements and the dipole antenna.

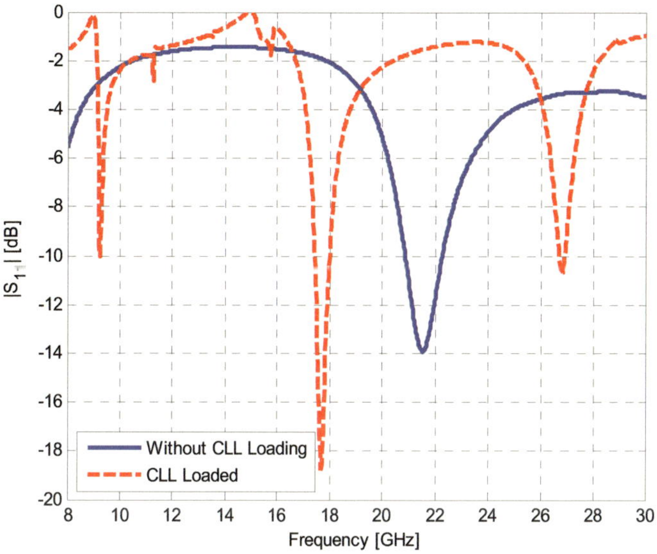

(a)

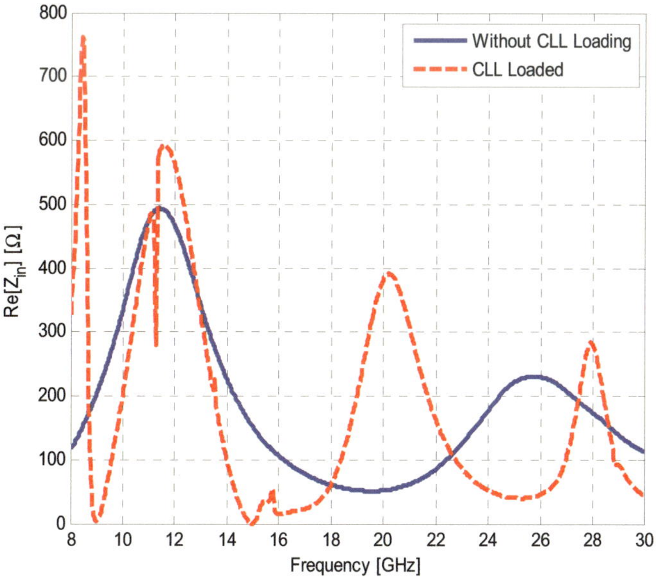

(b)

Fig. 13. (a) simulated reflection coefficient, and (b) Input Impedance of the CLL loaded and unloaded wire dipole antennas.

Fig. 14, shows the surface current densities on the CLL loaded and unloaded dipole antennas. As can be seen, the current in the CLL loaded region is the superposition of two currents oriented in opposite directions. However, the surface current caused by CLL cells is dominant, and thus the current in all parts of the dipole has the same phase. To clarify the operation mechanism of the CLL loaded dipole antenna. We assume that dipole antenna is surrounded by CLL cells, where the current direction of the dipole is reversed.

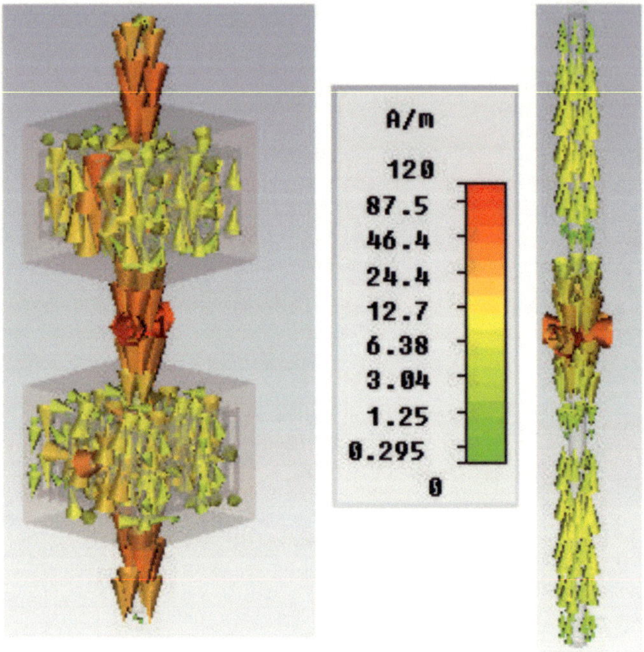

Fig. 14. Surface current densities on the CLL loaded dipole (left) and unloaded dipole (right) at f= 27GHz.

Fig. 15, conceptually explains the distribution of the surface current density on the CLL loaded dipole antenna. However, one must consider the complex interactions between the dipole arms and the CLL structures, such as the effects of the finite dimensions of the CLL structures on the current distribution. Consequently, full wave analysis methods have to be used in the antenna designs. However, to simplify the analysis, one can assume that the transverse dimensions of CLL loaded region are infinite in extent. Thus, as previous section, one can explain the concept based on the image theorem, i.e., when an electric current is vertical to a PMC (PEC) region, the current image has the reversed (same) direction.

For the current at the bottom of the dipole flowing into the CLL loaded region, the CLL loaded region acts as a PMC cover (Erentok et al., 2005). At the result, the direction of the image current is opposite to that of the original current, as shown in Fig. 15. In contrast, for the current at the top of dipole flowing into the CLL loaded region, the CLL loaded region acts as an AEC (Erentok et al., 2005), and thus the image current in the CLL loaded region has the same direction as the original current. The total surface current in the CLL loaded region is obtained as the sum of the two image currents and original current. At the result, the current phase of the CLL loaded dipole antenna remains unchanged through the antenna, as shown in Fig. 15.

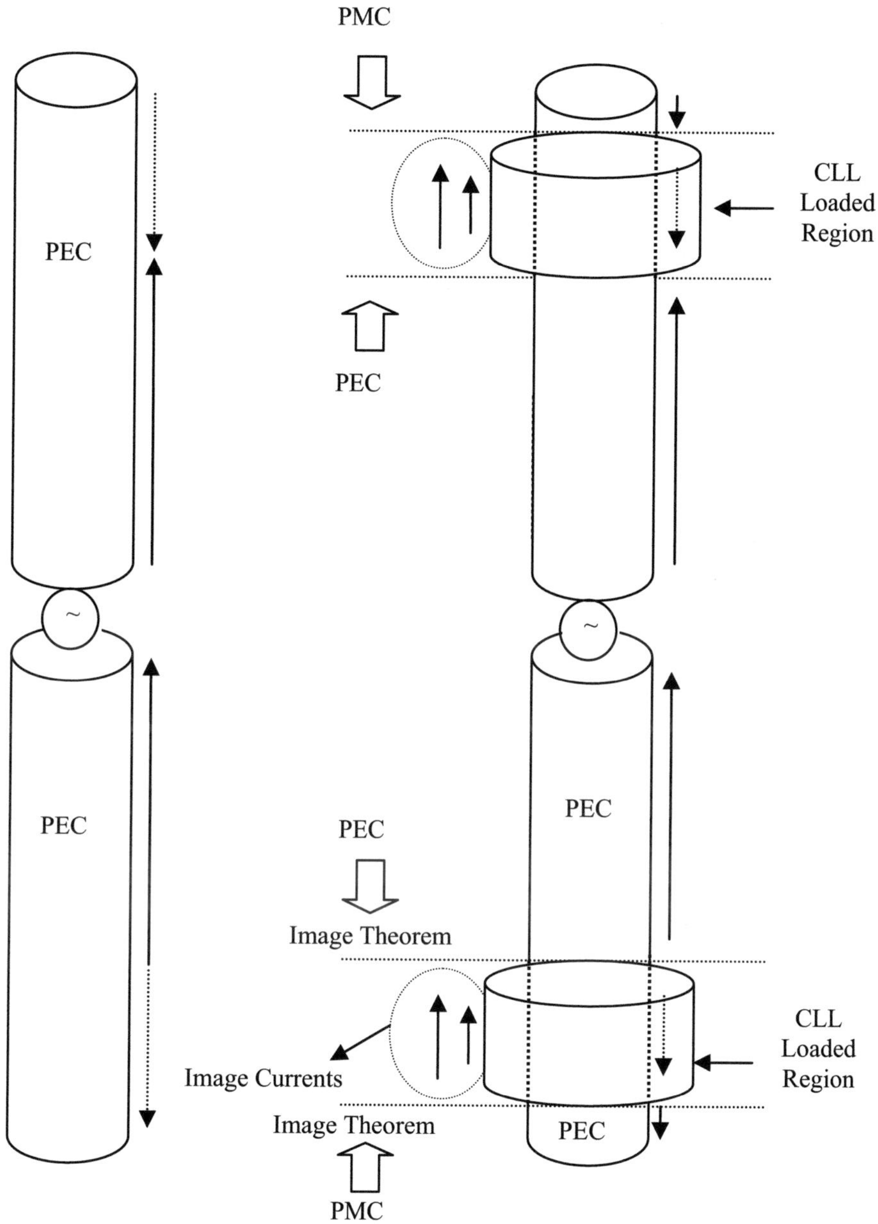

Fig. 15. Conceptual schematics of the conventional (left) and CLL loaded dipole (right).

The radiation patterns of the conventional and CLL loaded dipole antennas are also shown in Fig. 16, respectively. As can be seen, when the dipole antenna is loaded with CLL structure, the radiation patterns improve significantly, especially at the second harmonic (27GHz) of the main resonant frequency where antenna is matched well. The Simulation shows that above 20dB gain enhancement has been achieved in azimuth plane using CLL loading technique.

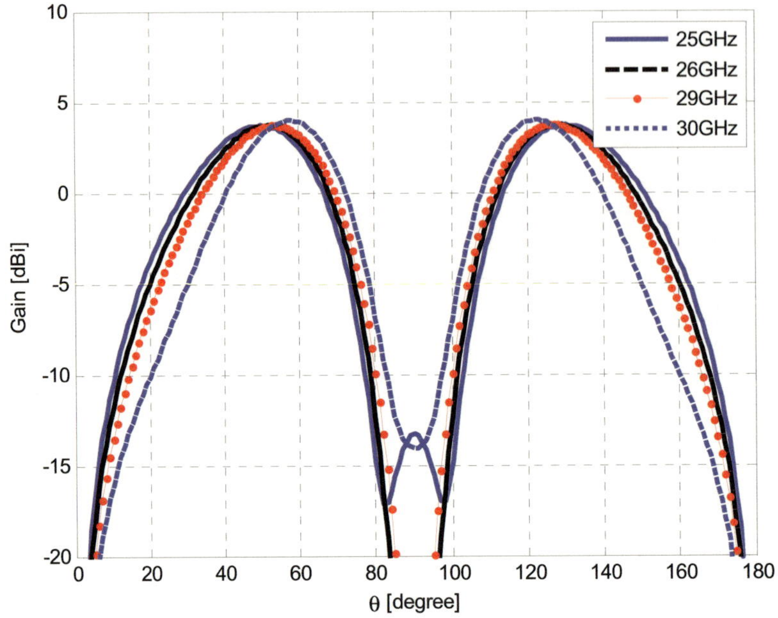

(a)

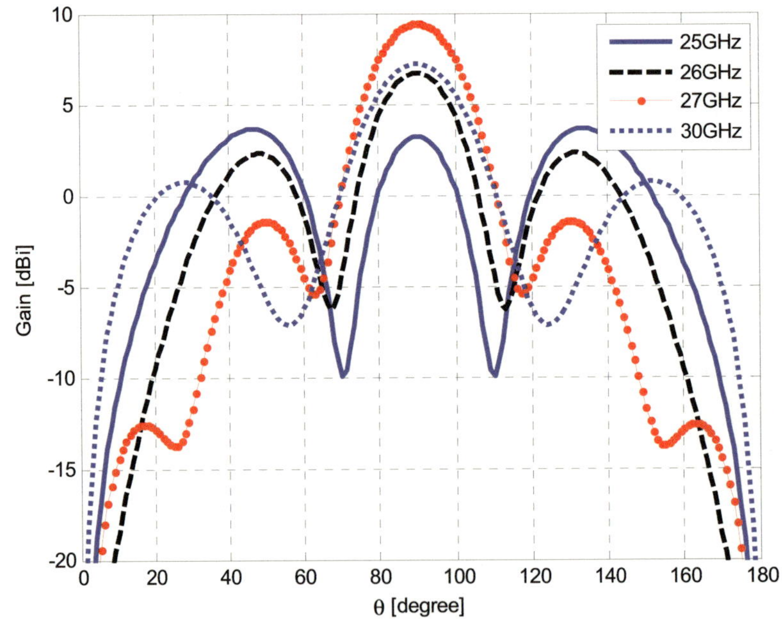

(b)

Fig. 16. Radiation patterns of (a) conventional wire dipole antenna, and (b) the CLL loaded dipole antenna versus frequency at $\varphi=0^0$.

2.4 Compact dual band loaded dipole antenna, incorporating artificial magnetic conductors

The interesting features of printed dipole antennas make it possible for them to be conformal with installation platforms such as automobiles and vessels, without affecting the aerodynamic or mechanical properties of the vehicles. In the recent years, due to unique electromagnetic properties, metamaterials have been widely considered in monopole and dipole antennas to improve their performance. However, the left-handed dipole antennas (Zhu et al., 2009; Iizuka et al., 2006; Iizuka et al., 2007), albeit compact, suffer from the losses in the loading components, namely, interdigitated capacitors and meander inductors, resulting in a very low gain.

In this section, we propose a new dual band printed dipole antenna in that CLL elements as reactive loads are placed close to the edge of the dipole. The losses associated with the CLL elements are very low in the frequency range of interest, resulting in both the acceptable gain and radiation pattern. When the CLL elements are incorporated, the resonant behavior of the unloaded printed dipole antenna changes. As a result, a new resonance is appeared with the frequency determined by the CLL dimensions. In fact, the CLL element can be easily described as an LC resonant circuit in which the resonant frequency is mainly determined by the loop inductance and the gap capacitor. It is worth noting that the resonant behavior of the CLL element starts to appear at a frequency in which the free space wavelength is much larger than its size. However, the second resonant frequency of the CLL-loaded dipole occurs at a frequency higher than the main resonant frequency of the unloaded dipole. To further reduce the second resonant frequency without increasing the area occupied by the antenna, chip capacitors are incorporated in the CLL elements. The chip capacitor provides a tuning capability of the second resonant frequency. Thus the frequency ratio between these two frequencies can be readily controlled by incorporating different chip capacitors into the capacitive gaps of the CLL elements. It is worthwhile to point out here that the CLL elements integrated with chip capacitors miniaturize the size of the printed dipole antenna.

The reason is that when the chip capacitor value is increased, the CLL resonant frequency decreases, and thus the second resonant frequency of the dipole shifts down to a lower frequency. The proposed dual band CLL loaded dipole antenna radiates effectively at both resonant frequencies with good return losses and gains as well as acceptable omnidirectional radiation patterns. The high-frequency structure simulator (Ansoft HFSS) is adopted for the simulations.

2.4.1 Dual band printed dipole antenna

In order to test the proposed approach, double-sided printed dipole antenna is loaded by CLL elements. Fig. 17, shows the CLL-loaded printed dipole antenna together with the CLL element. The dipole and CLL elements are printed on a FR4 substrate with a thickness of 0.8mm and a dielectric constant of 4.4 to reduce the cost of the antenna and to make it more rigid in construction. The CLL-loaded printed dipole has also been optimized to realize better performance. The optimized parameters of the proposed CLL-loaded dipole antenna are labeled in the Fig. 17. A prototype of the proposed CLL-loaded printed dipole is fabricated to validate the simulation results.

A photograph of the fabricated printed dipole antenna is shown in Fig. 18. Fig. 19, shows the reflection coefficient of the proposed dual band printed dipole antenna as well as the unloaded dipole antenna. As can be seen, the agreement between the simulation and measurement results is reasonably good. It is observed that when the CLL elements are added, two resonance frequencies become distinguishable from each other and thus two nulls are clearly observed in the reflection coefficient curve. The unloaded dipole antenna resonates at around 2.75GHz. In contrast, the fabricated CLL-loaded dipole resonates at 2.15 and 4.45GHz, as shown in Fig. 19. The lower resonant frequency corresponds to that of the original printed dipole and remains approximately unchanged while the higher resonant frequency is mainly due to the CLL loading. To further understand the performance of the printed dipole antenna near the CLL elements, Fig. 20, shows the magnitude of the S-parameters versus frequency for the CLL-based metamaterial.

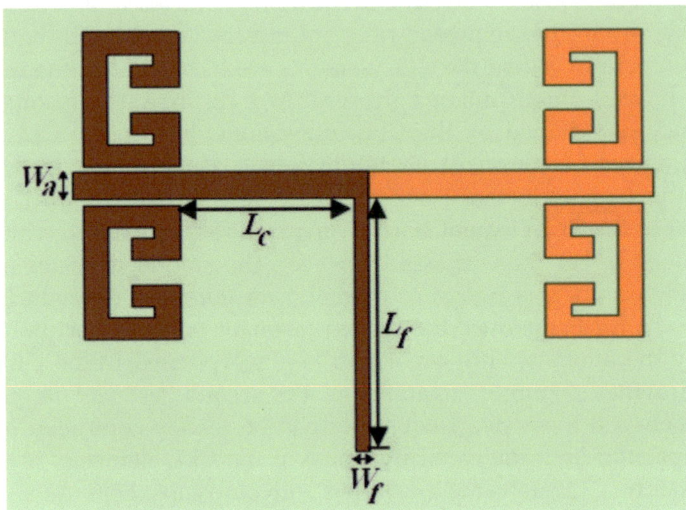

Fig. 17. Geometry of a CLL-loaded printed dipole antenna, L_f=23mm, L_c=13.67mm, W_f=1mm, W_a=2.5mm, L_1=3.73mm, L_2=4.95mm, L_3=7.62mm, G=2.28mm, W=2mm. From (Jafargholi et al., 2010), copyright © 2010 by Praise Worthy Prize, S. r. l.

Fig. 18. Photograph of a fabricated CLL-loaded printed dipole antenna. From (Jafargholi et al., 2010), copyright © 2010 by Praise Worthy Prize, S. r. l.

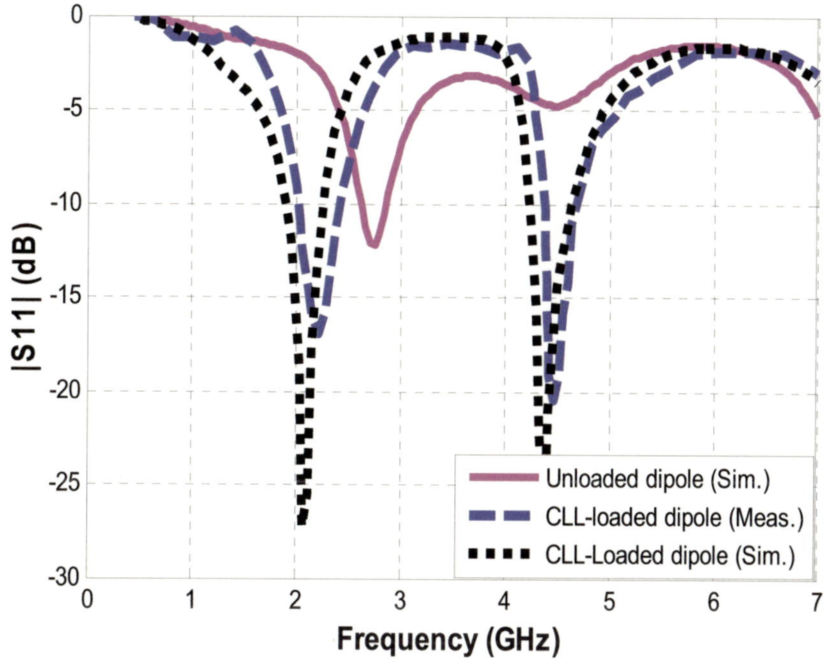

Fig. 19. Reflection coefficient comparison between the CLL-loaded and unloaded printed dipole antenna. From (Jafargholi et al., 2010), copyright © 2010 by Praise Worthy Prize, S. r. l.

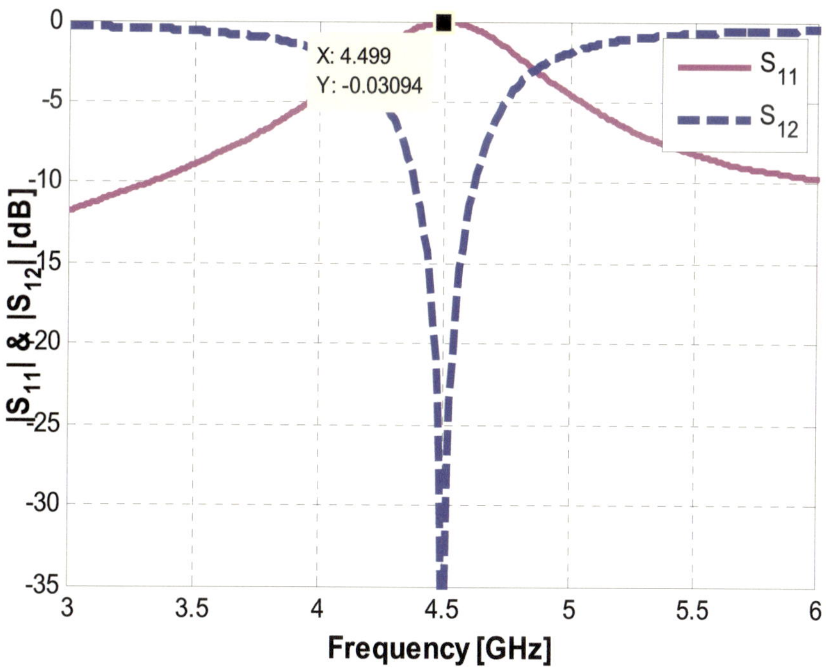

Fig. 20. Magnitude of the S-parameters versus frequency for the CLL-based metamaterial. From (Jafargholi et al., 2010), copyright © 2010 by Praise Worthy Prize, S. r. l.

It is observed that the CLL element effectively resonates at around 4.5GHz with small loss. This frequency coincides with the second resonant frequency of the CLL-loaded printed dipole (see Fig. 19). The radiation patterns of the proposed dual band printed dipole are measured at the resonant frequencies of 2.15 and 4.45GHz. Fig. 21, shows the measured and simulated E-plane radiation patterns at first and second resonant frequencies. The antenna gains at first and second resonant frequencies are 1.8dB, and 3.9dB, respectively. As a result, the losses introduced by the CLL elements are significantly low in the frequency range of interest. In other words, the proposed dual band dipole antenna has acceptable performance in both gain and radiation pattern.

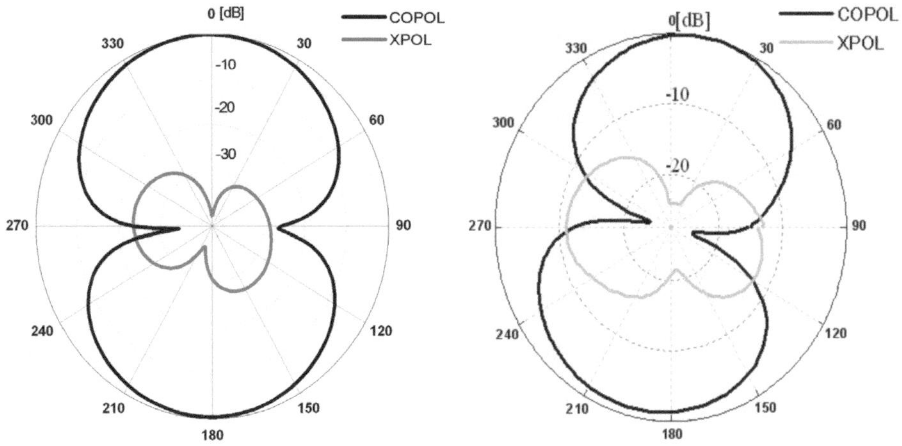

(a)

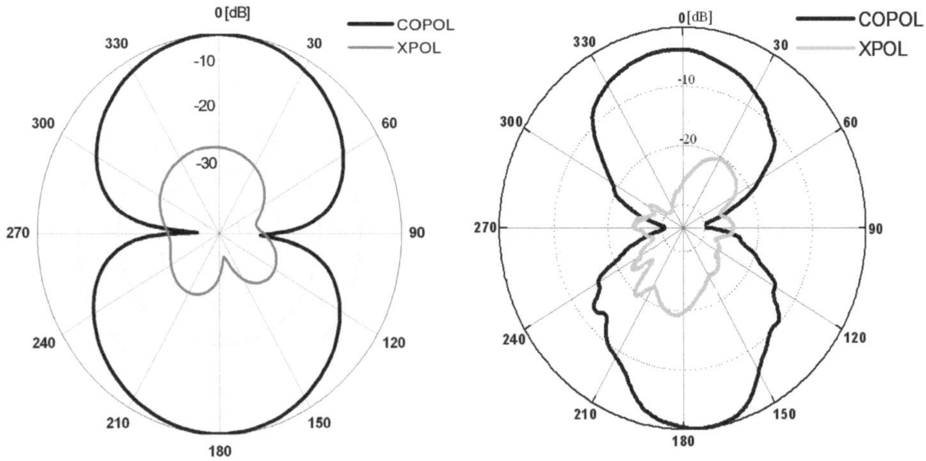

(b)

Fig. 21. E-plane radiation patterns of the CLL-loaded dipole antenna at (a) 2.15GHz, and (b) 4.45GHz. (Right hand figures are measurements). From (Jafargholi et al., 2010), copyright © 2010 by Praise Worthy Prize, S. r. l.

2.4.2 Miniaturized CLL-loaded printed dipole antenna incorporating chip capacitors

In principle, the size reduction can be arbitrarily achieved if it would be feasible to fabricate a proper metamaterial element that has a negative permeability at a frequency lower than the natural resonance frequency of the corresponding unloaded dipole antenna. For a metamaterial comprised of resonant CLL elements, this can be achieved by capacitive loading of the CLL elements. To verify and confirm the proposed approach, a prototype of a CLL-loaded dipole antenna, in which each CLL ring is loaded with a 0.68pF chip capacitor, is fabricated and measured.

A photograph of the fabricated miniaturized printed dipole antenna is shown in Fig. 22. The magnitude of the S-parameters for the CLL-based metamaterial loaded with a 0.68PF chip capacitor is shown in Fig. 23. As can be seen, the resonant frequency of the CLL element shifts down to 1.33GHz by incorporating 0.68PF chip capacitor (see Figs. 21, 23). In order to meet the specification of both the ISM system and mobile communication, the miniaturized printed dipole should radiate linearly polarized waves at 1.8GHz and 2.45GHz.

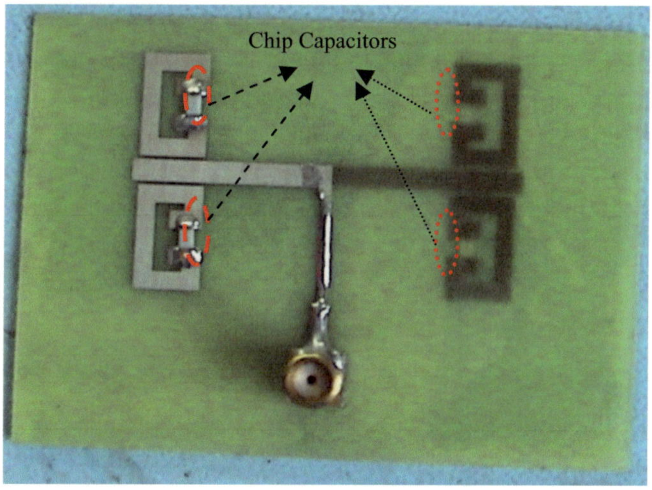

Fig. 22. Photograph of the miniaturized CLL-loaded printed dipole antenna (incorporating 0.68pF chip capacitors). From (Jafargholi et al., 2010), copyright © 2010 by Praise Worthy Prize, S. r. l.

Fig. 24, compares the measured and simulated reflection coefficient of the proposed miniaturized CLL-loaded printed dipole antenna. As can be seen, the second resonant frequency of the proposed CLL-loaded dipole of Section 1-2, considerably shifts down to a lower frequency by incorporating chip capacitors. The resonant frequencies of the proposed miniaturized CLL-loaded printed dipole are lower than the main resonant frequency of the unloaded dipole antenna. It should be pointed out that the antenna radiation patterns at both resonant frequencies are quite similar to that of a half wavelength dipole, as shown in Fig. 25. The antenna gains at first and second resonant frequencies are 0.62dB, and 2.26dB, respectively.

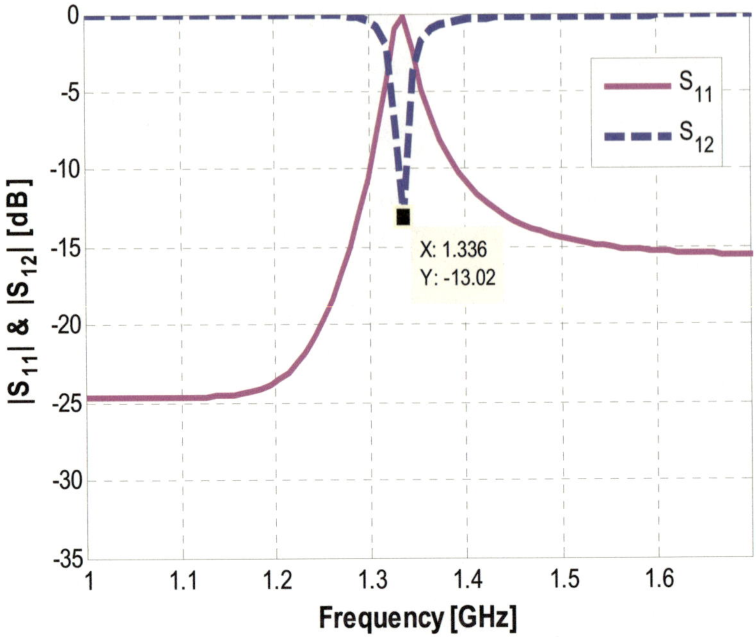

Fig. 23. Magnitude of the S-parameters versus frequency for the CLL based metamaterial loaded with a 0.68PF chip capacitor. From (Jafargholi et al., 2010), copyright © 2010 by Praise Worthy Prize, S. r. l.

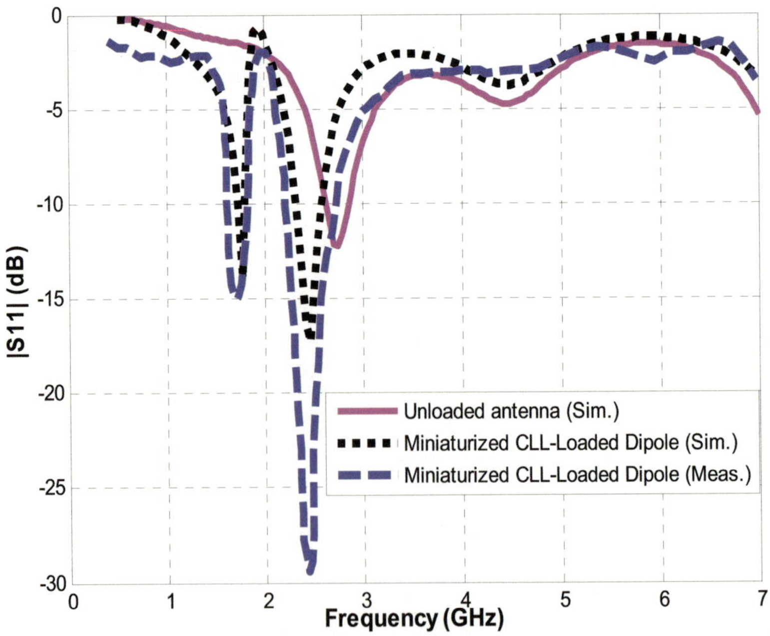

Fig. 24. Measured and simulated reflection coefficient of the miniaturized CLL-loaded printed dipole antenna, as compared to that of the unloaded printed dipole. From (Jafargholi et al., 2010), copyright © 2010 by Praise Worthy Prize, S. r. l.

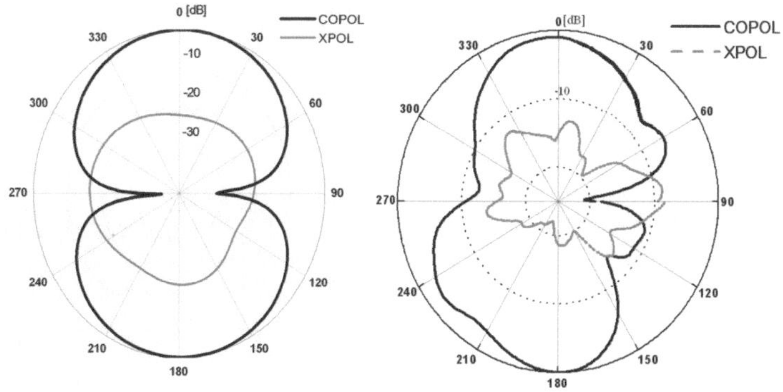

(a)

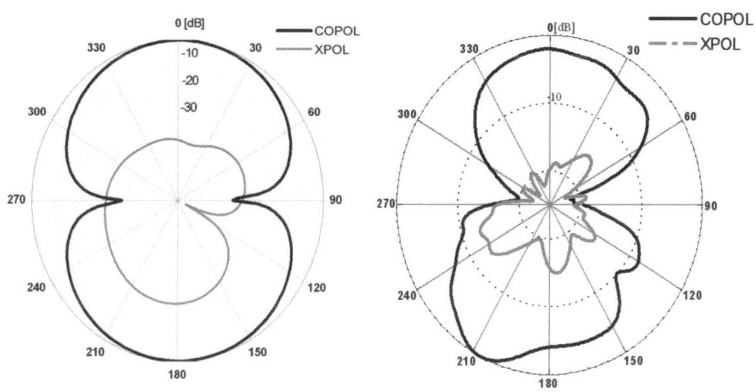

(b)

Fig. 25. E-plane radiation patterns of the miniaturized CLL-loaded printed dipole antenna of Section 1-3 at (a) 1.713GHz, and (b) 2.434GHz. (Right hand figures are measurements). From (Jafargholi et al., 2010), copyright © 2010 by Praise Worthy Prize, S. r. l.

3. Acknowledgment

The authors would like to thank Iran Telecommunication Research Centre (ITRC) for its financial supports.

4. References

[1] C. A. Balanis, *Advanced Engineering Electromagnetics*, John Wiley & Sons, New York, 1989.

[2] V. G. Veselago, "The electrodynamics of substances with simultaneously negative values of ε and μ," *Sov. Phys. Usp.*, vol. 10, no. 4, 509–514, 1968.

[3] M. Veysi, M. Kamyab, S. M. Mousavi, and A. Jafargholi "Wideband Miniaturized Polarization Dependent HIS Incorporating Metamaterials," *IEEE Antennas Wireless Propag. Letter*, vol. 9, 764-766, 2010.

[4] M. Rafaei Booket, M. Kamyab, A. Jafargholi and S. M. Mousavi, "Analytic Modeling and Implementation of The Printed Dipole Antenna loaded with CRLH Structures, " *Progressive In Electromagnetic Research B*, Vol. 20, 167-186, 2010.

[5] M. Veysi, M. Kamyab, J. Moghaddasi, and A. Jafargholi, "Transmission Phase Characterizations of Metamaterial Covers for Antenna Application," *Progressive In Electromagnetic Research Letter*, Vol.21, pp. 49-57, 2011.

[6] M. Rafaei Booket, A. Jafargholi, M. Kamyab, H. Eskandari, M. Veysi, and S. M. Mousavi, "A Compact Multi-Band Printed Dipole Antenna Loaded With Single-Cell MTM," Pending Publication in *IET Microwave Antenna Propag.*, 2011.

[7] A. Jafargholi, M. Kamyab, and M. Veysi, "PMC-based Waveguide-fed Slot Array," *ISRN Communications and Networking*, Hindawi, Vol. 2011, Article ID 941070, 5 Pages.

[8] A. Jafargholi, M. Kamyab, and M. Veysi, "Artificial Magnetic Conductor Loaded Monopole Antenna," *IEEE Antennas Wireless Propag. Letter*, vol. 9, 211-214, 2010.

[9] A. Jafargholi, M. Kamyab, M. Rafaei Booket, and M. Veysi, "A Compact Dual-band Printed Dipole Antenna Loaded with CLL-Based Metamaterials," *International Review of Electrical Engineering*, IREE, Vol. 5, No. 6, pp. 2710-2714, 2010.

[10] J. McVay, N. Engheta, and A. Hoorfar, "High-impedance metamaterial surfaces using Hilbert-curve inclusions," *IEEE Microwave Wireless Components Lett.*, vol. 14, no. 3, pp. 130–132, Mar. 2004.

[11] A. Erentok, P. Luljak, and R. W. Ziolkowski, "Antenna performance near a volumetric metamaterial realization of an artificial magnetic conductor," *IEEE Trans. Antennas Propagat.*, vol. 53, pp. 160–172, Jan. 2005.

[12] R. W. Ziolkowski and A. Kipple ,"Application of double negative metamaterials to increase the power radiated by electrically small antennas," *IEEE Trans. Antennas Propagat.*, vol. 51, no. 10, pp. 2626–2640, Oct. 2003.

[13] Q. Liu, P. S. Hall, and A. L. Borja ," Efficiency of Electrically Small Dipole Antennas Loaded With Left-Handed Transmission Lines," *IEEE Trans. Antennas Propagat.*, vol. 57, no. 10, pp. 3009–3017, Oct. 2009.

[14] S. D. Rogers, C. M. Butler, and A. Q. Martin, " Design and Realization of GA-Optimized Wire Monopole and Matching Network With 20:1 Bandwidth," *IEEE Trans. Antennas Propagat.*, vol. 51, no. 3, pp. 493–502, March. 2003.

[15] A. Erentok, and R. W. Ziolkowski, "Metamaterial-Inspired Efficient Electrically Small Antennas" *IEEE Trans. Antennas Propagat.*, vol. 56, pp. 691–707, March 2008.

[16] A. Erentok, *Metamaterial-Based Electrically Small Antennas*, Ph.D. dissertation at University of Arizona, 2007.

[17] J. Zhu, M. A. Antoniades, and G. V. Eleftheriades "A Compact Tri-Band Monopole Antenna With Single-Cell Metamaterial Loading" *IEEE Trans. Antennas Propagat.*, vol. 58, pp. 1031–1038, April. 2010.

[18] M. A. Antoniades and G. V. Eleftheriades, "A broadband dual-mode monopole antenna using NRI-TL metamaterial loading," *IEEE Antennas Wireless Propag. Lett.*, vol. 8, pp. 258–261, 2009.

[19] J. Zhu, M. A. Antoniades, and G. V. Eleftheriades, "A tri-band compact metamaterial-loaded monopole antenna for WiFi and WiMAX applications,"presented at the *IEEE Antennas and Propagation Society Int. Symp.*, Jun. 2009.

[20] J. Zhu and G. V. Eleftheriades, "Dual-band metamaterial-inspired small monopole antenna for WiFi applications," *Electron. Lett.*, vol. 45, no. 22, pp. 1104–1106, Oct. 2009.

[21] H. Iizuka, P. S. Hall, and A. L. Borja, "Dipole Antenna With Left-Handed Loading" *IEEE Antennas Wireless Propag. Lett.*, vol. 5, pp. 483–485, 2006.

[22] H. Iizuka, and P. S. Hall, "Left-Handed Dipole Antennas and Their Implementations" *IEEE Trans. Antennas Propagat.*, vol. 55, pp. 1246–1253, May 2007.

23

Compact Coplanar Waveguide Metamaterial-Inspired Lines and Its Use in Highly Selective and Tunable Bandpass Filters

Alejandro L. Borja[1], James R. Kelly[2], Angel Belenguer[1], Joaquin Cascon[1] and Vicente E. Boria[3]
[1]*Departamento de Ingeniería Eléctrica, Electrónica, Automática y Comunicaciones, Escuela, Politécnica de Cuenca, Universidad de Castilla-La Mancha*
[2]*School of Electronic, Electrical and Computer Engineering, University of Birmingham*
[3]*Instituto de Telecomunicaciones y Aplicaciones Multimedia, Universidad Politécnica de Valencia*
[1,3]*Spain*
[2]*UK*

1. Introduction

During the last years, the metamaterials field has grown rapidly due to the possibility of accomplishing a methodology to achieve negative effective parameters ε_{eff} Pendry et al. (1996) and μ_{eff} Pendry et al. (1999), and their experimental verification Smith et al. (2000) - Shelby, Smith & Schultz (2001). The main research work has been concentrated on the theoretical consequences of negative parameters, as well as techniques for practically realizing left-handed media for various optical/microwave concepts and applications. Among recent concepts, properties, and devices based on engineered metamaterials, we can mention frequency selective structures, which have opened the path to a new range of passive devices for guided applications. In this context, split ring resonator (SRR) loaded transmission lines represent the cutting edge of research in the field of one-dimensional (1D) planar left-handed structures. These structures were firstly proposed by Martin et al. in 2003 Martin et al. (2003) by magnetically coupling a shunted coplanar waveguide (CPW) and pairs of SRRs. These planar devices exhibit backward propagation in a narrow frequency band above the resonant frequency of the rings, with the necessary degree of flexibility to design compact low insertion losses filters. Thus, based on this former configuration different approaches have been proposed with the aim of improving performances and overcome possible drawbacks such as asymmetrical response shape or transmission bands with smooth edges. For instance, a combined right/left-handed CPW structure was implemented by cascading SRRs-wire and SRRs-gap stages Bonache et al. (2005), satisfactorily achieving a transmission upper band with a sharp cut-off. Furthermore, new CPW lines with extra loading elements have shown interesting properties in terms of improved out-of-band behavior and response selectivity Borja, Carbonell, Boria, Cascon & Lippens (2010) - Borja, Carbonell, Boria &

Lippens (2010b). By these means, it is possible to control several restrictive trade-offs by simply adjusting the loading elements. In this way it is possible to obtain quite a symmetric frequency response along with controllable bandwidths and compact dimensions.

In this chapter, the properties of a variety of metamaterial designs are analyzed and discussed in order to develop novel small planar metamaterial frequency selective structures. The study of these configurations is based on full-wave electromagnetic analysis, equivalent circuit simulations and measured responses of different prototypes designed for microwave operation. In particular, section 2 presents different CPW lines based on a split ring resonator SRR technology and loaded with metallic strips and gaps. The properties of such structures can be controlled by properly designing or adding loading elements. In this regard, the use of shunt wires permits to control frequency selectivity by means of an engineering of the electric plasma frequency, providing deeper upper band rejection levels. On the other hand, the addition of series capacitances to previous unit cell implementations provides a transmission response which is almost symmetric while exhibiting a right-handed character along the pass band, contrary to conventional left-handed lines. In sections 3 and 4 enhanced out-of-band rejection properties and reconfigurable responses can be obtained by the use of cascaded and varactor loaded basic cells, respectively. Finally, the main conclusions of the chapter are outlined in section 5.

2. Split ring resonators based coplanar waveguide lines

In this section, different arrangements corresponding to configurations of SRRs loaded CPW lines have been considered. Two loading elements, i.e. shunt strips, and series gaps have been successively included in the CPW line. These inclusions are much smaller than the electrical wavelength of the propagating wave. Therefore, effective medium considerations in the CPW and lumped equivalent circuit models have been used to understand and better explain the complex features of wave propagation inherent to these loaded lines. In this section numerical studies and experiments are also included.

2.1 Split ring resonators loaded coplanar waveguide with shunt strips

The first model is based on the combination of SRRs and shunt wires within a host CPW line. This resonant structure has already been deeply analyzed Martin et al. (2003), and therefore it will be used as a reference result. In summary, this unit cell behaves as a left-handed propagation uni-dimensional transmission line that, if its size is sufficiently small as compared to the electrical wavelength, can have a double negative effective medium behaviour. By using the SRRs it is possible to synthesize a negative value of effective permeability, whereas shunt wires in the CPW provide a negative effective permittivity. For this reason, the structure can be considered as a double negative effective medium, when operating at frequencies slightly above the resonance of the isolated SRR Aznar et al. (2008).

The proposed left-handed structure is depicted in Fig. 1. It consists of a host CPW loaded with SRRs and two shunt strips. SRRs are symmetrically placed on the rear of the substrate, while thin metal wires connect the signal line to the ground plane at positions coincident with the center of the SRRs. Table 1 defines the CPW geometry, unit cell characteristics, and common parameter values for this and subsequent configurations.

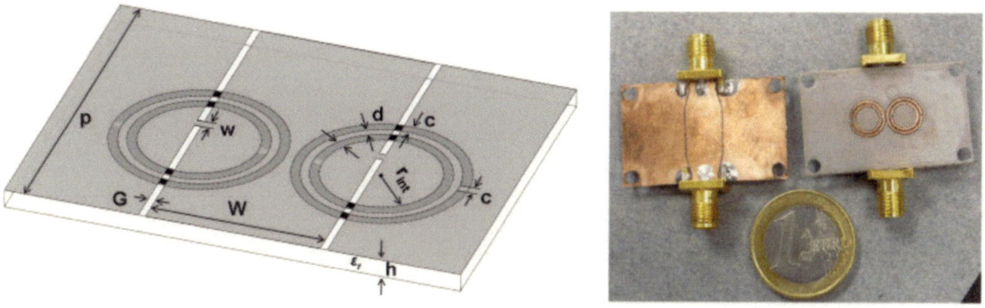

(a) (b)

Fig. 1. (a) SRRs loaded CPW with shunt strips, and (b) photograph of the prototype.

W(mm)	G(mm)	p(mm)	w(mm)	h(mm)	ϵ_r
7.7	0.3	10	0.4	0.508	2.2
tgδ	$t(\mu$m)	σ(S/m)	r_{int}(mm)	c(mm)	d(mm)
0.0009	35	$5.8 \cdot 10^7$	2.6	0.4	0.4

Table 1. Unit cell characteristics.

W is the line width, G is the gap between conductors, w is the strip width, and p the unit cell period. Substrate characteristics are height h, permittivity ϵ_r, and loss tangent $tg\,\delta$. Cooper metallization was utilized. This had a thickness t, and conductivity σ. Moreover, a prototype has been fabricated in order to verify the propagation behavior of the cell, see Fig. 1 (b). A taper section is added at both SMA connections to properly feed the device. The sample has been fabricated on a Neltec NY9220 dielectric substrate using a mechanical milling process. The milling was performed by a LPKF Protomat 93S machine. Thereafter, the fabricated device has been measured and characterized by means of a Rohde & Schwarz vector network analyzer ZVA-24, calibrated with a Through-Open-Short-Match kit, in the frequency band from 3 to 5 GHz.

In parallel, the structure proposed is analyzed and compared with the lumped element equivalent circuit of the unit cell, see Fig. 2. Due to physical symmetry properties, the magnetic wall concept has been used so that the equivalent circuit corresponds to one half of the basic cell. The equivalent circuit model can be transformed to an equivalent π-circuit type, as it was described by Aznar et al. (2008),

Each SRR can be represented by a simple LC parallel resonator circuit in the vicinity of resonance, with elements L_s and C_s. L and C are the per-section inductance and capacitance defined from the geometry of the CPW line and calculated as it is advised in Mongia et al. (1999). L_p is the equivalent inductance of connecting wires, which divides the inductances L_s and L into two parts, as proposed in Rogla et al. (2007) and thoroughly verified in Aznar et al. (2008). SRRs are modeled by parallel resonant circuits inductively coupled to the line through a coupling constant k. It is calculated by means of the fractional area theory explained in Martin et al. (2003). The values of the different lumped elements and parameters are summarized in Table 2.

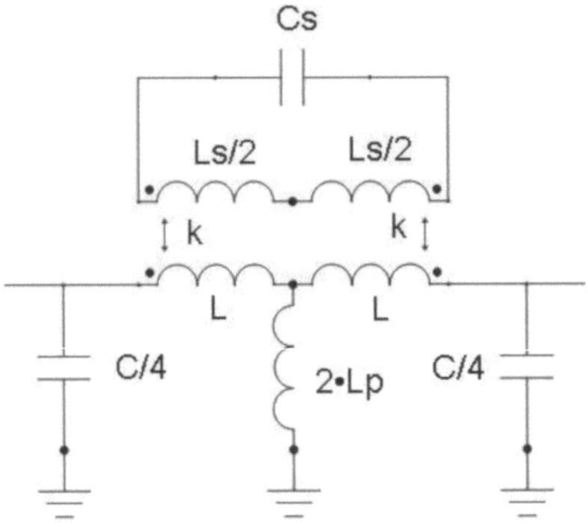

Fig. 2. Lumped equivalent circuit model of the SRRs loaded CPW with shunt strips, originally proposed in Rogla et al. (2007).

C_s (pF)	L_s (nH)	C (pF)	L (nH)	L_p (pH)	k
0.104	14.8	0.672	2.11	131	0.342

Table 2. Equivalent circuit parameters of the SRRs loaded CPW with shunt strips.

The transmission S_{21} and reflection S_{11} coefficients for the left-handed structure shown in Fig. 3 (a), have been obtained from three different sources, i. e. full-wave simulations, lumped equivalent circuit simulations, and experimental results. Agreement is found to be very good in all cases. Nonetheless, a frequency shift for the equivalent circuit response is observed. This shift could be minimized by simply tuning L_s and C_s element values, which control the SRR resonance. As it can be seen, the frequency response exhibits a pass band centred around 4.3 GHz with a transmission zero close to 3.6 GHz. According to the model of Fig. 2, the structure should exhibit a transmission zero (all injected power is returned back to the source) at that frequency where the series branch opens, and it occurs at the resonant frequency of the coupled SRRs.

The analysis of these transmission characteristics is performed by the extraction of the effective medium parameters of the uni-dimensional propagation structure, see Fig. 3 (b). The extraction is based on the well known Nicolson-Ross-Weir (NRW) procedure used in Smith et al. (2005), where the real parts of the permittivity and permeability are retrieved from the scattering parameters. The results obtained confirm the presence of a narrow pass band between 4.1 GHz and 4.5 GHz, corresponding to a double negative frequency band as it is expected. The maximum transmission is achieved when the matching condition $\epsilon \cong \mu$, and thus reduced impedance $\bar{Z} = \sqrt{\mu/\epsilon} \cong 1$, is satisfied. Henceforward, the transmission line becomes single negative as permeability reaches positive values after the magnetic plasma frequency f_{mp}. A double positive medium is subsequently obtained above the electric plasma frequency f_{ep}.

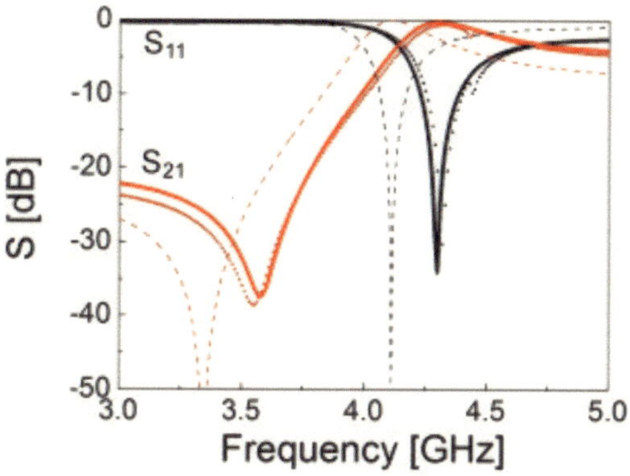

(a)

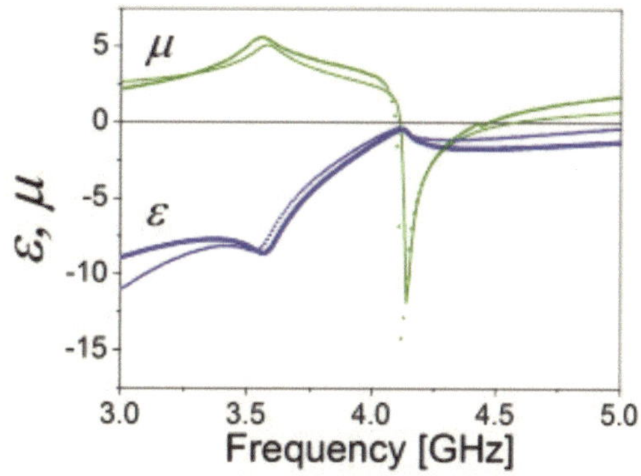

(b)

Fig. 3. (a) Simulated and experimental S_{11} and S_{21} parameters. (thick solid line: Full-wave simulation, symbol: Measurement, thin dashed line: lumped equivalent circuit simulation). (b) Simulated (line) and experimental (symbol) real parts of extracted permittivity (thick) and permeability (thin) according to the NRW method.

2.2 Split ring resonators loaded coplanar waveguide with series gaps

The next model, depicted in Fig. 4 (a), is based on the combination of SRRs and series gaps. In this configuration, the shunt wires located on the top side of the substrate have been substituted by series gaps. These elements are dual of the shunt wires and they are located symmetrically with regard to the center of the unit cell. The dimensions of the elements, defined in Table 1, are the same ones as those considered in the previous configuration. Moreover, the gap width g and separation p_1 are 0.25 mm and 5 mm, respectively. A prototype has been manufactured in order to verify the behaviour experimentally, see Fig. 4 (b). The sample has been fabricated using the same substrate and process described in section 2.1.

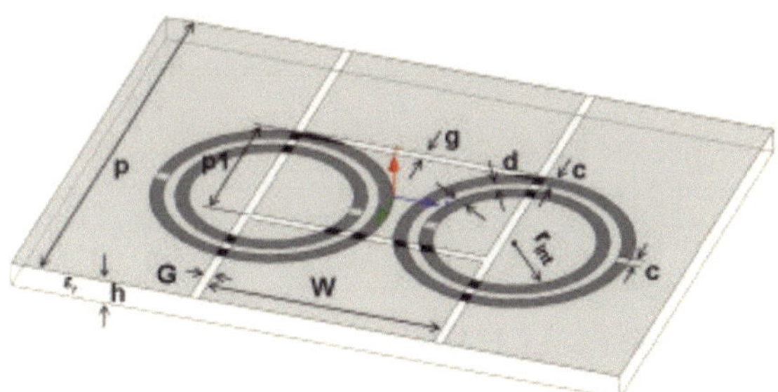

(a)

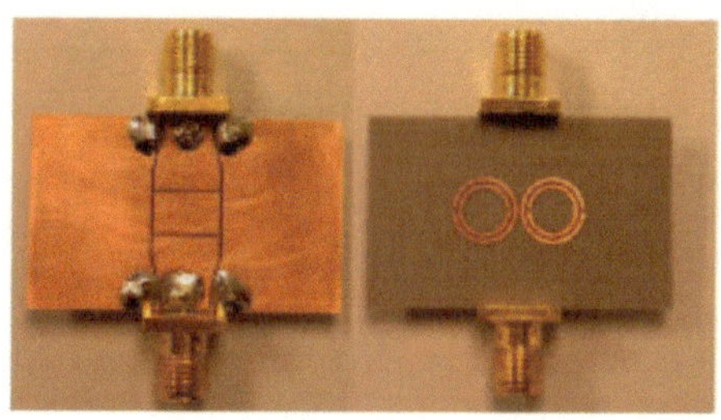

(b)

Fig. 4. (a) SRRs loaded CPW with series gaps and (b) photograph of the prototype.

Likewise, a lumped element equivalent circuit of the elemental cell, shown in Fig. 5, is used to asses the different properties of the new configuration under study. The CPW is modeled as described before, it includes two series capacitances which have been properly modified accordingly to the magnetic wall theory. A gap discontinuity in a CPW line can be represented by means of an equivalent two-port π-network, as presented by Deleniv et al. (1999). The

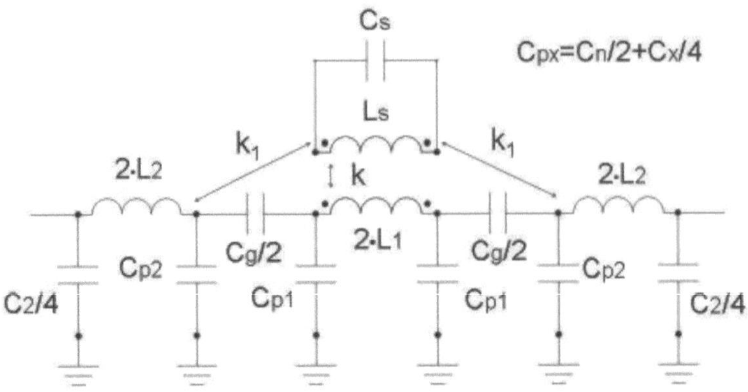

Fig. 5. Lumped equivalent circuit model of the SRRs loaded CPW with series gaps.

π-network comprises a series capacitance C_g together with a shunt capacitance C_n. The series capacitance C_g describes the reactance due to the gap discontinuity, whilst the shunt capacitance C_n accounts for grounding edge effects at the sides of the gap. L_1 and C_1 are the per-section inductance and capacitance of the line between the gaps. Also, L_2 and C_2 account for the per-section inductance and capacitance of the transmission line. C_{pi} (i= 1,2) is the equivalent capacitance of the two shunt capacitances, $C_{n/2}$ and $C_{i/4}$. The SRRs are inductively coupled to different parts of the line. Coupling to the central portion of the line, between gaps, is modeled by a coupling constant k. k is calculated applying the fractional area theory. The value of k has been adjusted slightly since the series gaps modify the coupling between the line and the rings. Following this adjustment, the bandwidth and location of the transmission zero can easily be tuned. Similarly, the coupling constant k_1 represents the interaction between the external portions of the CPW and the SRRs. The value of k_1 was adjusted by means of a curve fitting procedure, which ensured that the equivalent circuit simulations agreed well with the measurement results. Table 3 gives the final values of the circuit elements within the equivalent circuit.

C_s (pF)	C_g (pF)	C_n (pF)	C_1 (pF)	C_2 (pF)
0.104	0.3	0.0485	0.3192	0.1596
L_s (nH)	L_1 (nH)	L_2 (nH)	k	k_1
14.8	1	0.5	0.6	0.35

Table 3. Equivalent circuit parameters of the SRRs loaded CPW with series gaps.

Fig. 6 (a) shows the scattering parameters obtained through simulation and measurement. A good agreement between the three responses, i.e. full-wave, equivalent circuit, and measurements is observed. As a relevant feature, it can be mentioned that the structure exhibits a transmission band centred around 3.8 GHz and a transmission zero around 4.5 GHz. This is contrary to the performance of a CPW line loaded with SRRs and shunt strips, where the transmission zero is located below the pass band, see Fig. 3 (a). In this structure the transmission zero is located at high frequencies and the pass band appears before the anti-resonance. An important reminder here is that the SRRs are exactly the same in these two configurations, and only the other loading elements (series gaps and shunt wires, not resonant by themselves in this frequency range) are different. The parallel inductive contribution of the wires is replaced by a series capacitance. For this reason the device exhibits dual behaviour compared to a CPW line loaded with SRRs and shunt strips.

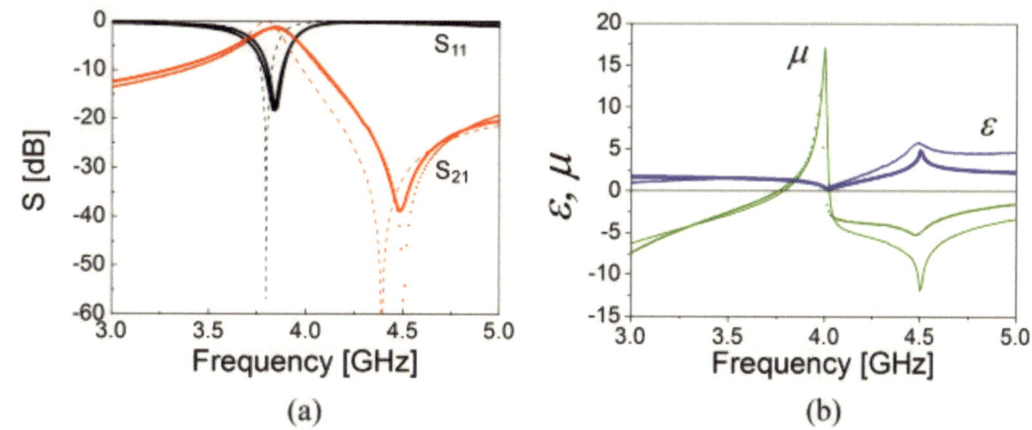

Fig. 6. (a) Simulated and experimental S_{11} and S_{21} parameters. (thick solid line: Full-wave simulation, symbol: Measurement, thin dashed line: lumped equivalent circuit simulation). (b) Simulated (line) and experimental (symbol) real parts of extracted permittivity (thick) and permeability (thin) according to the NRW method.

This performance can be explained by the interpretation of the retrieved permittivity and permeability of the structure, presented in Fig. 6 (b). The generation of a pass band is related to a double-positive condition. Permittivity remains positive in the whole measured frequency band. Permeability is also positive in a small frequency range, below the resonant frequency of the SRR (4 GHz), and negative with a monotonous variation outside it. In this case, the SRR is contributing to generate a right-handed transmission band. This is due to the series capacitance loaded in the line, which precludes transmission outside of the area where the SRR resonates. Permittivity is also affected by the presence of the SRR, but it remains positive as mentioned before. Note also, as expected, that peak transmissions correspond to the different crossings of the ϵ and μ curves ($\epsilon \cong \mu$ gives the matching condition and hence maximum transmission and minimum reflection). The transmission zero in Fig. 6 (b) is located coincident with the positions of the slope changes in the effective parameters. These slope changes are in turn due to the superposition of the 'monotonous' behaviours of the simple loading elements (series gaps) with the complex (quasi-Lorentz-type) behaviour of the SRR loaded in the line.

2.3 Split ring resonators loaded coplanar waveguide with shunt strips and series gaps

At last, the characteristics of transmission lines combining all the previous elements, shunt strips and series gaps in CPW technology are studied. The unit cell, described in Fig. 7 (a), consists of an SRRs based CPW loaded with two shunt metallic strips and two series capacitances.

The gaps are symmetrically placed with respect to the SRRs, while the thin shunt strips connecting the central line to the ground are placed at those positions coincident with the

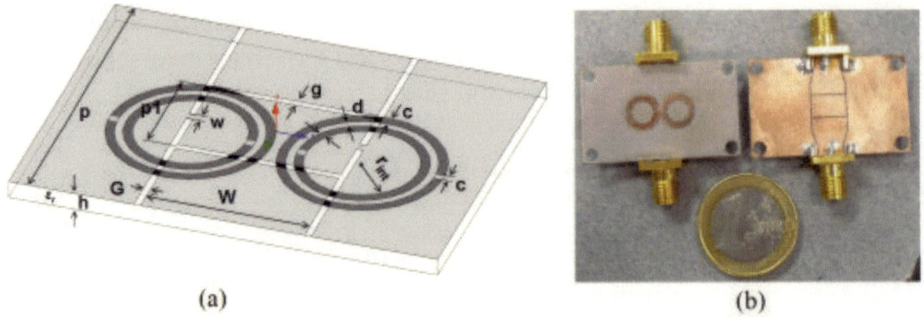

Fig. 7. (a) SRRs loaded CPW with shunt strips and series gaps, and (b) photograph of the prototype.

center of the SRRs. The series gap elements can be interpreted as the dual counterpart of the shunted inductive strips as it was discussed previously. These elements, shunt strips and series gaps, are in the coupling regions of the SRRs so that, as it is next explained, combined effects of previous properties take place. The configuration of the unit cell is the same as that employed in previous sections. The dimensions are given in Table 1. In Fig. 7 (b), the prototype of the unit cell is depicted.

The lumped equivalent circuit model of the basic cell, which will be used for the interpretation of the structure, is presented in Fig. 8. As it was reported previously, due to symmetry, the magnetic wall theory has been applied. In addition, the different elements of the circuit model are identical to those described in sections 2.1 and 2.2 (see also Table 4).

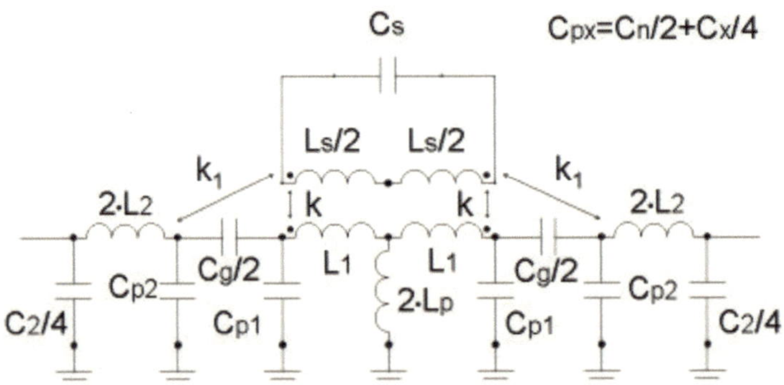

Fig. 8. Lumped equivalent circuit model of the SRRs loaded CPW with shunt strips and series gaps.

$C_s\ (pF)$	$C_g\ (pF)$	$C_n\ (pF)$	$C_1\ (pF)$	$C_2\ (pF)$
0.104	0.44	0.0485	0.588	1.3

$L_s\ (nH)$	$L_1\ (nH)$	$L_2\ (nH)$	$L_p\ (pH)$	k	k_1
14.8	1.84	4.08	131	0.6	0.1

Table 4. Equivalent circuit parameters of the SRRs loaded CPW with shunt strips and series gaps.

The simulated and measured frequency responses for the proposed structure are shown in Fig. 9 (a). An excellent agreement between full-wave simulation data and experimental results can be observed. There is also satisfactory agreement between the results obtained using the proposed equivalent circuit and those derived by other means.

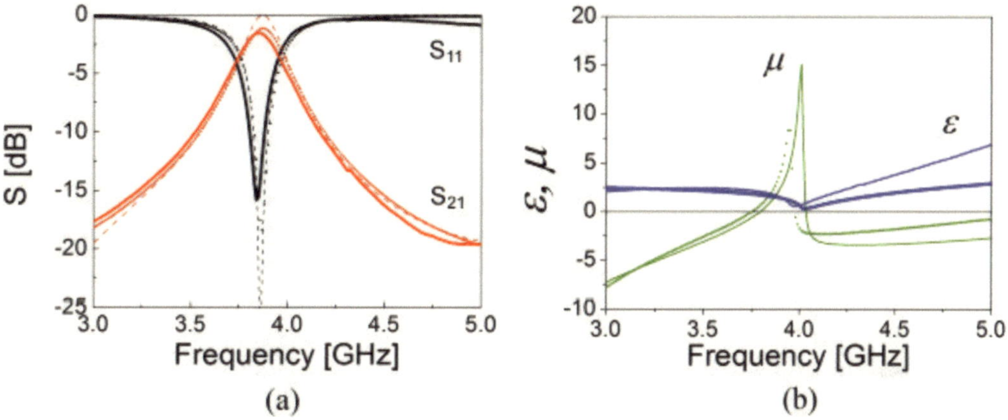

Fig. 9. (a) Simulated and experimental S_{11} and S_{21} parameters. (thick solid line: Full-wave simulation, symbol: Measurement, thin dashed line: lumped equivalent circuit simulation). (b) Simulated (line) and experimental (symbol) real parts of extracted permittivity (thick) and permeability (thin) according to the NRW method.

In the two cases previously studied, namely SRR loaded CPW with shunt strips and SRR loaded CPW with series gaps, a highly asymmetric frequency response with a characteristic anti-resonance effect (dip in the transmission) was obtained. In contrast, the frequency dependence of the cell with strips and gaps simultaneously exhibits an almost symmetrical pass band response centred around 3.9 GHz, just below the intrinsic SRR resonance. There are no transmission zeros in the vicinity of this pass band.

The third structure exhibits better selectivity than the previous ones, where it was only improved at frequencies above the pass band. The absence of transmission zeros is attributed to the transmission levels of each single element. In a CPW line loaded by SRRs, the shunt strips generate a transmission zero at lower frequencies but permit transmission above pass band. For this reason, the transmission zero introduced by the gaps disappears. In the same way, the transmission zero introduced by the strips is cancelled by the high transmission levels due to the series gaps.

The pass band is associated with the double positive condition (positive permittivity and permeability), see Fig. 9 (b). This condition is only achieved over a very narrow range of frequencies around, 3.9 GHz. The absence of transmission zeros can be attributed to the lack of slope changes in the effective parameters. For the models studied in sections 2.1 and 2.2, the transmission zeros in Fig. 3 and Fig. 6 are located coincident with the position of the slope changes in the effective parameters. These slope changes are due to a superposition of effects caused by the simple loading elements (series gaps and shunt wires) and the SRR. In common with the CPW line loaded with SRRs and gaps, this structure has a right-handed behaviour.

For this reason one can conclude that the gaps effect would dominate over that of the shunt strips. Whenever shunt strips are used there will be a transmission zero in the lower part of the frequency spectrum. This is the only effect using shunt strips. The frequency response will be symmetric and right-handed if the gaps are present. Additionally, as it can be expected, peak transmission corresponds to the matched condition $\epsilon \cong \mu$.

3. Compact and highly selective left-handed transmissions lines loaded with split ring resonators and wide strips

In the present section, it will be shown numerically and experimentally that problems related to out-of-band rejection can be alleviated by a proper arrangement of the loading elements responsible of the electrical response (shunt wires). Also, it is demonstrated that the selectivity of the transmission window can be improved by cascading basic cells. This opens up the possibility to fabricate band pass filters based on the SRR technology with excellent trade-offs between selectivity, insertion losses, and out-of-band rejection.

3.1 Split ring resonators loaded coplanar waveguide with wide shunt strips

The electromagnetic properties of left-handed materials, which are highly dispersive due to the transition between the single negative and double negative conditions, have shown interesting frequency filtering properties. These properties rely on the same physical principle, namely the magnetic coupling of a transmission medium to micro-resonators. This effect is responsible for producing a negative effective permeability above the resonant frequency of the resonators. On the other hand, an arrangement of shunt strips create a medium which exhibits negative values of the effective permittivity (ϵ_{eff}) and a high pass filter response. The overlap of the spectrum where ϵ_{eff} and μ_{eff} are simultaneously negative gives a frequency band in which the wave propagation is backward. This correspond to the so-called Left-Handed (LH) rule in terms of E, H and k trihedron. Indeed, it is now well established that the dispersion properties of the negative effective permittivity and permeability are very different, with a Drude-like and Lorentz-type frequency variation, respectively (see Fig. 10).

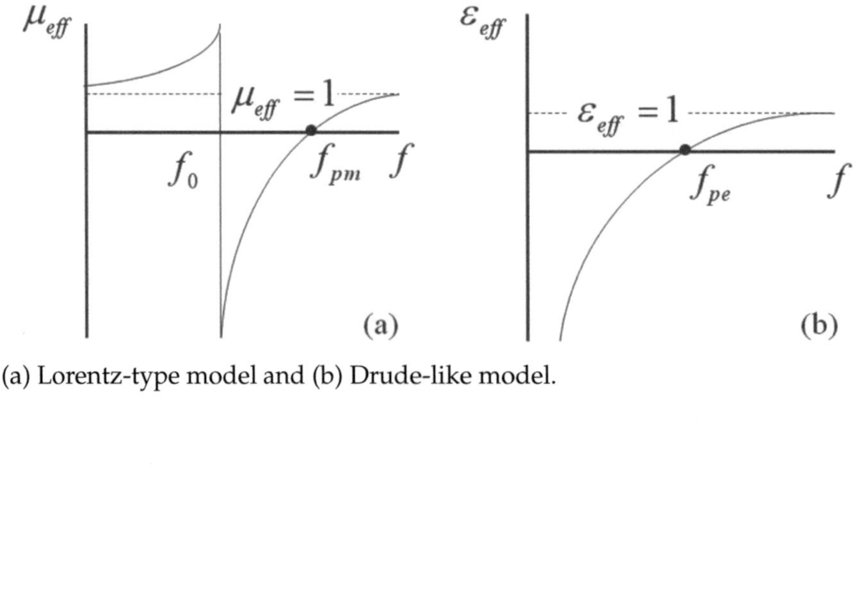

Fig. 10. (a) Lorentz-type model and (b) Drude-like model.

In short, whereas the permittivity increases continuously from negative values to positive ones, at the crossing point known as the electrical plasma frequency, the variations of the effective permeability versus frequency show a resonant response. The two frequencies involved are the resonance frequency of resonators and the magnetic plasma frequency. As a consequence, the asymmetric and double negative overall response is the superposition of a resonant transmission onto a baseline that increases with frequency. Under these conditions, the electrical plasma frequency which defines the transition between the negative and positive value of ϵ_{eff} is generally adjusted until it is slightly higher than the magnetic plasma frequency. By this choice it is generally believed that the impedance matching conditions can be met with comparable values of μ and ϵ and hence impedance $Z \approx 1$.

An alternative technique for improving the selectivity of the transmission window is proposed below, without any additional loading elements. It is based on the engineering of the electric plasma frequency f_{pe}. The idea is to increase f_{pe} with respect to the magnetic plasma frequency f_{pm}. In this way it is possible to ensure that the material is single negative over a broader range of frequencies. Consequently, the rejection level is increased in the upper part of the spectrum. In addition, the impedance matching conditions are not significantly degraded due to the fact that the effective permittivity is also influenced by the resonance effect of the SRRs. For this reason, the values of ϵ_{eff} are still quite comparable to those of μ_{eff} in the vicinity of the resonance. F_{pe} is tailored by enlarging the width of the shunt strip. This also has the effect of dramatically decreasing the coupling between the resonator and the CPW line. As a consequence, the selectivity of the left-handed pass band is enhanced. The model proposed in section 2.1 is further analyzed in order to improve its frequency response. Fig. 11 (a) shows a general schematic of the unit cell.

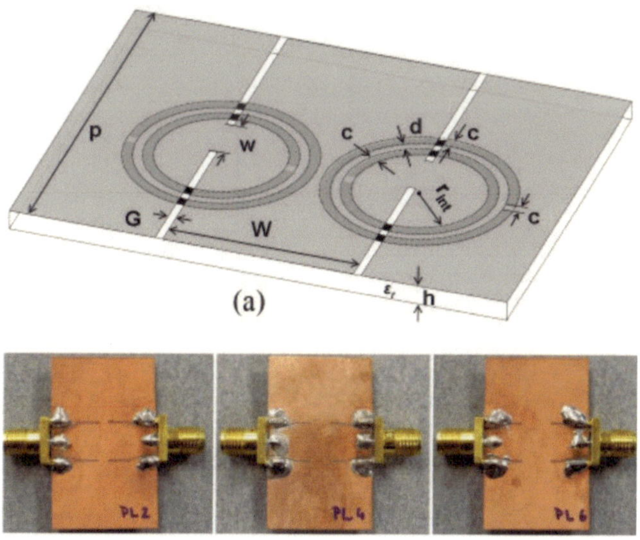

Fig. 11. (a) SRRs loaded CPW with wide shunt strips configuration, and (b) photograph of prototypes with widths w=2, 4 and 6 mm, respectively.

The dimensions and characteristics are given in Table 1. Additionally, Fig. 11 (b) shows three fabricated prototypes with identical dimensions, only w has been modified for the different devices. The width of connecting wires is varied from w=2 mm to w=6 mm, in 2 mm steps. In previous implementations, w was approximately equal to the SRR strip width c (where c=0.4 mm). In this case, w was increased by a factor of 10. With these SRR dimensions, the unloaded Q factor calculated using the eigenmode solver in HFSS, was found around 400. This assumes a copper conductivity of $\sigma = 5.8 \cdot 10^7$ S/m and dielectric loss tangent of tg δ=0.0009. Next, the simulated and measured responses are compared. Fig. 12 shows the scattering parameters S_{11} and S_{21} as a function of frequency for the four single-cell prototypes, which simply differ by the width of the shunt strip w as aforementioned.

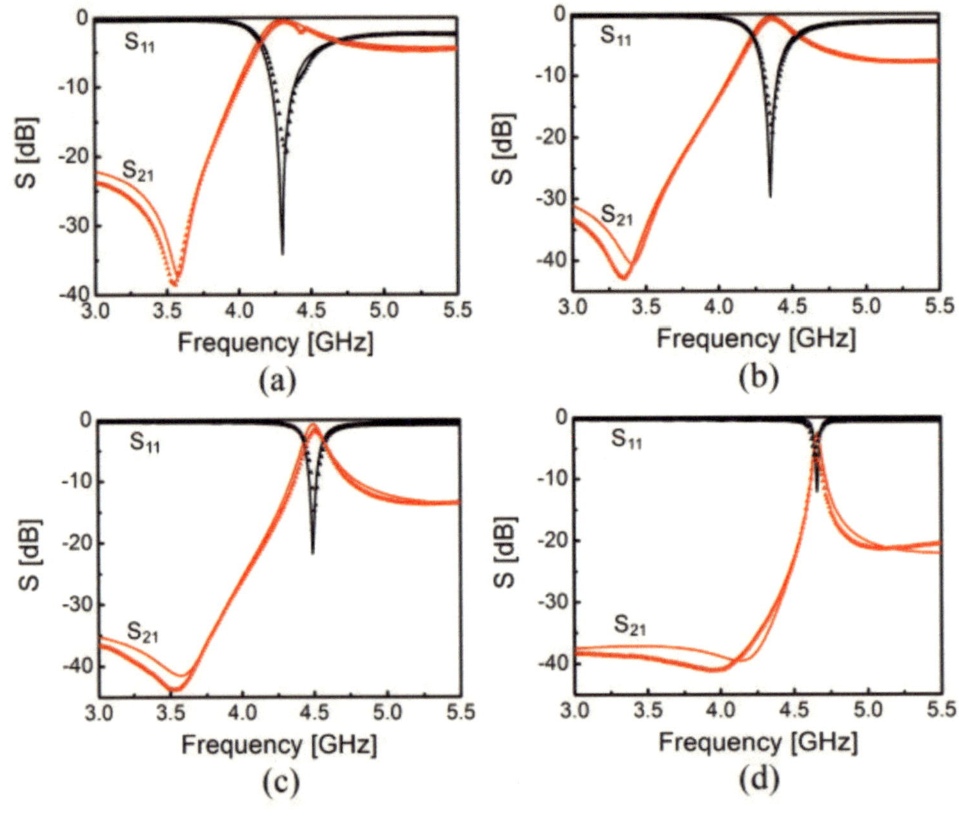

Fig. 12. Comparison between measured (symbol) and calculated (solid lines) scattering parameters for (a) w=0.4 mm, (b) w=2 mm, (c) w=4 mm and (d) w=6 mm.

In Fig. 12 (a) the width was set to w=0.4 mm, while in Fig. 12 (b), (c), and (d) it was increased to w=2 mm, w=4 mm, and w=6 mm respectively. In practice, the S_{ij} parameters were measured by means of a Vector Network Analyzer between 3 and 5.5 GHz. The calibration procedure is based on a Through-Open-Short-Match method. The tapered sections which interconnect the coaxial connector and the basic cell have been removed by a de-embedding process. Comparing Fig. 12 (a), (b), (c), and (d), it can be seen that enlarging the strip width by one order of magnitude, keeping the SRR geometry unchanged, has different major consequences:

an increase of the selectivity, a decrease of the bandwidth, a shift of the resonant frequency, and an increase of the insertion losses.

First of all, it can be noticed that the out-of-band rejection maximum at around 5 GHz which corresponds to the transmission level where the derivative of S_{21} is vanishing, shows a huge increase. Rejection levels are shown in Table 5. Let us keep in mind that such a high rejection level was obtained with a single cell. The measured return losses are equal to -20 dB for w=0.4 mm at around 4.25 GHz, while they reach -15 dB for w=4 mm at around 4.5 GHz, showing that the mismatch degradation is moderate, and compatible with real life applications where return loss around -10 dB can be enough.

	w=0.4 mm	w=2 mm	w=4 mm	w=6 mm
Out-of-band rejection (dB) at 5 GHz	−4.5	−7.6	−13	−22.7
FBW (%)	12.7	7.2	3.3	1.2

Table 5. Shunt strip effect on the out-of-band rejection and FBW.

Furthermore, it can be shown that the bandwidth at half maximum of the transmission is considerably narrower with a fractional bandwidth (FBW) decreasing from $FBW = 12.7\,\%$ for w=0.4 mm to $FBW = 1.2\,\%$ for w=6 mm, see Table 5.

Despite the fact that the SRR dimensions were kept unchanged between the four prototypes, a slight shift in the resonance frequency was observed. It is not due to fabrication tolerances, owing to the good fit with the full-wave simulation results. This shift is attributed to a slight alteration of the SRRs excitation, as a consequence of the modification of the strip. The interaction between SRRs and shunt strips is what determines in practice the position of the maximum transmission peak in the left-handed transmission band. Thus, as long as the width w is varied, the transmission band is shifted. Finally, it can be noted that there is a moderate increase in the insertion losses due to the increase in width of the shunt strip. It is for the reason that very selective frequency responses are deeply affected by the conductor and dielectric losses. Also, SRRs are weakly excited.

In order to have further insight into the electromagnetic properties in terms of dispersion characteristics, the frequency dependence of the effective parameters has been retrieved. To this aim, the Nicolson-Ross-Weir procedure, which has also been applied in section 2, is used. It is worthwhile to mention that the retrieval process is usually utilized when the dimensions of the unit cell are electrically small compared to the wavelength, generally $\lambda_g/10$. It means that when the working frequency increases the extracted parameters are less accurate. The applicability of such methods are strongly related to the frequency range where the device is considered to be used. Therefore, the results extracted from this analysis method should be considered as an estimation rather than an exhaustive study.

Fig. 13 shows the frequency dependence of the real parts of the effective permittivity and permeability for w=0.4 mm and w=4 mm. For simplicity, two values are shown, being enough to verify the effect of the shunt strip width increase. The key result is the observation of a shift of the frequency when ϵ_{eff} goes from the negative values to positive ones, this corner frequency corresponding to f_{pe}. Quantitatively it can be seen that f_{pe} is shifted from 6.6 GHz to 8.8 GHz for simulations, and from 5.2 GHz to 7.4 GHz for measurements, when the strip width is widened by an order of magnitude starting from w=0.4 mm. It is also important to note that a resonant feature is superimposed on the conventional Drude-like variation ($\epsilon = 1 - (\omega_{pe}/\omega)^2$) envelope, helping to keep good input impedances. ϵ_{eff} reaches values comparable to the values of μ_{eff} due to the resonant effect of the SRRs, thus having in both cases good impedance matching levels. On the contrary, the dispersion of the effective permeability is less affected by the increase of the strip width.

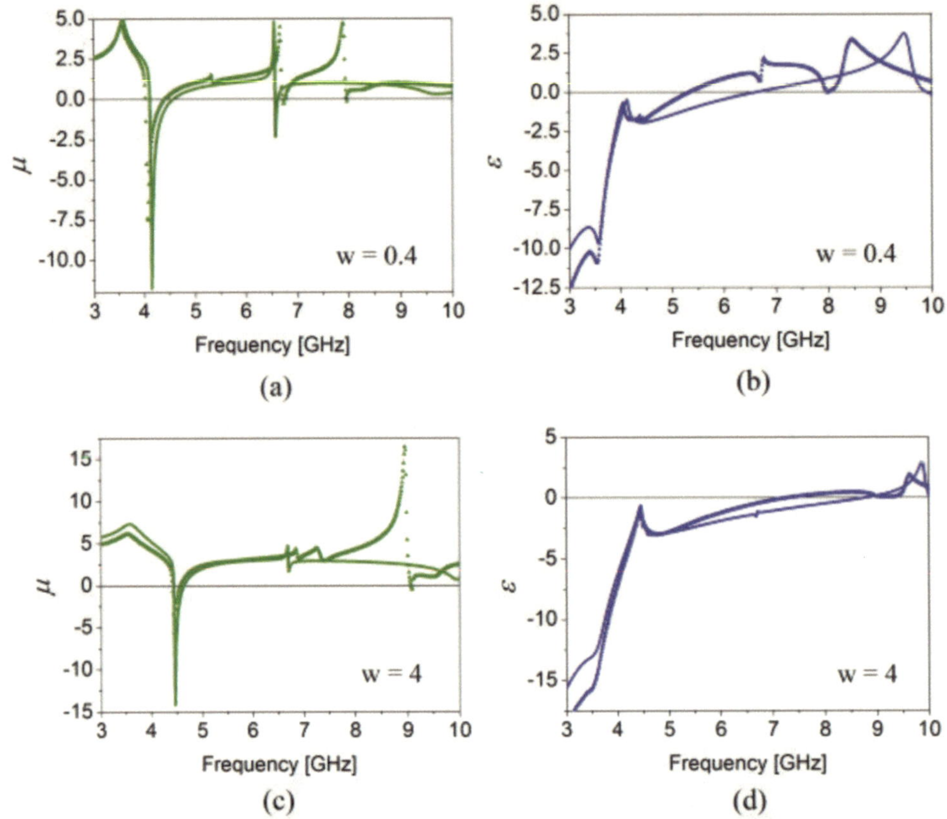

Fig. 13. Frequency dependence of the real parts of the permeability ((a),(c)) and effective permittivity ((b),(d)) for w=0.4 mm and w=4 mm, respectively. Simulation (line) and experimental (symbol) results.

The characteristic frequencies of a Lorentz dispersion law, f_o and f_{pm} respectively, remain practically unchanged by the increase of w. However, a slight frequency shift and narrower window with negative permeability values is observed, which is a consequence of the modified interaction between the SRRs and the wide shunt strip. Derived from these results, wide shunt strips present higher electric plasma frequencies, and consequently deeper rejections in the out-of-band are achieved. More to the point, through this shunt strip enlarging, the coupling between the CPW and the SRRs is reduced, providing narrower bandwidth and higher insertion losses. Both effects are interdependent, and the alteration of w brings always together a modified rejection in the out-of-band and coupling levels between the SRRs and the line. In both cases the comparison of the measured and calculated data, shows a relatively good agreement. However, some discrepancy can be noticed. This discrepancy is attributed to different reasons. Two taper sections are used in the fabricated devices, so in order to retrieve effective parameters they have to be de-embeeded. This is a possible source of errors. At the same time, finite size ground planes, fabrication tolerances, and the measurement procedure also introduce variations in the experimental response.

3.2 Array of split ring resonators loaded coplanar waveguide with wide shunt strips

The previous section has shown that there is a strong relationship between rejection in the out-of-band region, bandwidth, insertion losses, and selectivity as a function of the shunt strip width. Devices with a high selectivity and narrow bandwidth (high Q) generally present high values of insertion losses. In contrast, devices with moderate or low insertion losses do not have sufficiently selective responses to accomplish design requirements with restrictive out-of-band rejection or narrow bandwidth. Therefore, the key aspect in narrow band pass filters is the necessity of a trade-off between achievable insertion losses in the pass band and required frequency selectivity. The advantage of using the proposed SRRs loaded CPW line with wide strips is the possibility to fulfill this trade-off, as will be shown, when several unit cells are connected. The topology of the structure and a fabricated prototype are presented in Fig. 14.

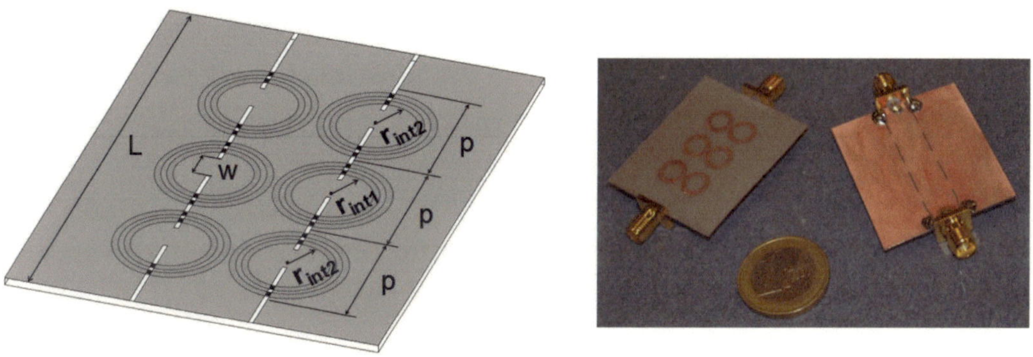

(a) (b)

Fig. 14. Model of the SRRs loaded CPW line with wide strips composed of 3 cascaded unit cells.

The proposed structure consists of three optimized stages, where the internal radius r_{int1} and r_{int2}, and the unit cell period p have been modified in order to achieve a reflection coefficient $S_{11} < -10$ dB along the pass band for two different strip widths w=3.5 mm and w= 4.5 mm. After optimization, parameter values used in the process have been set to r_{int1}=2.6 mm, r_{int2}= 2.58 mm, and p=10 mm for w=3.5 mm, while r_{int1}=2.61 mm, r_{int2}=2.59 mm, and p=9.9 mm for w= 4.5 mm. The total length of the device is L=30 mm. The rest of the cell parameters are the same as those given in Table 1. In Fig. 15 the frequency response of the two optimized filters is shown. In addition, Table 6 summarizes the main properties of the proposed filters. It can be seen that the response of both structures shows excellent insertion loss levels for the two different widths used even if narrow pass bands are achieved. Besides, good rejections levels, better than -30 dB, below and above the pass band are obtained with small ripple characteristics in the pass band of 0.4 dB and 0.6 dB. The main advantages of the filters proposed are the small insertion losses obtained in very selective filters with compact dimensions. In particular, the advantage of miniaturization when SRRs loaded CPW are implemented can be clearly appreciated in Fig. 16.

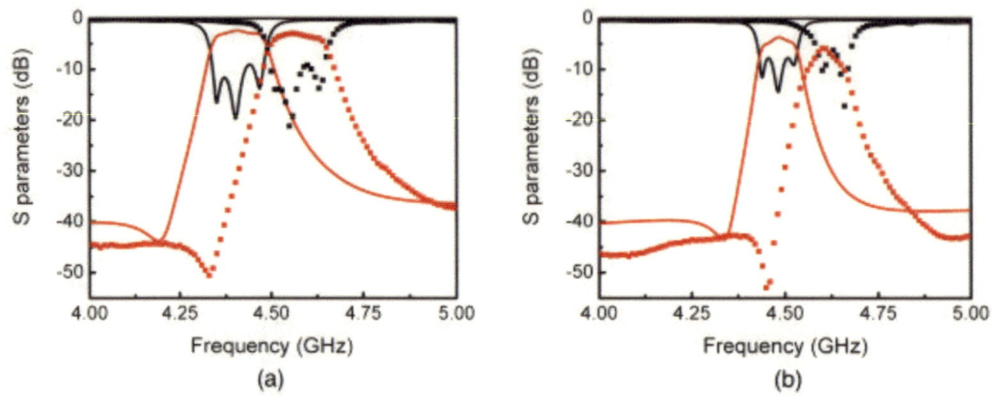

Fig. 15. Simulated (solid line) and experimental (symbol) S_{11} and S_{21} parameters for the optimized 3 stage SRRs loaded CPW line with (a) w=3.5 mm, and (b)w=4.5 mm.

	FBW (%)	IL (dB)	Ripple (dB)
w=3.5 mm	3.9	1.3	0.4
w=4.5 mm	1.8	2.2	0.6

Table 6. Characteristics of the optimized 3 stage SRRs loaded CPW simulated response.

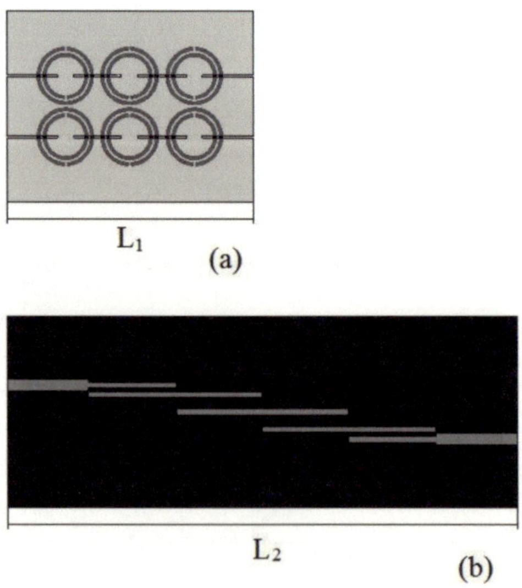

Fig. 16. Comparison of the order-3 layouts. (a) SRRs loaded CPW with wide shunt wires filter, and (b) edge-coupled line filter.

It compares the dimensions of two simulated models, namely the SRRs loaded CPW with wide shunt strips and a conventional edge-coupled lines filter. The total length L_1 is roughly 30 mm (three times the unit cell period), whereas the length L_2 is approximately 68 mm. The proposed filters are shortened by a factor of 2.3. The length of the edge-coupled filter could be further reduced by bending the coupled half-wavelength resonators in a U-shape. However, the order, and thus the length, of the filter cannot be reduced to keep good out-of-band rejection levels. On the contrary, the size of the proposed filter can still be reduced by enlarging the width of the shunt strips. Also, a comparison in terms of S-parameters is shown in Fig. 17. As it can be observed, both filters have a similar behavior. Attenuation levels outside pass band are comparable up to -30 dB for both cases. Also, a slight increment of approximately 1.1 dB in the insertion loss level can be observed for the edge-coupled line filter.

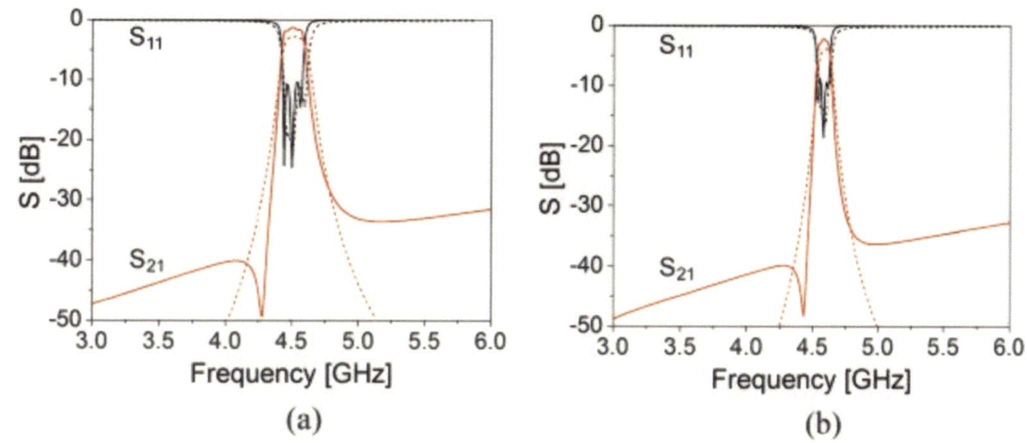

Fig. 17. Simulated S_{11} and S_{21} parameters for the optimized 3 stage SRRs loaded CPW line with (a) w=3.5 mm, and (b)w=4.5 mm. Also, the simulated S-parameters for a conventional order-3 edge-coupled filter with similar performance are depicted (dashed line).

4. Synthesis of compact and highly selective filters with tunable responses

This section presents a tunable filter based on the metamaterial transmission lines incorporating dispersive cells. Generally, in order to generate reconfigurable devices, individual reconfigurable components such as tunable capacitors and resonators are used. In particular, this reconfigurable capability is generated by an alteration of the SRRs' resonant frequency using reverse-biased varactors. Subsequently, the adjustable resonant frequency rings are employed to load the host transmission lines.

4.1 Tunable basic cell configuration

Fig. 18 shows the sketch of the varactor diode loaded SRR implemented in a CPW based configuration. The varactor is a reverse-biased semiconductor diode connected between the concentric rings of the SRR. Its capacitance can be tuned by changing the DC voltage applied to its pads. Fig. 19 shows the simulated scattering responses of the basic cell for different values of applied bias voltages (different capacitances and thus SRR resonant frequencies). The tuning range obtained is roughly 1.1 GHz with insertion losses of 2.7 dB for the higher

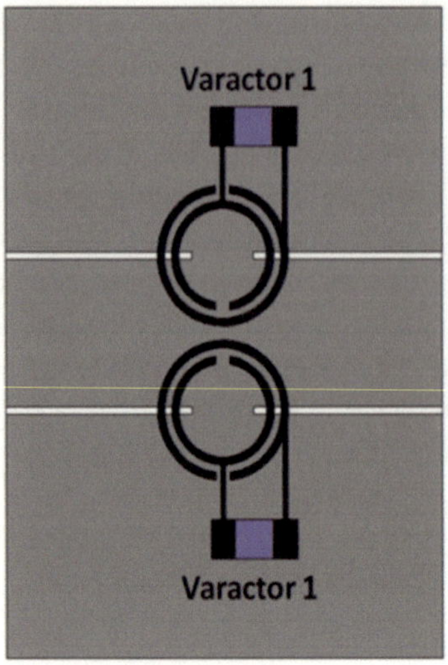

Fig. 18. Basic cell CPW line loaded with wide shunt strips and SRRs based diode varactors. Filter dimensions are the same as in figure 1, except w = 4 mm.

pass band response at around 5 GHz (f_{max}), and a maximum value of 3.8 dB when the tunable pass band is shift to 3.9 GHz ($0.78 f_{max}$). Moreover, as observed, rejection above the pass band increases as the operating frequency decreases. Compared to other tunable cell configurations, this approach effectively yields a broader frequency tuning range whilst preserving good insertion losses and a very narrow bandpass response.

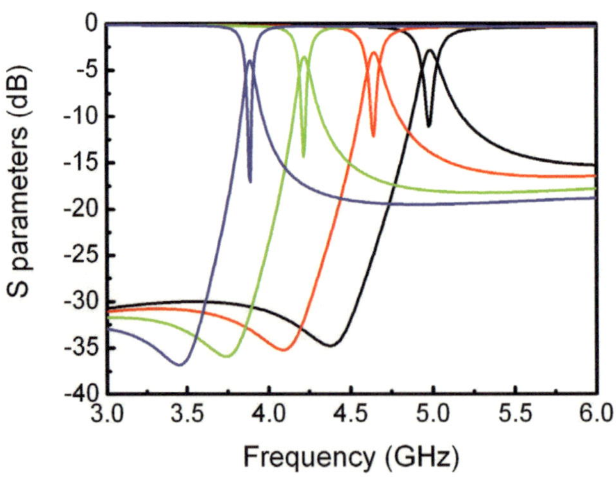

Fig. 19. Simulated scattering parameters of the tunable basic cell.

4.2 Three-cells array of split ring resonators loaded coplanar waveguide with varactors

Next, a synthesis technique for coupled resonators is applied to design a reconfigurable filter. We used the generic coupled resonators scheme of an N-order band pass filter structure by simply cascading the previous tunable basic cell. A third order reconfigurable band pass filter using varactor loaded SRRs, with a pass band centred between 4 and 5 GHz is considered. The layout of the filter is shown in Fig. 20.

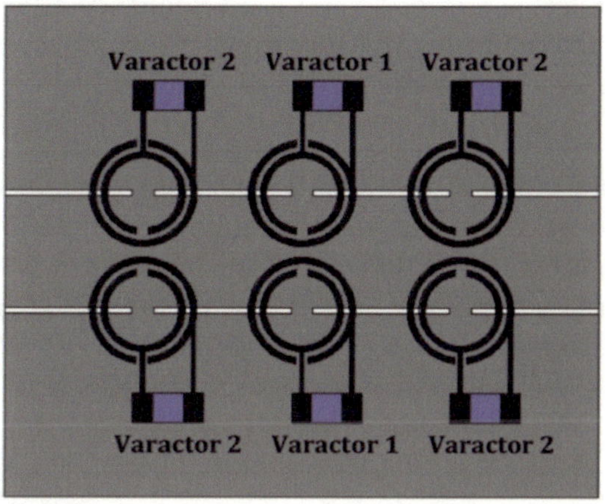

Fig. 20. Three-stage tunable filter loaded with wide shunt strips and diode varactors. Filter dimensions are the same ones as in figure 1, except w=2mm.

The design is performed by optimizing independently the central and side stages. In this regard, different loaded SRRs cells are used, with independent diode varactors loading elements as depicted in Fig. 20. The simulated S-parameter results of the third order tunable bandpass filter are shown in Fig. 21. The simulated filter can be tuned from 3.9 GHz to 4.9 GHz (approximately 25% variation), by changing the biasing voltage from 0 to 25 V. The simulated insertion losses at 3.9 GHz are 4.3 dB and 3.2 dB at 4.9 GHz. For the applications where such difference is acceptable, the bandwidth of the filter is calculated to be 2.6% and 5%, respectively.

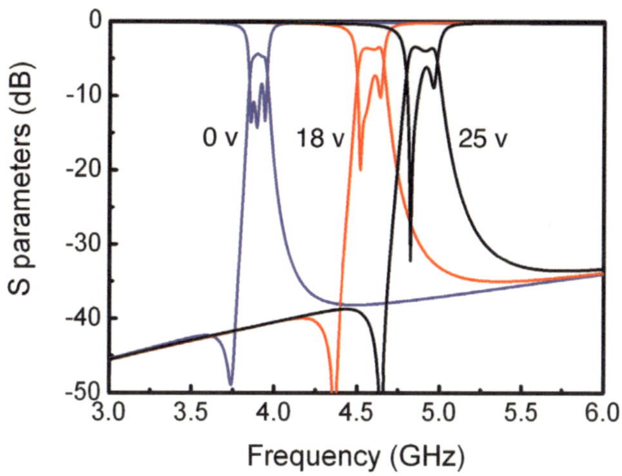

Fig. 21. Optimized simulated response of the 3-stage tunable filter.

Despite the slight increment of the insertion losses in the lower frequency band, the tuning range of the varactor loaded SRR filter is fairly good, and its size is roughly 2.5 times smaller than conventional edge-coupled filters working in the same frequency range. In addition, using a single tunable filter instead of several fixed-frequency filter bands can add system flexibility, which may warrant this slightly augmented insertion losses.

5. Summary

In this chapter, different types of coplanar transmission lines loaded with SRRs, shunt wires, series gaps, diode varactors and a combination of them have been analyzed. When different loading elements are added, the left-handed behavior originally predicted for these microstructures can be modified, and their frequency selectivity can be enhanced while maintaining the advantage of miniaturization. The analysis has been performed through full-wave electromagnetic simulations, lumped equivalent circuit models, and measurements of different prototypes.

In section 2, the propagation features of loaded CPW lines have been presented. Left-handed or right-handed propagation can be achieved depending on the loading elements included in the line, using SRRs as a common element. The use of series gaps, and their combination with shunt wires, increases the possibilities of generating narrow transmission bands in the vicinity of the SRR resonance. For instance, they can be tailored to even obtain symmetrical frequency responses. In conclusion, the proposed structures offer an alternative for designing planar frequency filtering structures in applications with severe restrictions in terms of rejection, selectivity, and size.

Section 3 has shown that it is possible to improve the selectivity of highly dispersive transmission lines, made of CPWs loaded with SRRs and shunt wires. This can be achieved by a proper engineering of the electric plasma frequency with respect to the magnetic plasma frequency. Beyond a higher rejection level, which was expected due to the deepening of the forbidden gap between the left- and right-handed dispersion branches, there is also a huge enhancement of the loaded Q quality factors. The structure also maintains low insertion losses. By cascading elementary cells it is possible to achieve a further increase of the steepness of the rejection with low insertion losses. Finally, section 4 has demonstrated reconfigurable filters having narrow bandpass responses, good insertion loss, and good frequency tuning range. The devices were based on varactor diode loaded SRRs and rigorous optimization processes.

Potential use of these miniaturized high-Q frequency selective cells can be foreseen in many modern microwave areas, notably in automotive, radar, wireless communication systems and biosensors.

6. References

Aznar, F., Bonache, J. & Martin, F. (2008). Improved circuit model for left-handed lines loaded with split ring resonators, *Applied Physics Letters* 92(4): 043512.

Bonache, J., Martin, F., Falcone, F., Garcia, J., Gil, I., Lopetegi T.and Laso, M. A. G., Marques, R., Medina, F. & Sorolla, M. (2005). Compact coplanar waveguide band-pass filter at the S-band, *Microwave and Optical Technology Letters* 46(1): 33–35.

Borja, A. L., Carbonell, J., Boria, V. E., Cascon, J. & Lippens, D. (2010). Synthesis of compact and highly selective filters via metamaterial-inspired coplanar waveguide line technologies, *IET Microwaves, Antennas and Propagation* 4(8): 1098–1104.

Borja, A. L., Carbonell, J., Boria, V. E. & Lippens, D. (2008). Synthesis of compact and highly selective filters via metamaterial-inspired coplanar waveguide line technologies, *Applied Physics Letters* 93: 203505.

Borja, A. L., Carbonell, J., Boria, V. E. & Lippens, D. (2010a). A 2% bandwidth C-band filter using cascaded split ring resonators, *IEEE Antennas and Wireless Propagation Letters* 9: 256–259.

Borja, A. L., Carbonell, J., Boria, V. E. & Lippens, D. (2010b). A compact coplanar waveguide metamaterial-inspired line and its use in tunable narrow bandpass filters, *40th European Microwave Conference (EuMC), 26 Sep. - 1 Oct., Paris.*

Borja, A. L., Carbonell, J., Boria, V. E. & Lippens, D. (2009). Highly selective left-handed transmission line loaded with split ring resonators and wires, *Applied Physics Letters* 94: 143503.

Carbonell, J., Borja, A. L., Boria, V. E. & Lippens, D. (2009). Duality and superposition in split ring resonator loaded planar transmission, *IEEE Antennas and Wireless Propagation Letters* 8: 886–889.

Deleniv, A., Vendik, I. & Gevorgian, S. (1999). Modeling gap discontinuity in coplanar waveguide using quasistatic spectral domain method, *2000 John Wiley & Sons, International Journal on RF and Microwave* 10: 150–158.

Martin, F., Bonache, J., Falcone, F., Sorolla, M. & Marques, R. (2003). Split ring resonator-based left-handed coplanar waveguide, *Applied Physics Letters* 83(22): 4652–4654.

Mongia, R., Bahl, I. & Bhartia, P. (1999). RF and microwave coupled-line circuits, *Artech House, Boston* .

Pendry, J. B., Holden, A. J., Robbins, D. J. & Stewart, W. J. (1999). Magnetism from conductors and enhanced nonlinear phenomena, *IEEE Transactions on Microwave Theory and Techniques* 47(11): 2075–2084.

Pendry, J. B., Holden, A. J., Stewart, W. J. & Youngs, I. (1996). Extremely low frequency plasmons in metallic mesostructures, *Physical Review Letters* 76: 4773–4776.

Rogla, L. J., Carbonell, J. & Boria, V. E. (2007). Study of equivalent circuits for open-ring and split-ring resonators in coplanar waveguide technology, *IET Microwaves, Antennas and Propagation* 1(1): 170–176.

Shelby, A., Smith, D. R. & Schultz, S. (2001). Experimental verification of a negative index of refraction, *Science* 292: 7779.

Shelby, R. A., Smith, D. R., Nemat-Nasser, S. C. & Schultz, S. (2001). Microwave transmission through a two-dimensional, isotropic, left-handed metamaterial, *Applied Physics Letters* 78(4): 489–491.

Smith, D. R. & Kroll, N. (2000). Negative refractive index in left-handed materials, *Physical Review Letters* 85: 2933–2936.

Smith, D. R., Padilla, W. J., Vier, D. C., Nemat-Nasser, S. C. & Schultz, S. (2000). Composite medium with simultaneously negative permeability and permittivity, *Physical Review Letters* 84(18): 4184–4187.

Smith, D. R., Vier, D. C., Koschny, T. & Soukoulis, C. M. (2005). Electromagnetic parameter retrieval from inhomogeneous metamaterials, *Physical Review E* 71(1): 036617.